APPLIED AND INDUSTRIAL MATHEMATICS IN ITALY II

Series on Advances in Mathematics for Applied Sciences – Vol. 75

APPLIED AND INDUSTRIAL MATHEMATICS IN ITALY II

Selected Contributions from the 8th SIMAI Conference

Baia Samuele (Ragusa), Italy 22 – 26 May 2006

Edited by

Vincenzo Cutello
Università di Catania, Italy

Giorgio Fotia
CRS4, Pula, Italy

Luigia Puccio
Università di Messina, Italy

World Scientific

NEW JERSEY • LONDON • SINGAPORE • BEIJING • SHANGHAI • HONG KONG • TAIPEI • CHENNAI

Published by

World Scientific Publishing Co. Pte. Ltd.
5 Toh Tuck Link, Singapore 596224
USA office: 27 Warren Street, Suite 401-402, Hackensack, NJ 07601
UK office: 57 Shelton Street, Covent Garden, London WC2H 9HE

British Library Cataloguing-in-Publication Data
A catalogue record for this book is available from the British Library.

Series on Advances in Mathematics for Applied Sciences — Vol. 75
APPLIED AND INDUSTRIAL MATHEMATICS IN ITALY II
Selected Contributions from the Eighth SIMAI Conference

ISBN-13 978-981-270-938-7
ISBN-10 981-270-938-X

Printed in Singapore by World Scientific Printers (S) Pte Ltd

PREFACE

Industrial mathematics is evolving into an important branch of mathematics. The scientific community is becoming increasingly aware of this new trend and it is actively engaged in bridging the gap between the highly specialized mathematical and computational research and the emerging demands of innovation coming from industry.

The present contributed volume intends to provide an overview of the most recent research activities currently being pursued in Italy in this direction and to illustrate possible ways of using results coming from pure and applied mathematical research in the resolution of problems coming outside mathematics. The field of applied and industrial mathematics in its broadest sense is covered, and as such, the mathematical content ranges from rigorous and formal analytical results to analysis of computational techniques, from modeling to engineering oriented computational simulations. We hope that, as a whole, the volume may provide to researchers working in industry a general view of existing skills in Academia, and to academics a selection of applications of mathematics to real world problems, to take inspiration from for new theoretical developments, and to discover new applications of existing mathematical ideas.

This book represents a constructive outcome of the activities of the Italian Society for Applied and Industrial Mathematics (SIMAI) to promote advances in applied and industrial mathematics. In particular, these activities include the 8th SIMAI Conference organized in 2006 which, by bringing together research and practitioners in mathematical modeling and numerical solutions, represents a key occasion to promote and stimulate research in applied mathematics and to foster interactions with industry. The 53 contributions collected here have been selected from a larger number of submitted papers by the Scientific Committee of SIMAI consisting of Ubaldo Barberis, Franco Brezzi, Enrico De Bernardis, Mario Primicerio, Fausto Saleri, Alessandro Speranza, Vanda Valente and the Editors of this book. We wish to thank all of them as well as the referees that have been involved in the selection procedure.

We thank especially Santa Agreste, Angela Ricciardello and Alberto Vocaturo, for the thorough editing which involved reformatting of some of the contributions, and for the careful preparation of the lay-out of this volume.

The Editors

Catania, Pula, Messina, March 15th 2007

CONTENTS

NUMERICAL APPROXIMATION OF A BGK-TYPE RELAXATION MODEL FOR REACTIVE MIXTURES

A. AIMI, M. DILIGENTI, M. GROPPI and C. GUARDASONI

Department of Mathematics, University of Parma
V.le G.P. Usberti, 53/A, 43100 Parma, Italy
alessandra.aimi,mauro.diligenti,maria.groppi@unipr.it
chiara.guardasoni@unimi.it

A consistent BGK-type approach to reacting gas mixtures, according to Boltzmann-like kinetics for a bimolecular reversible chemical reaction has been recently introduced. In this paper we apply a numerical strategy based on time splitting techniques to simulate the reactive BGK equations. These techniques have the advantage of simplifying the problem by treating separately the convection step and collision step. Numerical results of the time-dependent Riemann problem for the reactive BGK system are presented.

Keywords: Boltzmann equation, BGK model, reacting mixtures, Riemann problem, splitting methods

1. Introduction

In recent years, new suggestions and proposals concerning the mathematical modelling and applications of multi-component gaseous flows with chemical reactions have been published. In this context, several kinetic approaches have been developed in the last decades, starting from the pioneering work by Prigogine-Xhrouet.[1] The increasing interest on kinetic models is mainly motivated by the fact that they enable the macroscopic laws to be derived from elementary principles, providing consistent macroscopic theories in the hydrodynamic limit, and moreover they allow to deduce transport and structure coefficients, that are not directly obtainable from macroscopic approaches (the interested reader is referred to the comprehensive Ref. 2). The so called BGK equations[3,4] constitute a well known model of the nonlinear Boltzmann equation and a simpler tool of investigation in particular for reacting gaseous flows, for which the collision part of the kinetic equations becomes much heavier. A recent extension of a consistent BGK-type approach for inert gas mixtures to reacting gases, according to a Boltzmann-

like kinetic model developed in Ref. 5 for a bimolecular reversible chemical reaction of the type: $A^1 + A^2 \rightleftharpoons A^3 + A^4$, has been investigated in Ref. 6. This model is based on the simple idea of introducing only one suitable BGK collision operator for each species s, taking into account all interactions with whatever species r.
Here we propose a numerical strategy to simulate the reactive BGK equations in more general space-dependent situations. In particular we focus here on problems with axial symmetry, which are of interest in many applications like for instance the classical evaporation-condensation problem.[7] The method is based on time splitting techniques, which are widely used in the numerical analysis of the classical Boltzmann equation[8–10] but their application to kinetic systems describing reacting gas mixtures has not been yet discussed, to our knowledge. The time splitting approach has the advantage of simplifying the problem by treating separately the two steps, the convection or transport step, which solves the free-streaming equations along the characteristic lines, and the collision step, which solves the spatially homogeneous BGK equations. The numerical solution of this latter, which can be regarded as a Cauchy problem, is evaluated with Runge-Kutta explicit schemes of different order. Here we consider a splitting method, for which the truncation error per time step Δt is $O(\Delta t^3)$, but due to the accumulation of errors, the convergence rate is $O(\Delta t^2)$ coupled with an explicit Runge-Kutta method of order 2. Numerical results on time-dependent Riemann problem for reacting mixtures of four gases are presented.

2. Model equations

The BGK approximation introduced in Ref. 6 of the Boltzmann-type model worked out in Ref. 5 for chemical reactions is described by the following kinetic equations

$$\frac{\partial f^s}{\partial t} + \mathbf{v} \cdot \frac{\partial f^s}{\partial \mathbf{x}} = \nu_s \big(\mathcal{M}_s - f^s\big), \qquad s = 1, \ldots, 4, \tag{1}$$

where f^s is the general distribution function, $\mathcal{M}_s$ is an auxiliary local Maxwellian depending on velocity vector variable $\mathbf{v}$, molecular masses m^s, Boltzmann constant K and disposable parameters $n_s, \mathbf{u}_s, T_s$:

$$\mathcal{M}_s = n_s \left(\frac{m^s}{2\pi K T_s}\right)^{3/2} \exp\left[-\frac{m^s}{2KT_s}(\mathbf{v} - \mathbf{u}_s)^2\right] \qquad s = 1, \ldots, 4. \tag{2}$$

At last in (1), ν_s represents the inverse of the s-th relaxation time, possibly depending on macroscopic fields, but independent of $\mathbf{v}$. The above auxiliary fields n_s, $\mathbf{u}_s$, T_s are determined from the corresponding actual moments of the distribution functions f^s (namely number density n^s, mass velocity $\mathbf{u}^s$ and temperature T^s of each component) by requiring that the exchange rates for mass, momentum and total (kinetic plus chemical) energy prescribed by (1) coincide with those deduced from the reactive Boltzmann equations (see Ref. 5, 6 for a detailed derivation). In presence of chemical reactions, these exchange rates can be made explicit under the assumption of dominant elastic collisions and thus of "slow" chemical reactions (the so called tempered reaction regime), as for instance in the carbon-oxygen chain. We consider the application of BGK equations (1) to problems with axial symmetry with respect to an axis (say, $x_1 \equiv x$). In such cases, the distribution functions f^s depend on $\mathbf{v}$ only through its modulus and its latitudinal angle with respect to that axis, and all transverse components of the macroscopic velocities $\mathbf{u}^s$ vanish (i.e. $u_2^s = u_3^s = 0$). As well known in the literature,[11] in this case a reduction to a fully one-dimensional problem is possible, though describing still a three-dimensional velocity space, with a sensible simplification of the computational apparatus. This kind of problem is not only important for theoretical investigation, but also quite frequent in practical applications.[7]

Let us introduce the new unknowns

$$\phi_1^s(x,v,t) = \int_{\mathbb{R}^2} f^s \, \mathrm{d}v_2 \mathrm{d}v_3, \qquad \phi_2^s(x,v,t) = \int_{\mathbb{R}^2} (v_2^2 + v_3^2) f^s \, \mathrm{d}v_2 \mathrm{d}v_3, \quad (3)$$

each depending only on one space and one velocity variable. Starting from (1) and using Chu reduction[11] we obtain the following system of BGK equations for the unknown vector $\underline{\phi}^s = (\phi_1^s, \phi_2^s)^\top$, coupled with initial conditions

$$\begin{cases} \dfrac{\partial \underline{\phi}^s}{\partial t} + v \dfrac{\partial \underline{\phi}^s}{\partial x} = \nu_s(\underline{\phi}_s^e - \underline{\phi}^s) \quad t > 0, \ x \in \mathbb{R}, \ v \in \mathbb{R}, \qquad s = 1, \ldots, 4; \\ \underline{\phi}^s(x,v,0) = \underline{\phi}_0^s(x,v). \end{cases} \quad (4)$$

The BGK equations (4) describe a relaxation process towards the vector functions $\underline{\phi}_s^e = \left(\phi_{s,1}^e, \phi_{s,2}^e\right)^\top$, which is obtained by Chu transform of (2) and has the form

$$\begin{aligned} \left(\phi_{s,1}^e, \phi_{s,2}^e\right)^\top &= \left(M_s, \frac{2KT_s}{m^s} M_s\right)^\top, \\ M_s &= n_s \left(\frac{m^s}{2\pi K T_s}\right)^{1/2} \exp\left[-\frac{m^s}{2KT_s}(v - u_s)^2\right]. \end{aligned} \quad (5)$$

The Chu transform reduces the auxiliary velocity to a scalar parameter u_s, owing to the axial symmetry. To determine the auxiliary parameters $n_s = n_s(x,t)$, $u_s = u_s(x,t)$, $T_s = T_s(x,t)$ it is necessary first to compute the exchange rates for the Boltzmann reactive model described in Ref. 5; such rates are known analytically for Maxwell molecules,[12] even in the tempered reaction regime, and besides may be expressed in terms of mass m^s of each component of the mixture (with $m^1+m^2=m^3+m^4=M$), energies of chemical link E^s, and energy difference between reactants and products $\Delta E = -\sum_{s=1}^{4} \lambda^s E^s$ (with $\lambda^1 = \lambda^2 = -\lambda^3 = -\lambda^4 = 1$), conventionally assumed to be positive. Those rates (not reported here for brevity) involve fundamental macroscopic moments of distribution functions f^s, which are given in terms of ϕ_1^s and ϕ_2^s as

$$n^s = \int_{\mathbb{R}} \phi_1^s \mathrm{d}v, \;\; u^s = \frac{1}{n^s}\int_{\mathbb{R}} v\phi_1^s \mathrm{d}v, \;\; T^s = \frac{m^s}{3Kn^s}\int_{\mathbb{R}} [(v-u^s)^2\phi_1^s + \phi_2^s]\mathrm{d}v. \tag{6}$$

Here u^s denotes the first component of the mass velocity, since axial symmetry implies $u_2^s = u_3^s = 0$ (and for the same reason $u_{s,2} = u_{s,3} = 0$). We point out that the unknowns ϕ_1^s and ϕ_2^s provide a reduced description of the velocity distributions, if compared to f^s, but they suffice for our purposes. In (5), the auxiliary parameters: n_s, u_s, T_s are determined by requiring that the BGK scheme prescribes the same exchange rates of the Boltzmann model. The resulting expressions, as well as a clear derivation and a complete description may be found in Ref. 6. The discussion involves also the global macroscopic parameters: number density n, mass density ρ, mass velocity u, scalar pressure p or temperature T, which are expressed in terms of single component parameters by

$$\begin{gathered} n = \sum_{s=1}^{4} n^s, \qquad \rho = \sum_{s=1}^{4} \rho^s = \sum_{s=1}^{4} m^s n^s, \qquad u = \frac{1}{\rho}\sum_{s=1}^{4} m^s n^s u^s, \\ p = nKT = \sum_{s=1}^{4} n^s K T^s + \frac{1}{3}\sum_{s=1}^{4} m^s n^s \left(u^s - u\right)^2 . \end{gathered} \tag{7}$$

The macroscopic collision frequencies $\nu_s = \nu_s(x,t)$, which measure the strength at which BGK model equations push distributions towards equilibrium, can be evaluated by a suitable estimation of the actual average number of collisions taking place for each species.[6] It is remarkable that the consistency properties of the reactive BGK model, proved in Ref. 6, are independent from the choice of macroscopic collision frequencies. Anyway, a suitable evaluation is needed in order to avoid artificial acceleration or slowing down of the relaxation process.

3. Numerical approximation

We rewrite here the system (4) pointing out, in particular, the dependence of the vector functions $\underline{\phi}^e_s$ and of the macroscopic collision frequencies ν_s on the components of the unknown vector solution $\underline{\Phi} = (\underline{\phi}^1, \underline{\phi}^2, \underline{\phi}^3, \underline{\phi}^4)^\top$. In fact the auxiliary parameters n_s, u_s, T_s, appearing in (5), as well as the frequencies ν_s, follow from the definition of macroscopic moments of the solution through (7). Hence, we consider the following equivalent one-dimensional (in both space and velocity) initial-value nonlinear problem

$$\begin{cases} \dfrac{\partial \underline{\phi}^s}{\partial t} + v\dfrac{\partial \underline{\phi}^s}{\partial x} = \nu_s(\underline{\Phi}) \left[\underline{\phi}^e_s(\underline{\Phi}) - \underline{\phi}^s\right], \; s = 1, \cdots, 4; \quad t > 0,\, x \in \mathbb{R},\; v \in \mathbb{R}, \\ \underline{\phi}^s(x, v, 0) = \underline{\phi}^s_0(x, v), \end{cases} \tag{8}$$

for the unknowns $\underline{\phi}^s = (\phi^s_1, \phi^s_2)^\top$. Problem (8) can be rewritten in the form

$$\begin{cases} \dfrac{\partial \underline{\phi}^s}{\partial t} = A[\underline{\phi}^s] + B[\underline{\phi}^s], \; s = 1, \ldots, 4; \quad t > 0,\, x \in \mathbb{R},\; v \in \mathbb{R}, \\ \underline{\phi}^s(x, v, 0) = \underline{\phi}^s_0(x, v), \end{cases} \tag{9}$$

where: $A[\underline{\phi}^s] = -v\,\dfrac{\partial \underline{\phi}^s}{\partial x}$ is the convection operator, and $B[\underline{\phi}^s] = \nu_s(\underline{\Phi})\left[\underline{\phi}^e_s(\underline{\Phi}) - \underline{\phi}^s\right]$ is the collision operator. Therefore, in order to compute numerically the solution, it is usual to solve: $\dfrac{\partial \underline{\phi}^s}{\partial t} = A[\underline{\phi}^s]$ and $\dfrac{\partial \underline{\phi}^s}{\partial t} = B[\underline{\phi}^s]$, separately. This procedure is known as a splitting method and it is a common tool for the numerical analysis of the Boltzmann equation. The method consists of two steps, the convection step, which solves the collisionless equation (free-transport equation), and the collision step, which solves the space-homogeneous equation. Setting the solution $\underline{\Phi}(t)$ of the problem (9) as $\underline{\Phi}(t) = S^t_{A+B}(\underline{\Phi}_0)$, where: $\underline{\Phi}_0 = (\underline{\phi}^1_0, \underline{\phi}^2_0, \underline{\phi}^3_0, \underline{\phi}^4_0)^\top$, the conventional splitting method (CSM) is nothing more than the following approximation of the operator $S^{\Delta t}_{A+B}$

$$S^{\Delta t}_{A+B} \approx S^{\Delta t}_B S^{\Delta t}_A\,, \tag{10}$$

for which it holds[8]

$$S^{\Delta t}_{A+B}(\underline{\Phi}_0) = S^{\Delta t}_B[S^{\Delta t}_A(\underline{\Phi}_0)] + \frac{\Delta t^2}{2}[A'_{\underline{\Phi}_0} B(\underline{\Phi}_0) - B'_{\underline{\Phi}_0} A(\underline{\Phi}_0)] + O(\Delta t^3). \tag{11}$$

In spite of this result, because of the accumulation of errors, the convergence rate over $[0, N\Delta t]$ is $O(\Delta t)$. Note that the order of accuracy of this simple splitting does not improve even if we solve analytically both collision and convection steps. In the CSM, the collision step may also be performed before the convection step, having

$$S_{A+B}^{\Delta t}(\underline{\Phi}_0) = S_A^{\Delta t}[S_B^{\Delta t}(\underline{\Phi}_0)] - \frac{\Delta t^2}{2}[A'_{\underline{\Phi}_0} B(\underline{\Phi}_0) - B'_{\underline{\Phi}_0} A(\underline{\Phi}_0)] + O(\Delta t^3). \quad (12)$$

In this case, the leading term of the truncation error differs only in sign from that one obtained by (11) (see Ref. 8). From this last remark, we deduce a $O(\Delta t^3)$ accuracy per time step for the following approximation of the operator $S_{A+B}^{\Delta t}$

$$S_{A+B}^{\Delta t} \approx \frac{S_A^{\Delta t} S_B^{\Delta t} + S_B^{\Delta t} S_A^{\Delta t}}{2}, \quad (13)$$

i.e., the mean of two applications of CSM with inverted steps: convection-collision and collision-convection. In this way the leading errors of the above methods cancel each other. Moreover, the convergence rate over $[0, N\Delta t]$, due to the accumulation of errors, is $O(\Delta t^2)$, Ref. 8. Other higher order splittings can be found in Ref. 10.

For the numerical implementation of splitting techniques, at first we have to define a finite numerical domain in the phase space $[x_L, x_R] \times [v_L, v_R]$ dependent on the problem data: in particular the choice of $[v_L,\ v_R]$ is related to the initial velocity distributions while the choice of $[x_L,\ x_R]$ depends on the observation time interval. Consequently, we impose the following conditions (at infinity) for $s = 1, \cdots, 4$,

$$\underline{\phi}^s(x, v, t) = \underline{\phi}^s(x_L, v, t) \qquad x \le x_L, \quad \forall v, \forall t$$

$$\underline{\phi}^s(x, v, t) = \underline{\phi}^s(x_R, v, t) \qquad x \ge x_R, \quad \forall v, \forall t.$$

Let $x_r = x_L + r\Delta x$, $r = 0, \cdots, N_x$ and $v_q = v_L + q\Delta v$, $q = 0, \cdots, N_v$, be uniform grids defined in $[x_L, x_R]$ and in $[v_L, v_R]$, respectively.

Having set $\underline{\psi}^s(x_r, v_q, 0) = \underline{\phi}_0^s(x_r, v_q)$ and $t_i = i\,\Delta t$, for $i = 0, \cdots, N$-1, the realization of conventional splitting method (10) is as follows

Convection step:

$$\begin{cases} \dfrac{\partial \underline{\varphi}^s}{\partial t} + v\dfrac{\partial \underline{\varphi}^s}{\partial x} = 0, & t_i < t \le t_{i+1} \\ \underline{\varphi}^s(x_r, v_q, t_i) = \underline{\psi}^s(x_r, v_q, t_i) \end{cases}$$

with formal solution: $\underline{\varphi}^s(x_r, v_q, t_{i+1}) = \underline{\psi}^s(x_r - v_q\Delta t, v_q, t_i)$ and

Collision step:

$$\begin{cases} \dfrac{\partial \underline{\psi}^s}{\partial t} = \nu_s(\underline{\Psi})\left[\underline{\phi}_s^e(\underline{\Psi}) - \underline{\psi}^s\right], & t_i < t \le t_{i+1} \\ \underline{\psi}^s(x_r, v_q, t_i) = \underline{\varphi}^s(x_r, v_q, t_{i+1}) \end{cases}$$

where $\underline{\Psi} = (\underline{\psi}^1, \underline{\psi}^2, \underline{\psi}^3, \underline{\psi}^4)^\top$.
In the convection step, the problem is to evaluate the formal solution at time t_{i+1}, because it should be obtained from the initial condition evaluated at time t_i along the characteristic lines, but $\underline{\psi}^s$ is known only in the original nodes of the grid. To overcome this difficulty, we have considered the following algorithm:

- compute the nearest grid point from $x_r - v_q\Delta t$, named x_j;
- compute $\underline{\psi}^s(x_r - v_q\Delta t, v_q, t_i)$ using Taylor expansion around x_j truncated at a suitable order p, with derivatives approximated with centered finite differences at the same accuracy.[13]

To preserve the accuracy order per time step of the chosen splitting procedure, namely $k = 2$, $k = 3$ for CSM and (13), respectively, the order p is chosen such that: $(\Delta x)^{p+1} \leq (\Delta t)^k$. For this accuracy purpose, we have evaluated $\underline{\psi}^s(x_r - v_q\Delta t, v_q, t_i)$ with Taylor expansion rather than linear interpolation. Furthermore, for the same reason, it is useful to choose the space and time steps satisfying the following Courant-Friedrichs-Lewy (CFL) condition

$$\frac{\Delta x}{2} > \Delta t \max\{|v_L|, |v_R|\}. \tag{14}$$

Note that from (14) we have $\dfrac{\Delta x}{2} > \Delta t|v_q|$, with $\min\{|v_L|, |v_R|\} \leq |v_q| \leq \max\{|v_L|, |v_R|\}$ and in the foregoing algorithm, the nearest grid point from $x_r - v_q\Delta t$ is always x_r.
In the discretization process the time step Δt and the space step Δx are proportional to one another, i.e., $\Delta t = \lambda\Delta x$ for some positive constant λ.
For the numerical solution of collision steps, we compute a numerical approximation of the moments of $\underline{\psi}^s$ needed in $\underline{\phi}_s^e(\underline{\Psi})$ using composite Simpson rule over $[v_L, v_R]$. The velocity step Δv is chosen to assure the double precision stability to the numerical approximation of the moments related to the initial data. Then, time-advancing is carried out using classical explicit Runge-Kutta methods of order $k-1$, $k = 2, 3$. This choice maintains the overall accuracy order of CSM and of the method (13), respectively.

4. The Riemann problem for reacting mixtures of 4 gases

In this section we present some results related to one-dimensional time-dependent Riemann problem for reactive gaseous flows. This problem starts from piecewise-constant initial data, having a single discontinuity

$$\underline{\phi}_0^s(x,v) = \begin{cases} \underline{\phi}_{0L}^s(x,v) & x \le x^* \\ \underline{\phi}_{0R}^s(x,v) & x > x^* \, . \end{cases} \tag{15}$$

We consider elastic microscopic collision frequencies ν_k^{rs}, $k = 0, 1$, defined in Ref. 6, constant with respect to the impact speed and affected by a factor $1/\varepsilon = 10^l$, where ε is the Knudsen number, which corresponds to approach the fluid limit under the assumption that elastic scattering is the dominant process in the evolution (slow chemical reaction). All numerical values used in the simulations are to be considered as dimensionless; they have been chosen from existing literature (see e.g. Ref. 14 and Ref. 15), for illustrative and comparison purposes.

We consider two examples of Riemann problem with initial data reproducing macroscopic field such that $u_L = u_R = 0$, $\rho_L > \rho_R$, $T_L > T_R$. In both cases the structure of the exact solution for the inert case exhibits a shock wave propagating to the right, a contact shock wave propagating to the right and a rarefaction wave front moving to the left.[16] When chemical reactions are taken into account, the solution can be found only numerically and its structure resembles the one of the corresponding inert problem.[16]

Test 1. In this first numerical test we consider a mixture of four gases having the following different values of masses $m_1 = 0.018$, $m_2 = 0.001$, $m_3 = 0.017$, $m_4 = 0.002$; the symmetric matrix $\mathcal{N}$ of collision frequencies for elastic scattering for $k = 0$ is

$$\mathcal{N} = \begin{bmatrix} \nu_0^{11} = 5 & \nu_0^{12} = 6 & \nu_0^{13} = 2 & \nu_0^{14} = 7 \\ & \nu_0^{22} = 4 & \nu_0^{23} = 5 & \nu_0^{24} = 8 \\ & & \nu_0^{33} = 4 & \nu_0^{34} = 3 \\ & & & \nu_0^{44} = 6 \end{bmatrix} \tag{16}$$

and $\nu_1^{sr} = \nu_0^{sr}$ for $s, r = 1, \ldots, 4$. The initial data $\underline{\phi}_0^s$ are chosen as Maxwellians reproducing prescribed global mass density ρ_0, mean velocity u_0 and scalar pressure p_0. Correspondingly, the initial mass density, mean velocity and scalar pressure for the four gases are

$$(\rho_0, u_0, p_0) = \begin{cases} \rho_{01} = 1.08 \;\; \rho_{02} = 0.05 \;\; \rho_{03} = 2.55 \;\; \rho_{04} = 0.2 & \\ u_{0i} = 0, \;\; i = 1, \cdots, 4, \;\; p_0 = 2690 & x \le 0, \\ & \\ \rho_{01} = 0.54 \;\; \rho_{02} = 0.02 \;\; \rho_{03} = 1.7 \;\; \rho_{04} = 0.1, & \\ u_{0i} = 0, \;\; i = 1, \cdots, 4, \;\; p_0 = 500 & x > 0. \end{cases} \tag{17}$$

In this example we consider different values for the chemical collision frequency ν_{12}^{34}, defined in Ref. 6, and for the energy gap ΔE.
We began with the choice $\Delta E = 0$, which represents an important test for the consistency of our BGK model and of the relevant numerical approximation. It is known in fact that, starting from the reactive Boltzmann equations which our BGK model originated from, it is possible to derive, in the hydrodynamic limit, reactive Euler equations as zero order asymptotic approximation.[5] It is remarkable that, when $\Delta E = 0$, this system gives the classical Euler equations for global mass densities, mean velocity and temperature, whereas the single mass densities ρ_i may differ from the inert case. Numerical results shown below in Figure 1 at time $t = 0.015$ for global density and global mean velocity confirm this behavior also when we start from the BGK model and approach the fluid limit under the same assumption of dominant elastic scattering (the same happens to temperature, not reported here). The computational domain in space and velocity is: $[-1, 1] \times [-500, 500]$. Numerical computations were carried out until the final time instant $t = 0.015$ with $(\Delta t, \Delta x, \Delta v) = (10^{-5}, 10^{-2}, 5)$. Knudsen number is $\varepsilon = 10^{-1}$. Simulations have been performed using both CSM

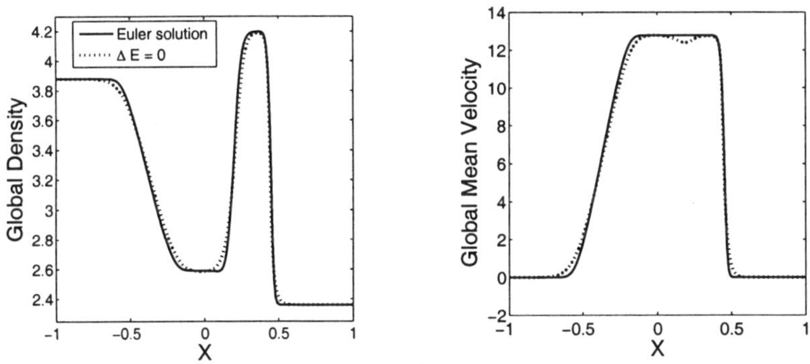

Fig. 1. Test of consistency for the reactive BGK model (with $\Delta E = 0$ at time $t = 0.015$).

coupled with Euler method and splitting (13) coupled with Heun method for the numerical solution of the collision steps. The variations of the profiles of the global density and global temperature for different values of the chemical collision frequency are shown in Figure 2 and for different values ΔE of energy difference between reactants and products are shown in Figure 3. We can notice that, in this example, the temperature increases either when the chemical reaction is faster or when ΔE is higher.

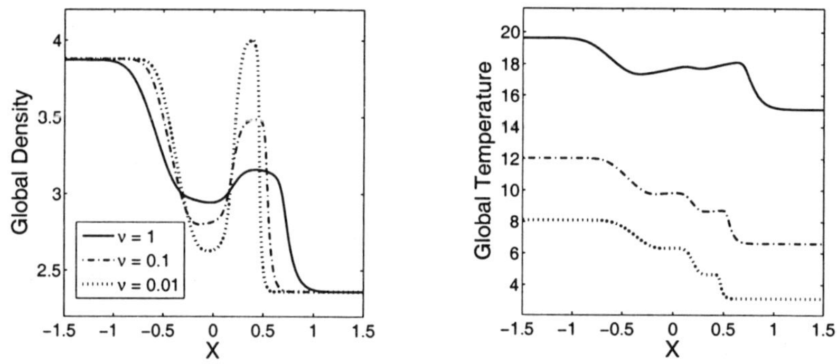

Fig. 2. Global density and global temperature for $\Delta E = 100$ at time $t = 0.015$ for different values of chemical collision frequency ν_{12}^{34}.

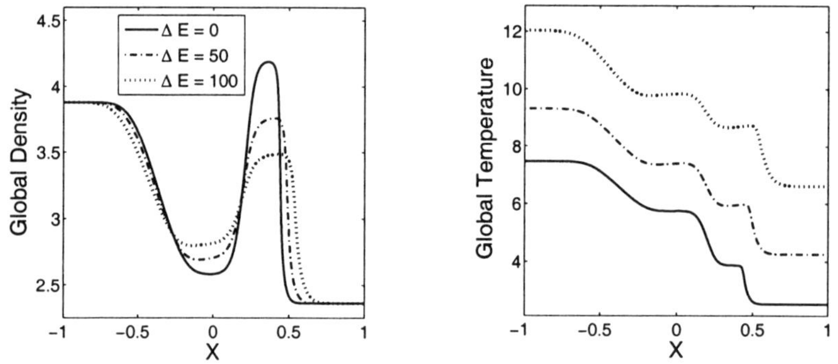

Fig. 3. Global density and global temperature for $\nu_{12}^{34} = 0.1$ at time $t = 0.015$ for different values of ΔE.

Test 2. In this second numerical text we consider a different mixture of four gases having the following different values of masses $m_1 = 58.5$, $m_2 = 18$, $m_3 = 40$, $m_4 = 36.5$, and the new symmetric matrix of collision frequencies for elastic scattering for $k = 0$ is $10^2\mathcal{N}$, where $\mathcal{N}$ is the matrix (16). The initial mass density, mean velocity and scalar pressure for the four gases are

$$(\rho_0, u_0, p_0) = \begin{cases} \rho_{0i} = \dfrac{i}{10},\ u_{0i} = 0,\ i = 1, \cdots, 4,\ p_0 = \dfrac{5}{3} & x \leq 0.5, \\ \\ \rho_{0i} = \dfrac{i}{80},\ u_{0i} = 0,\ i = 1, \cdots, 4,\ p_0 = \dfrac{1}{6} & x > 0.5. \end{cases} \tag{18}$$

These initial data reproduce the classical Sod problem[17] for Euler equations. The computational domain in space and velocity is: $[-0.5, 1.5] \times [-15, 15]$. Numerical computations were carried out until the final time instant $t = 0.2$ with $(\Delta t, \Delta x, \Delta v) = (5{\cdot}10^{-4}, 5{\cdot}10^{-3}, 5{\cdot}10^{-2})$. Knudsen number is $\varepsilon = 10^{-2}$. In this test we set the chemical collision frequency $\nu_{12}^{34} = 100$ and we consider different values of the energy gap ΔE. Simulations have been per-

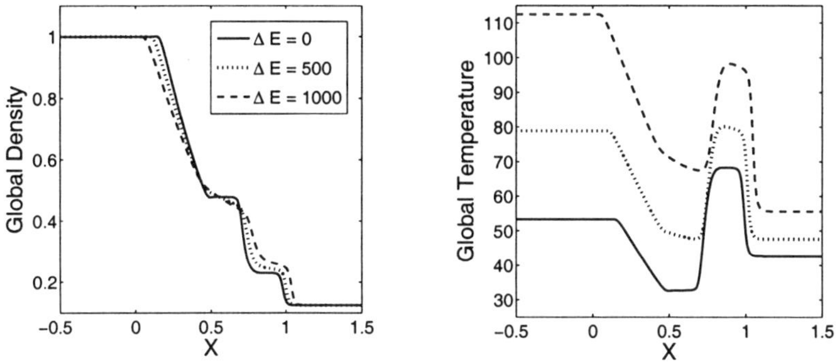

Fig. 4. Global density and global temperature at time instant $t = 0.2$ for different values of ΔE.

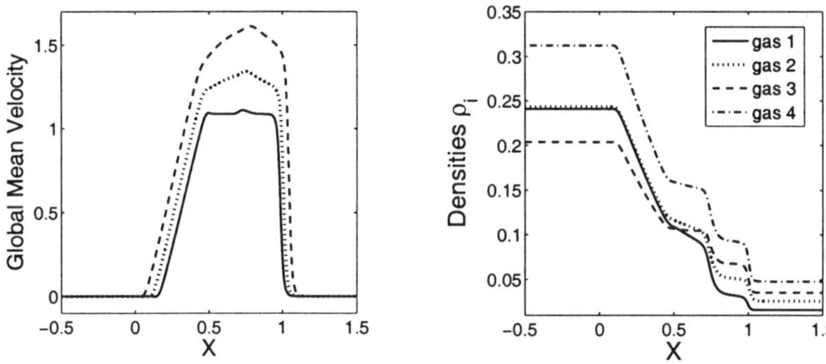

Fig. 5. Global mean velocity for different values of ΔE and densities $\rho_i, i = 1, \cdots, 4$ for $\Delta E = 500$ at time instant $t = 0.2$.

formed using splitting (13) coupled with Heun method for the numerical solution of the collision steps. The variations of the profiles of global density ρ, global temperature T and global mean velocity u for different values ΔE at time instant $t = 0.2$ are shown in Figure 4 and in Figure 5 (left). In Figure 5 (right) we report the profiles of densities ρ_i, $i = 1, \cdots, 4$, at the same time instant, for $\Delta E = 500$. The global moments of the inert mixture overlap the profiles obtained for $\Delta E = 0$, whereas this is not true

for the single components of the mixture. The higher ΔE, the greater the variations with respect to the inert case.
We notice moreover that the data used in Test 2 give rise to solutions in which the structure of density ρ and temperature T are somehow reversed with respect to Test 1.

References

1. I. Prigogine and E. Xhrouet, On the perturbation of Maxwell distribution function by chemical reaction in gases, *Physica* **XV**, 913 (1949).
2. V. Giovangigli, *Multicomponent Flow Modeling* (Birkhäuser, Boston, 1999).
3. P.L. Bhatnagar, E.P. Gross and K. Krook, A model for collision processes in gases, *Phys. Rev.* **94**, 511 (1954).
4. P. Welander, On the temperature jump in a rarefied gas, *Ark. Fys.* **7**, 507 (1954).
5. A. Rossani and G. Spiga, A note on the kinetic theory of chemically reacting gases, *Physica A* **272**, 563 (1999).
6. M. Groppi and G. Spiga, A Bhatnagar–Gross–Krook–type approach for chemically reacting gas mixtures, *Physics of Fluids* **16**, 4273 (2004).
7. Y. Sone, *Kinetic Theory and Fluid Dynamics* (Birkhäuser Verlag, Boston, 2002).
8. A.V. Bobylev and T. Ohwada, On the generalization of Strang's splitting scheme, *Riv. Mat. Univ. Parma* **(6) 2***, 235 (1999).
9. F. Filbet and G. Russo, High order numerical methods for the space non-homogeneous Boltzmann equation, *J. Comp. Phys.* **186**, 457 (2003).
10. T. Ohwada, Higher order approximation methods for the Boltzmann equation, *J. Comp. Phys.* **139**, 1 (1998).
11. C.K. Chu, Kinetic-theoretic description of the formation of a shock wave, *Phys. Fluids* **8**, 12 (1965).
12. M. Bisi, M. Groppi and G. Spiga, Grad's distribution functions in the kinetic equations for a chemical reaction, *Continuum Mech. Thermodyn.* **14**, 207 (2002).
13. W.G. Bickley, Formulae for numerical differentiation, *Math. Gazette* **25**, 19 (1941).
14. C. Baranger and S. Pieraccini, Numerical simulation of models for reacting polytropic gases, in *"WASCOM 2005"—13th Conference on Waves and Stability in Continuous Media* (World Sci. Publ., Hackensack, NJ, 2006), 28.
15. M. Groppi and M. Pennacchio, An IMEX finite volume scheme for reactive Euler equations arising from kinetic theory, *Commun. Math. Sci.* **1**, 449 (2003).
16. F. Conforto, A. Jannelli, R. Monaco, T. Ruggeri, On the Riemann problem for a system of balance laws modelling a reactive gas mixture, *Physica A* **373**, 67 (2007).
17. G.A. Sod, A survey of several finite difference methods for systems of non-linear hyperbolic conservation laws, *J. Comput. Phys.* **27**, 1 (1978).

ENERGY-TRANSPORT MODELS FOR SEMICONDUCTOR DEVICES AND THEIR COUPLING WITH ELECTRIC NETWORKS

G. ALÌ

Istituto per le Applicazioni del Calcolo "M. Picone", C.N.R., sez. di Napoli
via P. Castellino 111, 80131 Napoli, Italy
and INFN-Gruppo c. Cosenza, Italy,
E-mail: g.ali@iac.cnr.it

M. CARINI

Dipartimento di Matematica, Università della Calabria,
ponte P. Bucci cubo 30b, 87036 Arcavacata di Rende (Cosenza), Italy
and INFN-Gruppo c. Cosenza, Italy,
E-mail: carini@unical.it

We study the coupling of a semiconductor device, modeled by means of an energy-transport model, with an electric network, when thermal effects are taken into account. In this case, the network can be described by two concurrent topologies, one related to the transfer of electric current, and the other one related to the heat transfer. A mathematical model is proposed for this complex coupling.

Keywords: Semiconductor equations; Electric networks; Thermal network; PDAE modeling; Coupling effects.

1. Introduction

Microelectronics is an interesting applicative subject, with a pervasive impact in our society. Not only is it interesting for its important industrial applications, but also because mathematical modeling plays in it an indisputably central role, which is even more crucial when important technological changes take place.

Nowadays the increasing miniaturization of electronic devices is leading to a fast transition from microelectronics to nanoelectronics. In this new framework many second-order effects, which were previously neglected, become relevant. Thus, a renewed mathematical effort is needed for modeling such effects.

In this paper we concentrate on thermal effects. The inclusion of thermal effects in microelectronic modeling dates back to the seventies,[1] but it was considered systematically in applications only in recent times.[2,3]

Here, we present a refined model of electric networks with semiconductor devices, in the presence of temperature variations and heat transfer. This situation can be described by means of a companion thermal network.[4,5] At variance with other models, the temperature is includes as an additional unknown of the semiconductor devices, which are modeled by means of energy-transport equations. Therefore, the devices are directly coupled both to the electric and to the thermal network.

The paper is organized as follows. First we present the model for the electric network and the semiconductor devices. In the following section, we describe the thermal network. Finally, we discuss the thermal coupling of the above models.

2. Refined modeling of electric networks with devices

2.1. *The network equations*

An electrical network can be modeled by a directed graph to which electrical quantities are attached. Specifically, an electric voltage is applied to each node, and an electric current flows through each node. Moreover different electric component can be included in different branches, such as resistances, inductors, capacitors, and semiconductor devices.

If only these components are present, the network equations are given by Kirchhoff's laws and by appropriate constitutive equations. Using Modified Nodal Analysis[6,7] they can be written in the form:

$$\begin{pmatrix} \mathbf{A}_C\mathbf{C}\mathbf{A}_C^\top & 0 & 0 \\ 0 & \mathbf{L} & 0 \\ 0 & 0 & 0 \end{pmatrix} \frac{d\mathbf{x}}{dt} + \begin{pmatrix} \mathbf{A}_R\mathbf{G}\mathbf{A}_R^\top & \mathbf{A}_L & \mathbf{A}_V \\ -\mathbf{A}_L^\top & 0 & 0 \\ -\mathbf{A}_V^\top & 0 & 0 \end{pmatrix} \mathbf{x} + \begin{pmatrix} \mathbf{A}_I \boldsymbol{i} \\ 0 \\ \mathbf{v} \end{pmatrix} + \begin{pmatrix} \mathbf{A}_\lambda \boldsymbol{\lambda} \\ 0 \\ 0 \end{pmatrix} = 0, \qquad (1)$$

where the vector $\mathbf{x} = (\mathbf{u}, \boldsymbol{j}_L, \boldsymbol{j}_V)$ comprises the network unknowns, that is, the node potentials, $\mathbf{u} \in \mathbb{R}^n$, the currents through inductive branches, $\boldsymbol{j}_L \in \mathbb{R}^{n_L}$, and the currents through branches with voltage sources, $\boldsymbol{j}_V \in \mathbb{R}^{n_V}$.

The electric components included in the network are described by the matrices $\mathbf{G} \in \mathbb{R}^{n_R \times n_R}$, $\mathbf{L} \in \mathbb{R}^{n_L \times n_L}$, $\mathbf{C} \in \mathbb{R}^{n_C \times n_C}$, which are, respectively, the conductance, capacitance and inductance matrices. They are positive-definite and symmetric, and, generally they can be expressed as:[8]

$$\mathbf{G} = \frac{\partial \boldsymbol{j}_R}{\partial \mathbf{w}_R}(\mathbf{A}_R^\top \mathbf{u}), \quad \mathbf{L} = \frac{\partial \boldsymbol{\phi}_L}{\partial \mathbf{w}_L}(\boldsymbol{j}_L), \quad \mathbf{C} = \frac{\partial \mathbf{q}_C}{\partial \mathbf{w}_C}(\mathbf{A}_C^\top \mathbf{u}),$$

where the vectors $\boldsymbol{j}_R \in \mathbb{R}^{n_R}$, $\boldsymbol{\phi}_L \in \mathbb{R}^{n_L}$, $\mathbf{q}_C \in \mathbb{R}^{n_C}$, comprise, respectively, the current through each resistance, the magnetic flux through each inductor, and the charge stored by each capacitor. In (1), we are considering a linear RLC network, for which the above matrices are constant.

The topology of the electric network is described by means of incidence matrices, $\mathbf{A}_R \in \mathbb{R}^{n \times n_R}$, $\mathbf{A}_L \in \mathbb{R}^{n \times n_L}$, $\mathbf{A}_C \in \mathbb{R}^{n \times n_C}$, $\mathbf{A}_V \in \mathbb{R}^{n \times n_V}$, $\mathbf{A}_I \in \mathbb{R}^{n \times n_I}$, whose entries are 1, or 0, depending on whether a node is, or is not, connected to a branch with an electric component with the right label. The first three matrices select branches with resistances, inductors and capacitors, respectively. The next two incidence matrices, respectively, select branches with independent voltage sources, $\mathbf{v} \in \mathbb{R}^{n_V}$, and independent current sources, $\boldsymbol{i} \in \mathbb{R}^{n_I}$.

We also need an incidence matrix, $\mathbf{A}_\lambda \in \mathbb{R}^{n \times n_\lambda}$, for semiconductor devices included in the network. In this case, n_λ is the overall number of Ohmic contacts, and the vector $\boldsymbol{\lambda}$ comprises all currents through Ohmic contacts.

Finally, equation (1) must be supplemented by consistent initial data:[9]

$$\mathbf{x}(t_0) = \mathbf{x}_0. \tag{2}$$

2.2. *The semiconductor equations*

A semiconductor device can be described[10–12] by a domain $\Omega \subset \mathbb{R}$, which represents the region occupied by a doped semiconductor, where charge carriers move, under the action of external and self-induced electric potential. In general a semiconductor device reaches its equilibrium state much faster than the typical relaxation time for the electric network to which the device is connected. Therefore, in first approximation, we can limit ourselves to consider a steady-state model for the devices.

The most general energy-transport model for semiconductors, can be written in the following form:[13–15]

$$\begin{cases} -\mathrm{div}\, \mathbf{J}_n = -R, \quad -\mathrm{div}\, \mathbf{J}_p = R, \\ -\mathrm{div}\, \mathbf{J}_w = W, \\ -\mathrm{div}\, (\epsilon \nabla \phi) = N + p - n, \end{cases} \qquad \text{with } \mathbf{x} \in \Omega. \tag{3}$$

Here, $\mathbf{J}_n$, $\mathbf{J}_p$, $\mathbf{J}_w$ are the electron, hole and energy flux densities, respectively, n, p are the electron and hole number densities, respectively, and

ϕ is the electric field. Moreover, $\epsilon(\mathbf{x})$ is the dielectric constant, R is the recombination-generation term, and W is the energy relaxation term. Finally, $N(\mathbf{x})$ is the doping profile, which models the specific structure of the device.

System (3) is a nonlinear, elliptic system for the electric potential, the temperature, T, and the quasi-Fermi potentials, ϕ_n, ϕ_p, which are defined by the Maxwell-Boltzmann relations,

$$n = n_i \exp\left(\frac{\phi - \phi_n}{k_B T}\right), \quad p = n_i \exp\left(\frac{\phi_p - \phi}{k_B T}\right).$$

In fact, for the flux densities we can assign constitutive relations of the form:

$$\mathbf{J}_{\mathrm{a}} = \sum_{\mathrm{b}=n,p,w} D_{\mathrm{ab}}(\phi, \phi_n, \phi_p, T)\mathbf{Y}_b, \qquad \mathrm{a} = n, p, w. \tag{4}$$

Here, the thermodynamic forces[13,14] $\mathbf{Y}_b$ are given by

$$\mathbf{Y}_n = \nabla\left(-\frac{\phi_n}{k_B T}\right), \quad \mathbf{Y}_p = \nabla\left(-\frac{\phi_p}{k_B T}\right), \quad \mathbf{Y}_w = \nabla\left(-\frac{1}{k_B T}\right),$$

and we have $D_{\mathrm{ab}} = D_{\mathrm{ba}}$, by virtue of Onsager's principle, and $D_{np} = 0$.

The system is close once we know the recombination-generation term $R = R(\phi, \phi_n, \phi_p, T)$, and the relaxation term $W = W(\phi - \phi_n, \phi_p - \phi, T)$, which satisfies the inequality

$$W(\phi - \phi_n, \phi_p - \phi, T)(T - T_{\mathrm{env}}) \leq 0,$$

where T_{env} is the environment temperature.

Notice that in this model we identify the electron and hole temperature with a common temperature. It is also possible to consider more general models with two distinct temperatures which measure the thermal excitation of electrons and holes as independent carrier families.

For system (3), we need to assign appropriate boundary conditions. We split the boundary of the semiconductor domain in two distinct parts,

$$\partial\Omega = \Gamma_D \cup \Gamma_N, \quad \Gamma_D \cap \Gamma_N = \emptyset.$$

In the Dirichlet part, Γ_D, we assign Dirichlet conditions for the potentials and the temperature:

$$\phi_n = \phi_p = \phi - \phi_{\mathrm{bi}} = u_D, \tag{5}$$
$$T = T_D, \tag{6}$$

where $\phi_{\rm bi}$ is the built-in potential, depending on the doping profile, u_D is the applied external potential, and T_D is the external temperature at boundaries.

In the Neumann region, Γ_N, we assign insulating conditions for the fluxes,

$$\boldsymbol{\nu}\cdot\nabla\phi_n = \boldsymbol{\nu}\cdot\nabla\phi_p = \boldsymbol{\nu}\cdot\nabla\phi = 0, \tag{7}$$

$$\boldsymbol{\nu}\cdot\nabla T = 0, \tag{8}$$

where $\boldsymbol{\nu}$ is the exterior normal unitary vector to the boundary.

2.3. *Electric network-device coupling*

The electric network equations (1) and the semiconductor equations (3) are not independent. To model their electric coupling, we assume for simplicity a single device with n_λ Ohmic contacts. Thus, we assume that

$$\Gamma_D = \bigcup_{k=1}^{n_\lambda} \Gamma_k, \qquad \Gamma_i \cap \Gamma_j = \emptyset, \quad i \neq j.$$

The input coming from the network into the device, consists of the applied voltages at the Ohmic contacts, that is,

$$u_D\big|_{\Gamma_k} = u_k, \qquad k = 1, \ldots, n_\lambda, \tag{9}$$

with

$$\mathbf{u}_\lambda := \begin{pmatrix} u_1(t) \\ \vdots \\ u_{n_\lambda}(t) \end{pmatrix}, \qquad \mathbf{u}_\lambda = \mathbf{A}_\lambda^\top \mathbf{u}. \tag{10}$$

The output going from the device into the network, consists of the electrical currents flowing through the Ohmic contacts,

$$\boldsymbol{\lambda} := \begin{pmatrix} \lambda_1(t) \\ \vdots \\ \lambda_{n_\lambda}(t) \end{pmatrix}, \qquad \lambda_k \text{ current through } \Gamma_k,\ k = 1, \ldots, n_\lambda. \tag{11}$$

In terms of the device variables, λ_k is given by:

$$\lambda_k = \int_{\Gamma_k} (\mathbf{J}_n + \mathbf{J}_p)\cdot\boldsymbol{\nu}\, d\sigma. \tag{12}$$

The above conditions can be simplified a bit. First of all, it is simple to see that the currents are not independent, since

$$\int_{\Gamma_D} (\mathbf{J}_n + \mathbf{J}_p)\cdot\boldsymbol{\nu}\, d\sigma = \int_{\partial\Omega} (\mathbf{J}_n + \mathbf{J}_p)\cdot\boldsymbol{\nu}\, d\sigma = \int_{\Omega} \mathrm{div}\cdot(\mathbf{J}_n + \mathbf{J}_p)\, d\mathbf{x} = 0.$$

Then, we have

$$\sum_{k=1}^{n_\lambda} \lambda_k = 0, \tag{13}$$

and we can write $\boldsymbol{\lambda} = \boldsymbol{\mu}\mathbf{I}_\lambda$, with

$$\boldsymbol{\mu} := \begin{pmatrix} 1 & & 0 \\ & \ddots & \\ 0 & & 1 \\ -1 & \cdots & -1 \end{pmatrix}, \qquad \mathbf{I}_\lambda := \begin{pmatrix} \lambda_1 \\ \vdots \\ \lambda_{n_\lambda - 1} \end{pmatrix}. \tag{14}$$

It follows that the coupling term appearing in (1) simplifies as

$$\mathbf{A}_\lambda \boldsymbol{\lambda} = \mathbf{A}_D \mathbf{I}_\lambda, \qquad \mathbf{A}_D := \mathbf{A}_\lambda \boldsymbol{\mu}. \tag{15}$$

Also, the applied potential are redundant, since the energy-transport system senses differences of tension rather than potentials. Therefore, $\mathbf{I}_\lambda$ is invariant under the transformation

$$u_D\big|_{\Gamma_k} = u_k \;\rightarrow\; u_D\big|_{\Gamma_k} = u_k - u_{n_\lambda}, \quad k = 1, \ldots, n_\lambda,$$

and we can replace the coupling conditions (10) with

$$\mathbf{u}_\lambda := \begin{pmatrix} \mathbf{V}_\lambda(t) \\ 0 \end{pmatrix}, \quad \mathbf{V}_\lambda = \mathbf{A}_D^\top \mathbf{u}. \tag{16}$$

3. Companion thermal network

So far we described temperature variations only inside the devices contained in the electric network. To describe thermal effect also in the network and in its substrate, we need to introduce a companion thermal network[4,16] to the electric network.

The components of the thermal network can be divided in two categories:

- distributed thermal elements, that is, space-distributed elements whose electric characteristics depend on temperature, associated with a distributed temperature, $T(x,t)$;
- lumped thermal elements, that is, thermally relevant elements with small extension or large thermal conductivity, associated with a lumped temperature, $\hat{T}(t)$.

The lumped thermal elements (for instance, resistances or contact nodes) are usually elements of the electric network, while the distributed thermal element (for instance, heat-transferring structure, electric lines or devices) are not necessarily elements of the electric network.

3.1. *Distributed thermal components*

We have two kinds of distributed thermal components: semiconductor devices connected to the network, and thermal lines which connect two nodes of the network. For simplicity, we consider one-dimensional thermal lines. Each line is normalized to the space interval $[0, 1]$.

We have seen that the incidence matrix for semiconductor devices is $\mathbf{A}_\lambda \in \mathbb{R}^{n \times n_\lambda}$. We comprise all the temperature variables for each device in a single vector variable, $\mathbf{T}^{\mathrm{D}}(\mathbf{x}, t) \in \mathbb{R}^{n_D}$.

As for the second kind of distributed components, we assume to have m_d distributed thermal lines, identified by means of the incidence matrix $\mathbf{A}_d^{\mathrm{th}} = \mathbf{A}_{d,1}^{\mathrm{th}} - \mathbf{A}_{d,2}^{\mathrm{th}} \in \mathbb{R}^{n \times m_d}$. We comprise all the temperature variables for thermal lines in a single vector variable, $\mathbf{T}^{\mathrm{d}}(x, t) \in \mathbb{R}^{m_d}$.

In the i-th distributed thermal branch, the i-th temperature function $T_i^{\mathrm{d}} = T_i^{\mathrm{d}}(x, t)$ satisfies the one-dimensional heat equation:

$$\mathcal{M}_i \frac{\partial T_i^{\mathrm{d}}}{\partial t} = \frac{\partial}{\partial x}\left(\Lambda_i \frac{\partial T_i^{\mathrm{d}}}{\partial x}\right) - \mathcal{S}_i(T_i^{\mathrm{d}} - T_{\mathrm{env}}) + \mathcal{P}_i, \tag{17}$$

with $(x, t) \in [0, 1] \times [0, \infty)$, and $i = 1, \ldots, m_d$. In (17), $\mathcal{M}_i = \mathcal{M}_i(x)$ is the thermal mass, $\Lambda_i = \Lambda_i(x)$ is the heat conductivity, $\mathcal{S}_i = \mathcal{S}_i(x)$ is the transmission function, that is, the thermal radiation to the environment, and $\mathcal{P}_i$ is the electric-thermal source.

The i-th equation (17), is supplemented with initial-boundary conditions:

$$T_i^{\mathrm{d}}(x, 0) = T_{0,i}^{\mathrm{d}}(x), \quad x \in [0, 1], \tag{18}$$

$$T_i^{\mathrm{d}}(0, t) = T_{1,i}^{\mathrm{d}}(t), \quad T_i^{\mathrm{d}}(1, t) = T_{2,i}^{\mathrm{d}}(t), \quad t \in [0, \infty). \tag{19}$$

In conclusion, the distributed thermal branches (including devices) can be described by the vector function $\mathbf{T} = \left(\mathbf{T}^{\mathrm{d}}, \mathbf{T}^{\mathrm{D}}\right) \in \mathbb{R}^{m_d + n_d}$, depending on $(x, \mathbf{x}, t) \in [0, 1] \times \Omega \times [0, \infty)$.

3.2. *Lumped thermal components and thermal nodes*

We assume to have m_l branches with lumped thermal elements, identified by means of the incidence matrix $\mathbf{A}_l^{\mathrm{th}} \in \mathbb{R}^{n \times m_l}$. The temperature variable for lumped thermal branches is a vector $\hat{\mathbf{T}}^l(t) \in \mathbb{R}^{m_l}$.

In the i-th lumped thermal branch, the i-th temperature function $\hat{T}_i^l = \hat{T}_i^l(t)$ satisfies a lumped version of the heat equation (17):

$$\hat{\mathcal{M}}_i \frac{d\hat{T}_i^l}{dt} = \hat{\mathcal{F}}_i - \hat{\mathcal{S}}_i(\hat{T}_i^l - T_{\mathrm{env}}) + \hat{\mathcal{P}}_i, \tag{20}$$

where $\hat{\mathcal{M}}_i$ is the lumped thermal mass, $\hat{\Lambda}_i$ is the lumped heat conductivity, $\hat{\mathcal{S}}_i$ is the lumped transmission function, $\hat{\mathcal{P}}_i$ is the electric-thermal source, $\hat{\mathcal{F}}_i$ is the net heat flux coming from distributed thermal branches connected to the i-th lumped thermal branch.

Once we have identified the thermal branches, we need to identify the nodes of the thermal network. We assume that there are m thermal nodes, with the attached temperature variable $\hat{\mathbf{T}}(t) \in \mathbb{R}^m$.

To link the temperature $\hat{\mathbf{T}}$ of the thermal nodes to the temperature $\hat{\mathbf{T}}^l$ of the lumped thermal components, we introduce a matrix $\mathbf{M} \in \{0,1\}^{m \times m_l}$, with components m_{ij}, $i = 1, \ldots, m$, $j = 1, \ldots, m_l$. The matrix $\mathbf{M}$ maps lumped thermal branches to thermal nodes, that is, $m_{ij} = 1$ if the lumped thermal branch j corresponds to the thermal node i, and $m_{ij} = 0$ otherwise.

Next, we introduce the matrices

$$\hat{\boldsymbol{\mathcal{M}}} = \mathbf{M}\,\mathrm{diag}(\hat{\mathcal{M}}_1, \ldots, \hat{\mathcal{M}}_{m_l})\,\mathbf{M}^\top, \quad \hat{\boldsymbol{\mathcal{S}}} = \mathbf{M}\,\mathrm{diag}(\hat{\mathcal{S}}_1, \ldots, \hat{\mathcal{S}}_{m_l})\,\mathbf{M}^\top,$$

and the vectors

$$\hat{\boldsymbol{\mathcal{F}}} = \begin{pmatrix} \hat{\mathcal{F}}_1 \\ \vdots \\ \hat{\mathcal{F}}_m \end{pmatrix}, \quad \hat{\boldsymbol{\mathcal{P}}} = \begin{pmatrix} \hat{\mathcal{P}}_1 \\ \vdots \\ \hat{\mathcal{P}}_{m_l} \end{pmatrix}.$$

Then, we can write

$$\hat{\boldsymbol{\mathcal{M}}} \frac{d\hat{\mathbf{T}}}{dt} = \hat{\boldsymbol{\mathcal{F}}} - \hat{\boldsymbol{\mathcal{S}}}(\hat{\mathbf{T}} - T_{\mathrm{env}}\mathbf{I}) + \mathbf{M}\,\hat{\boldsymbol{\mathcal{P}}}. \tag{21}$$

This is the final version of the lumped heat equation.

The expressions of $\hat{\boldsymbol{\mathcal{F}}}$ and $\hat{\boldsymbol{\mathcal{P}}}$ are part of the coupling conditions.

4. Coupling conditions

So far, we have written four coupled systems of equations, namely: the (lumped) electric network equations (1), the (distributed) device equations (3), the lumped heat equations (or lumped thermal network equations) (20), and the distributed heat equations (17). We still need to clarify the mutual dependencies among these systems. We have already discussed the electric coupling between the network equations and the device equations. In a similar way, we can consider the thermal coupling between the thermal network equations and the distributed heat equations. Moreover, we have to describe how thermal properties affect electric properties and *viceversa*. Finally, we describe the coupling between the device equations and the thermal network.

4.1. *Lumped-distributed thermal coupling*

The equations (17), for the distributed temperature vector function, $\mathbf{T}^{\mathrm{d}}(x,t)$, and the equations (20) for the lumped temperature vector function, $\hat{\mathbf{T}}(t)$, are coupled, because:

- the boundary conditions for $\mathbf{T}^{\mathrm{d}}$ are given in terms of $\hat{\mathbf{T}}$;
- the heat fluxes, computed by the distributed temperature, enter the heat equation for the thermal nodes.

Let us consider the boundary conditions (19). We can write them as

$$\mathbf{T}^{\mathrm{d}}(0,t) = \mathbf{T}_1^{\mathrm{d}}(t), \quad \mathbf{T}^{\mathrm{d}}(1,t) = \mathbf{T}_2^{\mathrm{d}}(t), t \in [0,\infty). \tag{22}$$

We introduce the matrix $\mathbf{N} \in \{0,1\}^{m\times n}$, with components n_{ij}, $i = 1,\ldots,m$, $j = 1,\ldots,n$, which maps electric nodes to thermal nodes, that is, $n_{ij} = 1$ if the electric node j corresponds to the thermal node i, and $n_{ij} = 0$ otherwise. Then, with the help of $\mathbf{N}$, we can define the incidence matrix

$$\mathbf{A}_\lambda^{\mathrm{th}} = \mathbf{N}\left(\mathbf{A}_{d,1}^{\mathrm{th}}, \mathbf{A}_{d,2}^{\mathrm{th}}\right) \in \mathbb{R}^{m\times m_d}.$$

Then, the lumped-distributed thermal coupling conditions are

$$\begin{pmatrix} \mathbf{T}_1^{\mathrm{d}} \\ \mathbf{T}_2^{\mathrm{d}} \end{pmatrix} = \mathbf{A}_\lambda^{\mathrm{th}\top}\hat{\mathbf{T}}. \tag{23}$$

As for the heat fluxes, we can write $\hat{\mathcal{F}} = \hat{\mathcal{F}}^{\mathrm{d}} + \hat{\mathcal{F}}^{\mathrm{D}}$, where the two terms comprise heat fluxes from distributed thermal branches and devices, respectively. Then, we can express $\mathcal{F}^{\mathrm{d}}$ in terms of $\mathbf{T}^{\mathrm{d}}$:

$$\hat{\mathcal{F}}^{\mathrm{d}}(t) = \mathbf{A}_\lambda^{\mathrm{th}} \begin{pmatrix} \mathbf{\Lambda}(0)\dfrac{\partial}{\partial x}\mathbf{T}^{\mathrm{d}}(0,t) \\ \\ -\mathbf{\Lambda}(1)\dfrac{\partial}{\partial x}\mathbf{T}^{\mathrm{d}}(1,t) \end{pmatrix}, \tag{24}$$

where

$$\mathbf{\Lambda} = \mathrm{diag}(\Lambda_1,\ldots,\Lambda_{m_d}).$$

4.2. *Thermal-electric coupling*

First, let us consider the thermal-to-electric coupling. This coupling occurs because some components of the electric network may depend on temperature. The most important components which exhibit this behavior are semiconductor devices and resistors.[17,18] We will discuss later the case of semiconductor devices. Here, we concentrate briefly on resistors.

A resistance may have a quadratic dependence on temperature.[19] We can state this dependence by writing:

$$\boldsymbol{j}_R = \boldsymbol{j}_R(\mathbf{A}_R^\top \mathbf{u}, \mathbf{T}). \tag{25}$$

We recall that $\boldsymbol{j}_R$ defines the matrix $\mathbf{G} = \dfrac{\partial \mathbf{r}}{\partial \mathbf{w}_R}(\mathbf{A}_R^\top \mathbf{u})$, which appears in the network equations.

Next, let us consider electric-to-thermal coupling. This coupling occurs because the power dissipated by a resistance is the product of the current through the resistance times the applied voltage.[20]

The power dissipated by all resistances is:

$$\boldsymbol{\mathcal{P}}_R = \operatorname{diag}(\boldsymbol{j}_R)\mathbf{A}_R^\top \mathbf{u} \in \mathbb{R}^{n_R}. \tag{26}$$

We need to relate resistive branches with distributed and lumped thermal branches. Thus, we introduce a matrix $\mathbf{K} \in \{0,1\}^{m_d \times n_R}$, with components k_{ij}, and a matrix $\hat{\mathbf{K}} \in \{0,1\}^{m_l \times n_R}$, with components $\hat{k}_{ij}$, defined as follows:

- **K** maps resistor branches to distributed thermal branches, that is, $k_{ij} = 1$ if the resistor branch j corresponds to the thermal distributed branch i, and $k_{ij} = 0$ otherwise.
- $\hat{\mathbf{K}}$ maps resistor branches to lumped thermal branches, that is, $\hat{k}_{ij} = 1$ if the resistor branch j corresponds to the thermal distributed branch i, and $\hat{k}_{ij} = 0$ otherwise.

Then, we can write

$$\boldsymbol{\mathcal{P}} = \mathbf{K}\boldsymbol{\mathcal{P}}_R(\mathbf{u}), \qquad \hat{\boldsymbol{\mathcal{P}}} = \hat{\mathbf{K}}\boldsymbol{\mathcal{P}}_R(\mathbf{u}). \tag{27}$$

Note that the first of the last expression is a lumped quantity which appears in a distributed equation. It is possible to distribute this expression by using appropriate functions living on the branches.

4.3. *Thermal network-device coupling*

The coupling between the thermal network and the device takes place through the Dirichlet boundary $\Gamma_D = \bigcup_{k=1}^{n_D} \Gamma_k$.

The input coming from the thermal network to the device consists of the external temperatures at the Dirichlet boundaries: $T = T_D$ in .

$$T_D\big|_{\Gamma_k} = T_k, \qquad k = 1, \ldots, n_\lambda, \tag{28}$$

with

$$\mathbf{T}_\lambda := \begin{pmatrix} T_1(t) \\ \vdots \\ T_{n_\lambda}(t) \end{pmatrix}, \quad \mathbf{T}_\lambda = \mathbf{A}_\lambda^\top \mathbf{N}^\top \hat{\mathbf{T}}. \tag{29}$$

The output going from the device into the thermal network, consists of the heat fluxes through the Dirichlet boundaries,

$$\begin{pmatrix} \hat{\mathcal{F}}_1^{\mathrm{D}}(t) \\ \vdots \\ \hat{\mathcal{F}}_{n_D}^{\mathrm{D}}(t) \end{pmatrix}, \quad \hat{\mathcal{F}}_k^{\mathrm{D}} \text{ heat flux through } \Gamma_k,\ k = 1, \ldots, n_D. \tag{30}$$

In terms of the device variables, we can write

$$\mathbf{J}_w = \mathbf{J}_{\mathrm{th}} - \phi(\mathbf{J}_n + \mathbf{J}_p), \tag{31}$$

where $\mathbf{J}_{\mathrm{th}}$ is the thermal flux density. Then, we can write

$$\hat{\mathcal{F}}_k^{\mathrm{D}} = \int_{\Gamma_k} \mathbf{J}_{\mathrm{th}} \cdot \boldsymbol{\nu}\, d\sigma. \tag{32}$$

5. Conclusions

In this paper we have presented a refined modeling of networks with devices, including thermal effects, which can be described by means of a companion thermal network.

As for analysis of the resulting model, we note that after neglecting alla thermal effects, the model reduces to a well known PDAE model for an electric network with devices.[9] On the other end, if we replace all devices included in the electric network with equivalent circuits, the model reduces to a known PDAE model for thermal effects in an electric network.[4] For both models existence and well-posedness results are known. Thus, in principle, it should be possible to combine the analytical strategies devised for tackling these simpler models in order to study the well-posedness of the combined model. This analysis will be developed in a subsequent paper.

References

1. K. Fukahori, *Computer simulation of monolithic circuit performance in the presence of electro-thermal interactions*, PhD thesis, University of California, Berkeley (CA, USA, 1977).
2. J. Bielefeld, G. Pelz, H. B. Abel, G. Zimmer, *IEEE Trans. Elec. Dev.* **42**, 1968 (1995).

3. Ch. Deml, P. Türkes, *IEEE Transactions on Industry Applications* **35**, 657 (1999).
4. A. Bartel, *Partial Differential-Algebraic Models in Chip Design - Thermal and Semiconductor Problems* (VDI, 2004).
5. A. Bartel, M. Günther, *Math. Comp. Modell. Dyn. Syst* **9**, 25 (2003).
6. C. W. Ho, A. E. Ruehli, P. A. Brennan, *IEEE Trans. Circuits and Systems* **22**, 505 (1975).
7. W. J. McCalla, *Fundamentals of Computer Aided Circuit Simulation* (Kluwer Acad. Publ. Group, Dordrecht, 1988).
8. M. Günther, *Partielle differential-algebraische Systeme in der numerischen Zeitbereichsanalyse elektrischer Schaltungen* (VDI, 2001).
9. G. Alì, A. Bartel, M. Günther, C. Tischendorf, *Math. Model. Meth. Appl. Sc.* **13**, 1261 (2003).
10. S. Selberherr, *Analysis and Simulation of Semiconductor Devices* (Springer, 1984).
11. P. A. Markowich, C. A. Ringhofer and C. Schmeiser, *Semiconductor Equations* (Springer, 1990).
12. J. W. Jerome, *Analysis of charge transport. A mathematical study of semiconductor devices* (Springer, 1995).
13. S. R. de Groot, P. Mazur, *Nonequilibrium thermodynamics*, reprint of the original book published in 1962 by NorthHolland (Dover, NewYork, 1984).
14. H. Kreuzer, *Nonequilibrium thermodynamics and its statistical foundations* (Clarendon Press, Oxford, 1981).
15. P. Degond, S. Génieys, A. Jüngel, *Math. Meth. Appl. Sci.* **21**, 1399 (1998).
16. A. Bartel, U. Feldmann, *Modeling and Simulation for Thermal-Electric Coupling in an SOI-Circuit*, in *Scientific Computing in Electrical Engineering*, eds. A. M. Anile, G. Alì, G. Mascali, Series: Mathematics in Industry, Subseries: The European Consortium for Mathematics in Industry , Vol. 9 (Springer, 2006), pp. 27–32.
17. G. Massobrio, P. Antognetti, *Semiconductor Device Modeling with SPICE*, 2nd ed. (McGraw-Hill, New York, 1993).
18. L. T. Su, D. A. Antoniadis, N. D. Arora, B. S. Doyle, D. B. Krakauer, *IEEE Electron Device Letters* **15** 374 (1994).
19. A. Bartel, M. Günther, M. Schulz, *Modeling and Discretization of a Thermal-Electric Test Circuit*, in *Modeling, Simulation and Optimization of Integrated Circuits*, eds. K. Antreich, R. Bulirsch, A. Gilg, P. Rentrop, Int. Series Num. Math. Vol. 146 (Birkhäuser, Basel 2003), pp. 187-201.
20. A. Chryssafis, W. Love, *Solid-State Electron* **22**, 249 (1979).

DIMENSION REDUCTION FOR DISCRETE SYSTEMS

R. ALICANDRO

DAEIMI, Università di Cassino,
via Di Biasio 43, 03043 Cassino (FR), Italy
E-mail: alicandr@unicas.it

A. BRAIDES

Dipartimento di Matematica, Università di Roma 'Tor Vergata'
via della Ricerca Scientifica, 00133 Roma, Italy
E-mail: braides@mat.uniroma2.it

M. CICALESE

Dipartimento di Matematica e Applicazioni 'R. Caccioppoli', Università 'Federico II'
via Cintia, 80126 Napoli, Italy
E-mail: cicalese@unina.it

We review some results concerning the variational analysis of the dimension reduction problem for discrete thin films drawing a parallel with the analogous problem in the continuum and highlighting some new features.

1. Introduction

In this paper we review some of the results obtained in Ref. [1] concerning the description of the overall behaviour of variational pair-interaction lattice systems defined on 'thin' domains of $\mathbb{Z}^N$; *i.e.* on domains consisting on a finite number M of mutually interacting copies of a portion of a $(N-1)$-dimensional discrete lattice. After drawing a parallel with the analog theories for 'continuous' thin films about compactness and homogenization results, we show that new phenomena arise due to the different nature of the microscopic interactions in particular due to boundary layer effects that render the effective behaviour depend in a non-trivial way on the number M of layers.

In a more precise notation, we consider energies depending on functions parameterized on a portion of the lattice $\mathbb{Z}^N$ consisting of a 'cylindrical' set

$$\mathcal{Z}(\omega, M) = (\omega \cap \mathbb{Z}^{N-1}) \times \{1, \dots M\}$$

of the form

$$\sum_{\alpha,\beta\in\mathcal{Z}(\omega,M)} \varphi_{\alpha,\beta}(u_\alpha - u_\beta), \qquad u : \mathcal{Z}(\omega, M) \to \mathbb{R}^d$$

when the size of ω is large. Upon introducing a small positive parameter ε and scaling ω to a fixed size, obtaining the discrete thin domain

$$\mathcal{Z}_\varepsilon(\omega, M) = (\omega \cap \varepsilon\mathbb{Z}^{N-1}) \times \{\varepsilon, 2\varepsilon, \ldots M\varepsilon\}, \tag{1}$$

this problem can be reformulated as the description of the Γ-limit of energies of the form

$$F_\varepsilon(u) = \sum_{\alpha,\beta\in\mathcal{Z}_\varepsilon(\omega,M)} \varepsilon^{N-1} f^\varepsilon_{\alpha,\beta}\Big(\frac{u_\alpha - u_\beta}{\varepsilon}\Big), \qquad u : \mathcal{Z}_\varepsilon(\omega, M) \to \mathbb{R}^d \tag{2}$$

(at this point a more general dependence of the energy densities on ε is introduced for the sake of generality).

The energies above may be viewed as the discrete analog of thin-film energies of the form

$$\mathcal{F}_\varepsilon(u) = \frac{1}{\varepsilon}\int_{\omega\times(0,M\varepsilon)} W_\varepsilon(x, Du(x))\, dx \qquad u \in W^{1,p}(\omega \times (0, M\varepsilon); \mathbb{R}^d). \tag{3}$$

For such functionals a number of results in the framework of variational convergence have been obtained since the pioneering work of Le Dret and Raoult [10]; in particular if W_ε uniformly satisfy some p-growth conditions, a general compactness result by Braides, Fonseca and Francfort [7] shows that, upon subsequences, we can always suitably define a Γ-limit energy in the lower-dimensional set ω of the form

$$\mathcal{F}(u) = \int_\omega W(\hat{x}, Du(\hat{x}))\, d\hat{x}, \qquad u \in W^{1,p}(\omega; \mathbb{R}^d) \tag{4}$$

(here, $\hat{x} = (x_1, \ldots, x_{n-1})$. Particular cases are when $W_\varepsilon = W_0(Du)$ is independent of ε and x, in which the limit W is simply given (see Ref. [10]) by

$$W(A) = M\, Q\overline{W}_0(A), \qquad \overline{W}_0(A) = \inf_z W_0(A\,|\,z), \tag{5}$$

where Q stands for the operation of quasiconvexification (note the trivial dependence on M), and when $W_\varepsilon(x, A) = W(x/\varepsilon, A)$, in which suitable homogenization formulas hold (see Ref. [7]).

On the other hand, a general compactness theory for discrete systems with energy densities of polynomial growth defined on 'thick' domains by Alicandro and Cicalese [2] is also available. In particular, that theory can be applied in the case above when $M = 1$ and the energies F_ε are simply interpreted as defined on $\omega \cap \varepsilon\mathbb{Z}^{N-1}$. In this case again an energy of the same form as $\mathcal{F}$ above can be proven to be the Γ-limit in a suitable sense of such F_ε (discrete functions must be identified with piecewise-constant interpolations). Appropriate homogenization formulas apply as well if the discrete interactions possess some periodicity.

In the general case $M > 1$ we note that the energies F_ε can also be seen as defined on M copies of $\omega \cap \varepsilon\mathbb{Z}^{N-1}$ interacting through pair-interactions corresponding to $\beta - \alpha$ with a non-zero component in the N-th variable. We first show that functions on which the limit energy is finite, that are thus defined on M copies of ω, are actually equal on each of these copies, so that the limit energy can be defined on the only set ω. More precise homogenization formulas are given in the case when the energy densities are periodic; *i.e.* $f^\varepsilon_{\alpha,\beta} = f_{\alpha/\varepsilon,\beta/\varepsilon}$ and there exists an integer k such that $f_{i,j} = f_{i',j'}$ if $i - i' = j - j' \in k\mathbb{Z}^{N-1}$. As in Ref. [7] these formulas are defined through minimum problems on rectangles with boundary conditions on the lateral boundaries only. In the discrete case it must be noted that these formulas are necessary also in the 'trivial' case when $f_{i,j} = f_{j-i}$; *i.e.* the energy densities depend only on the distance of α and β in the unscaled reference lattice $\mathbb{Z}^N$, as already observed for 'thick' domains, except when only nearest-neighbour interactions are taken into account (see Ref. [5]). In the case of thin domains an additional scale effect must be taken into account, since long-range interactions (next-to nearest interactions and further) produce different effects close to the upper and lower free boundaries than in the interior (see Figure 1). For all the proofs of the results presented here we refer to Ref. [1].

Finally, we mention that our work has connections with a number of papers where variational methods for thin structures are dealt with but in a different perspectives, most notably that of Friesecke and James [9], some works by Blanc *et al.* [3] and more recently by Schmidt [11].

2. Notation and preliminaries

Given N, $d \in \mathbb{N}$, we denote by $\{e_1, e_2, \ldots, e_N\}$ the standard basis in $\mathbb{R}^N$, by $|\cdot|$ the usual euclidean norm and by $\mathcal{M}^{d\times N}$ the space of $d \times N$ matrices. If $B \subset \mathbb{R}^N$ is a Borel set, we will denote by $|B|$ its Lebesgue measure. We use standard notation for L^p and Sobolev spaces.

We also introduce a useful notation for difference quotient along any direction. Fix $\xi \in \mathbb{R}^N$; for $\varepsilon > 0$ and for every $u : \mathbb{R}^N \to \mathbb{R}^d$ we define

$$D^\xi_\varepsilon u(x) := \frac{u(x + \varepsilon\xi) - u(x)}{\varepsilon|\xi|}.$$

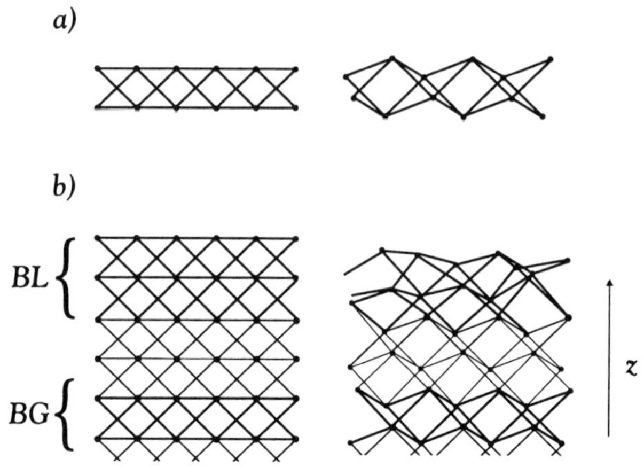

Fig. 1. Different effects in a simple model for thin films with nearest and next to nearest interactions: a) ground state geometry for a two layers thin film; b) bulk geometry (BG) and boundary layer effect (BL) for a multi-layer thin film subject to a vertical deformation gradient z.

Moreover we set

$$\mathcal{A}_\varepsilon(\Omega) := \{u : \mathbb{R}^N \to \mathbb{R}^d \ : \ u \text{ constant on } \alpha + [0,\varepsilon)^N \text{ for any } \alpha \in \varepsilon\mathbb{Z}^N \cap \Omega\}.$$

3. Compactness theorem

In what follows, for the sake of simplicity, ω will denote a bounded convex open set of $\mathbb{R}^{N-1}$ with Lipschitz boundary, the general case of a non-convex ω being dealt with similarly (see Ref. [2]). Given $M \in \mathbb{N}$, $A \subset \mathbb{R}^{N-1}$ and $\varepsilon > 0$, we set

$$A_\varepsilon := A \times [0, \varepsilon(M-1)].$$

In the following we will identify the set of functions $\mathcal{A}_\varepsilon(\Omega_\varepsilon)$ with $[\mathcal{A}_\varepsilon(\omega)]^M$ through the bijection $u \mapsto (u^0, u^1, \ldots, u^{M-1})$ where, for any $i \in \{0, 1, \ldots, M-1\}$, $u^i \in \mathcal{A}_\varepsilon(\omega)$ is defined by

$$u^i(\beta) = u(\beta, \varepsilon i) \quad \forall \beta \in \varepsilon\mathbb{Z}^{N-1} \cap \omega.$$

We consider the family of functionals $F_\varepsilon : [L^p(\omega; \mathbb{R}^d)]^M \to [0, +\infty]$ defined as

$$F_\varepsilon(u) = \begin{cases} \displaystyle\sum_{\xi \in \mathbb{Z}^N} \sum_{\alpha, \alpha+\varepsilon\xi \in \Omega_\varepsilon \cap \varepsilon\mathbb{Z}^N} \varepsilon^{N-1} f_\varepsilon^\xi \left(\alpha, D_\varepsilon^\xi u(\alpha)\right) & \text{if } u \in \mathcal{A}_\varepsilon(\Omega_\varepsilon) \\ +\infty & \text{otherwise,} \end{cases} \tag{1}$$

and $f_\varepsilon^\xi : (\varepsilon\mathbb{Z}^N \cap \Omega_\varepsilon) \times \mathbb{R}^d \to [0, +\infty)$ is a given function. Let $p > 1$, on f_ε^ξ we make the following assumptions:

$$f_\varepsilon^{e_i}(\alpha, z) \geq c_1(|z|^p - 1) \quad \forall(\alpha, z) \in (\varepsilon\mathbb{Z}^N \cap \Omega) \times \mathbb{R}^d,\ i \in \{1, \ldots, N\} \tag{2}$$

$$f_\varepsilon^\xi(\alpha, z) \leq C_\varepsilon^\xi(|z|^p + 1) \quad \forall(\alpha, z) \in (\varepsilon\mathbb{Z}^N \cap \Omega) \times \mathbb{R}^d,\ \xi \in \mathbb{Z}^N, \tag{3}$$

where $c_1 > 0$ and $\{C_\varepsilon^\xi\}_{\varepsilon,\xi}$ satisfies:

$$\limsup_{\varepsilon\to 0+} \sum_{\xi\in\mathbb{Z}^N} C_\varepsilon^\xi < +\infty; \tag{H1}$$

$$\forall\delta > 0 \quad \exists\, M_\delta > 0: \quad \limsup_{\varepsilon\to 0+} \sum_{|\xi|>M_\delta} C_\varepsilon^\xi < \delta. \tag{H2}$$

In the following, referring to Ref. [4] for all the definitions and properties of Γ-convergence, we will denote by $F'(u)$ and $F''(u)$ the $\Gamma\text{-}\liminf_{\varepsilon\to 0+} F_\varepsilon(u)$ and the $\Gamma\text{-}\limsup_{\varepsilon\to 0+} F_\varepsilon(u)$ with respect to the $[L^p(\omega; \mathbb{R}^d)]^M$-topology. Moreover if $u \in [L^p(\omega; \mathbb{R}^d)]^M$ is such that $u^0 = u^1 = \cdots = u^{M-1}$ we will simply write $u \in L^p(\omega; \mathbb{R}^d)$. The main result of this section is stated in the following theorem.

Theorem 3.1. ***(Compactness)*** *Let $\{f_\varepsilon^\xi\}_{\varepsilon,\xi}$ satisfy (2), (3) and let (H1)-(H2) hold. Then:*

(i) if $u \in [L^p(\omega; \mathbb{R}^d)]^M$ is such that $F'(u) < +\infty$, then

$$u^0 = u^1 = \cdots = u^{M-1} \in W^{1,p}(\omega; \mathbb{R}^d)$$

and

$$F'(u) \geq c\, \left(\|\nabla u\|^p_{L^p(\omega;\mathbb{R}^{d\times N-1})} - |\omega|\right) \tag{4}$$

for some positive constant c independent on u.

(ii) for every $u \in W^{1,p}(\omega; \mathbb{R}^d)$ there holds

$$F''(u) \leq C\, \left(\|\nabla u\|^p_{L^p(\omega;\mathbb{R}^{d\times N-1})} + |\omega|\right) \tag{5}$$

for some positive constant C independent on u.

(iii) for every sequence (ε_j) of positive real numbers converging to 0, there exists a subsequence (ε_{j_k}) and a Carathéodory function quasi-convex in the second variable $f : \omega \times \mathbb{R}^{d\times N-1} \to [0, +\infty)$ satisfying

$$c(|S|^p - 1) \leq f(x, S) \leq C(|S|^p + 1),$$

with $0 < c < C$, *such that* $(F_{\varepsilon_{j_k}}(\cdot))$ Γ*-converges with respect to the* $[L^p(\omega;\mathbb{R}^d)]^M$*-topology to the functional* $F : [L^p(\omega;\mathbb{R}^d)]^M \to [0,+\infty]$ *defined as*

$$F(u) = \begin{cases} \displaystyle\int_\omega f(x,\nabla u)\,dx & \text{if } u \in W^{1,p}(\omega;\mathbb{R}^d) \\ +\infty & \text{otherwise.} \end{cases} \tag{6}$$

Remark 3.1. (Convergence of minimum problems) In the case our functionals are subject to Dirichlet boundary conditions, a compactness theorem analogous to the previous one still holds and, by the properties of Γ-convergence, a convergence of minimum problems can be derived (see section 2.3 in Ref. [1]).

4. Homogenization

In this section we will give a homogenization result by presenting a Γ-convergence theorem for the energies F_ε in the case that the functions f_ε^ξ are obtained by rescaling by ε functions f^ξ periodic in the space variable.

Let $\mathbf{k} = (k_1, \ldots, k_{N-1}) \in \mathbf{N}^{N-1}$ be given and set

$$\mathcal{R}_{\mathbf{k}} := (0,k_1) \times \cdots \times (0,k_{N-1}).$$

For any $\xi \in \mathbf{Z}^N$, let $f^\xi : \mathbf{Z}^N \times \mathbf{R}^d \to [0,+\infty)$ be such that $f^\xi((\cdot,\alpha_N),z)$ is $\mathcal{R}_{\mathbf{k}}$-periodic for any $\alpha_N \in \mathbf{Z}$ and $z \in \mathbf{R}^d$. Then we consider f_ε^ξ of the following form

$$f_\varepsilon^\xi(\alpha,z) := f^\xi\left(\frac{\alpha}{\varepsilon}, z\right). \tag{1}$$

In this case, the growth conditions (2) and (3) and hypotheses (H1) and (H2) can be rewritten as follows:

$$f^{e_i}(\alpha,z) \geq c_1(|z|^p - 1), \quad \forall i \in \{1,\ldots,N\}, \tag{2}$$

$$f^\xi(\alpha,z) \leq C^\xi(|z|^p + 1), \tag{3}$$

where

$$\sum_{\xi\in\mathbf{Z}^N} C^\xi < +\infty. \tag{H3}$$

For every $r > 0$ we set $Q_r := (0,r)^{N-1}$ and, for $\varepsilon > 0$ and $S \in \mathcal{M}^{d\times N-1}$, we define

$$\mathcal{A}_{\varepsilon,S}(Q_r) := \left\{u \in \mathcal{A}_\varepsilon(\mathbf{R}^N) : u(\alpha) = S(\alpha_\pi) \text{ if } \left(\alpha_\pi + [-\varepsilon,\varepsilon]^{N-1}\right) \cap A^c \neq \emptyset\right\}.$$

The following homogenization theorem holds true.

Theorem 4.1. ***(Homogenization)*** *Let $\{f_\varepsilon^\xi\}_{\varepsilon,\xi}$ satisfy (1)-(3) and let (H3) hold. Then, (F_ε) Γ-converges with respect to the $[L^p(\omega;\mathbf{R}^d)]^M$-topology to the functional $F:[L^p(\omega;\mathbf{R}^d)]^M \to [0,+\infty]$ defined as*

$$F(u)=\begin{cases}\displaystyle\int_\Omega f_{hom}(\nabla u)\,dx & \text{if } u\in W^{1,p}(\omega;\mathbf{R}^d)\\ +\infty & \text{otherwise,}\end{cases} \tag{4}$$

where $f_{hom}:\mathcal{M}^{d\times N-1}\to[0,+\infty)$ is given by the following homogenization formula

$$f_{hom}(S):=\lim_{h\to+\infty}\frac{1}{h^{N-1}}\inf\Big\{\sum_{\xi\in\mathbf{Z}^N}\ \sum_{\substack{\beta,\ \beta+\xi\in\\ (Q_h)_1\cap\mathbf{Z}^N}} f^\xi(\beta,D_1^\xi v(\beta)),\ v\in\mathcal{A}_{1,S}(Q_h)\Big\} \tag{5}$$

Remark 4.1. (More explicit formula for the homogenized density) The formula defining f_{hom} can be simplified in some particular cases (see section 3.1 and 3.2 in Ref. [1]).

case 1. In the case $f^\xi(\alpha,\cdot)$ is convex for all $\alpha\in\mathbf{Z}^N$ and $\xi\in\mathbf{Z}^N$, the formula for f_{hom} reduces to a cell problem formula.

case 2. When f^ξ does not depend on α and $f^\xi(z)\not\equiv 0$ if and only if $\xi_N=0$ or $\xi_\pi=0$, that is only planar and vertical interactions are taken into account, then the formula for f_{hom} can be split in two independent homogenization formulas as follows

$$f_{hom}(S)=Mf^\pi_{hom}(S)$$
$$+\inf\{\sum_{k=1}^{M-1}\sum_{j=0}^{M-1-k} f^{ke_N}(j,\frac{u(j+k)-u(j)}{k}):\ u:\{0,\dots,M-1\}\to\mathbf{R}^d\},$$

where f^π_{hom} is given by a homogenization formula accounting only for planar interactions.

5. Discrete and continuous models for dimension reduction: layer dependence and asymptotic formulas

In this section we are interested in analyzing the asymptotic behavior of the function f_{hom} when the number of layers M tends to infinity and to draw a parallel with the existing theories on the continuum. To this aim, we will assume that the energy densities f_ε^ξ satisfy the hypotheses of Theorem 4.1 and, given $k_N \in \mathbf{N}$, the additional condition

$$f^\xi((\alpha_\pi,\cdot),z) \quad \text{is } (0,k_N)\text{-periodic } \forall z \in \mathbf{R}^d, \quad \forall \alpha_\pi \in \omega \cap \mathbf{Z}^N. \qquad (H4)$$

Under these assumptions, in Theorem 4.1 in Ref. [2] it has been proved that, given a convex bounded open set $\Omega \subset \mathbf{R}^N$, the family of functionals $F_\varepsilon^b : L^p(\Omega;\mathbf{R}^d) \to [0,+\infty]$ defined as

$$F_\varepsilon^b(u) = \begin{cases} \sum\limits_{\xi\in\mathbf{Z}^N} \sum\limits_{\alpha,\alpha+\varepsilon\xi\in\Omega} \varepsilon^N f^\xi\left(\frac{\alpha}{\varepsilon}, D_\varepsilon^\xi u(\alpha)\right) & \text{if } u \in \mathcal{A}_\varepsilon(\Omega) \\ +\infty & \text{otherwise,} \end{cases} \qquad (1)$$

Γ-converges with respect to the $L^p(\Omega;\mathbf{R}^d)$-topology to the functional $F^b : L^p(\Omega;\mathbf{R}^d) \to [0,+\infty]$ defined as

$$F^b(u) = \begin{cases} \displaystyle\int_\Omega f_{hom}^b(\nabla u)\,dx & \text{if } u \in W^{1,p}(\Omega;\mathbf{R}^d) \\ +\infty & \text{otherwise,} \end{cases} \qquad (2)$$

where f_{hom}^b is given by a homogenization formula analogous to (5).

Let us denote F_ε by F_ε^M and f_{hom} by f_{hom}^M thus highlighting the dependence on M in all the formulas we have obtained in the previous sections for the energy densities.

Given $\delta > 0$, suppose that in the definition (1) we replace M by $M_\varepsilon = \frac{\delta}{\varepsilon}$. Then $\frac{1}{M_\varepsilon}F_\varepsilon^{M_\varepsilon}(u) = F_\varepsilon^b(u)$ where F_ε^b is given by (1) with $\Omega = \omega \times (0,\delta)$. The homogenization result for thick domains implies that, for all $u \in W^{1,p}(\omega \times (0,\delta);\mathbf{R}^d)$,

$$\Gamma\text{-}\lim_{\varepsilon\to 0} \frac{1}{M_\varepsilon} F_\varepsilon^{M_\varepsilon}(u) = \frac{1}{\delta} \int\limits_{\omega\times(0,\delta)} f_{hom}^b(\nabla u(x))\,dx =: F_\delta(u).$$

On the other hand, in the pioneering paper by Le Dret and Raoult [10] it has been proved that F_δ Γ-converges, as δ tends to zero, to the functional

$$\int\limits_\omega Q\overline{f}_{hom}(\nabla u(x))\,dx \quad u \in W^{1,p}(\omega;\mathbf{R}^d),$$

where

$$\overline{f}_{hom}(S) := \inf\{f_{hom}^b(S|z),\ z \in \mathbf{R}^d\} \qquad (3)$$

and $Q\overline{f}_{hom}$ denotes the quasi-convex envelope of $\overline{f}_{hom}$. This considerations lead us to ask ourselves if

$$\lim_{M\to+\infty} \frac{1}{M} f_{hom}^M(S) = Q\overline{f}_{hom}(S). \qquad (4)$$

The following Theorem provides a partial answer to this problem (see section 4 in Ref. [1] for a more detailed analysis).

Proposition 5.1. *Under the hypotheses of Theorem 4.1 and (H4) , for all* $S \in \mathbf{R}^{d\times(N-1)}$ *there holds*

$$\lim_{M\to+\infty} \frac{1}{M} f^M_{hom}(S) \leq Q\overline{f}_{hom}(S),$$

where $\overline{f}_{hom}$ *is given by (3). If* $f^{\xi}(\cdot)$ *satisfies the hypotheses of cases* 1 *or* 2 *in Remark (4.1), then the equality holds in the previous formula.*

6. Remarks on the Cauchy-Born rule

In this section we find an example of 2D-1D family of dimension reduction problems (indexed by M) leading to a non trivial dependence of the Cauchy-Born rule on the number of layers.

Definition 6.1. Let $N = 2$ and let F_ε and f_{hom} be as in Theorem 4.1 with $\omega = (0,1)$. We say that $z \in \mathbf{R}^d$ is a *strong Cauchy-Born (sCB) state* or a *weak Cauchy-Born (wCB) state* if respectively, there exists

$$u_\varepsilon(t) \to zt \text{ such that } F_\varepsilon(u_\varepsilon) \to f_{hom}(z),$$

and such that

(i) $\#\{\alpha \in (0,1) \cap \varepsilon\mathbf{Z} :\ u_\varepsilon(\alpha+\varepsilon) - u_\varepsilon(\alpha) \neq \varepsilon z\} = o(\frac{1}{\varepsilon})$,
(ii) $\#\{\alpha \in (0,1) \cap \varepsilon\mathbf{Z} :\ u_\varepsilon(\alpha+2\varepsilon) - u_\varepsilon(\alpha) \neq 2\varepsilon z\} = o(\frac{1}{\varepsilon})$.

In the previous two cases we simply say that the Cauchy-Born (CB) rule holds at z or, shortly, that z is a CB state.

From a mechanical point of view, in this two different cases, to a macroscopic strain z it corresponds a micro-structure of

(i) uniformly displaced material points,
(ii) periodically displaced material points on the microscopic scale 2ε.

Example 6.1. Given $u \in \mathcal{A}_\varepsilon((0,1) \times [0, \varepsilon(M-1)])$ let us define

$$\begin{aligned} F^M_\varepsilon(u) = & \sum_{\alpha,\alpha+\varepsilon e_1 \in (0,1)_\varepsilon} \varepsilon f_1(D^{e_1}_\varepsilon u(\alpha)) + \sum_{\alpha,\alpha+\varepsilon(e_1+e_2) \in (0,1)_\varepsilon} \varepsilon f_2(D^{e_1+e_2}_\varepsilon u(\alpha)) \\ & + \sum_{\alpha,\alpha+\varepsilon(e_1-e_2) \in (0,1)_\varepsilon} \varepsilon f_3(D^{e_1-e_2}_\varepsilon u(\alpha)), \end{aligned} \tag{1}$$

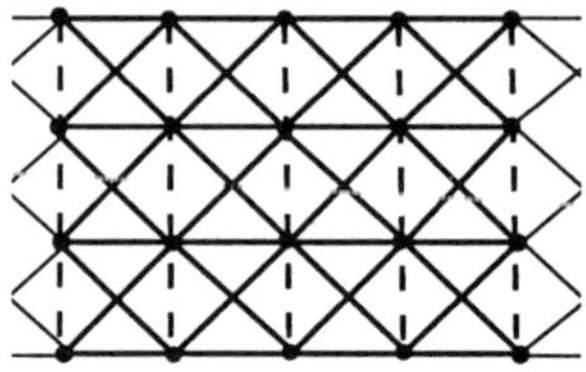

Fig. 2. Reference configurations of the discrete system (the dashed lines represent negligible interactions).

with $f_i : \mathbf{R}^d \to [0, +\infty)$ satisfying

$$c(|z|^p - 1) \le f_i(z) \le C(|z|^p + 1), \quad i = 1, 2, \qquad f_3(z) \le C(|z|^p + 1).$$

Note that for this energies the interactions along direction e_2 are not taken into account (see Figure 2 a)). Thus, hypotheses of section 4 are not satisfied. Anyway, thanks to the additional coerciveness condition on f_2, it is easy to see that again finite difference quotients along e_2 can be controlled and so the conclusion of Theorem (4.1) still holds true.

Proposition 6.1. *Let F_ε^M be defined by (1) and let f_{hom}^M be the density energy of its Γ-limit. Then it holds*

$$2\left((M-2)(\widetilde{f}\,)^{**} + (\widehat{f}\,)^{**}\right)(z) \le f_{hom}^M(z) \le 2\left((M-2)\widetilde{f} + \widehat{f}\,\right)^{**}(z) \qquad (2)$$

for any $z \in \mathbf{R}^d$, where $\widetilde{f} : \mathbf{R}^d \to [0, +\infty)$ and $\widehat{f} : \mathbf{R}^d \to [0, +\infty)$ are defined by the following formulas

$$\begin{aligned} \widetilde{f}(z) &= \frac{1}{4} \inf \{f_1(z_1) + f_1(z_2) : z_1 + z_2 = 2z\} \\ &\quad + \frac{1}{2} \inf \left\{ f_2\Big(\frac{z_1}{\sqrt{2}}\Big) + f_3\Big(\frac{z_2}{\sqrt{2}}\Big) : z_1 + z_2 = 2z \right\}, \\ \widehat{f}(z) &= \frac{1}{2} \inf \{f_1(z_1) + f_1(z_2) : z_1 + z_2 = 2z\} \\ &\quad + \frac{1}{2} \inf \left\{ f_2\Big(\frac{z_1}{\sqrt{2}}\Big) + f_3\Big(\frac{z_2}{\sqrt{2}}\Big) : z_1 + z_2 = 2z \right\}. \end{aligned} \qquad (3)$$

Remark 6.1. Note that, by (2) we deduce that

$$f_{hom}^2(z) = 2(\widehat{f}\,)^{**}(z) \text{ and } \lim_{M \to +\infty} \frac{f_{hom}^M(z)}{M} = 2(\widetilde{f}\,)^{**}(z),$$

with $\widehat{f}$ and $\widetilde{f}$ as in (3) and where f^{**} denotes the l.s.c. and convex envelope of f.

Set

$$C(M) := \{z \in \mathbf{R}^d : \ z \text{ is not a CB state for } F_\varepsilon^M\},$$

in Ref. [1] it has been proved, thanks to the non asymptotic formula for f_{hom}^M highlighted in the previous remark, that

$$C(2) = \{z : \ (\widehat{f}\,)^{**}(z) < \widehat{f}(z)\}$$

and

$$\lim_{M\to+\infty} C(M) = \{z : \ (\widetilde{f}\,)^{**}(z) < \widetilde{f}(z)\}.$$

With the particular choice

$$f_1(z) = (z+1)^2 \wedge (z-1)^2 \qquad f_2(z) = f_3(z) = z^2,$$

an explicit computation gives that

$$C(M) \subseteq C(2) = \Big(-\frac{5}{6}, -\frac{1}{6}\Big) \cup \Big(\frac{1}{6}, \frac{5}{6}\Big)$$
$$\lim_{M\to\infty} C(M) = \Big(-\frac{3}{4}, -\frac{1}{4}\Big) \cup \Big(\frac{1}{4}, \frac{3}{4}\Big) \subset C(2),$$

thus highlighting that the set $C(M)$ of all the deformations where the CB rule fails asymptotically decreases with the number of layers.

References

1. R. Alicandro, A. Braides and M. Cicalese, Continuum limits of discrete thin films with superlinear growth densities, submitted (download @ http://cvgmt.sns.it/).
2. R. Alicandro and M. Cicalese, Representation result for continuum limits of discrete energies with superlinear growth, *SIAM J. Math Anal.* **36** (2004), 1-37.
3. X. Blanc, C. Le Bris, Thomas-Fermi type models for polymer and thin films, *Adv. Diff. Eq.* **5** (2000), 977-1032.
4. A. Braides, *Γ-convergence for Beginners.* Oxford University Press, Oxford, 2002.
5. A. Braides, Non local variational limits of discrete systems, *Comm. Contemporary Math.* **2** (2000), 285-297.
6. A. Braides and M. Cicalese. Surface energies in nonconvex discrete systems, to appear on *Mathematical Models and Methods in Applied Sciences* (download @ http://cvgmt.sns.it/)
7. A. Braides, I. Fonseca and G. Francfort. 3D-2D asymptotic analysis for inhomogeneous thin films. *Indiana Univ. Math. J.* **49** (2000), 1367–1404.
8. A. Braides and G. Francfort, Bounds on the effective behaviour of a square conducting lattice, *Proc. R. Soc. London A.* **460** (2004), 1755–1769.

9. G. Friesecke and R.D. James, A scheme for the passage from atomic to continuum theory for thin films, nanotubes and nanorods, *J. Mech. Phys. Solids* **48** (2000), 1519-1540.
10. H. Le Dret and A. Raoult, The nonlinear membrane model as variational limit of nonlinear three-dimensional elasticity. *J.Math. Pures Appl.* **74** (1995), 549-578.
11. B. Schmidt, On the passage from atomic to continuum theory for thin films. Preprint (2005) (download @ http://www.mis.mpg.de/).

RECTANGULAR DUALIZATION OF BICONNECTED PLANAR GRAPHS IN LINEAR TIME AND RELATED APPLICATIONS

M. ANCONA, S. DRAGO and G. QUERCINI
Department of Computer Science (DISI), University of Genoa,
Via Dodecaneso 35, 16146 Genoa, Italy
E-mail: {ancona, drago, quercini}@disi.unige.it
http://www.disi.unige.it

A. BOGDANOVYCH
Faculty of Information Technology, University of Technology Sydney,
Sydney, NSW, Australia
E-mail: anton@it.uts.edu.au
http://www.it.uts.edu.au

Although rectangular dualization has been studied for several years in the context of floorplanning problems, its descriptive power has not been fully exploited for graph representation. The main obstacle is that the computation of a rectangular dual of any planar biconnected graph requires a sequence of non-trivial steps, some of which are still under investigation. In particular, the most tricky issue is the optimal *management* of separating triangles, for which no existing algorithm runs in linear time. In this paper we present our advances in rectangular dualization and we show two applications that, while very different, explain better than others its role.

Keywords: Rectangular Dual; Separating Triangles; Hierarchical Drawing; Structured Graphs; Electronic Institutions; Network Topologies

1. Introduction

Rectangular dualization is the computation of a rectangular dual of a planar graph. It was originally introduced to find rectangular topologies for floorplanning of integrated circuits:[1] by a floorplan, a rectangular chip area is partitioned into rectilinear polygons corresponding to the relative location of functional entities of the circuit. Subsequently, it found application in many other fields, in particular becoming effective in visualization problems. In this paper we briefly describe two possible applications. The first

one is visualization of network topologies, concerning the problem of engineering and optimizing large communication networks. This task is very challenging, as often real networks are huge, including hundreds and even thousands of nodes and links. In order to help human operators in maintaining and updating the description and documentation of its structure, a network is usually described in form of a hierarchy of subnetworks. Rectangular dualization is a very useful technique in graph drawing, especially when applied to the hierarchical drawing of a structured graph.
The second application domain which could benefit from rectangular dualization is the automatic design of Virtual Worlds whose development follows the 3D Electronic Institutions methodology,[2] which helps to separate the development into two independent phases: specification of the interaction rules and design of the 3D Interaction environment. During the specification phase not only the interactions rules are specified, but also the basic interaction components are determined. One part of the specification is the graph, which describes which scenes are required in the system (nodes of the graph), shows how the transitions between scenes are made (nodes) and which scenes can be reached from another scenes (arcs). By rectangular dualization of this graph, a two dimensional map of the Virtual Worlds is achieved and then scenes and transitions become 3-dimensional rooms, while the arcs of the graph determine which rooms have to be placed next to each other and have a shared door. The use of rectangular dualization is strongly limited by the fact that not all planar graphs admit a rectangular dual. However, it is possible to apply a minimal set of transformations to the original graph to obtain a graph admitting a rectangular dual representation. In the following sections we first describe an algorithm which computes the rectangular dual of any planar graph (section 2); then we outline the advantages of rectangular dualization in the two applications mentioned above (sections 3 and 4). In section 5 an overview of the paper and of future work is given.

2. OcORD: Optimal Constructor of a Rectangular Dual

A rectangular dual of a planar graph $G = (V, E)$ is a rectangle R partitioned into a set $\Gamma = R_1, ...R_n$ of non overlapping rectangles such that:

(1) no four rectangles meet at the same point;
(2) there is a one-to-one correspondence $f : V \rightarrow \Gamma$ such that two vertices u and v are adjacent in G if and only if their corresponding rectangles $f(u)$ and $f(v)$ share a common boundary.

Graphs not admitting a rectangular dual contain separating triangles, triangular regions of the graph that are not faces.[3–5] The idea behind OcORD, our tool aiming at constructing the rectangular dual of any planar graph in linear time, is to remove (we will use the term "break" from now on) all the separating triangles, if any, from the input graph, to obtain a graph admitting a rectangular dual. This can be accomplished by adding a *crossover* vertex on one edge of each separating triangle, as Lai and Leinwand proposed in.[6] However, their approach does not take care of adding a minimal set of crossover vertices, as they conjectured it was a NP-complete problem. In fact, it is possible to break two or more separating triangles by adding only one crossover vertex on an edge shared by all of them, instead of adding a vertex for each separating triangle. If a minimal set of crossover vertices is added, then we say that the breaking of separating triangles is optimal. In OCoRD this task is performed in five steps:

(1) four external vertices are added according to.[5]
(2) All the separating triangles are detected with the algorithm described in[7] and the geometrical dual of the resulting graph is computed.
(3) All the separating triangles are collapsed into macro-vertices, which creates a hierarchical structured graph (section 3.3).
(4) A crossover vertex is added on the duals of the edges belonging to a minimum covering. In[8] a formal proof that the number of crossover vertices added is minimum is given.
(5) The resulting graph is triangulated with the algorithm described in[9]

The graph obtained after these transformations admits a rectangular dual, that can be computed in linear time with several algorithms.[5,10] However, this procedure does not run in linear time and is fairly complicated, as it requires structuring the graph in step 3. Instead, we are investigating an algorithm that works on the plain graph and exploits matching in cubic graphs, that can be computed in linear time, as shown in.[11] Moreover, we have also developed a heuristic method that optimally breaks all the separating triangles in almost all the practical cases. We reserve to describe it in future work.

3. Visualization of Network Topologies

The plainest way to describe a communication network is to model the relationships among sites and links by means of a weighted undirected graph, but unless some more assumptions are taken, this approach can raise several problems. Problems are encountered when networks have a hierarchical topology, where nodes are classified in levels, usually two or three, according to their geographical position, number of users, and so on. Moreover, the optimization of large communication networks (for example the minimization of the number of message hops, i.e. the number of graph nodes traversed by a message) requires the use of special methods which need a most suitable graph representation. Many issues are encountered during network design phase, when repeated analysis and visualization of the network has to be performed: practical networks include hundreds or often thousands of nodes and links, so that even a simple description and documentation of the network structure is hard to maintain and update. In this case the network is usually described in form of a hierarchy of subnetworks that are represented by collapsing some subnetworks into single nodes or single links to be described in separated documents. Such a hierarchical approach to network (and graph) description can be formalized into a complex but flexible graph structure called *structured graph*. In the following sections we stress the importance of rectangular dualization in hierarchical graph drawing.

3.1. *Role of Rectangular Dualization in Network Drawing*

Using a rectangular dual in graph drawing offers several advantages:

(1) Its computational complexity is optimal: $O(n)$ time and $O(n^2)$ area, like other most efficient methods.
(2) It provides a symmetrical drawing with respect to x and y coordinates.
(3) It can be naturally extended to a hierarchical drawing of a structured graph G, by introducing the concept of *structured rectangular dual* (see section 3.3)
(4) The construction of a 2-visibility drawing from a rectangular dual is immediate (see Figure 1).
(5) A 2-bend[a] rectilinear drawing can be obtained from a rectangular dual in linear time.
(6) Also a 1-bend rectilinear drawing can be obtained from a rectangular dual in linear time.

The 2-visibility drawing method has been introduced by Kant.[12] In a 2-visibility drawing the vertices of a given graph are represented by rectangles (rectangular boxes) and the adjacency relationships are expressed by

[a] A bend is a point where the drawing of an edge changes its direction. A drawing is said to be k-bend if each edge has at most k bends

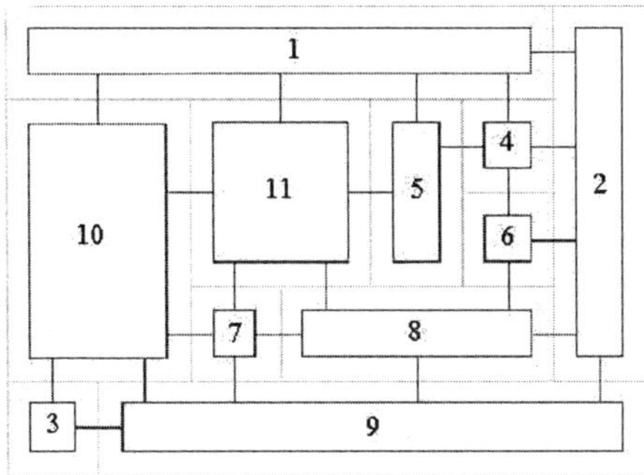

Fig. 1. A 2-visibility representation computed from a rectangular dual (depicted in the background)

horizontal and vertical lines drawn between boxes. The authors claim a high quality of the drawing. This kind of drawing can be trivially derived from the rectangular dual of a graph as follows. Two adjacent rectangles $R1$ and $R2$ of the dual graph share a portion of an edge that we call *window*. If inside $R1$ and $R2$ we draw two smaller rectangles large enough to be mutually visible through the window, they can be connected by a straight segment. The result of this procedure is shown in figure 1.

3.2. *Drawing Modes and Styles*

In this section we distinguish two concepts: *drawing mode* and *drawing style*. A drawing mode defines the form of edge and vertex representation adopted: in *straight-line* drawing mode edges are drawn as line segments connecting vertices; in *orthogonal drawing* the edges are drawn as sequences of segments parallel to x or y axes; in bus mode drawing (Figure 2) bundles of parallel edges are collected into a single path (or bus) until each edge emerges from the bus, and reaches its destination vertex (in some cases each bundle is identified by a label). Drawing modes for vertices include points, squares, rectangles or circles of fixed or variable size.

A drawing style is the combination of a drawing mode with a specific algorithm displaying data. In other words, a drawing style is concerned with relative positioning of vertices and edges in the plane, whereas a drawing mode with the form of representing edges and vertices. For example, in grid drawing style, vertices and edges are drawn only on a discrete grid in the plane; in the Kandinsky model (also known as *Podevs*), and in some of its variants, like the simple Kandinsky model (*Podevsnef*), vertices are represented by squares of equal size, placed on a coarse grid and edges are drawn, in orthogonal mode, on a finer grid.

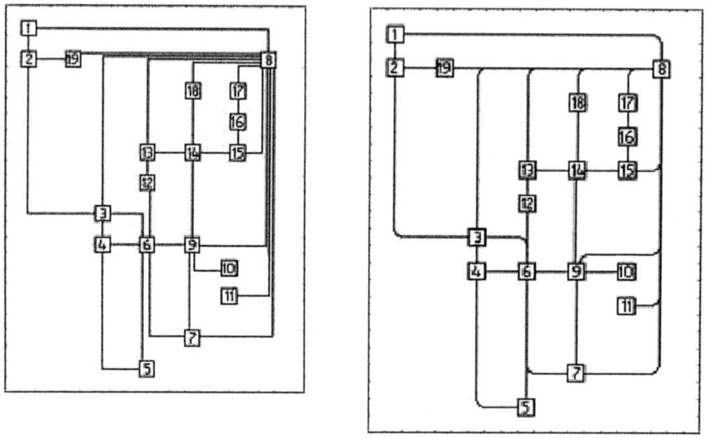

(a) A classical drawing (b) A bus mode drawing

Fig. 2. Classical drawing and bus mode drawing of a given graph

The bus drawing mode is useful for undirected graphs, since the arrow-heads would introduce ambiguities. In figure 2 Kandinsky drawing (figure 2(a)) and bus-mode drawing (figure 2(b)) of the same graph are compared. The latter, although it is 2-bend while the first is 1-bend, is more readable, as edges are not too close each other. To create a bus style drawing, we simply have to ignore the offsets of the edges on the finer grid, and transform each bend into a curved corner. Despite its evident utility (see schematics drawing tools like *Orcad*) this mode has not been considered by researchers in graph drawing. In this paper we use only bus-mode layouts because it is particularly useful for connecting vertices of high degree, a frequent case in clustered graphs. Bus-mode drawing is a powerful way for structuring and clustering edges of a graph. Buses represent a structured set of interconnections: while a sub-graph can be represented by a single macro-vertex, a set of edges contributing to the same logical function can be condensed into a single bus, that can be considered as a special form of hyper-edge. Our algorithm for constructing the dual graph assigns integer coordinates to each rectangle associated with a vertex. In this way, by scaling the rectangle sizes we can allocate space both for squares representing vertices and zones dedicated to interconnections. It is a trivial task to derive a bus-mode rectilinear drawing of a graph G from its rectangular dual. Such a drawing could be based on paths all composed by exactly two bends with a pleasant and clear visual effect. We call this kind of drawing the naive 2-bend bus-mode drawing. However, with a minimum effort, we can produce a 1-bend (maximum) drawing in linear time.

3.3. *Hierarchical Drawing of Structured Graphs*

Managing large networks requires the decomposition of the network in manageable units organized in a hierarchy (tree) describing how single parts contribute to form the original network. A *structured graph* (or *clustered graph*, henceforth referred to as "SG") is a form of abstraction applied to a large graph in order to make it modular and more manageable. The abstraction consists in collapsing a subgraph to a single vertex (called a *macrovertex*), or to a single link (called a *macrolink*), thus obtaining a simpler and hierarchically described graph. The structuring operation is usually iterated recursively until a large network is decomposed into relatively small and manageable components and subcomponents. Every subcluster is defined at several levels of nesting adopting a methodology that is usually applied to every large project (software and hardware design) involving hundreds or thousands of components: "modularity". Figure 3 shows an example of a structured graph. We will refer to the graph represented in figure as "gd94".

Formally, a SG is a pair $H = (G, T)$, where G is a connected simple (multi- or hyper-) graph and T is a tree describing the structure of H. The leaves of T are exactly the vertices of G and each node $t \in T$ represents a cluster $G_t = (V_t, E_t)$ of G such that $V_t \subseteq V$ is the subset of vertices that are leaves of the subtree of T rooted at t. G_t is the subgraph generated by V_t, while $H_t = (G_t, T_t)$ is the SG associated with t. Notice that the planarity of the underlying base graph does not imply the planarity of a structured graph if arbitrary subgraphs are collapsed. In order to preserve planarity, the collapsing operation should respect some conditions. Some researchers require H to be *c-planar*,[13] namely there exists a planar embedding of H in the plane, with a planar drawing such that no region R surrounding a macro-element is crossed by an edge having both vertices external to it. Instead, we give the following conditions:

(1) The original graph must be planar.
(2) Only complete and connected subgraphs are collapsed.
(3) Elements of the same kind (namely macro-vertices or macro-links) must be completely nested or independent, thus having an empty intersec-

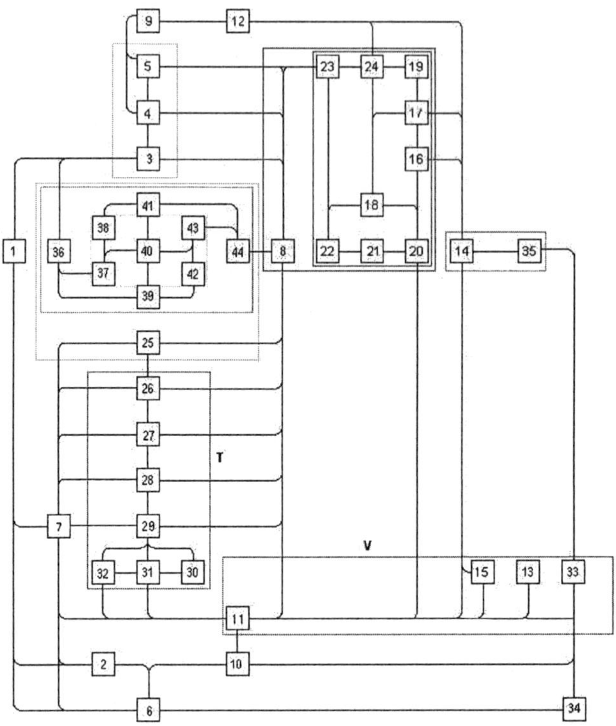

Fig. 3. Example of a structured graph (gd94), where each cluster is surrounded by a rectangle

tion.

(4) Different macro-elements (vertices and/or links) may share only vertices and not one or more links.

and we report the following well-known results.

Definition 3.1. A graph G' is said to be a subdivision of a graph G if G' is obtained from G by replacing some of its edges with paths having at most their endvertices in common.

Theorem 3.1 (Kuratowski, 1929). *A graph is planar if and only if it does not contain a subdivision of* K_5 *and* $K_{3,3}$.

Now we can prove the following:

Theorem 3.2. *In relation with the above definition of a structured graph,* *we have that if only connected subgraphs are collapsed, the resulting structured graphs are all planar.*

Proof. By absurd, suppose that there is a non planar clustered graph G' of a planar graph G. For Kuratowski's theorem G' contains a subdivision K of a K_5 or $K_{3,3}$. G' must contain at least a macro-vertex because G is planar. Let m be a macro-vertex of G' and Gm the corresponding clustered subgraph in G; by the connectivity of G_m follows that there is a path in G_m joining the two ports of m on which K is defined in G', that is, there exists a subdivision K'' of G, that is a contradiction. □

3.4. *Hierarchically Structured Rectangular Duals*

Drawing a structured graph, or even drawing a medium-sized graph, requires to fit the available page size for a drawing, i.e., to decompose the representation into sheets and to refer them to the original graph. Either for structured graphs, often the size of a single component does not comply with visualization requirements. In this case, a visualization-oriented decomposition should be preferred. A hierarchical structure of a graph can be extracted from its rectangular dual. To this purpose we introduce the concept of *hierarchically structured rectangular dual* or, more simply, *hierarchical rectangular dual* (HRD) of a graph G. The HRD of a graph G is a tree of rectangles (R, T) such that the rectangles of the dual graph represent either single vertices, like in a standard rectangular dual, or connected subgraphs of G, abstracted to a single vertex (see, for example, rectangles violet (V) and turquoise (T) of figure 3) and called *macro-rectangles.*

In figure 3 clusters are obtained from a rectangular dual computed on "gd94" using OcORD. Rectangles in the figure are rectangles in the rectangular dual that have other rectangles inside, meaning that they do not represent vertices, but subgraphs. We are far from saying that rectangular dualization is a method for clustering graphs, as clustering is not only a topological issue, but has to take into account of many other factors. However, if we restrict our attention only to drawing, clusters suggested by a rectangular dual seem to be very efficient.

3.4.1. *Computing a HRD*

In the previous section we have described how a HRD suggests a clusterization of the primal graph. It is also possible to perform the inverse step, namely computing a HRD from a planar structured graph fulfilling conditions listed in section 3.3. The idea is the following: we build the rectangular dual R, where each cluster C_i is depicted as a rectangle R_i; for each cluster C_i we compute its rectangular dual and we draw it inside R_i. This procedures preserves the adjacencies between the macro-vertices, without caring about adjacencies between simple vertices lying in different clusters. Thus, the adjacency between two vertices does not necessarily imply the adjacency of the two corresponding rectangles in the rectangular dual. In order to fulfil this second constraint, each cluster must specify an interface to the other clusters. More in details, if A and B are two adjacent clusters, a vertex i (called *interface vertex*) is added so that each edge (a, b), $a \in A$ and $b \in B$, is split into two edges (b, i) and (a, i). Each cluster is then surrounded by the interface vertices. Thus, we compute the HRD of this graph and we draw the rectangular dual of each cluster inside the corresponding rectangles. However, this procedure raises several problems, in terms of efficiency of the final drawing, and this is the main reason why here we give only a brief overview. First of all, it is not clear how many interface vertices have to be added: an interface vertex for each adjacency of a cluster does not seem to be an optimal choice. Next, when computing the rectangular dual of clusters, we consider it having its interface vertices on the external face. Thus, an interface vertex between cluster A and cluster B appears on the external face (and in the rectangular dual) of both clusters, meaning that they are replicated. This topic will be object of further analysis in our research.

4. Representation of 3D Electronic Institutions

In this section we only introduce in short an application that, while being very different from network visualization, demonstrates the power and versatility of rectangular dualization. For more details the interested reader can refer to.[14]

3D Electronic Institutions are a new method of software design of open systems based on the metaphor of 3D Virtual Worlds. One of the drawbacks of the Virtual Worlds technology is that its design and development has emerged as a phenomenon shaped by a home computer user, rather than by the research and development in universities or companies. Thus, Virtual Words do not have the means to enforce technological norms and rules of their inhabitants. The enforcement of organizational conventions in 3D Electronic Institutions methodology is achieved by separating different patterns of conversational activities into separate methodological entities (scenes), assigning different roles to different types of participants, specifying the rules (protocols) for inter-participant interactions and defining the role flow of participants between different scenes. The specification of scenes and the role flow are done in a form of a directed graph, where nodes represent scenes and arcs and their labels define the role flow. This graph, called *Performative Structure*, forms a basis for the visualization of the system. Rectangular dualization plays its central role in automatically transforming the Performative Structure into a 3D Virtual World. In fact, the rectangular dual of the Performative Structure is a 2D map of the institution, which is further transformed into a 3D Virtual World. In such a process, rectangular dualization offers a space optimal solution, as it can be used for minimizing the distance between two agents (entities in the virtual world) that are expected to have frequent interactions. The automatic generation of a 3D Virtual Word consists of the following steps:

(1) The redundant information contained in the Performative Structure is filtered out. If some nodes of the graph are connected with more than one arc, only one randomly selected arc is left and all the others are deleted.
(2) The Performative Structure is transformed into a format compatible with OcORD input and a rectangular dual is computed. In the rectangular dual, scenes and transitions are transformed into rooms and connections are visualized as doors.
(3) A 3D Virtual World is generated from the 2D map created at the previous step. The generated 3D Virtual World is visualized and connected to the infrastructure, which controls the correct behaviour of the participants.

5. Conclusions and Future Work

In this paper we presented the main reasons behind our interest in rectangular dualization, describing some useful applications that benefit from this technique. Some work is still to be done in developing an efficient and linear-time algorithm for computing the rectangular dual of every planar graph. Moreover, the computation of a HRD of a structured graph is far from being efficient. Finally, rectangular dualization is only a particular case of a more general class of rectangular representations, called rectangular layouts. Most of them show the same problems arising in rectangular dualization and have a similar impact in practical applications (think about cartograms, for example). Our aim is also to extend our results to these layouts.

References

1. J. Grason, An Approach to Computerized Space Planning Using Graph Theory, in *Proceedings of the 8th Workshop on Design Automation (DAC '71)*, (ACM Press, New York, NY, USA, June 1971).
2. A. Bogdanovych, M. Esteva, S. Simoff, C. Sierra and B. Helmut, A Methodology for 3D Electronic Institutions, in *Proceedings of the Sixth International Joint Conference on Autonomous Agents and Multi-Agent Systems (AAMAS '07)*, (Springer-Verlag, Honolulu, USA, May 2007).
3. K. Kozminski and E. Kinnen, An Algorithm for Finding a Rectangular Dual of a Planar Graph for Use in Area Planning for VLSI Integrated Circuits, in *Proceedings of the 21st Conference on Design Automation (DAC '84)*, (IEEE Press, Piscataway, NJ, USA, June 1984).
4. S. M. Leinwand and Y. T. Lai, An algorithm for Building Rectangular Floor-Plans, in *Proceedings of the 21st Conference on Design Automation (DAC '84)*, (IEEE Press, Piscataway, NJ, USA, June 1984).
5. G. Kant and X. He, Two Algorithms for Finding Rectangular Duals of Planar Graphs, in *Proceedings of the 19th International Workshop on Graph-Theoretic Concepts in Computer Science (WG93)*, (Springer-Verlag, London, UK, June 1994).
6. Y. T. Lai and S. M. Leinwand, Algorithms for Floorplan Design via Rectangular Dualization, *IEEE Trans. on CAD of Integrated Circuits and Systems* **7**, 1278 (1988).
7. N. Chiba and T. Nishizeki, Arboricity and Subgraph Listing Algorithms, *SIAM Journal on Computing* **14**, 210 (1985).
8. A. Accornero, M. Ancona and S. Varini, All Separating Triangles in a Plane Graph Can Be Optimally "Broken" in Polynomial Time, *International Journal of Foundations of Computer Science* **11**, 405 (2000).
9. T. Biedl, G. Kant and M. Kaufmann, On Triangulating Planar Graphs under the Four-Connectivity Constraint, *Algorithmica* **19**, 427 (1997).
10. J. Bhasker and S. Sahni, A Linear Algorithm to Find a Rectangular Dual of a Planar Triangulated Graph, in *Proceedings of the 23rd ACM/IEEE Conference on Design automation (DAC '86)*, (IEEE Press, Piscataway, NJ, USA, June 1986).
11. T. Biedl, B. Prosenjit, E. Demaine and A. Lubiw, Efficient Algorithms for Petersen's Matching Theorem, *Journal of Algorithms* **38**, 110 (2001).
12. U. Foßmeier, G. Kant and M. Kaufmann, 2-Visibility Drawings of Planar Graphs, in *Proceedings of the International Symposium on Graph Drawing (GD '96)*, (Springer-Verlag, London, UK, September 1997).
13. Q. Feng, Algorithms for Drawing Clustered Graphs, PhD Thesis, Department of Computer Science and Software Engineering, The University of Newcastle, Australia, (Newcastle, Australia, 1997).
14. A. Bogdanovych and S. Drago, Euclidean Representation of 3D Electronic Institutions: Automatic Generation, in *Proceedings of the Working Conference on Advanced Visual Interfaces (AVI '06)*, (ACM Press, New York, NY, USA, May 2006).

REMARKS ON CONTACT POWERS AND NULL LAGRANGIAN FLUXES

L. ANSINI and G. VERGARA CAFFARELLI

Dipartimento di Metodi e Modelli Matematici per le Scienze Applicate
Università 'La Sapienza'
Via A. Scarpa 16 - 00161 Roma, Italy

We consider grade-N elastic bodies (N=1,2) with stored energy function given by a null lagragian and we derive some identities concerning stress measures.

1. Introduction

We consider *grade-N* elastic bodies, that is, material bodies whose response to deformation processes is specified by a *stored energy function* of the deformation and its iterated gradients up to a fixed integer order N. We here concentrate on the case when the stored energies are null lagragians. A *null lagragian* is a real-valued set function defined over the collection of parts P of a given body having a two-fold integral representation:

$$P \mapsto \int_{\mathrm{P}} \sigma \qquad \text{and} \qquad \mathrm{P} \mapsto \int_{\partial \mathrm{P}} \tau$$

such that

$$\int_{\mathrm{P}} \sigma = \int_{\partial \mathrm{P}} \tau \tag{1}$$

for all deformations $\boldsymbol{f}$ and all body parts P with almost-everywhere-smooth boundary surface ∂P, positively oriented by the outer unit normal $\mathbf{n}$.
We take the densities σ, τ to depend on the deformation $\boldsymbol{f}$ and its iterated gradients up to the same order N and to be consistently represented in terms of a vector field $\boldsymbol{u}$:

$$\sigma = \mathrm{Div}\, \boldsymbol{u}, \tag{2}$$

$$\tau = \boldsymbol{u} \cdot \boldsymbol{n}. \tag{3}$$

We observe that the surface density τ has the form of a contact flux. Mutually compatible pairs of contact forces and stress distributions satisfy certain identities that have been derived by Podio-Guidugli & Vergara Caffarelli[3] by using the arbitrariness in the choice of body parts P.
Here we show that those identities can be also derived making use of test functions in the place of the arbitrary body parts.
We observe, in particular, that those identities hold for material bodies *with faded boundaries* in the sense that their periphery consists of a region where a "regularized characteristic" function assumes values between zero and one.
More precisely, for $\boldsymbol{f}(p)$ be a deformation vector field, let $\sigma(p, \boldsymbol{f}, {}_1\boldsymbol{F}, {}_2\boldsymbol{F})$, ${}_1\boldsymbol{F} \equiv \boldsymbol{F} = \nabla \boldsymbol{f}$ and ${}_2\boldsymbol{F} = \nabla(\nabla \boldsymbol{f})$, be a (grade-1 or) grade-2 null lagrangian over R^3; furthermore, let θ be a $C_0^\infty(R^3)$ test function, and consider the associated total energy

$$E_\theta = \int_{R^3} \theta(p)\ \sigma(p, \boldsymbol{f}, {}_1\boldsymbol{F}, {}_2\boldsymbol{F}).$$

Then, identity (1) can be rewritten as:

$$\int_{R^3} \theta(p)\ \sigma(p, \boldsymbol{f}, {}_1\boldsymbol{F}, {}_2\boldsymbol{F}) = -\int_{R^3} \nabla\theta(p) \cdot \boldsymbol{u}(p, \boldsymbol{f}, {}_1\boldsymbol{F}, {}_2\boldsymbol{F}). \tag{4}$$

Moreover, the contact power expended over the velocity field $\boldsymbol{v}$ is given by:

$$\begin{aligned} \delta\, E\,[\boldsymbol{v}] &:= \frac{d}{d\epsilon}\left(\int_{R^3} \theta(p)\,\sigma(p, \boldsymbol{f} + \epsilon\, \boldsymbol{v}, {}_1\boldsymbol{F} + \epsilon\, \nabla \boldsymbol{v}, {}_2\boldsymbol{F} + \epsilon\ {}_2\nabla \boldsymbol{v})\right)_{\epsilon=0} \qquad (5)\\ &= -\frac{d}{d\epsilon}\left(\int_{R^3} \nabla\theta(p) \cdot \boldsymbol{u}(p, \boldsymbol{f} + \epsilon\, \boldsymbol{v}, {}_1\boldsymbol{F} + \epsilon\, \nabla \boldsymbol{v}, {}_2\boldsymbol{F} + \epsilon\ {}_2\nabla \boldsymbol{v})\right)_{\epsilon=0}. \end{aligned}$$

Remark.

Let σ be a grade-1 null lagrangian. In view of (2),

$$\sigma(p, \boldsymbol{f}, {}_1\boldsymbol{F}) = \mathrm{Div}\, \boldsymbol{u}(p, \boldsymbol{f}, {}_1\boldsymbol{F}),$$

or rather

$$\sigma = \sum_i \big[(u_i)_{p_i} + \sum_j (u_i)_{f_j}\, f_{j/i} + \sum_{j,h} (u_i)_{{}_1F_{jh}}\, f_{j/hi}\big],$$

where the last term of the right side must be identically null, because σ cannot depend on ${}_2\boldsymbol{F}$:

$$\sum_{i,h} (u_i)_{{}_1F_{jh}}\, f_{j/hi} = 0. \tag{6}$$

This consequence is implied by the skew-symmetry condition

$$(u_i)_{{}_1F_{jh}} = -(u_h)_{{}_1F_{ji}}. \tag{7}$$

Let the left minor transpose ${}^t\boldsymbol{U}$ of a third-order tensor $\boldsymbol{U}$ be defined by ${}^tU_{i,j,h} = U_{j,i,h}$. By (6), the skew- symmetry condition (7) can be rewritten in the following form:

$${}^t\partial_{{}_1\boldsymbol{F}}\, \boldsymbol{u}\, [\boldsymbol{A}] = 0,$$

for all symmetric second-order tensors $\boldsymbol{A}$.

Quite similary, let σ be a grade-2 null lagrangian. In view of (2),

$$\sigma(p, \boldsymbol{f}, {}_1\boldsymbol{F}, {}_2\boldsymbol{F}) = \operatorname{Div} \boldsymbol{u}(p, \boldsymbol{f}, {}_1\boldsymbol{F}, {}_2\boldsymbol{F}),$$

whence

$$\sigma = \sum_i \Big[(u_i)_{p_i} + \sum_j (u_i)_{f_j}\, f_{j/i} + \sum_{j,h} (u_i)_{{}_1F_{jh}}\, f_{j/hi} + \sum_{j,h,k} (u_i)_{{}_2F_{jhk}}\, f_{j/hki}\Big];$$

here the term involving the third gradient of the deformation must be null:

$$\sum_{i,h,k} (u_i)_{{}_2F_{jhk}}\, f_{j/hki} = 0.$$

The related skew-symmetry condition:

$$(u_i)_{{}_2F_{jhk}} + (u_k)_{{}_2F_{jih}} + (u_h)_{{}_2F_{jki}} = 0$$

can be written as:

$${}^t\partial_{{}_2\boldsymbol{F}}\, \boldsymbol{u}\, [\boldsymbol{A}] = 0$$

for all third-order tensors $\boldsymbol{A}$ insensitive to the permutation of any pair of indexes h, k, i.

2. Grade-1 Null Lagrangians

Let σ be a grade-1 null lagrangian. We know that

$$\sigma(p, f, {}_1\boldsymbol{F}) = \operatorname{Div} \boldsymbol{u}(p, f, {}_1\boldsymbol{F}),$$

where $\boldsymbol{u}(p, f, {}_1\boldsymbol{F})$ is a suitable skew-symmetric vector field; by (5),

$$\int_{R^3} \theta(\partial_f\, \sigma \cdot \boldsymbol{v} + \partial_{{}_1\boldsymbol{F}}\, \sigma \cdot \nabla \boldsymbol{v}) = -\int_{R^3} \partial_f\, \boldsymbol{u} \cdot \nabla\theta \otimes \boldsymbol{v} + \partial_{{}_1\boldsymbol{F}}\, \boldsymbol{u} \cdot \nabla\theta \otimes \nabla \boldsymbol{v}.$$

Integrating by parts and using the skew-symmetry condition, we have the integral identity:

$$\int_{R^3} \theta\, (\partial_f\, \sigma - \operatorname{Div} \partial_{{}_1\boldsymbol{F}}\, \sigma) \cdot \boldsymbol{v} = \int_{R^3} \Big((\partial_{{}_1\boldsymbol{F}}\, \sigma)^T - \partial_f\, \boldsymbol{u} + \operatorname{Div} \partial_{{}_1\boldsymbol{F}}\, \boldsymbol{u}\Big) \cdot \nabla\theta \otimes \boldsymbol{v}.$$

The arbitrariness of $\boldsymbol{v}$, θ and $\nabla\theta$ yields the following two identities:

$$\partial_f \sigma - \mathrm{Div}\, \partial_{{}_1\boldsymbol{F}} \sigma = 0 \tag{8}$$

and

$$(\partial_{{}_1\boldsymbol{F}} \sigma)^T = \partial_f \boldsymbol{u} - \mathrm{Div}\, \partial_{{}_1\boldsymbol{F}} \boldsymbol{u}. \tag{9}$$

We may read identity (8) as an alternative representation of grade-1 null lagrangians, and identity (9) as the assertion that the stress associated to a grade-1 null lagrangian is the lagrangian derivative of the corresponding vector field $\boldsymbol{u}$.

3. Grade-2 Null Lagrangians

Let σ be a grade-2 null lagrangian. We know that

$$\sigma(p, \boldsymbol{f}, {}_1\boldsymbol{F}, {}_2\boldsymbol{F}) = \mathrm{Div}\, \boldsymbol{u}(p, \boldsymbol{f}, {}_1\boldsymbol{F}, {}_2\boldsymbol{F}),$$

where $\boldsymbol{u}(p, f, {}_1\boldsymbol{F}, {}_2\boldsymbol{F})$ is a suitable skew-symmetric vector field.
By (5), integrating by parts and using the skew-symmetry condition, we arrive at the following integral identity:

$$\begin{aligned}
\delta E\,[\boldsymbol{v}] :=& \int_{R^3} (\partial_f \sigma - \mathrm{Div}\, \partial_{{}_1\boldsymbol{F}} \sigma + \mathrm{Div}\,\mathrm{Div}\, \partial_{{}_2\boldsymbol{F}} \sigma) \cdot \boldsymbol{v}\, \theta \\
&+ \int_{R^3} (-\partial_{{}_1\boldsymbol{F}} \sigma + 2\,\mathrm{Div}\, \partial_{{}_2\boldsymbol{F}} \sigma) \cdot \boldsymbol{v} \otimes \nabla\theta + \partial_{{}_2\boldsymbol{F}} \sigma \cdot \boldsymbol{v} \otimes {}_2\nabla\theta \\
=& \int_{R^3} (-{}^t\partial_f \boldsymbol{u} + \mathrm{Div}\, {}^t\partial_{{}_1\boldsymbol{F}} \boldsymbol{u} - \mathrm{Div}\,(\mathrm{Div}\, {}^t\partial_{{}_2\boldsymbol{F}} \boldsymbol{u})) \cdot \boldsymbol{v} \otimes \nabla\theta \\
&+ \int_{R^3} ({}^t\partial_{{}_1\boldsymbol{F}} \boldsymbol{u} - 2\,\mathrm{Div}\, {}^t\partial_{{}_2\boldsymbol{F}} \boldsymbol{u}) \cdot \boldsymbol{v} \otimes {}_2\nabla\theta.
\end{aligned}$$

In view of the arbitrariness of $\boldsymbol{v}$, $\nabla\theta$ and ${}_2\nabla\theta$, we obtain the identities:

$$\partial_f \sigma - \mathrm{Div}\, \partial_{{}_1\boldsymbol{F}} \sigma + \mathrm{Div}\,\mathrm{Div}\, \partial_{{}_2\boldsymbol{F}} \sigma = 0, \tag{10}$$

$$\partial_{{}_1\boldsymbol{F}} \sigma - 2\,\mathrm{Div}\, \partial_{{}_2\boldsymbol{F}} \sigma = {}^t\partial_f \boldsymbol{u} - \mathrm{Div}\, {}^t\partial_{{}_1\boldsymbol{F}} \boldsymbol{u} + \mathrm{Div}\,(\mathrm{Div}\, {}^t\partial_{{}_2\boldsymbol{F}} \boldsymbol{u}) \tag{11}$$

and

$$\partial_{{}_2\boldsymbol{F}} \sigma[\boldsymbol{A}] = ({}^t\partial_{{}_1\boldsymbol{F}} \boldsymbol{u} - 2\,\mathrm{Div}\, {}^t\partial_{{}_2\boldsymbol{F}} \boldsymbol{u})[\boldsymbol{A}],$$

for all symmetric second-order tensors $\boldsymbol{A}$.
As before, the identities (10)-(12) give an alternative representation of grade-2 null lagrangians and of the stress measures associated to a grade-2 null lagrangian in terms of the corresponding vector field $\boldsymbol{u}$.

References

1. P. Podio-Guidugli and G. Vergara Caffarelli, Surface Interaction Potentials in Elasticity. *A.R.M.A* **109**, (1990) pp.343-383.
2. S. Carillo, P. Podio-Guidugli and G. Vergara Caffarelli, Second-Order Surface Interaction Potentials. In: M. Brocato, P. Podio-Guidugli (ed.), *Rational Continua, Classical and New: A Collection of papers dedicated to Gianfranco Capriz*, Springer-Verlag (2003) pp.19-38.
3. P. Podio-Guidugli and G. Vergara Caffarelli, On contact powers and null lagrangian fluxes. In: P. Fergola, F. Capone, M. Gentile, G. Guerriero (ed.), *Proceedings of the International Meeting New Trends in Mathematical Physics, Naples, Italy*, World Scientific (2004) pp.147-156.

INNOVATIVE FINANCIAL PRODUCTS FOR ENVIRONMENTAL PROTECTION IN AN EVOLUTIONARY CONTEXT

ANGELO ANTOCI*

DEIR, University of Sassari
Via Torre Tonda, 34 07100 Sassari, Italy
** E-mail: angelo.antoci@virgilio.it*
www.uniss.it

MARCELLO GALEOTTI*

DiMaD, University of Florence
Via Lombroso 6/7 50134 Florence, Italy
E-mail: marcello.galeotti@dmd.unifi.it

LUCIO GERONAZZO*

DiMaD, University of Florence
Via Lombroso 6/7 50134 Florence, Italy
E-mail: lucio.geronazzo@dmd.unifi.it

The main objective of the paper is to analyse the effects on economic agents' behaviour deriving from the introduction of financial activities aimed to environmental protection.

The environmental protection mechanism we study should permit exchange of financial activities among citizens, firms and Public Administration. Such a particular "financial market" is regulated by the Public Administration, but mainly fuelled by the interest of two classes of involved agents: firms and dwelling citizens. We assume that the adoption process of financial decisions is described by a two-population evolutionary game and we study the basic features of the resulting dynamics.

Keywords: Environment protection; Visitors and firms taxes vs environmental options; Evolutionary games; Dynamical systems.

1. Introduction

The main objective of the paper is to analyze the effects on economic agents' behavior deriving from the introduction of an environmental protection mechanism that can permit exchange of financial activities among visitors of a region R, firms operating in the region R and the Public Administration (PA). This *financial market* is regulated by the PA, but mainly fuelled by the interest of visitors and firms.

The context we analyze has the following features.

The PA offers to an individual who desires to spend a period of time in the region R the choice between:

(1) Purchasing a *call environmental option* sold by the PA at a given price, which implies a *cost* in the case of a *high environmental quality*, measured according to a properly defined quality index Q, but offers a *reimbursement* in the case of a low value of Q. Consequently, buying the financial option represents a self-assurance device against environmental degradation.

(2) Paying a fixed amount (entrance ticket) to the PA as a visitor's tax.

Analogously, the PA offers to a potentially polluting firm operating in the region R the choice between:

(3) Issuing a *put environmental option* bought by the PA: this financial activity implies a financial aid for the firm only if the quality index Q results higher than a properly defined threshold level.

(4) Paying a fixed amount as an environmental fine to carry on its activity in the region R.

Hence, in the case of a high value of Q, visitors bear a cost but find satisfaction in a high environmental quality, while the firms choosing (3) receive a financial support for investments aimed to environmental protection (hence such investments do not diminish their competitiveness). Furthermore, the PA attains the goal of improving the environmental quality at a low cost (if prices of financial activities are determined in such a way that the costs born by visitors compensate, at least partially, the financial aids to firms). In the case of a low value of Q, instead, the visitors choosing (1) receive a partial reimbursement for the environmental damage and the firms choosing (3) do not receive the financial aid.

The PA has to determine the prices of these financial activities taking into account, among other things, the number of visitors and firms respectively aiming to buy and sell the options, the costs of non polluting technologies, the transaction costs and so on.

As a consequence of the mechanism described, we can expect a strong interdependency between firms and visitors' payoffs. Our goal is to study the dynamics of such a *financial market*, arising from the interaction of economic agents and the PA. To this end, we represent the adoption process of choices by a two-population evolutionary game, where one population of firms strategically interacts with one population of visitors. The evolution of visitors and firms' behavior is modelled by the so called *replicator dynamics*, according to which the choices whose expected payoff is greater than the average one widespread in the populations at the expenses of the others.

The paper is structured as follows. In Section 2 we describe the model. Sections 3 and 4 are devoted to the analysis of the evolutionary dynamics. Section 5 concludes the paper.

2. The Model

The environmental protection mechanism we analyze is centered on the introduction of innovative financial activities issued with assistance of the Public Administration (PA), sold by polluting firms operating in the region R and bought by dwelling visitors or citizens. The motivations underlying the introduction of such mechanism are:

(1) Virtuous firms choosing a non polluting technology can obtain financial aids by selling the financial options.
(2) Visitors can protect themselves against environmental degradation by a self-assurance device (see [1]).
(3) The PA can attain the goal of improving the environmental quality at a low cost, since the costs born by visitors compensate, at least partially, the financial aids to firms.

The basic features of the "contracts" between the PA and visitors, and between the PA and firms, are the following.

The PA offers to an individual who desires to spend a period of time in the region R the choice between:

(V_1) purchasing an environmental option sold by the PA at a given price, that is, from the point of view of the visitor, a call option. It gives rise to a non-refundable cost in the case of high environmental quality, measured by a suitable pollution index Q. On the contrary, it gives the right, in the case of a low level of Q, to a defined positive payoff as an indemnity;

(V_2) paying a fixed amount (entrance ticket) to the PA as a visitor's tax.

The PA offers to a firm operating in R the choice between:

(F_1) issuing an environmental option bought by the PA, which, from the point of view of the firm, behaves as a put option, implying the payment of a fixed amount back if the quality index Q does not result sufficiently high (i.e. if a given environmental goal is not reached);

(F_2) paying a fixed amount as an environmental fine to carry on its activity in the region R.

We assume that the choice of F_2 implies the choice of a *polluting technology*. Furthermore, we don't consider (by assumption) the possibility of *free riding* (for a motivation of such hypothesis see [2]), i.e. firms have no incentive in issuing the financial asset if they are going to adopt a polluting technology.

2.1. *Our modelling choices*

As a consequence of the described mechanism a strong interdependence between firms and visitors' decisions is achieved. According to this, the profits of each firm strictly depend on the choices of both the other firms and the visitors. The same argument applies for visitors monetary payoffs.

A dynamical model arising as a natural choice from the above assumptions is a two-population evolutionary game, where the population of firms strategically interacts with the population of visitors.

With the aim of keeping the presentation at an understandable level, we study the interaction of firms and visitors in a simplified strategic context, preserving, nevertheless, the main features of the real dynamics. Namely, at each instant of time, two pairs of randomly chosen firms and visitors are assumed to match in order to play a one shot game. On occasion of each matching:

1) each firm has to choose (ex ante) between the two strategies: F_1 (selling the option described above) and F_2 (paying a fixed amount as an environmental fine);

2) each visitor has to choose (ex ante) between the two strategies: V_1 (buying the option described above) and V_2 (paying a fixed amount as an entrance ticket).

The analysis of such a context allows us to take into account all types of interdependence between economic agents (between two firms, between two visitors, between firms and visitors), working in a very simplified analytical setting.

2.2. *Visitors' payoff matrix*

To avoid trivial cases, we assume that, in each matching, the quality index Q results *high enough* (i.e. the predetermined environmental goal is reached) only if both firms choose F_1 (i.e. choose a non polluting technology).

We consider the following visitors' payoff matrix (where $\widetilde{p_2} = \overline{p} + p_2$ and

$\widetilde{p_1} = \overline{p} + p_1$):

	V_1, F_1, F_1	V_1, F_1, F_2	V_1, F_2, F_2	V_2, F_1, F_1	V_2, F_1, F_2	V_2, F_2, F_2
V_1	$-\widetilde{p_2}$	$-\widetilde{p_2} + \alpha_{21}$	$-\widetilde{p_2} + \alpha_{22}$	$-\widetilde{p_1}$	$-\widetilde{p_1} + \alpha_{11}$	$-\widetilde{p_1} + \alpha_{12}$
V_2	$-\overline{p}$	$-\overline{p}$	$-\overline{p}$	$-\overline{p}$	$-\overline{p}$	$-\overline{p}$

The entry a_{11} gives the payoff of a visitor choosing V_1, matched with another visitor and two firms choosing respectively V_1 and F_1, and so on.

The parameter $\overline{p} \geq 0$ represents the fixed amount paid as an entrance ticket when adopting the strategy V_2. The parameters $(\overline{p} + p_1)$ and $(\overline{p} + p_2)$, with $0 < p_1 < p_2$, are the prices (fixed by PA) of the financial activity described above in the case where, respectively, only one or both visitors choose V_1. So the price of the option is positively correlated to the number of visitors willing to buy it.

The parameters α_{ij} (the index $i = 1, 2$ indicating the number of visitors buying the self-assurance device and the index $j = 1, 2$ indicating the number of polluting firms, i.e. firms choosing F_2), fixed by the PA, are the *reimbursements*, which depend on the strategies adopted by firms and visitors according to the following rule: $\alpha_{1j} > \alpha_{2j}$, $\alpha_{i2} \geq \alpha_{i1}$, $i, j = 1, 2$.

Therefore, if the value of Q is low, the reimbursement is (coeteris paribus) inversely correlated to the number of visitors choosing V_1. For the sake of generality, we do not impose any restriction about the correlation between the reimbursement and the number of firms selling the option, i.e. adopting F_1.

We assume, furtherly, that the option provides actual insurance coverage, that is α_{12}, $\alpha_{22} > p_2$ (and consequently α_{11}, $\alpha_{21} > p_1$).

2.3. *Firms' payoff matrix*

We suppose the following payoff matrix holds for firms:

	F_1, V_1, V_1	F_1, V_1, V_2	F_1, V_2, V_2	F_2, V_1, V_1	F_2, V_1, V_2	F_2, V_2, V_2
F_1	$-c_{NP} + \beta_2$	$-c_{NP} + \beta_1$	$-c_{NP} + \beta_0$	$-c_{NP}$	$-c_{NP}$	$-c_{NP}$
F_2	$-c_P - \overline{q}$	$-c_P - \overline{q}$	$-c_P - \overline{q}$	$-c_P - \overline{q}$	$-c_P - \overline{q}$	$-c_P - \overline{q}$

where c_{NP} and c_P, $0 < c_P < c_{NP}$, represent the costs of the non polluting and the polluting technology, respectively. We assume that $\beta_0 < \beta_1 < \beta_2$, i.e. the financial aid given to firms choosing F_1 increases with the number of visitors subscribing the options offered by the PA. Furthermore we assume the condition $c_{NP} - c_P < \beta_0$, stating that the financial aid is higher than the cost difference.

2.4. *Expected payoffs and dynamics*

Let the variable $x(t)$ denote the proportion of visitors adopting strategy V_1 at the instant of time t, $0 \le x(t) \le 1$.

Analogously let $y(t)$ denote the proportion of firms choosing strategy F_1 at the instant of time t, $0 \le y(t) \le 1$.

The expected payoff of an available strategy (at each one shot matching) depends on the ratio of agents adopting each strategy, ratio that represents the probability to meet agents playing the corresponding strategy.

Let us indicate by:

- $EV_i(x, y)$ the expected payoff of strategy V_i, $i = 1, 2$.
- $EF_j(x, y)$ the expected payoff of strategy F_j, $j = 1, 2$.

The process of adopting strategies is modelled by the so called *replicator dynamics* [see, e.g., [3]], according to which the strategies whose expected payoffs are greater than the average payoff, spread in the populations at the expenses of the others:

$$\dot{x} = x\left(EV_1 - \overline{EV}\right) \qquad (1)$$
$$\dot{y} = y\left(EF_1 - \overline{EF}\right)$$

where:

$$\overline{EV} = x \cdot EV_1 + (1 - x) \cdot EV_2$$

$$\overline{EF} = y \cdot EF_1 + (1 - y) \cdot EF_2$$

are average payoffs of the populations of visitors and of firms, respectively.

Explicitly, we can write (1) as a polynomial system defined in the square $[0, 1]^2$:

$$\dot{x} = x(1 - x)F(x, y) \qquad (2)$$

$$\dot{y} = y(1 - y)G(x, y) \qquad (3)$$

where (posing $\gamma_1 = \alpha_{11} - \alpha_{21}$, $\gamma_2 = \alpha_{12} - \alpha_{22}$, $\delta = p_2 - p_1$):

$$F(x,y) = -x\left[\delta y^2 + (\gamma_1+\delta)\,y(1-y) + (\gamma_2+\delta)\,(1-y)^2\right] \\ +\alpha_{11}y(1-y) + \alpha_{12}(1-y)^2 - p_1(y^2-y+1)$$

$$G(x,y) = y\left[\beta_2 x^2 + \beta_1 x(1-x) + \beta_0(1-x)^2\right] - (c_{NP}-c_P)(x^2-x+1)$$

3. Mathematical results

3.1. *Fixed points*

Notice that the vertices $(x,y) = (0,0)$, $(0,1)$, $(1,0)$, $(1,1)$ of the square $[0,1]^2$ are always fixed points of the system (2), (3). The other fixed points are the intersections:

i) between the locus $F(x,y) = 0$ and the horizontal edges $y = 0$ and $y = 1$ of the square;

ii) between the locus $G(x,y) = 0$ and the vertical edges $x = 0$ and $x = 1$;

iii) between $F(x,y) = 0$ and $G(x,y) = 0$ in the interior, $(0,1)^2$, of the square.

Let us investigate, in particular, the fixed points of type iii).

Note that the expressions in square brackets of $F(x,y)$ and $G(x,y)$ are positive for any y and any x in $[0,1]$, respectively, and that $(c_{NP}-c_P)(x^2-x+1) > 0\ \forall\ x$.

In the open square $(0,1)^2$, the isocline $\dot{x} = 0$ is represented by the intersection between $(0,1)^2$ and the graph of a function

$$x = \widetilde{x}(y) = \frac{P_2(y)}{Q_2(y)} \tag{4}$$

Similarly, the isocline $\dot{y} = 0$ is represented by the intersection between $(0,1)^2$ and the graph of a function

$$y = \widetilde{y}(x) = \frac{S_2(x)}{T_2(x)}, \tag{5}$$

where P_2, Q_2, S_2 and T_2 are degree two polynomials, with $Q_2(y) > 0$ for $y \in [0,1]$ and $S_2(x) > 0$, $T_2(x) > 0$ for $x \in [0,1]$. Clearly (4) and (5) have horizontal asymptotes, respectively, as $y \to \pm\infty$ and as $x \to \pm\infty$. Furthermore the two functions exhibit at most one maximum and one minimum.

First of all, let us consider (4). Under our assumptions, we check that $\widetilde{x}(1) < 0$ and $\widetilde{x}(0) > 1$. Consequently the graph of $\widetilde{x}(y)$ in $[0,1]^2$ cannot present both increasing and decreasing tracts, since in that case some vertical line would have three intersections with it. Thus, in $[0,1]^2$, $\widetilde{x}(y)$ is decreasing.

With regard to the isocline $\dot{y} = 0$, it is easily checked that function (5) has one maximum for $x = x_1 < 0$ and one minimum for $x = x_2 > 1$. Since $\widetilde{y}(0) = \frac{c_{NP} - c_P}{\beta_0} < 1$ (by assumption), the intersection between $[0,1]^2$ and the graph of (5) is again that of a decreasing function.

Finally we observe that the graphs of (4) and (5) can have at most five intersections (eventually outside $[0,1]^2$), in that the number of intersections corresponds to the number of roots of a degree five polynomial.

Notice that the above analysis implies that there are no fixed points in the interior of the edges $y = 0$ and $y = 1$ of $[0,1]^2$, while one fixed point always exists in the interior of each vertical edge, i.e. $x = 0$ and $x = 1$.

3.2. *Stability of fixed points*

Straightforward calculations show that the vertices $(x, y) = (0,1)$, $(1,0)$ are always sinks and the vertices $(x, y) = (0,0)$, $(1,1)$ are saddles. The fixed points in the interior of the edges $x = 0$ and $x = 1$ always have a positive eigenvalue in direction of the edge, so they cannot be sinks.

Now we want to examine the stability of the fixed points in the interior of $[0,1]^2$. Let $(\overline{x}, \overline{y}) \in (0,1)^2$ be a fixed point of the system (2), (3) and let $J = J(\overline{x}, \overline{y})$ be the jacobian matrix evaluated at such a point. It holds:

$$sign\ Det\ J = sign\left(\frac{\partial F}{\partial x}\frac{\partial G}{\partial y} - \frac{\partial F}{\partial y}\frac{\partial G}{\partial x}\right)$$

Therefore (see (4) and (5)):

$$Det\ J(\overline{x}, \overline{y}) > 0 \Leftrightarrow \widetilde{y}'(\overline{x}) \cdot \widetilde{x}'(\overline{y}) > 1 \tag{6}$$

$$Det\ J(\overline{x}, \overline{y}) < 0 \Leftrightarrow \widetilde{y}'(\overline{x}) \cdot \widetilde{x}'(\overline{y}) < 1 \tag{7}$$

In case (6) the fixed point is a sink or a source, in case (7) the fixed point is a saddle.

Proposition 3.1. *There are at most three fixed points in* $(0,1)^2$ *and at most one of them is not a saddle.*

Proof. We bound ourselves to a sketch of the proof, since the required computations are rather lengthy. According what we have seen, the intersections of $F(x, y) = 0$ and $G(x, y) - 0$ with $[0, 1]^2$ can be viewed as the graphs of two decreasing functions, say, respectively, $y = f(x)$ and $y = g(x)$. We can consider, in fact without restriction, the generic case.

A sink or a source correspond to an intersection $(\overline{x}, \overline{y})$ of the two graphs with $f'(\overline{x}) < g'(\overline{x})$. If more than one such point existed, there should be x_1, x_2, x_3, $0 \leq x_1 < x_2 < x_3 \leq 1$, such that $f'(x_1) < g'(x_1)$, $f'(x_2) > g'(x_2)$, $f'(x_3) < g'(x_3)$. So in (x_1, x_3) the two derivatives $f'(x)$ and $g'(x)$ should intersect at least twice. But in fact it can be shown that, given $f'(x_1) < g'(x_1)$, $f'(x)$ and $g'(x)$ can intersect at most once in $(x_1, 1)$. □

Proposition 3.2. *If there is only one fixed point in $(0, 1)^2$ and it is a sink, then there exists a repelling limit cycle.*

Proof. In this case it is easily checked that the fixed points on the vertical edges of $[0, 1]^2$, i.e. $A = \left(0, \frac{c_{NP} - c_P}{\beta_0}\right)$ and $B = \left(0, \frac{c_{NP} - c_P}{\beta_2}\right)$, are saddles. As $\dot{x} < 0$ at A and $\dot{x} > 0$ at B, it follows that the unstable manifolds of the two saddles lie on the respective edges, while the α-limit sets of their stable manifolds must stay in $(0, 1)^2$. Since the only fixed point in $(0, 1)^2$ is assumed to be a sink, then, by the Poincaré Bendixson Theorem, the α-limit set, the same for the two stable manifolds, can only be (generically) a repelling limit cycle surrounding the internal equilibrium. □

3.3. *Some numerical examples*

We rename the coefficients of $F(x, y)$ and $G(x, y)$ as follows:

$$p_2 - p_1 = a,\ \alpha_{11} - \alpha_{21} = b,\ \alpha_{12} - \alpha_{22} = c,\ \alpha_{11} = d,$$
$$\alpha_{12} = e,\ p_1 = f,\ \beta_2 = g,\ \beta_1 = h,\ \beta_0 = i,\ c_{NP} - c_P = l$$

In the Examples below bifurcations in the phase portrait of system (2), (3) take place as a suitable parameter $\varepsilon > 0$ varies.

Set

$$l = 1,\ i = 1.5 - o(\epsilon),\ h = 1.5 + o(\epsilon),\ g = 3$$
$$a = 3 - 2\epsilon,\ f = 1.5 + \epsilon$$
$$d = e = 4.5,\ b = c = o(\epsilon)$$

At $\epsilon = 0$ there is a triple intersection of the isoclines in $\left(\frac{1}{2}, \frac{1}{2}\right)$ (improper saddle). For $\epsilon > 0$ small enough (e.g. $\epsilon = 0.1$, figure 1), posing $o(\epsilon) = 0$, there exist a sink at $\left(\frac{1}{2}, \frac{1}{2}\right)$ and two saddles. The basin of the sink is bounded by the union of the stable manifolds of the saddles. When $\varepsilon = 0.3$ (figure 2) the basin of the sink is bounded by a repelling cycle. Between those two values of ϵ a saddle-connection occurs: at the bifurcation value there appears a *polycycle* constituted by trajectories *connecting* the two saddles.

4. Conclusions

We have seen that the fixed points $(x, y) = (0, 1)$ (where environmental quality is *high*, no visitor purchases and all firms issue the environmental options, i.e. all firms are not polluting) and $(x, y) = (1, 0)$ (where environmental quality is *low*, all visitors purchase and no firm sells the environmental options) are always locally attractive under replicator equations. Therefore the dynamics is always path dependent whatever the parameters values are. The states $(0, 1)$ and $(1, 0)$ are Nash equilibria and can be interpreted as stable *social conventions*, i.e. as strategy distributions which are customary, expected and self-enforcing in the sense of Lewis [4]. Note that visitors and firms' expected payoffs evaluated at $(0, 1)$ and $(1, 0)$ are given respectively by:

$$EV_2(0, 1) = -\overline{p},\ EF_1(0, 1) = -c_{NP} + \beta_0$$

$$EV_1(1, 0) = -\overline{p} - p_2 + \alpha_{22},\ EF_2(1, 0) = -c_P - \overline{q},$$

where $EV_2(0, 1) < EV_1(1, 0)$ and $EF_2(1, 0) < EF_1(0, 1)$.

Therefore, in $(0, 1)$ firms' profits are higher than in $(1, 0)$ and the better performance of firms is obtained in a context of high environmental quality. In $(1, 0)$ visitors' (monetary) payoffs are higher than in $(0, 1)$; however, in $(1, 0)$ visitors' welfare is negatively affected by environmental degradation. So $(1, 0)$ and $(0, 1)$ cannot be ordered in the sense of Pareto, if the functional form of visitors' utility is not known.

In any case, buying the financial option gives visitors a self-assurance device against environmental deterioration, in that if firms' choices generate low environmental quality, subscribing the financial option allows visitors to alleviate welfare reduction due to environmental degradation. Consequently, the Public Administration, by introducing a *market* for environmental options, prevent visitors' welfare from reaching *low* values, which may inhibit individuals from visiting the region.

The fixed points $(0,1)$ and $(1,0)$ are *pure population fixed points*, in that only one strategy is chosen in each population. The above analysis has proved that there may exist fixed points where both strategies coexist in both populations. Among these fixed points, at most one is attractive. It is easy to check that the expected payoff of visitors (respectively, of firms), evaluated at such a fixed point, is higher than the payoff in $(0,1)$ and lower than that in $(1,0)$ (respectively, is higher than the payoff in $(1,0)$ and lower than that in $(0,1)$). When three attractive fixed points are present, the dynamics becomes highly path dependent. As showed above, the existence of an attracting fixed point in the interior of $[0,1]^2$ deeply complicates the morphology of the attraction basins of $(0,1)$ and $(1,0)$, giving rise to an *indeterminacy* result about dynamics.

If environmental protection is the main objective of the PA, obtained not at the expenses of firms' profits, the fixed point $(0,1)$ is the *best* outcome that can be reached by the economy, while the other attracting fixed points can be viewed as *poverty traps*, being characterized by lower profits and environmental quality. To reach the desirable outcome $(0,1)$ is paramount the correct setting of the prices of options. We leave to future research the study of dynamics in markets where prices of options (which are considered as parameters in our model) are *optimally* evaluated by the PA and the market.

References

[1] N. Hanley, J. F. Shogren and B. White. *Environmental economics in theory and practice*, MACMILLAN PRESS, Bristol (1997).

[2] R. Horesh. *Environmental policy bonds: Injecting market incentives into the achievement of society's environmental goals*, OECD Paper, Paris (2002).

[3] J. W. Weibull. *Evolutionary game theory*, MIT Press, Cambridge (1995).

[4] D. Lewis. *Convention*, Harvard University Press, Cambridge (1969).

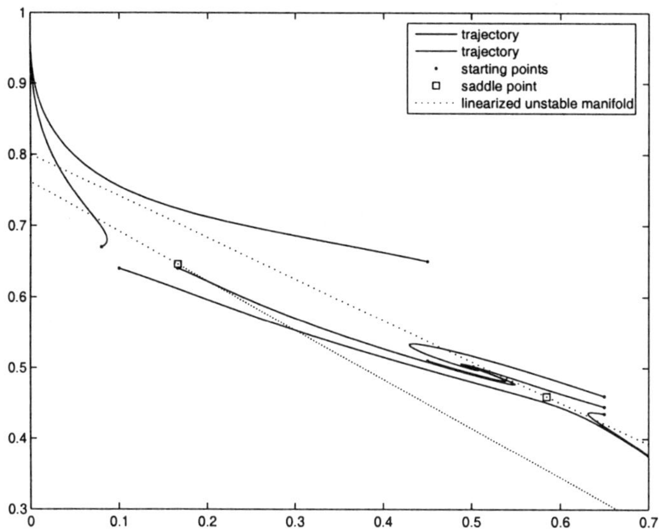

Fig. 1. one sink and two saddles from an improper saddle

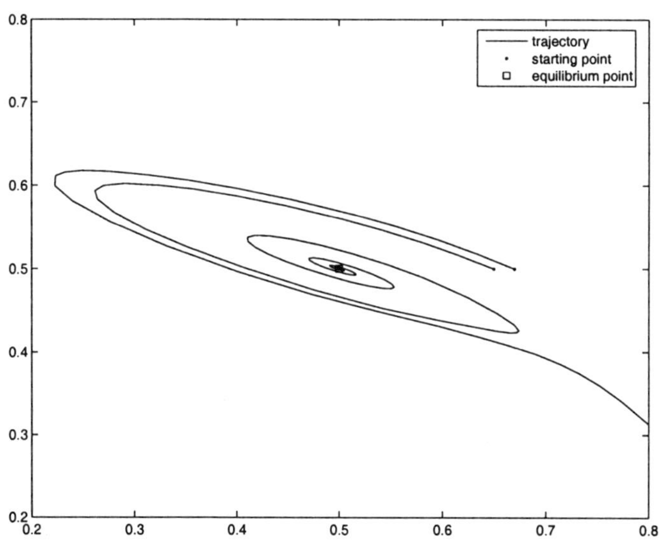

Fig. 2. one sink and two saddles with a repelling limit cycle

MULTIPLICATIVE SCHWARZ ALGORITHMS FOR SYMMETRIC DISCONTINUOUS GALERKIN METHODS

P. F. ANTONIETTI

Dipartimento di Matematica, Università di Pavia,
via Ferrata 7, 27100 Pavia, Italy
E-mail: paola.antonietti@unipv.it
http://www-dimat.unipv.it/antonietti

B. AYUSO

Istituto di Matematica Applicata e Tecnologie Informatiche–CNR,
via Ferrata 7, 27100 Pavia, Italy
E-mail: blanca@imati.cnr.it

We present some multiplicative non-overlapping Schwarz methods for discontinuous Galerkin approximations of second order elliptic problems. The construction and the analysis of two-level Schwarz methods are provided in a unified framework for a wide class of discontinuous Galerkin discretisations. Numerical experiments confirming the theoretical results are also included.

Keywords: Schwarz preconditioners; Discontinuous Galerkin methods.

1. Introduction

Since the earliest domain decomposition (DD) method of H.A. Schwarz was proposed, the development of Schwarz methods for classical conforming methods has been subject of studies and it is by now well understood (for comprehensive reviews see, e.g.,[1,2]). For discontinuous Galerkin (DG) discretisations, it is a completely new field and only few contributions can be found in the literature, see, e.g., Feng and Karakashian[3] , Lasser and Toselli[4] , Brenner and Wang[5] and Antonietti and Ayuso[6,7] .

Based on a totally discontinuous finite element space, the DG method was originally proposed for the numerical approximation of hyperbolic problems and, in recent years, has also become increasingly popular for the discretisation of elliptic equations. We refer to[8] for a unified presentation and analysis of all the DG methods for elliptic problems present in the literature. The reasons for this growth of interest in DG methods are numerous, but essentially lie on the fact that allowing for discontinuities in the finite element approximation gives tremendous flexibility in terms of mesh design and choice of shape functions. For example, DG methods easily handle non-conforming meshes, allow for approximations of various orders (facilitating so the *hp*-adaptivity), permit to approximate weakly the boundary conditions and to handle in a natural way possible discontinuities in the coefficients of the physical model.

In this paper, following the unified *flux formulation* framework proposed by Brezzi *et al.*[8] , we study some multiplicative Schwarz preconditioners for the algebraic linear systems arising from a wide class of symmetric DG approximations of elliptic problems. We prove that a simple Richardson iteration applied to the preconditioned systems converges and we show that the convergence estimates are of order $O(H/h)$. It is well known that for classical discretisation methods, the more sophisticated *substructuring* DD methods provide convergence bounds of order $O(\log(H/h)+1)$. The development of *substructuring* techniques for DG approximations will be subject of future research.

An outline of the paper is as follows. In Sec. 2 we recall the DG approximation of a diffusion problem. In Sec. 3 and Sec. 4 we provide the construction and the analysis of the non-overlapping Schwarz preconditioners. Two-dimensional numerical experiments on conforming and non-matching grids are presented in Sec. 5.

2. Discontinuous Galerkin Methods for Second Order Elliptic Problems

In this section, we set up some notation, introduce the model problem we will consider, and recall the DG formulation based on the *flux formulation.*

For a bounded domain B in $\mathbb{R}^d$, $d = 2,3$, we denote by $H^m(B)$ the standard Sobolev space of order $m \geq 0$. For $m = 0$, we write $L^2(B)$ instead of $H^0(B)$.

Let $\Omega \subset \mathbb{R}^d$, $d = 2,3$ be a convex open polygon or polyhedron and f a given function in $L^2(\Omega)$. We consider the following model problem

$$-\Delta u = f \quad \text{in } \Omega, \quad u = 0 \quad \text{on } \partial\Omega. \tag{1}$$

Remark 2.1. All the results we present are also valid for more general second order elliptic operators in divergence form, with possibly discontinuous coefficients, and other kinds boundary conditions.

In what follows C and c denote generic positive constants that may not be the same at different occurrences but that are always mesh-independent. Moreover, $x \approx y$ means that $cx \leq y \leq Cx$, with C, c positive constants.

2.1. *Meshes, Trace Operators and Finite Element Spaces*

Let $\mathcal{T}_h$ be a shape-regular (not necessarily matching) partition of Ω into disjoint open elements T such that $\overline{\Omega} = \cup_{T\in\mathcal{T}_h}\overline{T}$ where each $T \in \mathcal{T}_h$ is an affine image of a fixed master element $\widehat{T}$, and where $\widehat{T}$ is either the open unit d-simplex or the open unit hypercube in $\mathbb{R}^d$, $d = 2, 3$. Denoting by h_T the diameter of $T \in \mathcal{T}_h$, we define the mesh size h of $\mathcal{T}_h$ by $h = \max_{T\in\mathcal{T}_h} h_T$.

We define the $(d-1)$–dimensional faces (if $d = 2$, "face" means "edge") of $\mathcal{T}_h$ as follows. An *interior face* of $\mathcal{T}_h$ is the (non-empty) interior of $\partial T^+ \cap \partial T^-$, where $T^\pm$ are two adjacent elements of $\mathcal{T}_h$, not necessarily matching. Similarly, a *boundary face* of $\mathcal{T}_h$ is the (non-empty) interior of $\partial T \cap \partial\Omega$, where T is a boundary element of $\mathcal{T}_h$. We denote by $\mathscr{E}^I$ and $\mathscr{E}^B$ the sets of all interior and boundary faces of $\mathcal{T}_h$, respectively, and set $\mathscr{E} = \mathscr{E}^I \cup \mathscr{E}^B$. We will use the convention that

$$\int_{\mathscr{E}} \psi\, ds = \sum_{e\in\mathscr{E}} \int_e \psi\, ds.$$

We shall refer to $\mathcal{T}_h$ as the "fine" mesh and we assume that it satisfies the following assumptions: *i)* $\mathcal{T}_h$ is *locally quasi-uniform*; i.e., $h_{T^-} \approx h_{T^+}$, for all $T^\pm \in \mathcal{T}_h$ such that the interior of $T^+ \cap T^-$ is non empty; *ii)* $\mathcal{T}_h$ is *uniformly graded*; i.e., there exists a positive constant C such that for all $T \in \mathcal{T}_h$ and for all $e \in \mathscr{E}$, $h_T \leq Ch_e$, where h_e is the diameter of $e \in \mathscr{E}$.

We define the *local mesh size* $\mathtt{h}$ on $\mathscr{E}$ by setting $\mathtt{h}(\boldsymbol{x}) := \min\{h_{T^+}, h_{T^-}\}$ if $\boldsymbol{x}$ is in the interior of $\partial T^+ \cap \partial T^-$, and by $\mathtt{h}(\boldsymbol{x}) := h_T$ if $\boldsymbol{x} \in \partial T \cap \partial\Omega$.

Next we introduce some trace operators. Let $e \in \mathscr{E}^I$ be an interior face shared by two elements T^+ and T^- with outward normal unit vectors $\boldsymbol{n}^\pm$, respectively. For piecewise smooth vector-valued and scalar functions $\boldsymbol{\tau}$ and v, let $\boldsymbol{\tau}^\pm$ and $v^\pm$ be the traces of $\boldsymbol{\tau}$ and v on $\partial T^\pm$ taken from the interior of $T^\pm$, respectively. We define the *jump* and *weighted average* operators by

$$[\![\boldsymbol{\tau}]\!] = \boldsymbol{\tau}^+ \cdot \boldsymbol{n}^+ + \boldsymbol{\tau}^- \cdot \boldsymbol{n}^-, \qquad [\![v]\!] = v^+\boldsymbol{n}^+ + v^-\boldsymbol{n}^-,$$
$$\{\!\{\boldsymbol{\tau}\}\!\}_\delta = \delta\boldsymbol{\tau}^+ + (1-\delta)\boldsymbol{\tau}^-, \qquad \{\!\{\boldsymbol{v}\}\!\}_\delta = \delta\boldsymbol{v}^+ + (1-\delta)\boldsymbol{v}^-,$$

where the parameter $\delta \in [0, 1]$ (for $\delta = 1/2$ we drop the subindex and we simply write $\{\!\{\cdot\}\!\}_{1/2} = \{\!\{\cdot\}\!\}$). On a boundary face $e \in \mathscr{E}^B$, we set $[\![\boldsymbol{\tau}]\!] = \boldsymbol{\tau}\cdot\boldsymbol{n}$, $[\![v]\!] = v\,\boldsymbol{n}$, $\{\!\{\boldsymbol{\tau}\}\!\}_\delta = \boldsymbol{\tau}$ and $\{\!\{v\}\!\}_\delta = v$.

Let $\ell_h \geq 1$ be a given approximation order. For $\widehat{T}$ the open unit d-simplex, let $\widehat{\mathcal{P}}^{\ell_h}(\widehat{T})$ be the set of polynomials of total degree ℓ_h on $\widehat{T}$, and for $\widehat{T}$ the open unit hypercube in $\mathbb{R}^d$, let $\widehat{\mathcal{Q}}^{\ell_h}(\widehat{T})$ be the set of all tensor-product polynomials on $\widehat{T}$ of degree ℓ_h in each coordinate direction. For a given partition $\mathcal{T}_h$, we define the discontinuous finite element space V_h as

$$V_h = \{v \in L^2(\Omega) : v|_T \in \mathcal{M}^{\ell_h}(T) \quad \forall T \in \mathcal{T}_h\},$$

where $\mathcal{M}^{\ell_h}(T)$ is the mapping to T of $\widehat{\mathcal{M}}^{\ell_h}(\widehat{T})$, and $\mathcal{M}$ is either $\mathcal{P}$ or $\mathcal{Q}$.

Finally, we denote by $H^2(\mathcal{T}_h)$ the space of functions on Ω whose restriction to each element $T \in \mathcal{T}_h$ belongs to the Sobolev space $H^2(T)$.

2.2. *Discontinuous Galerkin Discretisation*

The *flux formulation* can be obtained by introducing the auxiliary variable $\boldsymbol{\sigma} = \nabla u$ and by rewriting problem (1) as a first order system of equations. By setting $\boldsymbol{\Sigma}_h = [V_h]^d$, we consider the following DG methods: find $(\boldsymbol{\sigma}_h, u_h) \in \boldsymbol{\Sigma}_h \times V_h$ such that, for all $(\boldsymbol{\tau}, v) \in \boldsymbol{\Sigma}_h \times V_h$, and for all $T \in \mathcal{T}_h$

$$\begin{aligned} &\int_T \boldsymbol{\sigma}_h \cdot \boldsymbol{\tau}\, dx + \int_T u_h \nabla \cdot \boldsymbol{\tau}\, dx - \int_{\partial T} \widehat{u}_h \boldsymbol{\tau} \cdot \boldsymbol{n}_T\, ds = 0, \\ &\int_T \boldsymbol{\sigma}_h \cdot \nabla v\, dx - \int_{\partial T} \widehat{\boldsymbol{\sigma}}_h \cdot \boldsymbol{n}_T\, v\, ds = \int_T f\, v\, dx, \end{aligned} \tag{2}$$

Here $\widehat{u} = \widehat{u}(u_h)$ and $\widehat{\boldsymbol{\sigma}} = \widehat{\boldsymbol{\sigma}}(\boldsymbol{\sigma}_h, u_h)$ are the scalar and vector *numerical fluxes*, respectively, and are nothing but approximations of the traces of u and $\boldsymbol{\sigma} = \nabla u$, respectively, on the boundary of T. Provided that $\widehat{u}$ is chosen independently of $\boldsymbol{\sigma}$, the variable $\boldsymbol{\sigma}$ can actually be eliminated, in an element-by-element manner, by using the first equation of (2), obtaining DG methods in primal form. Hence, denoting by ∇_h the elementwise application of the operator ∇, and defining, for $u, v \in H^2(\mathcal{T}_h)$,

$$\begin{aligned} A(u,v) = \int_\Omega \nabla_h u \cdot \nabla_h v\, dx + \int_{\mathcal{E}} [\![\widehat{u} - u]\!] \cdot \{\!\{\nabla_h v\}\!\}\, ds + \int_{\mathcal{E}^I} \{\!\{\widehat{u} - u\}\!\}\, [\![\nabla_h v]\!]\, ds \\ - \int_{\mathcal{E}} \{\!\{\widehat{\boldsymbol{\sigma}}\}\!\} \cdot [\![v]\!]\, ds - \int_{\mathcal{E}^I} [\![\widehat{\boldsymbol{\sigma}}]\!]\, \{\!\{v\}\!\}\, ds, \end{aligned} \tag{3}$$

the *primal* DG formulation reads: find $u_h \in V_h$ such that

$$A(u_h, v) = \int_\Omega f v\, dx,$$

for all $v \in V_h$. The definition of the *numerical fluxes* $\widehat{u}$ and $\widehat{\boldsymbol{\sigma}}$ as a suitable linear combination of averages and jumps of u_h and $\boldsymbol{\sigma}_h$ determines the different DG methods (see Table 1).

Table 1. Numerical fluxes on interior faces.

Method	$\widehat{u}$	$\widehat{\boldsymbol{\sigma}}$
SIPG[9]	$\{\!\{u_h\}\!\}$	$\{\!\{\nabla_h u_h\}\!\} - \alpha\,\mathrm{h}^{-1}\,[\![u_h]\!]$
SIPG(δ)[10]	$\{\!\{u_h\}\!\}_{(1-\delta)}$	$\{\!\{\nabla_h u_h\}\!\}_\delta - \alpha\,\mathrm{h}^{-1}\,[\![u_h]\!]$
LDG[11]	$\{\!\{u_h\}\!\} - \boldsymbol{\beta}\cdot[\![u_h]\!]$	$\{\!\{\boldsymbol{\sigma}_h\}\!\} + \boldsymbol{\beta}\cdot[\![\boldsymbol{\sigma}_h]\!] - \alpha\,\mathrm{h}^{-1}\,[\![u_h]\!]$

For instance, by choosing the numerical fluxes on internal and boundary faces, respectively, as

$$\widehat{u}|_e = \{\!\{u_h\}\!\}_{1-\delta}, \quad \widehat{\boldsymbol{\sigma}}|_e = \{\!\{\nabla_h u_h\}\!\}_\delta - \alpha\,\mathbf{h}\,[\![u_h]\!], \quad e \in \mathscr{E}^I,$$
$$\widehat{u}|_e = 0, \quad \widehat{\boldsymbol{\sigma}}|_e = \nabla_h u_h - \alpha\,\mathbf{h}\,u_h\,\boldsymbol{n}, \quad e \in \mathscr{E}^B,$$

we obtain the SIPG(δ) method of Stenberg[10] , and, for the particular choice $\delta = 1/2$, the SIPG method of Arnold[9] . The parameter $\alpha > 0$ (at our disposal) is independent of the mesh size. The local discontinuous Galerkin (LDG) method of Cockburn and Shu[11] can be obtained by choosing

$$\widehat{u}|_e = \{\!\{u_h\}\!\} + \boldsymbol{\beta}\cdot[\![u_h]\!], \quad \widehat{\boldsymbol{\sigma}}|_e = \{\!\{\boldsymbol{\sigma}_h\}\!\} - \boldsymbol{\beta}\,[\![\boldsymbol{\sigma}_h]\!] - \alpha\,\mathbf{h}\,[\![u_h]\!],$$

if e is an interior face, and $\widehat{u}|_e = 0$, $\widehat{\boldsymbol{\sigma}}|_e = \boldsymbol{\sigma}_h - \alpha\,\mathbf{h}u_h\,\boldsymbol{n}$, if e is a boundary face. The parameter $\boldsymbol{\beta} \in \mathbb{R}^d$ is taken such that $\|\boldsymbol{\beta}\|_{[L^\infty(\mathscr{E}^I)]^d} \leq C$.

From now on, since no confusion might arise, we drop the subindex h from the discrete functions.

3. Multiplicative Schwarz Preconditioners

In this section, we present our Schwarz methods for the DG discretisations introduced before. We refer to Antonietti and Ayuso[6,7] where Schwarz methods for a wider class of DG approximations are studied. We first introduce some notation, then we describe our algorithms in the variational form and from the algebraic point of view.

3.1. *Non-overlapping Partitions, Local and Coarse Solvers*

We consider three level of *nested* partitions of the domain Ω satisfying the previous assumptions: a subdomain partition $\mathcal{T}_N$ made of N non-overlapping subdomains Ω_i, a coarse partition $\mathcal{T}_H$ (with global mesh size H) and a fine partition $\mathcal{T}_h$ (with global mesh size h). For each Ω_i of $\mathcal{T}_N$ let $\mathscr{E}_i$ be the set of all faces of $\mathscr{E}$ (recall that $\mathscr{E}$ is the set of all faces of the fine partition) belonging to $\overline{\Omega}_i$, we define $\mathscr{E}_i^I = \{e \in \mathscr{E}_i : e \subset \Omega_i\}$, and

$$\mathscr{E}_i^B = \{e \in \mathscr{E}_i : e \subset \partial\Omega_i \cap \partial\Omega\}, \quad \Gamma_i = \{e \in \mathscr{E}_i : e \subset \partial\Omega_i \setminus \partial\Omega\}.$$

For $i = 1, \dots, N$, we next introduce our *local solvers*. We define the *local spaces* V_h^i as

$$V_h^i = \{u \in L^2(\Omega_i) \,:\, u|_T \in \mathcal{M}^{\ell_h}(T) \ \forall T \in \mathcal{T}_h, T \subset \Omega_i\},$$

and the *prolongation* operators $R_i^T : V_h^i \longrightarrow V_h$ as the classical injection operators[a] from the local space V_h^i to V_h. Our *local solvers* are defined by considering the DG approximation of the problem:

$$-\Delta u_i = f|_{\Omega_i} \ \text{ in } \partial\Omega_i, \quad u_i = 0 \ \text{ on } \Omega_i.$$

In view of (3), the local bilinear forms $A_i : V_h^i \times V_h^i \longrightarrow \mathbb{R}$ are defined by

$$\begin{aligned} A_i(u_i, v_i) &= \int_{\Omega_i} \nabla_h u_i \cdot \nabla_h v_i \, dx + \int_{\mathcal{E}_i} [\![\widehat{u}_i - u_i]\!] \cdot \{\!\{\nabla_h v_i\}\!\} \, ds \\ &+ \int_{\mathcal{E}_i^I} \{\!\{\widehat{u}_i - u_i\}\!\} \, [\![\nabla_h v_i]\!] \, ds - \int_{\mathcal{E}_i} \{\!\{\widehat{\boldsymbol{\sigma}}_i\}\!\} \cdot [\![v_i]\!] \, ds - \int_{\mathcal{E}_i^I} [\![\boldsymbol{\sigma}_i]\!] \, \{\!\{v_i\}\!\} \, ds, \end{aligned}$$

where $\widehat{u}_i$ and $\widehat{\boldsymbol{\sigma}}_i$ are the *local* numerical fluxes, and are defined, on $e \in \mathcal{E}_i^I$, as the numerical fluxes $\widehat{u}$, $\widehat{\boldsymbol{\sigma}}$ of the global DG method on interior faces, and, on $e \in \mathcal{E}_i^B \cup \Gamma_i$, as $\widehat{u}$ and $\widehat{\boldsymbol{\sigma}}$ on boundary faces.

Remark 3.1. From the definition of $A_i(\cdot, \cdot)$, it follows that we are using *approximate*[b] local solvers, that is $A(R_i^T u_i, R_i^T u_i) \neq A_i(u_i, u_i)$. In[6,7] it is shown that the following *local stability* property holds true

$$A(R_i^T u_i, R_i^T u_i) \leq \omega A_i(u_i, u_i) \quad \forall\, u_i \in V_h^i,$$

with $1 < \omega < 2$, provide that the penalty parameter is chosen big enough (see Remark 4.2 below).

Finally, we define our *coarse solver*. For a given approximation order $0 \leq \ell_H \leq \ell_h$, we define the coarse space V_h^0 as $V_h^0 = \{v_H \in L^2(\Omega) : v_H|_T \in \mathcal{M}^{\ell_H}(T) \ \forall T \in \mathcal{T}_H\}$, and the injection operator $R_0^T : V_h^0 \longrightarrow V_h$. The coarse solver is defined as the restriction to $V_h^0 \times V_h^0$ of $A(\cdot, \cdot)$, i.e.,

$$A_0(u_0, v_0) = A(R_0^T u_0, R_0^T v_0) \quad \forall\, u_0, v_0 \in V_h^0.$$

---

[a]For vector-valued functions R_i^T is defined componentwise.
[b]In all the previously proposed Schwarz methods (see, e.g.,[3,5]) *exact* local solvers were employed, i.e., $A(R_i^T u_i, R_i^T u_i) = A_i(u_i, u_i)$ for all $u_i \in V_h^i$.

3.2. *Variational and Algebraic Formulations*

For $i = 0, \dots, N$, we define the projection operator $P_i = R_i^T \widetilde{P}_i : V_h \longrightarrow V_h$, where $\widetilde{P}_i : V_h \longrightarrow V_h^i$ is defined as

$$A_i(\widetilde{P}_i u, v_i) = A(u, R_i^T v_i) \quad \forall\, v_i \in V_h^i.$$

The multiplicative Schwarz operator we consider is given by

$$P_{mu} = I - (I - P_N)(I - P_{N-1}) \cdots (I - P_0),$$

where $I : V_h \longrightarrow V_h$ is the identity operator. We also define the error propagation operator $E_N = (I - P_N)(I - P_{N-1}) \cdots (I - P_0)$ and observe that $P_{mu} = I - E_N$. Since even for symmetric DG approximations P_{mu} is not symmetric, we also consider the following *symmetrized* version of the multiplicative Schwarz operator

$$P_{mu}^{sym} = I - (I - P_0^*) \cdots (I - P_N^*)(I - P_N) \cdots (I - P_0),$$

where, for $i = 0, \dots, N$, P_i^* is the adjoint operator of P_i with respect to the inner product induced by $A(\cdot, \cdot)$, i.e., $A(P_i^* u, v) = A(u, P_i v)\ \forall\, u, v \in V_h$. Following[1,2] , the Schwarz method consists in replacing the discrete problem $Au = f$ by the equation $Pu = g$, where P can be either P_{mu} or P_{mu}^{sym}, and with the corresponding appropriate right-hand side g.

The Schwarz operators can be seen as preconditioned linear systems. The algebraic formulation problem of (2.2) is given by $\boldsymbol{A}\boldsymbol{u} = \boldsymbol{f}$. By taking into account that, for $i = 0, \dots, N$, the matrix representation of the projection-like operator P_i is $\boldsymbol{P}_i = \boldsymbol{R}_i^T \boldsymbol{A}_i^{-1} \boldsymbol{R}_i \boldsymbol{A}$, we have

$$\boldsymbol{P}_{mu} = \boldsymbol{I} - (\boldsymbol{I} - \boldsymbol{P}_N) \cdots (\boldsymbol{I} - \boldsymbol{P}_0) = \boldsymbol{B}_{mu} \boldsymbol{A},$$

The matrix $\boldsymbol{B}_{mu}$ is called Schwarz preconditioner. The preconditioned system to be solved is then $\boldsymbol{B}_{mu}\boldsymbol{A}\boldsymbol{u} = \boldsymbol{B}_{mu}\boldsymbol{f}$. Analogously, for the symmetrized version by taking into account that $\boldsymbol{P}_i^* = \boldsymbol{P}_i$, we get

$$\boldsymbol{P}_{mu}^{sym} = \boldsymbol{I} - (\boldsymbol{I} - \boldsymbol{P}_0) \cdots (\boldsymbol{I} - \boldsymbol{P}_N)(\boldsymbol{I} - \boldsymbol{P}_N) \cdots (\boldsymbol{I} - \boldsymbol{P}_0) = \boldsymbol{B}_{mu}^{sym} \boldsymbol{A}.$$

As already noticed, $\boldsymbol{P}_{mu}$ is not symmetric, and therefore a suitable iterative solver, as the *generalized minimal residual method* (GMRES) has to be used for solving the resulting linear system. For its symmetrized version, a linear solver designed for symmetric linear systems as the *conjugate gradient* (CG) method can be used as an acceleration method.

4. Convergence Results

In this section, following the abstract theory of Schwarz methods[1,2] , we present the convergence results. We refer to[7] for their proofs.

We introduce the norm induced by the bilinear form $A(\cdot,\cdot)$, that is

$$\|u\|_A^2 = A(u,u) \quad \forall\, u \in V_h,$$

and observe that $A(\cdot,\cdot)$ does indeed define an inner product and $\|\cdot\|_A$ is a norm provided the penalty parameter is taken so as to ensure the coercivity (in a suitable mesh-dependent norm) of $A(\cdot,\cdot)$. We denote by N_c the maximum number of adjacent subdomains a given subdomain can have, and define

$$C_0^2 = O\left(\max_{D\in\mathcal{T}_H} \frac{\operatorname{diam}(D)}{\min_{\substack{T\in\mathcal{T}_h \\ T\subset D}} \operatorname{diam}(T)} \right). \tag{4}$$

Remark 4.1. Whenever the fine and coarse partitions are globally quasi uniform, i.e, $H_D \approx H$ for all $D \in \mathcal{T}_H$ and $h_T \approx h$ for all $T \in \mathcal{T}_h$, we can write $C_0^2 = O(H/h)$.

In Theorem 4.1, we show that the energy norm of the error propagation operator

$$\|E_N\|_A^2 = \sup_{\substack{u\in V_h \\ u\neq 0}} \frac{A(E_N u, E_N u)}{A(u,u)},$$

is strictly smaller than one. This is a sufficient condition to ensure that a simple Richardson iteration applied to the preconditioned system converges.

Theorem 4.1. *Let $A(\cdot,\cdot)$ be the bilinear form of one of the DG methods given in Table 1. Then, there exists $\widetilde{\alpha} > 0$ such that if $\alpha \geq \widetilde{\alpha}$*

$$\|E_N\|_A^2 \leq 1 - \frac{2-\omega}{\alpha\, C_0^2(1 + 2\,\omega^2(N_c+1)^2)} < 1,$$

with $C_0^2 = O(H/h)$ (see (4) and Remark 4.1).

Theorem 4.1 also guarantees that the multiplicative Schwarz method can be accelerated with the GMRES linear solver.

As a direct consequence of Theorem 4.1 we can guarantee the convergence of the symmetrized multiplicative Schwarz method.

Corollary 4.1. *Let $A(\cdot,\cdot)$ be the bilinear form of one of the DG methods given in Table 1, and let $E_{N_s}^{sym}$ be the error propagation operator of the* *symmetrized multiplicative Schwarz method, i.e.,* $E_N^{sym} = E_N^* E_N$. *Then, there exists* $\widetilde{\alpha}$ *such that if* $\alpha \geq \widetilde{\alpha}$,

$$\|E_N^{sym}\|_A \leq 1 - \frac{2-\omega}{\alpha\, C_0^2(1+2\,\omega^2(N_c+1)^2)} < 1,$$

with $C_0^2 = O(H/h)$ *(see* (4) *and Remark 4.1).*

Since P_{mu}^{sym} is self-adjoint with respect to $A(\cdot,\cdot)$, we can use the Rayleigh quotient characterization of the extreme eigenvalues, i.e.,

$$\lambda_{\min}^A(P_{mu}^{sym}) = \min_{\substack{u\in V_h\\ u\neq 0}} \frac{A(P_{mu}^{sym}u,u)}{A(u,u)}, \quad \lambda_{\max}^A(P_{mu}^{sym}) = \max_{\substack{u\in V_h\\ u\neq 0}} \frac{A(P_{mu}^{sym}u,u)}{A(u,u)}.$$

The condition number $\kappa(P_{mu}^{sym})$ of P_{mu}^{sym} is given by

$$\kappa(P_{mu}^{sym}) = \frac{\lambda_{\max}^A(P_{mu}^{sym})}{\lambda_{\min}^A(P_{mu}^{sym})}.$$

We can use Corollary 4.1 to provide the following bound on the condition number $\kappa(P_{mu}^{sym})$ of the symmetrized multiplicative Schwarz method:

$$\kappa(P_{mu}^{sym}) \leq \frac{\alpha^* C_0^2\left(1+2\,\omega^2(N_c+1)^2\right)}{2-\omega}. \tag{5}$$

Remark 4.2. As in the classical Schwarz theory, our convergence analysis for P_{mu} and P_{mu}^{sym} relies upon the hypothesis that $\omega \in (1,2)$. Since we are using approximate local solvers, we need a technical assumption on the size of the penalty parameter to guarantee $\omega \in (1,2)$. Nevertheless, we wish to stress that the assumed size of $\widetilde{\alpha}$ is moderate (see[7] for details).

5. Numerical Experiments

We present some two-dimensional numerical experiments to illustrate the performance of the considered non-overlapping Schwarz methods. The performance of the proposed Schwarz methods with three-dimensional test cases will be the subject of future research. We take $\Omega = (0,1)\times(0,1)$, we choose an exact solution $u(x,y) = \exp(xy)$ and we adjust the load f and the (non-homogeneous) boundary conditions accordingly. The subdomain partitions consist of N squares, $N = 4, 16$ (see Fig. 1 for $N = 4$). We consider both matching and non-matching Cartesian grids and unstructured triangular grids (see Fig. 1 where the initial coarse and fine grids are depicted). We denote by H_0 and h_0 the corresponding initial coarse and fine mesh sizes, respectively, and we consider n successive global uniform refinements

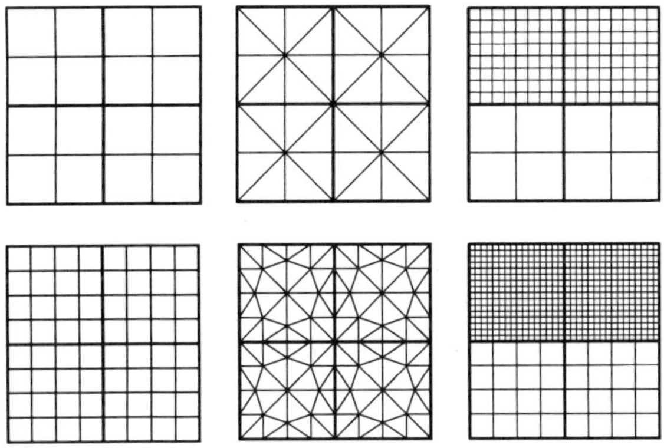

Fig. 1. Initial coarse (top) and fine (bottom) refinements on Cartesian grids, unstructured triangular grids and non-matching Cartesian grids, respectively, with $N = 4$.

of these initial grids so that the resulting mesh sizes are $H_n = H_0/2^n$ and $h_n = h_0/2^n$, with $n = 0, 1, 2, 3$. The (relative) tolerance is set to 10^{-9}.

We first address the scalability of the proposed multiplicative Schwarz method, i.e., the independence of the convergence rate of the number of subdomains. In Table 2 we compare the GMRES iteration counts for the SIP method ($\alpha = 10$) with $\ell_h = \ell_H = 1$ obtained on unstructured triangular grids (see Fig. 1) with $N = 4, 16$. The crosses in the last line of Table 2 (and below) indicate that we were not able to solve the non-preconditioned system due to the excessive GMRES memory storage requirements. As stated in Theorem 4.1, our preconditioner seems to be insensitive on the number of the subdomains. It can be also seen that, for fixed H we observe an asymptotic behaviour $O(1/\sqrt{h})$, and for fixed h, the computed convergence behaviour is slightly better than $O(\sqrt{H})$.

Table 2. $\boldsymbol{P}_{mu}\boldsymbol{u} = \boldsymbol{g}$: GMRES iteration counts.
SIPG method ($\alpha = 10$), $\ell_h = \ell_H = 1$, unstructured triangular grids.

	$N = 4$				$N = 16$			
$H \downarrow\ h \rightarrow$	h_0	$h_0/2$	$h_0/4$	$h_0/8$	h_0	$h_0/2$	$h_0/4$	$h_0/8$
H_0	11	18	30	50	11	17	28	47
$H_0/2$	-	11	19	35	-	10	17	30
$H_0/4$	-	-	12	24	-	-	11	18
$H_0/8$	-	-	-	17	-	-	-	12
# iter($\boldsymbol{A}$)	94	182	350	x	94	182	350	x

The rest of the numerical experiments of this section have been carried with $N = 16$. We address the performance of the multiplicative preconditioner for higher order polynomial approximations. In Table 3 we have reported the GMRES iteration counts obtained with the LDG method ($\alpha = 1$, $\boldsymbol{\beta} = (0.5, 0.5)^T$) on Cartesian grids, by choosing $\ell_h = 2$ and $\ell_H = 1, 2$. Notice that, by choosing $\ell_h = \ell_H = 2$ our preconditioner performs better than with $\ell_h = 2$ and $\ell_H = 1$. Moreover, our preconditioner performs well also for small values of the penalty parameter α, confirming that the hypothesis on the size of α, required in Theorem 4.1 is only technical and it is not needed in practice.

Table 3. $\boldsymbol{P}_{mu}\boldsymbol{u} = \boldsymbol{g}$: GMRES iteration counts.
LDG method ($\alpha = 1$, $\boldsymbol{\beta} = (0.5, 0.5)^T$), Cartesian grids.

	$\ell_h = \ell_H = 2$				$\ell_h = 2$, $\ell_H = 1$			
$H \downarrow$ $h \rightarrow$	h_0	$h_0/2$	$h_0/4$	$h_0/8$	h_0	$h_0/2$	$h_0/4$	$h_0/8$
H_0	12	16	21	28	22	30	40	53
$H_0/2$	-	9	12	16	-	17	23	32
$H_0/4$	-	-	7	8	-	-	16	21
$H_0/8$	-	-	-	5	-	-	-	13
# iter($\boldsymbol{A}$)	112	210	403	x	112	210	403	x

Now, we investigate the effect of the choice of a coarse space V_H made of piecewise constants ($\ell_H = 0$) on the performance of our multiplicative Schwarz method. In Table 4 we compare the GMRES iteration counts obtained with the SIPG method ($\alpha = 10$) on non-matching Cartesian grids with $\ell_h = 1$ and $\ell_H = 1, 0$. Clearly, a piecewise constant coarse solver deteriorates the performance of the multiplicative Schwarz preconditioner.

Table 4. $\boldsymbol{P}_{mu}\boldsymbol{u} = \boldsymbol{g}$: GMRES iteration counts.
SIPG method ($\alpha = 10$), non-matching Cartesian grids.

	$\ell_h = \ell_H = 1$				$\ell_h = 1$, $\ell_H = 0$			
$H \downarrow$ $h \rightarrow$	h_0	$h_0/2$	$h_0/4$	$h_0/8$	h_0	$h_0/2$	$h_0/4$	$h_0/8$
H_0	13	20	30	42	45	64	89	122
$H_0/2$	-	10	16	24	-	47	66	91
$H_0/4$	-	-	9	14	-	-	46	63
$H_0/8$	-	-	-	8	-	-	-	44
# iter($\boldsymbol{A}$)	277	538	x	x	277	538	x	x

Finally, we present some numerical computations obtained with the symmetrized multiplicative Schwarz preconditioner. In Table 5 we report the condition number estimates and the CG iteration counts computed with the SIPG ($\alpha = 10$) method by choosing $\ell_h = \ell_H = 1$ on Cartesian grids. Clearly, the numerical results reported in Table 5 confirm the theoretical condition number estimate given in (5): the condition number of the preconditioned systems behaves like $O(H/h)$.

Table 5. $\boldsymbol{P}_{mu}^{sym}\boldsymbol{u} = \boldsymbol{g}$: condition number estimates and CG iteration counts. SIPG method ($\alpha = 10$), $\ell_h = \ell_H = 1$, Cartesian grids.

	$\kappa(\boldsymbol{P}_{mu}^{sym})$					CG # iter$(\boldsymbol{P}_{mu}^{sym})$			
$H \downarrow$ $h \rightarrow$	h_0	$h_0/2$	$h_0/4$	$h_0/8$		h_0	$h_0/2$	$h_0/4$	$h_0/8$
H_0	5.17	10.42	21.16	43.15		13	21	29	37
$H_0/2$	-	4.77	9.38	18.65		-	11	17	23
$H_0/4$	-	-	4.73	9.40		-	-	10	15
$H_0/8$	-	-	-	4.54		-	-	-	9
$\kappa(\boldsymbol{A})$	265.29	1043.06	4155.47	16605.72	# iter$(\boldsymbol{A})$	77	155	303	592

References

1. B. F. Smith, P. E. Bjørstad and W. D. Gropp, *Domain decomposition: parallel multilevel methods for elliptic partial differential equations* (Cambridge University Press, Cambridge, 1996).
2. A. Toselli and O. Widlund, *Domain decomposition methods—algorithms and theory*, Springer Series in Computational Mathematics, Vol. 34 (Springer-Verlag, Berlin, 2005).
3. X. Feng and O. A. Karakashian, *SIAM J. Numer. Anal.* **39**, 1343 (2001).
4. C. Lasser and A. Toselli, *Math. Comp.* **72**, 1215 (2003).
5. S. C. Brenner and K. Wang, *Numer. Math.* **102**, 231 (2005).
6. P. F. Antonietti and B. Ayuso, *Schwarz domain decomposition preconditioners for discontinuous Galerkin approximations of elliptic problems: non-overlapping case*, *M2AN Math. Model. Numer. Anal.*, to appear.
7. P. F. Antonietti and B. Ayuso, *Multiplicative Schwarz methods for discontinuous Galerkin approximations of elliptic problems*, tech. rep., IMATI-CNR 10-PV (2006), submitted.
8. D. N. Arnold, F. Brezzi, B. Cockburn and L. D. Marini, *SIAM J. Numer. Anal.* **39**, 1749 (2001/02).
9. D. N. Arnold, *SIAM J. Numer. Anal.* **19**, 742 (1982).
10. R. Stenberg, Mortaring by a method of J. A. Nitsche, in *Computational Mechanics: New Trends and Applications*, (CIMNE, Barcelona, 1998).
11. B. Cockburn and C.-W. Shu, *SIAM J. Numer. Anal.* **35**, 2440 (1998).

TOPOLOGICAL CALCULUS: BETWEEN ALGEBRAIC TOPOLOGY AND ELECTROMAGNETIC FIELDS

W. ARRIGHETTI[†] and G. GEROSA[*]

Department of Electronic Engineering, "La Sapienza" Università di Roma, via Eudossiana 18, 00184, Rome, Italy
E-mail: arrighetti@die.uniroma1.it[†], gerosa@die.uniroma1.it[*]
www.die.uniroma1.it/strutture/labcem/

Topological behaviour of self-similar spectra for fractal domains is shown and applied to solve electromagnetic problems on fractal geometries, like for example the Sierpinski gasket.. Two different mathematical tools are employed: the Topological Calculus,[1,2] which frames a topology-consistent,[3] discrete counterpart to domains and operators and the Iterated Function Systems (IFSs)[4] to produce fractals as limit sets of simple recursion mappings. Topological invariants and Analytical features of a set can be easily extracted from such a discrete model, even for complex geometries like fractal ones.

One of the targets of this work is to show how recursion symmetries of a (pre-) fractal set, mathematically coded by "algebraic" relationships between its parts, are sole responsible for the self-similar distribution of its (laplacian) eigenvalues: no metric information is needed for this property to be observed.

Another primary target is to show how Topological Calculus easily allows for an almost instantaneous discretization of contoinuum equations of any (topological) field theory. Investigating the natural modes of self-similar domains is important to many applications whose core geometry is prefractal or at least highly irregular. Most recently, transport[5,6] and electromagnetic pehnomena were focused: IFS-generated waveguides, resonators[7,8] and antennas[9] exhibiting multi-band properties. Such complex domains need careful mathematical formulations in order to transfer traditional-geometric properties to them; Topological Calculus is one of such discrete formulations.

Keywords: Topological Calculus; simplicial complex; cochain; boundary; coboundary; cohomology; adjacency matrix; Serpinski gasket; self-similar spectrum; Maxwell's equations

1. Introduction to topological calculus

1.1. *Simplices and simplicial complexes*

Topological Calculus deals with topological field theories in the setting of Simplicial Cohomology on infinite fields (e.g. $\mathbb{R}$ or $\mathbb{C}$).[3]

Let $\mathbf{v}_0, \mathbf{v}_1, ..., \mathbf{v}_p \in \mathbb{R}^n$ be geometrically independent, i.e.

$$\dim_{\mathbb{R}}\text{span}(\mathbf{v}_0 - \mathbf{v}_k)_{1\leq k\leq p} = p.$$

The *p-simplex* formed by these $p+1$ vectors is the smallest convex closed set containing them, i.e.:

$$\sigma = (\mathbf{v}_0|\mathbf{v}_1|\ldots|\mathbf{v}_p) \stackrel{\text{def}}{=} \left\{ \sum_{i=0}^{p} t_i\mathbf{v}_i \;\middle|\; t_0, t_1, \ldots, t_p \in [0,1], \; \sum_{i=0}^{p} t_i = 1 \right\}. \quad (1)$$

A p-simplex is the p-dimensional analogue of a triangle (i.e. a 2-simplex): 0-simplices are points, 1-simplices segments, 3-simplices tetrahedra. The boundary $\partial\sigma$ of a simplex is the union of $p+1$ $(p-1)$-simplices (those with each one of its vertices removed). An *oriented* p-simplex is one which the permutation order of an ordering relation "$\prec$" among its vertices is set to: for $\mathbf{v}_0 \prec \mathbf{v}_1 \prec \ldots \prec \mathbf{v}_p$ the corresponding simplex is written as $\vec{\sigma} := [\mathbf{v}_0, \mathbf{v}_1, \ldots, \mathbf{v}_p]$. Even permutations correspond to the same oriented simplex $\vec{\sigma}$; odd permutations correspond to the other one: $\overleftarrow{\sigma} :\equiv -\vec{\sigma}$.

A *simplicial complex* $\Sigma \subseteq \mathbb{R}^n$ is the union set of a finite (or countable) number of simplices $\sigma_1, \sigma_2, \sigma_3, \ldots$ such that: $\Sigma = \sigma_1 \cup \sigma_2 \cup \sigma_3 \cup \ldots$; $\sigma \cap \sigma'$ is a simplex of Σ. Formally, $\sigma \cap \sigma' \in \Sigma$ and $\sigma^\circ \cap \sigma'^\circ = \emptyset$, $\forall \sigma, \sigma' \in \Sigma$. Two simplices $\sigma, \sigma' \in \Sigma$ are *adjacent* to each other if their intersection is non-empty (thus is a lower-dimensional simplex of Σ itself); that is written as $\sigma @_\Sigma \sigma'$. The *dimension* $\dim\Sigma$ of Σ is the maximum dimension of its simplices. Whenever an orientation is chosen for all its $\dim\Sigma$-simplices, Σ is *oriented* and written as $\vec{\Sigma}$.

A *p-adjacency matrix* $H_{\Sigma,p}$ can be defined, for any $1 \leq p \leq \dim\Sigma$ (and after a formal ordering among its p-simplices), as skew-symmetric such that its $(i,j)^{\text{th}}$ element is either ± 1 or 0 if the ith p-simplex is either adjacent to the jth one (sign given by compatible orientations) or not, respectively.[2] As Σ $\left[\vec{\Sigma}\right]$, for $\dim\Sigma = 1$ is algebraically equivalent to a simple [oriented] graph[a], $H_{\Sigma,0}$ coincides with the graph's incidence matrix H_Σ.

1.2. *Simplicial homology and cohomology*

Let $G(+)$ be an abelian group. A G-valued *p-chain* on $\vec{\Sigma}$ is a function $c_p : \vec{\Sigma} \longrightarrow G$ such that $c_p(\overleftarrow{\sigma}) = -c_p(\vec{\sigma})$, $\forall \sigma \in \Sigma$. The set of all the G-valued p-chains on $\vec{\Sigma}$ is the abelian group $\mathrm{C}_p(\Sigma; G)$; the elements of its

[a]skΣ, the *skeleton* of Σ, is the simplicial complex formed by Σ's 0- and 1-simplices.

conjugate group, $\mathrm{C}^p(\Sigma; G)$, are called *p-cochains*. Let $R(+, \cdot)$ be an abelian ring. R-valued chains (and cochains accordingly) can be represented via *elementary chains*: for every p-simplex $\sigma \in \Sigma$ an elementary p-chain $C_\sigma \in \mathrm{C}_p(\Sigma; R)$ exists such that $C_\sigma(\vec{\tau}) = \delta_{\sigma,\tau}$. This way every (co-)chain can be decomposed with coefficients $c_{\vec{\sigma}} \in R$, one for every simplex:

$$C_\sigma = \sum_{\sigma \in \mathrm{C}_p(\Sigma; R)} c_{\vec{\sigma}} \, \vec{\sigma} \, . \tag{2}$$

The *p-boundary* linear operator $\partial_p : \mathrm{C}_p(\Sigma; R) \longrightarrow \mathrm{C}_{p-1}(\Sigma; R)$ is defined to map every elementary p-chain to a linear combination (with alternating ± 1 unity's signs) of elementary $(p-1)$-chains associated to its boundarying $(p-1)$-simplices, i.e. (identifying a simplex's elementary chain with the simplex itself):

$$\partial_p \left[\mathbf{v}_0, \mathbf{v}_1, \ldots, \mathbf{v}_p\right] = \sum_{i=0}^{p} (-1)^i \left[\mathbf{v}_0, \ldots, \mathbf{v}_{i-1}, \mathbf{v}_{i+1}, \ldots, \mathbf{v}_p\right] \tag{3}$$

The *p-coboundary* is the dual operator $\delta_p : \mathrm{C}^p(\Sigma; R) \longrightarrow \mathrm{C}^{p+1}(\Sigma; R)$, acting by linearly combining all the elementary $(p+1)$-cochains adjacent to a given elementary p-cochain:[5,6]

$$\delta_p \, \vec{\sigma} = \sum_{\mathbf{u} @_\Sigma \sigma} [\mathbf{u}, \vec{\sigma}], \tag{4}$$

where $[\mathbf{u}, \vec{\sigma}]$ is the oriented $(p+1)$-simplex formed by vertex $\mathbf{u}$ and the vertices of σ, with the orientation (and ordering) inherited by $\vec{\sigma}$'s.

p-(co-)chains annihilated by the corresponding (co-)boundary operator are called *p-(co-)cycles* (because their values on consecutive simplices forming a closed rpath inside Σ sum to zero); with this definition in mind, the following groups are defined:

$$\mathrm{Z}_p(\Sigma; R) \stackrel{\text{def}}{=} \mathrm{Ker}\partial_p, \quad \mathrm{B}_p(\mathit{\Sigma}; R) \stackrel{\text{def}}{=} \mathrm{Im}\partial_{p+1}; \tag{5}$$

$$\mathrm{Z}^p(\mathit{\Sigma}; R) \stackrel{\text{def}}{=} \mathrm{Ker}\delta_p, \quad \mathrm{B}^p(\mathit{\Sigma}; R) \stackrel{\text{def}}{=} \mathrm{Im}\delta_{p-1}. \tag{6}$$

Finally, the quotient groups defined by them are called p^{th} *Homology* and p^{th} *Cohomology* groups, repsectively:

$$\mathrm{H}_p(\mathit{\Sigma}; R) \stackrel{\text{def}}{=} \frac{\mathrm{Z}_p(\mathit{\Sigma}; R)}{\mathrm{B}_p(\mathit{\Sigma}; R)}; \quad \mathrm{H}^p(\mathit{\Sigma}; R) \stackrel{\text{def}}{=} \frac{\mathrm{Z}^p(\mathit{\Sigma}; R)}{\mathrm{B}^p(\mathit{\Sigma}; R)}. \tag{7}$$

Elements in $\mathrm{H}_p(\mathit{\Sigma}; R)$ are (equivalence classes of) p-cycles being boundary to *no* $(p+1)$-chains; there are as many as the number of p-dimensional "holes", i.e. the pth *Betti number*[b] $\beta_p(\Sigma)$, of Σ. Elements in $\mathrm{H}^p(\Sigma; R)$ are p-cocycles being coboundary to *no* $(p-1)$-cochains as well.

1.3. *Spectral decomposition*

Infinite-field-valued p-chain and p-cochain groups[c] on finite simplicial complexes, such those used in Topological Calculus[d], are isomorphic vector spaces. This fact has both drawbacks and advantages.[2] As for the group-valued case, operations defined on the field get inherited by the (co-)chain spaces, whose bases are formed by the elementary (co-)chains. Aso, homology and cohomology spaces are isomorphic to each other, i.e. $\mathrm{H}_p(\Sigma; \mathbb{R}) \cong \mathrm{H}^p(\Sigma; \mathbb{R})$.

The pth *Laplace-Beltrami* operator (or *p-beltramian*) $\Delta_p : \mathrm{C}^p(\Sigma; \mathbb{R}) \longrightarrow \mathrm{C}_p(\Sigma; \mathbb{R})$ is defined as[e]:

$$\Delta_p \stackrel{\text{def}}{=} -\left(\partial_{p+1}\delta_p - \delta_{p-1}\partial_p\right). \tag{8}$$

Its action on elementary p-chains (i.e., formally, on every p-simplex) is proven to be:

$$\begin{aligned}\Delta_p\left[\mathbf{v}_0, \mathbf{v}_1, \ldots, \mathbf{v}_p\right] = &- \left[\mathbf{v}_0, \mathbf{v}_1, \ldots, \mathbf{v}_p\right] \mathrm{ad}_\Sigma \left[\mathbf{v}_0, \mathbf{v}_1, \ldots, \mathbf{v}_p\right] + \\ &- 2\sum_{i=0}^{p} \\ &\sum_{\mathbf{u} @_\Sigma [\mathbf{v}_0, \ldots, \not{\mathbf{v}}_i, \ldots, \mathbf{v}_p]} \left[\mathbf{v}_0, \ldots, \mathbf{v}_{i-1}, \mathbf{u}, \mathbf{v}_{i+1}, \ldots, \mathbf{v}_p\right]. \end{aligned} \tag{9}$$

Simpler cases are $p = 0$ and $p = n$, as either one of the I-order simplicial operators vanishes, so $\Delta_0 \equiv -\partial_1\delta_0$ and $\Delta_n \equiv \delta_{n-1}\partial_n$:

$$\Delta_0[\mathbf{v}] = -[\mathbf{v}]\,\mathrm{ad}_\Sigma\mathbf{v} + \sum_{\mathbf{u} @_\Sigma \mathbf{v}} [\mathbf{u}], \tag{10}$$

$$\begin{aligned}\Delta_n\left[\mathbf{v}_0, \mathbf{v}_1, \ldots, \mathbf{v}_n\right] = &- \sum_{i=0}^{p} \\ &\sum_{\mathbf{u} @_\Sigma [\mathbf{v}_0, \ldots, \not{\mathbf{v}}_i, \ldots, \mathbf{v}_n]} \left[\mathbf{v}_0, \ldots, \mathbf{v}_{i-1}, \mathbf{u}, \mathbf{v}_{i+1}, \ldots, \mathbf{v}_n\right]. \end{aligned} \tag{11}$$

---

[b]The sequence of Betti numbers for all p's is a topological invariant for a given simplicial complex (or topological space[3]).

[c]Algebraic Topology usually deals with integers Z or their finite moduli, Z_p.

[d]Real numbers' field $\mathbb{R}$ will be used here, but (hyper-)complex one would do too.

[e]$\mathrm{ad}_\Sigma\mathbf{v}$ is vertex $\mathbf{v}$'s *adjacency number*, i.e. the number of its adjacent vertices.

In a topological field theory, boundary, coboundary and Laplace-Beltrami operators are the analogue to vector calculus' ones[f] (grad, div, curl, ∇^2).

Lemma 1.1. *$\partial_{p+1}\delta_p$ and $\delta_{p-1}\partial_p$ operators of a simplicial complex $\vec{\Sigma}$ are self-adjoint, non-negative definite and self-commuting. Furthermore $\mathrm{Z}_p(\Sigma;\mathbb{R}) = \mathrm{Ker}(\partial_{p+1}\delta_p)$ and $\mathrm{Z}^p(\Sigma;\mathbb{R}) = \mathrm{Ker}(\delta_{p-1}\partial_p)$.*

Lemma 1.1 implies that the beltramian operator is self-adjoint and non-positive definite as well. Topological properties of (co-)homology spaces directly involve the following orthogonal decomposition for p-chains.

Theorem 1.1. $\mathrm{C}_p(\Sigma;\mathbb{R}) = \mathrm{B}_p(\Sigma;\mathbb{R}) \oplus \mathrm{B}^p(\Sigma;\mathbb{R}) \oplus \mathrm{H}^p(\Sigma;\mathbb{R})$, *i.e. for every p-chain x a $(p+1)$-chain ϕ, a $(p-1)$-chain ψ and a unique harmonic[g] p-chain h exist such that:*

$$x = \partial_{p+1}\phi + \delta_{p-1}\psi + h. \tag{12}$$

The presence of an harmonic p-chain into the decomposition, which is the analogue of Hodge's Decomposition Theorem,[10] depends on the presence of p-dimensional "holes" inside the simplicial complex: there are as many indipendent harmonic chains as $\beta_p(\Sigma)$ (as well as the TEM modes for a non-simply-connected waveguide[7]).

Using elementary p-chains as a basis for $\mathrm{C}_p(\Sigma;\mathbb{R})$, the representative matrix of Δ_p is (first equation is valid for $0 < p < n$):

$$\begin{cases} [\Delta_p] = -2\,|H_{\Sigma,p}| - \mathrm{diag}\,(\mathrm{ad}_\Sigma\sigma)_{\sigma\in\Sigma} \\ [\Delta_0] = -\,|H_{\Sigma,0}| - \mathrm{diag}\,(\mathrm{ad}_\Sigma\mathbf{v})_{\mathbf{v}\in\Sigma} \\ [\Delta_n] = -\,|H_{\Sigma,n}| \end{cases} \tag{13}$$

where $|H_\Sigma|$ is the matrix whose elements are the absolute values of the incidence matrix ones.

2. Simplicial electromagnetics

The linear equations of any classical field theory can be translated in the setting of Topological Calculus almost at sight, by also triangulating the continuum domain; for nonlinear equations one has to resort to simplicial analogue to inner and outer (exterior) products: the *cap* and *cup* products.[2]

---

[f] p-cochains are, in fact, the discrete counterpart to differential p-forms.[10]
[g] p-chains $h \in \mathrm{H}^p(\Sigma;\mathbb{R})$ are said to be *harmonic* because $\Delta_p h = 0$, which is the simplicial version of Laplace's equation.

For example, consider the Electromagnetics on a simplicial complex: force fields like the electric field E and the magnetic field H are represented by 1-cochains, since they are physically, directly measured along paths; the electric displacement field D and the magnetic flux density field B are represented by 2-cochains, since they are computed as fluxes, i.e. across surfaces (surface densities, like the electric density current, J). The electric charge density instead, being evaluated inside volumes, is represented by a 3-cochain ϱ. The Maxwell's equations are written in the time domain as:

$$\begin{cases} \delta_2 D = \varrho \\ \delta_2 B = 0 \\ \delta_1 E + \dot{B} = 0 \\ \delta_1 H - \dot{D} = J \end{cases} \tag{14}$$

By Fourier-transforming the Maxwell's equations in the frequency domain $\omega \in \mathbb{R}$, complex cochains are considered,

$$\begin{cases} \delta_2 \hat{D} = \hat{\varrho} \\ \delta_2 \hat{B} = 0 \\ \delta_1 \hat{E} + i\omega \hat{B} = 0 \\ \delta_1 \hat{H} - i\omega \hat{D} = \hat{J} \end{cases} \tag{15}$$

where, for example, the value of $E(\omega)$ on a side (1-simplex) is the value of the electric field vector along that direction within the simplicial complex, whereas the value of $\varrho(\omega)$ on a specific tetrahedron (3-simplex) is the electric charge contained within that three-dimensional region; both quantities referring to electrons (or plane waves' spectra) at frequency $\omega/(2\pi)$. The electrodynamic potentials A and V are defined such that, in the *Lorenz' gauge* $\partial_1 A - c^{-2}\dot{V} = 0$, the electromagnetic fields are governed by simplified Helmholtz's equations (where $\kappa = \omega\sqrt{\epsilon\mu}$), here written in the source-less case:

$$\begin{cases} \Delta_0 \hat{V} + \kappa^2 \hat{V} = 0 \\ \Delta_1 \hat{A} + \kappa^2 \hat{A} = 0 \end{cases} \tag{16}$$

As one can see, the solution of any electromagnetic problem in a bounded, triangulated region heavily depends on the behaviour of the Laplace-Beltrami operator on simplicial complexes, which will be analyzied in the next chapter for a special fractal case. A simple application of this simplicial formulation is the immediate recovering of topology-based features of the electromagnetic fields: the 1st cohomology space for example corresponds to the harmonic fields such that $\Delta_1 \hat{A} = 0$, i.e. to the static potentials existing only if the domain is not simply connected. In fact, the number of such independent static modes equals the numbers of holes inside the domain, which is exactly its 1st Betti number. For a two-dimensional triangulation of an electromagnetic waveguide's cross-section, the computation of the cohomology space recovers, for example, all the (finite) *TEM* modes of the waveguide.

3. Prefractal simplicial complexes

3.1. *Simplicial IFSs*

An *Iterated Function System* (IFS) is a finite set $\{\mathbf{w}_1, \mathbf{w}_2, \ldots, \mathbf{w}_p\}$ of contraction mappings on a complete metric space (e.g. $\mathbb{R}^n$),[4,11]). Let $E_0 \subset \mathbb{R}^n$ be a compact 'initiator', which a sequence $(E_N)_{N\in\mathbb{N}_0}$ of compact sets is generated from:

$$E_{N+1} = \mathbf{w}(E_N) := \bigcup_{j=1}^{p} \mathbf{w}_j(E_N), \qquad \forall N \in \mathbb{N}_0. \tag{17}$$

It is well known that map $\mathbf{w}$ is contracting in the Hausdorff's metric space of compact sets[11] and has one fixed point (a compact set $F \subset \mathbb{R}^n$), called the IFS' *attractor*: $\lim_N \mathbf{w}^N(E_0) = F$, for every compact set $E_0 \subset \mathbb{R}^n$. F is often self-similar and has non-integer fractal dimensions. The IFS is "*just-touching*" if $\mathbf{w}_i^\circ \cap \mathbf{w}_j^\circ = \emptyset$, always for $i < j$. Finite order iterate $E_N \subset \mathbb{R}^n$ is called the IFS' N^{th} *prefractal* (of initiator E_0).

An IFS is said to be *simplicial* if its contraction mappings are simplicial mappings too, i.e. if its iterates computed starting from a simplicial complex Σ_0 still form a sequence of simplicial complexes $(\Sigma_N)_{N\in\mathbb{N}}$. As the simplicial complexes are essentially algebraic objects, despite being defined as closed sets in an Euclidean space, any metric information can be neglected whenever the field theory which is to be modeled is, in fact, topological. In this context all the numerics are done with just Σ's adjacency matrices; a simplicial IFS is just specified by the way the adjacency matrices are recursively built (and grow in dimension, as the number of simplices increases).

Figure 1 shows an initiator $\check{\Sigma}_0$ and the first 3 iterations of a *simplicial Sierpinski gasket*. As the Sierpinski gasket's IFS is "just-touching", $\check{\Sigma}_n$ is made by three copies of $\check{\Sigma}_{n-1}$, linked with each other via 15 triangles, 5 for every mutual joints between the 3 copies of $\check{\Sigma}_{n-1}$. The incidence and adjacency matrices are iteratively built using the following renormalization scheme ($1 \le p \le 2$):[5]

$$H_{\check{\Sigma}_{n,p}} = L_{n,p} + \bigoplus^{3} H_{\check{\Sigma}_{n-1},p}, \tag{18}$$

Fig. 1. Simplicial Sierpinski gaskets $\check{\Sigma}_n$, $0 \leq n \leq 3$.

where the skew-symmetric matrix $L_{n,p}$ adds nonzero elements for every linkage p-simplices to the diagonal-block matrix whose blocks are all equal to $\check{\Sigma}_{n-1}$'s adjacency matrix (one for each copy). Algebraic Topology provides an easy way to compute $\check{\Sigma}_n$'s lacunarity:

$$\beta_1(\check{\Sigma}_n) = \dim_{\mathbb{Z}} \mathrm{H}_1(\check{\Sigma}_n) = \frac{3^n - 1}{2}. \tag{19}$$

3.2. *Self-similar spectrum of simplicial prefractals*

The combination of Topological Calculus and simplicial IFSs often allows the extraction of useful discrete topology-based parameters for complex domains (e.g. modeling porous or irregular-shaped materials). Once the studied domain is triangulated[h] (i.e. adjacency matrices are computed) many discretized differential operators have a simple closed-form representation in this frame. The triangulation can also be furtherly refined to embrace finer topological 'defects'. Geometric self-similarity of the domain is reflected by the self-similar diagonal-block structure of its adjacency matrices (18) and, via (9)-(13), to the beltramians' spectra.[7]

The simplicial Sierpinski gasket can be used as an example for diffusion (Fourier's equation) as well as electromagentic (Helmholtz's equation)

[h] A *triangulation* is a homeomorphism from a topological space to a simplicial complex, e.g. the discretization required to model a continuum euclidean domain to an unstructured simplicial mesh.

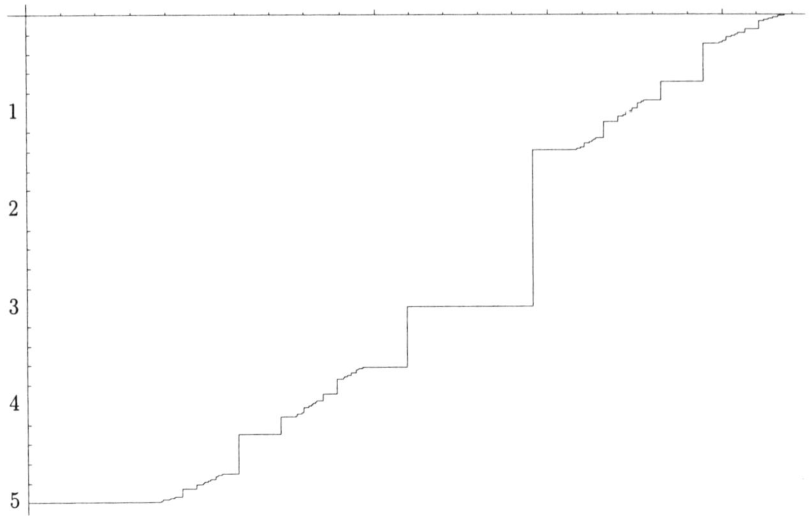

Fig. 2. Spectrum of 0-beltramian for simplicial Sierpinski gasket's 6th prefractal $\check{\Sigma}_6$.

problems on fractals. Matrices representing Δ_0, Δ_1 and Δ_2 operators depend on 0-, 1- and 2-simplex adjacencies and are computed accordingly. The following figures show the eigenvalues' number vs. eigenvalue plots of the Laplace-Beltrami operator on the Sierpinski gasket's 6th prefractal: Fig.2 is referred to the 0-chains' case (i.e. the simplicial analogue to scalar fields), Fig.3 to 1-chains' case (analogue to vector fields), Fig.4 to 2-chains' (discrete analogue to vector and pseudovector fields, respectively).

Horizontal "plateaux" on the plots are associtated to multiple eigenvalues (the broader the plateau, the greater is the degeneracy). The eigenvalues' distribution is self-similar, as well as eigenfunctions' patterns; this fact is connected to the presence of several eigenfunctions localized on distinct copies into the prefractal: $\check{\Sigma}_n$ admits 3 copies of $\check{\Sigma}_{n-1}$'s natural modes, which admits 3 copies of $\check{\Sigma}_{n-2}$'s natural modes itself (i.e. $\check{\Sigma}_n$ admits 9 copies of $\check{\Sigma}_{n-2}$'s ones), and so on. Such modes, whose patterns are directly inherited by previous-order prefractals (and whose degeneracy is reflected by multiple-eigenvalues' plateaux in Figs. 2,3,4), are called *diaperiodic* modes and naturally exist on continuous models of prefractals too:

Every new iteration also brings newer modes for $\check{\Sigma}_n$, due to the topological influence of the interconnections between the three copies of $\check{\Sigma}_{n-1}$, which admits 'interconnective modes' due to its three copies of $\check{\Sigma}_{n-2}$, and so on.

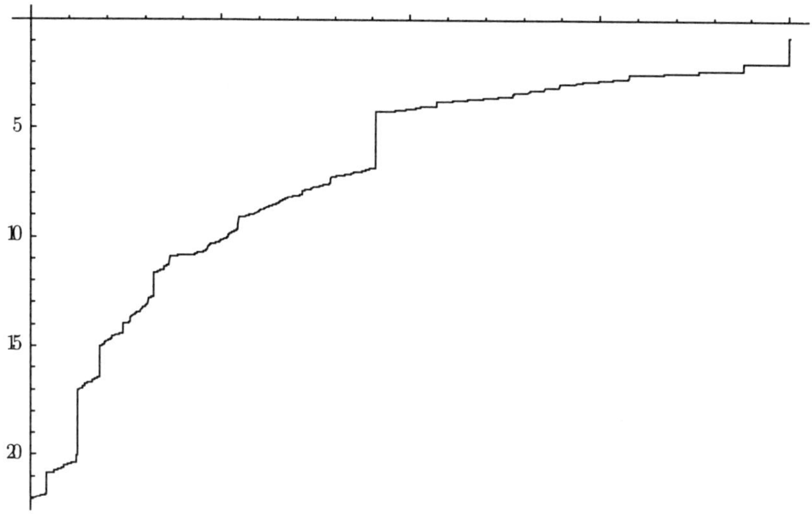

Fig. 3. Spectrum of 1-beltramian for simplicial Sierpinski gasket's 6th prefractal $\check{\Sigma}_6$.

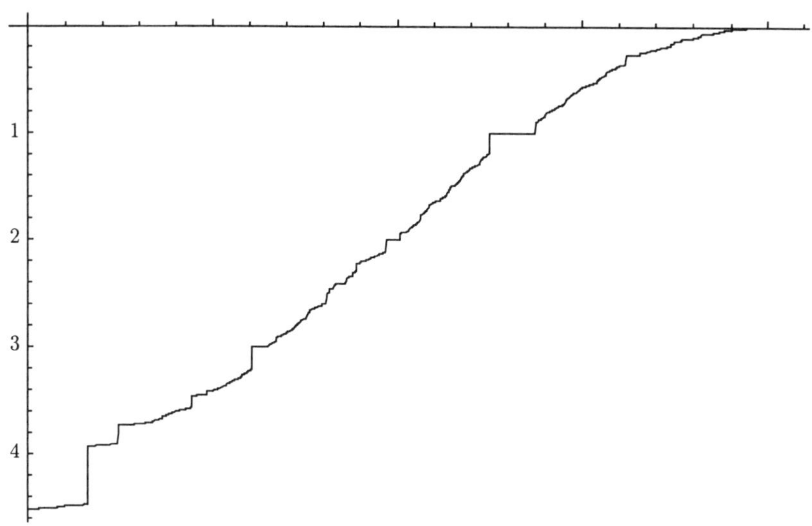

Fig. 4. Spectrum of 2-beltramian for simplicial Sierpinski gasket's 6th prefractal $\check{\Sigma}_6$.

The presence of self-similar modes – and diaperiodic ones above all – has been already observed in the 'continuum case'[7,8] (for prefractal waveguides' cross-sections and resonators), but it can be observed for prefractal simplicial complexes too. That also proves that "fractality" is not only a metric property of a set (like in the usual IFSs[11]), but also depends on the way the studied domain is built up from its very components. In a purely topological reasoning (where both the domain, i.e. the simplicial complex, and the operators do not bear metric information) the algebrics behind the construction of a simplicial IFS are enough to observe self-similar spectral properties.

References

1. W. Arrighetti, *Simplicial prefractals and Topological Calculus* (Fractals in Engineering V, Tours, 2004).
2. W. Arrighetti, *Analysis of Fractal Electromagnetic Structures*, Laurea thesis, "La Sapienza" University, Rome, (2002).
3. A. Hatcher, *Algebraic Topology* (Cambridge University Press, Cambridge, 2002), www.math.cornell.edu/~hatcher/.
4. K. Falconer, *Fractal Geometry: mathematical foundations and applications* (Wiley, Chichester, 1990).
5. W.A. Schwalm, B. Moritz, M. Giona, M.K. Schwalm, *Vector difference calculus for physical lattice models*, *Phys Rev E* **59**: 1217 (1999).
6. M. Giona, *Contour integrals and vector calculus on fractal curves and interfaces*, *Chaos Soliton Fract* **10**: 1349 (1999).
7. W. Arrighetti, G. Gerosa, *Spectral analysis of Šerpinskij carpet-like prefractal waveguides and resonators*, *IEEE Microw Wirl Comp* **15**: 30 (2005).
8. W. Arrighetti, G. Gerosa, *Can you hear the fractal dimension of a drum?*, in *Applied and Industrial Mathematics in Italy*, eds. M. Primicerio, R. Spigler, V. Valente, Series on Advances in Mathematics for Applied Sciences, Vol. 69, pp. 65–75 (World Scientific, 2005), www.arxiv.org/abs/math.SP/0503748.
9. J. Parrón, J. Romeu, J.M. Rius, J.R. Mosig, *Method of moments enhancement technique for the analysis of Sierpinski pre-fractal antennas*, *IEEE T Antenn Propag* **51**: 1872 (2003).
10. D. Huybrechts, *Complex Geometry* (Springer, Berlin, 2005).
11. M.F. Barnsley, *Fractals Everywhere* (Acedemic Press, London, 1993).

DETERMINISTIC SOLUTION OF BOLTZMANN EQUATIONS GOVERNING THE DYNAMICS OF ELECTRONS AND PHONONS IN CARBON NANOTUBES

CH. AUER and F. SCHÜRRER

Institute of Theoretical and Computational Physics,
Graz University of Technology,
8010 Graz, Austria
E-mail: auer@itp.tu-graz.ac.at, schuerrer@itp.tu-graz.ac.at

C. ERTLER

Institute of Theoretical Physics, University of Regensburg,
93053 Regensburg, Germany
E-mail: Christian.Ertler@physik.uni-regensburg.de

It is an interesting challenge to utilize carbon nanotubes for future electronic building blocks. We present a kinetic model for the dynamic description of the interplay of electrons and phonons in carbon nanotubes. Simulations based on a deterministic solution procedure to the Boltzmann-like equations lead to a better understanding of the influence of hot phonons on the temporal evolution of the electron current in metallic nanotubes on a substrate.

Keywords: Carbon nanotube; Kinetic theory; Hot-phonon effect

1. Introduction

Single-wall carbon nanotubes (SWNTs) are of great scientific and engineering interest due to their remarkable electrical properties. These features originate from their crystalline one-dimensional structure and the strong covalent carbon-carbon bonding. SWNTs are important candidates for interconnects and high performance field effect transistors[1–3] in future electronic devices.

SWNTs can be thought as a single layer of a graphite crystal rolled up to a seamless cylinder. From a crystallographical point of view, they are characterized by their wrapping vector $\boldsymbol{R}_{nm} = (n, m) = n\boldsymbol{a}_1 + m\boldsymbol{a}_2$ determining the folding of the graphite sheet, where $\boldsymbol{a}_1$ and $\boldsymbol{a}_2$ are the graphene lattice vectors.[4] Depending on the wrapping vector, various chiralities of nanotubes can be obtained which govern the electrical characteristics of nanotubes. We are interested in investigating metallic SWNTs for which $n = m$, so-called armchair nanotubes. At first glance, nanotubes are considered as one-dimensional quantum wires with ballistic electron transport. However, based on high-field transport measurements, it was found that the scattering of electrons with optical phonons destroys the ballistic behavior.[5,6] Moreover, the generation of optical phonons during high-field electron transport, which can be directly detected by a Raman scattering experiment as mentioned by Lazzeri et. al. in,[7] is essential to understand the reduction of the conductivity at high fields.

In the past, the high-field transport in metallic SWNTs was studied at a macroscopic level in[6,8–10] or by solving the semiclassical Boltzmann equation in.[5,6] In the latter case, the dynamics of electrons was treated in a kinetic way, while the phonons were kept in equilibrium at a fixed lattice temperature. In order to model the effect of hot phonons on the distribution of electrons, we introduced a kinetic model for both electrons and optical phonons.[11] With this model, we investigated the steady-state transport in metallic SWNTs lying on a substrate and found good agreement with measurements. A further study of the high-field transport in SWNTs was presented in.[12]

In this paper, we investigate the transient behavior of interacting electrons and phonons in carbon nanotubes. In section 2, we present our kinetic model, which is a powerful tool for accurately treating transport phenomena influenced by interactions between hot-phonons and hot-electrons. The applied deterministic solution procedure is focused on in section 3. We clarify the hot phonon effect in transients of microscopic and macroscopic quantities in section 4. Finally, important conclusions are drawn in section 5.

2. Transport Model

We concentrate on armchair nanotubes with equal chiral indices $n = m$. The allowed electronic states are then characterized by two equivalent points $\boldsymbol{K}$ and $\boldsymbol{K}' = 2\boldsymbol{K}$ in the reciprocal space.[4] In this case, the electronic energies are well approximated by the linear dispersion relations

$$\varepsilon_i(k) = \hbar v_i k, \quad i = 1, 2 \tag{1}$$

as plotted in Fig. 1, where $v_1 = +v_F$ and $v_2 = -v_F$ are the positive and negative Fermi velocities, respectively.[7] With $\hbar$ we denote the reduced Plank constant $h/2\pi$ and k stands for the component of the electron momentum

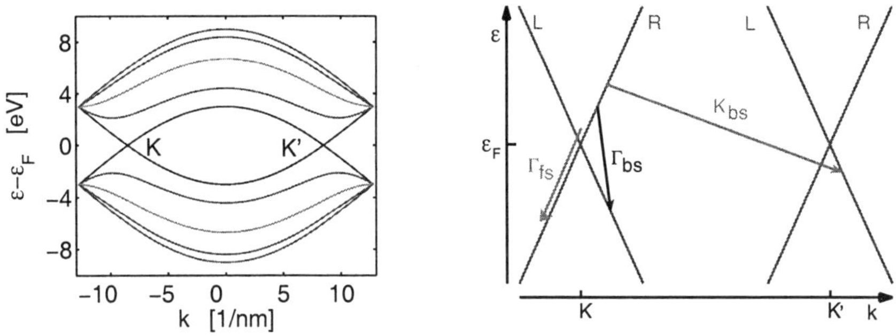

Fig. 1. Left plot: electronic band structure of a metallic SWNT. Right plot: linear approximation of the band structure around the Fermi energy ϵ_F at the symmetry points K and K′ and scattering processes for Γ- and K-phonons (bs=back scattering, fs=forward scattering).

along the tube axis. Electrons in corresponding states according to $\boldsymbol{K}$ and $\boldsymbol{K}'$ can be considered as equivalent in our transport model. Therefore, it is sufficient to introduce only two distribution functions $f_i(\varepsilon, x, t)$ for right ($i = 1$) and left moving ($i = 2$) electrons[5] with $\varepsilon = \varepsilon_i(k)$, position x along the tube axis and time t. We assume tube diameters of approximately $d = 2$ nm, which allows us to neglect higher energy subbands for electrons with energies <0.5 eV.

The Boltzmann equations

$$\partial_t f_i + v_i \partial_x f_i - e_0 v_i E \partial_\varepsilon f_i = \mathcal{C}_i, \tag{2}$$

govern the evolution of the distribution functions $f_i = f_i(\varepsilon, x, t)$ with E denoting the electric field along the tube axis and e_0 the electron charge. The collision operators

$$\mathcal{C}_i = \mathcal{C}_i^{ac} + \sum_{\eta=1}^{3} \mathcal{C}_i^{\eta} \tag{3}$$

include the interaction of electrons with acoustic phonons, described by

$$\mathcal{C}_i^{ac} = \frac{v_F}{l_{ac}}(f_j - f_i), \quad j \neq i. \tag{4}$$

The operators

$$\begin{aligned} \mathcal{C}_i^{\eta} = \gamma_\eta \Big\{ & g_\eta(q_i^-) f_j^- (1 - f_i) + [g_\eta(q_i^+) + 1] f_j^+ (1 - f_i) \\ & - g_\eta(q_i^+) f_i (1 - f_j^+) - [g_\eta(q_i^-) + 1] f_i (1 - f_j^-) \Big\}, \quad j \neq i, \end{aligned} \tag{5}$$

model back- and forward scattering with phonons of the modes $\eta = 1, 2$ and $\eta = 3$, respectively (Fig. 1). In Eq. (4), l_{ac} stands for the acoustic mean free path (MFP).[5] Electron scattering at impurities can be taken into account by inserting the elastic MFP $l_e = 1/(1/l_{ac} + 1/l_{im})$ instead of l_{ac} in (4). Further, in Eq. (5), γ_η denotes the electron-phonon coupling constants. We used the abbreviations $f_i = f_i(\varepsilon, x, t)$, $f_i^\pm = f_i(\varepsilon \pm \hbar\omega_\eta, x, t)$ and for $\eta = 1, 2$, $q_i^\pm = \mp(2\varepsilon \pm \hbar\omega_\eta)/\hbar v_i$ but for $\eta = 3$, $q_i^+ = q_i^- = q_i = \omega_3/v_i$. The mode indices $\eta = 1, 2, 3$ refer to K-phonons, longitudinal optical Γ-phonons and transverse optical Γ-phonons, respectively. The phonon energies for K- and Γ- phonons are given by $\hbar\omega_1 \sim 161.2$ meV and $\hbar\omega_2 = \hbar\omega_3 \sim 196.0$ meV.[7] It is important to note that elastic scattering with acoustic phonons mainly determines the conductance at low biases. However, in the case of high electric fields, the current is essentially limited by inelastic scattering with optical phonons.

In contrast to earlier studies,[5,6,8] we treat the phonon distribution functions $g_\eta(q, x, t)$, which is determined by

$$\partial_t g_\eta + \nu_\eta \partial_x g_\eta = \mathcal{D}_\eta, \tag{6}$$

as unknowns in our extended kinetic model for electrons and optical phonons. The phonon distribution functions depend on the one-dimensional wave vector q, position x and time t. In Eq. (6), nonzero phonon velocities ν_η are considered to incorporate the effect of spatial diffusion of optical phonons. The dispersion of Γ- and K-phonons in graphene is linear for small q[13] leading to $|\nu_1| \sim 7230$ m/s, $|\nu_2| \sim 2950$ m/s and $|\nu_3| \sim 0$. The phonon collision operator

$$\mathcal{D}_\eta = \mathcal{D}_\eta^{ep} + \mathcal{D}_\eta^{pp} \tag{7}$$

consists of two terms. The first one takes into account electron-electron interactions and the second one refers to phonon-phonon processes. For back-scattering of phonons, we obtain

$$\begin{aligned}\mathcal{D}_\eta^{ep} = 2\sum_{i=1}^{2} \gamma_\eta \Big\{ &(g_\eta + 1) f_i(\varepsilon_i^+)[1 - f_j(\varepsilon_i^-)] \\ &- g_\eta f_j(\varepsilon_i^-)[1 - f_i(\varepsilon_i^+)] \Big\}, , \quad j \neq i,\end{aligned} \tag{8}$$

where the abbreviations $g_\eta = g_\eta(q, x, t)$, $f_i(\varepsilon_i^\pm) = f_i(\varepsilon_i^\pm, x, t)$ with $\varepsilon_i^\pm = \hbar(v_i q \pm \omega_\eta)/2$ are used. For $\eta = 3$, the electron-phonon collision operator reads

$$\mathcal{D}_3^{ep} = \gamma_3 \sum_{i=1}^{2} J_i \delta_{q,q_i} \int_{\mathbb{R}} d\varepsilon \Big\{ (g_3 + 1) f_i(\varepsilon)[1 - f_i(\varepsilon^-)] - g_3 f_i(\varepsilon^-)[1 - f_i(\varepsilon)] \Big\} \tag{9}$$

with $\varepsilon^- = \varepsilon - \hbar\omega_3$ and $J_i = 4L/hv_F$ denoting the density of states for electrons of type i with respect to a tube of the length L. For brevity, we omitted the space and time variables x and t of the distribution functions in (8) and (9). By virtue of the Kronecker delta δ_{q,q_i} in the collision operator (9), only phonons with the wave vectors $q_i = \omega_3/\nu_i$ are emitted and absorbed by forward scattering of electrons as a consequence of the linear dispersion of electrons and the constant phonon frequency. Phonon-phonon interactions are dealt with by the operator

$$\mathcal{D}_\eta^{pp} = -\frac{1}{\tau_\eta}[g_\eta(q, x, t) - g_\eta^0], \tag{10}$$

where τ_η denotes the relaxation time determined by the anharmonic contributions to the interatomic potential, and g_η^0 is the Bose-Einstein distribution at a fixed lattice temperature T. Measurements of the linewidths of Raman G^+ peaks in metallic nanotubes imply a life time of $\tau^{an} \sim 3.3$ ps.[14]

We consider ohmically contacted SWNTs of finite length. Since the Boltzmann equations (2) and (6) are hyperbolic conservation laws, we must apply only inflow boundary conditions by fixing the values of the distribution functions for right moving particles at $x = 0$ and for left moving particles at $x = L$. For electrons at the left contact, e.g., this condition reads

$$f_1(\varepsilon, x = 0, t) = t_1^2 f^0(T) + (1 - t_1^2) f_2(\varepsilon, x = 0, t), \tag{11}$$

where t_1^2 is the transmission coefficient of the contact and $f^0(T) = [1 + \exp(\varepsilon/k_B T)]^{-1}$ denotes the Fermi-Dirac distribution. Concerning the optical phonons, we use the inflow boundary conditions $g_\eta(q, x = 0, t) = g_\eta^0$ for $\nu_\eta(q) > 0$ and $\eta = 1, 2$.

The introduced Boltzmann equations (2) and (6) represent a kinetic transport model which includes both the dynamics of electrons and optical phonons. It enables us to investigate the transient far-from-equilibrium behavior of the electron-phonon system in metallic SWNTs.

3. Numerical Approach

In contrast to classical semiconductor devices, the determination of the electron transport in SWNTs is a low-dimensional problem. The distribution functions $f_i(\varepsilon, x, t)$ and $g_\eta(q, x, t)$ depend on only two phase-space variables. Therefore, the kinetic equations can be solved very efficiently by means of a deterministic solver.

In our approach, we use a fixed uniform discretization of the phase-space variables x, ε and q. The discretization length $\Delta\varepsilon$ of the energy variable is chosen in such a way that $\hbar\omega_\eta = \sigma_\eta \Delta\varepsilon$ with $\sigma_\eta \in \mathbb{N}$. Based on this assumption, the energy grid is defined by $\varepsilon_n = -\hat{\varepsilon} + n\Delta\varepsilon$ for $n = 0, \ldots, N$. Hence, the maximal energy on this grid is $\hat{\varepsilon} = \Delta\varepsilon N/2$. Further, we fix the grid points $q_m^\eta = -\hat{q}^\eta + m\Delta q$ for $m = 0, \ldots, N - \sigma_\eta$ with $\Delta q = 2\Delta\varepsilon/\hbar v_F$ and $\hat{q}^\eta = \Delta q(N - \sigma_\eta)/2$ for the wave vector of the phonon modes $\eta = 1, 2$. These grids ensure that the energy and momentum relations $\varepsilon(k') = \varepsilon(k) \pm \hbar\omega_\eta$ as well as $k' = k \pm q$ are satisfied at the discrete level in each individual back scattering process. Consequently, the collision operators (3) and (8) can be evaluated exactly in terms of the discretized distribution functions $f_i(x, \varepsilon_n, t)$ and $g_\eta(x, q_m^\eta, t)$. Applying the midpoint rule approximates the integrals with respect to ε in the collision operator (9) very efficiently.

The left-hand sides of the kinetic equations (2) and (6) are hyperbolic conservation laws. Therefore, we use a conservative high-order finite-difference scheme to reconstruct the derivatives with respect to x, ε and q in these equations. The derivative $\partial_\varepsilon f_i$, for instance, is then approximated by

$$\partial_\varepsilon f_i(x, \varepsilon_n, t) \approx \frac{1}{\Delta\varepsilon}[F_{in+1/2}(x,t) - F_{in-1/2}(x,t)]. \tag{12}$$

The terms $F_{in+1/2}(x,t)$ are called numerical fluxes. By applying weighted essentially nonoscillatory (WENO) schemes,[15] we express these quantities in terms of the discretized distribution functions $f_i(x, \varepsilon_n, t)$. This fifth-order version of the WENO method has already been successfully applied to solve the semiconductor Boltzmann equation.[16]

The complete discretization of the coupled kinetic equations for electrons (2) and phonons (6) by means of this method leads to a system of ordinary differential equations, which governs the temporal evolution of the distribution functions evaluated at the grid points. The system of ODEs is solved by means of an explicit total variation diminishing (TVD) second-order Runge-Kutta type scheme.[17]

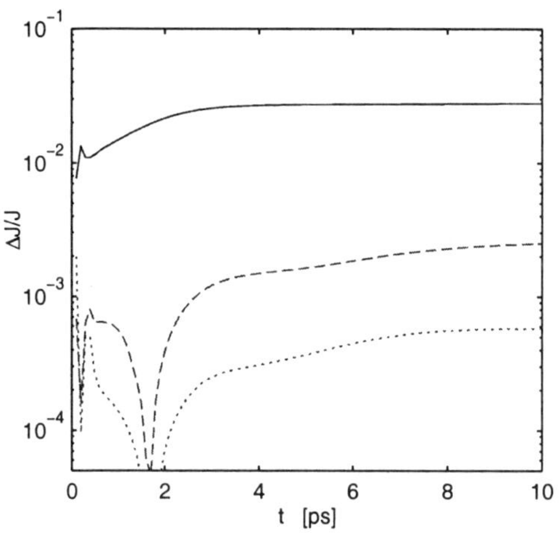

Fig. 2. Relative error $\Delta J/J$ of the mean current in a SWCNT (d = 2 nm, L = 300 nm, U = 1 V): errors of the calculations with the grids $N_z \times N_\epsilon = 25 \times 50$ (solid line), $N_z \times N_\epsilon = 50 \times 100$ (dashed line) and $N_z \times N_\epsilon = 100 \times 200$ (dotted line). (reference calculation: $N_z \times N_\epsilon = 200 \times 400$, basic energy range: $[-2, 2]$ eV)

4. Simulation Results

In this section, we study the dynamics of electrons and phonons in ohmically contacted metallic SWNTs under low- and high-field conditions. We proved already[11] that our model predicts steady-state properties of metallic SWNTs in agreement with experimental results presented in.[6]

The following simulations refer to a nanotube on a substrate with length L = 300 nm and diameter d = 2 nm. We assumed the phonon energies $\hbar\omega_1$ = 160 meV and $\hbar\omega_2 = \hbar\omega_3$ = 200 meV as well as the Fermi velocity $v_F = 8.4 \times 10^5$ m/s, which corresponds to the values reported in.[7] As group velocities of the optical phonon modes, we use ν_1 = 5000 m/s, ν_2 = 3000 m/s and $\nu_3 = 0$. The value ν_1 for the zone-boundary phonons is lower than the maximum of 7230 m/s obtained in[13] for $\boldsymbol{q} = \boldsymbol{K}$. We use ν_1 = 5000 m/s, which is a good approximation of the q-averaged phonon velocity. Further, we suppose τ_η = 3.5 ps for the relaxation times of the decay of optical phonons for $\eta = 1, 2, 3$. Concerning the elastic MFP of electrons, we found l_e = 750 nm[11] as a fitting parameter of our transport model regarding measured current-voltage characteristics. According to the density functional theory calculations in,[7] the coupling coefficients for the interaction of electrons with optical phonons depend on the tube diameter

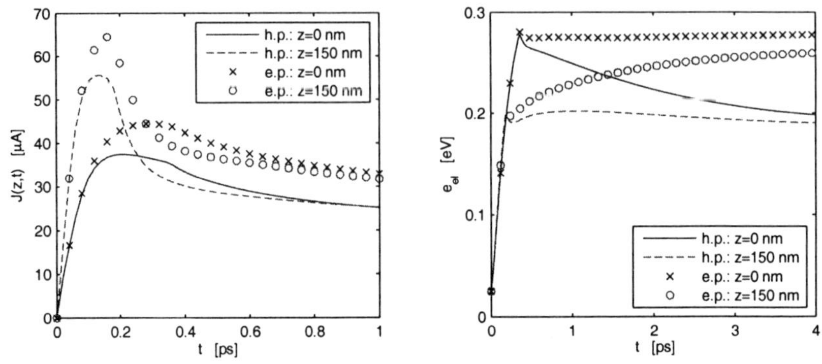

Fig. 3. Temporal change of the current (left) and the energy (right) of electrons in a SWCNT ($d = 2$ nm, $L = 300$ nm, $U = 1$ V): current J (left) and mean electron energy e_{el} (right) as a function of time t at fixed positions z for hot phonons (h.p.) and equilibrium phonons (e.p.) with adapted temperatures.

via

$$\gamma_\eta = \frac{v_F}{l_\eta}\frac{1}{d} \tag{13}$$

with the constants $l_1 = 92.0$ and $l_2 = l_3 = 225.6$. Finally, the transmission probabilities at the left ($i = 1$) and right ($i = 2$) contacts of the nanotube are assumed to be $t_i^2 = 0.92$.

To test the convergence behavior of the developed numerical scheme, we perform simulations of the carrier transport in the metallic nanotube based on several refinements of the two-dimensional phase space grid. According to the given length $L = 300$ nm of the nanotube and the basic energy range of $[-2, 2]$ eV, the grids $N_x \times N_\epsilon = 25 \times 50$, 50×100, 100×200 and 200×400 are chosen. As a measure of accuracy, we consider the relative error $\Delta\bar{J}_{N_x}(t)/\bar{J}_{N_x}(t)$ of the space-averaged current in the nanotube with respect to the mean reference current $\bar{J}_{ref}(t)$. The currents are defined by

$$\bar{J}_{N_x}(t) = \frac{1}{N_x}\sum_{i=1}^{N_x} J(x_i, t), \quad \Delta\bar{J}_{N_x}(t) = |\bar{J}_{N_x}(t) - \bar{J}_{ref}(t)|.$$

The reference current $\bar{J}_{ref}(t)$ results from simulations based on the grid $N_x \times N_\epsilon = 200 \times 400$. Figure 2 depicts the relative error of the mean current for the chosen phase space grids versus time. We observe that the relative error is smaller than 2 % even for the coarsest grid. Comparing the results for different grid sizes reveals the order of the numerical reconstruction to

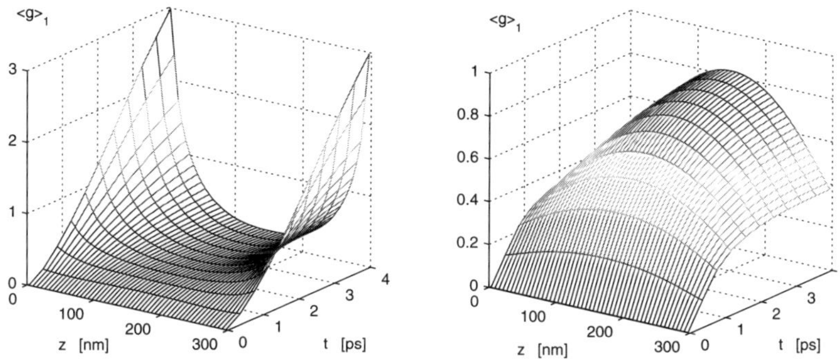

Fig. 4. Average number of K-phonons $\langle g \rangle_1(z,t)$ as functions of z and t in a SWCNT (d = 2 nm, L = 300 nm, U = 1 V): hot phonons (left) and equilibrium phonons at adapted temperature (right).

be in the range of 2–3. For the following investigations, we use the grid size $N_x \times N_\epsilon = 50 \times 100$. In this case, the relative error of the mean current is $< 0.3\,\%$.

The main aim of our study is to investigate the time-dependent influence of nonequilibrium phonons on the electron current in metallic nanotubes. For this purpose, we compare in Fig. 3 the temporal evolution of the current and the energy obtained in two different ways. The calculations are carried out at a bias of 1 V. In the first case, the phonons are kept in equilibrium at adapted temperatures. These temperatures are defined by the same number and energies of phonons as in the calculations in which we also solved the Boltzmann equation for optical phonons. In Fig. 3, the graphs in the left plot marked with crosses (circles) refer to the current in the nanotube at z = 0 nm (z = 150 nm). In the second case, the results plotted as solid ($z = 0$ nm) and dashed ($z = 150$ nm) lines refer to simulations based on nonequilibrium phonon distributions obtained from solving the coupled system of BE for electrons and optical phonons. During the first 0.1 ps, we observe a strong increase of the current mainly determined due to the ballistic transport of electrons. However, the current drops down with the onset of a strong emission of optical phonons. This effect is less pronounced at $z = 0$ nm as in the middle of the nanotube at $z = 150$ nm. The results obtained by the two different models coincide completely as long as the ballistic transport is dominant. After 0.1 ps, the plots reveal that the calculations based on thermalized phonons overestimate the electron current

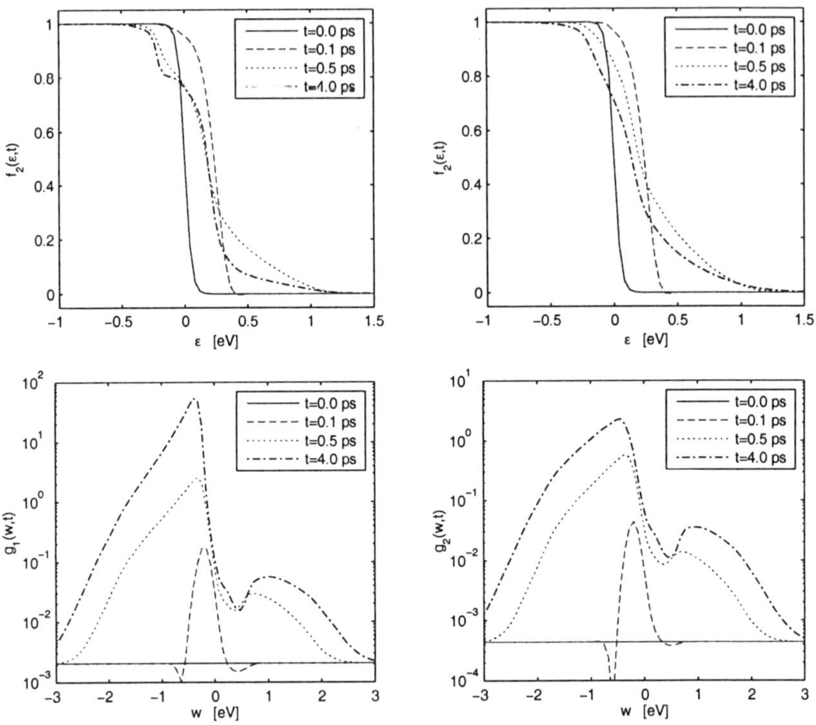

Fig. 5. Distribution functions of left-moving electrons $f_2(\varepsilon, z, t)$ (top) and optical phonons $g_\eta(w, z, t)$ (bottom) at $z = 0$ nm and different times t in a SWCNT ($d = 2$ nm, $L = 300$ nm, $U = 1$ V): electron distributions in the case of hot phonons (top left) and for equilibrium phonons at adapted temperature (top right); distribution of hot K-phonons g_1 (bottom left) and hot Γ-phonons g_2 (bottom right) as functions of $w = \hbar v_F q$.

significantly.

The right plot in Fig. 3 shows a comparison of the temporal change of the mean electron energy $e_{el} = E_{el}/N_{el}$ with

$$E_{el} = \int_0^\infty D\, \varepsilon\, [f_1(\varepsilon, z, t) + f_2(\varepsilon, z, t)] d\varepsilon,$$

$$N_{el} = \int_0^\infty D\, [f_1(\varepsilon, z, t) + f_2(\varepsilon, z, t)] d\varepsilon,$$

where $D = 2/\pi\hbar v_F$, obtained by means of the two different methods as defined before. Here, we also observe that during the period of ballistic transport the results for both models agree. However, as soon as the emission of optical phonons becomes effective, the mean electron energy is dramatically overestimated when phonons are kept in equilibrium.

In Fig. 4 we plotted the temporal evolution of the momentum averaged distribution of K-phonons obtained by considering hot phonons (left plot) and by keeping the phonons in equilibrium at adapted temperatures (right plot). The results are completely different. The thermalized phonons exhibit maxima at the half length of the nanotube in contrast to the much higher peaks of the nonequilibrium phonons at the boundaries.

The deviations in the time evolution of the distribution functions $f_2(\varepsilon, z, t)$ of left moving electrons based on calculations by considering hot (top left plot) and alternatively equilibrium phonons (top right plot) depicted in Fig. 5 are responsible for the different results obtained for the evolution of the electron current and the mean electron energy. The increasing redistribution of energetic electrons caused by hot phonon processes in course of the relaxation towards the steady state can be clearly observed in the top left plot of Fig. 5.

The plots at the bottom of Fig. 5 display the temporal evolution of the distributions of K- and Γ-phonons at the left end of the nanotube ($z = 0$). The distributions characterized by a sharp peak (it should be noted that the distribution functions are plotted in a logarithmic scale) are strongly asymmetric. Phonons with negative momentum are dominant. This means that mainly electrons with certain energies are scattered by optical phonons as can be seen in the top left plot of Fig. 5. We observe that the electron distribution functions at corresponding energies are increasingly depopulated in the period following the end of the ballistic transport.

5. Conclusion

We present a coupled system of semiclassical Boltzmann equations to investigate the hot phonon effect in metallic carbon nanotubes. Our simulations of transport in SWNTs rely on a direct solution to the transport equations. We demonstrate the high accuracy of the numerical scheme even for rather coarse grids and low computational effort. To clarify the effect of hot phonons in the transient transport regime, simulations based on two different models are carried out. In addition to the full dynamic description of the electron-phonon system by means of semiclassical Boltzmann equations, simulations based on equilibrium phonons at adapted temperatures corresponding to the heating of the phonon gas are performed. It turns out that the results of the two models coincide as long as the current is determined by the ballistic transport of electrons. However, in the case of the high-field transport accompanied by increasing phonon processes, we observe significant deviations. The investigated transients of the distribution functions of electrons and optical phonons as well as moments of them reveal that an accurate reconstruction of these quantities requires a complete kinetic description of electrons and optical phonons.

Acknowledgments

The authors are indebted to Prof. A. M. Anile of the Dipartimento di Matematica e Informatica, Università di Catania, Italy, for drawing their attention to the investigated problem. This work has been supported by the Fond zur Förderung der wissenschaftlichen Forschung, Vienna, under contract number P17438-N08.

References

1. A. Javey, J. Guo, D.B. Farmer, Q. Wang, D. Wang, R. G. Gordon, M. Lundstrom and H. Dai, *Nano Lett.* **4**, 447 (2004).
2. Y. M. Lin et al., *IEEE Elec, Dev. Lett.* **26**, 823 (2005).
3. H. Dai et al., *Nano: Brief Reports and Reviews* **1**, 1 (2006).
4. M. S. Dresselhaus and G. Dresselhaus, *Rev. Mater. Res.* **34**, 247 (2004).
5. Z. Yao, C. Kane, C. Dekker, *Phys. Rev. Lett.* **84**, 2941 (2000).
6. A. Javey et al., *Phys. Rev. Lett.* **92**, 106804 (2004).
7. M. Lazzeri, S. Piscanec, F. Mauri, A. C. Ferrari and J. Robertson, *Phys. Rev. Lett.* **95**, 236802 (2005).
8. J.-Y. Park, S. Rosenblatt, Y. Yaish, V. Sazonova, H. Üstünel, S. Braig, T. A. Arias, P. Brouwer and P. L. McEuen, *Nano Lett.* **4**, 517 (2004).
9. E. Pop, D. Mann, J. Cao, Q. Wang, K. Goodson and H. Dai, *Phys. Rev. Lett.* **95**, 155505 (2005).
10. D. Mann, E. Pop, J. Cao, Q. Wang, K. Goodson and H. Dai, *J. Phys. Chem. B* **110**, 1502 (2006).
11. Ch. Auer, F. Schürrer and C. Ertler, *Phys. Rev.* B **74**, 165409 (2006).
12. M. Lazzeri and F. Mauri, *Phys. Rev.* B **73**, 165419 (2006).
13. S. Piscanec, M. Lazzeri, F. Mauri, A. C. Ferrari and J. Robertson, *Phys. Rev. Lett.* **93**, 185503 (2004).
14. M. Lazzeri, S. Piscanec, F. Mauri, A. C. Ferrari and J. Robertson, cond-mat/0508700 (2005).
15. G. Jiang and C.-W. Shu, *J. Comput. Phys.* **126**, 202 (1996).
16. J. A. Carrillo, I. M. Gamba, A. Majorana and C.-W. Shu, *J. Comput. Phys.* **184**, 498 (2003).
17. C.-W. Shu and S. Osher *J. Comput. Phys.* **77**, 439 (2003).

DEVELOPMENT OF CURVE-BASED CRYPTOGRAPHY

ROBERTO M. AVANZI
Fakultät für Mathematik - Ruhr-Universität Bochum
Universitätsstraße 150 - 44801 Bochum - Germany
roberto.avanzi@ruhr-uni-bochum.de

In the last years, curve based cryptography has seen tremendous development. First proposed in 1985 by Koblitz and Miller, elliptic curve cryptography (ECC) slowly proved itself to be a valid alternative to RSA. Later, also hyperelliptic curves have been added to the arsenal of cryptographic primitives. Today curve based cryptography is a well established technology: In this survey we shall first very broadly review its development; we shall then move to a survey of recent results dealing specifically with Koblitz curves.

Keywords: Elliptic and Hyperelliptic Curve Cryptography, Koblitz Curves, Scalar Multiplication, Integer Expansions.

1. Introduction

This paper serves two purposes: to explain from a historical perspective why elliptic and hyperelliptic curves are a good choice for cryptography, and to review some recent results about the performance of a specific class of elliptic curves, Koblitz curves, as an example of the dramatic developments that characterize the whole field.

We begin with the situation of cryptography (§ 2) and of elliptic curves (§ 3) before the cryptographic applications of elliptic curves were envisioned. We then move to the early development of curve based cryptography (§ 4), and how it established itself a viable alternative to RSA and to ElGamal systems (§ 5). The second part (§ 6) describes recent new scalar multiplication methods for Koblitz curves.

2. Cryptography Until Elliptic Curves

Before 1975, the problem of establishing a secret key by communicating over an unsecured channel had no publicly known solution: a secret key had to be transmitted beforehand by other means.

In 1975 Diffie and Hellman [14] proposed a very elegant solution. *Let*

G be a cyclic group with generator g, both publicly known. Two users Alice and Bob want to establish a secret key known only to them. Alice chooses a secret integer a and sends the element $a \cdot g$ to Bob. Similarly, Bob chooses a secret integer b, and sends $b \cdot g$ to Alice. Alice then computes $a(b \cdot g) = ab \cdot g$, Bob computes $b(a \cdot g) = ab \cdot g$. The secret key is then (derived from) $\kappa = ab \cdot g$. We say that Alice and Bob established *the secret key κ.*

To decipher the communication an eavesdropper must compute κ solely from $G, g, a \cdot g$ and $b \cdot g$. This is the *Computational Diffie–Hellman problem* (CDHP). If the parameters G and g are chosen to make the CDHP *intractable*, then the key κ is a secret only known to Alice and Bob.

Diffie and Hellman suggested to use the multiplicative group G of a finite field and a primitive element α of the field for g. A necessary, but not always sufficient, condition for the CDHP to be intractable is that the discrete logarithm problem (DLP) in G be *hard.* Given $h \in G = \langle g \rangle$, the DLP consists in determining an integer a such that $h = a \cdot g$, which is called the *discrete logarithm* of h to the base of g. Diffie and Hellman in fact mentioned the idea of designing cryptosystems around the DLP in finite fields, and one such cryptosystem was proposed by ElGamal [16] in 1985. They also introduced the concept of *public key cryptosystem*, whereby a message sent, say, by Bob to Alice, can be encrypted using a *public key* that Alice has made available to everyone, but can only be deciphered by Alice using her own *secret key.* The first important public key cryptosystem, RSA [47], was introduced in 1978.

As other systems, such as Knapsack-based systems, were successfully attacked, RSA quickly took the lead, but alternatives have been proposed at a steady pace: In the 29 years since the introduction of RSA, cryptosystems based on elliptic [32,44] and hyperelliptic [33] curves, multivariate quadratic equations [41], polynomial factorization [29], as well as many others have been proposed. Many of these cryptosystems hold their own speed or bandwidth advantages over RSA. Standards regulate, for example, the use of Elliptic Curve Cryptography (ECC) [19,30] and of NTRU [13]. Cryptographic hardware for some of these systems has already hit the market. In the meantime, the RSA system has been shown to be weaker than initially believed. Still, RSA has remained unchallenged among the public key cryptosystems in terms of dissemination. The most important and successful alternative to RSA so far is provided by Elliptic Curves, especially in view of NSA's recent important endorsements: the acquisition of a special license on Certicom's elliptic curve-related patent portfolio and the inclusion of these curves in the Suite B [40] of recommended algorithms.

3. Elliptic Curves Before Cryptography

In layman's terms the subject of elliptic curves can be roughly described as the study of the solutions to equations

$$E \; : \; y^2 = x^3 + ax^2 + b \tag{1}$$

where the polynomial x^3+ax^2+b has no repeated roots. The equation (1) is said to be in *Weierstrass form*. The points on elliptic curves forms a group, and the group law can be expressed by explicit formulas. Since the DLP on these groups is in general a hard problem, they are natural candidates for cryptographic applications. Because of their structure, elliptic curves have important connections to number theory – such as to factorization of integers [39], which in turn is another connection to cryptography!

Elliptic integrals (related to elliptic curves) of the form $\int r\big(x, \sqrt{p(x)}\big)dx$, where the polynomial $p(x)$ has degree three or four and has no repeated roots, had been studied in 1655 by J. Wallis. In 1694 J. Bernoulli studied elliptic curves to determine of the length of arcs of a lemniscate. Many types of Diophantine equations, such as Pell equations or the problem of finding integers expressible as the sum of two rational cubes, lead to the problem of finding torsion points on suitable elliptic curves. A generalization of Ritt's second theorem [46] leads to a Pell equation in polynomials and thus to torsion points on elliptic curves [10].

Elliptic curves enjoy connections to topology, modular forms, zeta functions and analysis. They played a fundamental role in A. Wiles' recent proof of Fermat's Last Theorem, via a connection first seen by Y. Hellegouarch and G. Frey, and then proved by K. Ribet.

G. Frey also played an important role in the development of the cryptographic uses of elliptic curves, in particular of their security aspects.

4. The Early Years and the RSA Juggernaut

In 1985 V. Miller [44] and N. Koblitz [32] proposed to design cryptosystems around the DLP in the group of rational points of an elliptic curve over a finite field. This marked the beginning of ECC. It was soon noticed that while ECC offered greater potential security it was slow compared to RSA. In fact, the security of RSA relies on the difficulty of factoring a product of two large prime, and at that time algorithms for integer factorization were less efficient than those in use today. Hence RSA could use much smaller parameters than those it uses today, with better performance.

Certicom Corp., founded in 1985 by S. Vanstone, has been one of the first companies to seriously consider the applications of ECC. Working together with the University of Waterloo, Canada, Certicom played an important role in all kinds of ECC research. The performance of ECC has been improved by a few orders of magnitude in the subsequent years, also thanks to the contributions made by Certicom.

At the same time, few computational problems have seen successes as spectacular as those in integer factorization. (For a survey of the history of integer factoring methods and for DLP computations in finite fields as well as in varieties see, for example, Chapters 19, 20, 21, 22 and 25 of [4].) Until a few years ago 512-bit RSA moduli were considered secure. But, on Aug. 22, 1999 the 514-bit RSA-155 challenge was factored. This made 512 bit RSA keys no longer secure, and the minimum security requirement is now 1024 bits. On May 10, 2005, J. Franke and T. Kleinjung announced that they factored the RSA-200 number, a 662 bit integer. Moving from 512 to 1024 bits, de facto slows RSA down by a factor around 4.

1024-bit RSA keys are the next target. Several hardware designs have been proposed to factor them, such as the TWINKLE [49] ("The Weizmann Institute Key Locating Engine") sieving device, TWIRL [50], ("The Weizmann Institute Relation Locator"), Mesh-based Sieving [24] and the SHARK [21] device. SHARK might be realizable for 1 million dollars to factor 1024 bit RSA moduli in one year.

5. A Blossoming Field

In 1987, shortly after having proposed the use of elliptic curves, Koblitz [33] suggested the Jacobians of hyperelliptic curves (HEC) of higher genus.

On November 3-4, 1997 the first ECC Workshop was held in Waterloo. Around the same time the research groups of C. Paar at the Worchester Polytechnic Institute and of C. K. Koc at the Ohio State University made serious advancements in software and hardware implementations of ECC. They also initiated the Workshop on Cryptographic Hardware and Embedded Systems (CHES, now an IACR workshop), that became an ideal venue to present practical developments in curve based cryptography.

HECC have also been enjoying increasing attention in recent years. They have long been considered as not competitive with ECC because of the difficulty of constructing suitable curves and of their poor performance with respect to ECC [51]. But, in the subsequent years also this view changed.

Firstly, it is now possible to efficiently construct genus 2 and 3 HEC whose divisor class group has almost prime order of cryptographic relevance. For curves over prime fields, a genus 2 analogue of Schoof's point counting algorithm can be used: The first version [22] was too slow, but several improvements, in particular [23] made it possible to count points on curves which are sufficiently large. Another technique is the complex multiplication method [42,53]. For small characteristic, T. Satoh [48] proposed a fast point counting algorithm for elliptic curves, later improved and extended in several directions, including HEC [31,37,43].

Secondly, the performance of the HEC group operations has been considerably improved. For genus 2, from the first explicit formulae by R. Harley [28] we arrived to T. Lange's monumental work [36]. For genus 3, see for instance the formulae by J. Pelzl [45]. Thorough comparisons and new, improved formulae can be found in [18] and [9].

In Europe, Germany lead the research on curve cryptography.

G. Frey was one of the leading proponents of a DFG Graduate School "Mathematical and Engineering Method for Secure Data Transmission" that fostered the development of ECC and HECC. Weng's [53] Thesis on the complex multiplication method, and Lange's [35] Thesis on hyperelliptic curves were only a few of the important works written under Frey's supervision. This Graduate School was followed by the EU-funded project AREHCC (Advanced Research in Elliptic and Hyperelliptic Curve Cryptography). This european network connected universities and corporations. It produced an important corpus of knowledge, resources and patents. Among these results we mention the first investigation on countermeasures against side channel attacks on hyperelliptic cryptosystems [1], formulae for computing on genus 2 curves [36] and, following the research on EC performance by the Waterloo group [26], performance assessments [2] for HEC.

After the foundation of the Horst Görtz Institute for IT-Security (HGI) in Bochum, the Ruhr-University of Bochum, situated just 20 kilometers far from Essen in the small area of Germany known as Ruhrgebiet, became another important node in the development of these new technologies. The appointment of C. Paar in Bochum lead to a good synergy with Essen.

Cooperation between the researchers in the Ruhrgebiet and in Waterloo has always been intense. For instance, the first ECC Workshops have always been held in alternating years in Essen and Waterloo.

In the United Kingdom, N. Smart's research group in Bristol is a very important protagonist of the research in elliptic curve cryptography.

Many developments also take place inside the European Network of Excellence ECRYPT, encompassing 23 academic and 9 industrial partners. Coordinator is the Katholieke Universiteit Leuven, founder alongside the Ecole Normale Supérieure (Paris) and the Ruhr–University Bochum.

Several books now cover the field [4,11,12,27].

Security

The security of a curve-based cryptosystem depends on the hardness of the DLP in the Jacobian Variety of the underlying algebraic curve.

So far no subexponential algorithm for solving the DLP on elliptic and hyperelliptic curves of genus at most 4 was found; For curves defined over a fixed field and large, increasing genus the complexity of solving the discrete logarithm becomes subexponential in the group order [17] by using index calculus methods, whereas for fixed genus and increasing field size the complexity of the best methods is still exponential but with a lower exponent than Pollard's methods for genus at least three.

On the other hand the best algorithms for solving the factorization problem and the DLP in finite fields are subexponential. Therefore, to achieve a security increase equivalent to *doubling* the RSA key size, one needs to *add only a few bits* to an EC group. For example, according to [38] the security of 1323 bit RSA (or of a 137 bits subgroup of a 1024 bit finite field) is attained by an EC over a 157 bit field and with a group of prime order. For that same level of security, the field would have 79 bits for a HEC of genus two, 59 bits for genus three and 53 bits for genus four (the Jacobian of the curve must have a subgroup of prime order of at least 157 bits); In comparison, the security of 2048 bits RSA is (roughly) achieved by 200-bits curve groups, and that of 3072 bits RSA by 240-bits curve groups [15].

Systems based on curves of genus between 2 and 4 also offer much shorter key sizes compared to RSA for comparable security, genus 2 offering the same security as EC, and genus 3,resp. 4 requiring 12.5%, resp. 33.3% more bits – for instance, 180 and 213 bits instead of 160 for ECC).

There are thus obvious bandwidth and performance advantages in using curve based systems, in particular as security requirements increase.

6. Scalar Multiplication

In this Section we shall describe improvements to the scalar multiplication i.e. the computation of $n \cdot P$ for an integer n and an element P of the considered group, in Koblitz curves.

In order to compute $n \cdot P$ on an elliptic curve the first algorithm that had been used was the double-and-add algorithm, based on the binary expansion $\sum_{i=0}^{\ell} n_i 2^i \in \mathbb{N}$, $n_i = 0$ or 1 of the scalar n. This is given as Algorithm 1. Put $\lambda(n) = \log_2 n$. This method requires, on average, $\lambda(n)$ doublings and $\lambda(n)/2$ additions in the group G.

It can be generalised to any group G endowed with a suitable group

Algorithm 1. Double-and-add method for scalar multiplication

INPUT: An element P of a group $(G, +, 0)$, $n = \sum_{i=0}^{\ell} n_i 2^i \in \mathbb{N}$, $n_i = 0$ or 1.
OUTPUT: The element $n \cdot P \in G$.

```
1.  Q ← 0 ∈ G
2.  for i = ℓ downto 0 do
3.      Q ← 2·Q
4.      if n_i = 1 then Q ← Q + P
5.  return Q
```

automorphism ϕ satisfying a monic minimal polynomial over the rational integers. ϕ is identified with a root of the polynomial and then the scalars are written to the base of ϕ. Then the doubling is replaced in Algorithm 1 by an application of ϕ. Various digit sets can be used, as we shall see in the following example.

6.1. *Koblitz Curves and Digit Sets*

Consider *Koblitz curves* [34] defined by the equation

$$E_a\colon y^2 + xy = x^3 + ax^2 + 1 \qquad \text{with} \qquad a \in \{0, 1\} \tag{2}$$

over the field $\mathbb{F}_{2^n}$. The Frobenius endomorphism τ, is the map induced on $E_a(\mathbb{F}_{2^n})$ by the Frobenius automorphism of the field extension $\mathbb{F}_{2^n}/\mathbb{F}_2$, that maps field elements to their squares. Set $\mu = (-1)^{1-a}$. It is known [52] that τ permutes the points $P \in E_a(\mathbb{F}_{2^n})$, and $(\tau^2 + 2)P = \mu\tau(P)$. Identify τ with a root of $\tau^2 - \mu\tau + 2 = 0$. If we write an integer z as $\sum_{i=0}^{\ell} z_i \tau^i$, where the digits z_i belong to a suitably defined digit set $\mathcal{D}$, then we can compute zP as $\sum_{i=0}^{\ell} z_i \tau^i(P)$ via a Horner scheme. The resulting variant of Algorithm 1 [34,52] is called "τ-and-add" method. Since a Frobenius operation is much faster than a group doubling, scalar multiplication on Koblitz curves is more efficient than on generic elliptic curves.

The elements dP for all $d \in \mathcal{D}$ must be computed before the Horner scheme, the cost being roughly proportional to the size of the digit set. Larger digit sets permit the construction of representations $\sum_{i=0}^{\ell} z_i \tau^i$ with fewer non-zero coefficients and thus fewer group additions in the Horner scheme. Optimal performance is attained upon balancing digit set size and number of non-zero coefficients.

A *reduced residue set* modulo τ^w is a set of representatives of each residue class of $\mathbb{Z}[\tau]$ modulo τ^w which is coprime to τ. Solinas [52] constructs a digit set adjoining the zero to a reduced residue set modulo τ^w whose elements have minimal norm in their classes. This digit set is uniquely determined [8]. Solinas' recoding enjoys the *width-w non-adjacent property*

$$z_i \neq 0 \qquad \text{implies} \qquad z_{w+i-1} = \ldots = z_{i+1} = 0 \ , \tag{3}$$

and is called the τ-adic width-w non-adjacent form (or τ-w-NAF for short). Every integer admits a unique τ-w-NAF.

Of course, we can construct other digit sets without requiring the digits to have minimal norm. In this case it is possible that not all integers admit finite expansions. Digit sets that always guarantee a finite expansion are called w-NADS (*width w non adjacent digit sets*). They have been first characterised in [8]. Particularly interesting is the following w-NADS:

Theorem 6.1. *Let $w \geq 2$. Then $\{\pm\bar{\tau}^k : 0 \leq k < 2^{w-2}\}$ is a reduced residue system modulo τ^w. ($\bar{\tau}$ is the complex conjugate of τ.)*

Let $\mathcal{D} := \{0\} \cup \{\pm\bar{\tau}^k : 0 \leq k < 2^{w-2}\}$. If $w \in \{2,3,4,5,6\}$ then $\mathcal{D}$ is a w-NADS. If $w \in \{7,8,9,10,11,12\}$ then $\mathcal{D}$ is not a w-NADS.

In [8] it is also shown how to use $\mathcal{D}$ with large values of w and still obtaining finite recodings for all inputs. Before showing how to use this family of digit sets, we need to briefly talk about point halving.

6.2. *Point Halving on Koblitz Curves and Sublinear Scalar Multiplication*

For any given point P on an EC defined over a field of characteristic two, *point halving* consists in computing a point Q such that $2Q = P$. According to [20], halving is about two times faster than doubling. Thus, developing the scalar to the base of $1/2$ modulo the group order one can devise a halve-and-add scalar multiplication method which is faster than the double-and-add method for generic elliptic curves. It is not useful for Koblitz curves because halving is slower than a Frobenius operation.

In [3] a halving is inserted in the "τ-and-add" method to speed up Koblitz curve scalar multiplication. A refined version of this approach [6,7], is equivalent to using the digit set $\{0, \pm 1, \pm\bar{\tau}\}$ modulo τ^3.

For $n \in \mathbb{Z}[\tau]$ we compute zP using an expansion $y = \sum_{i=0}^{\ell} y_i \tau^i$ of the integer $y := \bar{\tau}^{2^{w-2}-1} z$ where the digits y_i belong to the digit set of Theorem 6.1. Write $y_i = \epsilon_i \bar{\tau}^{k_i}$ with $\epsilon_i \in \{0, \pm 1\}$ and define

$$y^{(k)} = \sum_{i\,:\,0 \leq i \leq \ell,\ y_i = \pm\bar{\tau}^k} \epsilon_i \tau^i \ .$$

Now $y = \sum_{k=0}^{2^{w-2}-1} y^{(k)} \bar{\tau}^k$ and it is easy to verify that

$$zP = \bar{\tau}^{-(2^{w-2}-1)} yP = \sum_{m=0}^{2^{w-2}-1} \left(\frac{\tau}{2}\right)^{2^{w-2}-1-m} (y^{(m)})P \; .$$

The last expression is evaluated by a Horner scheme in $\frac{\tau}{2}$, i.e. by repeated applications of τ and a point halving (since $2 = \tau\bar{\tau}$) interleaved with additions of $y^{(0)}P$, $y^{(1)}P$, etc. The elements $y^{(k)}P$ are computed by a τ-and-add loop as usual, i.e. by a second, nested Horner scheme. We can thus compute any scalar multiple of a base point with most of the speed advantages of a large digit set but without the disadvantage of having to store many multiples of the base point first. Under reasonable assumptions [8]:

Theorem 6.2. *There exists a sublinear scalar multiplication algorithm on Koblitz Curves with constant input-dependent memory consumption.*

Sublinear refers to the number of group operations, and "constant memory consumption" refers to the number of points we need to store during the computation – each one taking $O(n)$ bits. A similar result can be found in [5]. Usual windowed methods have, of course, similar time complexity but use storage for $2^{w-2} - 1$ points [52] and thus $O(n2^w) = O(n^2/\log n)$ bits of memory. Now a comparison of the scalar multiplication methods using this idea with classical methods for field sizes actually used in standards.

bits	τ-and-add		Solinas's w-NAF		This paper	
	Cost	Mem	Cost	Mem	Cost	Mem
163	596.97	1	406.55	8	359.59	1
233	996.86	1	523.13	8	500.23	1
283	1773.14	1	605.06	16	585.94	1
409	5833.29	1	832.23	16	801.94	1
571	17725.39	1	1311.20	32	1058.10	1

The second and third columns report speed and memory consumption of a simple scalar multiplication on Koblitz curves using affine coordinates and the τ-NAF, i.e. a τ-adic recoding using the digit set $\{0, \pm 1\}$ and a width-2 non-adjacency property. The next column pair refers to a state-of-the art implementation of a windowed τ-adic recoding with mixed coordinates and optimal window width. The last two columns refers to the method based on the techniques reviewed here. "Cost" means cost of a full scalar multiplication expressed in field multiplications (cf. [8]), and "Mem" refers to the number of points we need to store.

The newest methods reduce memory footprint and improves performance with respect to the previous state of the art techniques. In particular, speed improvements ranging from 2 to more than 10 in a few years, as seen in the table above, are in fact not uncommon in curve-based cryptography. The paper [9] shows similar improvements for other classes of curves, such as genus 3 and 4 hyperelliptic Jacobians.

Acknowledgements. *This paper could not have been written without the direct and indirect help of many other fellow scientists, too many to mention. In particular the author wishes to express his gratitude to all his coauthors for having accompanied him in the study of this research field.*

References

1. R. M. Avanzi. *Countermeasures against differential power analysis for hyperelliptic curve cryptosystems.* CHES 2003. LNCS 2779, 366–381. Springer–Verlag, 2003.
2. R. M. Avanzi. *Aspects of hyperelliptic curves over large prime fields in software implementations.* In *Cryptographic Hardware and Embedded Systems – CHES 2004*, LNCS 3156, 148–162, Springer–Verlag, 2004.
3. R. M. Avanzi, M. Ciet, and F. Sica. *Faster Scalar Multiplication on Koblitz Curves combining Point Halving with the Frobenius Endomorphism.* PKC 2004, LNCS 2947, 28–40. Springer–Verlag, 2004.
4. R. M. Avanzi, H. Cohen, C. Doche, G. Frey, T. Lange, K. Nguyen, and F. Vercauteren. *Handbook of Elliptic and Hyperelliptic Curve Cryptography.* Chapman and Hall / CRC Press. 2005.
5. R. M. Avanzi, V. Dimitrov, C. Doche, and F. Sica. *Extending Scalar Multiplication using Double Bases.* ASIACRYPT 2006, LNCS 4284, 130–144.
6. R. M. Avanzi, C. Heuberger, and H. Prodinger. *Minimality of the Hamming Weight of the τ-NAF for Koblitz Curves and Improved Combination with Point Halving.* SAC 2005. LNCS 3897, 332-344. Springer–Verlag, 2006.
7. R. M. Avanzi, C. Heuberger, and H. Prodinger. *Scalar Multiplication on Koblitz Curves Using the Frobenius Endomorphism and its Combination with Point Halving: Extensions and Mathematical Analysis.* Algorithmica **46** (2006), 249–270
8. R. M. Avanzi, C. Heuberger, and Prodinger. *On Redundant τ-adic Expansions and Non-Adjacent Digit Sets.* Proceedings of SAC 2006 (to appear).
9. R. M. Avanzi, N. Thériault, and Z. Wang. *Rethinking Low Genus Hyperelliptic Jacobian Arithmetic over Binary Fields: Interplay of Field Arithmetic and Explicit Formulæ.* Submitted. (CACR Technical Report 2006-07.)
10. R. M. Avanzi and U. M. Zannier, *Genus one curves defined by separated variable polynomials*, Acta Arith. **99** (2001), 227–256.
11. I. F. Blake, G. Seroussi, and N. P. Smart. *Elliptic curves in cryptography*, volume 265 of *London Mathematical Society Lecture Note Series.* Cambridge University Press, 1999.
12. I. F. Blake, G. Seroussi, and N. P. Smart. *Advances in Elliptic Curve Cryptography*, volume 317 of *London Mathematical Society Lecture Note Series.* Cambridge University Press, 2005.

13. Consortium for Efficient Embedded Security. *Efficient embedded security standards #1: Implementation aspects of NTRU and NSS, Version 1*, 2002.
14. W. Diffie and M. E. Hellman, *New directions in cryptography*, IEEE Trans. Inform. Theory **22**(6) ,644–654 (1976).
15. ECRYPT. *Ecrypt yearly report on algorithms and keysizes.* Technical report, ECRYPT, 2005.
16. T. ElGamal, *A public key cryptosystem and a signature scheme based on discrete logarithms*, Advances in Cryptology – Crypto 1984, LNCS 196, 10–18. Springer–Verlag, Berlin, 1985.
17. A. Enge. *Computing discrete logarithms in high-genus hyperelliptic Jacobians in provably subexponential time.* Math. Comp. **71** 238, 729–742.
18. X. Fan, T. Wollinger and Y. Wang. *Efficient Doubling on Genus 3 Curves over Binary Fields.* IACR ePrint 2005/228.
19. National Institute of Standards and Technology. *Digital Signature Standard.* FIPS Publication 186-2, February 2000.
20. K. Fong, D. Hankerson, J. López, A. Menezes. *Field Inversion and Point Halving Revisited.* IEEE Trans. Computers 53(8), 1047–1059, 2004.
21. J. Franke, T. Kleinjung, C. Paar, J. Pelzl, C. Priplata, and C. Stahlke. *SHARK - a realizable special hardware sieving device for factoring 1024-bit integers.* Workshop on Special Purpose Hardware for Attacking Cryptographic Systems (SHARCS) 2005, February 2005.
22. P. Gaudry, and R. Harley, *Counting points on hyperelliptic curves over finite fields*, Algorithmic Number Theory Symposium – ANTS IV. LNCS 1838, 313–332. Springer–Verlag, 2000.
23. P. Gaudry and E. Schost. *Construction of Secure Random Curves of Genus 2 over Prime Fields.* EUROCRYPT 2004. LNCS 3027, 239–256. Springer–Verlag, 2004.
24. W. Geiselmann, R. Steinwandt. *Yet Another Sieving Device.* In *Proceedings of CT-RSA 2004.* LNCS 2964, 278–291. Springer–Verlag, 2004.
25. M. Gonda, K. Matsuo, K. Aoki, J. Chao, and S. Tsuji. *Improvements of addition algorithm on genus 3 hyperelliptic curves and their implementations.* SCIS 2004, 995–1000.
26. D. Hankerson, J. López Hernandez, and A. Menezes. *Software Implementation of Elliptic Curve Cryptography over Binary Fields.* Proceedings of CHES 2000, LNCS 1965, 1-24. Springer–Verlag, 2000.
27. D. Hankerson, A. J. Menezes, and S. A. Vanstone. *Guide to elliptic curve cryptography.* Springer–Verlag, 2003.
28. R. Harley. *Fast Arithmetic on Genus Two Curves.*
Available at `http://cristal.inria.fr/~harley/hyper/`
29. J. Hoffstein, J. Pipher and J. H. Silverman. NTRU: a ring-based public key cryptosystem. *Algorithmic number theory – ANTS III*, LNCS 1423, 267–288. Springer–Verlag, 1998.
30. IEEE Std 1363-2000. *IEEE Standard Specifications for Public-Key Cryptography.* IEEE Computer Society, August 29, 2000.
31. K.S. Kedlaya. *Counting points on hyperelliptic curves using Monsky–Washnitzer cohomology.* J. Ramanujan Math. Soc. **16** (2001), 323–338.

32. N. Koblitz. *Elliptic curve cryptosystems.* Mathematics of computation **48** (1987), 203–209.
33. N. Koblitz. *Hyperelliptic Cryptosystems.* J. of Cryptology **1** (1989), 139–150.
34. N. Koblitz. *CM-curves with good cryptographic properties.* In *Advances in Cryptology - Crypto 1991*, LNCS 576, 279–287. Springer-Verlag, Berlin, 1992.
35. T. Lange. *Efficient Arithmetic on Hyperelliptic Curves*, PhD thesis, Essen 2001.
36. T. Lange. *Formulae for Arithmetic on Genus 2 Hyperelliptic Curves.* Appl. Algebra Eng. Commun. Comput. **15** (5), 295–328 (2005)
37. A.G.B. Lauder and D. Wan. *Counting points on varieties over finite fields of small characteristic.* In: Algorithmic Number Theory: Lattices, Number Fields, Curves and Cryptography. Cambridge University Press, 2002.
38. A. K. Lenstra, E. R. Verheul. *Selecting Cryptographic Key Sizes.* J. Cryptology, **14** (4), 255–293, 2001.
39. H.W. Lenstra, Jr. *Factoring integers with elliptic curves.* Ann. of Math, **126** (1987), 649–673.
40. NSA Suite B of Recommended Algorithms. See: `http://www.nsa.gov/ia/industry/crypto_suite_b.cfm`
41. T. Matsumoto and H. Imai. *Public quadratic polynomial-tuples for efficient signature verification and message-encryption.* In *Advances in Cryptology — EUROCRYPT 1988.* LNCS 330, 419–545. Springer–Verlag, 1988.
42. J.-F. Mestre. *Construction des courbes de genre 2 a partir de leurs modules.* Progr. Math. **94** (1991), 313–334.
43. J.-F. Mestre. J.-F. Mestre. Lettre adressée à Gaudry et Harley, December 2000. In French. Available at `http://www.math.jussieu.fr/~mestre/`
44. V.S. Miller. *Use of elliptic curves in cryptography.* In: *Proceedings of CRYPTO '85.* LNCS 218, 417–426. Springer–Verlag, 1986.
45. J. Pelzl. *Fast Hyperelliptic Curve Cryptosystems for Embedded Processors.* Master's Thesis. Ruhr-University of Bochum, 2002.
46. J.F. Ritt, *Prime and composite polynomials*, TAMS **23** (1922), 51–66.
47. R. Rivest, A. Shamir, and L. Adleman. *A Method for Obtaining Digital Signatures and Public Key Cryptosystems. Communications of the ACM*, **21**, 120–126, 1978.
48. T. Satoh. *The canonical lift of an ordinary elliptic curve over a finite field and its point counting.* J. Ramanujan Math. Soc. **15** 4 (2000), 247–270.
49. A. Shamir. *Factoring large numbers with the TWINKLE device.* In: *Cryptographic Hardware and Embedded Systems (CHES) 1999*, LNCS 1717, 2–12. Springer–Verlag, 1999.
50. A. Shamir, E. Tromer. *Factoring large numbers with the TWIRL device.* In *Proceedings of Crypto 2003*, LNCS 2729, 1–26. Springer-Verlag, 2003.
51. N.P. Smart. *On the Performance of Hyperelliptic Cryptosystems.* EUROCRYPT '99, LNCS 1592, 165–175. Springer–Verlag, 1999.
52. J.A. Solinas. *Efficient Arithmetic on Koblitz Curves.* Designs, Codes and Cryptography **19** (2/3), 125–179, 2000.
53. A. Weng. *Konstruktion kryptographisch geeigneter Kurven mit komplexer Multiplikation.* PhD thesis, Universität Gesamthochschule Essen, 2001.

MODEL ORDER REDUCTION: AN ADVANCED, EFFICIENT AND AUTOMATED COMPUTATIONAL TOOL FOR MICROSYSTEMS

T. BECHTOLD, A. VERHOEVEN, E. J. W. TER MATEN

NXP Semiconductors, Research, DMS - Physical Design Methods,
Hich Tech Campus 48, 5656 AE Eindhoven, The Netherlands
E-mail: {Tamara.Bechtold,Arie.Verhoeven,Jan.ter.Maten}@NXP.com

T. VOSS

Delft university of Technology,
Mekelweg 2, 2628 CD Delft, The Netherlands
E-mail: {T.Voss@tudelft.nl

The goal of mathematical model order reduction (MOR) is to replace the non-automatic compact modeling, which is the state of the art in simulation flow of microelectronic and micro-electro- mechanical systems (MEMS). MOR offers a possibility of automatically creating small but very accurate models which can be used within system level simulation. The main challenges in integrating model order reduction as a standard tool in the current simulation flow are to be able to reduce non-linear ordinary differential equation systems (ODEs) and differential algebraic equation systems (DAEs), which arise from either spatial discretization of partial differential equations (PDEs) or from electronic circuit equations. We present a methodology for applying MOR to linear and non-linear ODEs and DAEs and numerical results for several MEMS and microelectronic devices.

Keywords: Model Order Reduction; Microsystem Simulation

1. Introduction

The decreasing size and growing complexity of micro-electronic systems impose new challenges on the designers and simulation tools.[1] The main requirement on modern simulation tools for microsystems is the automatic macromodeling with very high accuracy. In Fig. 1 (left) an "ideal" design flow is shown, which is at present hindered by the lack of modeling continuity.

Luckily, modern mathematical methods, known as model order reduction,[2] show most promising results to-wards filling this "gap". The main advantage of mathematical model order reduction over "classical compact modeling" is that it is formal, robust and can be automatized. Model order reduction starts with either an ODE or an DAE system (see Fig. 1 (right)). Numerical solution of time-dependent

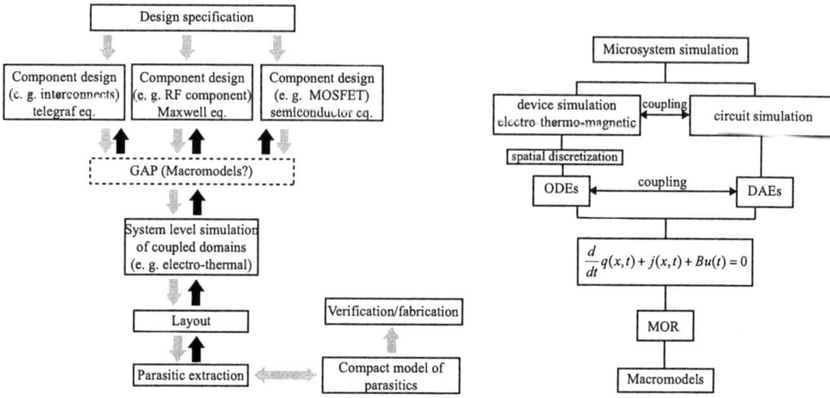

Fig. 1. Microsystem design flow is hindered by the lack of modeling continuity (left). The gap may be filled with macromodels obtained by MOR (right).

partial differential equations, like e. g. heat transfer equation, semiconductor equations and Maxwel equations includes spatial discretization of the computational domain. There are several discretization methods which can be used like, finite element method, boundary element method etc. They transform an PDE into an ODE system, whose dimension depends on the fineness of the discretization grid. Exactly at this moment, model order reduction can take over to significantly speed up the transient simulation of resulting large-scale ODE system.[4]

A similar approach can be applied to the other part of the microsystem, the circuitry. Due to the lumped nature of electrical current, the dynamics of the electrical circuits can be generally described by a differential-algebraic equation system. It is composed out of Kirchhoff's laws and characteristics branch equations of the networks elements. There are several schemes to set up the network equations like, modified nodal analysis (MNA), sparse tableau, etc. The dimension of the resulting DAE system is of the order of the number of elements in the circuit, which means that it can be extremely large, as today's VLSI circuits have hundreds of millions of elements. Again, model order reduction can be used to speed up the transient simulation.[3]

2. Principle of Model Order Reduction

Both, the ODE systems, which arise from spatial discretization of 1st order time-dependent PDEs, and the DAE systems, which describe the dynamics of electrical circuits at time t can be described by:

$$\frac{d}{dt}\mathbf{q}(\mathbf{x},t) + \mathbf{j}(\mathbf{x},t) + B\mathbf{u}(t) = \mathbf{0}, \tag{1}$$

with the difference that for "real" DAEs the partial derivative q_x is singular. In case of circuit equations for example, the vector-valued functions $\mathbf{q}(\mathbf{x},t)$ and $\mathbf{j}$ represent the contributions of respectively, reactive elements (such as capacitors and inductors) and of nonreactive elements (such as resistors) and all time-dependent sources are stored within $\mathbf{u}(t)$. In case of linear or linearized models, (1) simplifies to:

$$\begin{aligned} C \cdot \dot{\mathbf{x}} + G \cdot \mathbf{x} &= B \cdot \mathbf{u}(t) \\ \mathbf{y} &= L \cdot \mathbf{x} \end{aligned} \tag{2}$$

where $\mathbf{x}(t) \in \mathbf{R}^n$ is the state vector, $\mathbf{u}(t) \in \mathbf{R}^m$ is the input excitation vector and $\mathbf{y}(t) \in \mathbf{R}^p$ is the output measurement vector. G, $C \in \mathbf{R}^{n \times n}$ are the symmetric and sparse system matrices and for DAEs, C is singular. $B \in \mathbf{R}^{n \times m}$ and $L \in \mathbf{R}^{p \times n}$ are the user-defined input and output distribution arrays. n is the dimension of the system and m and p are the numbers of inputs and outputs. The idea of model order reduction is to replace (2) by the system of the same form but with much smaller dimension $r \ll n$:

$$\begin{aligned} C_r \cdot \dot{\mathbf{z}} + G_r \cdot \mathbf{z} &= B_r \cdot \mathbf{u}(t) \\ \mathbf{y_r} &= L_r \cdot \mathbf{z} \end{aligned} \tag{3}$$

which can be solved by suitable DAE or ODE solver and will approximate the input/output characteristics of (2). A transition from (2) to (3) is formal and is done in two steps. The first is the transformation of the state vector $\mathbf{x}$ to the vector of generalized coordinates $\mathbf{z}$ and truncation of a number of those generalized coordinates, which leads to some (hopefully) small error ϵ):

$$\mathbf{x} = V \cdot \mathbf{z} + \epsilon \tag{4}$$

The time-independant $V \in \mathbf{R}^{n \times r}$ is called the transformation or projection matrix, as (4) can be seen as the projection of the state vector onto some low-dimensional subspace defined by V. Note that the spatial and physical meaning of $\mathbf{x}$ is lost during such projection. In the second step, (2) is multiplied from the left hand side with another matrix $W^T \in \mathbf{R}^{r \times n}$, so that $C_r = W^T C V_r, G_r = W^T G V_r, B_r = W^T B$ and $L_r = EV$. Note that the number of inputs and outputs in the reduced system (3) is the same as in (2). The model order reduction process is schematically shown in Fig. 2.

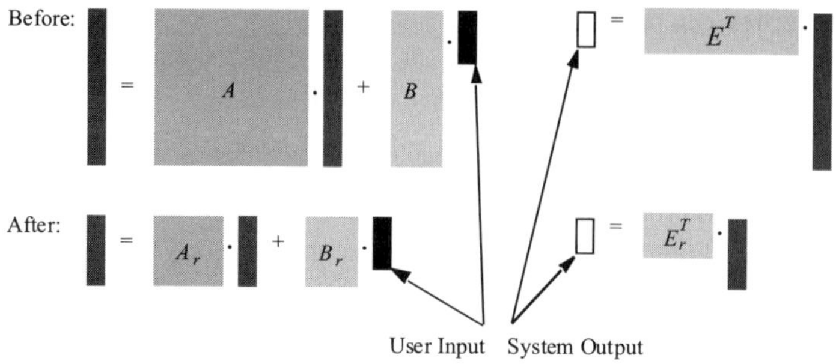

Fig. 2. Schematics of the system before and after model reduction step, (courtesy of J. G. Korvink, IMTEK, Germany).

The above principle of projection can also be applied directly to the second order ODE systems,[19] which may arise from spatial discretization of the 2nd order time-dependent PDEs (e. g. the equation of motion, which is often solved in MEMS simulation). Furthermore, it can be applied to the nonlinear system (1). This, however, is much more complicated and does not necessarily result in reduction of computational time, as will be shown in the following chapters.

3. Methods

In Fig. 3 a number of state-of-the-art methods for model order reduction of (1) is displayed.

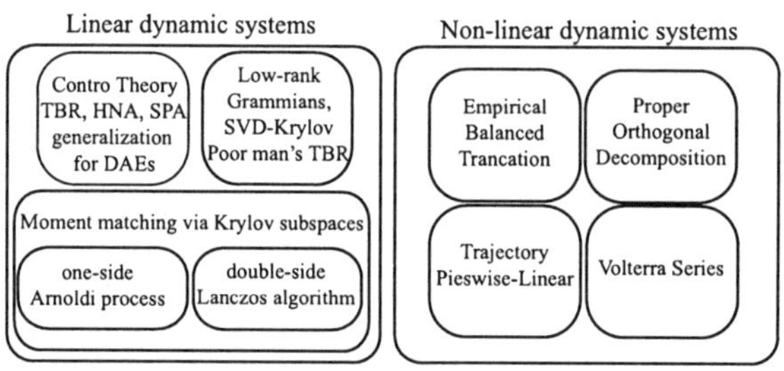

Fig. 3. Overview of model reduction methods for linear and non-linear dynamical systems.

3.1. *Linear MOR*

The control theory methods (mostly used are Truncated Balanced Realization, Hankel Norm Approximation and Singular Perturbation Approximation) offer a good mathematical basis and a global error estimate. For MOR of smaller-size linear ODE systems they have been successfully used since many years.[18] Unfortunately, their computational complexity is of n^3, as for the construction of V and W the singular value decomposition (SVD) of the large-scale system matrices is required. Their generalization to DAEs has been explored in.[7]

In order to overcome the pore scaling of SVD-based methods, Krylov subspace methods (also known as moment matching methods)[5] are mostly used for MOR of large-scale linear ODEs and DAEs. They are based on writing down the transfer function of (2) in the Laplace-domain, developing it into Taylor series around a frequency point of interest (single or multiple frequency points might be chosen for "classical" or rational Krylov methods) and truncating the terms of the higher order. The left over terms define a reduced model's rational transfer function (r terms define a reduced model of order r for one-sided projections and of order $\frac{r}{2}$ for two-sided projections). The Taylor coefficients are called moments of the transfer function and can not be explicitly computed in a stable way. Rather, the bases for the Krylov subspaces are computed and stored within V and W. The main disadvantage of Krylov methods is the lack of global error estimate and the fact that the preservation of stability and passivity of the reduced model is not guaranteed in general. In order to overcome this bottle-necks without loosing the speed-up, a research is going on, on how to combine the advantages of the SVD-based and Krylov-based methods.

An SVD-Krylov method[9] is based on iteratively updating the interpolation frequencies (and so the transformation matrices V and W) for the rational Krylov method by using the eigenvalues of the reduced model in each iteration. After convergence, a stability of the reduced model is guaranteed, as well as it's error minimality. The Poor man's TBR has been proposed in.[8] It is based on the singular value decomposition of the projection matrix V, which is computed by rational Krylov. In this way, an additional reduction of model size compared to rational Krylov is achieved and at the same time the error estimate property of TBR is inherited. Low-rank Gramian approximation aims at speeding up the solution of the sparse generalized Lyapunov equations. Hereby, the Gramians are approximated through the low-rank matrices. In[6] the applicability of balanced truncation on parallel computers was extended to sparse systems with up to $O(10^5)$ states. The method inherits the preservation of stability and passivity and the global error bound of "classical" TBR. Fig. 4 summarizes the properties of the available algorithms.

Linear MOR methods	Advantages	Disadvantages
Control theory methods (SVD-based)	global error estimate preservation of stability and passivity fully auomatic	computational effort $O(n^3)$
Moment maching (Krylov-based)	low computational effort	no global error estimate manual selection of r preservation of stbility and passivity only in some cases
Hybrids (SVD-Krylov poor man's TBR low rank gramians)	global error estimate computational effort $O(n^2)$ preservation of stability and passivity	still under development

Fig. 4. Methods for model order reduction of linear dynamic systems.

3.2. *Nonlinear MOR*

We said that the goal of model order reduction is to produce a lower dimensional system that has approximately the same response characteristics for all the inputs as the original system. For the linear systems of the form (2) and their reduced systems (3) this was the case. However, the direct projection of the general non-linear system (1) leads at present mostly to the reduced models, which approximate the original one only for a certain single input. Furthermore, the extraction of the reduced order model requires the simulation of the original one. Sometimes, the simulation of the reduced model might require even longer CPU time than the simulation of the full-scale model, which is of course, not what we want. Hence, the reduced non-linear models are at best, meaningfull for re-use.
The idea behind Proper Orthogonal Decomposition (POD)[11] is to directly project the original nonlinear system (1) onto some subspace with smaller dimension. As this, however, does not lead to the reduction of the computational time, Missing Point Estimation technique (MPE)[12] is used to speed up the simulation.

Empirical balanced truncation[15] is an extended version of a POD method. Instead of creating the reduced subspace with only one relevant input and initial state, several training trajectories are created and the reduced subspace is built in a similar way as in the BTR method.

The idea behind the Trajectory Piecewise Linear (TPWL)[10] method is to linearise (1) several times along a training trajectory (corresponding to some typical input). The local linear reduced systems (can be created with arbitrary linear MOR method) are then used to create a global reduced subspace. The final TPWL model is constructed as a weighted sum of all local linearised reduced systems.

The idea behind Volterra series[14] is to construct a bilinear system, which approximates the first moments of the nonlinear system. Then linear model reduction techniques are used to create a reduced bilinear system which matches as many moments of the original system as possible.
Fig. 5 summarizes the properties of the available algorithms. In[16] a review of some additional methods is given.

Nonlinear MOR methods	Advantages	Disadvantages
POD	high accuracy	no speed up (MPE can help) no global error estimate
Empirical Balanced Truncation	god approximation different inputs/initial values possible	most expensive model extraction no speed-up (MPE can help) no global error estimate
TPWL	cheap reduced model evaluations	high memory consuming bad accuracy for highly nonlinear systems
Volterra Series	moment matching full-system simulation is not necessary	increased dimension of the state vector not applicable to DAEs

Fig. 5. Methods for model order reduction of nonlinear dynamic systems.

4. Applications

The test-cases below are academic examples of electrical circuits, which closely resemble the industry-relevant models. They have been defined as MOR case studies for the European project COMSON.[13]

4.1. *Transmission Line (linear DAE system)*

Fig. 6 shows an academic model of transmission line. This very simple model has been chosen, because it resembles the interconnect modeling and can be effectively used for testing new MOR algorithms. It consists of scalable number of RLC ladders and after modified nodal analysis results in a linear DAE system of the form (2).

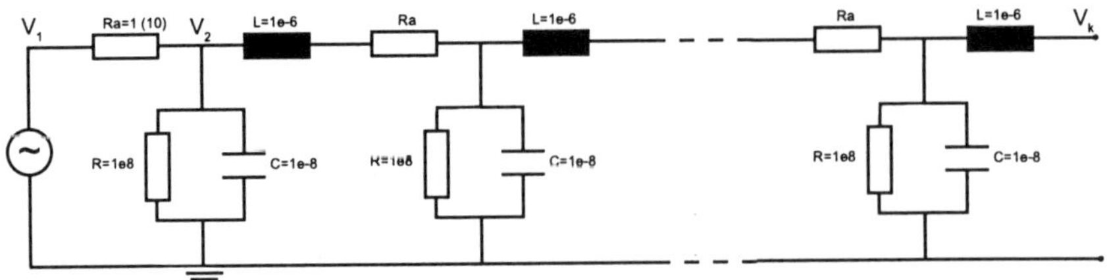

Fig. 6. Structure of transmission line test case

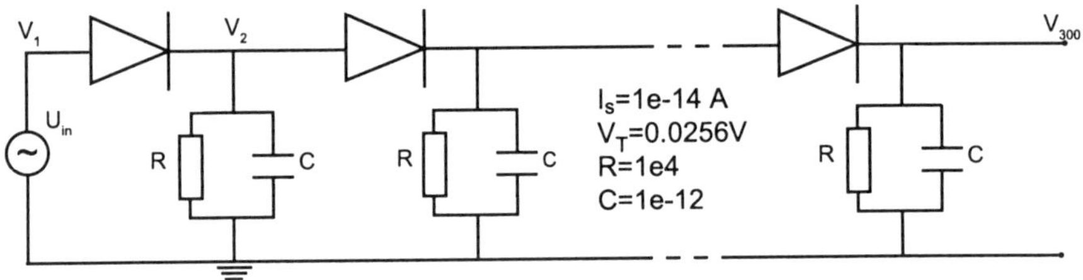

Fig. 7. Structure of diode chain circuit

4.2. *Diode Chain (nonlinear DAE system)*

Fig. 7 shows an academic, highly non-linear model of a diode chain. It consists of scalable number of diodes and is described by the following equations:

$$\begin{aligned} V_1 - U_{\text{in}}(10^9 t) &= 0, \\ i_E - g(V_1, V_2) &= 0, \\ g(V_1, V_2) - g(V_2, V_3) - C\dot{V}_2 - \tfrac{1}{R}V_2 &= 0, \\ &\vdots \\ g(V_{N-1}, V_N) - g(V_N, V_{N+1}) - C\dot{V}_N - \tfrac{1}{R}V_N &= 0, \\ g(V_N, V_{N+1}) - C\dot{V}_{N+1} - \tfrac{1}{R}V_{N+1} &= 0, \end{aligned}$$

$$g(V_a, V_b) = \begin{cases} (I_s e^{\frac{V_a - V_b}{V_T}} - 1) & \text{if } V_a - V_b > 0.5 \\ 0 & \text{otherwise} \end{cases}$$

$$U_{\text{in}}(t) = \begin{cases} 20 & \text{if } t \leq 10 \\ 170 - 15t & \text{if } 10 < t \leq 11 \\ 5 & \text{if } t > 11 \end{cases}$$

4.3. *Pyrotechnical Microthruster (linear ODE system)*

The Pyrotechnical microthruster is a MEMS device, which was fabricated within a European project Micropyros (founded under IST-99047). It is based on the integration of solid fuel with a silicon micro-machined structure (see Fig. 8).

The heat transfer within the hotplate is described through the following equations:

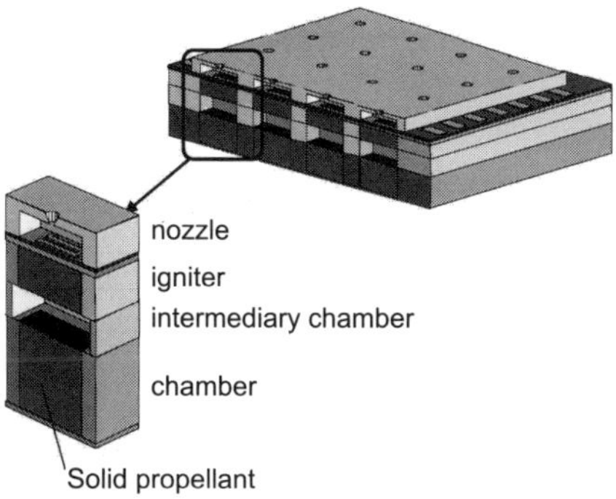

Fig. 8. A structure of pyrotechnical microthruster array (courtesy of C. Rossi LAAS-CNRS, France).

$$\nabla \bullet (\kappa \nabla T) + Q - \rho C_p \frac{\partial T}{\partial t} = 0, Q = \mathrm{j}^2 R \tag{5}$$

where κ is the thermal conductivity in W/mK and C_p is the specific heat capacity in J/kgK. Assuming that both are temperature independent around the working point, which is realistic, the finite element based spatial discretization of (5) leads to a large linear ODE system (2).

5. Numerical Results

Transmission line model of order 6002 has been reduced with Krylov-based block Arnoldi algorithm from[3] down to 150, with the relative error less than 0.02% (not shown). We have experienced problems, as the MNA system matrices C and K coming from Pstar (NXP in-house circuit simulator) were indefinite. In Fig. 9 (left) the instable reduced order model is shown. After the multiplication of the equations which correspond to the inductor branches and the voltage source branch with -1 and so making the system matrices positive semi-definite, stability of the reduced model was gained (see Fig. 9 (right)), as well as the excellent approximation (the relative error, which is less than 0.02% is not shown). This example shows however, that the engineers may easily experience difficulties in an attempt to automatize model order reduction process.

Our experiments show, that the above problems do not appear for reducing the linear ODE systems, which arise from spatial discretization of PDEs. Fig. 10 (left) shows the relative error of the BTA and Arnoldi-based reduction of the microthruster device from order 1071 down to order 7. As expected, the BTA shows smaller error in the transient phase but performs worse for the steady-state.

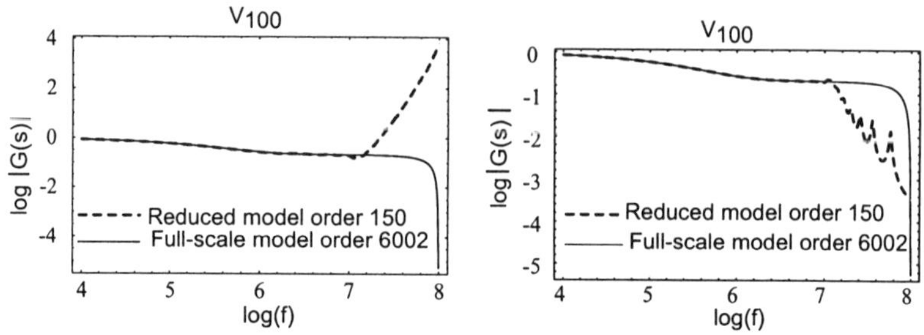

Fig. 9. Instable frequency response of a reduced transmission line model at V_100 (left) and the stable reduced model, after conversion of the original model into positive definite DAE system (right).

Fig. 10 (right) shows, how relative error between two successive reduced models can be effectively used as an error estimate[17] for Arnoldi algorithm. In this way, compact model extraction becomes completely automatic.

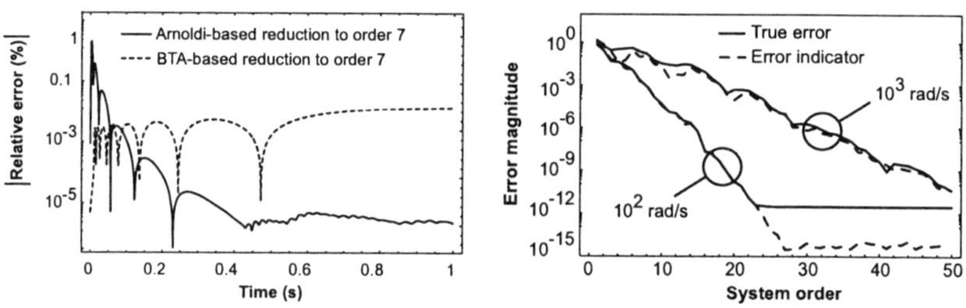

Fig. 10. Relative Error of the BTA and Arnoldi-based reduction (expansion around $s_0 = 0$)in a single output node of the microthruster model of order 1071 (left). Convergence of the relative error between the two successive reduced models for the microthruster (right).

Lastly, in Fig. 11 we show the relative errors over all output nodes of the diode chain model with order 302. The POD models are, as expected, more accurate, but much slower to simulate than the TPWL models (see the corresponding extraction and simulation times in[20]). A significant speed up has been achieved by combining the POD with MPE.

6. Conclusion and Outlook

It has been shown, that the lack of automatic compact modeling is the main bottleneck in the design flow of today's micro- and nanosystems. Mathematical model order reduction appears to be a "perfect" tool to solve this problem. In this paper, we have described and demonstrated the methodology for applying model order reduction to microelectronic and MEMS models. Furthermore, we have reviewed

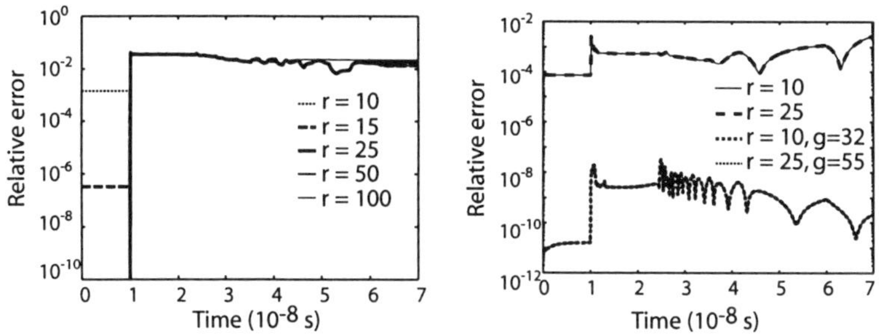

Fig. 11. Relative errors over all nodes of diode chain model, left-reduction by TPWL, right-reduction by POD combined with MPE.

the most promising methods for linear and nonlinear MOR. Although, there are still difficulties to completely atomize the process, which are due to necessity of extraction a full-scale model in a proper form and/or due to missing a reliable global error estimate for Krylov-based methods, it looks like that model order reduction for linear ODE and DAE models is mature enough for common engineering use. This is not the case for more realistic and thus much more complicated nonlinear models. Although, we have presented promising numerical results for an academic model of diode-chain, model order reduction of industry-relevant nonlinear models, which might be subjected to a broad spectrum of excitations, remains a research topic.

Furthermore, as model reduction in its original form does not allow us to preserve parameters within the system, which is essential for a quick design iteration, a parametric model order reduction (see e. g.[21]) should be researched.

References

1. G. Denk, "Circuit Simulation for Nanoelecronics", in Scientific Computing in Electrical Engineering, Editors: A. M. Anile, G. Ali, G. Mascali, *Springer*, pp. 13-20,(2006).
2. A. C. Antoulas, "Approximation of Large-Scale Dynamical Systems", *SIAM*, (2005).
3. R. W. Freund, "Krylov-subspace methods for reduced order modeling in circuit simulation", *J. of Comput. and Appl. Math.*, 123, pp. 395-421,(2000).
4. T. Bechtold, E. B. Rudnyi, J. G. Korvink, "Fast Simulation of Electro-Thermal MEMS", *Springer*,(2006).
5. R. W. Freund, "Model reduction methods based on Krylov-subspaces", *Acta Numerica*, pp. 267-319,(2003).
6. J. M. Badia, P. Benner, R. Mayo, E. S. Quintana-Orti, G. Quintana-Orti, A. Remon,"Balanced Truncation Model Reduction of Large and Sparse Generalized Linear Systems", *submitted to Elsevier Science*.
7. T. Stykel, "Gramian-based model reduction for descriptor systems", *Math. Control Signals Sys.*, 16, pp.297-319, (2004).
8. J. R. Phillips, M. Silveira, "Poor Man's TBR: A Simple Model Reduction Scheme",

IEEE Trans. on Comp. Aid. Desig. Integr. Circ. Syst., 24(1), pp. 43-55, (2005).
9. S. Gugercin, A. C. Antoulas, "Model reduction of large-scale systems by least squares", *Linear Alg. and its Applics.*, 415, pp. 290-321, (2006).
10. M. Rewienski, J. White, "A Trajectory Piecewise-Linear Approach to Model Order Reduction and Fast Simulation of Nonlinear Circuits and Micro-machined Devices", *Proc. of the Int. Conf. on CAD*, pp. 252-257, (2001).
11. P. Astrid, "Reduction of process simulation models: a proper orthogonal decomposition approach", *PhD thesis*, Eindhoven University of Technology, (2004).
12. P. Astrid, A. Verhoeven, "Application of Least Squares MPE technique in the reduced order modeling of electrical circuits", *Proceedings MTNS*,(2006).
13. *www.comson.org*.
14. Z.Bai, D.Skoogh, "Krylov Subspace Techniques for Reduced-Order Modeling of Nonlinear Dynamical Systems", *University of California report*, (2001).
15. S. Lall, J.E. Marsden, S. Glavaski, "A subspace approch to balanced truncation for model reduction of nonlinear systems", *International Journal of robust and nonlinear controll*, pp. 519-535, (2002).
16. L. Feng, "Review of model reduction methods for numerical simulation of nonlinear circuits", *Applied Mathematics and Computation*, 167(N 1), pp. 576-591. (2005).
17. T. Bechtold, E. B. Rudnyi, J. G. Korvink, "Error indicators for fully automatic extraction of heat-transfer macromodels for MEMS", *Journal of Micromechanics and Microengineering*, 15(N 3), pp. 430-440, (2005).
18. G. Obinata, B. D. O. Anderson, "Model Reduction for Control System Design", *Springer*, (2004).
19. S. B. Salimbahrami, "Structure Preserving Order Reduction of Large Scale Second Order Models", *PhD thesis, Technical University of Munic*, (2005).
20. T. Voss, A. Verhoeven, T. Bechtold, J. ter Maten, "Model Order Reduction for Nonlinear Differential Algebraic Equations in Circuit Simulation", *Proc. ECMI06 to appear at Springer.*
21. G. Shi, B. Hu, C. J. R. Shi, "On Symbolic Model Order Reduction", *IEEE Transactions on Computer-Aided Design of Integrated Circuits and Systems*, 25(7), pp. 1257-1272, (2006).

Acknowledgment: We would like to acknowledge the EU Marie Curie RTN project COMSON grant n. MRTN-CT-2005-019417.

A NEW FINITE ELEMENT METHOD FOR KIRCHHOFF PLATES

LOURENÇO BEIRÃO DA VEIGA*

Dipartimento di Matematica "Federigo Enriques"
via Saldini 50, 20133 Milano, Italy
** E-mail: beirao@mat.unimi.it*
www.mat.unimi.it/users/beirao

JARKKO NIIRANEN AND ROLF STENBERG

Institute of Mathematics, Helsinki University of Technology
P.O. Box 1100, 02015 TKK, Finland
E-mail: {jarkko.niiranen, rolf.stenberg}@tkk.fi

Based on the ideas from Refs. 1 and 2 we present a new finite element method for the Kirchhoff plate bending model.[3] The method uses C^0 basis functions for the deflection and the rotation, i.e. the same approach as used for the Reissner–Mindlin model. To account for the effective shear force at the free boundary a stabilization term is added. We prove optimal a-priori and a-posteriori error estimates. The corresponding results from benchmark problems are also reported.

Keywords: finite elements, Kirchhoff plate model, biharmonic problem, free boundary, a-priori error analysis, a-posteriori error analysis

1. Introduction

We consider the classical Kirchhoff plate bending problem. The natural variational space for this biharmonic problem is the second order Sobolev space. Thus, a conforming finite element approximation requires globally C^1-continuous elements which imply a high polynomial order. A way to avoid using high order polynomial spaces is to write the Kirchhoff problem as the limit of the Reissner–Mindlin problem written in mixed form. On the other hand, in the presence of the free boundary conditions, this leads to a method which is not consistent: The solution of the Kirchhoff problem does not coincide with the solution of the mixed Reissner–Mindlin problem with the thickness set equal to zero.[3,4]

In this paper we present for the Kirchhoff plate bending problem a family of (low order) finite elements for which the optimal convergence rate holds — even in the presence of free boundaries.[3] This method is a modification of the stabilized method for the Reissner–Mindlin plates presented in Ref. 2.

In the next section we describe the plate bending problem and introduce the new family of finite elements. In Section 3 we state optimal a-priori error estimates and also the reliability and efficiency results for a local a-posteriori error indicator. In the final section we present numerical results on benchmark computations, in order to verify the theoretical a-priori error estimates and illustrate the robustness of the a-posteriori error estimator.

2. The finite element method for the Kirchhoff plate problem

2.1. *The Kirchhoff plate bending model*

We consider the bending problem of an isotropic linearly elastic plate under the transverse loading $F = Gt^3 f$, where t denotes the thickness of the plate and G is the shear modulus for the material. We assume that the midsurface of the undeformed plate is described by a convex polygonal domain $\Omega \subset \mathbb{R}^2$. The plate is considered to be clamped on the part Γ_{c} of its boundary $\partial\Omega$, simply supported on the part $\Gamma_{\mathrm{s}} \subset \partial\Omega$ and free on $\Gamma_{\mathrm{f}} \subset \partial\Omega$. We indicate with $\mathcal{V}$ the set of all the corner points in Γ_{f} corresponding to an angle of the boundary Γ_{f}.

Then, following the Kirchhoff plate bending model and assuming that the load is sufficiently regular, the deflection of the plate can be found as the solution of the following well known biharmonic problem:

For the scaled loading f find the deflection w such that

$$\begin{aligned}
&D\Delta^2 w = Gf && \text{in } \Omega\,,\\
&w = 0\,, \quad \nabla w \cdot \boldsymbol{n} = 0 && \text{on } \Gamma_{\mathrm{c}}\,,\\
&w = 0\,, \quad \boldsymbol{n}^T \boldsymbol{M} \boldsymbol{n} = 0 && \text{on } \Gamma_{\mathrm{s}}\,,\\
&\boldsymbol{n}^T \boldsymbol{M} \boldsymbol{n} = 0\,, \quad \tfrac{\partial}{\partial \boldsymbol{s}} \boldsymbol{s}^T \boldsymbol{M} \boldsymbol{n} + (\mathbf{div}\, \boldsymbol{M}) \cdot \boldsymbol{n} = 0 && \text{on } \Gamma_{\mathrm{f}}\,,\\
&\boldsymbol{s}_1^T \boldsymbol{M} \boldsymbol{n}_1(c) = \boldsymbol{s}_2^T \boldsymbol{M} \boldsymbol{n}_2(c) && \forall c \in \mathcal{V}\,,
\end{aligned} \tag{1}$$

where $\boldsymbol{n}$ and $\boldsymbol{s}$, respectively, are the unit outward normal and the unit counterclockwise tangent to the boundary. The indices 1 and 2 indicate the sides of the boundary angle at the corner point c. The scaled bending modulus and the bending moment are

$$D = \frac{E}{12(1-\nu^2)}\,, \tag{2}$$

$$\boldsymbol{M} = \frac{G}{6}\big(\boldsymbol{\varepsilon}(\nabla w) + \frac{\nu}{1-\nu}\operatorname{div}\nabla w \boldsymbol{I}\big)\,, \tag{3}$$

with the Young modulus E and the Poisson ratio ν for the material, and the symmetric gradient operator $\boldsymbol{\varepsilon}$. The shear modulus can be written in the form

$$G = \frac{E}{2(1+\nu)}\,. \tag{4}$$

Next we introduce, respectively, the spaces for the deflection, rotation and the shear stress (Lagrange multiplier)

$$W = \left\{v \in H^1(\Omega) \mid v = 0 \text{ on } \Gamma_c \cup \Gamma_s\right\}\,, \tag{5}$$

$$\boldsymbol{V} = \left\{\boldsymbol{\eta} \in [H^1(\Omega)]^2 \mid \boldsymbol{\eta} = 0 \text{ on } \Gamma_c\,,\ \boldsymbol{\eta}\cdot\boldsymbol{s} = 0 \text{ on } \Gamma_s\right\}\,, \tag{6}$$

$$\boldsymbol{H} = \{\boldsymbol{u} \in L^2(\Omega) \mid \operatorname{rot}\boldsymbol{u} \in L^2(\Omega)\ ,\ \boldsymbol{u}\cdot\boldsymbol{s} = 0 \text{ in } \Gamma_c \cup \Gamma_s\}\,, \tag{7}$$

$$\boldsymbol{Q} = \boldsymbol{H}'\,. \tag{8}$$

Now the mixed variational formulation reads:

Find $(w, \boldsymbol{\beta}, \boldsymbol{q}) \in W \times \boldsymbol{V} \times \boldsymbol{Q}$ such that

$$a(\boldsymbol{\beta}, \boldsymbol{\eta}) + \kappa\langle \boldsymbol{q}, \nabla v - \boldsymbol{\eta}\rangle = (f, v) \qquad \forall (v, \boldsymbol{\eta}) \in W \times \boldsymbol{V}\,, \tag{9}$$

$$\langle \nabla w - \boldsymbol{\beta}, \boldsymbol{r}\rangle = 0 \qquad \forall \boldsymbol{r} \in \boldsymbol{Q}\,, \tag{10}$$

where κ is the shear correction factor and a is the following bilinear form:

$$a(\boldsymbol{\phi}, \boldsymbol{\eta}) = \frac{1}{6}\big((\boldsymbol{\varepsilon}(\boldsymbol{\phi}), \boldsymbol{\varepsilon}(\boldsymbol{\eta})) + \frac{\nu}{1-\nu}(\operatorname{div}\boldsymbol{\phi}, \operatorname{div}\boldsymbol{\eta})\big)\ \ \forall \boldsymbol{\phi}, \boldsymbol{\eta} \in \boldsymbol{V}\,. \tag{11}$$

The brackets $\langle\cdot,\cdot\rangle$ above indicate the duality product between functions of $\boldsymbol{H}$ and $\boldsymbol{Q}$.

2.2. *The new finite element method*

For simplicity, we assume that the boundary of the plate is at least partly clamped or simply supported. Let a regular family of triangular meshes on Ω be given. Given an integer $k \geq 1$, we then define the discrete spaces for the deflection and the rotation, respectively,

$$W_h = \{v \in W \mid v_{|K} \in P_{k+1}(K)\ \ \forall K \in \mathcal{C}_h\}\,, \tag{12}$$

$$\boldsymbol{V}_h = \{\boldsymbol{\eta} \in \boldsymbol{V} \mid \boldsymbol{\eta}_{|K} \in [P_k(K)]^2\ \ \forall K \in \mathcal{C}_h\}\,, \tag{13}$$

where $\mathcal{C}_h$ represents the set of all the triangles K of the mesh and $P_k(K)$ is the space of polynomials of degree k on K.

In the sequel, we will indicate with h_K the diameter of each element K, while by h we denote the maximum size of all the elements in the mesh. Moreover, we will use the notation e for a general edge of the triangulation and h_e for the length of e. Let also two positive stability constants γ and α be assigned.[3] Then, the discrete problem reads (with $\kappa = 1$):

Method 2.1. Find $(w_h, \boldsymbol{\beta}_h) \in W_h \times \boldsymbol{V}_h$ such that

$$\mathcal{A}_h(w_h, \boldsymbol{\beta}_h; v, \boldsymbol{\eta}) = (f, v) \qquad \forall (v, \boldsymbol{\eta}) \in W_h \times \boldsymbol{V}_h \,, \tag{14}$$

where the bilinear form $\mathcal{A}_h$ is

$$\mathcal{A}_h(z, \boldsymbol{\phi}; v, \boldsymbol{\eta}) = \mathcal{B}_h(z, \boldsymbol{\phi}; v, \boldsymbol{\eta}) + \mathcal{D}_h(z, \boldsymbol{\phi}; v, \boldsymbol{\eta}) \,, \tag{15}$$

with

$$\begin{aligned} \mathcal{B}_h(z, \boldsymbol{\phi}; v, \boldsymbol{\eta}) &= a(\boldsymbol{\phi}, \boldsymbol{\eta}) - \sum_{K \in \mathcal{C}_h} \alpha h_K^2 (\boldsymbol{L\phi}, \boldsymbol{L\eta})_K \\ &+ \sum_{K \in \mathcal{C}_h} \frac{1}{\alpha h_K^2} (\nabla z - \boldsymbol{\phi} - \alpha h_K^2 \boldsymbol{L\phi}, \nabla v - \boldsymbol{\eta} - \alpha h_K^2 \boldsymbol{L\eta})_K \end{aligned} \tag{16}$$

and

$$\begin{aligned} \mathcal{D}_h(z, \boldsymbol{\phi}; v, \boldsymbol{\eta}) = \sum_{e \in \Gamma_{\mathrm{f},h}} \Big(&(M_{ns}(\boldsymbol{\phi}), [\nabla v - \boldsymbol{\eta}] \cdot \boldsymbol{s})_e \\ &+ ([\nabla z - \boldsymbol{\phi}] \cdot \boldsymbol{s}, M_{ns}(\boldsymbol{\eta}))_e + \frac{\gamma}{h_e} ([\nabla z - \boldsymbol{\phi}] \cdot \boldsymbol{s}, [\nabla v - \boldsymbol{\eta}] \cdot \boldsymbol{s})_e \Big) \end{aligned} \tag{17}$$

for all $(z, \boldsymbol{\phi}) \in W_h \times \boldsymbol{V}_h$, $(v, \boldsymbol{\eta}) \in W_h \times \boldsymbol{V}_h$, where $\Gamma_{\mathrm{f},h}$ represents the set of all the boundary edges in Γ_{f} and $M_{ns} = \boldsymbol{s}^T \boldsymbol{M} \boldsymbol{n}$.

The bilinear form $\mathcal{B}_h$ constitutes essentially of the original method of Ref. 2 with the thickness t set equal to zero, while the additional bilinear form $\mathcal{D}_h$ is introduced in order to avoid the convergence deterioration in the presence of free boundaries.[3,4]

3. A-priori and a-posteriori error estimates

3.1. *A-priori estimates*

For the deflection and the rotation, $(v, \boldsymbol{\eta}) \in W_h \times \boldsymbol{V}_h$, we introduce the following mesh dependent norms:

$$|(v, \boldsymbol{\eta})|_h^2 = \sum_{K \in \mathcal{C}_h} h_K^{-2} \|\nabla v - \boldsymbol{\eta}\|_{0,K}^2 , \tag{18}$$

$$\|v\|_{2,h}^2 = \|v\|_1^2 + \sum_{K \in \mathcal{C}_h} |v|_{2,K}^2 + \sum_{e \in \mathcal{T}_h} h_K^{-1} \| [\![\frac{\partial v}{\partial \boldsymbol{n}}]\!] \|_{0,e}^2 , \tag{19}$$

$$|||(v, \boldsymbol{\eta})|||_h = \|\boldsymbol{\eta}\|_1 + \|v\|_{2,h} + |(v, \boldsymbol{\eta})|_h . \tag{20}$$

With the space

$$\boldsymbol{V}_* = \{\boldsymbol{\eta} \in [H^1(\Omega)]^2 \mid \boldsymbol{\eta} = \boldsymbol{0} \text{ on } \Gamma_c, \ \boldsymbol{\eta} \cdot \boldsymbol{s} = 0 \text{ on } \Gamma_f \cup \Gamma_s\} \tag{21}$$

we introduce the norm

$$\|\boldsymbol{r}\|_{-1,*} = \sup_{\boldsymbol{\eta} \in \boldsymbol{V}_*} \frac{\langle \boldsymbol{r}, \boldsymbol{\eta} \rangle}{\|\boldsymbol{\eta}\|_1} , \tag{22}$$

which bounds above the classical norm $\| \cdot \|_{-1}$. We then have the following a-priori error estimate:[3]

Theorem 3.1. *Let $(w, \boldsymbol{\beta})$ be the solution of the problem (1) and $(w_h, \boldsymbol{\beta}_h)$ the solution of the problem (14). Then it holds*

$$|||(w - w_h, \boldsymbol{\beta} - \boldsymbol{\beta}_h)|||_h + \|\boldsymbol{q} - \boldsymbol{q}_h\|_{-1,*} \leq C h^s \|w\|_{s+2} \tag{23}$$

for all $1 \leq s \leq k$, where the exact and discrete shear stresses are

$$\boldsymbol{q}_{|K} = \frac{1}{\alpha h_K^2} (\nabla w - \boldsymbol{\beta} - \alpha h_K^2 \boldsymbol{L}\boldsymbol{\beta})_{|K} , \tag{24}$$

$$\boldsymbol{q}_{h|K} = \frac{1}{\alpha h_K^2} (\nabla w_h - \boldsymbol{\beta}_h - \alpha h_K^2 \boldsymbol{L}\boldsymbol{\beta}_h)_{|K} \quad \forall K \in \mathcal{C}_h . \tag{25}$$

For the shear stress $\boldsymbol{r} \in L^2(\Omega)$, the same result holds in the auxiliary norm

$$\|\boldsymbol{r}\|_{-1,h} = \Big(\sum_{K \in \mathcal{C}_h} h_K^2 \|\boldsymbol{r}\|_{0,K}^2 \Big)^{1/2} . \tag{26}$$

We also note that the following norm equivalence holds:

Lemma 3.1. *There is a positive constant C such that*

$$C|||(v, \boldsymbol{\eta})|||_h \leq \|\boldsymbol{\eta}\|_1 + |(v, \boldsymbol{\eta})|_h \leq |||(v, \boldsymbol{\eta})|||_h \quad \forall (v, \boldsymbol{\eta}) \in W_h \times \boldsymbol{V}_h . \tag{27}$$

3.2. *A-posteriori estimates*

We now present the reliability and the efficiency results for an a-posteriori error estimator of our method. To this end, we introduce

$$\tilde{\eta}_K^2 := h_K^4 \|f_h + \operatorname{div} \boldsymbol{q}_h\|_{0,K}^2 + h_K^{-2} \|\nabla w_h - \boldsymbol{\beta}_h\|_{0,K}^2 , \tag{28}$$

$$\eta_e^2 := h_e^3 \|[\![\boldsymbol{q}_h \cdot \boldsymbol{n}]\!]\|_{0,e}^2 + h_e \|[\![\boldsymbol{M}(\boldsymbol{\beta}_h)\boldsymbol{n}]\!]\|_{0,e}^2 , \tag{29}$$

$$\eta_{s,e}^2 := h_e \|M_{nn}(\boldsymbol{\beta}_h)\|_{0,e}^2 , \tag{30}$$

$$\eta_{f,e}^2 := h_e \|M_{nn}(\boldsymbol{\beta}_h)\|_{0,e}^2 + h_e^3 \|\frac{\partial}{\partial s} M_{ns}(\boldsymbol{\beta}_h) - \boldsymbol{q}_h \cdot \boldsymbol{n}\|_{0,e}^2 , \tag{31}$$

where f_h is some approximation of the load f and $[\![\cdot]\!]$ represents the jump operator (which is assumed to be equal to the function value on boundary edges).

Then for any element $K \in \mathcal{C}_h$, the local error indicator is

$$\eta_K := \Big(\tilde{\eta}_K^2 + \frac{1}{2} \sum_{e \in \Gamma_{\mathrm{i},h} \cap \partial K} \eta_e^2 + \sum_{e \in \Gamma_{\mathrm{s},h} \cap \partial K} \eta_{s,e}^2 + \sum_{e \in \Gamma_{\mathrm{f},h} \cap \partial K} \eta_{f,e}^2 \Big)^{1/2} , \tag{32}$$

where $\Gamma_{\mathrm{i},h}$ represents the set of all the internal edges, while $\Gamma_{\mathrm{c},h}, \Gamma_{\mathrm{s},h}$ and $\Gamma_{\mathrm{f},h}$ represent the sets of all the boundary edges in $\Gamma_{\mathrm{c}}, \Gamma_{\mathrm{s}}$ and Γ_{f}, respectively. Finally, the global error indicator is defined as

$$\eta := \Big(\sum_{K \in \mathcal{C}_h} \eta_K^2 \Big)^{1/2} . \tag{33}$$

For the error analysis we assume that the following classical saturation assumption holds:

Assumption 3.1. *Given a mesh $\mathcal{C}_h$, let $\mathcal{C}_{h/2}$ be the mesh obtained by splitting each triangle $K \in \mathcal{C}_h$ into four triangles connecting the edge midpoints. Let $(w_{h/2}, \boldsymbol{\beta}_{h/2}, \boldsymbol{q}_{h/2})$ be the discrete solution corresponding to the mesh $\mathcal{C}_{h/2}$. We assume that there exists a constant ρ, $0 < \rho < 1$, such that*

$$\begin{aligned} &|||(w - w_{h/2}, \boldsymbol{\beta} - \boldsymbol{\beta}_{h/2})|||_{h/2} + \|\boldsymbol{q} - \boldsymbol{q}_{h/2}\|_{-1,*} \\ &\le \rho\big(|||(w - w_h, \boldsymbol{\beta} - \boldsymbol{\beta}_h)|||_h + \|\boldsymbol{q} - \boldsymbol{q}_h\|_{-1,*}\big) , \end{aligned} \tag{34}$$

where by $|||\cdot|||_{h/2}$ we indicate the norm $|||\cdot|||_h$ with respect to the new mesh $\mathcal{C}_{h/2}$.

We then have the following reliability and efficiency results for the error estimator:[3]

Theorem 3.2. *There exists a positive constant C such that*

$$|||(w - w_h, \boldsymbol{\beta} - \boldsymbol{\beta}_h)|||_h + \|\boldsymbol{q} - \boldsymbol{q}_h\|_{-1,*} \le C\Big(\sum_{K \in \mathcal{C}_h} \eta_K^2 + h_K^4 \|f - f_h\|_{0,K}^2 \Big)^{1/2} .$$

Theorem 3.3. *It holds*

$$\eta_K \le |||(w - w_h, \boldsymbol{\beta} - \boldsymbol{\beta}_h)|||_{h,\omega_K} + \|\boldsymbol{q} - \boldsymbol{q}_h\|_{-1,*,\omega_K} + h_K^2 \|f - f_h\|_{0,\omega_K} ,$$

where $|||\cdot|||_{h,\omega_K}$, $\|\cdot\|_{-1,*,\omega_K}$ *and* $\|\cdot\|_{0,\omega_K}$ *represent, respectively, the norms* $|||\cdot|||_h$, $\|\cdot\|_{-1,*}$ *and* $\|\cdot\|_0$ *restricted to the domain* ω_K*, the set of all the triangles sharing an edge with* K*.*

4. Numerical results

In all the cases the material constants are the Young modulus $E = 1$ and the Poisson ratio $\nu = 0.3$. The stability constants we have chosen for the polynomial degree $k = 1$ are $\alpha = 0.1$ and $\gamma = 100$. For $k = 2$ we have used the local stability constants α_K and γ_e, see the discussion in Ref. 3.

4.1. *A-priori test: semi-infinite plate*

Our first numerical computations are performed for a test problem for which an analytical solution has been obtained in Ref. 5. The domain is the semi-infinite region $\Omega = \{(x, y) \in \mathbb{R}^2 \mid y > 0\}$ and the loading is $f = \frac{1}{G} \cos x$. The shear modulus is $G = 1/(2(1 + \nu))$ and the shear corrector factor is $\kappa = 1$. The boundary $\Gamma = \{(x, y) \in \mathbb{R}^2 \mid y = 0\}$ is assumed to be free.

Due to the smoothness of the solution, from Theorem 3.1 and Lemma 3.1 it follows immediately the convergence rate

$$\|\boldsymbol{\beta} - \boldsymbol{\beta}_h\|_1 + |(w - w_h, \boldsymbol{\beta} - \boldsymbol{\beta}_h)|_h = \mathcal{O}(h^k) \ . \tag{35}$$

On the contrary, according to the observations in Ref. 3,4, the convergence rate for the original method, without the additional bilinear form $\mathcal{D}_h$, should be of the order $\mathcal{O}(h^{1/2})$.

We discretize the domain $\overline{D} = [0, \pi/2] \times [0, 3\pi/4]$ and set the symmetry conditions on the vertical boundaries $\{x = 0, 0 \le y \le 3\pi/4\}$, $\{x = \pi/2, 0 \le y \le 3\pi/4\}$, while on the upper horizontal boundary $\{y = 3\pi/4, 0 \le x \le \pi/2\}$ we use the non-homogeneus Dirichlet conditions adopting the exact solution as a reference. The computations are performed for five meshes corresponding to the uniform refinements for $N = 2, 4, 8, 16, 32$, where N is the number of elements on the boundary, see Figure 1.

Let D_b represent the boundary domain $[0, \pi/2] \times [0, \pi/4]$, see Figure 1.

In Figure 2 we depict the error in the norm of (35) equivalent to the norm $|||\cdot|||_h$ –see Lemma 3.1, for the polynomial degrees $k = 1, 2$.

The dashed line is the convergence graph for the original method (i.e. without the correction $\mathcal{D}_h$ in (16)) while the solid line refers to the new one, Method 2.1. As predicted by the theory, the convergence rate for the original method is $\mathcal{O}(h^{1/2})$, while the modified method follows the rate $\mathcal{O}(h^k)$.

4.2. *A-posteriori tests*

In this section we consider the case $k = 1$, i.e. the finite element approximation is quadratic for the deflection and linear for the rotation.

4.2.1. *Rectangle with simply supported boundaries*

First we consider the simply supported rectangle $\Omega = (0,1) \times (-1,1)$ with the uniform loading $f = 1$. The exact solution for the problem can be found by writing the load as a trigonometric series.

According to Ref. 6 the regularity in the corners of the plate is $w \in H^3$. This implies by Theorem 3.1 and Lemma 3.1 the theoretical convergence rate $\mathcal{O}(h^\sigma)$, $\sigma = \min\{1, k\} = 1$.

The convergence graphs for the adaptively refined meshes are shown in Figure 3. The two upper graphs (solid lines) represent, respectively, the global error estimator (red) and the global true error (blue), while the lower ones (dashed lines) are, respectively, for the maximum local estimator (red) and for the maximum of the local true errors (blue). Also the theoretical convergence rate $\mathcal{O}(h)$ is indicated in the same figure (black dashed line).

We note that for the global error the convergence rate in Figure 3 is the same as the theoretical value, i.e. the convergence rate for the uniform refinements. This holds for both the true and the estimated error.

The effectivity index for the adaptive error estimator, i.e. the ratio between the estimated and true error, is shown in Figure 6 for this test problem. The reported steps are taken from the adaptive refinements. As can be seen in the figure, the effectivity index lies between 0.6 and 1.1; the black dashed line represents the value 1. The effectivity index first decreases (between 8 and 200 elements) but then settles down near the value 0.7 (between 200 and 4209 elements).

The numerical computations show that the effectivity index remains in a certain, almost constant, level uniformly in the mesh size. We emphasize that although we have used here the norm in (35) the effectivity index with respect to the norm $|||\cdot|||_h$ is different only up to a constant level, because of the equivalence of these two norms. These facts confirm that the error estimate can be used as an reliable and efficient error measure.

4.2.2. *L-shaped domain with simply supported boundaries*

Here we use the error estimator as the only error measure, because of the lack of the exact solution for the following benchmark problem — but also due to the fact that the behavior of the error estimator was confirmed to be uniformly at the same level as the true error in the previous test problem.

We study the L-shaped nonconvex domain with the corners $(0,0)$, $(2,0)$, $(2,1)$, $(1,1)$, $(1,2)$ and $(0,2)$. The plate is uniformly loaded, $f = 1$, and all the boundaries are simply supported.

According to Ref. 6 the regularity in the critical L-corner is now $w \in H^{7/3}$. This implies by Theorem 3.1 and Lemma 3.1 the theoretical convergence rate $\mathcal{O}(h^\sigma)$, $\sigma = \min\{1/3, k\} = 1/3$.

The convergence graphs for the uniformly and adaptively refined meshes are shown in Figure 4. The two upper graphs (solid lines) represent the global error estimator, while the lower ones (dashed lines) indicate the maximum local estimator. Moreover, we show in the same figure also the theoretical convergence rates $\mathcal{O}(h)$ and $\mathcal{O}(h^{1/3})$ (black dashed lines). Finally, two example meshes from the adaptive process, with the distribution of the error estimator, are depicted in Figure 5.

For the uniform refinements, the convergence rate of the error estimator in Figure 4 (blue graphs) clearly follows the theoretical value $\mathcal{O}(h^{1/3})$. Differently, after the first adaptive steps, the method shows its robustness in finding the corner singularity of the solution and refining locally near the L-corner. This is seen in both the convergence graphs (red ones), especially in the local one.

References

1. P. Destuynder and T. Nevers, Une modification du modèle de Mindlin pour les plaques minces en flexion présentant un bord libre. *RAIRO Modél. Math. Anal. Numér.*, **22**, 217–242, 1988.
2. R. Stenberg, A new finite element formulation for the plate bending problem. P. G. Ciarlet, L. Trabucho, and M. Viaño eds. *Asymptotic Methods for Elastic Structures*, Proceedings of the International Conference, Lisbon, October 4–8, 1993, Walter de Gruyter & Co., Berlin - New-York, 209–221, 1995.
3. L. Beirão da Veiga and J. Niiranen and R. Stenberg, A family of C^0 finite elements for Kirchhoff plates. Helsinki University of Technology, Institute of Mathematics, Research Reports, A 483, January, 2006 (`http://www.math.tkk.fi/reports`).
4. L. Beirão da Veiga, Finite element methods for a modified Reissner–Mindlin free plate model. *SIAM J. Num. Anal.*, **42**, 1572–1591, 2004.
5. D. N. Arnold and R. S. Falk, Edge effects in the Reissner–Mindlin plate

theory. A. K. Noor, T. Belytschko and J. C. Simo eds. *Analytic and Computational Models of Shells*, ASME, New York, 71–90, 1989.
6. H. Melzer and R. Rannacher, Spannungskonzentrationen in Eckpunkten der vertikal belasteten Kirchhoffschen platte. *Bauingenieur*, **55**, 181–189, 1980.

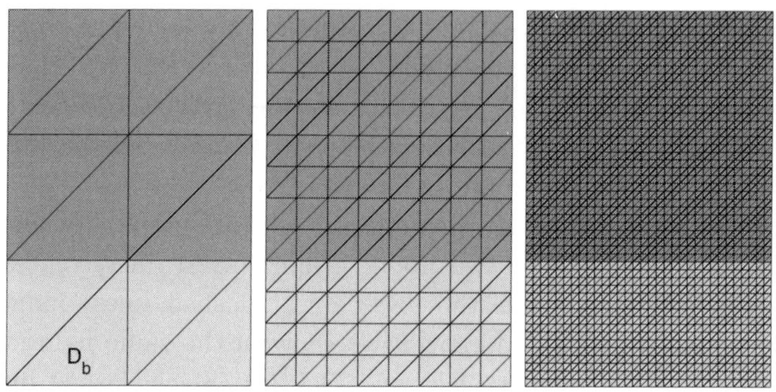

Fig. 1. Uniform meshes with the boundary region D_b for $N = 2, 8, 32$.

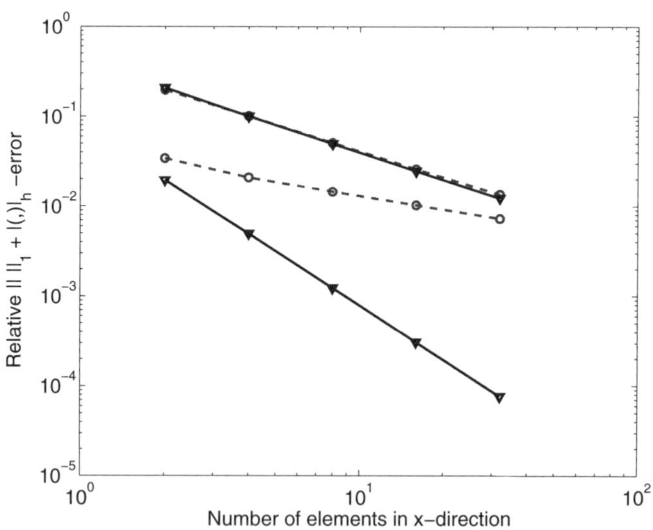

Fig. 2. Free boundary; Convergence of the error in the norm $||\boldsymbol{\beta} - \boldsymbol{\beta}_h||_{1,D_b} + |(w - w_h, \boldsymbol{\beta} - \boldsymbol{\beta}_h)|_{h,D_b}$ with $k = 1, 2$ (dashed line for the original, solid line for the new method).

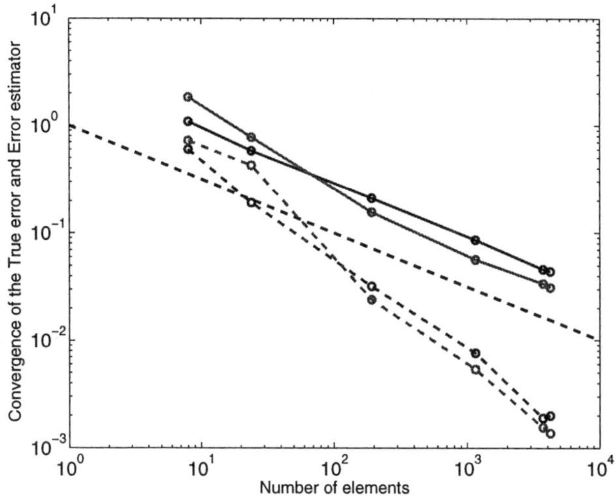

Fig. 3. Simply supported boundary; Convergence of true error and the error estimator for adaptive refinements; Solid lines for the global error and estimator, dashed lines for the maximum local error and estimator; straight dashed line for the theoretical convergence rate $\mathcal{O}(h)$.

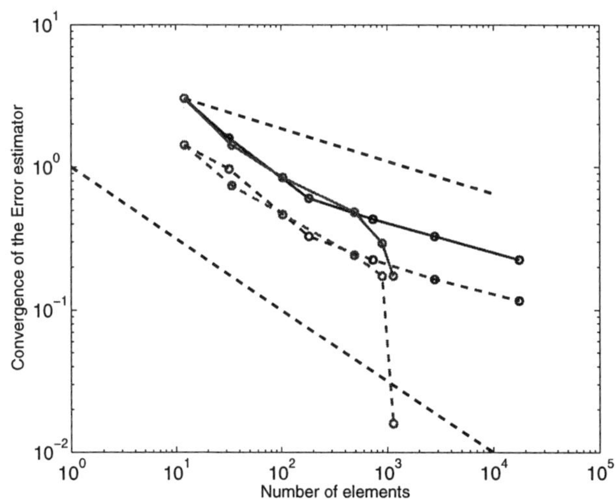

Fig. 4. Simply supported boundary and L-corner; Convergence of the global estimator (solid lines) and the maximum local estimator (dashed lines); uniform refinements and adaptive refinements are shown; Straight dashed lines for the theoretical convergence rates $\mathcal{O}(h)$ and $\mathcal{O}(h^{1/3})$.

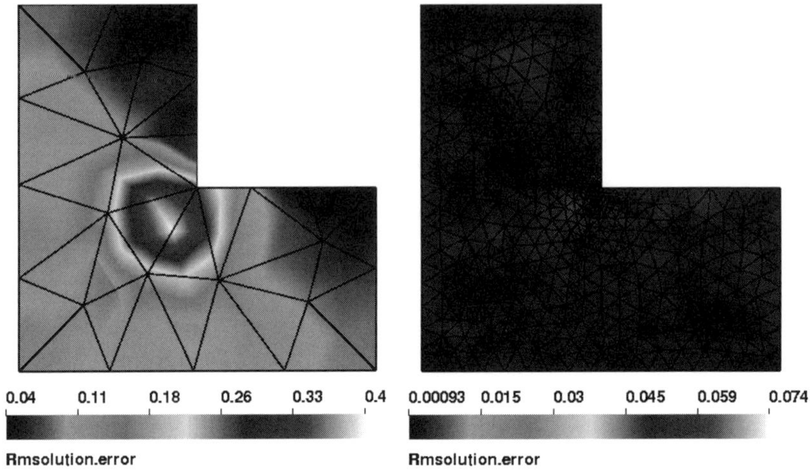

Fig. 5. Simply supported boundary and L-corner; Distribution of the error estimator for two different refinement steps.

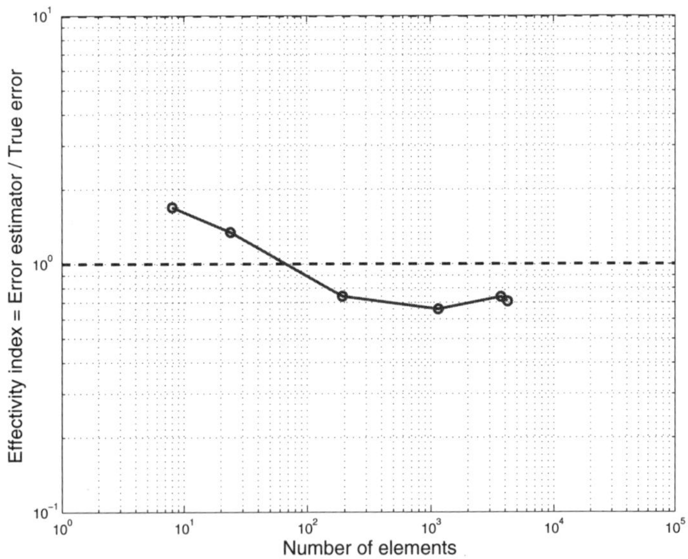

Fig. 6. Simply supported boundary; Effectivity index for the adaptive refinements.

A TWO-DIMENSIONAL TRUST-REGION METHOD FOR LARGE SCALE BOUND-CONSTRAINED NONLINEAR SYSTEMS *

S. BELLAVIA, M. MACCONI and B. MORINI

Dipartimento di Energetica "S.Stecco", Università degli Studi di Firenze
via C. Lombroso 6/17, 50134 Firenze, Italia
E-mail: stefania.bellavia@unifi.it, maria.macconi@unifi.it, benedetta.morini@unifi.it

A class of trust-region methods for solving large scale bound-constrained nonlinear systems has been proposed in a recent paper by the authors. These methods combine a Newton-Krylov method with a subspace trust-region strategy and with a strategy for handling the bounds. In this paper, an algorithm from that class is described and its numerical performance is illustrated by numerical experiments. The results of a comparison with an efficient approach in literature are shown too.

Keywords: Inexact Newton methods, forcing terms, Krylov subspace methods, subspace trust-region problems, bound constraints.

1. Introduction

In scientific and engineering computing areas it is often needed to solve bound-constrained nonlinear systems[1,2]

$$F(x) = 0, \qquad l \le x \le u, \tag{1}$$

where $x \in \mathbb{R}^n$, $l \in (\mathbb{R} \cup \{-\infty\})^n$, $u \in (\mathbb{R} \cup \{\infty\})^n$, $F : X \to \mathbb{R}^n$ with $X \subseteq \mathbb{R}^n$. The function F is assumed to be continuously differentiable and X is assumed to be an open set containing the feasible region $\Omega = \{x \in \mathbb{R}^n : l \le x \le u\}$.

In the last few years, numerical methods for (1) have been proposed and most of them are suitable for medium problems, see e.g. Ref. 3–5. In this work we are interested in solving large scale problems and focus on a general scheme for large bound-constrained problems presented by Bellavia and Morini in Ref. 6.

The class of procedures considered is in the framework of Newton-Krylov methods.[7] This means that, at the current iterate x_k, a Krylov method[8] is applied to the Newton equation

$$F'(x_k)p = -F(x_k), \tag{2}$$

in order to compute an Inexact Newton step p_k^I satisfying

$$\|F'(x_k)\, p_k^I + F(x_k)\| \le \eta_k \|F(x_k)\|, \tag{3}$$

where $\eta_k \in [0,1)$ is a scalar usually called forcing term.

To manage constrained nonlinear systems and obtain global convergence, the Newton-Krylov method is augmented with a strategy for handling the bounds and with a trust-region strategy defined over a small dimensional space. Thus, the work to solve the subspace trust-region problem is trivial while the dominant computational work has been shifted to the determination of the subspace. A general formulation for these methods is given in Ref. 6 along with the theoretical study showing global and fast local convergence under standard assumptions.

The preliminary computational results reported in Ref. 6 encouraged the development of a practical trust-region algorithm for large scale bound-constrained nonlinear systems. The purpose of this paper is to illustrate the numerical potential of a particular implementation of one algorithm in the framework. Also, the procedure outlined is compared with an alternative optimization technique suited for the least-squares problem $\min_{x\in\Omega} \|F(x)\|^2/2$ and contained in the `Matlab` Optimization Toolbox.[9] The results confirm the favorable behavior of the proposed algorithm and encourage further experiments on a larger spectrum of applications.

1.1. *Some notations*

The symbol $\|\cdot\|$ indicates the Euclidean vector norm and the subordinate matrix norm. The symbol $(x)_i$ represents the ith component of the vector $x \in \mathbb{R}^n$. When clear from the context, the brackets are omitted. For any function, the symbol f_k denotes the value of f at the point x_k of the sequence $\{x_k\}$.

Let $int(\Omega) = \{x \in \mathbb{R}^n : l < x < u\}$. For $x \in int(\Omega)$, $D(x)$ denotes the diagonal matrix

$$D(x) = diag((D(x))_{1,1}, (D(x))_{2,2}, \dots, (D(x))_{n,n}) \tag{4}$$

with $(D(x))_{i,i}$ given by

$$(D(x)_{i,i}) = \begin{cases} u_i - x_i & \text{if} \quad (\nabla f(x))_i < 0 \;\; \text{and} \;\; u_i < \infty \\ x_i - l_i & \text{if} \quad (\nabla f(x))_i > 0 \;\; \text{and} \;\; l_i > -\infty \\ \min\{x_i - l_i,\, u_i - x_i\} & \text{if} \quad (\nabla f(x))_i = 0 \;\; \text{and} \;\; l_i > -\infty \\ & \qquad\qquad\qquad\qquad\qquad \text{or } u_i < \infty \\ 1 & \text{otherwise} \end{cases} .$$

2. The two-dimensional subspace interior trust-region method

In this section, we present one iteration of the method implemented and discuss the computational effort required to perform it. The current iterate x_k is assumed to be strictly feasible, i.e. $x_k \in int(\Omega)$.

Let m_k be the following local model of F around x_k

$$m_k(p) = \frac{1}{2}\|F(x_k) + F'(x_k)p\|^2, \tag{5}$$

where F' is the Jacobian matrix of F and $p \in \mathbb{R}^n$. Assume that m_k is an adequate representation of the merit function

$$f(x) = \frac{1}{2}\|F(x)\|^2, \tag{6}$$

in a trust-region defined as a sphere of radius Δ_k centered at x_k. Then, the trust-region problem takes the form

$$\min_{p \in \mathbb{R}^n} \{m_k(p) : \|p\| \leq \Delta_k\}. \tag{7}$$

Several algorithms[10] exist for solving (7). Some of them look for a nearly exact solution and typically involve the computation of eigenvalues and a Newton process applied to the secular equation. Other approaches find approximate solutions by simple strategies and are less costly to implement.

Since we are interested in large scale problems, we follow the approach proposed in Ref. 6 where the local model m_k is minimized on a two-dimensional subspace S_k, i.e. problem (7) is replaced by

$$\min_{p \in S_k} \{m_k(p) : \;\; \|p\| \leq \Delta_k\}. \tag{8}$$

The subspace S_k is computed cheaply with the aid of the Krylov method used to solve the Newton equation (2). In fact, we assign $S_k = span\{p_k^I, \nabla f_k\}$ where p_k^I is given in (3) and ∇f_k is the gradient of the merit function f at x_k. The theoretical analysis carried out in Ref. 6 shows that the exact solution of (8) is not required to ensure global and locally fast convergence. Thus, in the sequel we assume that problem (8) is solved approximately and let p_k^S be an approximate solution to (8).

The step p_k^S may not be well suited for being used since the point $x_k^S = x_k + p_k^S$ may be infeasible. Therefore, we consider the projection of $x_k + p_k^S$ onto Ω followed by a small step toward the interior of Ω, i.e.

$$p_k^{TR} = \alpha_k (P(x_k + p_k^S) - x_k), \tag{9}$$

where $\alpha_k \in (0,\, 1)$ and $(P(x))_i = \max\{l_i,\, \min\{x_i,\, u_i\}\}$, $i = 1, \ldots, n$.

The strategy for computing the new iterate requires the generalized Cauchy point p_k^C defined as follows. First, we consider the following scaled gradient of f at x_k:

$$d_k = -D_k \nabla f_k, \tag{10}$$

where $D_k = D(x_k)$ is defined by (4). Second, we compute the minimizer of m_k along d_k subject to satisfy the trust-region bounds and let

$$\begin{aligned} \tau_k &= \underset{\tau > 0}{\operatorname{argmin}} \{m_k(\tau d_k) : \|\tau d_k\| \le \Delta_k\} \\ &= \min\left\{ \frac{\nabla f_k^T D_k \nabla f_k}{\|F_k' D_k \nabla f_k\|^2}, \frac{\Delta_k}{\|D_k \nabla f_k\|} \right\}. \end{aligned} \tag{11}$$

Then, the generalized Cauchy point is such that

$$p_k^C = c_k\, d_k, \qquad c_k = \begin{cases} \tau_k & \text{if} \quad x_k + \tau_k d_k \in int(\Omega) \\ \theta \lambda_k & \text{otherwise} \end{cases} \tag{12}$$

with $\theta \in (0, 1)$ and $\lambda_k = \operatorname{argmin}\{l > 0, x_k + l d_k \in \Omega\}$.

To make the method globally convergent, the step p_k used to update x_k is required to satisfy the following decrease condition on m_k

$$\rho_c(p_k) = \frac{m_k(0) - m_k(p_k)}{m_k(0) - m_k(p_k^C)} \ge \beta_1, \tag{13}$$

with $\beta_1 \in (0, 1)$. We force this condition and let the vector p_k^{TR} given in (9) play the role of the trial step. Hence, we set $p_k = p_k^{TR}$ if it satisfies (13). Otherwise, we let

$$p_k = t p_k^C + (1 - t) p_k^{TR}, \tag{14}$$

where $t \in (0, 1)$ is fixed so that $\rho_c(p_k) = \beta_1$. It is easy to see that the value of t is given by

$$t = \frac{z^T u - w}{\|u\|^2}, \tag{15}$$

where $u = F'_k(p_k^C - p_k^{TR})$, $\quad z = -(F_k + F'_k p_k^{TR})$,
$w = \left((z^T u)^2 - 2\,\|u\|^2\,(F_k^T F'_k(p_k^{TR} - \beta_1 p_k^C) + \|F'_k p_k^{TR}\|^2/2 - \beta_1 \|F'_k p_k^C\|^2/2)\right)^{\frac{1}{2}}$.

Finally, the step p_k must provide a good agreement between m_k and the merit function f. Thus, the condition

$$\rho_f(p_k) = \frac{f(x_k) - f(x_k + p_k)}{m_k(0) - m_k(p_k)} \geq \beta_2, \tag{16}$$

is tested for a given constant $\beta_2 \in (0,1)$. If this condition doesn't hold, then the step p_k is rejected, the trust-region radius Δ_k is reduced and a new trial step is computed. If (16) holds, then the new iterate $x_{k+1} = x_k + p_k$ is formed and the trust-region radius is updated. In practice, given a positive constant β_3 such that $\beta_2 < \beta_3 < 1$, if

$$\rho_f(p_k) \geq \beta_3, \tag{17}$$

we allow a possible increase in Δ_k. Otherwise the trust region radius is kept the same.

The above considerations lead to the following Two-Dimensional Subspace Interior Trust-Region (SITR-2D) method:

SITR-2D METHOD

Given: $x_0 \in int(\Omega)$, $\Delta_0 > 0$, $\beta_1 \in (0,1)$, $0 < \beta_2 < \beta_3 < 1$, $\theta \in (0,1)$.

For $k = 0, 1, 2, \ldots$

1. Choose $\alpha_k \in (0,1)$, $\eta_k \in [0,1)$.
2. Apply a Krylov method to (2) to find a vector p_k^I satisfying (3).
3. Compute an approximate solution p_k^S to (8).
4. Form p_k^{TR} by (9) and p_k^C by (12).
5. If $\rho_c(p_k^{TR}) \geq \beta_1$
   set $p_k = p_k^{TR}$
   otherwise
   compute p_k by (14) with t given by (15).
6. If $\rho_f(p_k) < \beta_2$ reduce Δ_k and go to Step 3.
7. Set $x_{k+1} = x_k + p_k$.
8. Choose $\Delta_{k+1} \geq \Delta_k$ according to (17).

From the computational point of view, our description of the method is complete if we specify the algorithm to compute p_k^S. In our implementation we approximately solve (8) with a very low computational effort by the following dogleg strategy:

DOGLEG ALGORITHM FOR PROBLEM (8)

Given: $x_k \in int(\Omega)$, $p_k^I \in \mathbb{R}^n$, $\Delta_k > 0$.

1. Compute ∇f_k and $w_1 = \nabla f_k / \|\nabla f_k\|$.
2. Compute $w_2 = p_k^I - (w_1^T p_k^I)\, w_1$. Set $w_2 = w_2/\|w_2\|$.
3. Set $W = [w_1, w_2] \in \mathbb{R}^{n\times 2}$.
4. Compute $\tau = \min\{\|W^T\nabla f_k\|^2/\|F_k' W W^T \nabla f_k\|^2,\ \Delta_k/\|W^T\nabla f_k\|\}$.
5. Compute the two-dimensional Cauchy step $s^c = -\tau W^T \nabla f_k$.
6. Solve the linear system

$$\begin{pmatrix} \|F_k' w_1\|^2 & w_2^T {F_k'}^T F_k' w_1 \\ w_2^T {F_k'}^T F_k' w_1 & \|F_k' w_2\|^2 \end{pmatrix} \begin{pmatrix} s_1^m \\ s_2^m \end{pmatrix} = \begin{pmatrix} \|\nabla f_k\| \\ 0 \end{pmatrix},$$

   and set $s^m = (s_1^m, s_2^m)^T$.
7. Determine the two-dimensional dogleg step s_k:

$$s_k = \begin{cases} s^m & \text{if } \|s^m\| \le \Delta_k; \\ s^c & \text{if } \|s^c\| = \Delta_k; \\ (1-\sigma)s^c + \sigma\, s^m & \text{otherwise} \end{cases}$$

   where $\sigma \in (0,1)$ is uniquely determined so that $\|s_k\| = \Delta_k$.
8. Compute $p_k^S = W s_k$.

The above algorithm works in the following way. First of all, an orthonormal basis $W = [w_1, w_2] \in \mathbb{R}^{n\times 2}$ for S_k is computed. Thus, for each vector $p \in S_k$ there exists $s \in \mathbb{R}^2$ such that $p = Ws$ and problem (8) can be written as

$$\min_{s\in\mathbb{R}^2} \{\|F_k + F_k' W\, s\|^2 : \ \|s\| \le \Delta_k\}. \tag{18}$$

Then, a standard dogleg method[10] is applied to solve (18). The Cauchy point $s^c \in \mathbb{R}^2$ for problem (18) is found in Steps 4-5 and the solution $s^m \in \mathbb{R}^2$ to the unconstrained minimization problem $\min_{s\in\mathbb{R}^2} \|F_k + F_k' W s\|^2$ is evaluated in Step 6 by solving the 2×2 linear system. Finally, the dogleg solution s_k to (18) is formed (Step 7) and p_k^S is computed (Step 8). We remark that the vector s^m could be computed also via the QR decomposition of the $n \times 2$ matrix $F_k' W$. Anyway, the computational effort of the above dogleg procedure is very low.

It is worthy noting that, if the linear system (2) is solved up to the full accuracy, i.e. the Krylov solver is applied with a tight stopping criterion, the two dimensional subspace dogleg strategy reduces to the standard dogleg strategy.

We conclude this section with some comments on the computational cost of each iteration of the SITR-2D method. The main cost is given by the computation of the Inexact Newton step p_k^I by a Krylov method while the other steps can be cheaply performed. In particular, computing ∇f_k calls for a matrix-vector product and the evaluation of the generalized Cauchy step p_k^C involves one matrix-vector product to compute the scalar τ_k in (11). The main computational effort of the dogleg algorithm for problem (8) consists of one step of the Gram-Schmidt procedure and two matrix-vector products needed to form the linear system of dimension two. We note that the computation of τ does not require any further matrix-vector product as $WW^T\nabla f_k = \nabla f_k$ and $\|W^T\nabla f_k\| = \|\nabla f_k\|$. Finally, the computation of the actual step p_k requires two matrix-vector products for obtaining the scalar t given in (15).

3. Some numerical experiments

We implemented the `SITR-2D` method in a Double Precision FORTRAN code. The numerical experiments were performed with an Intel Pentium M Processor 1700 MHz and 1 Gb of RAM, with machine precision $\epsilon_m = 2 \cdot 10^{-16}$.

The application of our method requires some parameters. We fixed $\Delta_0 = 1$, $\beta_1 = 0.1$, $\beta_2 = 0.25$, $\beta_3 = 0.75$, $\theta = 0.99995$, see Ref. 3.

The Jacobian of F was computed analytically while the inexact Newton step p_k^I was computed by GMRES(20)[8] with a maximum of 49 restart, for a total of 1000 GMRES iterations. If after 1000 GMRES iterations condition (3) is not met, the step p_k^I given by the last computed GMRES iterate is used. Regarding the choice of the forcing terms, we implemented[7]

$$\eta_0 = 0.9, \qquad \eta_k = 0.9\,\frac{\|F_k\|^2}{\|F_{k-1}\|^2}, \quad k > 0,$$

and used the safeguards suggested in [11, p. 305]. This choice along with $\alpha_k = \max(0.95, 1 - \|F_k\|)$ ensures quadratic convergence.[6] Concerning the trust-region update, if the step fails to satisfy (16) the trust-region radius is reduced setting $\Delta_k = \min\{\Delta_k/4, \|p_k\|/2\}$. If (17) holds we set $\Delta_{k+1} = \max\{\Delta_k, 2\|p_k\|\}$, otherwise we let $\Delta_{k+1} = \Delta_k$.

Convergence is declared when

$$\|F_k\| \leq 10^{-6}. \tag{19}$$

Failure is declared if one of the following situations occurs: either 150 iterations or 1000 F-evaluations are performed; the trust-region size is reduced below $\sqrt{\epsilon_m}$; $\|F_{k+1} - F_k\| \leq 100\epsilon_m \|F_k\|$; $\|D_k^{-1} \nabla f_k\| < 100\epsilon_m$. The last two occurrences indicate that the sequence $\{x_k\}$ is approaching a minimum of f in Ω.[3]

The aim of this section is to prove the computational feasibility of our method and to allow a comparison with the well-assessed function `lsqonlin` contained in the `Matlab` Optimization Toolbox. The function `lsqonlin` addresses the problem

$$\min_{x \in \Omega} f(x),$$

where f is the merit function in (6) and by default it chooses the large-scale algorithm. This algorithm is a subspace trust-region method and it is based on the interior-reflective Newton method described in Ref. 12, 13. In particular, at each iteration, the Newton equation is formulated as a symmetric linear system and solved by the method of preconditioned Conjugate Gradients (PCG). Then, the trust region problem is restricted to a two-dimensional space spanned by the resulting inexact Newton step and the direction d_k.

We run `lsqnonlin` using the analytical form of the Jacobian. We chose the stopping tolerances `TolFun`$=10^{-10}$, `TolX`$=0$, in order to make differences in final precision negligible with respect to `SITR-2D`. The remaining settings are the default ones. The method terminates successfully when the change in the value of f is less than the tolerance `TolFun` while it is considered to fail when it converges to a minimum of f that is not a zero of F. Further, a failure is declared if 1000 F-evaluations or 150 nonlinear iterations were not enough to satisfy the stopping criteria.

3.1. *The test problems*

We next give a summary of the test problems used. These problems have more than one solution and the bounds allow to select specified solutions.

`P1. The Chandrasekhar H-equation.` We consider the discrete analogue of the Chandrasekhar H-equation, [14, p. 87]. It depends on a parameter $c \in (0, 1]$. We set $n = 400$ and $c = 0.9999$. We use the bounds $l_i = 0$, $u_i = 10^{300}$, $i = 1 \ldots, n$.

`P2. Integral equation.` This problem is a discrete analogue of the integral equation given in Ref. 15. The discretization used is the same of problem `P1`. The dimension is $n = 10^3$ and the bounds are $l_i = 0$, $u_i = -10^{300}$, $i = 1 \ldots, n$.

`P3. Five diagonal system`. This is the nonlinear system [16, Problem 4.8]. We set $n = 5 \cdot 10^3$, $l_i = 1$, $u_i = 10^{300}$, $i = 1 \ldots, n$.

`P4. Seven diagonal system`. This is the nonlinear system [16, Problem 4.9]. We set $n = 5 \cdot 10^3$, $l_i = 0$, $u_i = 10^{300}$, $i = 1 \ldots, n$.

`P5. Poisson problem`. This problem is the standard five-points finite differences analogue of the nonlinear partial differential equation given in [16, Problem 4.25]. We let $n = 10^4$, $l_i = -5$, $u_i = 5$, $i = 1 \ldots, n$.

`P6. Bratu problem`. This is the standard five-points finite differences analogue of the nonlinear partial differential equation given in [16, Problem 4.24]. We let $n = 10^4$, $l_i = -10^{300}$, $u_i = 1.5$, $i = 1 \ldots, n$.

`P7. A chemical equilibrium system`. A nonlinear system of $n = 11 \cdot 10^3$ equations was obtained augmenting System 1 given in Ref. 2. We use the bounds $l_i = 0$ and $u_i = 10^{300}$ for $i = 1 \ldots, n$.

`P8. The Bratu NCP problem`. The Bratu nonlinear complementarity problem[17] was reformulated as a system of $n = 5 \cdot 10^4$ nonlinear equations with $l_i = 0$ and $u_i = 10^{300}$ for $i = 1 \ldots, n$. It depends on a parameter λ and we set $\lambda = 6$.

All tests from problems P1-P4 and P7 were solved without preconditioning, while the incomplete LU factorization ILU(0) was used in the solution of the tests from problems P5, P6 and P8.

3.2. *Numerical results*

We applied `SITR-2D` and `lsqnonlin` algorithms to problems `P1-P8`. For each problem, we fixed four values for the initial guess x_0 and set $x_0 = 10^{\gamma-2}$, $\gamma = 0, 1, 2, 3$, in problems `P1,P4,P7,P8`, $x_0 = -10^{\gamma-2}$, $\gamma = 0, 1, 2, 3$, in problems `P2` and `P6`, $x_0 = 1 + 10^{\gamma-2}$, $\gamma = 0, 1, 2, 3$, in problem `P3`, $x_0 = l + (\gamma + 1)(u - l)/5$, $\gamma = 0, 1, 2, 3$, in problem `P5`.

To evaluate and compare the performance of the procedures `SITR-2D` and `lsqnonlin`, we use the performance profile approach proposed by Dolan and Moré.[18] The profile of each code is measured considering the ratio of its computational effort versus the best computational effort of the codes and here the number of performed function evaluations is adopted as a measure of the computational effort. Specifically, for each test t and solver s, we let $\mathtt{nvf}_{t,s}$ denote the number of F-evaluations required to solve test t by solver s and $\mathtt{nvf}_t^*$ be the lowest number of F-evaluations required by the codes to solve test t. Then, the ratio

$$r_{t,s} = \frac{\mathtt{nvf}_{t,s}}{\mathtt{nvf}_t^*},$$

measures the performance on test t by solver s with the best performance by the two solvers on such test. Clearly, $r_{t,s} \geq 1$ and $r_{t,s} = 1$ means that the solver s was the most convenient in solving test t. Finally, for the code s the performance profile is defined as

$$pf_s(\delta) = \frac{\text{no. of tests s.t. } r_{t,s} \leq \delta}{\text{total no. of tests}}, \quad \delta \geq 1.$$

Using this approach, there is no need to discard solver failures from the data. In fact, if a code does not solve test t, $r_{t,s}$ is assigned a large number, say r_m. In our experiments setting $r_m = 15$ we capture the overall performance of the solvers. We underline that the left side of the picture gives the percentage of the tests for which a solver is most convenient, whereas the right side gives the percentage of the tests that were successfully solved by each solver.

Table 1. Performance of SITR-2D method

SITR-2D method									
Pb#	γ	itnl	nvf	itl	Pb#	γ	itnl	nvf	itl
1	0	12	13	22	5	0	13	14	159
	1	12	13	23		1	10	11	147
	2	11	12	23		2	9	10	167
	3	–				3	13	14	182
2	0	8	9	10	6	0	11	12	280
	1	8	9	9		1	12	13	266
	2	4	5	4		2	13	14	298
	3	13	14	14		3	17	18	322
3	0	5	6	20	7	0	27	30	177
	1	7	8	40		1	24	25	170
	2	12	13	50		2	31	36	185
	3	70	82	333		3	–		
4	0	21	22	157	8	0	15	16	369
	1	61	65	1284		1	17	18	369
	2	6	7	65		2	16	17	343
	3	33	38	112		3	–		

The results with the SITR-2D method are displayed in Table 1. The column with heading γ indicates the scalar used to form the initial guess x_0. The columns with headings itnl, nvf, itl indicate the number of nonlinear iterations, function evaluations and linear iterations performed, respectively. If the method fails to solve a test, this is denoted by the symbol "–". The results provided in Table 1 show the computational feasibility of our method. The SITR-2D algorithm solved successfully 29 tests on a total of 32 tests. Except two runs, i.e. P3 with $\gamma = 3$ and P4 with $\gamma = 1$, the numbers `itnl` and `nvf` are small and this shows the computational efficiency of the method. Moreover, the number `itl` indicates the effectiveness of the preconditioning techniques used in solving tests from P5, P6 and P8.

Figure 1 shows the performance profiles of the codes in the interval $[0, 15]$. From the values of $pf_s(1)$, it is clear that `SITR-2D` had the most wins, i.e. it solved about 75% of the tests with the greatest efficiency. Also, focusing on the ability of completing a run successfully we have again that `SITR-2D` compares very favorably with the `lsqnonlin` code. In fact, our method solved about 90% of the tests while `lsqnonlin` solved about 75% of the tests. Finally, the performance profile of `SITR-2D` readily flattens and our algorithm has the ability to solve over 90% of the tests within a factor 4 of `lsqnonlin` solver. On the other hand, the worst run of `lsqnonlin` is within a factor 8 of `SITR-2D` algorithm.

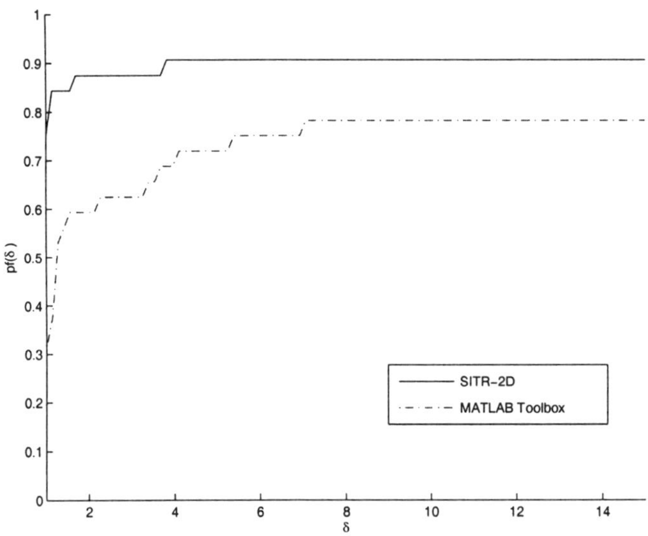

Fig. 1. Performance profile

Acknowledgments

Work supported by MIUR, Rome, Italy, through "Cofinanziamenti Programmi di Ricerca Scientifica di Interesse Nazionale" and GNCS-INDAM.

References

1. C.A. Floudas et al., *Handbook of test problems in local and global optimization*, Kluwer Academic Publishers, Nonconvex Optimization and its Applications, 33, 1999.
2. K. Meintjes, A.P. Morgan, *Chemical Equilibrium systems as numerical tests problems*, ACM Trans. Math. Soft., 16 (1990), pp. 143-151.
3. S. Bellavia, M. Macconi, B. Morini, *STRSCNE: A scaled trust-region solver for constrained nonlinear equations*, Comput. Optim. and Appl., 28 (2004), pp. 31-50.
4. C. Kanzow, A. Klug, *An interior-point affine-scaling trust-region method for semismooth equations with box constraints*, Comput. Optim. and Appl., to appear.
5. X.J. Tong, L.Qi, *On the convergence of a trust-region method for solving constrained nonlinear equations with degenerate solutions*, J. Optim. Theory Appl., 123 (2004), pp. 187-212.
6. S. Bellavia, B. Morini, *Subspace trust-region methods for large bound-constrained nonlinear equations*, SIAM J. Numer. Anal., 44 (2006), pp. 1535-1555.
7. S.C. Eisenstat, H.F. Walker, *Choosing the forcing term in an inexact Newton method*, SIAM J. Sci. Comput., 17 (1996), pp. 16-32.
8. Y. Saad, *Iterative methods for sparse linear systems*, SIAM, 2003.
9. *Matlab 7*, The Math Works, Natick, MA.
10. A.R. Conn, N.I.M. Gould, Ph.L. Toint, *Trust-region methods*, SMPS/SIAM Series on Optimization, 2000.
11. M. Pernice, H.F. Walker, *NITSOL: a new iterative solver for nonlinear systems*, SIAM J. Sci Comput., 19 (1998), pp. 302-318.
12. T.F. Coleman, Y. Li, *On the convergence of interior-reflective Newton methods for nonlinear minimization subject to bounds*, Math. Programming, 67 (1994), pp.189-224.
13. T.F. Coleman, Y. Li, *An interior trust region approach for nonlinear minimization subject to bounds*, SIAM J. Optim, 6 (1996), pp. 418-445.
14. C.T. Kelley, *Iterative Methods for Linear and Nonlinear Equations*, Frontiers in Applied Mathematics, SIAM, 1995.
15. C.T. Kelley, J.I. Northrup, *A pointwise Quasi-Newton method for integral equations*, SIAM J. Numer. Anal., 25 (1988), pp. 1138-1155.
16. L. Luksan, J. Vlcek, *Sparse and partially separable test problems for unconstrained and equality constrained optimization*, Institute of Computer Science, Academy of Sciences of the Czech Republic, Technical Report No. 767, 1999.
17. S.P. Dirkse, M.C. Ferris, *MCPLIB: A Collection of Nonlinear Mixed Complementarity Problems*, Technical Report, Computer Sciences Department, University of Wisconsin, Madison, Wisconsin, 1994.
18. E.D. Dolan, J.J. Moré, *Benchmarking optimization software with performance profiles*, Math. Programming 91 (2002), 201-213.

An adaptive finite element semi-Lagrangian Runge-Kutta-Chebyshev method for combustion problems

R. BERMEJO and J. CARPIO

Universidad Politécnica de Madrid
Dpto de Matemática Aplicada, E.S.T.I.Industriales
C/ José Gutiérrez Abascal 2 , 28006 Madrid, Spain.
E-mail: rbermejo@etsii.upm.es, jaime.carpio@upm.es

We present in this paper an adaptive semi-Lagrangian Runge-Kutta-Chebyshev method in a finite element framework to integrate the combustion equations.

Keywords: Combustion, semi-Lagrangian scheme, Runge-Kutta-Chebyshev method, adaptivity, finite element

1. Introduction and governing equations

This paper presents a fully adaptive scheme to perform numerical simulations of planar lifted flames. To do so, we consider the systems of equations composed by the compressible Navier-Stokes equations at low Mach number and the convection-diffusion-reaction equations for temperature and species plus the equation of state. The formulation of the problem is as follows. Let $D \subset \mathbb{R}^2$ be a domain with appropriately smooth boundary $\partial D = \Gamma^D \cup \Gamma^N$, $\Gamma^D \cap \Gamma^N = \emptyset$, where Γ^D and Γ^N are the pieces of ∂D for Dirichlet and Neumann boundary conditions, respectively. On $D \times (0,T]$ we define the following variables: density of fluid ρ, the hydrodynamic correction of pressure p, the flow velocity $\mathbf{v} = (v_1, v_2)$, the temperature T (expressed in Kelvin degrees) and the species mass fractions $(Y_i)_{i=F,O_2,N_2,P}$, where F, O_2, N_2 and P stand for fuel, oxygen, nitrogen and products of combustion respectively. The system of equations of the model is

Navier-Stokes equations:

$$\begin{cases} \dfrac{\partial \rho}{\partial t} + \nabla \cdot (\rho \mathbf{v}), \\ \rho \left(\dfrac{\partial \mathbf{v}}{\partial t} + \mathbf{v} \cdot \nabla \mathbf{v} \right) = \nabla \cdot (\mu \nabla \mathbf{v}) - \nabla p. \end{cases} \tag{1a}$$

Combustion equations:

$$\begin{cases} \rho\left(\dfrac{\partial T}{\partial t}+\mathbf{v}\cdot\nabla T\right)=\nabla\cdot(\rho D_T\nabla T)-\dfrac{H_F}{c_p}w_F, \\ \rho\left(\dfrac{\partial Y_i}{\partial t}+\mathbf{v}\cdot\nabla Y_i\right)=\nabla\cdot(\rho D_i\nabla Y_i)+w_i, \\ Y_{N_2}=1-Y_F-Y_{O_2}-Y_P,\ \rho=\dfrac{M}{T}\dfrac{\rho_0T_0}{M_0}, \end{cases} \tag{1b}$$

plus the boundary conditions

$$\begin{cases} \mathbf{v}=\mathbf{v}_D \text{ on } \Gamma_{\mathbf{v}}^D,\ \mu\dfrac{\partial \mathbf{v}}{\partial n}-p\mathbf{n}=\mathbf{0} \text{ on } \Gamma_{\mathbf{v}}^N, \\ T=T_D \quad \text{on } \Gamma_T^D,\ \varrho D_T\dfrac{\partial T}{\partial n}=0 \text{ on } \Gamma_T^N, \\ Y_i=Y_i^D \quad \text{on } \Gamma_{Y_i}^D,\ \varrho D_i\dfrac{\partial Y_i}{\partial n}=0 \text{ on } \Gamma_{Y_i}^N, \end{cases}$$

and the initial conditions in D

$$\begin{cases} \mathbf{v}(x,0)=\mathbf{U},\quad T(x,0)=T_0 \\ Y_F(x,0)=0,\quad Y_{O_2}(x,0)=Y_{O_2,A} \text{ and } Y_P(x,0)=0, \end{cases} \tag{2a}$$

where the subscript A stands for air. To make the model more realistic we shall consider that the dynamic diffusion coefficients (namely, the viscosity μ, the thermal diffusivity ρD_T, the mass diffusivity of each species ρD_F, ρD_{O_2}, ρD_P) depend upon the temperature according to the power law.

$$\frac{\mu}{\mu_0}=\frac{\rho D_i}{(\rho D_i)_0}=\left(\frac{T}{T_0}\right)^{\sigma} \qquad \text{with } \sigma=0.7, \tag{3}$$

where μ_0, $(\rho D_i)_0$ and T_0 denote a referent or initial value in D.

It is assumed the overall one step chemical kinetics and the local rate at which the overall combustion process takes place is modeled by the Arrhenius law of the form [1]

$$w_F=\frac{w_{O_2}}{r}=-\frac{w_P}{r+1}=-\rho BY_FY_{O_2}\exp\left[-\frac{E}{RT}\right] \tag{4}$$

where w_i is the mass of species i produced per unit volume per unit time, $R=8.314J/molK$ is the universal gas constant, B denotes the so called pre-exponential frequency factor of the chemical reaction given by

$$B=\frac{S_L^2}{D_{TA}}\frac{\beta^3}{4}\left(\frac{T_s}{T_0}\right)^{1-\sigma}\frac{S+1}{S}\frac{1}{Y_{O_2,A}}\exp\left[\frac{E}{RT_s}\frac{(T-T_s)}{T}\right], \tag{5}$$

where S_L denotes the planar flame velocity; $S=\frac{rY_{F,O}}{Y_{O_2}}$ is the stoichiometric ratio, r being the mass of oxygen burnt; T_s denotes the stoichiometric flame temperature given by $T_s=T_0+H_F\frac{Y_{F,O}}{c_p(1+S)}$, H_F being the amount of heat released by unit mass of fuel consumed; and E is the activation energy modeled by the expression

$$\frac{E(\phi)}{E_0} = \begin{cases} 1 + 2(0.7 - \phi)^2 & \text{if} \quad \phi \leq 0.7 \\ 1 & \text{if} \quad 0.7 \leq \phi \leq 1.0 \\ 1 + 1.472\,(\phi - 1)^2 & \text{if} \quad 1.0 \leq \phi. \end{cases} \tag{6}$$

Here, E_0 is the activation energy for the stoichiometric mixture, and ϕ is the local equivalence ratio defined in terms of the mass fractions of fuel and oxygen in the upstream fresh mixture as $\phi = rY_{F,u}/Y_{O_2,u}$. ϕ is approximated as

$$\begin{cases} \phi = S\dfrac{Z}{1-Z} \quad \text{with} \\ Z = (SY_F/Y_{F,0} - Y_{O_2}/Y_{O_2,A} + 1)\,/(1+S) \end{cases} \tag{7}$$

The parameter $\beta = \frac{E(T_s - T_o)}{RT_s^2}$ is the Zeldovich number, which is moderately large in combustion processes. Figure 1 is a schematic representation of the physics of the phenomenon modeled by (1a)-(2a).

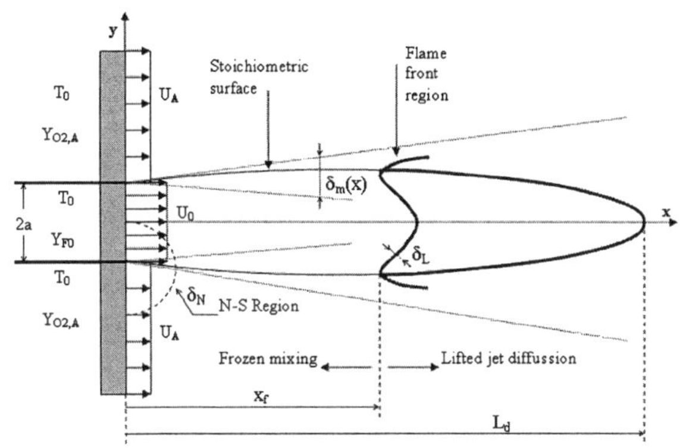

Fig. 1. A sketch of a laminar lifted flame in a planar diffusion jet.

The air stream is flowing in with velocity U_A and temperature T_0 through a porous wall, whereas the fuel stream flows in with velocity U_0 and temperature T_0 through a hole of diameter $2a$. As we move into the interior of D we first encounter a region of length scale $\delta_N = D_{TA}/U_0$ known as the Navier-Stokes region, it is in this region where the mixing layers originate. Further downstream, at a distance $x \gg \delta_N$, the Navier-Stokes region evolves to a slender mixing layer of thickness $\delta_m = \left(\frac{D_{TA}x}{U_0}\right)^{1/2}$, $\delta_m \ll x$. We may ignite a flame in the mixing layer by an external source such that the flame front becomes rapidly elongated by the action of the flow and a quasi-planar premixed flame is generated. When the jet Reynolds number is large the configuration of laminar jet diffusion flames is a slender jet with a developed length $L_d \gg a$, and the characteristic thickness of the quasi-planar premixed flame is $\delta_L = D_{TA}/S_L$, with $\delta_L \ll a$. For typical hydrocarbons $\delta_L \simeq 10^{-4}m$. The premixed flame moves upstream and downstream along the stoichiometric surface. When the ratio U_0/S_L is below a critical value, the diffusion flame will be anchored in the Navier-Stokes region near the injector. As U_0/S_L grows the flame will be lifted at a distance x_f such that $\delta_N < x_f < L_d$. When U_0/S_L goes beyond a critical value the flame will be blown-off and leave the domain D. After the description of the phenomenology of lifted flames, it is clear that a good choice for numerical simulations should be an adaptive approach.

The numerical method presented in this paper is a fully adaptive -time and space- formulation of semi-Lagrangian **I**mplicit **E**xplicit **R**unge **K**utta **C**hebyshev (IMEX RKC) methods specifically devised for the integration of combustion equations. Thus, to ease the formulation of the numerical method we choose as a model for the combustion problem

$$\begin{cases} \dfrac{Du}{Dt} = \nabla \cdot (k\nabla u) + f(u) \ \text{ in } \ D \times (0,T), \\ u(x,0) = u_0(x) \ \text{ in } \ D, \\ u = u_D \ \text{ on } \ \Gamma^D \ \text{ and } \ k\dfrac{\partial u}{\partial n} = 0 \ \text{ on } \ \Gamma^N, \end{cases} \tag{8}$$

where $\frac{D}{Dt} = \frac{\partial}{\partial t} + \mathbf{v} \cdot \nabla$ is the material derivative operator, k is the diffusion coefficient that depends on u, x and t, such that there exist positive constants k_1 and k_2, $k_1 \leq k \leq k_2$, with $\varkappa = \frac{k_2}{k_1} = O(1)$. We further assume that if $|r_{\max}|$ and $|r_{\min}|$ are the largest and smallest (in modulus) eigenvalues of $\frac{\partial f}{\partial u}$ then $r = \frac{|r_{\max}|}{|r_{\min}|} \gg 1$, $r \gg \varkappa$ and $\frac{|\mathbf{v}|}{|k_1|} \gg 1$.

2. The Finite element Semi-Lagrangian Implicit-Explicit-Runge-Kutta Chebyshev method.

Given the real parameter h_0, $0 < h_0 < 1$, let h be the space discretization parameter such that $0 < h < h_0$; we assume that a regular triangulation $\mathcal{T}_h$ composed by elements $\overline{T}_j$ with Lipschitz boundary Γ_j is generated in $\overline{D}$, and a conforming family of finite element spaces

$$\begin{cases} V_h = \{v_h \in C^0(\overline{D}) : v_h \mid_{T_j} \in P(\overline{T}_j) \ \forall \overline{T}_j \in \mathcal{T}_h\} \\ V_{hD} = \{v_h \in V_h : v_h \mid_{\Gamma_D} = 0\} \end{cases} \tag{9}$$

is associated with $\mathcal{T}_h$. Here, $P(\overline{T}_j)$ denotes the set of polynomials like functions defined in $\overline{T}_j$. The degree m of the the polynomials we use in this paper is 2 unless otherwise stated. If M is the number of mesh points, then any element of V_h is expressed as

$$v_h = \sum_{i=1}^{M} V_i \Phi_i(x), \tag{10}$$

where $V_i = v_h(x_i)$, x_i being the $i-th$ mesh point, and $\{\Phi_i\}$ is the set of global nodal basis functions of V_h characterized by the property $\Phi_i(x_j) = \delta_{ij}$.

2.1. *The IMEX RKC method of second order*

The explicit Runge Kutta Chebyshev methods proposed by van der Houwen and coworkers [2] have a stability region that is a narrow strip extending along the negative real axis in the complex plain with the real stability boundary $\beta \simeq s^2$, s being the number of stages; so that, these methods are designed to enlarge the real stability boundary increasing the number of stages. Explicit RKC methods work well for reaction-diffusion equations with moderate stiffness and Jacobian matrix close to a normal matrix; however, when the reaction terms are very stiff and the diffusion terms are moderately stiff, explicit RKC methods are not efficient because they require a very large number of stages in order to be stable. To overcome this problem [3] proposes an implicit-explicit extension of the explicit RKC methods. The new IMEX RKC methods treat implicitly the reaction terms, which are responsible of the stiffness, leaving explicit the diffusion terms. But in convection dominated reaction-diffusion problems IMEX RKC methods do not work well or are not sufficiently efficient because the Jacobian matrix of the equations may have complex eigenvalues with imaginary part large enough to make them lie outside the stability region unless one chooses a small time step length.

For a presentation of the IMEX RKC method we follow [3] and consider the *ode* system $y' = F_D(t,y) + F_R(t,y)$, which is obtained from (8) by application of the method of lines. Here, $F_D(t,y)$ stands for the diffusion terms of the equations that in the numerical formulation are treated explicitly, and $F_R(t,y)$ denotes the reaction terms which are treated implicitly. Assuming that at time t_n a numerical solution y^n is known and taking the time step length Δt and the number of the stages $s \geq 2$, the IMEX RKC computes the solution y^{n+1} at time t_{n+1} as follows:

$$\begin{cases} W_0 = y^n, \\ W_1 = W_0 + \widetilde{\mu}_1 \Delta t F_{D,0} + \widetilde{\mu}_1 \Delta t F_{R,1}, \\ W_j = (1-\mu_j-\nu_j)W_0 + \mu_j W_{j-1} + \nu_j W_{j-2} + \mu_j \Delta t F_{D,j-1} + \widetilde{\gamma}_0 \Delta t F_{D,0} + \\ \qquad \left[\widetilde{\gamma}_j - (1-\mu_j-\nu_j)\widetilde{\mu}_1\right] \Delta t F_{R,0} - \nu_j \widetilde{\mu}_1 \Delta t F_{R,j-2} + \widetilde{\mu}_1 \Delta t F_{R,j} \quad (2 \le j \le s) \\ y^{n+1} = W_s, \end{cases}$$

where $F_{D,j} = F_D(t_n + c_j \Delta t, W_j)$ and $F_{R,j} = F_R(t_n + c_j \Delta t, W_j)$. All coefficients are available in analytical form for arbitrary $s \ge 2$ in [3].

To calculate a numerical approximation to the weak solution of (8) by the semi-Lagrangian IMEX RKC method we divide the interval $[0,T]$ into subintervals $[t_n, t_{n+1}]$ of length $\Delta t = t_{n+1} - t_n$ and consider the sequence of problems

$$\begin{cases} \dfrac{Du}{Dt} = \nabla \cdot (k \nabla u) + f(u) \text{ in } D \times (t_n, t_{n+1}), \\ u(x,t_n) = \overline{u}(x,t_n) \text{ in } D, \\ u(x,t_{n+1}) = u_D \text{ on } \Gamma^D \text{ and } k\dfrac{\partial u(x,t_{n+1})}{\partial n} = 0 \text{ on } \Gamma^N, \end{cases} \tag{11}$$

where

$$\overline{u}(x,t_n) = u(X(x,t_{n+1};t_n),t_n), \tag{12}$$

and $X(x,t_{n+1};t_n) \in \overline{D}$ is the foot at time $t = t_n$ of the characteristic curve of the operator $\frac{D}{Dt}$ that goes by (x,t_{n+1}). Assuming $\mathbf{v} \in L^\infty(0,T;W^{1,\infty}(D))$, $X(x,t_{n+1};t)$ is the unique solution of

$$\begin{cases} \dfrac{dX(x,t_{n+1};t)}{dt} = \mathbf{v}(X(x,t_{n+1};t),t) \\ X(x,t_{n+1};t_{n+1}) = x. \end{cases}$$

The application of IMEX RKC to integrate (11) along the characteristics $X(x,t_{n+1};t)$ in the framework of finite elements yields the following scheme:

Assuming that at time t_n the solution $u_h^n \in V_{hD}$ is known, calculate $\overline{u}_h^n \in V_{hD}, W_{h0}, \cdots, W_{hs} \in V_{hD}$ and $u_h^{n+1} \in V_{hD}$ such that for all $v_h \in V_{hD}$

$$\int_D W_{h0} v_h dx = \int_D \overline{u}_h^n v_h dx, \tag{13a}$$

$$\int_D W_{h1} v_h dx - \widetilde{\mu}_1 \Delta t \int_D F_{R,1}(W_{h1}) v_h dx = \int_D W_{h0} v_h dx - \widetilde{\mu}_1 \Delta t \int_D K(\overline{u}_h^n) \nabla \overline{u}_h^n \cdot \nabla v_h dx, \tag{13b}$$

for $2 \leq j \leq s$

$$\begin{cases} \int_D W_{hj} v_h dx - \widetilde{\mu}_1 \Delta t \int_D F_{R,j}(W_{hj}) v_h dx = \\ (1-\mu_j-\nu_j)\int_D W_{h0} v_h dx + \int_D (\mu_j W_{hj-1} + \nu_j W_{hj-2}) v_h dx - \\ \widetilde{\gamma}_0 \Delta t \int_D K(\overline{u}_h^n)\nabla \overline{u}_h^n \cdot \nabla v_h dx - \widetilde{\mu}_j \Delta t \int_D K(W_{hj-1})\nabla W_{hj-1} \cdot \nabla v_h dx + \\ \Delta t \left[\widetilde{\gamma}_j - (1-\mu_j-\nu_j)\widetilde{\mu}_1\right] \int_D F_{R,0}(\overline{u}_h^n) v_h dx \widetilde{\gamma}_j - \nu_j \widetilde{\mu}_1 \Delta t \int_D F_{R,j-2}(W_{hj-2}) v_h dx, \end{cases} \tag{13c}$$

and finally

$$u_h^{n+1} = W_{hs}. \tag{13d}$$

In these formulae, $F_{D,0} = \nabla \cdot (K(\overline{u}_h^n)\nabla \overline{u}_h^n(x))$, $F_{D,j} = \nabla \cdot (K(W_j)\nabla W_j(x))$ $(1 \leq j \leq s)$ and $F_{R,j} = \sum_{i=1}^M f(W_j)_i \Phi_i(x)$ $(0 \leq j \leq s)$; $f(W_j)$ denotes the value of $f(W_j)$ at the mesh-node x_i and $\Phi_i(x)_i$ is the set of global basis functions of V_{hD}. It still remains to be said how we calculate $\overline{u}_h^n(x)$. Following [4] we set

$$\overline{u}_h^n(x) = \sum_{i=1}^M \overline{U}_i^n \Phi_i(x), \tag{14a}$$

where, using the shorthand notations $X^n(x_i) = X(x, t_{n+1}; t_n)$, $U^{n+} = \max_l\{U_l^n\}_{T_K}$ and $U^{n-} = \min_l\{U_l^n\}_{T_k}$, here T_k is the element of the mesh where the point $X^n(x_i)$ is located,

$$\overline{U}_i^n = \begin{cases} U^{n+} & \text{if } U^{n+} > \sum_{i=1}^M U_j^n \Phi_j(X^n(x_i)), \\ U^{n-} & \text{if } U^{n-} < \sum_{i=1}^M U_j^n \Phi_j(X^n(x_i)), \\ \sum_{i=1}^M U_j^n \Phi_j(X^n(x_i)), & \text{otherwise.} \end{cases} \tag{14b}$$

3. The adaptive method

This section is devoted to the description of the implementation of the semi-Lagrangian IMEX RKC method in an adaptive framework. The idea is to start with an initial Δt_0 and an initial mesh $\mathcal{T}_h^0$ fine enough to represent well the initial condition, and then at every time step to adapt both of them according to the features of the solution, such that an error criterion must be satisfied. Denoting by V_h^n and V_{hD}^n the finite element spaces associated with the mesh $\mathcal{T}_h^n$ at time t_n, we formulate the semi-Lagrangian IMEX RKC adaptive method as follows.

Suppose that at time t_n, $\mathcal{T}_h^n$, u_h^n, V_h^n, V_{hD}^n and Δt_n are known, then execute the following procedure:

(1) *Mesh adaptation and semi-Lagrangian step: using the information supplied by the semi-Lagrangian step in the calculation of* $\overline{u}_u^n$ *generate the mesh* $\mathcal{T}_h^{n+1}$ *from the initial mesh* $\mathcal{T}_h^0$ *by successive refinements* $\mathcal{T}_{h,p}^0$, $0 \leq p \leq p_{\max}$.
(2) *IMEX RKC step: apply (13a)-(13d) to calculate* $u_h^{n+1} \in V_{hD}^{n+1}$ *with time step* Δt_n.
(3) *Time adaptation: accept or reject* u_h^{n+1} *and consequently adapt the length of the time step*
(4) *Set* $n = n+1$ *and go to 1.*

3.1. *Mesh adaptation*

In time dependent problems there are several strategies to obtain the meshes $\mathcal{T}_h^n$. One of such strategies, that fits very well (in the sense that it is fast and efficient) in the framework of the semi-Lagrangian methods, is the one that at every time step starts with the initial mesh $\mathcal{T}_h^0$ and generates the mesh $\mathcal{T}_h^n$ by successive local mesh refinements $\mathcal{T}_{h,p}^0$ $(0 \leq p \leq p_{\max})$ based on the error indicators. Thus:

At $p = 0$, *choose the parameters* θ, ν $(0 < \theta,\ \nu < 1)$, Tol_h, p^*, p^{**} *and the macro triangulation* $(\mathcal{T}_{h,p}^0, \{x_{j,p}\})$, *where* $\{x_{j,p}\}$ *is the set of mesh nodes in* $\mathcal{T}_{h,p}^0$.

- **For** $p = 0, 1, .., p^*$ **do**
  (1) *Use the semi-Lagrangian step to calculate* $\{X^n(x_{j,p})\}$ *and* $\{\overline{U}_{j,p}^n\}$.
  (2) Set

$$\widetilde{u}_{h,p}(x) = \sum_j \overline{U}_{j,p}^n \Psi_j(x), \tag{15a}$$

where

$$\Psi_j(x)\mid_{\overline{T}_k} \in P_1(\overline{T}_k) \ \ \forall \overline{T}_k \in \mathcal{T}_{h,p}^0, \tag{15b}$$

and compute local and global error indicators for the triangulation $\mathcal{T}_{h,p}^0$ *as:*

$$\begin{cases} \eta_k^2 = \frac{h_k^2}{2}\left([\partial_n \widetilde{u}_{h,p}]\right)^2_{\partial \overline{T}_k \setminus \partial\Omega} \\ \eta^2 = \sum_k \frac{h_k^2}{2}\left([\partial_n \widetilde{u}_{h,p}]\right)^2_{\partial \overline{T}_k \setminus \partial\Omega} \end{cases} \tag{16}$$

where h_k *is the diameter of the element* $\overline{T}_k$, $[\partial_n \widetilde{u}_{h,p}]$ *denotes the jump of* $\partial_n \widetilde{u}_{h,p}$ *across the inter-element edges.*
  (3) *Mark the elements of* $\mathcal{T}_{h,p}^0$ *to be refined and obtain a new triangulation* $(\mathcal{T}_{h,p+1}^0, \{x_{j,p+1}\})$

if $\eta < Tol_h$

set $p = p^*$ *and continue*

end if

- *Enlarge the partition* $(\mathcal{T}^0_{h,p^*+1},\ \{\{x_{j,p^*+1}\})$ *to improve the approximation of* $F_{R,0}(\widetilde{u}_{h,p})$

  For $p = p^*, ..., p^{**}$ **do**

  Repeat the steps from 1. to 3. but with the reaction rate $F_{R,0}(\widetilde{u}_{h,p})$ instead of $\widetilde{u}_{h,p}$.
- *Set* $(\mathcal{T}^{n+1}_h,\ \{x_j\}) = (\mathcal{T}^{(0)}_{h,p^{**}},\ \{x_{j,p^{**}}\})$

To mark the elements that have to be refined we use algorithm GERS (Guaranteed Error Reduction Strategy) [5]. The refinement of marked triangles is made bisecting the largest edge by joining its mid-point with the opposite vertex and taking the vertices thus created as the vertices of a new refinement.

3.2. *Time step adaptation and number of stages*

Defining at time t_{n+1} the error estimator Est^{n+1} as [4]

$$\left\|Est^{n+1}\right\|^2_w := \sum_{\overline{T}_k \in \mathcal{T}^{n+1}_h} \frac{|\overline{T}_k|}{m_k} \sum_{i=1}^{m_k} \left(\frac{Est^{n+1}_i}{w_i}\right)^2, \qquad w_i = \mathsf{atol}_i + \mathsf{rtol}\left|U^{n+1}_i\right| \tag{17}$$

where m_k is the number of mesh points per element, $|\overline{T}_k|$ denotes the area of $\overline{T}_k$, atol_i and rtol being prescribed tolerances (in the numerical example both are 10^{-3} for all i) and Est^{n+1}_i (see [3]) is given by

$$Est^{n+1}_i = \frac{1}{15}\left[12\left(\overline{u}^n_h - u^{n+1}_h\right)_i + 6\Delta t_n \left(F_h(t_n, \overline{u}^n_h) + \left(F_h(t_{n+1}, u^{n+1}_h\right)_i\right]\,.$$

The adjustment of the size of the time step is made by the formula [4]

$$\Delta t_{new} = \min\left(10, \max\left(0.1, \mathsf{fac}\right)\right)\Delta t_{old}, \tag{18}$$

where

$$\mathsf{fac} = \begin{cases} 0.8\left(\dfrac{\|Est^n\|^{1/3}_w}{\|Est^{n+1}\|^{1/3}_w}\dfrac{\Delta t_n}{\Delta t_{n-1}}\right)\dfrac{1}{\|Est^{n+1}\|^{1/3}_w} & \text{when } u^{n+1}_h \text{ is accepted,} \\[2ex] \dfrac{0.8}{\|Est^{n+1}\|^{1/3}_w} \text{ when } u^{n+1}_h \text{ is rejected.} \end{cases} \tag{19}$$

We still have to fix the number of stages s for which this time step is absolutely stable as far as the explicit part of the RKC method is concerned. This number is given by the formula [4]

$$s = 1 + Integer\ part \left[\sqrt{1 + \frac{2.6\Delta t_n}{0.653} \max_i \left(\frac{\sum_j |s_{ij}|}{m_{ii}} \right)} \right], \qquad (20)$$

where $s_{ij} = \int_D k(\overline{u}_h^n) \nabla \phi_i \cdot \nabla \phi_j dx$ and $m_{ii} = \int_D \phi_i^2 dx$.

4. Numerical examples

To illustrate the behavior of the method in solving (1a)-(2a) to simulate the generation of planar lifted flames, we show here the results for the case when a jet of air is flowing into the domain $D = (0, 2L) \times (-L, L), L = 500\delta_L$, with the same velocity $\mathbf{U} = (U_0, 0)$ of a jet of methane (the fuel) of mass fraction $Y_{F,O} = 0.23$, flowing into the domain through a hole of width $2a = 20\delta_L$. The upper, lower and right hand side boundaries are all Γ^N boundaries. The values of other parameters used in this experiment are the following: $E_0 = 1.25 \cdot 10^5 J/mol$, $Y_{O_2,A} = 0.23$, $r = 4$, $S = 4$, $H_F/c_p = 4 \cdot 10^4 K \frac{Kg\ Fluid}{Kg\ CH_4}$, $T_0 = 300K$, $T_s = 2140K$, $\beta_0 = 6$. For the ideal gas law we have:

$$\frac{1}{M} = \left[\frac{Y_F}{16} + \frac{Y_{O_2}}{32} + \frac{Y_P}{80} + \frac{(1 - Y_F - Y_{O_2} - Y_P)}{28} \right] \times 10^3, \qquad \frac{\rho_0 T_0}{M_0} = 11000.$$

We apply the semi-Lagrangian IMEX RKC method to solve (1a)-(2a) via the following scheme:

Assuming that at time t_n, $n \geq 0$, u^n, T^n, Y_F^n, $Y_{O_2}^n$, Y_P^n and ρ^n are known, we proceed to the step $n+1$ as follows:

(1) *Mesh adaptation plus semi-Lagrangian step.*
(2) *Calculate T^{n+1}, $Y_{O_2}^{n+1}; Y_F^{n+1}$, Y_P^{n+1} applying the adaptive semi-Lagrangian-IMEX RKC scheme with P_2 elements for these variables.*
(3) *Time adaptation: adapt the time step size.*
(4) *Calculate ρ^{n+1} by using the state equation. At this stage we use P_1 elements for ρ.*
(5) *Calculate the solution of the compressible Navier-Stokes equations at low Mach numbers. For this stage we extend the method of [6] to low Mach number flows and use positive interpolation for the semi-Lagrangian step.*
(6) *Set n=n+1 and go to 1.*

Figure 2 show a series of snapshots of the solution. The pictures on the right panels represent the distributions of temperature (upper half) and details of the mesh (lower half). The left panels are the plots of the stoichiometric line $\Phi = 1$ with the reaction velocity w_F. The results are obtained for a ratio $U_0/S_L = 2$ and the steady state is reached after 4000 time steps. The final mesh has 11628 elements with diameter h in the range $0.9\delta_L \leq 44\delta_L$. A simulation of this example with a fixed mesh would have required about 2000.000 elements and, therefore, a significant increase in CPU time.

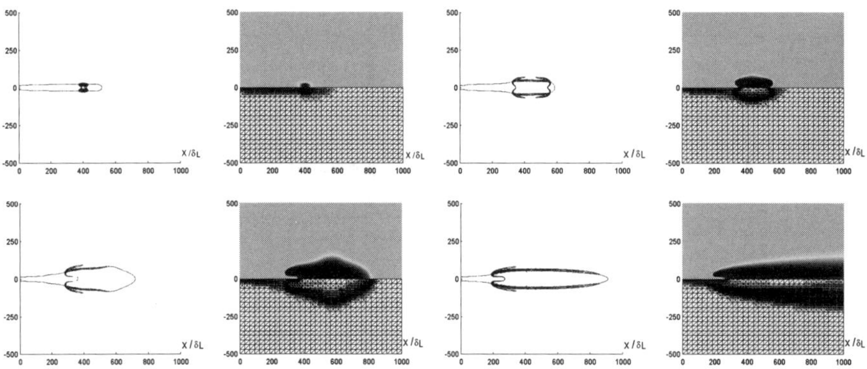

Fig. 2. Snapshots of a planar jet with $U_A = U_0$ at time steps 45, 345, 1200 and 4000.

When the steady state is reached the flame front will stay at a distance x_f from the porous wall and the jet will have a developed length L_d. To compare our results with those obtained by other researchers in the field, we also show in Figure 3 the results of x_f/δ_L when $a \to \infty$ and $S = 4$; this situation is the so called mixing layer. The full line corresponds to the results obtained in [1] by calculating the numerical solution of the stationary model by finite differences without adaptivity, whereas the results of our model are the points denoted by asterisks.

Figure 4 shows the evolution of the number of stages s and the values of $\Delta t S_L^2/D_{TA}$ versus the number of time steps.

The distribution of CPU time is the following: Navier-Stokes problem, 48%; semi-Lagrangian adaptive stage, 5%; IMEX RKC, 47%.

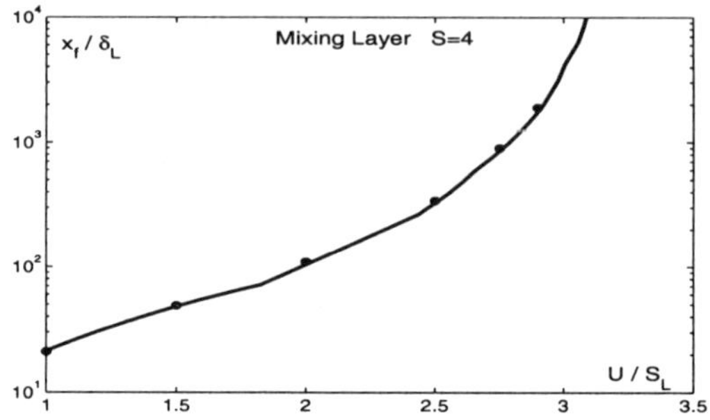

Fig. 3. Mixing layer.

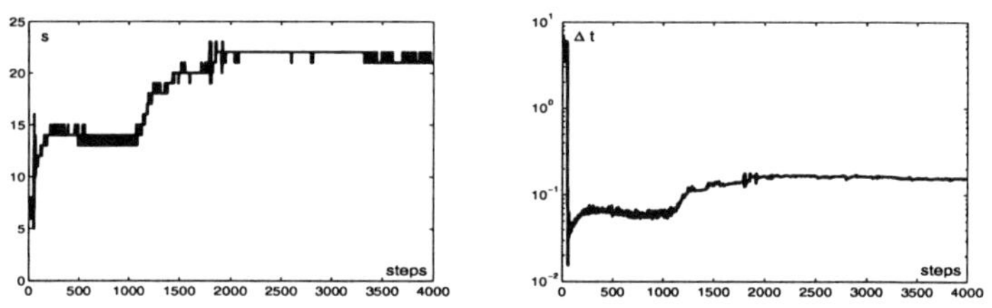

Fig. 4. Evolution of number of stages and variation of time step.

References

1. E, Fernandez-Tarrazo, M. Vera, A. Liñán, *Liftoff and blowoff of a diffusion flame between parallel streams of fuel and air.* Combustion and Flame 144 (2006) 261-276.
2. P.J. van der Houwen, B. P. Sommeijer, *On the internal stability of explicit m-stage Runge-Kutta methods for large m-values.* Z. Augew. Math. Mech. 60. (1980) 479-485.
3. J.G. Verwer, B. P. Sommeijer, *An implicit-explicit Runge-Kutta-Chebyshev scheme for diffusion-reaction equations.* SIAM J. Sci. Comput. 25 (2004) 1824-1835.
4. R. Bermejo, J. Carpio, *An adaptive finite element-semi-Lagrangian implicit-explicit Runge-Kutta-Chebyshev method for convection dominated reaction-diffusion problem.* Applied Numerical Mathematics (2006), doi:10.1016/ j.apnum.2006.10.008.
5. A. Schmidt, K. G. Siebert, *ALBERT: Designe of Adaptive Finite Element Software. The Finite Elemnt Toolbox ALBERTA.* Springer Lectures Notes in Computational Science and Engineering. Springer Berlin (2005).
6. A. Allievi, R. Bermejo, *Finite elemnt modified method of characteristics for Navier-Stokes equations.* Int. J. Numer. Meth. Fluids, 32 (2000) 439-464.

ACTIVE INFRARED THERMOGRAPHY IN NON-DESTRUCTIVE EVALUATION OF SURFACE CORROSION 2: HEAT EXCHANGE BETWEEN SPECIMEN AND ENVIRONMENT

P. BISON

CNR-ITC, Padova, Italy
E-mail: paolo.bison@itc.cnr.it

M. CESERI

Dipartimento di Matematica Università di Firenze, Firenze, Italy
E-mail: ceseri@math.unifi.it

D. FASINO

Dipartimento di Matematica e Informatica, Università di Udine, Udine, Italy
E-mail: fasino@dimi.uniud.it

G. INGLESE

CNR-IAC, Firenze, Italy
E-mail: gabriele@fi.iac.cnr.it

A thin plate Ω has an inaccessible side in contact with aggressive external agents. On the other side we are able to heat the plate and take temperature maps (thermal data) in laboratory conditions. Detecting and evaluating damages on the inaccessible side from thermal data requires the solution of a nonlinear inverse problem for the heat equation. To do this, it is extremely important to assign correct boundary conditions, in particular on the inaccessible boundary of Ω. In several cases the boundary conditions must take account of heat exchange between Ω and the environment. Here we discuss, from the quantitative point of view, the relation between the physical constants of the system (conductivity, width of the plate, ...) and the heat transfer through the boundary of Ω.

Keywords: Infrared thermography, Non-Destructive Evaluation, Robin Boundary Conditions, Inverse Problems.

1. Introduction

Nondestructive Evaluation (NDE) via Infrared Thermography (IRT) is aimed at detection and evaluation of hidden defects of conductors from the analysis of temperature maps obtained with an infrared camera (see for example the introduction of Ref. 3). IRT is particularly effective in the detection of subsurface anomalies (see for example Ref. 12). For this reason we assume that the conductor Ω is a thin plate (of width $a \ll d$ where $d = \min_{P \in \partial\Omega}(\max_{Q \in \Omega} |P - Q|))$ with one accessible side S_{Top}. The other one (S_{Bot}) is assumed to be inaccessible. Ω is an isotropic and homogeneous conductor with conductivity κ_0.

In practice, we heat the conductor by means of a lamp that induces a heat flux of density ϕ through S_{Top}. Temperature maps U at S_{Top} are obtained by means of a IR camera.

The goal is to detect and evaluate anomalies on S_{Bot} from the knowledge of ϕ and U on the opposite side. Here, heat flux density ϕ is assumed periodic (of period τ) in time and spatially constant. It is known that the so called *thermal diffusion length* is

$$\mu = \sqrt{\frac{\kappa_0 \tau}{\pi C_0}}$$

(C_0 being the volumetric heat capacity). It means that if $a > \mu$ we have no chances to investigate anomalies on S_{Bot}. Since a is usually given and the thermal parameters are at least approximately known, the flux ϕ must be suitably modulated acting on τ.

Since we cannot describe in details the energy balance in the space between lamp and specimen, ϕ is only known by a factor. If necessary, this factor can be regarded as an additional unknown in anomalies evaluation (see for example Ref. 1 where a non constant flux is recovered together with a specimen damage).

Anomalies must be detected before they become really dangerous. Hence, their size ϵ is assumed to be small with respect to the width a of Ω. For this reason it is straightforward to adopt perturbative reconstruction methods based on expansion in powers of ϵ like the one described in Ref. 3. This paper intends to contribute in reducing the gap still existing between mathematical approach (see for example Ref. 4 and references therein) and application-oriented research (see for example the reports listed in http://ntrs.nasa.gov/search.jsp) in thermal nondestructive evaluation.

In particular, we carry on a fine modelization of the direct problem in the damaged domain, taking account of radiative and convective boundary conditions that include heat exchange with the environment. It is known that heat exchange is not negligible when the Biot number (the ratio between the thermal resistance of our specimen and the one of the outer environment) is large in a sense that will be specified and analised in Sec. 3.

The implementation of a reconstruction algorithm is still in progress.

2. The mathematical model

We model the specimen Ω by means of the open parallelepiped

$$\Omega = (0,1) \times (0,1) \times (0,a)$$

where the *thickness* a is assumed to be lower than 1. Ω is an isotropic and homogeneous conductor of heat so that its *conductivity tensor* is $\kappa_0 I$ (I is the identity matrix 3×3). Hence, κ_0 and its *volumetric heat capacity* C_0 are constant in space and we assume also that they do not change in time. Moreover, during the test, the temperature field varies in a range small enough to make negligible any dependence of κ_0 and C_0 from the temperature itself.

Let $u^0(x,y,z,t)$ be the temperature in Ω for $t \in (0,T)$. The function u^0 solves the heat equation

$$u_t^0 = \frac{1}{C_0} \nabla \cdot \kappa_0 \nabla u^0 \qquad \text{in } \Omega \times (0,T), \tag{1}$$

with initial data $u^0(x,y,z,0) = g^0(x,y,z)$ in Ω and Robin boundary conditions:

$$u_x^0(0,y,z,t) = u_x^0(1,y,z,t) = 0 \tag{2}$$

in $(0,1) \times (0,a) \times (0,T)$,

$$u_y^0(x,0,z,t) = u_y^0(x,1,z,t) = 0 \tag{3}$$

in $(0,1) \times (0,a) \times (0,T)$,

$$u_z^0(x,y,0,t) - \gamma_{\text{bot}}[u^0(x,y,0,t) - u_{\text{bot}}] = 0 \tag{4}$$

and

$$u_z^0(x,y,a,t) + \gamma_{\text{top}}[u^0(x,y,a,t) - u_{\text{top}} - \frac{1}{h_{\text{top}}}\phi(x,y,t)] = 0 \tag{5}$$

in $(0,1)\times(0,1)\times(0,T)$. Here γ_{bot} and γ_{top} are constant positive coefficients defined as

$$\gamma_{\text{bot}} = \frac{h_{\text{bot}}}{\kappa_0} \qquad \gamma_{\text{top}} = \frac{h_{\text{top}}}{\kappa_0}$$

h_{bot} and h_{top} being the *heat exchange coefficients* between the specimen surfaces and the environment on bottom and top sides rispectively. u_{bot} and u_{top} are external temperature values (see for example Ref. 6), while the function ϕ in Eq. (5) describes the method adopted for heating our specimen through the accessible side $z = a$ of Ω.

Observe that the minus sign behind γ_{bot} is due to the fact that u_z^0 in $z = 0$ is the opposite of the outward normal derivative. In what follows, we will refer to u^0 as *background solution.* Furthermore, let $f^0(x, y, t_k) = u^0(x, y, a, t_k)$ be the (background) temperature maps collected at instants $0 = t_0 < t_1 < \ldots < t_N = T$.

2.1. *The perturbed domain*

Assume that the only effect of the external aggression to the specimen, is some material loss from the inaccessible side. The damaged specimen is represented as

$$\Omega_{\epsilon\theta} = \{(x, y, z) : 0 < x, y < 1,\ \epsilon\theta(x, y) < z < a\}$$

with $\epsilon \ll a$. Here θ is a sufficiently smooth (say C^∞) function such that $\theta(0, y) = \theta(1, y) = \theta(x, 0) = \theta(x, 1) = 0$ and $0 \le \theta(x, y) \le 1$. The support of θ (i.e., the open set where $\theta > 0$) is strictly included in $(0,1)\times(0,1)$. A transverse section of $\Omega_{\epsilon\theta}$ is shown in Fig. 1.

Loss of matter is described by the nonnegative function $\epsilon\theta$. It is reasonable to assume that corrosion or any other damaging process act slowly in the time scale of our experiments. Hence, $\epsilon\theta$ is assumed to be time-independent in what follows.

Temperature in $\Omega_{\epsilon\theta}$ still satisfies heat equation Eq.(1) but the boundary conditions on the bottom side of the domain are clearly modified. We have

$$\nabla u(x, y, \epsilon\theta(x, y), t) \cdot \mathbf{n}(x, y, \epsilon\theta(x, y)) + \gamma_{\text{bot}}[u(x, y, \epsilon\theta(x, y), t) - u_{\text{bot}}] = 0$$

in $(0,1)\times(0,1)\times(0,T)$ where $\mathbf{n}$ is the outward normal unit vector on the boundary of $\Omega_{\epsilon\theta}$. The remaining boundary conditions are analogous to Eqs. (2), (3), (5).

As for the initial condition, we can assume that the value of u for $t = 0$ is not dependent on the perturbation $\epsilon\theta$. This assumption should require some more comment if we did not assume a periodic ϕ in what follows.

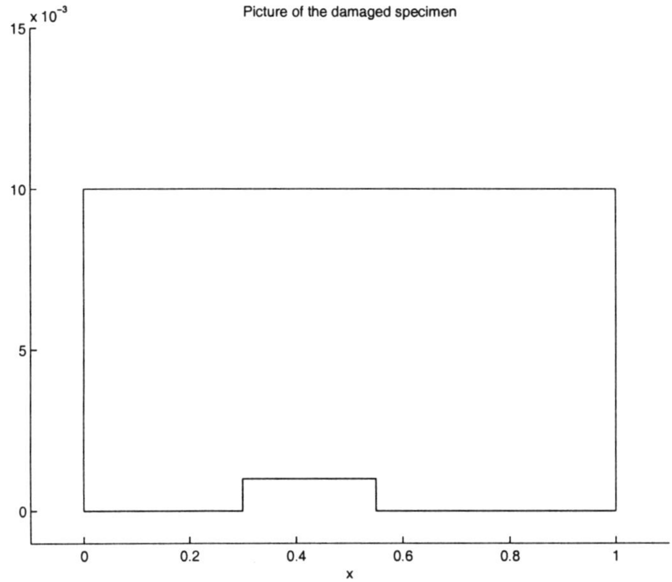

Fig. 1. A transverse section of the damaged specimen: the heigth of the damage is $\frac{a}{10}$, where $a = 0.01$.

The temperature maps $f(x, y, t_k) = u(x, y, a, t_k)$, collected at instants $0 = t_0 < t_1 < \ldots < t_N = T$ are modified as an effect of corrosion. The discrepancies $f(x, y, t_k) - f^0(x, y, t_k)$ (*contrast* in temperature response) and the flux data ϕ will be used in the following to recover $\epsilon\theta$.

3. Comments about Robin boundary conditions in NDE via IRT models

The goal of the present communication is to show relevance of Robin boundary conditions depending on the physical constants. We recall (see Ref. 6) that the boundary conditions (4) and (5) are derived from Fourier's law inside the domain and Newton's law outside, taking account of conservation of energy. More precisely, we have

$$-\kappa_0 \cdot \nabla u^0 = Flux$$

in Ω ($Flux$ is the pointwise specific thermal flux). On the other hand, the outward normal flux through the upper boundary is (linearized in the temperature)

$$Flux_N(x, y, a, t) = h_{top}(u^0((x, y, a, t) - u_{top})).$$

Nonlinear flux (as a function of temperature) are introduced in Ref. 6 page 21 (standard examples) and are studied for example in Ref. 11 (glass cooling) and Ref. 9 (virtual fluxes and homegenization of domain perturbation).

Since the normal component of the gradient is clearly u_z^0, we have (5) with

$$\gamma_{\rm top} = \frac{h_{\rm top}}{\kappa_0}.$$

In the same way we can derive $\gamma_{\rm bot}$.

3.1. *Biot number and heat exchange*

The heat flux through the top boundary of Ω is produced by a system of lamps at a finite distance from the specimen. Periodicity can be modulated by suitably operating on potentiometers. It is assumed that light passes through the air without noticeable absorption, while it is completely absorbed by Ω. Actually, the light is absorbed by the specimen in a layer whose thickness is much lower than a. Hence, temperature on the heated surface raises and generates a flux that partly diffuses inside the material by pure conduction. It happens that some heat is exchanged towards the environment by radiation and convection, provided the temperature of the environment is lower than that of the heated surface. How much the heat is diffused inside the material and how much it is exchanged with the environment depends mainly on the thermal conductivity k_0 and on the heat exchange coefficient $h_{\rm bot}$ (or $h_{\rm top}$). In ordinary conditions (as for example smooth sample surface in contact with still air) this coefficient assumes values around $10[Wm^{-2}K^{-1}]$. On the other hand thermal conductivity may assume a wide range of values depending on the material: It ranges from 10^{-2} to $10^{2}[Wm^{-1}K^{-1}]$. Conductivity is therefore the main parameter affecting the heat exchange with the environment and viceversa.

To be more precise, rewrite the boundary condition (4) as follows

$$u_\zeta^0(P) - \frac{ah_{\rm bot}}{\kappa_0}[u^0(P) - u_{\rm bot}] = u_\zeta^0(P) - \frac{R_{\rm con}}{R_{\rm ext}}[u^0(P) - u_{\rm bot}] = 0$$

where $P = (x, y, 0, t)$, $\zeta = z/a$, $R_{\rm con} = a/\kappa_0$ and $R_{\rm ext} = h_{\rm top}^{-1}$. $R_{\rm con}$ and $R_{\rm ext}$ are the thermal resistances of the sample (resistance to conduction) and of the environment (resistance to exchange) respectively. The ratio between those two thermal resistances is widely known as *Biot number*. When the Biot number is low, heat exchange with the environment is negligible and boundary condition (4) is in practice equivalent to the adiabatic one

$$u_z^0(x, y, 0, t) = 0.$$

In mathematical terms, Robin condition can be substituted by a Neumann one. When Biot number is large enough (practitioners say $> .1$) it is necessary to use (4).

3.2. *Modeling by means of Robin boundary conditions*

In the light of our discussion on Biot number, there are different physical situations in which the exchange term is relevant. Some examples follow:

Example 3.1. The specimen is plastic or ceramic i.e its conductivity is very low. These kind of material are corrosion resistant , but they can be anyhow damaged in other ways.

Example 3.2. Sometimes on the surface of metallic conductor a skin of a poor conductor (grease or oxide) grows up. The presence of such a skin, affetcs the energy exchange between the metallic body and the exterior. This interaction can be summarized (in first approximation) by means of a Robin condition (see Carslaw and Jager[6] pag 20). A similar situation is also observed in some homogenization models of coating (see for example Ref. 5).

Example 3.3. Consider a metallic pipe. Assume that the metal is relatively not much conductive (for example a kind of steel that can be used in parts of boilers, called AISI304) and that the width of the metal is small with respect to the diameter of the pipe. If some fluid (water or else) flows inside the pipe, the exchange coefficient at the interface between metal and fluid becomes large (from 100 to 1000). Corrosion detection in this case is clearly useful for ordinary maintenance.

We use Ex. (3.3) to illustrate our discussion. In presence of a periodic heat flux (of period τ) through the accessible side $z = a$, we observe that $R_{\text{con}} = \mu/\kappa_0$. Here μ is the thermal diffusion length or *characteristic penetration depth*

$$\mu = \sqrt{\frac{\kappa_0 \tau}{\pi C_0}}$$

already seen in the introductive section. Hence, τ must be chosen so that $a \leq \mu$.

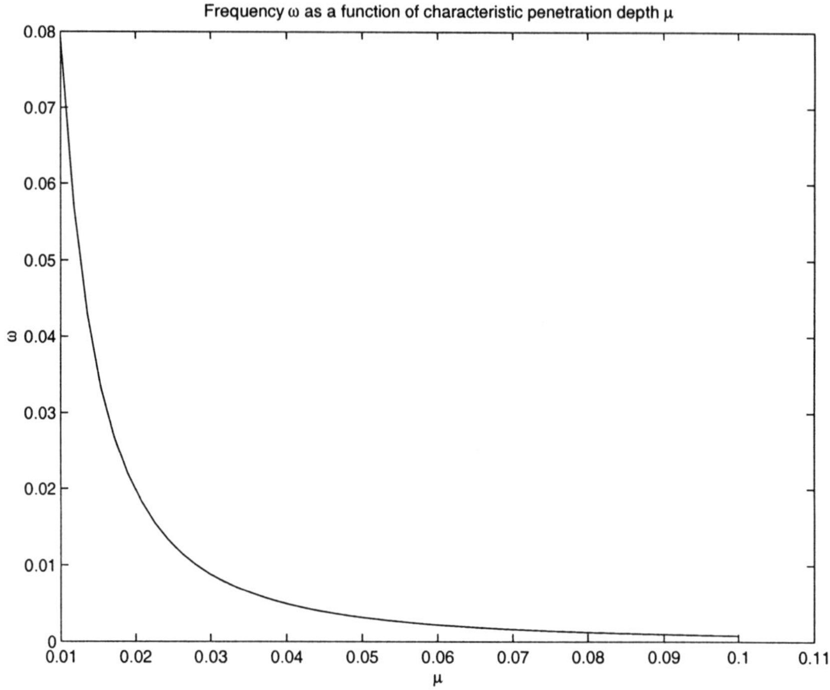

Fig. 2. Frequency ω vs. depth μ: note that for little length a frequency of order 10^{-2} is sufficient.

In Fig. 2, we show the graph of the frequency $\omega = \tau^{-1}$ as a function of μ for the material AISI304 (see Ex. (3.3) above). In a pipe in which the ratio between the width of the metal and the diameter of the section is $a = .01$, we have that $\omega \approx .08$ is a suitable frequency for the investigation of damages on the inaccesible surface ($\tau \approx 12.5$).

Furthermore, it is easy to observe that the Biot number behaves like the square root of the period τ so that we expect that for large τ the exchange term in the inaccessible boundary condition becomes relevant.

We recall that heat equation in presence of a periodic flux ϕ can be Fourier transformed in a sequence (parametrized by $j \in \mathrm{Z}$) of complex Helmoltz equations. When $\phi(x, y, t) = \phi_0(1 + \sin(\frac{2\pi}{\tau}t))$, it is easy to check that only the equations for $j = -1, 0, 1$ are non trivial (see for example Ref. 3 section 4).

Consider a transverse section of the pipe (ring) and regard it for simplicity as a rectangle. Assumed that $a = .01$ and, consequently, $\tau \approx 12.5$, we solved the Helmoltz equation for the damaged specimen $\Omega_{\epsilon\theta}$ when the physical constants are the ones typical of AISI304:

- thermal conductivity $k_0 = 15W/mK$;
- density $\rho = 7900kg/m^3$;
- specific heat $c_p = 480J/kgK$.

The size of the damage is $a/10$. We considered a heat flux ϕ constant in space. We fixed the coefficient h_{top} (laboratory side) to $10W/m^2K$, while we varied only the h_{bot} (inaccessible side).

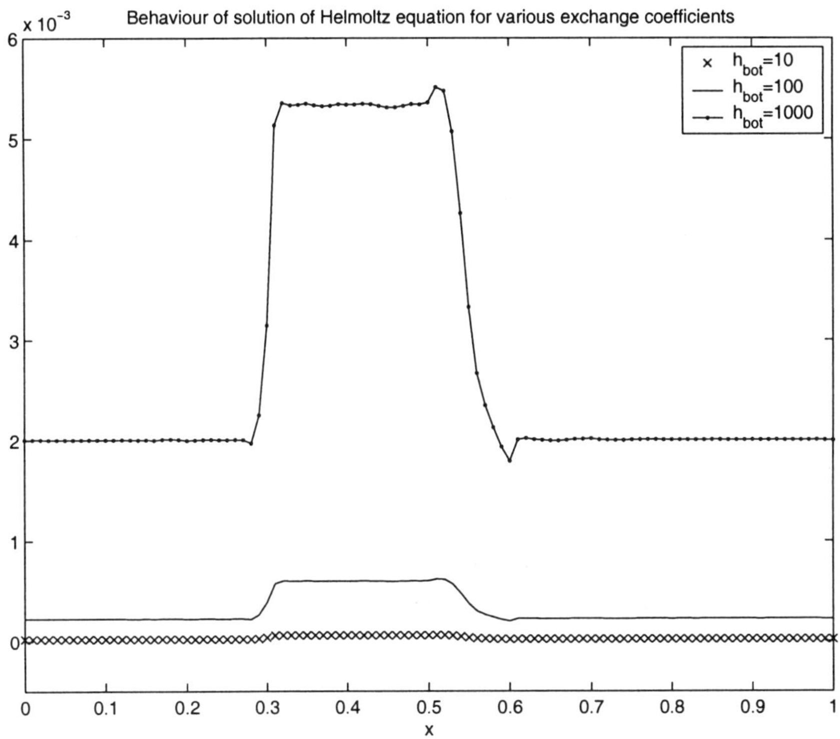

Fig. 3. Differences of the solution in case of exchange ($h_{bot} > 0$) and adiabatic ($h_{bot} = 0$) boundary conditions.

We chose the values $h_{bot} = 10, 100, 1000$ in order to compare the solutions (for $z = a$) with the case $h_{bot} = 0$. The results of this computations are plotted in Fig. 3: as one can see the solution with $h_{bot} = 10$ does not give additional information with respect to the adiabatic case. As we increase the exchange coefficient the differences in the temperature maps increase so that the Robin boundary condition on bottom surface must be retained.

In summary, on one hand a sufficiently large period allows us to go deeper in the specimen, but doing so, the use Robin condition becomes reliable for a better description of the heat exchange process through the inaccessible boundary.

Finally, straightforward calculations shows the essence of the analytical relation between the inaccessible exchange parameter γ_{bot} and the solution u of the following BVP for Laplace's equation:

$$\Delta u = 0$$

$$u_y(x,0) - \gamma_{\text{bot}} u(x,0) = 0$$
$$u_y(x,a) + \gamma_{\text{top}} u(x,a) - \gamma_{\text{top}}/h_{\text{top}}\phi = 0$$
$$u_x(0,y) = u_x(1,y) = 0$$
for $x \in (0,1)$ and $y \in (0,a)$.

The solution takes the form $u(x,y) = my + q$. We have

$m = \gamma_{\text{top}}(\phi/h_{\text{top}} - u(a))$, $u_y = m$, $q = u(0) = u(a) - am$ (no dependence on x!).

Hence, $\gamma_{\text{bot}} = m/q = \gamma_{\text{top}}(\phi/h_{\text{top}} - u(a))/q$ (if $\gamma_{\text{bot}} = 0$ we have $u = \phi/h_{\text{top}}$.

Since $q = u(a) - a\gamma_{\text{top}}(\phi/h_{\text{top}} - u(a)) = u(a)(1 + a\gamma_{\text{top}}) - a\gamma_{\text{top}}\phi/h_{\text{top}}$ we have $\gamma_{\text{bot}} = -\gamma_{\text{top}}(u(a) - \phi/h_{\text{top}})/(u(a) + a\gamma_{\text{top}}(u(a) - \phi/h_{\text{top}}))$

or

$u(a) = \phi/h_{\text{top}}(\gamma_{\text{top}} + a\gamma_{\text{top}}\gamma_{\text{bot}})/(\gamma_{\text{top}} + a\gamma_{\text{top}}\gamma_{\text{bot}} + \gamma_{\text{bot}})$.

Example 3.4. $\gamma_{\text{top}} = 10$, $a = .01$, $\phi/h_{\text{top}} = \alpha$ ($u(a) = \alpha$ when $\gamma_{\text{bot}} = 0$).

$u(a) = \alpha(10 + 0.1\gamma_{\text{bot}})/(10 + 1.1\gamma_{\text{bot}})$. If $\gamma_{\text{bot}} = 500$ we have $u(a)/\alpha \approx .1$ which means a significant relative difference.

References

1. H. T. Banks, F. Kojima and W. P. Winfree, Boundary estimation problems arising in thermal tomography. *Inverse Problems* **6**, 897–921 (1990).
2. P. Bison, Some application of IR thermography to thermal non destructive testing. *Minicourse on Applied Inverse Problems*, CNR-IAC Firenze 2002.
3. P. Bison, D. Fasino and G. Inglese, Active infrared thermography in non-destructive evaluation of surface corrosion, in *Series on Advances in Mathematics for Applied Sciences Vol 69 - Proceedings of the 7th SIMAI Conference - Venice 2004* (World Scientific: New Jersey), 143–154 (2005).
4. K. Brian and L. F. Caudill, Reconstruction of an unknown boundary portion from Cauchy data in n dimensions. *Inverse Problems* **21**, 239–255 (2005).

5. G. Buttazzo and R. V. Khon Reinforcement by a thin layer with oscillating thickness. *Appl Math Optim* **16**, 246–261 (1987).
6. H. S. Carslaw and J. C. Jaeger, *Conduction of Heat in Solids.* Oxford University Press (1959).
7. D. J. Crowther, L. D. Favro, P. K. Kuo and R. L. Thomas, Inverse Scattering Algorithm Applied to Infrared Thermal Wave Images. *J. Appl. Phys.* **73**, 1714–1723 (1993).
8. H. W. Engl, M. Hanke and A. Neubauer, *Regularization of Inverse Problems.* Kluwer Accademic Publisher (2000).
9. D. Fasino and G. Inglese Recovering unknown terms in a nonlinear boundary condition for Laplace's equation *IMA Journal of Applied Mathematics* 1–21 (2006). doi: 10.1093/imamath/hxl021
10. P. Kaup and F. Santosa, Nondestructive evaluation of corrosion damage using electrostatic measurements. *J. Nondestructive Evaluation* **14**, 127–136 (1995).
11. F. T. Lentes and N. Siedow Three-dimensional radiative heat transfer in glass cooling processes *Glass Sci. Technol. Glastech. Ber.* **72**, 188–196 (1999).
12. X. Maldague, Applications of Infrared Thermography to Nondestructive Evaluation. Invited chapter in *Trends in Optical Nondestructive Testing.* Pramod Rastogi editor, 591–609 (2000).
13. L. E. Payne, *Improperly Posed Problems in Partial Differential Equations* SIAM (1975).

LEXSEGMENT IDEALS AND SIMPLICIAL COHOMOLOGY GROUPS

V. BONANZINGA

DIMET, University of Reggio Calabria, Faculty of Engineering,
via Graziella (Feo di Vito),
Reggio Calabria, 89100, Italy
E-mail: vittoria.bonanzinga@unirc.it,
vittoria.bonanzinga@tin.it.

L. SORRENTI

Department of Mathematics, University of Messina,
C.da Papardo, Salita Sperone 31,
Messina, 98166, Italy
E-mail: sorrenti@dipmat.unime.it,
sorrenti.loredana@tiscali.it.

Let V be a $k-$vector space with basis $e_1, \ldots, e_n$ and let E be the exterior algebra over V. For any subset $\sigma = \{i_1, \ldots, i_d\}$ of $\{1, \ldots, n\}$ with $i_1 < i_2 < \ldots < i_d$ we call $e_\sigma = e_{i_1} \wedge \ldots \wedge e_{i_d}$ a monomial of degree d and we denote the set of all monomials of degree d by M_d. We order the monomials lexicographically so that $e_1 > e_2 > \ldots > e_n$. Then a lexsegment ideal is an ideal generated by a subset of M_d of the form $L(u, v) = \{w \in M_d : u \geq w \geq v\}$, where u, $v \in M_d$ and $u \geq v$. We describe all lexsegment ideals with linear resolution in the exterior algebra. Then we study the vanishing and non vanishing of reduced simplicial cohomology groups of a simplicial complex Δ and of certain subcomplexes of Δ with coefficients in a field k. Finally we give an idea of the applicative aspects of our results.

Keywords: lexsegment ideal; linear resolution; simplicial cohomology groups.

1. Introduction and formulation of the problems

Let V be a $K-$vector space with basis $e_1, ..., e_n$ and let E be the exterior algebra over V. For any subset $\sigma = \{i_1, ..., i_d\}$ of $\{1, ..., n\}$ with $i_1 < i_2 < ... < i_d$ we call $e_\sigma = e_{i_1} \wedge ... \wedge e_{i_d}$ a monomial of degree d and we denote the set of all monomials of degree d by M_d. In order to simplify notation, we set uv for the product $u \wedge v$ for any two elements u and v. An ideal $I \subset E$ generated by monomials is called a monomial ideal. We order the monomials lexicographically so that $e_1 > e_2 > \ldots > e_n$. A lexsegment of degree d is a subset of M_d of the form $L(u,v) = \{w \in M_d : u \geq w \geq v\}$. A lexsegment ideal is an ideal generated by a lexsegment. Lexsegment ideals in the polynomial ring were first introduced by Hulett and Martin [11]. In extremal combinatorics one usually considers more special lexsegment ideals, which we call initial lexsegment ideals. These are ideals which, in each degree, are generated by initial lexsegments, that is, sets of the form: $L^i(v) = \{w \in M_d : w \geq v\}$, called simple lexsegments in usual terminology. A final lexsegment is a set of the form: $L^f(u) = \{w \in M_d : w \leq u\}$. A final lexsegment ideal is an ideal which is generated by a final lexsegment. A lexsegment L is called completely lexsegment if all the iterated shadows of L are again lexsegment. We recall that the shadow of a set S of monomials in M_d is the set of the non-zero monomials: $\mathrm{Shad}(S) = \{se_i : \forall s \in S, \forall i \notin \mathrm{supp}(s)\}$, where the support of a monomial $u \in E$ is $\mathrm{supp}(u) = \{i : e_i \textit{ divides } u\}$. We define the $i-$th shadow recursively by $\mathrm{Shad}^i(S) = \mathrm{Shad}(\mathrm{Shad}^{i-1}(S))$. In Ref. [4] completely lexsegment ideals have been characterized and the sufficient conditions for lexsegment ideals to have linear resolution have been determined. In the first section of this paper we describe all completely lexsegment ideals in the exterior algebra with a linear resolution. In Sec. 3 we study the vanishing and non vanishing of reduced simplicial cohomology groups of a simplicial complex Δ and of certain subcomplexes of Δ with coefficients in a field k and we compute the reduced simplicial cohomology groups of a certain subcomplex of a lexsegment simplicial complex. In Sec. 4 we give an idea of the applicative aspects of the results obtained in Sec. 3.

2. Completely lexsegment ideals in the exterior algebra

In Ref. [4] the first author characterized completely lexsegment ideals in the exterior algebra and gave some sufficient conditions for lexsegment ideals to have a linear resolution. The following theorem shows that the problem of studying minimal resolutions of lexsegment ideals in the exterior algebra can be reduced to studying minimal resolutions of squarefree lexsegment ideals.

Theorem 2.1. *Let V be a vector space with basis $e_1, \ldots, e_n$ over a field k and let E be the exterior algebra over V. Let $I \subseteq E$ be an ideal in the exterior algebra and let J be the corresponding squarefree monomial ideal in the polynomial ring. Then:*

(1) J is completely lexsegment in R if and only if I is completely lexsegment in E.

(2) J has a linear free resolution over R if and only if the corresponding ideal I in E has a linear free resolution

Proof. The assertion (1) follows from the definition of completely lexsegment ideals in the exterior algebra [4]. The assertion (2) is due to Aramova, Avramov, Herzog (see Corollary 2.2 in Ref. [1]). □

In Ref. [5] the problem of describing squarefree lexsegment with a linear resolution is solved. By applying 2.1 and the results obtained in Ref. [5] we obtain a full description for lexsegment ideals with a linear resolution in the exterior algebra. In order to quote the main result we introduce some notation which we will use throughout this paper. For $w \in M_d$ we set

$$m(w) = min\{i : i \in \text{supp} w\},\ M(w) = max\{i : i \in \text{supp} w\}$$

$$w'' = w/e_{m(w)},\ w' = w/e_{M(w)}.$$

Theorem 2.2. *Let u, $v \in M_d$ be monomials with $u \geq v$ and $I = (L(u,v))$ be a completely lexsegment ideal. Let k be the smallest integer such that $(i_k, j_k) \neq (k,k)$ and $B = \{w \in M_d : w < v, x_k w' > u\}$. Then I has a linear resolution if, and only if, one of the following conditions holds:*

(a) $u = e_1 \cdots e_{k-1} e_{i_k} \cdots e_{i_d}$ with $e_{i_k} \cdots e_{i_d} \geq e_{k+2} \cdots e_{d+2}$ and $v = e_1 \cdots e_{k-1} e_{n-d+k} e_{n-d+k+1} \cdots e_n$

(b) $i_k = k$ and the following condition holds: for every $(w_1, w_2) \in B \times B$ with $w_1 \neq w_2$ there exists an index l, $m(w_1) \leq l < M(w_2)$ such that $\frac{e_k w_1}{e_{m(w_1)}} \leq u$, $\frac{e_k w_2}{x_l} \leq u$ and $\frac{w_1}{e_{m(w_1)}} \neq \frac{w_2}{e_l}$.

This theorem gives us a description of completely lexsegment ideals with a linear resolution. Arguing as in Ref. [5] and applying 2.2 we obtain the procedure to determine all arbitrary lexsegment ideals with a linear resolution.

In the following propositions we use the term "completely lexsegment ideal" in the (more general) sense, that, each homogeneous component of the ideal is spanned by a lexsegment, but the ideal is not necessarily generated in one degree.

Proposition 2.1. *Let $u \geq v$ be monomials of degree d in E, $u \neq e_1 \ldots e_d$, $v \neq e_{n-d+1} \ldots e_n$. Let $I = (L(u,v))$, $J = (L^i(v))$ and $K = (L^f(u))$. Then* *(a)* $I = J \cap K$ *if, and only if,* I *is completely lexsegment.*

(b) $J \cap K$ *is completely lexsegment if and only if* K *is completely lexsegment, that is, if, and only if,* $u \geq e_3 \dots e_{d+2}$.

Proof.

(a) If I is completely lexsegment, then by Lemma 1.5 in Ref. [4], $I = J \cap K$. Suppose $I = J \cap K$ and let $u = e_{i_1} \dots e_{i_d}$. If $i_1 \neq 1$, then by Theorem 1.1 I is not completely lexsegment. Moreover, if $w = e_{r_1} \dots e_{r_d}$ is a monomial with $w < v \leq u$ then $r_1 \neq 1$. It is easy to see that $we_1 \in J \cap K$, but $we_1 \notin I$. Thus in this case $I \neq J \cap K$. We assume now $i_1 = 1$. It follows from the persistence theorem 1.6 in Ref. [4] that K is completely lexsegment. J is obviously completely lexsegment because it is an initial lexsegment. From Lemma 1.4 in Ref. [4] $J \cap K$ is generated in degree d and in degree $d+1$. Then I is completely lexsegment if, and only if, $(J \cap K)_d$ and $(J \cap K)_{d+1}$ are lexsegment. $(J \cap K)_d = (L^i(v) \cap L^f(u))$ is obviously lexsegment. Since $(J \cap K)_{d+1} = (\text{Shad}(L^i(v)) \cap \text{Shad}(L^f(u)))$ and J and K are completely lexsegment, it follows that $(J \cap K)_{d+1}$ is a lexsegment, therefore I is completely lexsegment.

(b) From persistence Theorem 1.6 in Ref. [4] K is completely lexsegment if, and only if, $u \geq e_3 e_4 \dots e_{d+2}$. It is clear that if K is completely lexsegment then also $J \cap K$ is completely lexsegment. Hence to conclude the proof it is enough to show that if $J \cap K$ is completely lexsegment then $u \geq e_3 e_4 \dots e_{d+2}$. Assume that $e_3 e_4 \dots e_{d+2} > u \geq v$. Then $e_2 e_3 e_4 \dots e_{d+2} > e_2 v > ve_{M(c(v))}$ and, since $m(c(u)) = 1$, $e_2 e_3 e_4 \dots e_{d+2} < ue_{m(c(u))} = ue_1$. It follows that $e_2 e_3 \dots e_{d+2} \in \text{Shad}(L(u, v)) = L(ue_1, ve_{M(c(v))})$.
Moreover $e_2 e_3 \dots e_{d+2} \notin K$. Hence ue_1, $ve_{M(c(v))}$ are in $J \cap K$, but $L(ue_1, ve_{M(c(v))}) \nsubseteq J \cap K$. This implies that $(J \cap K)_{d+1}$ is not lexsegment. Then $J \cap K$ is not completely lexsegment. □

3. The reduced simplicial cohomology groups of a simplicial complex Δ

Let Δ be a simplicial complex on the set of vertices $[n]$. We write I_Δ for the ideal of E generated by all monomials e_F, with $e_F \notin \Delta$ and we denote by $k\{\Delta\}$ the quotient algebra E/I_Δ. When T is a subset of $[n]$, let Δ_T denote the simplicial complex consisting of all faces $\sigma \in \Delta$ with $\sigma \subset T$. If $a = (a_1, \dots, a_n) \in \mathbb{N}^n$, $|a| = a_1 + \dots + a_n$ and $\text{supp}(a) = \{i : a_i \neq 0\}$.

The resolution of I_Δ has a multigraded structure, so that the $k-$vector spaces $\mathrm{Tor}_i^E(k\{\Delta\};k)$ are $\mathbb{Z}^n-$graded $k-$ vector spaces. The next theorem (see Ref. [3], Theorem 6.4) shows that the $\mathbb{Z}^n-$ graded components of $\mathrm{Tor}_i^E(k\{\Delta\},k)$ are isomorphic to the reduced simplicial cohomology of certain subcomplexes of Δ. This result is the precise analogue to Hochster's formula [12] which describes the Z^n-graded components of $\mathrm{Tor}_i^R(k[\Delta],k)$ when R is the polynomial ring.

Theorem 3.1. *Let Δ be a simplicial complex and $a \in \mathbb{N}^n$. Set $d = |a|$ and $T = \mathrm{supp}(a)$. Then, for all $i \geq 0$, we have*

$$\mathrm{Tor}_i^E(k\{\Delta\},k)_a \cong \tilde{H}^{d-i-1}(\Delta_T;k)$$

where $\tilde{H}^*(\cdot;k)$ denotes the reduced simplicial cohomology groups with the coefficient field k.

Corollary 3.1. *Let Δ be a simplicial complex and $a \in \mathbb{N}^n$. Set $T = \mathrm{supp}(a)$ and $|a| = i+j$. Then, for all $i \geq 0$, we have*

$$\mathrm{Tor}_i^E(k\{\Delta\},k)_{i+j} = \bigoplus_T \tilde{H}^{j-1}(\Delta_T,k)$$

where the sum is taken over all $T \subset [n]$.

Proof. It follows immediately from the previous theorem. □

In the following results we study the vanishing and non vanishing of reduced simplicial cohomology groups of a simplicial complex Δ when I_Δ has a linear resolution or when I_Δ is a completely lexsegment ideal.

Proposition 3.1. *If I_Δ is a monomial ideal generated in degree $d \leq n$ and with a linear resolution, then*

$$\tilde{H}^i(\Delta;k) = 0 \quad for\ all\ i \neq d-2$$

Proof. Since I_Δ has a linear resolution, one has

$$\mathrm{Tor}_i^E(k\{\Delta\},k)_{i+j} = 0 \quad j \neq d-1.$$

If $a = (1,\ldots,1)$ then $T = [n]$ and $\Delta = \Delta_T$ and from Theorem 3.1

$$\mathrm{Tor}_i^E(k\{\Delta\},k)_{(1,\ldots,1)} \cong \tilde{H}^{n-i-1}(\Delta,k).$$

Since

$$\mathrm{Tor}_i^E(k\{\Delta\},k)_{(1,\ldots,1)} \subset \mathrm{Tor}_i^E(k\{\Delta\},k)_n$$

it follows

$$\tilde{H}^{n-i-1}(\Delta, k) \subset \mathrm{Tor}_i^E(k\{\Delta\}, k)_n.$$

If $i+j=n$ and $j \neq d-1$ then $j = n-i \neq d-1$ and $\mathrm{Tor}_i^E(k\{\Delta\},k)_n = 0$. Therefore

$$\tilde{H}^{n-i-1}(\Delta, k) = 0 \quad \text{if} \quad n-i \neq d-1$$

and the assertion holds. □

Definition 3.1. A simplicial complex Δ is called lexsegment if the corresponding ideal I_Δ is a lexsegment ideal.

In the next theorem we compute the reduced simplicial cohomology groups of a certain subcomplex of a lexsegment simplicial complex.

Theorem 3.2. *If I_Δ is a completely lexsegment ideal generated in degree $d \leq n$, then*

$$\tilde{H}^i(\Delta_T, k) = 0 \quad \textit{for all } i \neq d-2, d-3 \textit{ and all } \emptyset \neq T \subset [n]$$

Proof. Let $u \geq v$ be monomials in M_d. If $I = (L(u,v))$ is completely lexsegment then from Lemma 1.5. in Ref. [4] $I = J \cap K$, where $J = (L^i(v))$ and $K = (L^f(u))$. The short exact sequence

$$0 \longrightarrow I \longrightarrow J \oplus K \longrightarrow L \longrightarrow 0$$

where $L = J + K = (M_d)$ gives rise to a long exact sequence

$$\ldots \longrightarrow \mathrm{Tor}_{i+1}(L,k)_{i+j} \longrightarrow \mathrm{Tor}_i(I,k)_{i+j} \longrightarrow \mathrm{Tor}_i(I,k)_{i+j} \oplus \mathrm{Tor}_i(K,k)_{i+j}$$

$$\longrightarrow \mathrm{Tor}_i(L,k)_{i+j} \longrightarrow \mathrm{Tor}_{i-1}(I,k)_{i+j} \longrightarrow \ldots$$

If I is generated in degree d, since J, K and L have a linear resolution $\mathrm{Tor}_i(J,k)_{i+j} = \mathrm{Tor}_i(K,k)_{i+j} = \mathrm{Tor}_i(L,k)_{i+j} = 0$ for $j \neq d$ and $\mathrm{Tor}_{i+1}(L,k)_{i+1+j} = 0$ for $j \neq d$, hence $\mathrm{Tor}_{i+1}(L,k)_{i+j} = 0$ for $j \neq d-1$. Therefore $\mathrm{Tor}_i(I,k)_{i+j} = 0$ for $j \neq d-1, d$. From

$$0 \longrightarrow I \longrightarrow E \longrightarrow k\{\Delta\} \longrightarrow 0$$

it follows that

$$\mathrm{Tor}_{i+1}(k\{\Delta\},k)_{i+1+j} = \mathrm{Tor}_i(I,k) = 0 \Leftrightarrow j+1 \neq d-1, d,$$

hence

$$\mathrm{Tor}_{i+1}(k\{\Delta\},k)_{i+1+j} = 0 \text{ for } j \neq d-1, d-2$$

and $i \geq 0$. Therefore

$$\mathrm{Tor}_i(k\{\Delta\}, k)_{i+j} = 0 \text{ for } j \neq d-1, d-2 \text{ and } i > 0. \tag{1}$$

From Corollary 3.1

$$\mathrm{Tor}_i^E(k\{\Delta\}, k)_{i+j} = \bigoplus_T \tilde{H}^{j-1}(\Delta_T; k), \tag{2}$$

with $T = \mathrm{supp}(a)$ and $|a| = i + j$. From (1) and (2)

$$\tilde{H}^{j-1}(\Delta_T, k) = 0 \;\; for \;\; j \neq d-1, d-2$$

for all $\emptyset \neq T \subset [n]$, hence the assertion holds. □

4. Applicative aspects

The computation of homology and cohomology simplicial groups is a useful tool in computational topology, an active research field which deals with the study of topological properties of an object that can be computed to some finite accuracy. There is a growing literature on the formalization and representation of topological questions for computer applications, see Ref. [6] for a survey of this field. Application areas include digital image processing, solid modelling for computer design, topology preserving morphisms in computer graphics and 3-d models of protein molecules. The earliest applications work on extracting topological information from data targeted digital images. These are typically represented by binary data on a fixed regular grid in two or three dimensions, pixels and voxels. This field has many applications, for example, in computing the boundaries of a drainage basin from satellite data [20]. Much work in this area focuses on algorithms for the labeling of components [13], boundaries [20] and other features of digital images.
In order to give an idea of these applications we consider the following.

Problem 4.1. *Detecting some information about the structure of a digital image, given a finite set S of data of the image.*

One of the procedures usually used to solve this problem can be summarized in three steps:

(1) Modeling of X by a topological space ($X \subset \mathbb{R}^2$, $X \subset \mathbb{R}^3$).
(2) Construction of a triangulation Δ associated with S.
(3) Computation of the simplicial homology groups of Δ.

The homology groups characterize the number and the type of holes and the number of connected components of X.
An approach of this kind is used by many authors. In Ref. [9] M. Grandis develops combinatorial homotopical tools consisting essentially of intrinsic homotopic theory for simplicial complexes, and in Ref. [10] he applies these tools to explore mathematical models representing images. In Ref. [16] and in Ref. [17] V. Robins considers the problem of determining information about the topology of a subset $X \subseteq \mathbb{R}^n$ given only a finite point approximation to X. The computability of homology groups from a given triangulation is well-known and the algorithm uses simple linear algebra [14]. However, these algorithms have run times at best cubic in the number of simplices. So, the problem of computing homology groups is an active area of research. In order to solve problems related to cost times it would be useful to build algorithms with a lower computational cost.
Our approach in computing homology is theoretical. In the previous sections we have studied the combinatoric properties of the monomial ideals associated with certain simplicial complexes. In order to show how our results can be applied to detect some information about a topological space we first give some preliminary remarks.

Let Δ be a simplicial complex. We embed Δ in the topological space $\mathbb{R}^n$.

Definition 4.1. We define the geometric realization of the simplicial complex Δ by

$$|\Delta| = \bigcup_{F \in \Delta} |F|.$$

Definition 4.2. Let X be a topological space, and $\varphi : X \to |\Delta|$ a homeomorphism. The pair (Δ, φ) is called triangulation of X.

The following is a fundamental result in algebraic topology [18]:

Theorem 4.1. *Let X be a topological space with triangulation Δ. Then*

$$\tilde{H}_q(\Delta; A) \cong \tilde{H}_q(X, A)$$

for all q.

Since we consider k field, the problem of computing homology is equivalent to the problem of computing cohomology as the following [18] shows:

Remark 4.1. If k is a field and Δ is a finite simplicial complex, then:

$$\tilde{H}_q(\Delta, k) \cong \tilde{H}^q(\Delta, k).$$

In many applications all that is needed is the rank of homology groups, that is, the Betti numbers. In fact, for a topological space X, there is the following interpretation of Betti numbers [15]: the first Betti number $\beta_0 = \text{rank}\tilde{H}_0(X, k)$ is equal to the number of connected components minus 1, the second Betti number $\beta_1 = \text{rank}\tilde{H}_1(X, k)$ is equal to the number of independent tunnel minus 1, the third Betti number $\beta_2 = \text{rank}\tilde{H}_2(X, k)$ is equal to the number of enclosed voids minus 1. In other words, the second Betti number gives the number of holes in dimension 1 and the third Betti number gives the number of holes in dimension 2. For example, if X is the planar image of Fig. 1 we can detect that X is connected with two holes by constructing a triangulation Δ of a finite set of X and by computing the ranks of the homology groups of Δ.

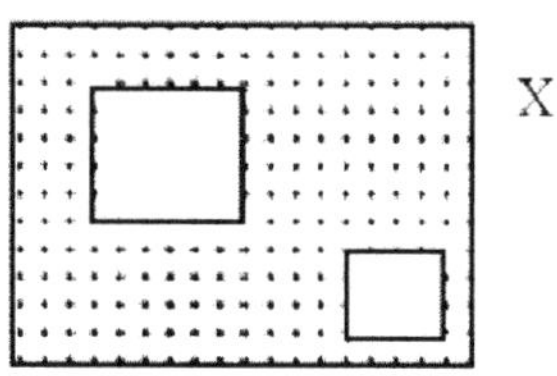

Fig. 1.

In the following examples we apply the results obtained in Sec. 3 to detect information about some topological spaces consisting of geometric surfaces.

Example 4.1. Let X be the topological space consisting of a closed cilindric surface (Fig. 2). X is homeomorphic to a prism surface. Then the simplicial complex Δ of Fig. 2 is a triangulation of Δ. The monomial ideal associated with Δ is $I_\Delta = (e_1e_3, e_2e_6, e_3e_6)$. The corresponding ideal of I_Δ in R is $J = (x_1x_3, x_2x_6, x_3x_6) \subset R$, and it has the following linear free resolution[a]:

$$0 \to R^2(-3) \to R^3(-2) \to J.$$

[a]The minimal free resolution of J is computed by using CoCoA[7].

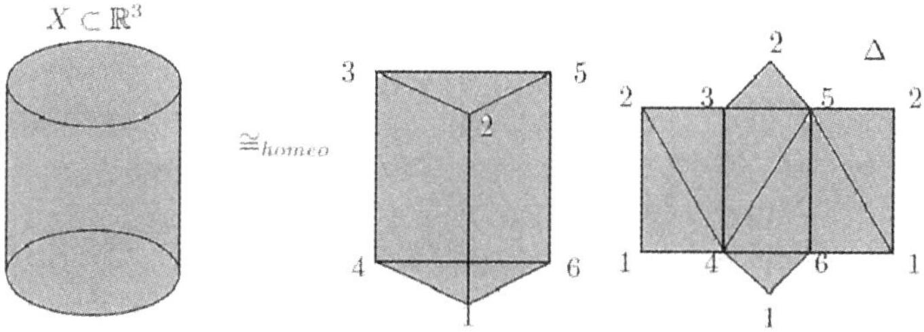

Fig. 2.

Then, it follows from Proposition 3.1 that $\tilde{H}^i(\Delta, k) = 0$, for all $i \neq d-2 = 0$. In particular, we obtain

$$\text{rank}\tilde{H}^1(\Delta, k) = \text{rank}\tilde{H}_1(\Delta, k) = \text{rank}\tilde{H}_1(X, k) = 0$$

and, that X has one indipendent tunnel,

$$\text{rank}\tilde{H}^2(\Delta, k) = \text{rank}\tilde{H}_2(\Delta, k) = \text{rank}\tilde{H}_2(X, k) = 0$$

and this means that X has one enclosed void.

Example 4.2. Let us consider the ideal $I_\Delta = (e_1e_2, e_1e_3, e_1e_4, e_1e_5) \subset E$. Since I is an initial lexsegment ideal then I is completely lexsegment. The simplicial complex Δ has vertex set $[n] = \{1, 2, \ldots, 5\}$. We have:
$\Delta = \{\{\emptyset\}, \{1\}, \{2\}, \{3\}, \{4\}, \{5\}, \{2,3\}, \{2,4\}, \{2,5\}, \{3,4\}, \{3,5\}, \{4,5\}, \{2,3,4\}, \{2,3,5\}, \{2,4,5\}, \{3,4,5\}\}$. If $T = \{2, 3, \ldots, 5\} \subset \{1, 2, \ldots, 5\}$ then $\Delta_T = \{\{\emptyset\}, \{2\}, \{3\}, \{4\}, \{5\}, \{2,3\}, \{2,4\}, \{2,5\}, \{3,4\}, \{3,5\}, \{4,5\}, \{2,3,4\}, \{2,3,5\}, \{2,4,5\}, \{3,4,5\}\}$. Note that Δ is a triangulation of the topological space X of Fig. 3, consisting of a closed conic surface. It follows

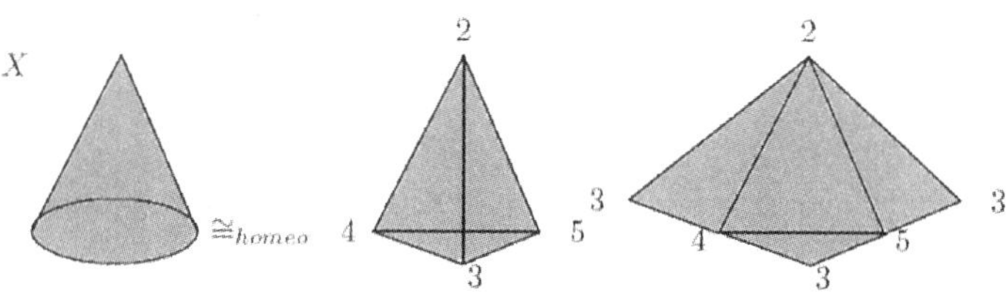

Fig. 3.

from Theorem 3.2 that

$$\tilde{H}^i(\Delta) = 0, \forall i \neq d-2, d-3.$$

Then

$$\text{rank}\tilde{H}_1(X) = \text{rank}\tilde{H}_1(\Delta, k) = \text{rank}\tilde{H}^1(\Delta, k) = 0,$$

$$\text{rank}(\tilde{H}_2(X)) = \text{rank}\tilde{H}_2(\Delta, k) = \text{rank}\tilde{H}^2(\Delta, k) = 0,$$

and this means that X has one independent tunnel and one enclosed void.

At the moment it is not clear how Proposition 3.1 and Theorem 3.2 could be applied to detect some information when X is a digital image. Our goal in the solution of Problem 4.1 is to give a contribution to the computation of homology (third step) in some special cases, coming from triangulations associated with completely lexsegment ideals or with ideals with a linear resolution. Concerning the first and the second step of 4.1, we believe that, under certain hypotheses, it is possible to associate the triangulations coming from these ideals to a finite set of data of a digital image. It would be interesting to find these hypotheses.

References

1. A. Aramova, L. L. Avramov and J. Herzog, Resolutions of monomial ideals and cohomology over exterior algebras, in *Trans. AMS.* **352** (2) (2000), pp. 579-594.
2. A. Aramova, J. Herzog and T. Hibi, Squarefree lexsegment ideals, *Math. Z.*, **228** (1998) 2, pp. 353-378.
3. A. Aramova, J. Herzog and T. Hibi, Gotzmann theorems for exterior algebras and combinatorics, *J. Algebra* **191** (1997), **26**, pp. 171-223
4. V. Bonanzinga, Lexsegment ideals in the exterior algebra, in *Geometric and Combinatorial aspects of commutative algebra: Lect. Notes in Pure and Appl. Math.*, eds. J. Herzog and G. Restuccia, **4**, (Dekker, New York, 1999), pp. 43-56.
5. V. Bonanzinga and L. Sorrenti, Squarefree lexsegment ideals with linear resolution, Preprint (2005).
6. T.K. Dey, H. Edelsbrunner and S. Guha, *Computational topology*, in *Advances in Discrete and Computational Geometry*, **223**, (Contemporary mathematics., AMS, 1999).
7. CoCoA Team, *CoCoA: a system for doing Computations in Commutative Algebra*, available at http://cocoa.dima.unige.it.
8. S. Eliahou and M. Kervaire, Minimal resolutions of some monomial ideals, *J. Algebra* **129** (1990), pp. 1-25.
9. M. Grandis, An intrinsec homotopy theory for simplicial complexes, with applications to image analysis, *Appl. Cat. Structures* **10** (2002), pp. 99-155.

10. M. Grandis, Ordinary and directed combinatorial homotopy, applied to image analisys and cuncurrency, *Homology, Homotopy and Applications*, **5**(2), (2003), pp. 211-231.
11. H. A. Hulett and H. M. Martin, Betti numbers of lexsegment ideals, *J. Algebra* **275** (2004), 2, pp. 629-638.
12. M. Hochster, Cohen-Macaulay rings, combinatorics and simplicial complexes, in Ring theory II, McDonald, eds. B. R. and Morris, R. A., *Lecture Notes in Pure and Appl. Math.* **26**, Marcel Dekker, (1977).
13. T. Y. Kong and A. Rosenfeld, Digital Topolgy: Introduction and survey, *Computer Vision, Graphics and Image Processing*, **48**, pp. 357-393, (1989).
14. J. R. Munkres, *Elements of Algebraic Topology*, Benjamin Cummings, (1984).
15. E. M. Patterson, *Topology*, University mathematical texts, Oliver and Bold LTD, (1996).
16. V. Robins, Computational Topology at Multiple Resolutions, PhD thesis, June (2000).
17. V. Robins, Computational Topology for Point Data: Betti Numbers of Alpha-Shapes, in *Morphology of Condensed Matter: Physics and Geometry of Spatially Complex Systems, Lecture Notes in Physics*, **600**, Springer (2002), pp. 261-275.
18. R. P. Stanley, *Combinatorics and commutative algebra*, Progress in Mathematics, **41**, Birkhäuser (1996).
19. J. K. Udupa, *Surface connectedness in digital spaces*, Topological Algorithms for Digital Images Processing, Elsevier, Amsterdam, (1993).
20. J. Williams, *Geographic information from Space: processing an applications of geocoded satellite images*, Wiley, (1995).

NONLINEAR ELECTRONIC TRANSPORT IN SEMICONDUCTOR SUPERLATTICES

L. L. BONILLA
Modelización, Simulación y Matemática Industrial, Universidad Carlos III de Madrid, 28911 Leganés, Spain
E-mail: bonilla@ing.uc3m.es

L. BARLETTI
Dipartimento di Matematica "Ulise Dini"; Università di Firenze, 50134 Firenze, Italy
E-mail: barletti@math.unifi.it

R. ESCOBEDO and M. ALVARO
Modelización, Simulación y Matemática Industrial, Universidad Carlos III de Madrid, 28911 Leganés, Spain

Nonlinear charge transport in strongly coupled semiconductor superlattices is described by single or two-miniband Wigner-Poisson kinetic equations with BGK collision terms. Balance equations for miniband populations and electric field are derived using the Chapman-Enskog method. Numerical solutions show stable self-oscillations of the current through a voltage biased superlattice.

Keywords: Quantum drift-diffusion equations, Chapman-Enskog method, Rashba spin-orbit interaction, modified Kane model.

1. Introduction

Semiconductor superlattices are essential ingredients in fast nanoscale oscillators, quantum cascade lasers and infrared detectors. A superlattice (SL) is a quasi-one-dimensional crystal originally proposed by Esaki and Tsu to observe Bloch oscillations, i.e., the periodic coherent motion of electrons in a miniband when an electric field is applied. Once the materials were grown, many interesting nonlinear phenomena were observed, such as self-oscillations of the current through the SL due to charge dipole motion, multistability of stationary charge and field profiles, etc. See the review 1.

Nonlinear charge transport in SLs has been widely studied in the last decade using balance equations for electron densities and electric field. These equations are either proposed using phenomenological arguments or derived ad hoc from kinetic theories.[1,2] Systematic derivations are scarce. For a single-miniband SL, the Chapman-Enskog (CE) method applied to a semiclassical Boltzmann-Poisson system whose collision term is of Bhatnagar-Gross-Krook (BGK) type yields a generalized drift-diffusion equation (GDDE),[3] and a quantum drift-diffusion equation (QDDE) when applied to a Wigner-Poisson-BGK (WPBGK) system.[4] For a semiclassical parabolic-band BGK-Poisson semiconductor system, the CE method had earlier been used to obtain balance equations.[5] The quantum WPBGK system contains two pseudo-differential operators, involving the band dispersion relation and the electric potential. The leading order approximation in the hyperbolic limit balances collisions and electric potential, and its solution is not obvious because the potential is an a priori unknown solution of the Poisson equation. SLs are simpler because their Wigner functions are periodic in the reciprocal lattice, the potential terms become multiplication operators in Fourier space, and the leading order approximation is straightforward to solve.[4]

For sufficiently high applied electric fields, electrons may populate higher minibands, then be scattered to the lowest, etc. Moreover, SLs with diluted magnetic impurities subject to a magnetic field may present spin polarization effects whose understanding is crucial to develop spintronic devices.[6] Even without magnetic impurities, spin polarization could appear due to Rashba spin-orbit interaction.[7] Once we consider electron spin, each miniband is split in two and single-miniband SLs become two-miniband SLs. We shall systematically derive quantum balance equations by the CE method.

2. Single miniband superlatice

The Wigner-Poisson-Bhatnagar-Gross-Krook (WPBGK) system for 1D electron transport in the lowest miniband of a strongly coupled SL is:

$$\begin{aligned}\frac{\partial f}{\partial t} &+ \frac{i}{\hbar}\left[\mathcal{E}\left(k+\frac{1}{2i}\frac{\partial}{\partial x}\right) - \mathcal{E}\left(k-\frac{1}{2i}\frac{\partial}{\partial x}\right)\right] f \\ &+ \frac{ie}{\hbar}\left[W\left(x+\frac{1}{2i}\frac{\partial}{\partial k}, t\right) - W\left(x-\frac{1}{2i}\frac{\partial}{\partial k}, t\right)\right] f \\ &= Q[f] \equiv -\nu_e\left(f - f^{FD}\right) - \nu_i\,\frac{f(x,k,t)-f(x,-k,t)}{2}, \end{aligned} \tag{2.1}$$

$$\varepsilon\,\frac{\partial^2 W}{\partial x^2} = \frac{e}{l}\,(n - N_D), \tag{2.2}$$

with

$$n = \frac{l}{2\pi}\int_{-\pi/l}^{\pi/l} f(x,k,t)dk = \frac{l}{2\pi}\int_{-\pi/l}^{\pi/l} f^{FD}(k;n)dk, \tag{2.3}$$

$$f^{FD}(k;n) = \frac{m^* k_B T}{\pi^2\hbar^2}\ln\left[1+\exp\left(\frac{\mu - \mathcal{E}(k)}{k_B T}\right)\right]. \tag{2.4}$$

Here f, n, N_D, $\mathcal{E}(k)$, d_B, d_W, $l = d_B + d_W$, W, ε, m^*, k_B, T, ν_e, ν_i and $-e < 0$ are the one-particle Wigner function, the 2D electron density, the 2D doping density, the miniband dispersion relation, the barrier width, the well width, the SL period, the electric potential, the SL permittivity, the effective mass of the electron, the Boltzmann constant, the lattice temperature, the frequency of the inelastic collisions responsible for energy relaxation, the frequency of the elastic impurity collisions and the electron charge, respectively.

The left-hand side of Eq. (2.1) can be straightforwardly derived from the Schrödinger-Poisson equation for the wave function in the miniband using the definition of the 1D Wigner function:

$$f(x,k,t) = \frac{2l}{S}\sum_{j=-\infty}^{\infty}\int \langle\psi^\dagger(x+\frac{jl}{2},y,z,t)\psi(x-\frac{jl}{2},y,z,t)\rangle e^{ijkl}dxdy. \tag{2.5}$$

The second quantized wave function ψ is a superposition of the Bloch states corresponding to the miniband and S is the SL cross section.[4] The right hand side in Eq. (2.1) is the sum of $-\nu_e\left(f - f^{FD}\right)$, which represents energy relaxation towards a 1D effective Fermi-Dirac (FD) distribution $f^{FD}(k;n)$ (local equilibrium, which is the 3D Fermi-Dirac distribution integrated over the lateral components of the wave vector (k, k_y, k_z)), and $-\nu_i[f(x,k,t)-f(x,-k,t)]/2$, which accounts for impurity elastic collisions.[3] For simplicity, the collision frequencies ν_e and ν_i are fixed constants. Exact and FD distribution functions have the same electron density, thereby preserving charge continuity. $\mu = \mu(n)$ results from solving (2.3) with (2.4).

The WPBGK system (2.1) to (2.4) should have a $2\pi/l$-periodic (in k) solution satisfying appropriate initial and boundary conditions:

$$f(x,k,t) = \sum_{j=-\infty}^{\infty} f_j(x,t)\, e^{ijkl}. \tag{2.6}$$

Defining $F = \partial W/\partial x$ (*minus* the electric field) and the average

$$\langle F\rangle_j(x,t) = \frac{1}{jl}\int_{-jl/2}^{jl/2} F(x+s,t)\,ds, \tag{2.7}$$

it is possible to obtain the following equivalent form of the Wigner equation[4]

$$\frac{\partial f}{\partial t} + \sum_{j=-\infty}^{\infty} \frac{ijl}{\hbar} e^{ijkl} \left(\mathcal{E}_j \frac{\partial}{\partial x} \langle f \rangle_j + e \langle F \rangle_j f_j \right) = Q[f]. \tag{2.8}$$

Integrating (2.8) over k yields the charge continuity equation

$$\frac{\partial n}{\partial t} + \frac{\partial}{\partial x} \sum_{j=1}^{\infty} \frac{2jl}{\hbar} \langle \mathrm{Im}(\mathcal{E}_{-j} f_j) \rangle_j = 0. \tag{2.9}$$

Here we can eliminate the electron density by using the Poisson equation and then integrate over x, thereby obtaining the nonlocal Ampère's law for the total current density $J(t)$:

$$\varepsilon \frac{\partial F}{\partial t} + \frac{2e}{\hbar} \sum_{j=1}^{\infty} j \langle \mathrm{Im}(\mathcal{E}_{-j} f_j) \rangle_j = J(t). \tag{2.10}$$

To derive the QDDE, we shall assume that the electric field contribution in (2.8) is comparable to the collision terms and that they dominate the other terms (*the hyperbolic limit*).[3] Let v_M and F_M be the electron velocity and field positive values at which the (zeroth order) drift velocity reaches its maximum. In this limit, the time t_0 it takes an electron with speed v_M to traverse a distance $x_0 = \varepsilon F_M l/(eN_D)$, over which the field variation is of order F_M, is much longer than the mean free time between collisions, $\nu_e^{-1} \sim \hbar/(eF_M l) = t_1$. We therefore define the *small parameter* $\lambda = t_1/t_0 = \hbar v_M N_D/(\varepsilon F_M^2 l^2)$ and formally multiply the first two terms on the left side of (2.1) or (2.8) by λ.[3,4] The result is

$$\lambda \left(\frac{\partial f}{\partial t} + \sum_{j=-\infty}^{\infty} \frac{ijl}{\hbar} e^{ijkl} \mathcal{E}_j \frac{\partial}{\partial x} \langle f \rangle_j \right) = Q[f] - \sum_{j=-\infty}^{\infty} \frac{iejl}{\hbar} e^{ijkl} \langle F \rangle_j f_j. \tag{2.11}$$

The solution of Eq. (2.11) for $\lambda = 0$ is calculated in terms of its Fourier coefficients as

$$f^{(0)}(k; F) = \sum_{j=-\infty}^{\infty} \frac{(1 - ij\mathcal{F}_j/\tau_e)\, f_j^{FD}}{1 + j^2 \mathcal{F}_j^2} e^{ijkl}, \tag{2.12}$$

where $\mathcal{F}_j = \langle F \rangle_j / F_M$, $F_M = \hbar\sqrt{\nu_e(\nu_e + \nu_i)}/(el)$ and $\tau_e = \sqrt{(\nu_e + \nu_i)/\nu_e}$.

The CE ansatz for the Wigner function is:[4]

$$f(x, k, t; \lambda) = f^{(0)}(k; F, n) + \sum_{m=1}^{\infty} f^{(m)}(k; F, n)\, \lambda^m, \tag{2.13}$$

$$\varepsilon \frac{\partial F}{\partial t} + \sum_{m=0}^{\infty} J^{(m)}(F, n)\, \lambda^m = J(t). \tag{2.14}$$

The coefficients $f^{(m)}(k;F,n)$ depend on the 'slow variables' x and t only through their dependence on F and n, which obey (2.2) and (2.14). The functionals $J^{(m)}(F,n)$ are chosen so that the $f^{(m)}(k;F,n)$ are bounded and $2\pi/l$-periodic in k. Keeping the desired number of terms and setting $\lambda = 1$ in (2.14) yields the sought QDDE. Inserting (2.13) - (2.14) in (2.11), we find the hierarchy:

$$\mathcal{L}f^{(1)} = -\left.\frac{\partial f^{(0)}}{\partial t}\right|_0 + \sum_{j=-\infty}^{\infty} \frac{ijl\mathcal{E}_j e^{ijkl}}{\hbar} \frac{\partial}{\partial x}\langle f^{(0)}\rangle_j \tag{2.15}$$

$$\mathcal{L}f^{(2)} = -\left.\frac{\partial f^{(1)}}{\partial t}\right|_0 + \sum_{j=-\infty}^{\infty} \frac{ijl\mathcal{E}_j e^{ijkl}}{\hbar} \frac{\partial}{\partial x}\langle f^{(1)}\rangle_j - \left.\frac{\partial}{\partial t} f^{(0)}\right|_1, \tag{2.16}$$

and so on. Here

$$\mathcal{L}u(k) \equiv \frac{ie}{\hbar}\sum_{-\infty}^{\infty} jl\langle F\rangle_j u_j e^{ijkl} + \left(\nu_e + \frac{\nu_i}{2}\right) u(k) - \frac{\nu_i}{2} u(-k), \tag{2.17}$$

and the subscripts 0 and 1 in the right hand side of these equations mean that $\varepsilon\, \partial F/\partial t$ is replaced by $J - J^{(0)}(F)$ and by $-J^{(1)}(F)$, respectively.

Inserting the expansion (2.13) into (2.3), we obtain the *compatibility condition* $f_0^{(m)} = 0$ (for $m > 0$), which implies that $\langle \mathcal{L}f^{(m)}\rangle = 0$, for $m > 0$. These solvability conditions yield $J^{(m)} = 2e\sum_{j=1}^{\infty} j\langle \mathrm{Im}(\mathcal{E}_{-j} f_j^{(m)})\rangle_j/\hbar$, which can also be obtained by insertion of Eq. (2.12) in (2.10).

We shall particularize our results to the tight-binding dispersion relation $\mathcal{E}(k) = \Delta(1 - \cos kl)/2$ (Δ is the miniband width and $v(k) = (\Delta l \sin kl)/(2\hbar)$ is the group velocity), having nonzero Fourier coefficients $\mathcal{E}_0 = \Delta/2$, $\mathcal{E}_{\pm 1} = -\Delta/4$. The leading order of Ampère's law (2.14) is

$$\varepsilon\frac{\partial F}{\partial t} + \frac{ev_M}{l}\left\langle n\mathcal{M}\frac{2\mathcal{F}_1}{1+\mathcal{F}_1^2}\right\rangle_1 = J(t), \tag{2.18}$$

where $v_M = \Delta l\, \mathcal{I}_1(M)/[4\hbar\tau_e \mathcal{I}_0(M)]$, $\mathcal{M} = \mathcal{I}_1(\tilde{\mu})\mathcal{I}_0(M)/[\mathcal{I}_1(M)\mathcal{I}_0(\tilde{\mu})]$, $\mathcal{I}_m(s) = \int_{-\pi}^{\pi} \cos(mk) \ln\left(1 + e^{s-\delta+\delta\cos k}\right) dk$, $\delta = \Delta/(2k_BT)$, $\tilde{\mu} = \mu/(k_BT)$. $\mu = Mk_BT$ (calculated graphically in Fig. 1 of Ref. 3) solves (2.3) with $n = N_D$.

The solution of (2.15) yields $J^{(1)}$ in (2.14), which is the first correction to the QDDE (2.18). The details can be found in References 4 and 8 (for the numerical procedure). An important point is that the nonlocal terms in the QDDE require that boundary conditions be imposed on the intervals $[-2l, 0]$ and $[Nl, Nl + 2l]$ for a N-period SL. Fig. 1 shows the current self-oscillations that appear when the QDDE is solved with boundary conditions $\varepsilon\partial F/\partial t + \sigma F = J$ at each point of the intervals $[-2l, 0]$ and $[Nl, Nl + 2l]$

and appropriate σ and dc voltage bias. Parameter values correspond to a 157-period 3.64 nm GaAs/0.93 nm AlAs SL at 5K, with $N_D = 4.57 \times 10^{10}$ cm^{-2}, $\nu_i = 2\nu_e = 18\times10^{12}$ Hz under a dc voltage bias of 1.62 V, which yield $x_0 = 16$ nm, $t_0 = 0.24$ ps, $J_0 = ev_M N_D/l = 1.10 \times 10^5$ A/cm^2. Cathode and anode contact conductivities σ are 2.5 and 0.62 $\Omega^{-1}\text{cm}^{-1}$, respectively.

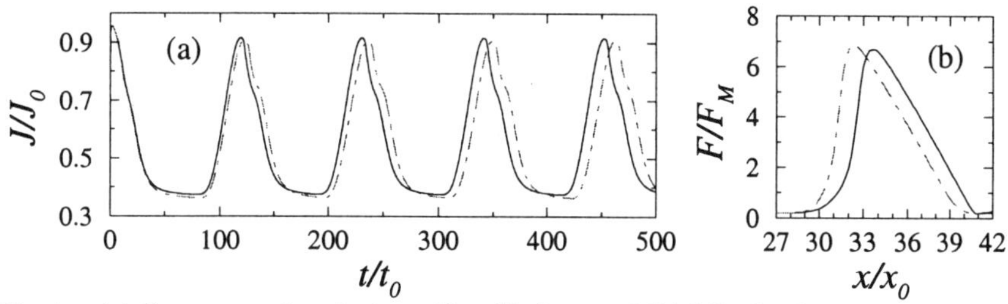

Fig. 1. (a) Current vs. time during self-oscillations, and (b) fully developed dipole wave. Solid line: QDDE, dashed line: GDDE.

3. Wigner description of a two-miniband superlattice

We shall consider a 2×2 Hamiltonian $\mathbf{H}(x, -i\partial/\partial x)$, in which

$$\mathbf{H}(x,k) = [h_0(k) - eW(x)]\boldsymbol{\sigma}_0 + \vec{h}(k)\cdot\vec{\boldsymbol{\sigma}}] \tag{3.1}$$
$$\equiv \begin{pmatrix} (\alpha+\gamma)(1-\cos kl) - eW(x) + g & -i\beta\sin kl \\ i\beta\sin kl & (\alpha-\gamma)(1-\cos kl) - eW(x) - g \end{pmatrix}.$$

Then $h_0 = \alpha(1-\cos kl)$, $h_1 = 0$, $h_2 = \beta\sin kl$, $h_3 = \gamma(1-\cos kl) + g$, and

$$\boldsymbol{\sigma}_0 = \begin{pmatrix} 1 & 0 \\ 0 & 1 \end{pmatrix}, \boldsymbol{\sigma}_1 = \begin{pmatrix} 0 & 1 \\ 1 & 0 \end{pmatrix}, \boldsymbol{\sigma}_2 = \begin{pmatrix} 0 & -i \\ i & 0 \end{pmatrix}, \boldsymbol{\sigma}_3 = \begin{pmatrix} 1 & 0 \\ 0 & -1 \end{pmatrix}. \tag{3.2}$$

The Hamiltonian (3.1) corresponds to the simplest 2×2 Kane model in which the quadratic and linear terms $(kl)^2/2$ and kl are replaced by $(1-\cos kl)$ and $\sin kl$, respectively. For a SL with two minibands, $2g$ is the miniband gap and $\alpha = (\Delta_1+\Delta_2)/4$ and $\gamma = (\Delta_1-\Delta_2)/4$, provided Δ_1 and Δ_2 are the miniband widths. In the case of a lateral SL, $g = \gamma = 0$, and $h_2\boldsymbol{\sigma}_2$ corresponds to the precession term in the Rashba spin-orbit interaction.[7] The other term, the intersubband coupling, depends on the momentum in the y direction and we have not included it here. Small modifications of (3.1) represent a single miniband SL with dilute magnetic impurities in the presence of a magnetic field B: $g = \gamma = h_2 = 0$, and $h_1 = \beta(B)$.[6] As in the case of a single miniband SL, $W(x)$ is the electric potential.

The energy minibands $\mathcal{E}^{\pm}(k)$ are the eigenvalues of the free Hamiltonian $\mathbf{H}_0(k) = h_0(k)\boldsymbol{\sigma}_0 + \vec{h}(k)\cdot\vec{\boldsymbol{\sigma}}$ and are given by

$$\mathcal{E}^{\pm}(k) = h_0(k) \pm |\vec{h}(k)|. \tag{3.3}$$

The corresponding spectral projections are $\mathbf{P}^{\pm}(k) = (\boldsymbol{\sigma}_0 \pm \vec{\nu}(k)\cdot\vec{\boldsymbol{\sigma}})/2$, with $\vec{\nu} = \vec{h}/|\vec{h}(k)|$, so that we can write $\mathbf{H}_0(k) = \mathcal{E}^{+}(k)\mathbf{P}^{+}(k) + \mathcal{E}^{-}(k)\mathbf{P}^{-}(k)$.

We shall now write the WPBGK equations for the Wigner matrix written in terms of the Pauli matrices $\boldsymbol{\sigma}_i$:

$$\mathbf{f}(x,k,t) = \sum_{i=1}^{3} f^i(x,k,t)\boldsymbol{\sigma}_i = f^0(x,k,t)\boldsymbol{\sigma}_0 + \vec{f}(x,k,t)\cdot\vec{\boldsymbol{\sigma}}. \tag{3.4}$$

The Wigner components are real and can be related to the coefficients of the Hermitian Wigner matrix by $f_{11} = f^0 + f^3$, $f_{12} = f^1 - if^2$, $f_{21} = f^1 + if^2$, $f_{22} = f^0 - f^3$. Hereinafter we shall use the equivalent notations

$$f = \begin{pmatrix} f^0 \\ \vec{f} \end{pmatrix} = \begin{pmatrix} f^0 \\ f^1 \\ f^2 \\ f^3 \end{pmatrix}. \tag{3.5}$$

The populations of the minibands with energies $\mathcal{E}^{\pm}$ are the moments:

$$n^{\pm}(x,t) = \frac{l}{2\pi}\int_{-\pi/l}^{\pi/l} \left[f^0(x,k,t) \pm \vec{\nu}\cdot\vec{f}(x,k,t)\right] dk, \tag{3.6}$$

and the total electron density is $n^+ + n^-$.

We shall restrict ourselves to the Rashba case, $g = \gamma = h_3 = 0$, from now on. Then $\vec{\nu} = (0,1,0)$ and $n^{\pm}$ are the densities of electrons having spin $\pm$. After some algebra, we can obtain the following WPBGK equations for the Wigner components

$$\frac{\partial f^0}{\partial t} + \frac{\alpha}{\hbar}\sin kl\,\Delta^- f^0 + \frac{\beta\cos kl}{\hbar}\,\Delta^- f^2 - \Theta f^0 = Q^0[f], \tag{3.7}$$

$$\frac{\partial \vec{f}}{\partial t} + \frac{\alpha\sin kl}{\hbar}\Delta^- \vec{f} + \frac{\beta}{\hbar}[\vec{\nu}\Delta^- f^0 \cos kl + \Delta^+(\vec{\nu}\times\vec{f})\sin kl]$$
$$-\Theta\vec{f} = \vec{Q}[f], \tag{3.8}$$

$$\varepsilon\,\frac{\partial^2 W}{\partial x^2} = \frac{e}{l}\,(n^+ + n^- - N_D), \tag{3.9}$$

$$\Theta f^i(x,k,t) = \sum_{j=-\infty}^{\infty} \frac{ejl}{i\hbar}\langle F(x,t)\rangle_j e^{ijkl} f^i_j(x,t). \tag{3.10}$$

Our collision model contains two terms: a BGK term which tries to send $f^0 \pm f^2$ to its local equilibrium (approximated by Boltzmann statistics at $T = 300\text{K}$) and a scattering term which tries to equalize n^+ and n^-:[6]

$$Q^0[f] = -\frac{f^0 - \Omega^0}{\tau}, \quad \vec{Q}[f] = -\frac{\vec{f} - \vec{\Omega}}{\tau} - \frac{\vec{f}}{\tau_{\text{sc}}}, \tag{3.11}$$

$$\Omega^0 = \frac{\phi^+ + \phi^-}{2}, \quad \vec{\Omega} = \frac{\phi^+ - \phi^-}{2}\vec{\nu}, \tag{3.12}$$

$$\phi^\pm(k; n^\pm) = \frac{m^* k_B T}{\pi^2 \hbar^2} \exp\left(\frac{\mu^\pm - \mathcal{E}^\pm(k)}{k_B T}\right), \tag{3.13}$$

$$\frac{l}{2\pi}\int_{-\pi/l}^{\pi/l} \phi^\pm(k; n^\pm)\, dk = n^\pm. \tag{3.14}$$

In (3.13), $\mu^\pm = \mu^\pm(n^\pm)$ solve (3.14). Our collision model satisfies charge continuity. In fact, (3.7) to (3.9) yield:

$$\begin{aligned} &\frac{\partial n^\pm}{\partial t} + \frac{l\Delta^-}{2\pi\hbar}\int_{-\pi/l}^{\pi/l} [\alpha \sin kl\,(f^0 \pm f^2) + \beta \cos kl\,(f^2 \pm f^0)]\, dk \\ &= \frac{l\Delta^-}{2\pi}\int_{-\pi/l}^{\pi/l} (Q^0[f] \pm Q^2[f])\, dk = \mp\frac{n^+ - n^-}{\tau_{\text{sc}}}, \end{aligned} \tag{3.15}$$

where we have employed $\int \Theta f^{0,2} dk = 0$. Then we obtain:

$$\frac{\partial}{\partial t}(n^+ + n^-) + \Delta^-\left[\frac{l}{\pi\hbar}\int_{-\pi/l}^{\pi/l} (\alpha\, \sin kl\, f^0 + \beta\, \cos kl\, f^2)\, dk\right] = 0. \tag{3.16}$$

Since $\Delta^- u(x) = l\,\partial\langle u(x)\rangle_1/\partial x$, (3.16) provides charge continuity. From (3.9) and (3.16), we get Ampère's law ($J(t)$ is the total current density):

$$\varepsilon\frac{\partial F}{\partial t} + \left\langle \frac{el}{\pi\hbar}\int_{-\pi/l}^{\pi/l} (\alpha\, \sin kl\, f^0 + \beta\, \cos kl\, f^2)\, dk \right\rangle_1 = J(t). \tag{3.17}$$

4. Quantum drift-diffusion equations

In the simpler case of a lateral SL with the precession term of Rashba spin-orbit interaction (but no intersubband coupling), we can obtain explicit rate equations for $n^\pm$ by means of the CE method. The general case (3.1) will be treated elsewhere. First of all, we should decide the order of magnitude of the terms in the WPBGK equations (3.7) and (3.8) in the hyperbolic limit. Recall that in this limit, the collision frequency $1/\tau$ and the Bloch frequency $eF_M l/\hbar$ are of the same order, about 10 THz for the SL of Section 2. The scattering time τ_{sc} is much longer than the collision time τ, and we shall consider $\tau/\tau_{\rm sc} = O(\lambda) \ll 1$. From (3.7) and (3.8), we can write the scaled WPBGK equations as follows:

$$\mathbb{L}f \equiv f - \tau\Theta f = \Omega - \lambda\left(\tau\frac{\partial f}{\partial t} + \Lambda f\right), \tag{4.1}$$

$$\Lambda f = \frac{\tau}{\tau_{\rm sc}}\begin{pmatrix} 0 \\ \vec{f} \end{pmatrix} + \frac{\alpha\tau}{\hbar}\sin kl\,\Delta^- f + \frac{\beta\tau}{\hbar}\cos kl\,\Delta^- \begin{pmatrix} f^2 \\ 0 \\ f^0 \\ 0 \end{pmatrix} + \frac{\beta\tau}{\hbar}\sin kl\,\Delta^+ \begin{pmatrix} 0 \\ f^3 \\ 0 \\ -f^1 \end{pmatrix}. \tag{4.2}$$

To derive the reduced balance equations, we use the following CE ansatz:

$$f(x,k,t;\epsilon) = f^{(0)}(k;n^+,n^-,F) + \sum_{m=1}^{\infty} f^{(m)}(k;n^+,n^-,F)\,\lambda^m, \tag{4.3}$$

$$\varepsilon\frac{\partial F}{\partial t} + \sum_{m=0}^{\infty} J_m(n^+,n^-,F)\,\lambda^m = J(t), \tag{4.4}$$

$$\frac{\partial n^\pm}{\partial t} = \sum_{m=0}^{\infty} A_m^\pm(n^+,n^-,F)\,\lambda^m. \tag{4.5}$$

$A_m^\pm$ and J_m are related through the Poisson equation (3.9), so that

$$A_m^+ + A_m^- = -\frac{l}{e}\frac{\partial J_m}{\partial x}. \tag{4.6}$$

Inserting (4.3) to (4.5) into (4.2), we get

$$\mathbb{L}f^{(0)} = \Omega, \tag{4.7}$$

$$\mathbb{L}f^{(1)} = -\tau\left.\frac{\partial f^{(0)}}{\partial t}\right|_0 - \Lambda f^{(0)}, \tag{4.8}$$

$$\mathbb{L}f^{(2)} = -\tau\left.\frac{\partial f^{(1)}}{\partial t}\right|_0 - \Lambda f^{(1)} - \tau\left.\frac{\partial f^{(0)}}{\partial t}\right|_1, \tag{4.9}$$

and so on. The subscripts 0 and 1 in the right hand side of these equations mean that we replace $\varepsilon\,\partial F/\partial t|_m = J\delta_{0m} - J_m$, $\partial n^\pm/\partial t|_m = A_m^\pm$. Moreover, inserting (4.3) into (3.6) yields the following compatibility and solvability conditions:

$$f_0^{(m)\,0} = f_0^{(m)\,2} = 0 \Longrightarrow (\mathbb{L}f^{(m)\,0})_0 = (\mathbb{L}f^{(m)\,2})_0 = 0, \quad m \geq 1. \tag{4.10}$$

To solve (4.7) for $f^{(0)} \equiv \varphi$, we first note that

$$-\tau\,\Theta\varphi = \sum_{j=-\infty}^{\infty} i\vartheta_j \varphi_j e^{ijkl}, \quad \vartheta_j \equiv \frac{\tau e j l}{\hbar}\,\langle F\rangle_j. \tag{4.11}$$

Then (4.7) and (3.12) yield

$$\varphi_j^0 = \frac{\phi_j^+ + \phi_j^-}{2}\,\frac{1 - i\vartheta_j}{1+\vartheta_j^2}, \quad \varphi_j^1 = \varphi_j^3 = 0, \quad \varphi_j^2 = \frac{\phi_j^+ - \phi_j^-}{2}\,\frac{1 - i\vartheta_j}{1+\vartheta_j^2}, \tag{4.12}$$

where we have used that the Fourier coefficients $\phi_j^\pm$ are real because $\phi^\pm$ are even functions of k. Similarly, the solution of (4.8) is $f^{(1)} \equiv \psi$ with

$$\psi_j^m = r_j^m\,\frac{1 - i\vartheta_j}{1+\vartheta_j^2} \quad (m = 0, 2), \quad \psi_j^1 = \psi_j^3 = 0. \tag{4.13}$$

Here r is the right hand side of (4.8). The balance equations can be found by calculating $A_m^\pm$ for $m = 0, 1$ from the solvability conditions for (4.8) and (4.9), or by simply inserting the solutions (4.12) and (4.13) in (3.15) and (3.17). In both cases, the result is:

$$\frac{\partial n^\pm}{\partial t} + \Delta^- D_\pm(n^+, n^-, F) = \mp R(n^+, n^-, F), \tag{4.14}$$

$$\varepsilon\,\frac{\partial F}{\partial t} + e\,\langle D_+ + D_-\rangle_1 = J, \tag{4.15}$$

$$D_\pm = -\frac{\alpha}{\hbar}\,\mathrm{Im}(\varphi_1^0 \pm \varphi_1^2 + \psi_1^0 \pm \psi_1^2) \pm\frac{\beta}{\hbar}\,\mathrm{Re}(\varphi_1^0 \pm \varphi_1^2 + \psi_1^0 \pm \psi_1^2), \quad R = \frac{n^+ - n^-}{2\tau_{sc}}. \tag{4.16}$$

A straightforward calculation of (4.16) yields

$$\begin{aligned} D_\pm = {} & \frac{(\alpha\vartheta_1 \pm \beta)\phi_1^\pm}{\hbar\,(1+\vartheta_1^2)} \mp \frac{\tau\,(\phi_1^+ - \phi_1^-)\,[2\alpha\vartheta_1 \pm \beta(1-\vartheta_1^2)]}{2\hbar\tau_{\rm sc}(1+\vartheta_1^2)^2} \\ & + \frac{[2\alpha\vartheta_1 \pm \beta(1-\vartheta_1^2)]\,\tau}{\hbar\,(1+\vartheta_1^2)^2}\,\frac{\partial\phi_1^\pm}{\partial n^\pm}\left[\Delta^-\left(\frac{\alpha\vartheta_1 \pm \beta}{\hbar\,(1+\vartheta_1^2)}\phi_1^\pm\right) \pm \frac{n^+ - n^-}{2\tau_{sc}}\right] \\ & + \frac{\alpha\,(3\vartheta_1^2 - 1) \pm \beta\vartheta_1(3-\vartheta_1^2)}{\hbar(1+\vartheta_1^2)^3}\,\frac{l\tau^2\phi_1^\pm}{\hbar\varepsilon}\left(\frac{J}{e} - \left\langle\left\langle\frac{\alpha\,(\phi_1^+ + \phi_1^-)\vartheta_1}{\hbar(1+\vartheta_1^2)}\right\rangle_1\right\rangle_1\right. \\ & \left. - \left\langle\left\langle\frac{\beta\,(\phi_1^+ - \phi_1^-)}{\hbar(1+\vartheta_1^2)}\right\rangle_1\right\rangle_1\right) - \frac{(\alpha^2+\beta^2)\tau}{2\hbar^2(1+\vartheta_1^2)}\,\Delta^- n^\pm \\ & + \frac{\tau}{2\hbar^2(1+\vartheta_1^2)}\left[(\alpha^2 - \beta^2 \mp 2\alpha\beta\vartheta_1)\,\Delta^-\left(\frac{\phi_2^\pm}{1+\vartheta_2^2}\right)\right. \\ & \left. + [(\beta^2-\alpha^2)\vartheta_1 \mp 2\alpha\beta]\,\Delta^-\left(\frac{\vartheta_2\phi_2^\pm}{1+\vartheta_2^2}\right)\right]. \end{aligned} \tag{4.17}$$

The following values of the parameters are typical of a GaAs/AlGaAs SL: $\alpha = 10$ meV, $\beta = 2.1$ meV, $l = 5$ nm, $T = 300$ K, $\tau = 10^{-13}$ s, $\tau_{sc} = 10^{-12}$ s, $N_D = 10^{10}$ cm^{-2}. Figure 2 shows the electron velocity, $v = Jl/(eN_D)$ (measured in units of $v_M = \alpha l/\hbar = 7.6 \times 10^6$ cm/s), as a function of field (measured in units of $F_M = 13.2$ kV/cm at which the electron velocity reaches its maximum) for a homogeneous solution of Eq. (4.15) with constant F and $n^+ = n^- = N_D/2$. We observe that there is a local maximum followed by a region of negative slope (negative differential velocity), which suggests a Gunn-type instability as in Section 2: self-sustained oscillations of the current through the SL due to motion of charge dipole waves under sufficiently high dc voltage bias.[1,2,4]

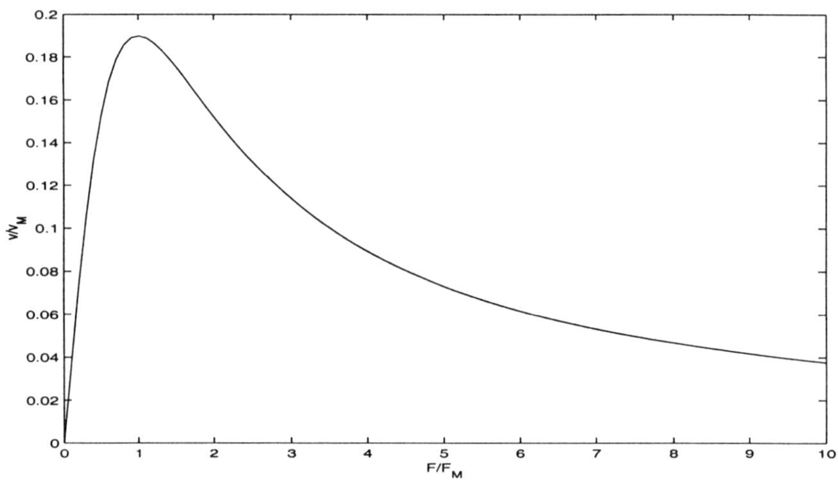

Fig. 2. Electron velocity vs. field for a homogeneous solution of (4.14) - (4.17).

5. Conclusions

For strongly coupled SLs having only one populated miniband, we have written a Wigner-Poisson-BGK system of equations and derived a quantum drift-diffusion equation for the field by using the Chapman-Enskog perturbation method. With appropriate voltage bias, a numerical solution of this equation yields self-sustained oscillations of the current due to recycling and motion of charge dipole domains. For SLs having two populated minibands coupled through a Rashba spin-orbit interaction, we have introduced a periodic version of the Kane Hamiltonian and derived the corresponding WPBGK system of equations. By using the CE method, we have derived quantum drift-diffusion equations for the miniband populations which contain generation-recombination terms. The spatially homogeneous solution of these equations provides an electron velocity which has a region of negative slope as a function of field. This hints to the possible existence of oscillatory instabilities and self-oscillations of the current due to motion of charge dipole waves under sufficiently high dc voltage bias.

Acknowledgements

This research was supported by the Spanish MECD grant MAT2005-05730-C02-01 and by the INDAM-GNFM project "Mathematical Models for Microelectronics".

References

1. L. L. Bonilla and H. T. Grahn, *Rep. Prog. Phys.* **68**, 577 (2005).
2. A. Wacker, *Phys. Rep.* **357**, 1 (2002).
3. L.L. Bonilla, R. Escobedo and A. Perales, *Phys. Rev. B* **68**, 241304(R) (2003).
4. L.L. Bonilla and R. Escobedo, *Math. Mod. Meth. Appl. Sci.* **15**(8), 1253 (2005).
5. C. Cercignani, I. M. Gamba and C. D. Levermore, *SIAM J. Appl. Math.* **61**, 1932 (2001).
6. D. Sánchez, A.H. MacDonald and G. Platero, *Phys. Rev. B* **65**, 035301 (2002).
7. P. Kleinert, V. V. Bryksin and O. Bleibaum, *Phys. Rev. B* **72**, 195311 (2005).
8. R. Escobedo and L. L. Bonilla, *J. Math. Chem.* **40**, 3 (2006).

BIOMASS GROWTH IN UNSATURATED POROUS MEDIA: HYDRAULIC PROPERTIES CHANGES

I. BORSI, A. FARINA, A. FASANO, M. PRIMICERIO

Dipartimento di Matematica - Università di Firenze
Viale G. Morgagni, 67/A
50134 Firenze, Italy
E-mail: borsi@math.unifi.it

We present a model to describe the biomass growth process taking place in an unsaturated porous medium during a bioremediation process. We focus on the so-called column experiment. At the initial time biomass and polluted water is inoculated in the column. The subsequent changes of hydraulic properties are analyzed. We also show some preliminary simulations.

Keywords: Microbial growth; Porous media; Bioremediation.

1. Introduction

The effect of the microbial growth on hydraulic properties of porous media is a topic studied in the framework of many applications, e.g. oil recovery, wastewater treatment, bioremediation, etc. (see [1]).
Studies on flow through porous media in presence of biomass growth are presented in the papers by Rockhold et al. [2–4]. As stated there, additional work is needed for modelling unsaturated condition.
The objective of this study is thus to analyze the flow through a contaminated unsaturated porous medium in presence of biomass growth processes which induce changes in the hydraulic properties of the medium itself.
In this paper we focus only on *anaerobic* processes, namely the model we develop does not account for the O_2 consumption and diffusion.

2. Problem description and physical assumptions

We consider a vertical column (whose high is $L \sim 1\,m$) of an unsaturated contaminated soil which represents a "laboratory scale" of a real vadose zone (the so-called "column experiment", see Fig. 1). The physical model is developed considering a 1-D approximation, so that x denotes the vertical

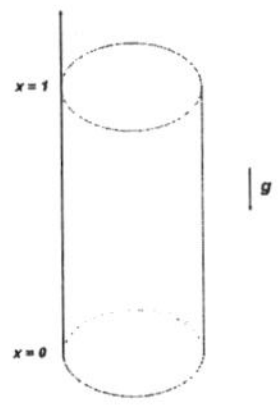

Fig. 1. A schematic of the column experiment.

coordinate of the column, pointing upwards.
At the initial time ($t = 0$) the saturation degree of the medium corresponds to the steady state. Then we inoculate biomass and (possibly) polluted water through the top surface. Our goal is to model the evolution of the biomass, pollutant concentration and hydraulic properties of the soil as well.

Hereafter we list the most significant physical assumptions (see also [5]):

A.1 The soil is a homogeneous, unsaturated, rigid porous medium.

A.2 The liquid phase which shares the empty space with air is composed by water (main component). We shall neglect the liquid density variation.

A.3 The pollutant is dissolved in water and adsorbed onto the soil grains. Moreover, the dissolved pollutant (below a certain concentration) acts as nutrient for the bacteria (*bio-reduction*), but above a certain threshold may become a toxic agent (see [6]).

A.4 The biomass is distributed either in water as suspension ("*free biomass*") or attached on the soil grains ("*attached biomass*"). In particular, there is no cluster formation in the free biomass. The mass of the single bacterium is known and denoted by $\mathfrak{m}_b$.

A.5 The attached biomass forms *porous clusters* so that the liquid can diffuse through them. The clusters porosity is a known constant denoted by ε_b. The pores are saturated at all times. Moreover, the concentration of the pollutant in the "entrapped water" equals the one of the "free water" (see [3], for instance). The number of bacteria that forms the unit mass of the attached biomass is a known constant denoted by $N^\star$, $[N^\star] = Kg^{-1}$. Of course, $N^\star$, ε_b and $\mathfrak{m}_b$ depend on the type of bacteria which are present inside the column.

A.6 We consider the attachment of free biomass on the clusters, but we neglect the inverse process (i.e. we neglect detachment). Indeed the experiments show that detachment is mainly due to the mechanical action caused by the "fast" water flux, [10].

A.7 The concentration of pollutant and bacteria in the liquid phase is low (few ppm).

Free biomass and attached biomass are responsible for different effects changing hydraulic properties. More precisely:

E.1 The free biomass causes essentially viscosity and surface tension variations.

E.2 The attached biomass growth causes medium porosity variations and affects the contact angle.

E.3 The above variations induce, in turn, changes in the permeability and in the relative saturation of the medium.

3. Notations and basic equations

We introduce the following quantities:

- ε_0, $[\varepsilon_0] = [-]$, initial porosity of the column (known parameter).
- ε_b, $[\varepsilon_b] = [-]$, clusters porosity (considered a known constant).
- ϕ^f, volume fraction occupied by the liquid and the gaseous (air) phase.
- ϕ^c, volume fraction occupied by the clusters.
- σ, $[\sigma] = [-]$ liquid phase saturation.
- $\theta_{lm} = \dfrac{\text{volume of "mobile" liquid}}{\text{porous medium volume}}$, $[\theta_{lm}] = [-]$. In particular, $\theta_{lm} = \sigma\phi^f$.
- $\theta_{lb} = \varepsilon_b\phi^c = \dfrac{\text{volume of "clusters – stored" liquid}}{\text{porous medium volume}}$, $[\theta_{lb}] = [-]$.

We thus have that

$$\phi^f + \phi^c = \varepsilon_0.$$

So, the volume fraction occupied by the liquid (accounting for mobile and stored) is

$$\theta_{lm} + \theta_{lb} = \sigma\phi^f + \varepsilon_b\phi^c = (\sigma - \varepsilon_b)\phi^f + \varepsilon_b\varepsilon_0. \tag{1}$$

The dependent variables which have to be determined are:

- $\sigma(x,t)$, the liquid phase saturation.
- $\phi^c\,(x,t)$, or alternatively $\phi^f = \varepsilon_0 - \phi^c$.
- $N_l(x,t) = \dfrac{\text{number of free bacteria in the liquid phase}}{\text{unit mass of free liquid}}$, $[N_l] = Kg^{-1}$.
- $w_A(x,t) = \dfrac{\text{mass of pollutant dissolved in the liquid phase}}{\text{unit mass of free liquid}}$.

- $w_s(x,t) = \dfrac{\text{mass of pollutant adsorbed onto soil grains}}{\text{unit mass of solid matrix}}$.

Darcy's law and Richards equation

We define the water pressure P and introduce the *capillary pressure* P_c and the *pressure head* ψ, setting

$$P_c = P_{\text{air}} - P = -P, \qquad \psi = \frac{-P}{\rho g} = \frac{P_c}{\rho g},$$

since, as usual, P_{air} has been rescaled to 0.

The well-known Darcy's law describes the *specific discharge* q,

$$q = -\frac{\rho g \mathcal{K}_{\text{sat}}}{\mu} k_{\text{rel}} \left(\frac{\partial \psi}{\partial x} + 1 \right),$$

where ρ water density, g gravity acceleration and

- $\mathcal{K}_{\text{sat}} = \mathcal{K}_{\text{sat}}(\phi^f)$, *saturated permeability*, $[\mathcal{K}_{\text{sat}}] = m^2$. For instance, we mention the Kozeny-Carman formula [11].
- $k_{\text{rel}} = k_{\text{rel}}(\psi)$, $[k_{\text{rel}}] = [-]$, *relative permeability* (see [14]).
- μ is the liquid phase viscosity. We assume the following law (based on experimental observations [2])

$$\mu = \tilde{\mu}(N_l) = \mu_0 \left(1 + b_3 N_l\right), \tag{2}$$

  with b_3 empirical parameter and μ_0 water viscosity (i.e. the liquid viscosity in absence of biomass).

Next, the Richards' equation [14] reads as

$$\frac{\partial}{\partial t} \left[\theta_{lm} + \theta_{lb}\right] = -\frac{\partial q}{\partial x}. \tag{3}$$

Now, introducing the *saturation curve* $\sigma = \tilde{\sigma}(\psi)$, we have $\theta_{lm} = \phi^f \tilde{\sigma}(\psi)$ so that, exploting (1), the mass conservation (3) rewrites as

$$\frac{\partial}{\partial t} \left[(\sigma - \varepsilon_b)\phi^f\right] = -\frac{\partial q}{\partial x}. \tag{4}$$

Evolution equation for the attached biomass phase

According to the literature (see for instance, [7] § 2.3 and [12]) we set

$$\frac{\partial \phi^c}{\partial t} = \underbrace{(c_1 N^\star) \left[\varepsilon_0 f \left(w_A \varepsilon_b \phi^c\right) - \phi^c\right] \phi^c}_{\text{biomass "bulk growth"}} + \underbrace{\lambda \mathfrak{H} \left(w_s - w_A\right) N_l \theta_{lm}}_{\text{attachment}}, \tag{5}$$

where:

- $[c_1] = Kg\, s^{-1}$.
- $w_A \varepsilon_b \phi^c$ is the amount of nutrient available for the attached biomass.
- $\varepsilon_0 f\,(w_A \varepsilon_b \phi^c)$, is a modified form of the *carrying capacity* (see [13] § 1.2). Actually, ε_0 is the maximum volume fraction allowed for the attached biomass and it is "modulated" by the function f, $0 \leq f \leq 1$, which accounts of both amount of nutrient and toxic effects (see assumption A.3).
- $\mathfrak{H}(\cdot)$ is the Heaviside function. We notice that the attachment term could be multiplied also by a function of ϕ^c, i.e. an "effective surface" term modeling the so-called *collector* and *collision* (or *sticking*) efficiencies.

Evolution equation for the free biomass

The free biomass is a component of the liquid phase. Therefore, following [8,9], the evolution equation for N_l is

$$\frac{\partial}{\partial t}\left(N_l \theta_{lm}\right) = \underbrace{-\frac{\partial}{\partial x}\left[q\, N_l\right] + \frac{\partial}{\partial x}\left[a_L\, |q|\, \frac{\partial N_l}{\partial x}\right]}_{\text{advection/dispersion}}$$
$$+ \underbrace{c_1 \left[N_{\max} f(w_A) - N_l\right] N_l \theta_{lm}}_{\text{free biomass growth}} \tag{6}$$
$$- \underbrace{\lambda \mathfrak{H}\left(w_s - w_A\right) N_l \theta_{lm}}_{\text{attachment}} .$$

The first term in (6) is the divergence of the advective flux $\mathbf{J}_{\text{adv}} = -qN_l$. The second term represents the dispersive flux $\mathbf{J}_{\text{disp}}$, which in 1–D setting has the following form

$$\mathbf{J}_{\text{disp}} = -\sigma \phi^f D, \quad D = \frac{a_L\,|q|}{\sigma \phi^f},$$

with a_L *longitudinal dispersion coefficient* (see [14]).
The "bulk growth" of the free biomass is modeled as in the case of the attached biomass, i.e. by means of a logistic–type dynamics. Here $N_{\max}$ is the equilibrium value. We set

$$N_{\max} = \mathfrak{n} N^\star, \qquad 0 < \mathfrak{n} < 1.$$

Equation for the adsorbed pollutant.

We describe the dynamics of w_s by the following equation

$$\frac{\partial w_S}{\partial t} = \underbrace{h_A\, w_A \varepsilon_b \phi^c (w^\star - w_S)}_{\text{adsorption term}} \qquad (7)$$

$$- \underbrace{h_{DB}\, N^\star \phi^c\, w_S}_{\text{bio-reduction term}},$$

stipulating, essentially, that only two effects are important: adsorption (desorption) and bio-reduction. In particular,

- $w_A \varepsilon_b \phi^c = \varepsilon_b \theta_{lb}$ is the amount of pollutant dissolved inside the biomass clusters.
- $w^\star$ is maximum concentration of pollutant (known parameter) which can be adsorbed by the soil.
- h_A, $[h_A] = s^{-1}$, is the *adsorption/desorption rate per unit concentration.*
- h_{DB}, $[h_{DB}] = Kg\, s^{-1}$, *bio–reduction rate per unit mass.*

Equation for w_A.

The total amount of pollutant (per unit volume of porous medium) dissolved in the free and "entrapped" water is $w_A(\theta_{lm}+\theta_{lb})$, i.e. $w_A(\theta_{lm}+\varepsilon_b\phi^c)$. According to [9], we write for it the following equation

$$\frac{\partial}{\partial t}\left[w_A(\theta_{lm} + \varepsilon_b \phi^b)\right] = \underbrace{-\frac{\partial}{\partial x}\left[q\, w_A\right] + \frac{\partial}{\partial x}\left[a_L\, |q|\, \frac{\partial w_A}{\partial x}\right]}_{\text{advection/dispersion term}}$$

$$- \underbrace{h_A w_A \varepsilon_b \phi^c (w^\star - w_S)}_{\text{adsorption term}} \qquad (8)$$

$$\underbrace{- h_{DB}\left[\theta_{lm} N_l + \phi^c N^\star\right] w_A\left(\theta_{lm} + \varepsilon_b \phi^c\right)}_{\text{bio-reduction term}}.$$

The bio–reduction term in (8) depends on the amount of bacteria which are present either in the liquid and in the clusters.

Summarizing, we have to solve the system of the governing equations (4), (5), (6), (7) and (8), which has to be endowed with initial and boundary conditions.

4. A simplified approach: biofilm and fluid media scaling

We now illustrate the basic idea of an approach to simplify the problem. The key point is to consider a porous medium constituted by a network of capillary tubes distributed uniformly in space. Next, we assume that the attached biomass phase forms a uniform layer (*biofilm*) completely coating the internal surfaces of the capillary tubes.

Focusing now on a single capillary tube, we compare two scenarios: capillary tube partially filled with "pure" water and capillary tube whose walls are coated by the biofilm and partially filled by a liquid whose components are water, bacteria and pollutant.

Denoting by p_c and $p_{c,\text{bio}}$ the capillary pressures which refer to the above scenarios, we may write the following Laplace formulas

$$p_c = \frac{2\gamma\, cos\alpha}{R}, \qquad p_{c,\text{bio}} = \frac{2\gamma_{\text{bio}}\, cos(\alpha_{\text{bio}})}{R_{\text{bio}}},$$

where R and R_{bio} denote the capillary radii, γ and γ_{bio} are the surface tensions and α, α_{bio} are the contact angles. Therefore,

$$\frac{p_{c,\text{bio}}}{p_c} = \frac{\gamma_{\text{bio}}}{\gamma}\frac{\cos\alpha_{\text{bio}}}{\cos\alpha}\frac{R}{R_{\text{bio}}}.$$

We now assume that the above formula holds true also for the averaged quantities, i.e.

$$\frac{P_{c,\text{bio}}}{P_c} = \frac{\gamma_{\text{bio}}}{\gamma}\frac{\cos\alpha_{\text{bio}}}{\cos\alpha}\frac{\langle R\rangle}{\langle R_{\text{bio}}\rangle}, \tag{9}$$

where $P_{c,\text{bio}} = \langle p_{c,\text{bio}}\rangle$, $P_c = \langle p_c\rangle$, since $\langle\cdot\rangle$ denotes the R.E.V. average. Now, selecting appropriate constitutive equations[a] for

$$\frac{\cos\alpha_{\text{bio}}}{\cos\alpha}, \quad \frac{\gamma_{\text{bio}}}{\gamma} \quad \text{and} \quad \frac{\langle R\rangle}{\langle R_{\text{bio}}\rangle},$$

we can define the parameter $\beta = \beta\left(\phi^f, N_l\right)$, (called *scaling factor* for the capillary pressure), such that

$$\frac{P_{c,\text{bio}}}{P_c} = \beta\left(\phi^f, N_l\right), \quad \Rightarrow \quad \psi = \beta\left(\phi^f, N_l\right)\psi_0, \tag{10}$$

where $\psi_0 = -\dfrac{P_c}{\rho g}$ is the pressure head in absence of biomass. Therefore,

$$\theta_{lm} = \tilde{\sigma}\left(\beta\left(\phi^f, N_l\right)\psi_0\right)\left(\varepsilon_0 - \phi^c\right).$$

[a] Such equations strongly depend on the intrisic geometry of the medium, see [5].

Such an approach (often called *fluid media scaling*, see [15]), presents an evident advantage: once the flow problem is specified, we determine ψ_0 using the "classical" Richards' equation, i.e. equation (4) where the term due to porosity changes is absent. As second step we evaluate ψ exploiting (10) and q through (2). Therefore, the mathematical problem is strongly simplified and its numerical solving turns out easier. Of course, the fluid media scaling suffers from an evident drawback: ψ and q do not fulfill the Richards' equation, i.e. mass conservation. Hence, such a property needs to be tested a posteriori.

5. Numerical simulations

In this section we present few numerical simulations we worked out considering the simplified model. A deeper analysis of the results can be found in the forthcoming paper.[5] Our main goals are: (i) to show that the found solution satisfies (within a suitable tolerance) the Richards' equation; (ii) to show that the results obtained agree, at least qualitatively, with the experimental data; (iii) to put in evidence that, in certain circumstances, variation of porosity and hydraulic properties is significant.

5.1. *Problem setting*

The PDEs system was solved in a 1D domain. Concerning the soil, we used the well-known van-Genuchten and Mualem forms for the saturation and permeability curve, respectively (see [14]). Moreover, a Cozeny-Karman function for the saturated permeability $k_{\text{sat}}(\phi^f)$ has been selected (see [11]). Finally, we run the simulation considering $T_{\max} = 7\ day$ as maximum time of the process.

Initial conditions

The initial stage of the experiment is characterized by absence of attached biomass and a given pollutant concentrations. Hence we set

$$w_A(x,0) = w_A^0, \quad w_S(x,0) = w_S^0, \quad N_l(x,0) = 0, \quad \phi^c(x,0) = 0,$$

with $w_A^0 = 0.3$ and $w_S^0 = 0.7$.

Boundary conditions

Following [4] we set on the column bottom, $x = 0$,

$$\left.\frac{\partial N_l}{\partial x}\right|_{x=0} = 0, \quad \left.\frac{\partial w_A}{\partial x}\right|_{x=0} = 0.$$

On the column top, $x = 1$ water (with pollutant and bacteria) is inoculated in the medium. We stipulate that pollutant and bacteria concentration in thc inflow watcr arc known, hcncc,

$$N_l(1,t) = N_1(t), \quad w_A(1,t) = w_{A\,1}(t)\,.$$

5.2. *Simulation results*

First, we mention the computed value of the quantity

$$\max_{(x,t)\in[0,1]\times[0,t_c]} \left| \frac{\partial(\theta_{lm} + \theta_{lb})}{\partial t} + \frac{\partial q}{\partial x} \right| \sim 3.8 \times 10^{-7}$$

which definitively shows that the computed solution $\psi = \beta\psi_0$ satisfies the Richards' equation.
Moreover, in Fig. 2-5 we report the plot of the most sgnificant quantities computed during the simulation. All the values are plotted at initial, intermedium and final time step.
Further comments on the obtained results will be reported in the forthcoming work [5].

References

1. Holden P.A., Fierer N., *Microbial processes in the vadose zone*,Vadose Zone J. **4** (2005), 1-21.
2. Rockhold M.L., Yarawood R.R., Niemet M.R., Bottomley P.J., Selker J.S., *Consideration for modelling bacterial-induced changes in hydraulic properties of variably saturated porous media*, Adv. Wat. Res. **25** (2002), 477-495.
3. Rockhold M.L., Yarwood R.R., Seker J.S., *Coupled microbial and transport processes in soils*, Vadose Zone J. **3** (2004), 368-383.
4. Rockhold M.L., Yarawood R.R., Niemet M.R., Bottomley P.J., Selker J.S., *Experimental observations and numerical modeling of coupled microbial and transport processes in variably saturated sand*, Vadose Zone J. **4** (2005), 407-417.
5. Borsi I., Farina A., Fasano A., Primicerio M., *Modelling changes of hydraulic properties induced by biomass growth in unsaturated porous media*, to appear.
6. Gallo C., Hassanizadeh S.M., *Modeling NAPL dissolution and biodegradation interactions: effect of toxicity and biomass growth limitations*,Computational Methods in Water Resources, S.M. Hassanizadeh et al. (Eds.), vol. **1** (2002), 859-866, Elsevier, Amsterdam.
7. Rajagopal K.R., Tao L., Mechanics of Mixtures, World Scientific, Singapore, 1995.
8. Schäfer A. et al., *Transport of bacteria in unsaturated porous media*, J. Contam. Hydrol. **33** (1998), 149-169.

9. Hassanizadeh S.M., Derivation of basic equations of mass transport in porous media, Part 2. Generalized Darcy's and Fick's laws, Adv. Water Resources, **9**, 207–222 (1986).
10. MacDonald T.R., et al., *Effects of shear detachment on biomass growth and in situ bioremediation*, Ground Water, **37** (1999), 555–563.
11. Carman, P.C, *Fluid flow through granular beds*, Transactions of Chemical Engineering, **15**, 150-166, London (1937).
12. Kingsland, S.E., Modeling Nature, University of Chicago press, Chicago, 1985.
13. Brauer, F., Castillo–Chavez, C., Mathematical Models in Population Biology and Epidemiology, Texts in Applied Mathematics, Vol. 40, Springer, New York, 2001.
14. Bear J., Verruijt A., Modelling Groundwater Flow and Pollution, Theory and Applications of Transport in Porous Media, Bear J. (Ed.), vol. 1, Kluwer, Dordrecht, 1987.
15. Miller, E.E., Miller R.D., *Physical theory for capillary flow phenomena*, J. Appl. Phys., **27** (1956), 324–332.
16. Coirier J., Mécanique des Milieux Continus, Dunod, Paris 1997.

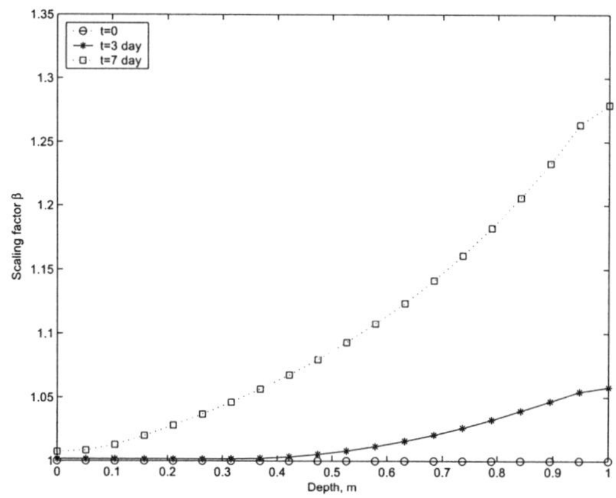

Fig. 2. Plot of the scaling factor β at different time values.

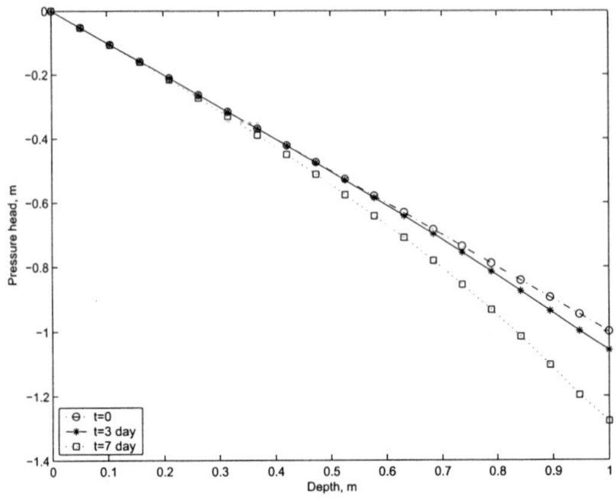

Fig. 3. Pressure head ψ at different times.

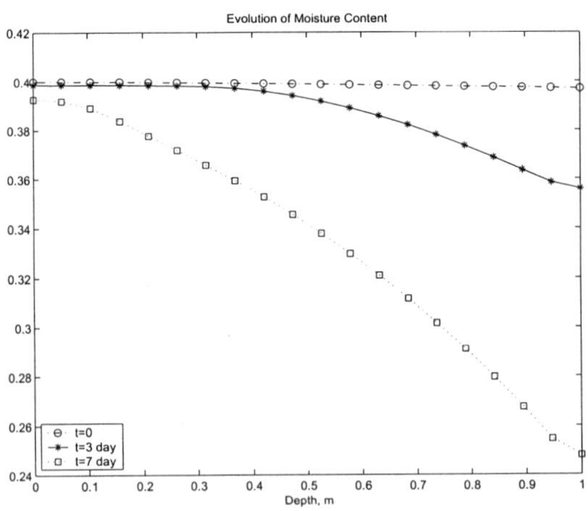

Fig. 4. Moisture content $(\theta_{lm} + \theta_{lb})$ at different times.

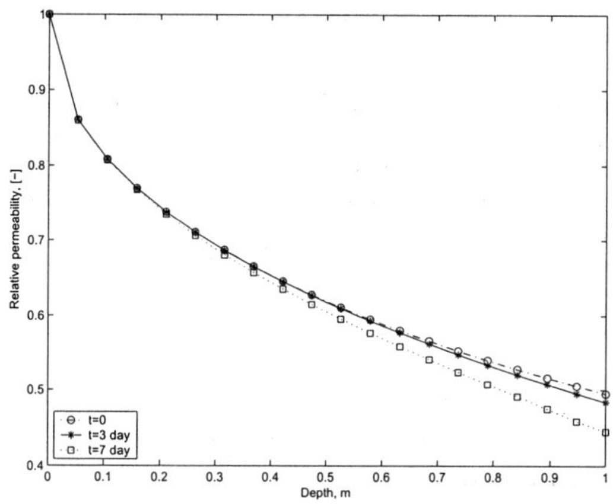

Fig. 5. Relative permeability k_{rel} at different times.

RECOVERING TRIVARIATE FUNCTIONS BY DATA ON TRACKS

M. BOZZINI* and M. ROSSINI**

Dipartimento di Matematica e Applicazioni, University of Milano Bicocca, via Cozzi 53, Milano, 20125, Italy
** E-mail: mira.bozzini@unimib.it*
*** E-mail: milvia.rossini@unimib.it*

In this paper we present a technique for recovering functions of three variables when scattered data on tracks are given. The aim is to obtain a good quality of reproduction with a contained computational cost. In order to do this we exploit the data structure which allows to solve lower dimension problems.

Keywords: Scattered data, trivariate functions, track data.

1. Introduction

The numerical approximation of scattered data is a very active area of research. In the literature we find several papers relevant to the bivariate case. In the last years a special attention has been addressed to the problem of recovering 3D objects while the problem of approximating functions of three variables $f : \Omega \subset \mathbb{R}^3 \to \mathbb{R}$ hasn't been considered so often even if many real phenomena can be described by scattered data in $\mathbb{R}^4$

$$X = \{X_1, \ldots, X_N\} \subset \Omega \subset \mathbb{R}^3, \quad Y = \{f(X_1), \ldots, f(X_N)\}. \tag{1}$$

Examples include

- Measurements of temperature at various location in a furnace
- EEG (electroencephalogram): measurements from electrodes attached to a scalp
- meteorological problems
- geological problems

We assume to have accurate data and then a popular choice is to use radial basis functions (RBFs) that are particularly well-suited to scattered data interpolation and their theory is developed in any dimension d. Let us denote the space of d-variate polynomials of order not exceeding m by P_m^d. We can interpolate the data (1) by conditionally positive definite radial functions $\phi : \mathbb{R}_+ \to \mathbb{R}$ of order $m \geq 0$. This means that for all possible choices of sets $X = \{X_1, \ldots, X_N\} \subset \Omega$ of N distinct points the quadratic form induced by the $N \times N$ matrix

$$A = (\phi(\|X_j - X_k\|_2))_{1 \leq j,k \leq N}, \tag{2}$$

is positive definite on the subspace

$$V := \left\{ \alpha \in \mathbb{R}^N \ : \ \sum_{j=1}^{N} \alpha_j p(X_j) = 0 \text{ for all } p \in P_m^d \right\}.$$

Note that $m = 0$ implies $V = \mathbb{R}^d$ because of $P_m^d = \{0\}$, and then the matrix A in (2) is positive definite. The most prominent examples of conditional positive definite radial basis functions of order m on $\mathbb{R}^d$ are

$$\begin{array}{ll}
\phi(r) = (-1)^{\lceil \beta/2 \rceil} r^{\beta}, \beta > 0, \beta \notin 2\mathbb{N}_0 & m \geq \lceil \beta/2 \rceil \\
\phi(r) = (-1)^{k+1} r^{2k} \log(r), \quad k \in \mathbb{N} & m \geq k+1 \\
\phi(r) = (c^2 + r^2)^{\beta/2}, \quad \beta < 0 & m \geq 0 \\
\phi(r) = (-1)^{\lceil \beta/2 \rceil} (c^2 + r^2)^{\beta/2}, \quad \beta > 0, \beta \notin 2\mathbb{N}_0 & m \geq \lceil \beta/2 \rceil \\
\phi(r) = e^{-\alpha r^2}, \quad \alpha > 0 & m \geq 0 \\
\phi(r) = (1 - r)_+^4 (1 + 4r) & m \geq 0, d \leq 3,
\end{array}$$

where $r = \| \cdot \|$, being $\| \cdot \|$ the Euclidean norm. See e.g. [19] for a comprehensive derivation of the properties of these functions.

When we interpolate three-variate functions, as already remarked in [4], we need samples of large (order of few thousands) or very large dimension (tens of thousands) in order to achieve a good accuracy. This means that we have to provide a technique that takes into account of the problems arising when dealing with large samples. The first question is the computational cost. In [17], Schaback defines precisely the properties that a methods should ideally have: a method is effective if the representation of the approximation is highly faithful to the actual function and if the computational cost is $O(N)$ for the construction and if it takes $O(1)$ operations to evaluate the solution at a single point.
Secondly, there is the still open problem of the severe ill conditioning of the interpolation matrix depending on the sample size N. To overcome this situation, for instance, adaptive techniques [18] or local methods [10] are suggested. In the first case, the approximation may exhibits undue oscillations [18]. While in the second case, samples with an irregular distribution may give a poor quality of reproduction.

In this paper we consider the problem of recovering the unknown function $f(x), \quad x \in \Omega = [0,1]^3$ when N scattered data on N_r straight lines of Ω are given (see Fig. 1). This situation happens in several applications as for instance the study of mineral concentrations in the subsoil where we collect the information by drillings, or in the measurements of the brain answers to depth electromagnetic impulses.
We want to remark that in the case we are considering, the data my present an irregular point distribution. Our goal is then to provide a technique with low computational cost that gives an accurate solution both in terms of approximation error and of graphical reproduction.
To this purpose we propose a three step method that, in part, overcomes the problem of the severe ill-conditioning.

In the next section we describe the method and discuss its computational cost. In §3 we provide a numerical example which shows the effectiveness of our method.

2. The method

The idea of the method is to exploit the data structure in order to

- keep the computational cost low;
- to use validated techniques for the low dimensions,
- obtain an approximation $g(x)$, $x \in \Omega$ with optimal theoretical approximation error.

Before going on, we make some assumptions and give some notations.
We indicate with r_k, $k = 1, \ldots, N_r$, the straight lines and we suppose that any r_k intersects the same two opposite faces of the cube, for instance the faces on $z = 0$ and on $z = 1$. On each line we have n_k scattered points and without loss of generality we set $n_k = n$, for $k = 1, \ldots, N_r$.

We assume that the unknown function f is smooth, $f(x) \in C^p(\Omega)$ $p > 1$. The theory tells that when the function we interpolate is highly regular,the higher accuracy is reached by the multiquadric $\phi(r) = \sqrt{\|r\|^2 + c^2}$. Unfortunately, the choice of a suitable scaling parameter c can give instability. In [5], it is shown how its variability range for achieving stability, becomes smaller as the space dimension grows. Moreover, numerical experiments point out than when the sample dimension grows, any value of the parameter can give a bad conditioning (see [8]). In addition we know (see [16]) that a local and variable choice of the parameter c gives better results.

2.1. *Step 1*

Taking into account of the above considerations and having scattered data on tracks, the natural choice is to interpolate the data on the lines by a multiquadric with variable values of the parameter c [3]. The first step consists in the solution of N_r univariate scattered data interpolation problems. In this way, we obtain the approximations $g_{r_k}(x)$, $x \in r_k$, of the restrictions, $f_{r_k}(x)$, of $f(x)$ to each line.
The interpolation error at each point $x \in r_k$ is

$$|g_{r_k}(x) - f_{r_k}(x)| = O(h_r^\rho), \quad k = 1, \ldots, N_r,$$

where h_r is the maximum of the center density on the lines, ρ depending on the function regularity.

2.1.1. *Computational cost of step 1*

The computational cost is relevant to the construction of the N_r interpolants. Since n is small (typically $n = 30$), it is given by the direct solution of N_r systems of size $n \times n$ an then it is $O(N_r n^3/3) = O(Nn^2/3)$.
Case 1). N large, N_r standard (few hundreds). In this case we have that $N_r = O(\sqrt{N})$ and $n^2/3 = O(N_r)$. Then the cost is

$$O(N\frac{n^2}{3}) \simeq O(N\sqrt{N}). \tag{3}$$

Case 2) N very large, $n \ll N$ then we have a cost that is $O(N)$.

2.2. *Step 2*

In the second step, we consider a number N_π of parallel planes to $z = 0$. Let π_j, $j = 1, \ldots, N_\pi$, be the planes. Their number N_π has to be chosen so that the planes describe correctly the phenomenon evolution represented by the variable z.
We intersect each plane π_j with the straight lines and, using the interpolants $g_{r_k}(x)$ of the first step, we get on π_j a new set of N_r scattered data (X_{π_j}, Y_{π_j})

$$X_{\pi_j} = \{X_1^j, \ldots, X_{N_r}^j\}, \quad Y_{\pi_j} = \{g_{r_1}(X_1^j), \ldots, g_{r_{N_r}}(X_{N_r}^j)\}, \tag{4}$$

which describe, with a good accuracy, the restriction $f_{\pi_j}(x)$ of $f(x)$ to π_j. The cost of the construction of each set (4) is $O(nN_r)$.
At this point, for $j = 1, \ldots, N_\pi$, we interpolate the data (4) by an interpolation method that has to preserve the accuracy of the first step. Then

we will use again the multiquadric basis. In addition we have to take into account of the sample dimension on π_j, i.e. N_r and of the point distribution on the planes. Generally we are in one of these two situations

(1) The original sample (1) has large size N. Then we have samples of standard dimension N_r on the planes π_j, $j = 1, \ldots, N_\pi$.
(2) The original sample (1) has very large dimension N. In this case N_r is large and in addition the points could be very irregularly distributed.

Case 1). We use again the classical multiquadric interpolant and we obtain $g_{\pi_j}(x)$, $x \in \pi_j$, $j = 1, \ldots, N_\pi$.

Case 2). It is convenient to use an "ad hoc method" that on one hand has to provide the solution with low computational cost and on the other has to work well when the points have irregular locations.
A good strategy is those proposed in [6], where the authors present a local interpolant constructed via a RBF modified Shepard method that in the construction of the nodal functions takes into account of the irregular point distribution. Namely, the interpolating function of the data (4) is given by

$$g_{\pi_j}(x) = \sum_{k=1}^{N_r} w_k(x) R_k(x), \quad x \in \pi_j, \tag{5}$$

where

$$R_k(x) = \sum_{i \in I_k} \alpha_i^k \phi(\|x - X_i^j\|) + p(x), \quad p(x) \in P_0^d, \quad x \in \pi_j, \tag{6}$$

is the radial basis function interpolant relative to the subset $\mathcal{X}_k = \{X_k^j, X_i^j, i \in I_k\}$, I_k being the set of indexes of n_q neighbours of X_k^j and $\phi(\cdot) = \sqrt{\|\cdot\| + c^2}$.
The nodal functions $R_k(x)$ are jointed together by an interpolating weight with compact support

$$w_k(x) = \frac{\left[(r_{w_k} - r_k)_+\right]^2}{\sum_{k=1}^{N_r} \left[\frac{(r_{w_k} - r_k)_+}{r_{w_k} r_k}\right]^2},$$

where $r_k = \|x - X_k^j\|$ and r_{w_k} is a radius of influence about the point X_k^j chosen just large enough to include n_w points and

$$(r_{w_k} - r_k)_+ = \begin{cases} r_{w_k} - r_k, \, r_k < r_{w_k}; \\ 0, \qquad\qquad r_k \geq r_{w_k}. \end{cases}$$

For each $x \in \pi_j$, the sum in (5) is only on n_w nodal functions.

The n_q points, needed to compute each nodal function are selected according to an adaptive algorithm that takes into account of the irregular distribution of the point on π_j. More precisely we consider the Delaunay triangulation, $D(X_{\pi_j})$, of X_{π_j} and, for each k, the points that are contiguous to the points of $\mathcal{X}_k$, within $D(X_{\pi_j})$, at a distance greater than $r_q = \max_{i \in I_k} \|X_i^j - X_k^j\|$ but less than $3\,\text{median}_{i \in I_k} \|X_i^j - X_k^j\|$.
We indicate with $\bar{I}_k$ the set of the corresponding indexes. We check now the ability of the initial nodal function $R_k(x)$ to reproduce the given functional values and we force the reproduction at those locations where it did not work well. That is

• we compute $R_k(X_i^j)$ with $i \in \bar{I}_k$ and

$$me := \text{median}_{i \in \bar{I}_k} |R_k(X_i^j) - g_{r_i}(X_i^j)|;$$

• we update I_k by adding those indexes $i \in \bar{I}_k$ such that

$$|R_k(X_i^j) - g_{r_i}(X_i^j)| > 2me;$$

• we construct $R_k(x)$ on the points corresponding to the updated indexes I_k.

This algorithm requires a cost of $O(N_r \log N_r)$ for the Delaunay triangulation and of of $O(N_r n_{av}^3/6)$ for the computation of the nodal functions $R_k(x)$ (n_{av} is the average number of data points involved for the nodal functions, typically $n_{av} < 2n_q$), The evaluation at a single point requires $O(n_w n_{av})$ operations.

2.2.1. *Computational cost of step 2*

In both cases we have the cost of obtaining on each π_j the sets set of N_r scattered data (X_{π_j}, Y_{π_j}) and it is $O(N_\pi N_r n) = O(N_\pi N)$. Generally $N_\pi < n$.

Case 1. The computational cost is

$$O(N_\pi N) + O(N_\pi \frac{N_r^3}{3}) \simeq O(\frac{N^2}{3}). \tag{7}$$

Case 2. The cost is

$$O(N_\pi N) + O(N_\pi N_r \log N_r) + O(N_\pi N_r n_{av}^3/3). \tag{8}$$

Now N is very large and $n, n_{av} \ll N$, then (8) becomes

$$O(N) + O(N \log N_r) \simeq O(N \log N). \tag{9}$$

2.3. *Step 3*

Using the approximations $g_{\pi_j}(x)$, $j = 1, \ldots, N_\pi$, it is possible to compute a set of gridded data (X_G, Y_G) of dimension $N_\pi\, N_G$. We construct the final approximation g by the polyharmonic B-spline interpolating (X_G, Y_G) via a discrete convolution product.

It is well-known that the polyharmonic splines have been studied by several authors because of their good properties (see, for instance, [9] for a wide description and references). Here we consider the class $SH_m(\mathbb{R}^d)$ of m-harmonic *cardinal* splines, $m > d/2$, which, according to [12], are defined as the subspace of $S'(\mathbb{R}^d)$ (the usual space of d-dimensional tempered distributions) whose elements f are in $C^{2m-d-1}(\mathbb{R}^d)$ and enjoy the property $\Delta^m f = 0$ on $\mathbb{R}^d \backslash \mathbb{Z}^d$, where $\Delta := \sum_{j=1}^d \partial^2/\partial x_j^2$ is the Laplacean operator and Δ^m is defined iteratively by $\Delta^m f = \Delta(\Delta^{m-1} f)$.
General properties of the space $SH_m(\mathbb{R}^d)$ can be founded for example in [11–13,15]. In [15] it is proved that, the space of polyharmonic splines can be generated by translates of a particular polyharmonic spline named *elementary B-splines* defined as $\Phi_m(x) := \Delta_1^m v_m(x)$, where v_m is the fundamental solution of Δ^m given by

$$v_m(x) = \begin{cases} c(d,m)\|x\|^{2m-d} & \text{if } d \text{ is odd,} \\ c(d,m)\|x\|^{2m-d}\log\|x\| & \text{if } d \text{ is even.} \end{cases}$$

The constant $c(d,m)$ depends only on d and m and is chosen so that $\Delta^m v_m = \mathit{Dirac}$ (the unit Dirac distribution at the origin). Δ_1 is the usual discrete version of Δ: $\Delta_1 f = \sum_{j=1}^d (f(\bullet - e_j) - 2f(\bullet) + f(\bullet + e_j))$.
We can express Φ_m, see [2], as convolution between a vector γ of $\mathbb{Z}^d$ and v_m, that is ($j \in \mathbb{Z}^d$):

$$\gamma_j := \begin{cases} -2d & \text{if } j = 0 \\ 1 & \text{if } |j| = 1 \\ 0 & \text{else,} \end{cases} \quad x \quad \gamma^m := \gamma * \gamma^{m-1} \quad \Phi_m = \gamma^m * v_m.$$

The important fact is that we can efficiently interpolate a set of gridded data $\{X_G, Y_G\}$. In fact, if we consider the Lagrangean m-harmonic spline L_m, i.e. $L_m(k) = \delta_{0,k}$ (the usual Kroneker symbol), we have that $L_m = a^m * \Phi_m$, where the vector $a^m \in l^1(\mathbb{Z}^d)$ can be computed explicitly according to [1]. The consequence is that the coefficients λ of the interpolant are obtained by the discrete convolution

$$\lambda = a^m * \bar{Y}_G,$$

where $\bar{Y}_G$ is a suitable extension of the data Y_G on a domain $\Omega_1 \supset \Omega$. Generally the cardinality of $\bar{Y}_G$ is less than two times the cardinality of Y_G.

Another advantage of interpolating gridded data by polyharmonic splines is that we achieve an optimal interpolation error. In fact (see [7]) if u is a function with compact support $\Omega_1 \supset \Omega$ belonging to the Sobolev space of order $2m$, we have that the interpolant I_u of u on $h\mathbb{Z}^d$ is such that

$$\|I_u(x) - u(x)\|_{\infty,\Omega} = O(h^{2m}).$$

2.3.1. *Computational cost of step 3*

The computational cost of the third step is given by

- the cost of the evaluation of the two-dimensional interpolants of the second step on a grid of size N_G. Generally $N_G = N_\pi \times N_\pi$.
- the cost of the interpolant coefficient computation via discrete convolution that is $O(N_\pi N_G)$.

Note that usually $N_G < N_r$.
Case 1). We have a cost of $O(N_\pi N_G N_r)$ for the gridded data computation. Now the number N_G of the gridded point on each plane doesn't exceed N_r and, as already said, $N_\pi \leq n$. Then the cost is

$$O(NN_r) + O(N) \simeq O(N\sqrt{N}). \tag{10}$$

Case 2). We have a cost of $O(N_\pi N_G n_w n_{av})$ for the gridded data computation. In this case $n, n_{av}, n_w \ll N$. Then we have a cost of $O(N)$.

The overall computational cost for the construction of $g(x)$ is
Case 1) $O(N^2/3)$.
Case 2) $O(N \log N)$.

The evaluation of $g(x)$ is done on a grid of Ω with step-size $2^{-liv}H$, being H the step-size of the uniform grid X_G and $liv > 1$. This allows to perform a fast evaluation by a discrete convolution and then to evaluate at each point with $O(1)$ operations.

3. Numerical experiments

The method has been tested on the functions of [14]. Here we present the results obtained for $F1$ that is a trivariate version of the well-known Franke's function.

$$F1 = .75\exp\left[-\frac{(9x-2)^2+(9y-2)^2+(9z-2)^2}{4}\right]$$
$$+ .75\exp\left[-\frac{(9x+1)^2}{49}-\frac{(9y+1)^2}{10}-\frac{(9z+1)^2}{10}\right]$$
$$+ .5\exp\left[-\frac{(9x-7)^2+(9y-3)^2+(9z-5)^2}{4}\right]$$
$$- .2\exp\left[-(9x-4)^2-(9y-7)^2-(9z-5)\right].$$

The results obtained for the case 1) (large size N) and case 2) (very large size N) are shown in Figs. 2 and 3 where the graphical visualization is relevant to the variation of one independent variable because usually, in a real problem, the variation of one independent variable represents the evolution of some phenomenon (see [4]). Then we have picked up the variable describing the evolution (for instance z), we have taken some values of it $z_1, \dots, z_l$, with $z_1 = 0$ and $z_l = 1$ and we have represented the evolution of the surfaces $g(x, y, z_k)$, $k = 1, \dots, l$. As shown by the figures, our technique give a good graphical accuracy in the whole domain and at the boundary. We have computed also the maximum errors e_∞, the root mean square errors e_2 (on a grid $33 \times 33 \times 33$) and the worst spectral condition index K_2 relevant to the different steps (see Tables 1 and 2).

Case 1) We have considered a sample of scattered data with large dimension. Namely $N = 3000$, $n = 30$ and $N_r = 100$. We have compared the results of our technique (M1) with those given by the global multiquadric interpolant (Mq). From Table 1 we conclude that we achieve the same accuracy but if we look at Fig. 2, we have also a better quality of reproduction. For completeness, we report the results obtained for $N = 2000$, in order

Method	e_2	e_∞	K_2
$M1$	7.1e-04	1.2e-02	10^{11}
Mq	6.4e-04	1.7e-02	10^{14}

to have a comparison with those in [10] where they have $e_\infty = 8.2e-2$, $e_2 = 2.9e-3$ with a computational time $t = 1.75$ while our results are $e_\infty = 1.4e-2$, $e_2 = 7.6e-4$ obtained with $t = 1$. **Case 2).** We have considered a set of very irregularly (see Fig. 1) scattered points with very

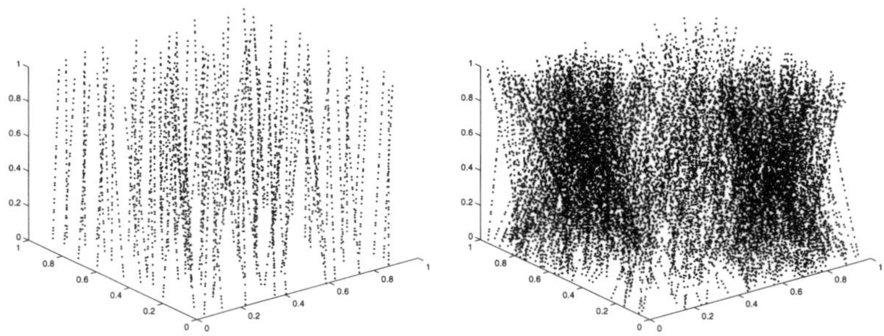

Fig. 1. Point locations. Left: case 1. Right: case 2

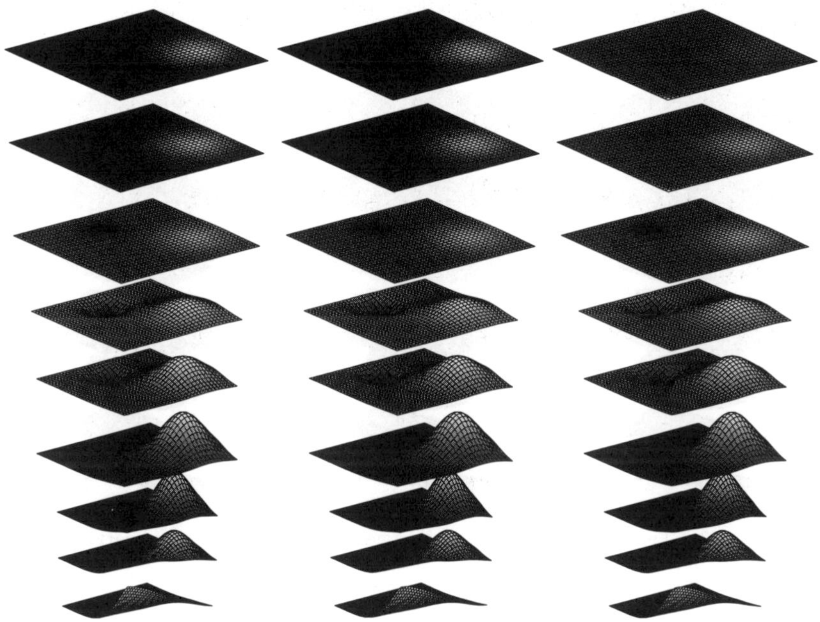

Fig. 2. Left: the true function, center: our approximation, right: the global multiquadric

large size: $N = 18600$, $n = 31$, $N_r = 600$. We have compared the results of the local technique (M1) with those given using in the second step the global multiquadric for interpolating the two-dimensional data on the planes (M2). Table 2 shows that using the local method described in §2 give a better accuracy, a better graphical reproduction of the function and "saves" also the conditioning without using ad hoc techniques that reduce the ill-conditioning due to the point locations. In this case it is not possi-

Method	e_2	e_∞	K_2
$M1$	3.4e-03	2.3e-02	10^{14}
$M2$	1.3e-02	6.6e-01	10^{20}

ble to compare the results with [10] because here we have a very irregular distribution of the locations.

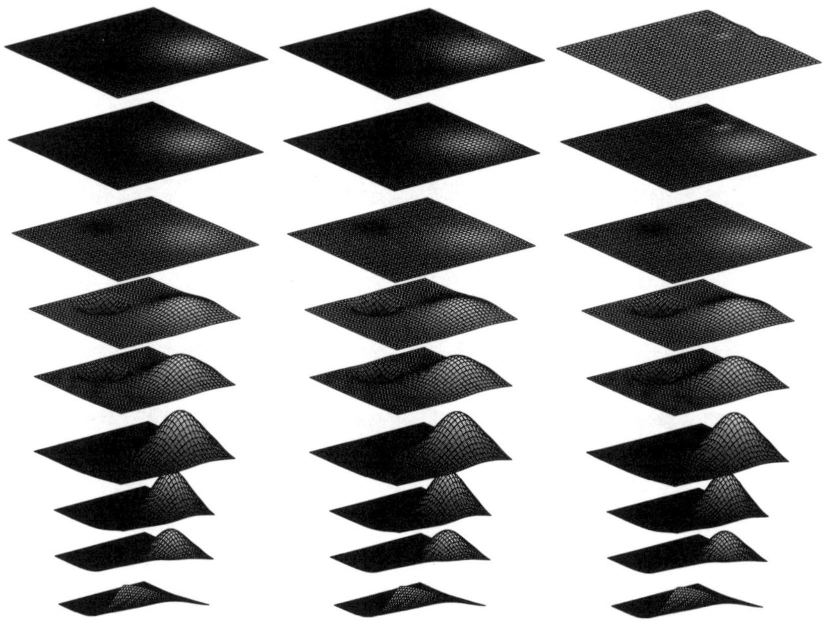

Fig. 3. Left: $F1$, center: M1 approximation, right: M2 approximation

References

1. B. Bacchelli, M. Bozzini, C. Rabut: A fast wavelet algorithm for multidimensional signals using polyharmonic splines. In *Curve and Surface Fitting: Saint-Malo 2003*, A. Cohen, J. L. Merrien and L. Schumaker eds., Nashboro Press, 2003, pp. 21-30
2. B. Bacchelli, M. Bozzini, C. Rabut and M. L. Varas: Decomposition and Reconstruction of Multidimensional Signals using polyharmonic pre-wavelets. Appl. Comput. Harmon. Anal. **18** (2005) pp. 282-299
3. M. Bozzini, L. Lenarduzzi, R. Schaback, Adaptive interpolation by scaled multiquadrics, Adv. Comp. Math. 16 (2002) pp. 375-387

4. M. Bozzini, M. Rossini, Testing methods for 3D scattered data interpolation, Monografias de la Academia de Ciencias de Zaragoza **20** (2002) pp. 111-135
5. M. Bozzini, L. Lenarduzzi, M. Rossini and R. Schaback, Interpolation by basis functions of different scales and shapes, Calcolo **41** (2004) pp. 77-87
6. M. Bozzini, L. Lenarduzzi, On the construction of surfaces from a very large sample with irregularly distributed locations, Preprint (2006)
7. M. D. Buhmann, Radial basis functions. Acta Numer., (2000) pp. 1-38
8. M. D. Buhmann, *Radial basis functions: theory and implementations.* Cambridge Monographs on Applied and Computational Mathematics, 12. Cambridge University Press, Cambridge, 2003.
9. A. Iske, *Multiresolution Methods in Scattered Data Modelling*, Springer Verlag, 2004
10. D. Lazzaro, L. Montefusco, Radial basis functions for the multivariate interpolation of large scattered data sets, J. Computational Applied Maths. 140 (2002) pp. 521–536
11. W. R. Madych, Spline type summability for multivariate sampling. Analysis of divergence (Orono, ME, 1997), in *Appl. Numer. Harmon. Anal.*, Birkhauser Boston, Boston, MA, 1999, pp. 475-512
12. W. R. Madych, S.A. Nelson, Polyharmonic cardinal splines. J. Approx. Theory **60** (1990) pp. 141-156
13. C. Micchelli, C. Rabut, and F. Utreras: Using the refinement equation for the construction of pre-wavelets, III: Elliptic splines. Numerical algorithms **1** (1991) pp. 331-352
14. G. M. Nielson. Scattered data modeling, IEEE Computer Graphics & Applications, **1**, (1993) pp. 60–70
15. C. Rabut, Elementary polyarmonic cardinal B-splines. Numer. Algorithms **2** (1992) pp. 39-46
16. S. Rippa, An algorithm for selecting a good value for the parameter c in radial basis function interpolation, Adv. Comput. Math, **11**, vol. 2–3 (1999) pp. 193–210
17. R. Schaback, Remarks on meshless local construction of surfaces. In *The Mathematics of Surfaces IX*, Proceedings of the Ninth IMA Conference on the Mathematics of Surfaces, Springer 2000
18. R. Schaback, H. Wendland: Adaptive greedy techniques for approximate solution of large RBF systems. Numer. Algorithms **24** (2000) pp. 239-254
19. R. Schaback, H. Wendland. Characterization and construction of radial basis functions. In *Eilat proceedings*, N. Dyn, D. Leviatan, and D Levin, eds, Cambridge University Press, (2000)

ON THE USE OF AN APPROXIMATE CONSTRAINT PRECONDITIONER IN A POTENTIAL REDUCTION ALGORITHM FOR QUADRATIC PROGRAMMING

S. CAFIERI, M. D'APUZZO, V. DE SIMONE, D. DI SERAFINO

Department of Mathematics, Second University of Naples,
81100 Caserta, Italy
E-mail: {sonia.cafieri,marco.dapuzzo,
valentina.desimone,daniela.diserafino}@unina2.it

We focus on the reuse of Constraint Preconditioners in the iterative solution of the augmented systems arising in Interior Point methods. We analyze different strategies for choosing the outer iterations in which the preconditioner is recomputed, in the context of a Potential Reduction algorithm for convex Quadratic Programming. The performance of these strategies is illustrated through a set of numerical experiments.

Keywords: Approximate Constraint Preconditioner; Potential Reduction Method; Quadratic Programming.

1. Introduction

The solution of linear systems of the form

$$\begin{pmatrix} Q+E & -A^T \\ -A & -F \end{pmatrix} \begin{pmatrix} \delta x \\ \delta y \end{pmatrix} = \begin{pmatrix} u \\ v \end{pmatrix}, \tag{1}$$

where $Q \in \Re^{n\times n}$ is symmetric positive semidefinite, $E \in \Re^{n\times n}$ and $F \in \Re^{m\times m}$ are diagonal with positive entries, $A \in \Re^{m\times n}$ and $m \leq n$, arises at each iteration of primal-dual Interior Point (IP) methods for the following convex Quadratic Programming problem:

$$\begin{aligned} &\text{minimize} \;\; p(x) \equiv \frac{1}{2}x^T Q x + c^T x, \\ &\text{subject to } Ax \geq b, \quad x \geq 0. \end{aligned}$$

When dealing with large-scale problems, the cost for solving System (1) with direct methods may become prohibitive in terms of memory and time.

Therefore, in recent years, there has been an increasing activity devoted to the development of suitable iterative techniques to solve such systems. Due to the ill conditioning of the system matrix, that increases as the IP iterate progresses toward the optimal solution, the use of effective preconditioners is mandatory. In the last decade large attention has been devoted to the class of Constraint Preconditioners (CPs), which have revealed their effectiveness when applied to systems of the type (1) (see Refs. 1–6). A CP has the same 2×2 block structure as the matrix to be preconditioned, with the same off-diagonal blocks, i.e. for the matrix in (1) it has the form

$$P = \begin{pmatrix} C & -A^T \\ -A & -D \end{pmatrix}, \tag{2}$$

where C and D are suitable approximations of $Q + E$ and F, respectively. A CP is often applied through its sparse direct factorization. Although this factorization is generally less expensive than the factorization of the matrix of System (1), in large-scale problems it may still account for most of the execution time of a single IP iteration. This has motivated the interest toward CP approximations aimed at reducing the computational cost of the preconditioner application.

Bergamaschi et al.[7] propose an approximate CP using a sparse approximation of the matrix A instead of the original matrix; Benzi and Simoncini[8] propose an inexact CP based on an incomplete factorization of the Schur complement $F + AC^{-1}A^T$; Dollar and Wathen[9] build approximate CPs by using incomplete Schilders' factorizations. An alternative approach consists in using for multiple IP iterations the CP that has been computed at a certain iteration. A similar approach is applied, e.g., in the numerical integration of ordinary differential equations, when the nonlinear equations arising in implicit methods are solved by the Newton method and the Jacobian matrix is kept fixed for some steps. In the context of IP methods, this idea has already been applied to incomplete Cholesky factorization preconditioners, coupled with the Conjugate Gradient (CG) method for the solution of the normal equations.[10,11]

The use of an approximate preconditioner is expected to increase the number of inner iterations; therefore, to have an effective approximation strategy, the time saved in the factorization of the preconditioner should pay off the time increase due to the larger iteration count. Furthermore, the approximate preconditioner should allow the computation of solutions accurate enough to guarantee the convergenge of the IP method.

In this paper we investigate the CP reuse in the context of an IP method for Quadratic Programming, the infeasible primal-dual Potential Reduction

(PR) method by Mizuno et al.[12] The behaviour of the exact CP coupled with CG, within this PR method, has been analyzed in Ref. 13. We present here a computational study of different strategies for selecting the PR iterations in which the preconditioner is recomputed. In this case, a simplified QMR is applied with the approximate preconditioner. A comparison with the behaviour of the exact CP is also performed.

The paper is organized as follows. In Sec. 2 we briefly describe the generic iteration of the PR method, focusing on the iterative solution of the KKT system, on the choice of the CP and on its approximation. In Sec. 3 we analyze different strategies for selecting the PR iterations in which the preconditioner is recomputed and discuss the results of numerical experiments carried out to evaluate their effectiveness. Concluding remarks are given in Sec. 4.

2. An Approximate Constraint Preconditioner

We focus on the reuse of CPs in the iterative solution of linear systems of the form (1), that arise in the infeasible primal-dual Potential Reduction method presented in Ref. 12. A full description of the inexact PR method and its convergence properties is beyond the scope of this paper; the reader is referred to Refs. 13,14 for more details. Here we only outline the generic k-th PR iteration, which requires the application of a Newton step to the following perturbed KKT equations:

$$\begin{pmatrix} Qx^k - A^T y^k - s^k + c \\ Ax^k - z^k - b \\ X^k S^k e \\ Y^k Z^k e \end{pmatrix} = \begin{pmatrix} 0 \\ 0 \\ (\Delta^k/\rho)e \\ (\Delta^k/\rho)e \end{pmatrix}, \tag{3}$$

where x^k and y^k are the primal and dual variables, z^k and s^k are the corresponding slack variables, e is the vector of all ones of appropriate dimension, Δ^k is the duality gap, $W = diag(w)$ for any vector w, and $\rho \geq n + m + \sqrt{n+m}$. This leads to the solution of the KKT linear system

$$\begin{pmatrix} Q & -A^T & -I & 0 \\ A & 0 & 0 & -I \\ S^k & 0 & X^k & 0 \\ 0 & Z^k & 0 & Y^k \end{pmatrix} \begin{pmatrix} \delta x^k \\ \delta y^k \\ \delta s^k \\ \delta z^k \end{pmatrix} = \begin{pmatrix} -r_d^k \\ -r_p^k \\ (-X^k S^k + (\Delta^k/\rho))e \\ (-Y^k Z^k + (\Delta^k/\rho))e \end{pmatrix}, \tag{4}$$

where I is the identity matrix of appropriate dimension and

$$r_p^k = Ax^k - z^k - b,$$
$$r_d^k = Qx^k - A^T y^k - s^k + c$$

are the primal and the dual infeasibility, respectively. By eliminating δs^k and δz^k, the system is reduced to the augmented system

$$\begin{pmatrix} Q+E^k & -A^T \\ -A & -F^k \end{pmatrix} \begin{pmatrix} \delta x^k \\ \delta y^k \end{pmatrix} = \begin{pmatrix} u \\ v \end{pmatrix}, \tag{5}$$

where $E^k = (X^k)^{-1}S^k$ and $F^k = (Y^k)^{-1}Z^k$. We focus on the iterative solution of the previous system in the case Q and A are large and sparse. We observe that, as the PR iterate approaches the solution of the optimization problem, some entries of E^k and F^k may become very large, thus producing an increasing ill conditioning in the system matrix. Ad hoc preconditioning technique are therefore required to obtain effective search directions.

We consider a CP of the form (2) where $C = diag(Q+E^k)$ and $D = F^k$. By applying this preconditioner to the matrix of System (5), all the eigenvalues become positive with at least m of them equal to 1 (see Refs. 3,4,15). Furthermore, with a proper choice of the starting guess, the preconditioned CG applied to System (5) is equivalent, in exact arithmetic, to a suitably preconditioned CG applied to the normal equations.[6]

Using a CP with the CG algorithm usually requires a sparse factorization of the preconditioner at each PR iteration. In order to reduce the factorization time, we propose to keep the preconditioner fixed for some PR iterations. This allows to reuse the CP factorization computed at a certain PR iteration, say $\bar{k}$, for a number $l_{\bar{k}} \geq 0$ of successive iterations. Hence the preconditioner is defined as follows:

$$P^{\bar{k}} = \begin{pmatrix} diag(Q+E^{\bar{k}}) & -A^T \\ -A & -F^{\bar{k}} \end{pmatrix}, \quad P^{\bar{k}+j} = P^{\bar{k}} \quad j = 1, ..., l_{\bar{k}}.$$

The number $l_{\bar{k}}$ of iterations for which the preconditioner is kept fixed should be chosen taking into account both the increasing ill conditioning in System (5), as the iterates approach the boundary of the feasible set, and the accuracy requirement in the solution of the system. Therefore, a suitable approach should not allow $l_{\bar{k}}$ to increase as k increases.

We observe also that the choice of the iterative solver to be used at a certain outer iteration has to be made according to the selected preconditioner. If the approximate preconditioner is used, the CG algorithm cannot be applied anymore and another Krylov subspace method has to be chosen.

In the next section we analyze some strategies for choosing $l_{\bar{k}}$ and present results obtained by using them in the PR algorithm.

3. Computational Experiments

In Ref. 13 we analyzed the performance of the infeasible PR method using CG with the exact CP to solve System (5). The preconditioner was implemented through a sparse LDL^T factorization. To avoid unnecessary CG iterations, adaptive stopping criteria were devised, that relate the accuracy of the solution of the augmented system to the quality of the current PR iterate. Numerical experiments carried out on several large-scale problems, either from the CUTEr collection or randomly generated, showed that the use of the preconditioned CG method generally reduces the execution time. Nevertheless, for some problems the time required by the factorization of the preconditioner may dominate the overall PR time.

Thus, we have implemented the approach described in Sec. 2, in which the preconditioner is held constant for some consecutive PR iterations. A simplified QMR algorithm with no look ahead[16] has been applied when the CP is reused, whereas the CG algorithm has been applied with the exact preconditioner. The algorithmic framework for the solution of the augmented systems by using this approach can be summarized as follows:

$k = 0$
repeat
  $k = k + 1$
  build and factorize CP
  solve the system by CG+CP
  $\bar{k} = k$
  choose $l_{\bar{k}} \geq 0$
  for $i = 1, l_{\bar{k}}$ **do**
    $k = k + 1$
    solve the system by QMR + CP
  endfor
until (PR convergence test satisfied)

The PR method has been stopped using the following condition, which involves the relative duality gap and the relative primal and dual infeasibilities:

$$\min\{\delta^k, \sigma_p^k, \sigma_d^k\} \leq tol_{PR},$$

where

$$\delta^k = \frac{\Delta^k}{1 + |p(x^k)|}, \quad \sigma_p^k = \frac{\|r_p^k\|}{1 + \|b\|}, \quad \sigma_d^k = \frac{\|r_d^k\|}{1 + \|c\|},$$

and tol_{PR} is a tolerance specified by the user, set to 10^{-7} in our experiments. To terminate the inner iterations the following adaptive criterion has been used:

$$tol_{CG/QMR} = \begin{cases} \min\left\{\tau\frac{\Delta^k}{\rho}, tol_{PR}\right\} & \text{if } \delta^k \leq \zeta \cdot tol_{PR} \\ \tau\frac{\Delta^k}{\rho} & \text{otherwise} \end{cases}.$$

This is based on the idea of adapting the accuracy requirement in the solution of the systems to the current value of the duality gap, and of requiring higher accuracy when the relative duality gap is close to tol_{PR}, but the infeasibility is not. In our experiments we have set $\tau = 10$ and $\zeta = 10$. Further details on the choice of the PR parameters can be found in Ref. 13.

Nine test problems from the CUTEr collection have been considered, modified to have only linear inequalities constraints and nonnegative variables without upper bounds. Details on these problems are given in Table 1. The tests have been performed on a personal computer with a Pentium IV

Table 1. Dimensions and number of nonzero entries of the Hessian (upper or lower triangle) and constraint matrices of the test problems.

Problem	n	m	$nnz(Q)$	$nnz(A)$
CVXQP1-M-a	1000	500	3984	1498
CVXQP1-M-b	10000	5000	39984	14998
CVXQP2-M-a	1000	250	3984	749
CVXQP2-M-b	10000	2500	39984	7499
CVXQP3-M-a	1000	750	3984	2247
CVXQP3-M-b	10000	7500	39984	22497
STCQP1-M-a	1025	510	5615	2805
STCQP1-M-b	4097	2052	26603	13338
STCQP1-M-c	8193	4095	57333	28665

processor (2.53 GHz), a memory of 1.256 GB and a L2 cache of 512 MB, running Linux Red Hat 9.0.

Firstly, we analyze the results obtained by setting $l_{\bar{k}} = 1$ for each $\bar{k}$, that is by performing the factorization of the preconditioner every other PR iteration. The number of inner and outer iterations, the time for the factorization of the preconditioners, the time for the solution of the linear systems, and the total execution time for PR are reported in Table 2 (bottom). For comparison purposes, the results obtained by recomputing CP at each PR iterations (CP-CG) are also presented in Table 2 (top). We see that CP-CG/QMR leads to a significant time reduction on those problems for which CP-CG requires factorization times much larger than the solution times. Furthermore, the number of outer iterations of the two approaches results almost the same, thus showing that CP-CG/QMR does not deteriorate the convergence behaviour of the PR method.

Table 2. Number of inner and outer iterations and execution times (sec.) of CP-CG and CP-CG/QMR

CP-CG

Problem	*it*	*CG it*	*fact. time*	*solve time*	*total time*
CVXQP1-M-a	24	243	8.62e−2	8.81e−2	1.97e−1
CVXQP1-M-b	30	415	1.45e+1	3.95e+0	1.88e+1
CVXQP2-M-a	25	269	1.70e−2	5.65e−2	8.90e−2
CVXQP2-M-b	27	407	1.96e−1	1.12e+0	1.55e+0
CVXQP3-M-a	26	290	3.34e−1	2.00e−1	5.61e−1
CVXQP3-M-b	30	503	9.65e+1	1.05e+1	1.07e+2
STCQP1-M-a	23	70	5.40e−1	6.12e−2	6.24e−1
STCQP1-M-b	21	73	8.18e+1	9.56e−1	8.28e+1
STCQP1-M-c	26	100	1.07e+3	4.63e+0	1.07e+3

CP-CG/QMR

Problem	*it*	*CG/QMR it*	*fact. time*	*solve time*	*total time*
CVXQP1-M-a	23	134/272	4.18e−2	1.57e−1	2.19e−1
CVXQP1-M-b	29	208/397	6.92e+0	6.08e+0	1.33e+1
CVXQP2-M-a	26	154/382	8.73e−3	8.73e−1	1.53e−1
CVXQP2-M-b	28	211/525	1.01e−1	2.49e+0	2.84e+0
CVXQP3-M-a	26	145/425	1.58e−1	4.07e−1	5.94e−1
CVXQP3-M-b	30	258/958	4.63e+1	2.31e+1	6.98e+1
STCQP1-M-a	24	36/528	2.54e−1	4.95e−1	7.73e−1
STCQP1-M-b	22	38/387	4.27e+1	5.30e+0	4.81e+1
STCQP1-M-c	26	54/287	5.15e+2	1.56e+1	5.31e+2

A more detailed analysis has shown that the number of QMR iterations significantly increases in the last PR steps, because either the augumented system matrix gets more and more ill-conditioned or the accuracy requirement in the solution of the system grows up. Therefore, the effectiveness of the approximate preconditioner deteriorates and the use of the exact CP results to be more effective. This behaviour is shown in Fig. 1 for the problem CVXQP2-M-a, where the number of inner iterations at each outer iteration is plotted for CP-CG and CP-CG/QMR. Thus, we have modified

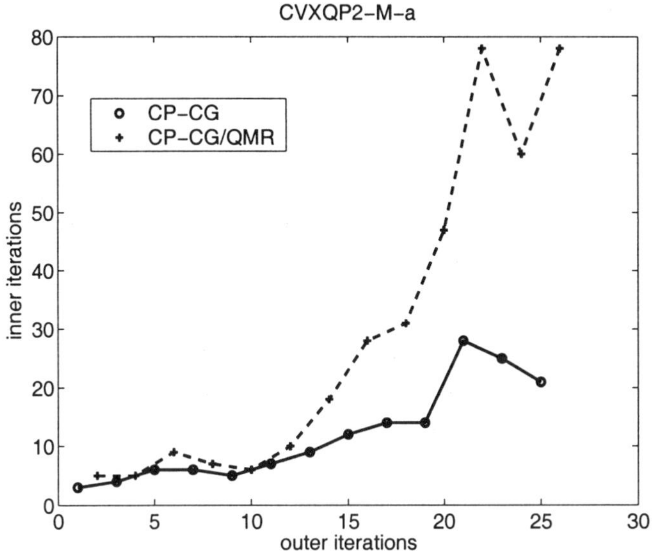

Fig. 1. Number of inner iterations performed at each outer iteration by CP-CG and CP-CG/QMR on the problem CVXQP2-M-a.

the previous approach by recomputing CP at each outer iteration once the algorithm gets close enough to the solution. More precisely, we have set

$$l_{\bar{k}} = \begin{cases} 1 & \text{if } \delta^{\bar{k}} > 10 \cdot tol_{PR} \\ 0 & \text{otherwise} \end{cases}.$$

This approach is referred to as CP-CG/QMR-v2 and the corresponding results are shown in Table 3. We see that CP-CG/QMR-v2 outperforms CP-CG/QMR on all the problems but STCQP1-M-b and STCQP1-M-c, according to the fact that these problems require factorization times much larger than the corresponding solution times. We also observe that CP-CG/QMR-v2 outperforms CP-CG on six problems.

Table 3. Number of inner and outer iterations and execution times (sec.) of CP-CG/QMR-v2.

CP-CG/QMR-v2					
Problem	*it*	*CG/QMR it*	*fact. time*	*solve time*	*total time*
CVXQP1-M-a	23	162/189	4.45e−2	1.32e−1	2.00e−1
CVXQP1-M-b	29	298/231	6.69e+0	5.08e+0	1.21e+1
CVXQP2-M-a	26	223/166	1.09e−2	8.69e−2	1.15e−1
CVXQP2-M-b	28	288/233	1.17e−1	1.65e+0	2.02e+0
CVXQP3-M-a	26	209/164	1.84e−1	2.87e−1	4.96e−1
CVXQP3-M-b	30	360/352	4.73e+1	1.48e+1	6.24e+1
STCQP1-M-a	24	45/213	3.50e−1	2.27e−1	6.01e−1
STCQP1-M-b	22	47/166	5.97e+1	2.42e+0	6.21e+1
STCQP1-M-c	26	60/166	6.08e+2	9.81e+0	6.18e+2

Finally, we have analyzed an adaptive approach in which the same CP factorization is reused until its effectiveness deteriorates in terms of QMR iterations required to solve the system. More precisely, the value of $l_{\bar{k}}$ has not been fixed *a priori*, but dynamically set as shown below:

$k = 0$; $qmrit_{max} = 0$; $l_{max} = 0$
repeat
  $k = k + 1$
  build and factorize CP
  solve the system by CG+CP
  $\bar{k} = k$; $qmrit_k = 0$; $l_{\bar{k}} = 0$
  while ($\delta^k > 10 \cdot tol_{PR}$ and $qmrit_k \leq qmrit_{max}$ and $l_{\bar{k}} \leq l_{max}$) **do**
    $k = k + 1$; $l_{\bar{k}} = l_{\bar{k}} + 1$
    solve the system by QMR + CP
    $qmrit_k$ = number of QMR iterations performed
    if ($l_{\bar{k}} = 1$) **then** $qmrit_{max} = qmrit_k$
  endwhile
  $l_{max} = l_{\bar{k}}$
until (PR convergence test satisfied)

The value of $qmrit_k$ is the number of QMR iterations performed at the

k-th PR step and $qmrit_{max}$ is the number of iterations required by QMR the first time the current approximate CP has been applied; l_{max} is the number of PR steps in which the previous CP has been reused. The results obtained with this adaptive approach (CP-CG/QMR-v3) are reported in Table 4. By comparing CP-CG/QMR-v3 with the other approaches, we see that CP-CG/QMR-v3 is generally more efficient on the largest problems, such as CVXQP3-M-b and STCQP1-M-c. Furthermore, CP-CG/QMR-v3 is faster than or comparable to CP-CG on six out of the nine problems.

Table 4. Number of inner and outer iterations and execution times (sec.) of CP-CG/QMR-v3.

		CP-CG/QMR-v3			
Problem	*it*	*CG/QMR it*	*fact. time*	*solve time*	*total time*
CVXQP1-M-a	22	127/408	2.95e−2	2.01e−1	2.51e−1
CVXQP1-M-b	31	252/520	4.13e+0	7.75e+0	1.22e+1
CVXQP2-M-a	27	199/349	8.50e−3	1.29e−1	1.55e−1
CVXQP2-M-b	27	246/623	9.44e−2	2.85e+0	3.19e+0
CVXQP3-M-a	25	165/428	1.10e−1	4.49e−1	5.85e−1
CVXQP3-M-b	30	303/572	2.92e+1	1.90e+1	4.87e+1
STCQP1-M-a	25	31/411	2.42e−1	3.68e−1	6.35e−1
STCQP1-M-b	22	38/295	4.74e+1	3.57e+0	5.10e+1
STCQP1-M-c	26	44/375	4.58e+2	1.70e+1	4.76e+2

From the above results we conclude that fixing the preconditioner for some outer iterations is potentially competitive with using the exact CP. Furthermore, an adaptive approach for deciding when CP has to be recomputed appears to be promising to achieve the best tradeoff between the reduction of the time for the factorization of the preconditioner and the increase of the time for solving the augmented system.

4. Conclusions

We focused on the use of approximate CPs in the iterative solution of the linear systems that arise in a PR algorithm for Quadratic Programming. We analyzed an approximation approach that consists in reusing for multiple outer iterations the preconditioner that has been computed at a certain iteration, with the aim of reducing the cost of the preconditioner application. We presented a computational study where different strategies for choosing when the preconditioner has to be recomputed were investigated. Numerical results showed that the proposed approach is potentially competitive with the exact preconditioner and deserves further investigation.

Acknowledgments

This work was partially supported by the Italian MIUR, FIRB Project *Large Scale Nonlinear Optimization*, grant no. RBNE01WBBB, and PRIN Project *Innovative Problems and Methods in Nonlinear Optimization*, grant no. 2005017083.

References

1. L. Luksan and J. Vlcek, *Numerical Linear Algebra with Applications* **5**, 219 (1998).
2. C. Keller, N. Gould and A. Wathen, *SIAM Journal on Matrix Analysis and Application* **21**, 1300 (2000).
3. C. Durazzi and V. Ruggiero, *Numerical Linear Algebra with Applications* **10**, 673 (2003).
4. L. Bergamaschi, J. Gondzio and G. Zilli, *Computational Optimization and Applications* **28**, 149 (2004).
5. M. Benzi, G. Golub and J. Liesen, *Acta Numerica* **14**, 1 (2005).
6. S. Cafieri, M. D'Apuzzo, V. De Simone and D. di Serafino, *Computational Optimization and Applications*, to appear (2007a) (Preprint 09/2004, Department of Mathematics, Second University of Naples).
7. L. Bergamaschi, J. Gondzio, M. Venturin and G. Zilli, *Computational Optimization and Applications*, to appear (2007).
8. M. Benzi and V. Simoncini, *Numerische Mathematik* **103**, 173 (2006).
9. H. S. Dollar and A. J. Wathen, *SIAM Journal of Scientific Computing* **27**, 1555 (2006).
10. N.K. Karmarkar and K.G. Ramakrishnan, *Mathematical Programming* **52**, 555 (1991).
11. T.J. Carpenter and D.F. Shanno, *Computational Optimization and Applications* **2**, 5 (1993).
12. S. Mizuno, M. Kojima and M.J. Todd, *SIAM Journal on Optimization* **5**, 52 (1995).
13. S. Cafieri, M. D'Apuzzo, V. De Simone and D. di Serafino, *Computational Optimization and Applications*, DOI 10.1007/s10589-006-9007-7 (2007b).
14. S. Cafieri, M. D'Apuzzo, V. De Simone , D. di Serafino and G. Toraldo, *Journal of Optimization Theory and Applications* **133** (2007).
15. H. S. Dollar, *Tech. Rep. NA-05-02, Numerical Analysis Group, Oxford University* (2005).
16. R.W. Freund and N.M. Nachtigal, A new Krylov-subspace method for symmetric indefinite linear system, in *Proceedings of the 14th IMACS World Congress on Computational and Applied Mathematics*, 1253 (1994).

OPTION HEDGING FOR HIGH FREQUENCY DATA MODELS

C. CECI

Dipartimento di Scienze, Facolta' di Economia
Universita' di Chieti-Pescara, I-65127-Pescara, Italy
E-mail: ceci@sci.unich.it

Hedging strategies for contingent claims are studied in a general model for high frequency data. The dynamics of the risky asset price is described through a marked point process Y, whose local characteristics depend on some hidden state variable X. The two processes Y and X may have common jump times, which means that the trading activity may affect the law of X and could be also related to the presence of catastrophic events. Since the market considered is incomplete one has to choose some approach to hedging derivatives. We choose the local risk-minimization criterion. When the price of the risky asset is a general semimartingale, if an optimal strategy exists, the value of the portfolio is computed in the terms of the so-called minimal martingale measure and may be interpreted as a possible arbitrage-free price. In the case where the price of the risky asset is modeled directly under a martingale measure, the computation of the risk-minimizing hedging strategy is given. By using a projection result, we also obtain the risk-minimizing hedging strategy under partial information when the hedger is restricted to observing only the past asset prices and not the exogenous process X which drives their dynamics.

Keywords: High-frequency data, option hedging, marked point processes, jump-diffusions.

Introduction

In models for intraday stock price movements asset prices are used to be described by marked point processes. In fact on a very small time scale, as in high frequency data, real asset prices are piecewise constant and jump in reaction to trades or to significant new information.

In many papers, (see, for instance [8, 9, 10, 12, 15, 14]) the asset price process is modeled as a double stochastic Poisson process with marks. In some of them, the local characteristics of this process depend on an unobservable state variable, which may describe the intraday market activity, the activity of other markets, macroeconomics factors or microstructure rules that drive the market.

In this paper we consider a more general model as that introduced in [4]. The behaviour of the asset prices is described via a general marked point process Y, whose local characteristics, in particular the jump-intensity, depend on an exogenous state variable X, which is modeled by a Markov jump-diffusion process. Moreover, the dynamics of Y and X may be strongly dependent, in particular the two processes may have common jump times. Hence our model could take into account also the possibility of catastrophic events. This kind of events, in fact, influence both the asset prices and the hidden state variable which drives their dynamics. We assume that the pair (X, Y) is a solution of a system of stochastic differential equations driven by a Browian motion and a Poisson random measure as a natural way to describe its dynamics.

In this note we are concerned with the hedging of contingent claims. When the given financial market is complete, every claim can be replicated by a self-financing dynamic portfolio strategy which only makes use of the existing assets. In this case, one can reduce to zero the risk of the claim by a suitable strategy. On the other hand, markets modeled by marked point processes, where infinite number of marks are allowed, are incomplete. Then one has to choose some approach to hedging derivatives. Since one cannot ask simultaneously for a perfect replication of a given claim by a portofolio strategy and the self-financing property of this strategy, one has to relax one of these conditions. In this paper we choose the local risk minimization, which keeps the replicability and relax the self-financing condition. In [7] the authors dealt with the study of hedging of contingent claims under market incompleteness by introducing the criterion of risk minimization in the case where the price process is a martingale under the real world probability measure. They proved that the optimal strategy can be obtained by the Kunita-Watanabe decomposition. In the general semimartingale case, since it does not exist in general any risk-minimizing strategy, the weaker concept of locally risk-minimizing strategy was introduced ([16]). In [6], under the further assumption that the risky asset price has continuous trajectories, it has been proved that an optimal strategy exists and that it can be computed by the Kunita-Watanabe decomposition under the minimal martingale measure.

In this paper we first consider the general case where the risky asset price is a semimartingale, but since it is not continuous, the results proved in [6] cannot be applied. However we prove that the value process of an optimal strategy (when it exists) can be again computed in terms of the minimal martingale measure. The explicit expression of the density of the minimal martingale measure is provided for our model where the filtration is generated by the Wiener process and the random Poisson measure. In [13] this has been done in the case of marked point processes with respect to their internal filtration.

In the last section we recall the main results obtained in [5] in the case where the price of the risky asset is a local martingale under the real world probability measure. The risk-minimizing strategy is computed by the Kunita-Watanabe decomposition. Moreover, by using a projection result ([17]), the risk-minimizing strategy when agents have access only to the information contained in the past asset prices (they have not knowledge of the latent state process) can be obtained by solving a filtering problem.

We do not discuss here this filtering problem since it is exhaustively studied in [4] when Y is a discrete valued process and in [5] when Y is a real-valued process.

1. The Model

We consider the same model studied in [4] and [5]. On some underlying filtered probability space $(\Omega, \mathcal{F}, \mathcal{F}_t, P)$ we consider a market with two traded assets: a riskless money market account and a risky asset. The risky asset price S is supposed having the form

$$S_t = S_0 e^{Y_t} \tag{1}$$

where

$$S_0 \in I\!R^+, \quad Y_t = \sum_{n=0}^{N_t} Z_n, \quad Z_0 = 0, \quad N_t = \sum_{n \geq 1} \mathbb{I}_{\{T_n \leq t\}},$$

the random times $\{T_n\}$ represent instants at which a large trade occurs or at which a market maker updates his quotes in reaction to significant new information, Y represents the logreturn process, $Z_n = Y_{T_n} - Y_{T_{n-1}}$ is the size of the nth logreturn change and N is the point process which counts the total number of changes.

Besides the risky asset, there is a risk-free asset traded in our market, whose price is taken equal to 1. This simply means that S is the discounted price of the risky asset and this helps to avoid more complicated notations. We will consider the case in which the $(P, \mathcal{F}_t)$-local characteristics ([2]) $(\lambda_t, \Phi_t(dz))$ of the marked point process Y may depend on some exogenous process X.

In [8, 9, 10], the possibility that the jump-times of N and X coincide has been excluded. In this note, we allow common jump times between N and X.

A natural way to describe this kind of behaviour is to suppose that the pair (X,Y) takes values in $\mathbb{R} \times \mathbb{R}$, and that it is a global solution to the following system

$$X_t = x_0 + \int_0^t b(X_s)\, ds + \int_0^t \sigma(X_s)\, dW_s + \int_0^t \int_Z K_0(s, X_{s^-}; \zeta)\, \mathcal{N}(ds, d\zeta) \tag{2}$$

$$Y_t = \int_0^t \int_Z K_1(s, X_{s^-}, Y_{s^-}; \zeta)\, \mathcal{N}(ds, d\zeta) \tag{3}$$

where $x_0 \in \mathbb{R}$, W_t is a $(P, \mathcal{F}_t)$-standard Brownian motion, $\mathcal{N}(dt, d\zeta)$ is a $(P, \mathcal{F}_t)$-Poisson random measure on $\mathbb{R}^+ \times Z$, independent of W_t, with mean measure $dt\ \nu(d\zeta)$, with $\nu(d\zeta)$ a σ-finite measure on a measurable space $(Z, \mathcal{Z})$. The $\mathbb{R}$-valued functions $b(x)$, $\sigma(x)$, $K_0(t, x; \zeta)$ and $K_1(t, x, y; \zeta)$ are jointly measurable functions of their arguments. Overall this paper we assume existence and uniqueness (at least weak uniqueness) to the system (2), (3) ([4, 5] for a discussion on this topic).

In Proposition (1.1) and (1.2) below we recall some results proved in [5]. At first, the $(P, \mathcal{F}_t)$-local characteristics $(\lambda_t, \Phi_t(dz))$ of Y are derived taking into account the representation (3). The time-dependency of $(\lambda_t, \Phi_t(dz))$ incorporate seasonality effects, which are typical for high frequency data. In particular λ_t, corresponds to the rate at which new economic information is absorbed by the market.

First, we introduce the sequence of jump times of Y

$$T_1 = inf\{t > 0 : \int_0^t \int_Z K_1(s, X_{s^-}, 0; \zeta)\, \mathcal{N}(ds, d\zeta) \neq 0\}$$

$$T_{n+1} = inf\{t > T_n : \int_{T_n}^t \int_Z K_1(s, X_{s^-}, Y_{T_n}; \zeta)\, \mathcal{N}(ds, d\zeta) \neq 0\}$$

and the sequence of the marks

$$Z_n = Y_{T_n} - Y_{T_{n-1}} = \int_Z K_1(T_n, X_{T_n^-}, Y_{T_{n-1}}; \zeta)\, \mathcal{N}(\{T_n\}, d\zeta).$$

Let us define

$$D_1(t, x, y) = \{\zeta \in Z : K_1(t, x, y; \zeta) \neq 0\} \tag{4}$$

and

$$D_0(t, x, y) = \{\zeta \in Z : K_0(t, x; \zeta) \neq 0, \quad K_1(t, x, y; \zeta) \neq 0\}. \tag{5}$$

Proposition 1.1.

Let $\forall T > 0$, $\forall t \in [0, T]$, $\forall A \in \mathcal{B}(I\!R)$ *(where* $\mathcal{B}(I\!R)$ *denotes the family of Borel subsets of* $I\!R$*)*

$$D_1^A(t, x, y) = \{\zeta \in Z : K_1(t, x, y; \zeta) \in A \setminus \{0\}\} \subseteq D_1(t, x, y), \tag{6}$$

and denote by m *the integer valued random measure associated to* Y *([2, 11])*

$$m(dt, dz) = \sum_{n \geq 1} \delta_{\{T_n, Z_n\}}(dt, dz) 1\!I_{\{T_n < \infty\}}. \tag{7}$$

Then, under the assumption

$$E[\int_0^T \nu(D_1(s, X_s, Y_s))\, ds] < \infty \qquad (\int_0^T \nu(D_1(s, X_s, Y_s))\, ds < \infty \quad P\text{--}a.s.) \tag{8}$$

the $(P, \mathcal{F}_t)$*-predictable projection of* m *is given by*

$$m^p(dt, dz) = \lambda_t \Phi_t(dz)dt = \lambda(t, X_{t^-}, Y_{t^-})\Phi(t, X_{t^-}, Y_{t^-}, dz)dt \tag{9}$$

where

$$\lambda_t = \lambda(t, X_{t^-}, Y_{t^-}) = \nu(D_1(t, X_{t^-}, Y_{t^-})) \tag{10}$$

provides the $(P, \mathcal{F}_t)$*-predictable intensity of the point process* $N_t = \sum_{n \geq 1} 1\!I_{\{T_n \leq t\}}$ *and on* $\{T_n < \infty\}$

$$\Phi_{T_n}(A) = \frac{\nu(D_1^A(T_n, X_{T_n^-}, Y_{T_n^-}))}{\nu(D_1(T_n, X_{T_n^-}, Y_{T_n^-}))}. \tag{11}$$

Moreover, whenever there exists a transition function $\mu(t, x, y, A)$ *such that,* $\forall A \in \mathcal{B}(I\!R)$

$$P\big(Z_n \in A \mid \mathcal{F}_{T_n -}\big) = \mu(T_n, X_{T_n^-}, Y_{T_n^-}, A)$$

then on $\{T_n < \infty\}$

$$\Phi_{T_n}(A) = P\big(Z_n \in A \mid \mathcal{F}_{T_n^-}\big). \tag{12}$$

By applying Itô formula to (3) we derive the joint dynamics of the pair (X, S):

$$X_t = x_0 + \int_0^t b(X_s)\, ds + \int_0^t \sigma(X_s)\, dW_s + \int_0^t \int_Z K_0(s, X_{s^-}; \zeta)\, \mathcal{N}(ds, d\zeta) \tag{13}$$

$$S_t = S_0 + \int_0^t \int_Z S_{r^-}(e^{K_1(r, X_{r^-}, log(S_{r^-}/S_0); \zeta)} - 1)\mathcal{N}(dr, d\zeta). \tag{14}$$

The pair (X, S) is a Markov process whose generator is given in the next proposition.

Proposition 1.2.
Under the assumptions, $\forall T > 0$

$$E[\int_0^T \sigma^2(X_s)\, ds\ ] < \infty \tag{15}$$

$$E[\int_0^T \nu(D_i(s, X_s, Y_s))\, ds] < \infty \quad i = 0, 1 \tag{16}$$

for real-valued, bounded functions $f(t, x, s)$ *such that* $\frac{\partial f}{\partial t}$, $\frac{\partial f}{\partial x}$, $\frac{\partial^2 f}{\partial x^2}$ *are bounded and continuous, the process*

$$f(t, X_t, S_t) - f(0, x_0, S_0) - \int_0^t Lf(r, X_r, S_r)\, dr \tag{17}$$

is a $(P, \mathcal{F}_t)$*-martingale , where*

$$\begin{aligned} Lf(t, x, s) &= \frac{\partial f}{\partial t}(t, x, s) + L_t f(t, x, s) = \\ &= \frac{\partial f}{\partial t}(t, x, s) + b(x)\, \frac{\partial f}{\partial x}(t, x, s) + \frac{1}{2}\, \sigma^2(x)\, \frac{\partial^2 f}{\partial x^2}(t, x, s) \\ &+ \int_Z \Big(f\big(t, x + K_0(t, x; \zeta), s e^{K_1(t, x, log(\frac{s}{S_0}); \zeta)}\big) - f(t, x, s)\Big)\, \nu(d\zeta). \end{aligned} \tag{18}$$

In [5] it has been studied the case where S is a $(P, \mathcal{F}_t)$-local martingale. Instead of this, here we will consider the more general case where S is a $(P, \mathcal{F}_t)$-semimartingale.

Proposition 1.3.
Under (8) and the following condition

$$\int_0^T \int_Z (e^{K_1(t, X_t, Y_t; \zeta)} - 1)^2 \nu(d\zeta) dt < +\infty \quad P - a.s. \tag{19}$$

S is a special semimartingale ([11]) with the decomposition

$$S_t = S_0 + M_t + A_t \tag{20}$$

where

$$A_t = \int_0^t \int_Z S_{r-} (e^{K_1(r, X_{r-}, Y_{r-}); \zeta)} - 1) \nu(d\zeta) dr$$

is a predictable process with paths locally of bounded variation,

$$M_t = \int_0^t \int_Z S_{r^-}(e^{K_1(r,X_{r^-},Y_{r^-};\zeta)} - 1)(\mathcal{N}(dr, d\zeta) - \nu(d\zeta)dr)$$

is a local martingale, locally square-integrable whose angle process is given by

$$< M >_t = \int_0^t \int_Z S_{r^-}^2 (e^{K_1(r,X_{r^-},Y_{r^-};\zeta)} - 1)^2 \nu(d\zeta)dr. \tag{21}$$

Proof.

First notice that (8) and (19) imply

$$\int_0^T \int_Z | e^{K_1(t,X_t,Y_t;\zeta)} - 1 | \nu(d\zeta)dt < +\infty \quad P-a.s. \tag{22}$$

hence

$$R_t = \int_0^t \int_Z (e^{K_1(r,X_{r^-},Y_{r^-};\zeta)} - 1)\mathcal{N}(dr, d\zeta)$$

is a semimartingale and by (19) square integrable. By (14) S is a semimartingale being the stochastic exponential of the semimartingale R. To conclude observe that, since S^2 is also a semimartingale being the stochastic exponential of the semimartingale

$$\int_0^t \int_Z (e^{2K_1(r,X_{r^-},Y_{r^-};\zeta)} - 1)\mathcal{N}(dr, d\zeta),$$

S is locally square-integrable. □

Let us observe that the following representations in terms of the integer valued measure m associated to Y hold

$$S_t = S_0 + \int_0^t \int_{\mathbb{R}} S_{r^-}(e^z - 1)m(dr, dz) \tag{23}$$

$$M_t = \int_0^t \int_{\mathbb{R}} S_{r^-}(e^z - 1)(m(dr, dz) - \lambda_r \phi_r(dz)) \tag{24}$$

$$A_t = \int_0^t \int_{\mathbb{R}} S_{r^-}(e^z - 1)\lambda_r \phi_r(dz), \tag{25}$$

and condition (19) can be written as

$$\int_0^T \int_{\mathbb{R}} (e^z - 1)^2 \lambda_r \phi_r(dz) < +\infty \quad P-a.s.$$

2. Hedging of a contingent claim

2.1. *Problem formulation*

Since our market is incomplete we have to choose some approach to hedging derivatives. In this paper we will use the criterion of risk minimization. This approach has been proposed in [7] in the martingale case and weakened in local sense in [16] for the general semimartingale case.

We consider a European contingent claim with maturity T whose payoff is given by $H(S_T)$ and such that $E[H^2(S_T)] < \infty$. The simplest example is given by a call option with strike price k where

$$H(S_T) = (S_T - k)^+.$$

We look for a trading strategy which generates the required payoff $H(S_T)$ and at the same time minimizes some measure of riskiness.

A $\{\mathcal{F}_t\}$-**trading strategy** is a pair $(\xi, \eta) = \{(\xi_t, \eta_t) : t \in [0, T]\}$, where ξ_t is an $\{\mathcal{F}_t\}$-predictable process and η_t is a process $\{\mathcal{F}_t\}$-adapted; ξ_t is the number of shares of the risky asset to be held at time t, while η_t is the amount invested in the riskless asset.

The **value** at time t of such a portfolio is given by

$$V_t = V_t(\xi, \eta) = \xi_t S_t + \eta_t.$$

We shall concentrate on strategies, (ξ, η), which are H-admissible in the sense that

$$V_T(\xi, \eta) = H(S_T) \qquad P - a.s.$$

and satisfies

$$E(\int_0^T \xi_t^2 d < S >_t) < \infty \tag{26}$$

$$E((sup_{t\in[0,T]} \mid V_t \mid)^2) < \infty. \tag{27}$$

The **cost process** of (ξ, η) is defined by

$$C_t(\xi, \eta) = V_t(\xi, \eta) - \int_0^t \xi_r dS_r \tag{28}$$

and provides the cumulative cost up to time t as current value of the portfolio minus total gains from trade. Under (26) and (27) C is a square integrable process. Moreover a strategy (ξ, η) is called self-financing if its cost process $C_t(\xi, \eta)$ is constant and it is called mean-self-financing if $C_t(\xi, \eta)$ is a martingale.

In an incomplete market perfect duplication is, in general, impossible and so the cost process will not be constant but fluctuate randomly over time. Hence we need a criterion to compare different strategies.

As a measure of riskiness, we introduce for each strategy (ξ, η) the conditional mean square error process

$$R_t(\xi,\eta) = E\Big((C_T(\xi,\eta) - C_t(\xi,\eta))^2 \mid \mathcal{F}_t\Big) \tag{29}$$

and the **problem of risk minimization** is formulated as follows

Given $H = H(S_T)$ with $E[H^2(S_T)] < \infty$, we have to find an H-admissible $\{\mathcal{F}_t\}$-strategy minimizing the $\{\mathcal{F}_t\}$-risk process, R_t, over the class of H-admissible $\{\mathcal{F}_t\}$-strategy. This strategy will be called $\{\mathcal{F}_t\}$-risk minimizing strategy.

In [7], in the martingale case, this problem was completely solved by using the Kunita-Watanabe decomposition. While, in the general case of a semimartingale there cannot exist any risk-minimizing strategy hence in [16] the weaker concept of locally risk-minimizing strategy was introduced.

It has been also proved that this definition is equivalent to the following

Definition 2.1.

An H-admissible strategy (ξ^*, η^*) is called optimal if the associated cost process $C(\xi^*, \eta^*)$ defined in (28) is a square-integrable $(P, \mathcal{F}_t)$-martingale orthogonal to M under P, that is the angle process $< C(\xi^*, \eta^*), M >= 0$ P-a.s..

This concept of optimal strategy is related to the existence of the minimal martingale measure as we will see in Proposition (2.2) and (2.3).

2.2. *The minimal martingale measure*

We recall that absence of arbitrage opportunities is related to the existence of risk-neutral probability measures. That is probability measures Q, equivalent to P, such that S is a local $(Q, \mathcal{F}_t)$-martingale. We concentrate our attention to the minimal martingale measure

Definition 2.2. A martingale measure P^* equivalent with respect to P is called minimal if any square-integrable $(P, \mathcal{F}_t)$-martingale which is orthogonal to M under P is still a martingale under P^*.

Existence and uniqueness of the minimal martingale measure for general semimartingales satisfying the structure condition, (SC), has been discussed in [1]. The (SC) condition requires that S assumes the form

$$S_t = S_0 + M_t + \int_0^t c_r d < M >_r$$

where M is a $(P, \mathcal{F}_t)$-local square integrable martingale and the predictable process c is such that

$$\int_0^t c_r^2 d < M >_r < \infty \quad P - a.s.$$

In our context, taking into account (20) and (19), the (SC) condition is fulfilled with

$$M_t = \int_0^t \int_Z S_{r^-} \left(e^{K_1(r, X_{r^-}, Y_{r^-}; \zeta)} - 1 \right) (\mathcal{N}(dr, d\zeta) - \nu(d\zeta)dr),$$

$$c_r = \frac{\int_Z (e^{K_1(r, X_{r^-}, Y_{r^-}; \zeta)} - 1)\nu(d\zeta)}{S_{r^-} \int_Z (e^{K_1(r, X_{r^-}, Y_{r^-}; \zeta)} - 1)^2 \nu(d\zeta)},$$

and under (8), the mean-variance tradeoff process is such that

$$K_t = \int_0^t c_r^2 d < M >_r = \int_0^t \frac{(\int_Z (e^{K_1(r, X_{r^-}, Y_{r^-}; \zeta)} - 1)\nu(d\zeta))^2}{\int_Z (e^{K_1(r, X_{r^-}, Y_{r^-}; \zeta)} - 1)^2 \nu(d\zeta)} dr \leq$$

$$\leq \int_0^t \nu(D_1(r, X_r, Y_r))dr < \infty \quad P - a.s.$$

where we recall that D_1 is defined in (4).

Hence by the result proved [1], we get the following proposition

Proposition 2.1.
Under (8), (19) if

$$c_t \Delta M_t < 1 \tag{30}$$

the minimal martingale measure P^* exists and is defined on $(\Omega, \mathcal{F}_T)$ by

$$\frac{dP^*}{dP} = L_T^*$$

where L^* is the Doleans-Dade exponential martingale associated to the $(P, \mathcal{F}_t)$-martingale $m_t = -\int_0^t c_r dM_r$.

Let us observe that condition (30) can be written as

$$\frac{\int_Z (e^{K_1(r, X_{r^-}, Y_{r^-}; \zeta)}) - 1)\nu(d\zeta)}{\int_Z (e^{K_1(r, X_{r^-}, Y_{r^-}; \zeta)} - 1)^2 \nu(d\zeta)} \int_Z (e^{K_1(r, X_{r^-}, Y_{r^-}; \zeta)}) - 1)\mathcal{N}(\{t\}, d\zeta) < 1$$

or equivalently, in terms of the local characteristics of Y, as

$$\frac{\int_{\mathbb{R}}(e^z-1)\Phi_t(dz)}{\int_{\mathbb{R}}(e^z-1)^2\Phi_t(dz)}(e^{\Delta Y_t}-1)<1.$$

2.3. *Existence of optimal strategies*

In [6] it is proved that an optimal strategy corresponds to the Follmer-Schweizer decomposition, more precisely

Proposition 2.2.

The existence of an optimal strategy is equivalent to a decomposition

$$H(S_T)=H_0+\int_0^T \xi_r^H dS_r+L_T^H \tag{31}$$

with H_0 square-integrable $\mathcal{F}_0$-measurable random variable, ξ^H predictable and satisfying (26), L^H square-integrable martingale orthogonal to M.

For such decomposition, the associated optimal strategy (ξ^*,η^*) is given by

$$\xi^*=\xi^H,\quad \eta^*=V(\xi^*,\eta^*)-\xi^*S$$

with

$$V_t(\xi^*,\eta^*)=H_0+\int_0^t \xi_r^H dS_r+L_t^H.$$

In [6], when S has continuous paths, it has been proved that the above decomposition is uniquely determined and coincides with the Kunita-Watanabe decomposition under the minimal martingale measure. Hence the optimal strategy exists and can be computed in terms of the minimal martingale measure. This result is obtained by using the property that the minimal martingale measure preserves orthogonality (see Theorem 3.5 of [6]), property which is not satisfied in the case where S has discontinuous paths. But, even if S is not continuous, if an optimal strategy exists, the value process associated to it can be computed again as the conditional expectation of the contingent claim $H(S_T)$ under the minimal martingale measure, as we will prove in the following proposition.

Proposition 2.3.

Assume (15), (16), (19) and (30).

If there exists an optimal strategy (ξ^*,η^*), the value process is given by

$$V_t(\xi^*,\eta^*)=E^{P^*}(H(S_T)\mid\mathcal{F}_t)=l(t,X_t,S_t) \tag{32}$$

where if $l \in C_b^{1,2}([0,T] \times \mathbb{R} \times \mathbb{R}^+)$ *it is a solution of the following integro-differential equation*

$$L^* l(t,x,y) = \frac{\partial l}{\partial t}(t,x,y) + b(x)\ \frac{\partial l}{\partial x}(t,x,y) + \frac{1}{2}\ \sigma(x)^2\ \frac{\partial^2 l}{\partial x^2}(t,x,y) + \quad (33)$$

$$\int_Z \Big(l\big(t, x + K_0(t,x;\zeta), ye^{K_1(t,x,log(y/S_0);\zeta)}\big) - l(t,x,y)\Big)(1 + U^*(t,x,y;\zeta))\nu(d\zeta) =$$

$$l(T,x,y) = H(y)$$

with

$$U^*(r,x,y;\zeta) = -(e^{K_1(r,x,log(y/S_0);\zeta)} - 1)\frac{\int_Z (e^{K_1(r,x,log(y/S_0);\zeta)} - 1)\nu(d\zeta)}{\int_Z (e^{K_1(r,x,log(y/S_0);\zeta)}) - 1)^2 \nu(d\zeta)}. \quad (34)$$

Proof.

Since $C(\xi^*, \eta^*)$ is a square-integrable $(P, \mathcal{F}_t)$-martingale orthogonal to M under P we have that it is a $(P^*, \mathcal{F}_t)$-martingale hence we get

$$E^{P^*}(H(S_T) \mid \mathcal{F}_t) = E^{P^*}(V_T(\xi^*, \eta^*) \mid \mathcal{F}_t) =$$

$$= E^{P^*}(C_T(\xi^*, \eta^*) + \int_0^T \xi_r^* dS_r \mid \mathcal{F}_t) = V_t(\xi^*, \eta^*).$$

By a suitable version of Girsanov Theorem ([3]), since

$$-\int_0^t c_r dM_r = \int_0^t U^*(r, X_{r^-}, S_{r^-}; \zeta)(\mathcal{N}(dr, d\zeta) - \nu(d\zeta)dr)$$

where $U^*(r,x,y;\zeta)$ is given in (34), we get that the $(P^*, \mathcal{F}_t)$- compensator of the integer-valued random measure $\mathcal{N}(dr, d\zeta)$ is given by

$$\nu^{P^*}(dr, d\zeta) = (1 + U^*(r, X_{r^-}, S_{r^-}; \zeta))\nu(d\zeta)dr.$$

Finally, by Itô formula we get that for any $f \in C_b^{1,2}([0,T] \times \mathbb{R} \times \mathbb{R}^+)$

$$f(t, X_t, S_t) = f(0, x_0, S_0) + \int_0^t L^* f(r, X_r, S_r)dr + m_t \quad (35)$$

where L^* is given in (33). By (15) and (16)

$$m_t = \int_0^t \sigma(X_r)\ \frac{\partial f}{\partial x}(r, X_r, S_r)dW_r +$$

$$\int_0^t \int_Z \Big(f\big(r, X_{r^-} + K_0(r, X_{r^-}; \zeta), S_{r^-} e^{K_1(r, X_{r^-}, Y_{r^-}; \zeta)}\big) - f(r, X_{r^-}, S_{r^-})\Big) \cdot$$

$$\cdot(1 + U^*(r, X_{r^-}, S_{r^-}; \zeta))(\mathcal{N}(dr, d\zeta) - \nu(d\zeta)dr)$$

is a $(P^*, \mathcal{F}_t)$-martingale. To this end it is sufficient to observe that

$$\int_Z \mid \Big(f(r, x+K_0(r, x; \zeta), ye^{K_1(r,x,log(y/S_0);\zeta)}) - f(r, x, y)\Big)(1+U^*(r, x, y; \zeta)) \mid \nu(d\zeta) \le$$

$$2\|f\|\Big(\nu\big(D_0(t, x, log(y/S_0))\big) + \nu\big(D_1(t, x, log(y/S_0))\big) + \int_Z \mid U^*(r, x, y; \zeta) \mid \nu(d\zeta)\Big)$$

and

$$\int_Z \mid U^*(r, x, y; \zeta) \mid \nu(d\zeta) \le \frac{(\int_Z \mid e^{K_1(r,x,log(y/S_0);\zeta)} - 1 \mid \nu(d\zeta))^2}{\int_Z (e^{K_1(r,x,log(y/S_0);\zeta)} - 1)^2 \nu(d\zeta)} \le$$

$$\le \nu\big(D_1(r, x, log(y/S_0))\big).$$

Hence under P^* the Markovianity of the pair (X, S) is preserved and L^* provides its generator.

Now, for $f = l$ in (35), since $l(t, X_t, S_t)$ is a $(P^*, \mathcal{F}_t)$-martingale all finite variations terms have to vanish and this leads to equation (33). □

Notice that analytical solutions to equation (33) are difficult to find but one could search approximating solutions. Otherwise one could compute the expectation in (32) by Monte Carlo simulations. This problem has been mentioned in [10] where related references are given.

2.4. *The martingale case*

In the sequel we shall assume (8), (19) and

$$\forall t \in [0, T], x \in I\!R, y \in I\!R \qquad \int_Z (e^{K_1(t,x,y;\zeta)} - 1)\nu(\zeta) = 0 \tag{36}$$

which ensure that P is a martingale measure for S.

Taking into account (23) condition (36) means, when (12) holds, that

$$E[e^{Z_n} - 1 \mid \mathcal{F}_{T_n^-}] = \int_{I\!R} (e^z - 1)\Phi_{T_n}(dz) = 0,$$

and this condition should be compared with that given in [8], where it is assumed that $E[e^{Z_n} - 1] = 0$.

The martingale case is discussed in [5]. In the sequel we will recall the main results there proved. Since S is a martingale we have that the decomposition (31) is given by the Kunita-Watanabe one and the following hold

Proposition 2.4.

Under the hypotheses (15), (16) and (19), let us define

$$g(t, X_t, S_t) := E(H(S_T) \mid \mathcal{F}_t). \tag{37}$$

If $g \in C_b^{1,2}([0,T] \times I\!R \times I\!R^+)$ *it is a solution of the following integro-differential equation*

$$Lg(t,x,y) = \frac{\partial g}{\partial t}(t,x,y) + b(x)\,\frac{\partial g}{\partial x}(t,x,y) + \frac{1}{2}\,\sigma(x)^2\,\frac{\partial^2 g}{\partial x^2}(t,x,y) + \tag{38}$$

$$\int_Z \Big(g\big(t, x + K_0(t,x;\zeta), y e^{K_1(t,x,log(y/S_0);\zeta)}\big) - g(t,x,y)\Big)\,\nu(d\zeta) = 0$$

$$g(T,x,y) = H(y).$$

Furthermore, the risk-minimizing hedging strategy (ξ^*, η^*) *is given by*

$$\xi_t^* = \frac{h(t, X_{t^-}, S_{t^-})}{S_{t^-}\Sigma(t, X_{t^-}, S_{t^-})} \tag{39}$$

$$\eta_t^* = g(t, X_{t^-}, S_{t^-}) - \xi_t^* S_t$$

where

$$h(t, X_{t^-}, S_{t^-}) = \int_Z \Big(e^{K_1(t,X_{t^-},Y_{t^-};\zeta)} - 1\Big)\cdot$$

$$\cdot\Big(g\big(t, X_{t^-} + K_0(t, X_{t^-};\zeta), S_{t^-} e^{K_1(t,X_{t^-},Y_{t^-};\zeta)}\big) - g(t, X_{t^-}, S_{t^-})\Big)\,\nu(d\zeta) \tag{40}$$

$$\Sigma(t, X_{t^-}, S_{t^-}) = \int_Z \big(e^{K_1(t,X_{t^-},Y_{t^-};\zeta)} - 1\big)^2 \nu(d\zeta) \tag{41}$$

and $Y_{t^-} = log(S_{t^-}/S_0)$.

The criterion of risk minimization is also well suited to deal with restricted information. We assume now that the hedger has access only to the information given by the past asset price, that is the filtration generated by S, $\mathcal{F}_t^S = \sigma\{S_r : r \le t\}$ which coincides with the filtration generated by Y, $\mathcal{F}_t^Y = \sigma\{Y_r : r \le t\}$.

In this framework we restrict our attention to $\{\mathcal{F}_t^S\}$-strategy and, as in [17,6], we consider the $\{\mathcal{F}_t^S\}$-risk process of an $\{\mathcal{F}_t^S\}$- strategy defined by

$$R_t^S(\xi,\eta) = E\Big((C_T(\xi,\eta) - C_t(\xi,\eta))^2 \mid \mathcal{F}_t^S\Big). \tag{42}$$

In [17] is proved that there exists a unique H-admissible, $\{\mathcal{F}_t^S\}$-risk minimizing strategy (ξ', η'), where ξ' is given by the Radon-Nikodym derivative of $< V(\xi^*, \eta^*), S >^{p,\mathcal{F}_t^S}$, with respect to $< S >^{p,\mathcal{F}_t^S}$:

$$\xi'_t = \frac{d < V(\xi^*, \eta^*), S >_t^{p,\mathcal{F}^S}}{d < S >_t^{p,\mathcal{F}^S}}$$

where (ξ^*, η^*) is the H-admissible, $\{\mathcal{F}_t\}$-risk minimizing strategy and, for any locally integrable process A of finite variation, $A^{p,\mathcal{F}^S}$ denotes the $\{\mathcal{F}_t^S\}$-predictable projection (see [11] for details).

Moreover η' is given by

$$\eta'_t = V_t(\xi', \eta') - \xi'_t S_t$$

and the value process is such that

$$V_t(\xi', \eta') = E[H(S_T) \mid \mathcal{F}_t^S] = E[g(t, X_t, S_t) \mid \mathcal{F}_t^S].$$

Hence, since

$$< V(\xi^*, \eta^*), S >_t = \int_0^t S_{r^-} h(r, X_{r^-}, S_{r^-}) dr, \quad < S >_t = \int_0^t S_{r^-}^2 \Sigma(r, X_{r^-}, S_{r^-}) dr$$

we get

$$\xi'_t = \frac{E[h(t, X_{t^-}, S_{t^-}) \mid \mathcal{F}_{t^-}^S]}{S_{t^-} E[\Sigma(t, X_{t^-}, S_{t^-}) \mid \mathcal{F}_{t^-}^S]} = \frac{\pi_{t^-}(h(t, \cdot, S_{t^-}))}{S_{t^-} \pi_{t^-}(\Sigma(t, \cdot, S_{t^-}))}$$

$$\eta' = \pi_t(g(t, \cdot, S_t)) - \xi'_t S_t$$

where π_{t^-} denotes the left-continuous version of the filter π_t. The filter π_t is the probability measure-valued $\mathcal{F}_t^S$-adapted process such that for any f bounded measurable function on $\mathbb{R}$

$$\pi_t(f(\cdot)) = E(f(X_t) \mid \mathcal{F}_t^S).$$

Thus the knowledge of the filter allows us to compute our strategy under restricted information.

For a discussion of the filtering problem see [4, 5].

References

1. J.P. Ansel, C. Stricker, *Unicité et existence de la loi minimale* in Séminaire de Probabilité XXVII, Lecture Notes in Mathematicas, Springer, 22-29.
2. P.Brémaud, *Point Processes and Queues*, Springer-Verlag (1980).

3. T. Bjork, Y. Kabanov and W. Runggaldier, *Bond market structure in presence of marked point processes*, Mathematical Finance **7** (2) (1997) 211-223.
4. C. Ceci and A. Gerardi, *A model for high frequency data under partial information: a filtering approach*, International Journal of Theoretical and Applied Finance **9** (4) (2006) 1-22.
5. C. Ceci, *Risk minimizing hedging for a partially observed high frequency data model*, Stochastics, **78** (1) (2006) 13-31.
6. H. Follmer and M. Schweizer, *Hedging of contingent claims under incomplete information*, in Applied Stochastic Analysis, Stochastic Monographs, 5, Gordon & Breach, London/New York, (1991) 389-414.
7. H. Follmer and D. Sondermann, *Hedging of non-redundant contingent claims*, in Contributions to Mathematical Economics, eds. W. Hilderbrand and A. Mas-Colell, North Holland, (1986) 205-223.
8. R. Frey, *Risk minimization with incomplete information in a model for high-frequency data*, Mathematical Finance **10** (2) (2000) 215-22.
9. R. Frey and W. Runggaldier, *Risk-minimization hedging strategies under restricted information: The case of stochastic volatility models observed only at discrete random times*, Mathematical Methods of Operations Research **50** (1999) 339-350.
10. R. Frey and W. Runggaldier, *A nonlinear filtering approach to volatility estimation with a view towards high frequency data*, International Journal of Theoretical and Applied Finance **4** (2) (2001) 199-210.
11. J. Jacod, *Calcul stochastique et problemes de martingales*, Springer-Verlag (1979).
12. M. Minozzo and S. Centanni, *Nonlinear Filtering using Reversible Jump Markov Chain Monte Carlo in a Model for High Frequency Data*, Proceedings of the "Convegno S.CO.", Treviso (2003), pp. 290-295.
13. J. L. Prigent, *Option pricing with a General Marked Point Process*, *Mathematics of Operations Research* **26** (1) (2001) 50-66.
14. L. C. G. Rogers and O. Zane, *Designing models for high-frequency data*, Preprint, University of Bath (1998).
15. T. Rydberg and N. Shephard, *A modelling framework for prices and trades made at the New York stock exchange*, Nuffield College working paper series 1999-W14, Oxford, *Nonlinear and Nonstationary Signal Processing*, eds. W.J. Fitzgerald *et al.*, Cambridge University Press (2000), pp. 217-246.
16. M. Schweizer, *Option hedging for semimartingales*, Stochastic Processes and their Applications **37** (1991) 339-363.
17. M. Schweizer, *Risk minimizing hedging strategies under restricted information*, Mathematical Finance **4** (1994) 327-342.
18. M. Schweizer, *A guided tour through quadratic hedging approaches*, in *Option Pricing, Interest Rate and Risk Management*, E. Jouini, J. Cvitanic, M. Musiela (eds.), Cambridge University Press (2001) 538-574.

MATHEMATICAL MODELS AND METHODS IN MICRO-NANO-TECHNOLOGIES

C. CERCIGNANI*, M. LAMPIS and S. LORENZANI
Dipartimento di Matematica, Politecnico di Milano, Milano, Italy 20133
**E-mail: carcer@mate.polimi.it*

Keywords: Boltzmann equation; slider-bearing; squeezed-film dampers.

1. Introduction

Microdevices are often operated in gaseous environments (typically air), and thus their performances are affected by the gas around them. In gas film lubrication problems, typical examples are the start/stop operations of hydrodynamic gas-lubricated bearings or flying head sliders employed in magnetic disk storage devices. Beyond the lubrication problems, microstructures that move in the direction parallel or perpendicular to their surfaces are used in surface-micromachined inertial sensors, resonating filter structures for signal processing and in micromachined capacitive accelerometers. At low pressures or in ultra-thin films, the mean free path of the gas molecules is not negligible compared with the film thickness. Thus, in this flow regime, the continuum equations are no longer valid and the Boltzmann equation must be considered. Since the basic constituent of the MEMS devices is the microchannel, where the gas flow is usually at low Mach number, the complicated structures typical of real microdevices can be investigated by solving highly idealized problems between parallel plates, such as plane Couette and Poiseuille flows.

2. The Poiseuille-Couette problem

Let us consider two plates separated by a distance h and a gas flowing parallel to them in the x direction due to a pressure gradient. The lower boundary (placed at $z = -h/2$) moves to the right with velocity U, while the upper boundary (placed at $z = h/2$) is fixed. Both boundaries are held at a constant temperature T_0. If the pressure gradient is taken to be small as well as the velocity U, it can be assumed that the velocity distribution of the flow is nearly the same as that occurring in an equilibrium state. This means that the Boltzmann equation can be linearized about a Maxwellian f_0 by putting:

$$f = f_0(1 + \tilde{h}) \tag{1}$$

where $f(x, z, \mathbf{c})$ is the distribution function for the molecular velocity $\mathbf{c}$ expressed in units of $(2RT_0)^{1/2}$ (R being the gas constant), z is the coordinate normal to the plates and $\tilde{h}(z, \mathbf{c})$ is the small perturbation upon the basic equilibrium state. Using Eq. (1), which defines the unknown perturbation function $\tilde{h}(z, \mathbf{c})$, the linearized Boltzmann equation reads:

$$kc_x + c_z \frac{\partial \tilde{h}}{\partial z} = L\tilde{h} \tag{2}$$

where $L\tilde{h}$ is the linearized collision operator and $k = 1/p\partial p/\partial x = 1/\rho\partial\rho/\partial x$, with p and ρ being the gas pressure and density, respectively. Since the collision integral has a rather complex form, simpler models have been introduced and are commonly considered to reduce the numerical costs of the Boltzmann equation solution. The simplest and most widely used model is the so-called BGK model introduced by Bhatnagar, Gross and Krook.[1] This model has the advantage of describing the right fluid limit but it leads to the wrong Prandtl number. Hence, even in the frame of a linearized analysis, one is induced to suspect that the incorrect Prandtl number can be influent in the transitional regime. Therefore, it appears worthwhile to investigate the Poiseuille-Couette problem, as an issue of relevance for applications, through a model equation more refined than the BGK one, that is the ellipsoidal statistical (ES) model.[2] The ES model, which allows the Prandtl number to take on its correct value, reads:

$$L\tilde{h} = \frac{1}{\theta'}\left[\nu + 2\mathbf{c}\cdot\mathbf{q} + \frac{2}{3}\tau(c^2 - 3/2) - \lambda c_i c_j P_{ij} + \frac{\lambda}{2}(\nu + 2/3\tau)c^2 - \tilde{h}\right] \tag{3}$$

where θ' is a suitable mean free time; ν and τ are the perturbations of the density ρ_0 and temperature T_0, respectively:

$$\nu = \pi^{-3/2} \int e^{-c^2} \tilde{h}(z, \mathbf{c}) d\mathbf{c} \tag{4}$$

$$\tau = \pi^{-3/2} \int (c^2 - 3/2) e^{-c^2} \tilde{h}(z, \mathbf{c}) d\mathbf{c} \ ; \tag{5}$$

$\mathbf{q}$ is the perturbation of the bulk velocity (in $(2RT_0)^{1/2}$ units)

$$\mathbf{q} = \pi^{-3/2} \int \mathbf{c}\, e^{-c^2} \tilde{h}(z, \mathbf{c}) d\mathbf{c} \tag{6}$$

and P_{ij} is the stress tensor (in $2\rho_0 RT_0$ units)

$$P_{ij} = \pi^{-3/2} \int c_i c_j e^{-c^2} \tilde{h}(z, \mathbf{c}) d\mathbf{c}. \tag{7}$$

Integrations are extended to the whole velocity space. In Eq. (3) λ is a constant to be chosen in such a way as to have the correct Prandtl number: λ is equal to 0 for the BGK model (Pr $=$ 1) and 1 for a Maxwell gas (Pr $= 2/3$).

Multiplying Eq. (2), with $L\tilde{h}$ given by Eq. (3), by $(c_x/\pi)\exp[-(c_x^2 + c_y^2)]$ and integrating with respect to c_x and c_y, it turns out:

$$\frac{1}{2}k + c_z \frac{\partial Z}{\partial z} = \frac{1}{\theta'} \left[\pi^{-1/2} \int_{-\infty}^{+\infty} e^{-c_{z1}^2} Z(z, c_{z_1})\, dc_{z_1} - \lambda c_z \pi^{-1/2} \cdot \int_{-\infty}^{+\infty} e^{-c_{z1}^2} c_{z_1} Z(z, c_{z_1})\, dc_{z_1} - Z(z, c_z) \right] \tag{8}$$

where by definition:

$$Z(z, c_z) = \pi^{-1} \int_{-\infty}^{+\infty} \int_{-\infty}^{+\infty} e^{-(c_x^2 + c_y^2)} c_x \tilde{h}(z, \mathbf{c})\, dc_x\, dc_y \ .$$

From Eq. (8) one can obtain the momentum conservation equation:

$$\frac{k}{2} + \frac{\partial}{\partial z} P_{xz} = 0 \tag{9}$$

The integration of Eq. (9) gives

$$P_{xz}(z) = -\frac{k}{2} z + \Pi \tag{10}$$

where Π is the integration constant. Therefore Eq. (8) can be rewritten as

$$\frac{1}{2}k + c_z\frac{\partial Z}{\partial z} = \frac{1}{\theta'}\left[\pi^{-1/2}\int_{-\infty}^{+\infty} e^{-c_{z_1}^2} Z(z, c_{z_1})\, dc_{z_1} + \frac{k}{2}\lambda c_z z - \lambda c_z \Pi \right.$$
$$\left. - Z(z, c_z)\right] \tag{11}$$

Appropriate boundary conditions on the two plates must be supplied for the Boltzmann equation (11) to be solved. If one assumes that the z component of the bulk velocity of the gas, defined by Eq. (6), is a known quantity, the integrodifferential Boltzmann equation (11) can be formally handled as an ordinary inhomogeneous differential equation whose solution reads:

$$Z(z, c_z) = \exp(-(z + \tfrac{h}{2}\text{sgn}c_z)/(c_z\theta'))Z(-\tfrac{h}{2}\text{sgn}c_z, c_z) + \int_{-\frac{h}{2}\text{sgn}c_z}^{z}$$
$$\exp\left(\tfrac{-|z-t|}{|c_z|\theta'}\right) \times [q(t) - k\theta'/2 - \lambda c_z \Pi + \lambda k c_z t/2]/(c_z\theta')\, dt \tag{12}$$

The values of the Z function at the boundary, $Z(-\frac{h}{2}\text{sgn}c_z, c_z)$, depend on the model of boundary conditions chosen. In the following, we will consider the Maxwell boundary conditions and specialize the analysis to walls having different physical properties so that two accommodation coefficients (α_1, α_2) must be used. In this case, the boundary conditions can be written as[3]

$$Z^+(h/2, c_z) = (1 - \alpha_1)Z^-(h/2, -c_z) \tag{13}$$

$$Z^+(-h/2, c_z) = \alpha_2 U + (1 - \alpha_2)Z^-(-h/2, -c_z) \tag{14}$$

where U is expressed in units of $(2RT_0)^{1/2}$; $Z^-(-h/2, c_z)$, $Z^-(h/2, c_z)$ are the distribution functions of the molecules impinging upon the walls and $Z^+(-h/2, c_z)$, $Z^+(h/2, c_z)$ the distribution functions of the molecules reemerging from them. Once the function at the boundary, $Z(-\frac{h}{2}\text{sgn}c_z, c_z)$, has been evaluated following the analytical procedure reported in Ref. 4, the substitution of the integral formula (12) in the definition of $q(z)$ gives the equation for the bulk velocity of the gas:

$$q^\lambda(z) = \frac{1}{2}k\theta'[1 - \psi_p^\lambda(u')] + U\psi_c^\lambda(u') \tag{15}$$

where the explicit form of the non-dimensional functions $\psi_p^\lambda(u')$ and $\psi_c^\lambda(u')$, giving the Poiseuille and Couette contributions, respectively, has been reported in a manuscript under revision and $u' = z/\theta'$.

Until now we have not mentioned the relation between θ' and the collision time θ defined in the BGK model. In order to get the same viscosity coefficient from the BGK model and the present one, we must put: $\theta' = (\lambda + 2)/2\theta$. Therefore, the rarefaction parameter $\delta' = h/\theta'$ can be rewritten in terms of the inverse Knudsen number $\delta = h/\theta$, appearing in the BGK solution of the Poiseuille-Couette problem, as follows: $\delta' = (2\delta)/(2+\lambda)$.

Using Eq. (15), the flow rate (per unit time through unit thickness) defined by:

$$F = \rho \int_{-h/2}^{h/2} q(z)dz \tag{16}$$

can be expressed as the sum of the Poiseuille flow (F_p^λ) and the Couette flow (F_c^λ) as follows:

$$F^\lambda = F_p^\lambda + F_c^\lambda = -\frac{\partial p}{\partial x} h^2 Q_p^\lambda(\delta, \alpha_1, \alpha_2) + \frac{\rho U h}{2} Q_c^\lambda(\delta, \alpha_1, \alpha_2) \tag{17}$$

where

$$Q_p^\lambda(\delta, \alpha_1, \alpha_2) = -\frac{1}{\delta'} + \frac{1}{\delta'^2} \int_{-\delta'/2}^{\delta'/2} \psi_p(u')du'$$

$$Q_c^\lambda(\delta, \alpha_1, \alpha_2) = \frac{2}{\delta'} \int_{-\delta'/2}^{\delta'/2} \psi_c(u')du'$$

are the non-dimensional volume flow rates.

3. The Reynolds equation and the flow of a rarefied gas in a microchannel

The analysis developed in the previous section can be applied to model gas flows in very small gaps between the moving surfaces of micromechanical structures. Two typical kinds of motions experienced by the planar surfaces of a microchannel are:

(a) the lateral motion of the lower boundary with slight inclination of the upper plate (slider bearing);

(b) normal oscillations between the two bounding surfaces (squeezed film).

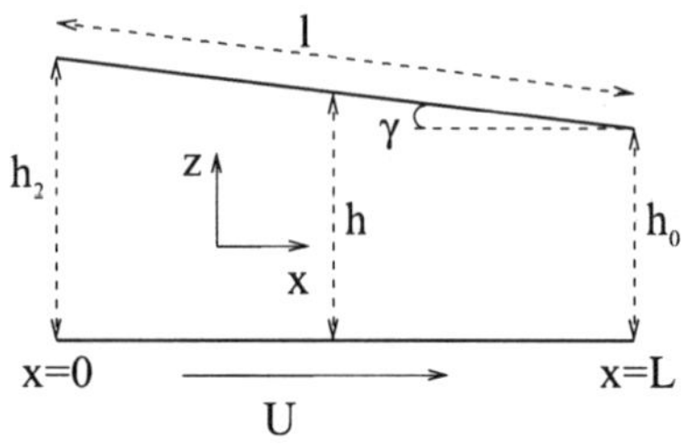

Fig. 1. Geometry of a slider bearing.

All of these situations can be analyzed using the Reynolds equation for a thin gas film.

3.1. *Slider-bearing problem*

In lubrication theory, gaseous layers are usually found between two solid bodies which are acted upon by forces (such as gravity) tending to push them together. To carry this load, the gas layer must develop normal stresses, largely dominated by pressure. Since the variations of the gas film thickness are very slow, the pressure distribution can be obtained by solving idealized problems between parallel plates, such as the plane Poiseuille-Couette flow considered in the previous section.[5]

The starting point to obtain the rarefied version of the Reynolds equation for lubrication is the mass balance equation. This equation is considerably simplified by the fact that the variations of density do not show up for slow motion in the steady flow case, which is an assumption usually fulfilled in the most important MEMS applications:

$$\frac{\partial u}{\partial x} + \frac{\partial v}{\partial y} + \frac{\partial w}{\partial z} = 0 \tag{18}$$

where the three velocity components are denoted by u, v, w. Let us consider a two-dimensional layer of gas between two walls located at $z = 0$ and $z = h$; the lower wall moves to the right with velocity U (see Fig. 1).

Integrating Eq. (18) across the layer (the y-direction has been suppressed) and taking into account that the normal velocity component at the boundaries is given by: $w = 0$ at $z = 0$, $w = u\partial h/\partial x$ at $z = h$, one obtains

$$\frac{\partial}{\partial x}\int_0^h u dz = 0 \tag{19}$$

Since the problem is linear and the pressure gradient is assumed to be constant across the layer, u is proportional to the sum of the velocities given by a Poiseuille flow with pressure gradient dp/dx and a Couette flow with the lower wall moving with velocity component U. Therefore, Eq. (19) turns out to be: $\partial F/\partial x = 0$, with the flow rate F given by Eq. (17).

Thus, the non-dimensional generalized Reynolds equation, modified to take into account the gas rarefaction effects, reads

$$\frac{d}{dX}\left(\tilde{Q}_p^\lambda(\delta_o PH, \alpha_1, \alpha_2) PH^3 \frac{dP}{dX} - Q_c^\lambda(\delta_o PH, \alpha_1, \alpha_2)\Lambda PH\right) = 0 \qquad (20)$$

where the following dimensionless quantities have been introduced: $X = x/l$, $P = p/p_o$, $H = h/h_o$; $\tilde{Q}_p^\lambda$ is the Poiseuille relative flow rate: $\tilde{Q}_p^\lambda = Q_p^\lambda/Q_{con}$, with $Q_{con} = \delta/6$ being the continuum flow limit; Λ is the bearing number defined as: $\Lambda = (6\mu U l)/(p_o h_o^2)$ with μ being the dynamic viscosity of the gas and p_o the ambient pressure. Furthermore, the rarefaction parameter δ has been expressed as: $\delta = \delta_o PH$, with δ_o being the characteristic inverse Knudsen number defined by the minimum film thickness, h_o, and the ambient pressure p_o: $\delta_o = (p_o h_o)/(\mu\sqrt{2RT_o})$. If the continuum limit ($\delta \to \infty$) is taken for any fixed $\alpha_1 = \alpha_2$, then the limiting solution for the Poiseuille and Couette flow rates is given by: $\tilde{Q}_p^\lambda \to 1$, $Q_c^\lambda \to 1$, so that Eq. (20) reduces to the classical Reynolds equation used in standard hydrodynamic lubrication theory.

Equation (20), solved numerically using relaxation methods, gives the pressure field in the gas film as a function of the longitudinal coordinate X. A comparison between the Reynolds equation solutions, obtained using the ES ($\lambda = 1$) and BGK models ($\lambda = 0$), and the numerical findings obtained from DSMC (Direct Simulation Monte Carlo) simulations (Alexander et al. 1994) and IP (Information Preservation) method (Jiang et al. 2005) in the case of Maxwell's boundary conditions on two physically identical walls, is shown in Fig. 2. The parameters describing the gas film geometric configuration were fixed at the following values: $h_2/h_o = 2$, $L/h_o = 100$.

Figure 2 shows that the present Reynolds equation solutions, obtained using the ES and BGK models, are in good agreement with the DSMC data presented by Alexander et al. (1994) and the IP results reported by Jiang et al. (2005). It is worth noting that the results of the IP method given by Jiang et al. are closer to the Reynolds equation numerical solutions than the DSMC data obtained previously by Alexander et al. (1994). Furthermore, the picture shows that the differences between the results given by the BGK and ES models are extremely small suggesting that in isothermal

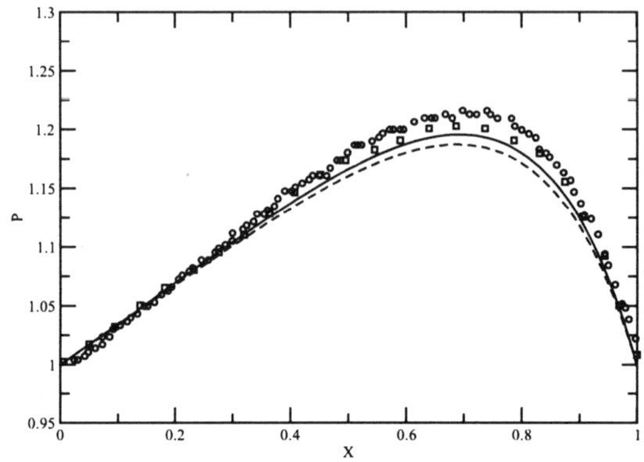

Fig. 2. Pressure profile versus X. Comparison between the Reynolds-BGK results (solid line), the Reynolds-ES results (dashed line), DSMC data (Alexander et al. 1994) (open circles) and IP data (Jiang et al. 2005) (open squares). The parameters are: $\delta_o = 0.7$, $\Lambda = 61.6$, $\alpha_1 = \alpha_2 = 0.7$.

conditions and at low Mach numbers the corrections introduced by a more refined kinetic model of the collisional Boltzmann operator are extremely small. Finally, it is worth stressing that the Reynolds equation solutions, obtained using the ES and BGK models, reveal the existence of inverted pressure profiles in the free-molecular flow regime, when two different accommodation coefficients (α_1, α_2) for the bounding surfaces are considered. Fig. 3 reveals that, for small δ_o, if one keeps the accommodation coefficient of the slider (α_1) fixed and varies the other one (α_2), the pressure distribution in the gas film, at fixed bearing number, increases with increasing α_2, as it always happens in the continuum region, while at fixed α_2, the pressure distribution decreases by increasing α_1. The origin of this kind of inverted pressure profiles can be traced back to the Couette contribution to the lubrication flow rate.[3]

3.2. *Squeezed-film dampers*

Micromechanical accelerometers, characterized by very small gaps between the moving elements and the fixed electrodes, often use a gas as damping medium. The quality factor of these oscillating microstructures damped by a squeezed gas film is dominated by the viscosity of the gas and the desired frequency response of the micromechanical device can be achieved by carefully controlling the gas pressure, which is modelled by using the Reynolds equation.

Let us consider two planar surfaces, parallel to the x-axis, separated

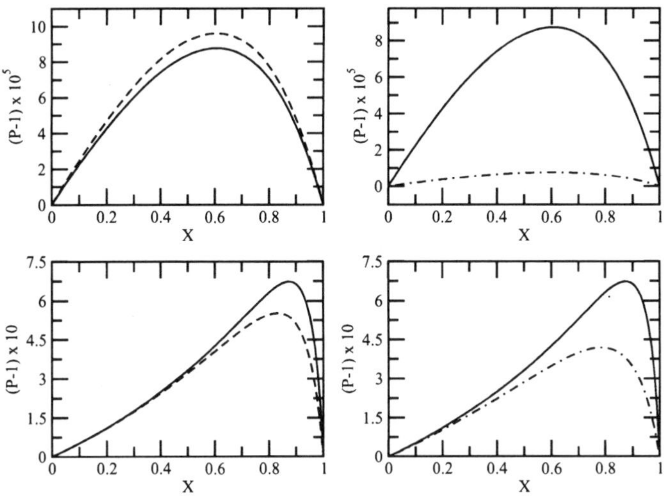

Fig. 3. Pressure profiles, from the Reynolds-BGK equation, versus X for $\Lambda = 50$. The line styles indicate $\alpha_1 = 0.1$ $\alpha_2 = 0.8$ (dashed), $\alpha_1 = 0.8$ $\alpha_2 = 0.8$ (solid), $\alpha_1 = 0.8$ $\alpha_2 = 0.1$ (dot dashed). The inverse Knudsen number δ_o is 10^{-3} (top panels) and 10 (bottom panels). The same trend can be also obtained through the Reynolds-ES equation.

by a distance h with a thin gas film between them. If the plates oscillate in a direction perpendicular to their surfaces, i.e. along the z direction, a Poiseuille-like flux is induced in the gap. The general continuity condition for the two-dimensional gas layer reads:

$$\frac{\partial \rho}{\partial t} + \frac{\partial}{\partial x}(\rho u) + \frac{\partial}{\partial z}(\rho w) = 0 \tag{21}$$

where ρ is the gas density, u and w are the x and z components of the bulk velocity of the gas, respectively. Integrating the mass flow conservation equation (21) across the film thickness and taking into account that the plates are parallel and the normal gas velocity component equals the wall velocity, one obtains

$$\frac{\partial}{\partial t}(\rho h) + \frac{\partial F}{\partial x} = 0 \tag{22}$$

where F is the gas flow rate (per unit time through unit thickness). In deriving Eq. (22) we have assumed that the variations of the gas density ρ in the z direction are negligible compared with the gradients along the longitudinal coordinate of the channel. The Reynolds equation, modified to take into account the gas rarefaction effects, can be derived starting from

Eq. (22) with the flow rate F given by the first term on the right hand side of Eq. (17). Assuming that the heat generation in the gas is negligible so that an isothermal process can be considered $(p - \rho/2)$, one obtains:

$$\frac{\partial}{\partial t}(ph) = \frac{h^3}{12\mu}\frac{\partial}{\partial x}\left[\frac{\partial p}{\partial x} p\tilde{Q}_p^{\lambda}(\delta, \alpha_1, \alpha_2)\right] \tag{23}$$

which reduces to the classical Reynolds equation in the continuum limit. The term on the right hand side of the Reynolds equation (23) represents the flow due to the pressure gradient, while the left hand side shows the flow induced by normal (squeeze) motions of the bounding surface.

If one assumes that the variation of plate spacing, δz, is small compared with the gap h, and that the pressure variations are small compared with the static pressure level p_a, Eq. (23) can be linearized

$$\frac{h^2 p_a}{12\mu_{\text{eff}}}\frac{\partial^2}{\partial x^2}\left(\frac{p}{p_a}\right) - \frac{\partial}{\partial t}\left(\frac{p}{p_a}\right) = \frac{\partial}{\partial t}\left(\frac{\delta z}{h}\right) \tag{24}$$

where p_a is the static pressure, $p(x,t)$ is the small variation in the pressure distribution, h is the initial gap between the plates, δz is the small variation in the plate gap. μ_{eff} is an effective viscosity expressed as a function of the rarefaction parameter δ and the Poiseuille flow rate coefficient $Q(\delta)$ as follows: $\mu_{\text{eff}} = (\mu\delta)/(6Q(\delta))$

In the continuum limit $Q(\delta) \longrightarrow \delta/6$, so that $\mu_{\text{eff}} \longrightarrow \mu$. Thus, the gas rarefaction effects along with the influence of the accommodation coefficients, are included in Eq. (24) by replacing the static gas viscosity μ with a pressure-dependent effective viscosity. In order to assess the viability of the effective viscosity approach, the numerical results obtained by using Eq. (24) are compared with the experimental data collected on a silicon biaxial accelerometer produced by STMicroelectronics. In spite of its apparently complex structure, a real micromechanical accelerometer usually has a highly repetitive layout whose basic units consist of two or three-dimensional microchannels where different sets of bounding walls move in the direction perpendicular or parallel to their surfaces (see Fig. 4).

Since the analysis performed on the general two-dimensional configuration of the real accelerometer of Fig. 4, by means of the solution of the Boltzmann equation based on the BGK model, has shown that squeeze film flow due to the Poiseuille flow provides the most important contribution to damping forces, one could suspect to be able to predict correctly the global damping, restricting the survey to the longest branches of the

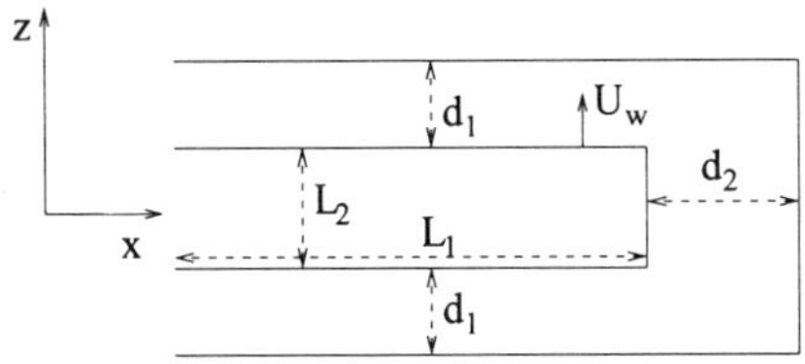

Fig. 4. Geometry of a two-dimensional microchannel. The fixed parameters of the apparatus are: $d_1 = 2.6\ \mu m$; $d_2 = 4.2\ \mu m$; $L_1 = 15\ \mu m$; $L_2 = 3.9\ \mu m$. The central 'rotor' moves with velocity U_w in the z direction while the external boundaries are fixed.

channel of Fig. 4 and applying there the one-dimensional Reynolds equation (24). In the quasi-static configuration of the experimental apparatus, the force exerted by the gas is directly proportional to the viscosity coefficient. Therefore, if F^a is the damping force evaluated at 1 bar, and F the damping force at decreasing pressures, the following formula holds:

$$F/F^a = \mu_{\text{eff}}/\mu_{\text{eff}}^a \tag{25}$$

where μ_{eff}^a is the modified viscosity at ambient pressure. Rather than directly presenting forces, the comparison with experiments is performed in terms of modified viscosity employing Eq. (25). In Fig. 5 the plot of $\log(\mu_{\text{eff}}/\mu_{\text{eff}}^a)$ versus $-\log \mathrm{p}$ is displayed where p is the pressure in bar, and $-\log \mathrm{p} = \log(\mathrm{Kn}/\mathrm{Kn}^{\mathrm{a}})$, with Kn^{a} being the reference Knudsen number at 1 bar for the structure at hand:

$$\mathrm{Kn}^{\mathrm{a}} = \ell/\mathrm{d}_1 = 0.064/2.6 = 0.0246 \tag{26}$$

and ℓ being the molecular mean free path. Unfortunately the agreement between the effective viscosity computed with the one-dimensional kinetic model and the experimental one is not satisfactory and errors explode at low pressures. However, it is worth recalling that formulas for μ_{eff} have been obtained under several hypotheses which are not completely fulfilled here. In particular, it is assumed that the flow is exactly one-dimensional and the length of the channel is infinite in the direction of the flow. Indeed, in the real MEMS configuration presented in Fig. 4 end effects cannot be neglected and a secondary flow also develops along the length of the plates. Actually, the comparison of the two data-set reported in Fig. 5 suggests that it is as if the reference Knudsen number for the flow between parallel plates were different from the one computed in Eq. (26). Since in the slip-flow pressure range, damping forces are accurately evaluated using a continuum numerical code corrected with slip boundary conditions, a new

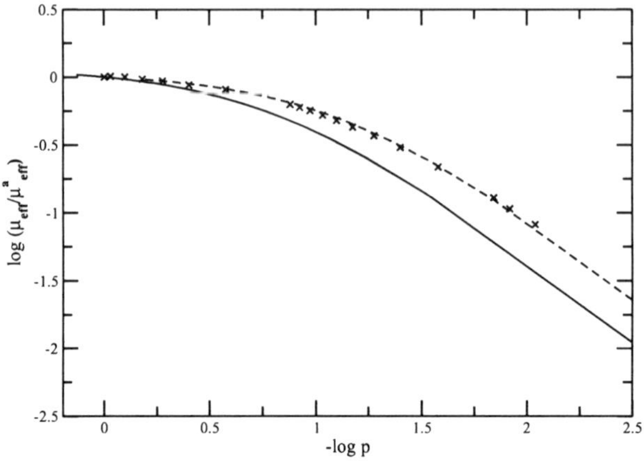

Fig. 5. Comparison between the normalized damping forces measured experimentally (crosses) and the one-dimensional approximation based on the Reynolds equation: the solid line is obtained using as reference Knudsen number the value given by Eq. (26), while the dashed line is obtained using the slip-corrected reference Knudsen number given by $Kn^a = 0.012$.

reference Knudsen number can be computed on the basis of simple slip flow simulations.[8] Figure 5 shows that the numerical results obtained using the one-dimensional Reynolds equation on the basis of a slip-corrected reference Knudsen number are in good agreement with the experimental results in the transitional regime.

References

1. P.L. Bhatnagar, E.P. Gross and M. Krook, *Phys. Rev.* **94**, 511 (1954).
2. C. Cercignani and G. Tironi, *Il Nuovo Cimento* **43**, 64 (1966).
3. C. Cercignani, M. Lampis and S. Lorenzani, *Phys. Fluids* **18**, 087102 (2006).
4. C. Cercignani, M. Lampis and S. Lorenzani, *Phys. Fluids* **16**, 3426 (2004).
5. C. Cercignani, *Slow Rarefied Flows. Theory and Application to Micro-Electro-Mechanical Systems* (Birkhäuser, Basel, 2006).
6. F.J. Alexander, A.L. Garcia and B.J. Alder, *Phys. Fluids* **6**, 3854 (1994).
7. J.Z. Jiang, C. Shen and J. Fan, *in Rarefied Gas Dynamics*, 24th Int. Symp., M. Capitelli Ed., AIP Conf. Proc. 762, 180-185, New York (2005).
8. C. Cercignani, A. Frangi, S. Lorenzani and B. Vigna, *Engineering Analysis with Boundary Elements*, in press (2006).

A GENERAL MODEL FOR WAX DIFFUSION IN CRUDE OILS UNDER THERMAL GRADIENT

E. COMPARINI* and F. TALAMUCCI

Dipartimento di Matematica "U. Dini", Università di Firenze, Firenze, I50134, Italy

**E-mail: elena.comparini@math.unifi.it*

We consider a general model for the complex phenomenon of wax deposition in crude oils. Wax is present either as dissolved in oil or suspended as a crystallized phase. The solubility of wax decreases very sharply with temperature. The presence of a thermal gradient induces both a dynamics of transfer from dissolved to solid phase and the formation of a gel–like deposit layer at the cold wall. The process is described including different stages of evolution: we start from the fully saturated system, then, after the onset of an unsaturated front we deal with the simultaneous presence of saturated and unsaturated regions up to the complete unsaturation of the system.

Keywords: Waxy crude oils; Molecular diffusion; Heat and mass transfer; Wax deposition; Free boundary problem

1. Introduction

Waxy crude oils are mixtures of mineral oils, paraffins, aromatics and other impurities.

The presence of wax components in oil, under particular thermal conditions, causes problems during transport in subsea pipelines, like wax precipitation with the consequent formation of wax crystals, deposition of the crystals on internal walls of pipeline, increasing of viscosity, diminuition of capacity of crude oil, removal of paraffin deposit.

The formation of a solid deposit in pipeline walls is a phenomenon of crucial importance in oil industry because it can cause the blockage of a line, so that this complex phenomenon has been the object of a large number of papers, see the survey paper [7].

We refer to [11], [10] for the description of the mechanism of diffusion in non-isothermal solutions that causes wax deposition. A one-dimensional model describing the phenomenon of thermally induced mass transport in partially saturated solutions under thermal gradient, including the displacement of all species (solvent, solute and segregated phase), is formulated in [10], where a mathematical analysis of the problem and some qualitative results have been obtained. In [11] the authors include in their analysis the case in which, because of a sufficiently low temperature, wax crystals aggregate in a gel-like structure that can be considered as immobile and not subject to diffusion.

In this paper we deal with a theoretical investigation of the process of thermally induced mass transport in a general situation, taking into account both the molecolar diffusion of wax and the displacement of crystals suspended in oil saturated with wax.

The mechanism at the origin of the formation of a deposit at the wall of the pipeline can be summarized remarking that solubility of wax depends on temperature, then dissolved wax under a certain temperature (*cloud point*) precipitates in form of wax crystals. Thermal gradient induces a concentration gradient that in its turn causes migration of wax towards the cold wall where it precipitates. Wax adheres to the wall forming a gel-like layer of increasing thickness.

The mathematical model is based on the assumption that the three components of the system, oil (solvent), dissolved wax (solute) and wax crystals (segregated phase) have the same density (supposed constant in the range of temperature considered). That implies that gravity has no effect and that the segregation/dissolution process does not change volume.

We start from a situation of full saturation of the solution in a three-dimensional domain assuming no flux condition at the boundaries analyzing the formation of the solid wax deposit at the cold wall and the possible desaturation of the solution starting from the warm wall.

2. Mass balance, mass flux

Let be $\Omega \subseteq \mathbb{R}^3$ a connected domain.
We define ϱ_α as the mass concentration (mass per unit volume of the system) of each component, referring to $\alpha = \gamma$ for oil, c for dissolved wax and G for segregated wax. It is also convenient to introduce the mass concentration of the liquid part, say Γ, made of dissolved wax and oil: $\varrho_\Gamma = \varrho_\gamma + \varrho_c$. The general conservation principle writes for each component

$$\frac{\partial \varrho_\alpha}{\partial t} + \nabla \cdot \mathbf{J}_\alpha = I_\alpha \tag{1}$$

where $\mathbf{J}_\alpha$ is the mass flux of the species α and I_α is the rate of mass production or loss.

Mass conservation requires $\sum_{\alpha=\gamma,c,G} I_\alpha = 0$. Furthermore, by the physics of the process is evident that the only mass transfer processes are dissolution or segregation of wax, so that $I_G = -I_c$, $I_\gamma = 0$, and then from (1):

$$\frac{\partial \varrho}{\partial t} + \nabla \cdot \mathbf{J} = 0, \qquad \varrho = \varrho_\gamma + \varrho_c + \varrho_G, \quad \mathbf{J} = \mathbf{J}_\gamma + \mathbf{J}_c + \mathbf{J}_G. \tag{2}$$

A particular attention must be paid to modelling the mass fluxes $\mathbf{J}_\alpha = \varrho_\alpha \mathbf{v}_\alpha$, $\alpha = \gamma, c, G$, where $\mathbf{v}_\alpha$ denotes the velocity of the α species ($\mathbf{J}_\alpha = \varrho_\alpha \mathbf{v}_\alpha$). Following a general scheme encompassing both dilute and concentrated solutions, the mass transfer of each species is due to the additive effects of diffusion and convection.

Let us say that $\mathbf{V}^A$ is some convective reference velocity and write

$$\mathbf{J}_\alpha = \varrho_\alpha(\mathbf{v}_\alpha - \mathbf{V}^A) + \varrho_\alpha \mathbf{V}^A \tag{3}$$

so that mass flux of each component is the sum of the convective flux $\mathbf{J}^A = \varrho_\alpha \mathbf{V}^A$ and the diffusive flux $\mathbf{J}^*_\alpha = \varrho_\alpha(\mathbf{v}_\alpha - \mathbf{V}^A)$.

According to [5], multicomponent diffusion of a N–species system, where the Nth species is designated as the solvent, can be described as

$$\mathbf{J}^*_i = C_i \left(\mathbf{v}_i - \mathbf{V}^A\right) = -\sum_{j=1}^{N-1} D^{A,C}_{i,j} \nabla C_j, \quad i = 1, \ldots, N-1 \tag{4}$$

where C_i denotes the i–concentration (not necessarily mass concentration) and the superscripts A, C in the molecular diffusivities $D^{A,C}_{i,j}$ stress the dependence on the choice of the concentrations (bulk density, mass fraction, volume fraction, relative concentration,...) and on the reference velocity.

The multicomponent diffusion coefficients $D^{A,C}_{i,j}$ are generally not symmetric and the diagonal terms $D^{A,C}_{i,i}$ (main terms) are similar in magnitude to binary values, see [5]. Each cross term $D^{A,C}_{i,j}$, $i \neq j$ is the contribution to the flux of the i solute by the concentration gradient of the j solute. Commonly, off–diagonal diffusion coefficients are much smaller than the main terms diffusion coefficients.

We have to remark that (4) demands no forced pressure and no thermal diffusion: actually, the range of temperature of the process we are studying is so restricted that (4) can be assumed with good approximation.

Let us examine now mass flux $\mathbf{J}_\alpha$ for each specific component.

It must be said that (see [5]) there is no evidence for what convective reference velocity $\mathbf{V}^A$ should be: the choice of the mass average velocity or the velocity of the solvent or other reference velocities depends on the phenomenology of the process.

As for segregated wax, it seems to be reasonable to assume a process of binary diffusion ($N = 2$) where G is the solute and the liquid part Γ (oil + dissolved wax) is the solvent. The reference velocity is in that case the mass average velocity defined by (see also (2)) $\mathbf{V}^* = \sum_{\alpha=\gamma,c,G} \varrho_\alpha \mathbf{v}_\alpha / \varrho = \mathbf{J}/\varrho$ and we write (4) (referring to mass concentrations) simply as $\mathbf{J}_G^* = \varrho_G(\mathbf{v}_G - \mathbf{V}^*) = -D_G \nabla \varrho_G$ where D_G is the binary diffusivity of wax crystals in the mixture.

Therefore, we get from (3)

$$\mathbf{J}_G = -D_G \nabla \varrho_G + \varrho_G \mathbf{J}/\varrho. \tag{5}$$

It is likely to assume that D_G depends on ϱ_G. If the concentration of solid wax is sufficiently high, we may think that D_G reduces to zero and wax crystals are simply transported via convection by the average mass flux. On the other hand, if the liquid component is present in eccess, the mass average velocity is approximately the solvent velocity, which can be assumed as the reference velocity $\mathbf{V}^A$.

Diffusion of dissolved wax c requires a special comment, since two different possibilities can be imagined:

A. dissolved wax diffuses in the mixture,

B. the process of diffusion of c occurs only in the liquid part whose concentration is $\varrho_\Gamma = \varrho_\gamma + \varrho_c$.

In the first case, the situation is identical to what we have just concluded for G in (5), so we write $\mathbf{J}_c = -D_c \nabla \varrho_c + \varrho_c \mathbf{J}/\varrho$, where D_c is the binary diffusivity of dissolved wax in the whole system and with respect to the mass average velocity.

In the latter case, we have to imagine a binary diffusion of c in the liquid part Γ only, disregarding wax crystals displacement.

Following such a point of view, (4) is used in order to link the relative flux of dissolved wax (i. e. with respect to the liquid part only) with the relative concentration of c in Γ. Hence, we write (4) as

$$\mathbf{J}_{c,rel}^* = \varrho_{c,rel}(\mathbf{v}_c - \mathbf{V}^A) = -\widetilde{D}_c \nabla \varrho_{c,rel}$$

where the appropriate choice of the reference velocity is $\mathbf{V}^A = \mathbf{v}_\Gamma = (\varrho_\gamma \mathbf{v}_\gamma + \varrho_c \mathbf{v}_c)/\varrho_\Gamma$ and the relative concentration has to be $\varrho_{c,rel} = \varrho_c/\eta_\Gamma$ with η_Γ volumetric fraction of Γ in a unit volume of the mixture. The coefficient $\widetilde{D}_c$ is the binary diffusivity of dissolved wax in the liquid part.

We finally get

$$\mathbf{J}_c = -\eta_\Gamma \widetilde{D}_c \nabla \frac{\varrho_c}{\eta_\Gamma} + \varrho_c \mathbf{v}_\Gamma = -\eta_\Gamma \widetilde{D}_c \nabla \frac{\varrho_c}{\eta_\Gamma} + \frac{\varrho_c}{\varrho_\Gamma} D_G \nabla \varrho_G + \frac{\varrho_c}{\varrho} \mathbf{J}. \tag{6}$$

The flux of oil is obtained by $\mathbf{J}_\gamma = \mathbf{J} - (\mathbf{J}_c + \mathbf{J}_G)$. Namely, we obtain in case A and B respectively

$$\mathbf{J}_\gamma = D_G \nabla \varrho_G + D_c \nabla \varrho_c + \frac{\varrho_\gamma}{\varrho}\,\mathbf{J}, \tag{7}$$

$$\mathbf{J}_\gamma = \frac{\varrho_\gamma}{\varrho_\Gamma} D_G \nabla \varrho_G + \eta_\Gamma \widetilde{D}_c \nabla \frac{\varrho_c}{\eta_\Gamma} + \frac{\varrho_\gamma}{\varrho}\,\mathbf{J}. \tag{8}$$

Obviously, different mechanisms describing the mass transfer process can be introduced, as, for istance, assuming a ternary diffusion of c, G in the solvent γ and the validity of (4) with $N = 3$, $\mathbf{V}^A = \mathbf{V}^*$ (in that case, the diffusion coefficients $D_{c,c}$, $D_{G,G}$, $D_{c,G}$ and $D_{G,c}$ must be specified).
One can use a volumetric approach defining the volumetric contents as $\gamma = \varrho_\gamma/d_\gamma$, $c = \varrho_c/d_c$, $G = \varrho_G/d_G$, $\Gamma = \gamma + c$, and the volumetric velocities as $\mathbf{J}_\alpha/d_\alpha = \alpha \mathbf{v}_\alpha$, where d_α, $\alpha = \gamma, c, G$ are the specific densities.
If all the components have the same constant specific density d, that is $d_\alpha = d$, $\alpha = \gamma$, c, G, and if $G + c + \gamma = 1$ (volume saturation), we have

$$\begin{cases} \mathbf{J}_G = -d\, D_G \nabla G + G\mathbf{J}, \\ \mathbf{J}_c = -d\, D_c \nabla c + c\mathbf{J}, \\ \mathbf{J}_\gamma = d\,(D_G \nabla G + D_c \nabla c) + \gamma \mathbf{J} \end{cases} \tag{9}$$

or $\mathbf{J}_c = -d\,\Gamma \widetilde{D}_c \nabla(c/\Gamma) + d\,(c/\Gamma) D_G \nabla G + c\mathbf{J}$ depending on the choice of assumption either A or B, respectively, for $\mathbf{J}_c$.
The corresponding balance equations for G and c are

$$\frac{\partial G}{\partial t} - \nabla \cdot [D_G \nabla G] + \nabla G \cdot \mathbf{J}/d = I_G/d \tag{10}$$

$$\frac{\partial c}{\partial t} - \nabla \cdot [D_c \nabla c] + \nabla c \cdot \mathbf{J}/d = -I_G/d \qquad \text{CASE A} \tag{11}$$

$$\frac{\partial c}{\partial t} - \nabla \cdot [\Gamma \widetilde{D}_c \nabla (c/\Gamma)] + \nabla \cdot [(c/\Gamma) D_G \nabla G] \tag{12}$$
$$+ \nabla\, c \cdot \mathbf{J}/d = -I_G/d \qquad \text{CASE B}$$

The species γ is computed simply by means of $\gamma = 1 - c - G$.
In the following analysis we will consider only case B, probably more adherent with the physical process.
In that case, the flux of total wax $c_{tot} = c + G$ is

$$\mathbf{J}_{c_{tot}} = -d\widetilde{D}_c \nabla c - d\widehat{D} \nabla G + c_{tot}\mathbf{J} \tag{13}$$

where $\widehat{D} = (1 - c/\Gamma) D_G + (c/\Gamma)\widetilde{D}_c$ and the mass balance can be written as

$$\frac{\partial c_{tot}}{\partial t} - \nabla \cdot (\widetilde{D}_c \nabla c + \widehat{D} \nabla G) + \nabla c_{tot} \cdot \mathbf{J}/d = 0. \tag{14}$$

3. Thermal balance, full or partial saturation

The dynamics of transition, i. e. the specification of the term I_G, depends on the thermodynamical assumptions we make for the model. Indeed, let us assume that the temperature T can be defined in each point of the system (i. e. the components are in thermal equilibrium in a representative elementary volume). It is known that for each value of T the concentration of dissolved wax in the solvent cannot exceed a saturation threshold, say $\varrho_{c,sat}$. Thus, we have $\varrho_{c,rel} \leq \varrho_{c,sat}(T)$. Typically $\varrho'_{c,sat}(T) > 0$.
If the components have the same specific densities we can write $c/\Gamma \leq c_S(T), \quad c_S(T) = \varrho_{c,sat}(T)/d, \, c'_S(T) > 0$.
Concerning with the total content of wax $c_{tot} = c + G$, either dissolved or crystallized, we can adopt two different points of view (see [10]):

- (E) thermodynamical equilibrium is istantaneously reached between dissolved wax and segregated wax, so that $c = \Gamma c_S(T)$, $G = c_{tot} - \Gamma c_S(T)$ in case of full saturation, $c = c_{tot} < \Gamma c_S$, $G = 0$, in case of partial saturation,
- (NE) segregated wax is present even if the solution is partially saturated.

In case of full saturation, that is $c = \Gamma c_S(T)$ each component can be expressed in term of G and T, since $c = \Gamma c_S(T) = (1 - G)c_S(T)$, $\gamma = [1 - c_S(T)](1 - G)$.
Thus, the dynamics of transition is automatically assigned and the mass balance for G is

$$
\begin{aligned}
&(1 - c_S(T))\frac{\partial G}{\partial t} - \nabla \cdot [(1 - c_S(T))D_G \nabla G] \\
&+(1 - G)c'_S(T)\frac{\partial T}{\partial t} - \nabla \cdot [\widetilde{D}_c(1 - G)c'_S(T)\nabla T] \\
&+[(1 - c_S(T))\nabla G + (1 - G)c'_S(T)\nabla T] \cdot \mathbf{J}/d = 0.
\end{aligned} \tag{15}
$$

If the total flux $\mathbf{J}$ can be in some way specified (as, for istance, whenever it is identically zero), then (15) is in terms of G and T only, since the diffusion coefficients are expected to be functions of G, c, (possibly of the sum c_{tot}) and of T.
In case of partial saturation $c/\Gamma < c_S(T)$, equation (15) cannot be used. Actually, an additional statement must be added in order to describe the mechanism of transition.
In that case, which we refer to as nonequilibrium models (NE), one can directly specify the term I_G as a function of c, G, c_S and T. A general law could be $I_G = -\Phi(\Gamma c_S - c)\ \Psi(G)$ where Φ, Ψ are empirical functions

satisfying

$$\Phi(0) = 0, \quad \Phi(y) > 0 \text{ if } y > 0, \quad \Phi'(y) > 0, \tag{16}$$

$$\Psi(0) = 0, \quad \Psi(G) > 0 \text{ if } G > 0 \tag{17}$$

In [9] $\Phi = \beta y$, with β positive constant, and $\Psi = H$, Heavyside function, are considered. More generally, it can be imagined that $\Psi(G) \sim 1$ starting from a value $G_0 > 0$ close to zero, while for very low concentration of G, corresponding to few isolated wax crystals, dissolution is more rapid, so that $\Psi(G) > 1$, $\Psi' < 0$ for $0 < G < G_0$.
The heat balance of the system, assuming that all the constituents have locally a common temperature T is (see [12])

$$\left[K(T) + \sum_{\alpha=\gamma,c,G} \varrho_\alpha \frac{\partial E_\alpha}{\partial T}\right] \frac{\partial T}{\partial t} + \nabla \cdot (-k\nabla T) + \tag{18}$$

$$\sum_{\alpha=\gamma,c,G} \left[\eta_\alpha + \frac{\partial E_\alpha}{\partial T}\right] \mathbf{J}_\alpha \cdot \nabla T + \sum_{\alpha=\gamma,c,G} I_\alpha \left[\int_0^T \eta_\alpha(y)dy + E_\alpha\right] = 0,$$

where $K(T) = \varrho\eta_\gamma + \varrho_c\eta_c + \varrho_G\eta_G$ is the equivalent heat capacity (η_α are the specific heats $\alpha = \gamma, c, G$), E_α latent energy per unit mass, $\alpha = \gamma, c, G$, k thermal conductivity of the mixture.
According to (18), the quasi–steady equation $(-k\nabla T) = 0$ is consistent if one makes the assumptions:

- heat released or absorbed during segregation or dissolution process is neglected,
- heat flux by conduction is no relevant,
- wax diffusivity is much less than the thermal diffusivity, so that the thermal equilibrium is istantaneously reached,
- the specific heat η_c and η_G are nearly the same.

Nevertheless, it must be remarked that thermal conductivities of different zones (saturated region, deposit of segregated wax,...) may be very different. In several cases, the range (T_1, T_2) of temperature is quite small, so that a linear solubility curve $c_s''(T) = 0$ can be used in order to describe the real situation.

4. Conditions at the external boundary and at the interfaces

At the boundary $\partial\Omega$ of Ω we assume that there is no mass exchange with the exterior of the species γ and c_{tot}:

$$\mathbf{J}_\gamma \cdot \mathbf{n} = \mathbf{J}_{c_{tot}} \cdot \mathbf{n} = 0, \quad \mathbf{x} \in \partial\Omega \tag{19}$$

($\mathbf{n}$ outward normal of $\partial\Omega$) and that the temperature is assigned:

$$T|_{\partial\Omega_1} = T_1 \quad T|_{\partial\Omega_2} = T_2 > T_1, \tag{20}$$

where $\partial\Omega_1$ ("cold" wall) and $\partial\Omega_2$ ("warm" wall) are the subsets of $\delta\Omega$ where the thermal flux outcoming or incoming, respectively (see [9]).
A special care must be devoted to writing the conditions on the internal boundaries of the system. Call Σ any interface in Ω (namely any internal boundary) and assume that $\dfrac{\partial g}{\partial t} + \nabla \cdot \mathbf{j} = 0$ holds for a quantity g with current flow $\mathbf{j}$. Then, the possible discontinuities across the interface Σ follow to the Rankine–Hugoniot condition (see, for istance, [6])

$$[[g]]\mathbf{V} \cdot \mathbf{n} = [[\mathbf{j}]] \cdot \mathbf{n} \tag{21}$$

where $\mathbf{V}$ is the velocity of the front Σ, $\mathbf{n}$, $[[*]]$ stand for the jump $[*]^+ - [*]^-$, with $+$ and $-$ denoting the two regions separated by Σ and $\mathbf{n}$ is the unit normal, say pointing $+$.
Applying (21) for the total wax and for oil and using the volumetric contents one gets

$$[[\gamma]]\, \mathbf{V} \cdot \mathbf{n} = [[\mathbf{J}_\gamma / d]] \cdot \mathbf{n}, \quad [[c + G]]\, \mathbf{V} \cdot \mathbf{n} = [[(\mathbf{J}_c + \mathbf{J}_G)/d]] \cdot \mathbf{n}. \tag{22}$$

We remark that $[[\mathbf{J}]] \cdot \mathbf{n} = 0$ and that the two conditions in (22) are non independent, since $c + G = 1 - \gamma$, $\mathbf{J}_c + \mathbf{J}_G = \mathbf{J} - \mathbf{J}_\gamma$.
Let us apply (22) for the two key situations:

1. $\Sigma = \Sigma_1$ is the interface between the deposit of segregated wax, say $\mathcal{D}$, and the rest of the mixture $\mathcal{R}$ (let us choose $-$ for $\mathcal{D}$, $+$ for $\mathcal{R}$),
2. $\Sigma = \Sigma_2$ is the front between the saturated region $\mathcal{S}$ and the unsaturated part $\mathcal{U}$ ($-$ for $\mathcal{S}$, $+$ for $\mathcal{U}$).

With respect to point 1, we postulate a pure mechanical criterion for the deposit growth, which is engendered by

- ▷ capture by adhesion of suspensions of a fraction $c_{tot}^{\mathcal{D}}$ of total wax c_{tot}^{+} at the front,
- ▷ contribution of the incoming mass flux, according to a part $\mathbf{J}_{tot}^{\mathcal{D}}$ of the total flux of wax $\mathbf{J}_{c_{tot}}^{+} = \mathbf{J}_c^{+} + \mathbf{J}_G^{+}$.

Assuming (by experimental evidence) that everything is at rest in the deposit $\mathcal{D}$, we formulate such as mechanism by stating

$$d[c_{tot}^{\mathcal{D}} - (1-\gamma^-)]\mathbf{V}_1 \cdot \mathbf{n} = \mathbf{J}_{c_{tot}}^{\mathcal{D}} \cdot \mathbf{n} \tag{23}$$

where γ^- is the oil content in the deposit.
Condition (23) is consistent with the total wax balance (22) only if the complementary fractions, which do not take part to deposition, are arranged in the right way beyond the front, namely

$$d[c_{tot}^{+} - c_{tot}^{\mathcal{D}}]\mathbf{V}_1 \cdot \mathbf{n} = [\mathbf{J}_{c_{tot}}^{+} - \mathbf{J}_{c_{tot}}^{\mathcal{D}}] \cdot \mathbf{n}. \tag{24}$$

Both (23) and (24) have to be considered at Σ_1 in order to formulate the free boundary problem. It may be of use to consider (23) and (22) in place of (24): as a matter of fact, (22) allows us to relate the velocity $\mathbf{v}_\gamma^+ = \mathbf{J}_\gamma^+/\gamma^+$ of the solvent at Σ_1 with the front speed:

$$(1-\gamma^-/\gamma^+)\mathbf{V}_1 \cdot \mathbf{n} = \mathbf{v}_\gamma^+ \cdot \mathbf{n}, \quad \gamma^+ = 1 - c_{tot}^+. \tag{25}$$

Remark 4.1. It must be said that if $c_{tot}^{\mathcal{D}} = c_{tot}^+$ and $\mathbf{J}_{c_{tot}}^{\mathcal{D}} = \mathbf{J}_{c_{tot}}^+$, then condition (23) coincides with the total balance (22) and (24) is of no use. In that case, an additional assumption must be introduced: one possibility is to state that $c^+ = (1-G^+)c_S(T)$. Actually, if all the dissolved wax is captured by the front, we may think that c attains saturation.

Whenever $c_{tot}^{\mathcal{D}} \neq c_{tot}^+$, we can eliminate $\mathbf{V}_1 \cdot \mathbf{n}$ from (23), (24) and get the following implicit condition at Σ_1:

$$[c_{tot}^{\mathcal{D}} - (1-\gamma^-)]\mathbf{J}_{c_{tot}}^{+} \cdot \mathbf{n} = [c_{tot}^{+} - (1-\gamma^-)]\mathbf{J}_{c_{tot}}^{\mathcal{D}} \cdot \mathbf{n}. \tag{26}$$

In (23) the fractions $c_{tot}^{\mathcal{D}}$ and $\mathbf{J}_{c_{tot}}^{\mathcal{D}}$ which build on the deposit must be specified.
A simple but reasonable possibility consists in setting

$$c_{tot}^{\mathcal{D}} = \lambda_c c^+ + \lambda_G G^+, \quad \mathbf{J}_{c_{tot}}^{\mathcal{D}} = \mu_c \mathbf{J}_c^+ + \mu_G \mathbf{J}_G^+ \tag{27}$$

where λ_c, λ_G, μ_c, μ_G are constant values in $[0,1]$ to be specified.
We remark that if one assumes that only the diffusive flux of dissolved wax contributes to the formation of the deposit, then $\mathbf{J}_{c_{tot}}^{\mathcal{D}} = \mu \mathbf{J}_{c,diff}^+ = -\mu d\, \Gamma \widetilde{D}_c \nabla(c/\Gamma)$, and conditions (23) and (24) become

$$\begin{cases} d[c_{tot}^{\mathcal{D}} - (1-\gamma^-)]\mathbf{V}_1 \cdot \mathbf{n} = \mu \mathbf{J}_{c,diff}^+, \\ d[c_{tot}^{+} - c_{tot}^{\mathcal{D}}]\mathbf{V}_1 \cdot \mathbf{n} = (1-\mu)\mathbf{J}_{c,diff}^+ - d(1-c/\Gamma)D_G \nabla G. \end{cases}$$

Let us point out some cases of (27), which have been considered in specific models in literature.

1. $\lambda_c = \lambda_G = 1$, that is $c_{tot}^{\mathcal{D}} = \lambda c_{tot}^{+}$ (both dissolved and segregated wax *in loco* are entrapped in the deposit).
   Conditions (23), (24) reduce to

$$\begin{cases} d[(c+G)^{+} - (1-\gamma^{-})]\mathbf{V}_1 \cdot \mathbf{n} = [\mu_c \mathbf{J}_c^{+} + \mu_G \mathbf{J}_G^{+}] \cdot \mathbf{n}, \\ [(1-\mu_c)\mathbf{J}_c^{+} + (1-\mu_G)\mathbf{J}_G^{+}] \cdot \mathbf{n} = 0. \end{cases} \tag{28}$$

   In particular, if $\mu_c = 1$ (i. e. the entire liquid wax flux contributes to the deposit), then (28) is, whenever $\mu_G \neq 1$

$$d[(c+G)^{+} - (1-\gamma^{-})]\mathbf{V}_1 \cdot \mathbf{n} = \mathbf{J}_c^{+} \cdot \mathbf{n}, \quad \mathbf{J}_G^{+} \cdot \mathbf{n} = 0, \tag{29}$$

   which corresponds to $\mu_G = 0$ (see [9]).
   On the contrary, if $\mu_G = 1$, we run into the case of Remark 4.1.
2. $\lambda_c = 0$, $\lambda_G = 1$, $\mu_c = 1$, $\mu_G = 0$, stating that wax entering the deposit is only the total segregated wax and flux flowing into the deposit is only the dissolved wax flux. In this case, (23), (24) are

$$[G^{+} - (1-\gamma^{-})]\mathbf{V}_1 \cdot \mathbf{n} = \mathbf{J}_c^{+} \cdot \mathbf{n}, \quad c^{+}\mathbf{V}_1 \cdot \mathbf{n} = \mathbf{J}_G^{+} \cdot \mathbf{n}. \tag{30}$$

3. $\lambda_c = 0$, $\lambda_G = \eta \leq 1$, $\mu_c = \chi \leq 1$, $\mu_G = 0$, wich corresponds to postulating the deposit of a fraction η of wax crystals and a part χ of dissolved wax flux. This time, (23) and (24) are

$$\begin{cases} d[\eta G^{+} - (1-\gamma^{-})]\mathbf{V}_1 \cdot \mathbf{n} = [\chi \mathbf{J}_c^{+}] \cdot \mathbf{n}, \\ d[c^{+} + (1-\eta)G^{+}]\mathbf{V}_1 \cdot \mathbf{n} = [(1-\chi)\mathbf{J}_c^{+} + \mathbf{J}_G^{+}] \cdot \mathbf{n}, \end{cases}$$

   Case 3 corresponds to the model formulated in [10].

Remark 4.2. Following a different point of view, if one conjectures to assign the rates of deposition as $c_{tot}^{\mathcal{D}} = \lambda(c,G,T)c_{tot}^{+}$, $0 < \lambda \leq 1$, $\mathbf{J}_{c_{tot}}^{\mathcal{D}} = \mu(c,G,T)\mathbf{J}_{c_{tot}}^{+}$, $0 < \mu \leq 1$, we deduce from (23), (24) that $\lambda = 1$ if and only if $\mu = 1$ and $\lambda = \mu$ if and only if they are both 1.
Whenever $\lambda \neq \mu$, condition (26) allows us to calculate the amount of total wax at the front: $c_{tot}^{+} = 1 - \mu(1-\gamma^{-})/(\lambda - \mu)$, which makes sense only if the functions λ, μ are such that $\lambda > \mu + (1-\gamma^{-})(1-\mu)$.

Whenever the deposit is contiguous to a fully saturated region, then, assuming (27), condition (23) writes

$$\begin{aligned} &d[\lambda_G G + \lambda_c(1-G)c_S - (1-\gamma^{-})]\mathbf{V}_1 \cdot \mathbf{n} = \\ &d[\mu_c c_S - \mu_G]D_G \nabla_G - \mu_c(1-G)c_S'(T)\widetilde{D}_c \nabla T, \end{aligned}$$

(we omitted the symbol +, for simplicity) and (24) similarly.

On the other hand, if $\mathcal{R}$ is unsaturated and thermodynamical equilibrium is assumed to hold, then G is identically zero in $\mathcal{R}$. Thus, according to (27), one should make the assumption $c_{tot}^{\mathcal{D}} = \lambda_c c^+$, $\quad \mathbf{J}_{c_{tot}}^{\mathcal{D}} = \mu_c \mathbf{J}_c^+$.
Arguing as in Remark 4.2 and considering that we expect $c = c_S(T)$ at Σ_1, we conclude that it must be $\lambda_c = \mu_c = 1$ and that

$$d(c^+ - (1-\gamma^-))\mathbf{V}_1 \cdot \mathbf{n} = \mathbf{J}_c^+ \cdot \mathbf{n}, \quad c = c_S(T). \tag{31}$$

We finally discuss the conditions on the desaturation front Σ_2, which separates the saturated region $\mathcal{S} = \{\mathbf{x} \in \Omega \mid c = \Gamma c_S(T)\}$ (say $-$) from the partially saturated part $\mathcal{U} = \{\mathbf{x} \in \Omega \mid c < \Gamma c_S(T)\}$ (say $+$).
It makes sense to assume that both c and G are continuous at the desaturation front:

$$[[c]] = 0, \quad [[G]] = 0 \quad \mathbf{x} \in \Sigma_2, \tag{32}$$

so that $c = (1-G)c_S(T)$ at the front (we have implicitly assumed $[[T]] = 0$). Hence, condition (22) at Σ_2 reduces to $\mathbf{J}_{c_{tot}} \cdot \mathbf{n} = 0$, which can be written, according to (13) and (32)

$$\begin{aligned} &\widetilde{D}_c \left((\nabla c)^+ - (1-G)c_S'(T)(\nabla T)^- \right) = \\ &-(1-c_S(T))D_G[[\nabla G]] - c_S(T)\widetilde{D}_c(\nabla G)^+. \end{aligned} \tag{33}$$

Note that the diffusion coefficients are continuous at Σ_2, since they are expected to depend (at most) on c, G and T.
If we assume thermodynamical equilibrium (model E), we have $G \equiv 0$ in $\mathcal{U}$, $c|_{\Sigma_2} = c_S$, $G|_{\Sigma_2} = 0$ and (33) reduces to

$$\widetilde{D}_c \left((\nabla c)^+ - c_S'(T)(\nabla T)^- \right) = -(1-c_S(T))D_G(\nabla G)^-. \tag{34}$$

It is clear that a wide set of situations can be described by varying the initial *scenario* of the model. Nevertheless, as we stated in Section 1, we focus our attention on the process which is expected to evolve according to the sequential stages:

- initial full saturation of Ω,
- formation of the solid wax deposit at the cold wall,
- desaturation process starting from the warm wall.

In order to follow such a scheme, we choose initially a total content of wax $c_{tot}(\mathbf{x}, 0) = c_{tot}^* > c_S(T_M)$ where $T_M = \max_{\mathbf{x} \in \Omega_2} T_2(\mathbf{x}, 0)$ (see (20)).
Assuming that the thermodynamical equilibrium is instantaneously reached, we have $(c/\Gamma)(\mathbf{x}, 0) = c_S(T(\mathbf{x}, 0))$, hence

$$c(\mathbf{x}, 0) = (1 - c_{tot}^*) \frac{c_S(T(\mathbf{x}, 0))}{[1 - c_S(T(\mathbf{x}, 0))]}, \quad G(\mathbf{x}, 0) = c_{tot}^* - c(\mathbf{x}, 0). \tag{35}$$

The initial profile of T is obtained by solving $-k\nabla T(\mathbf{x},0) = 0$ together with (see (20)) $T(\mathbf{x},0)|_{\Omega_1} = T_1(\mathbf{x},0)$, $T(\mathbf{x},0)|_{\Omega_2} = T_2(\mathbf{x},0)$.
Although we specify particular initial conditions, the problem offers formidable difficulties from the mathematical point of view: as to cite one, even assuming thermodynamical equilibrium at any time t, the onset of Σ_2 has to be ascribed to a time $\bar{t}$ such that $G(\mathbf{x},\bar{t})|_{\mathbf{x}\in\Omega_2} = 0$ and such as check seems to be very difficult in a generic domain Ω.

References

1. J. R. Cannon, A. Fasano, *Boundary value multidimensional problems in fast chemical reactions*, Arch. Rat. Mech. Anal., **53** (1973) 1-13.
2. S. Correra, A. Fasano, L. Fusi, M. Primicerio, *Modelling wax diffusion in crude oils: the cold finger device*, Appl. Math. Model., (Elsevier 2006), to appear.
3. S. Correra, A. Fasano, L. Fusi, M. Primicerio, F. Rosso, *Wax diffusivity under given thermal gradient: a mathematical model.* ZAMM (2006), to appear.
4. J. Bear *Dynamics of Fluids in Porous Media*, Environmental science serier (Elsevier 1972).
5. E. L. Cussler *Diffusion: Mass transfer in fluid systems*, (Cambridge University Press 1984).
6. Evans *Partial differential Equations*, AMS, (Providence 1999).
7. A. Fasano, S. Correra, L. Fusi, *Mathematical models for waxy crude oils.* Meccanica **39**, (2004) 441-483.
8. A. Fasano, M. Primicerio, *General free-boundary problems for the heat equation, I*, J. Math. Anal. Appl. , **57** (1977), 694-723.
9. A. Fasano, M. Primicerio, *Heat and mass transport in non-isothermal partially saturated oil-wax solutions.* New Trends in Math. Phys., P. Fergola, F. Capone, M. Gentile & G. Guerriero editors, (World Scientific 2005)
10. A. Fasano, M. Primicerio, *Temperature driven mass transport in concentrated saturated solutions.* Progress in Nonlinear Differential Equations and their Applications **61** (2005), 91-108.
11. A. Fasano, M. Primicerio, *Wax deposition in crude oils: a new approach.* Rendiconti Lincei, Matematica e Applicazioni, **16** Serie 9, Issue 4, Mathematical Physics (2005) 251-263
12. F. Talamucci, *Some problems concerning with mass and heat transfer in a multi–component system*, MAT Conferencias, seminarios y trabajos de Matemàtica, serie A **6** (2002) .

WAX DIFFUSION AND DEPOSITION IN THE PIPELINING OF WAXY OILS

S. CORRERA, D. MERINO-GARCIA

EniTecnologie, Via F. Maritano 26
San Donato Milanese, 20097 Milano, Italy

A. FASANO, L. FUSI*

Università Degli Studi di Firenze,
Dipartimento di Matematica "U. Dini",
Viale Morgagni 67/a, 50134, Firenze.
** E-mail: fusi@math.unifi.it*

Wax deposition from a waxy crude oil is modelled in turbulent flow in a pipeline. Molecular diffusion in a thin boundary layer near the pipe wall is taken as the only driving force for deposition. The model takes into account the ablation, which is a flow-related phenomenon that limits the growth of the deposit. The effect of desaturation of the oil is analyzed as well.

Keywords: Turbulent flow; Molecular diffusion; Wax deposition.

1. Introduction

In this work we study the deposition of wax on the walls of a pipeline during the transport of the so-called waxy crude oils (a class of mineral oils containing high molecular weight hydrocarbons). This problem is of crucial importance for petroleum industries, since the presence of a solid layer can lead, in the most dramatic scenario, to the total blockage of the line.

Waxy oils are characterized by the presence of heavy hydrocarbons compounds, primarily n-alkanes with a carbon number greater than 18. Iso-paraffins and naphtenes are also found in the deposits. These compounds tend to solidify at low temperatures, that is when the external temperature is lower than the Cloud Point T_{cloud}, i.e. the temperature at which solid crystals begin to appear. Solidification will preferentially occur on the wall, leading to deposition and causing undesirable effects, such as an increase in the pressure requirements, a decrease in flow rate and, in the worst situation, the complete blockage of the line.

Deposition may depend on various mechanisms which include both the transport of liquid waxes by molecular diffusion and the transport of solid material by Brownian diffusion and shear dispersion. There is a general accordance in considering that molecular diffusion towards the wall is the main responsible for the transport of waxes from the bulk to the wall.

In this work, wax deposition is modelled in the turbulent flow of a waxy crude oil in a cylindrical pipeline, taking into account molecular diffusion only. Deposition is contrasted by turbulence, by means of a mechanism termed "ablation". Possible desaturation is taken into account. For the sake of simplicity, waxes are lumped into a single pseudo-component. Moreover, following experimental evidence, the thermal properties and densities for solid and liquid waxes and for the solvent are considered to be the same.

At the beginning the oil is injected in the pipeline at a temperature greater than the cloud point. Because of turbulence the temperature in the bulk is homogeneized over cross sections except for a thin boundary layer, where a thermal gradient is produced. The latter induces a dissolved wax concentration gradient (molecular diffusion), as long as the oil is saturated by wax. Thus, in the boundary layer, dissolved wax diffuses towards the cold wall where the deposit layer begins to appear. Of course this occurs only when the temperature of the oil at wall has drop below the cloud point.

The concentration profile is the driving force of the diffusion of wax molecules towards the cold wall, which is assumed to be the only responsible of wax deposition. The deposit thickness and consistency are also affected by ablation, i.e. the removal of solid wax stripped from the deposit by the turbulent flow.

The performance of the model in the prediction of the deposition rate obviously relies on the physical properties of waxes. Particularly, the derivative of wax solubility with respect to temperature is of crucial importance, because it determines the concentration of liquid waxes in the radial direction. The second critical parameter is the diffusivity of dissolved wax, because it determines the rate at which waxes are driven to the wall.

"The model is based on the assumption that at all times the thickness of both the boundary and deposit layers is small compared to the radius of the pipe"; in fact, this is a condition that must be maintained for safe operation of the pipeline. For a more detailed description of the model we refer the reader to [1].

2. The thermal field

The flow occurs in a cylindrical pipe of radius R and length L in turbulent regime. The thermal field is homogeneized over cross sections except in a thin thermal boundary layer where it has the form

$$T(r,z) = (T_o - T_e)\exp\left\{-\frac{2\pi hR}{\rho cQ}z\right\}\cdot\left\{1-\frac{hR}{k}\ln\left(\frac{r}{R}\right)\right\} + T_e. \quad (1)$$

In (1) h is the heat transfer coefficient, k is heat conductivity, ρ is density, c is heat capacity, Q is the volumetric flow rate, T_o is the oil inlet temperature and T_e the temperature of the surroundings. Expression (1) is compatible with the assumption that heat transfer is quasi-steady and takes place mainly in the radial direction. Accordingly, we will assume that temperature variations induced by the external temperature T_e are slow enough, as it is confirmed by field experience. The thickness of the boundary layer can be obtained from the knowledge of the momentum boundary layer using classical correlations. The momentum boundary layer is determined imposing balance between the drag and propulsive forces in a unit length portion of the pipe. Indeed, denoting with σ_m the momentum boundary layer thickness and introducing the ratio $\epsilon_m = \sigma_m/R$, such a balance is expressed by

$$F_D = \frac{2\eta Q}{\epsilon_m R} = F_P = \Delta P \pi R^2, \quad (2)$$

where η is the viscosity of the oil and P is pressure. Relation (2) allows us to calculate ϵ_m. Then we use the correlation [2] $\epsilon_T = \epsilon_m \times 0.41$ to evaluate the thermal boundary layer thickness $\sigma_T = \epsilon_T R$.

Since mass deposition rate is proportional to the radial temperature gradient in the thermal boundary layer it is useful to write

$$\frac{\partial T(r,z)}{\partial r} = -(T_o - T_e)\exp\left\{-\frac{2\pi hR}{\rho cQ}z\right\}\cdot\frac{hR}{kr}. \quad (3)$$

The temperature of the turbulent core T_c is obtained writing the energy balance

$$\rho c\pi R^2 V\left(T_c(z) - T_c(z+dz)\right) = -\int_z^{z+dz} 2\pi Rk\frac{\partial T}{\partial r}(R,z)dz, \quad (4)$$

where $V = Q/R^2\pi$ is the velocity in the turbulent core. In the limit $dz \to 0$

$$\frac{dT_c}{dz} = \frac{2\pi hR(T_o - T_e)}{\rho cQ}\exp\left\{-\frac{2\pi hR}{\rho cQ}z\right\}. \quad (5)$$

Integrating with $T_c(0) = T_o$ we get

$$T_c = (T_o - T_e)\exp\left\{-\frac{2\pi h R}{\rho c Q} z\right\} + T_e. \tag{6}$$

3. The deposition equation

We denote by σ_d the deposit thickness and by $\nu = R - \sigma_d$ the reduced pipe radius. A fundamental assumption in our model is that the deposit thickness is small compared to the pipe radius, i.e. $\nu \approx R$. The deposition equation is written for the unknown ν. Deposition is the result of two contrasting mechanisms: molecular diffusion and ablation (removal of deposit by shear stripping). For the moment we suppose that the bulk of the turbulent flow contains so much wax that not only the oil is saturated with it, but also segregated crystals are present. This segregated phase acts as a wax reservoir, providing the mass progressively lost in the deposition phenomenon, thus keeping the oil saturated at each time. In this situation the concentration of dissolved wax is an increasing function of temperature $C_s(T)$ (wax solubility), linear in T (i.e. $dc_s/dT = \beta = const > 0$). The latter assumption is reasonable, since the range of temperature considered are not large. The deposition rate is proportional to the temperature radial gradient evaluated at the deposition front $r = \nu$:

$$j_{dep} = -\frac{D}{\psi}\frac{dC_s}{dT}\frac{\partial T}{\partial r}\bigg|_{r=\nu} =: -\frac{D\beta}{\psi}\frac{\partial T}{\partial r}\bigg|_{r=\nu}, \tag{7}$$

where ψ is the solid fraction of the deposit, D is dissolved wax diffusivity. For the moment we assume that ψ is constant, that is we do not take into account ageing. In section 5 we release this assumption and we propose a possible kinetics for the evolution of ψ. The rate of removal by ablation is proportional to the shear stress at the deposit front

$$j_{abl} = -\frac{A\tau}{\psi}, \tag{8}$$

where A is the ablation coefficient. Since

$$\tau = \frac{\eta V}{\epsilon_m \nu} = \frac{\eta Q}{\pi R^2 \epsilon_m \nu}, \tag{9}$$

$$j_{abl} = -\frac{A\eta Q}{\pi R^2 \psi \epsilon_m \nu}. \tag{10}$$

The deposition equation is obtained writing

$$\frac{d}{dt}\int_z^{z+dz} \rho\pi(R^2 - \nu^2(\xi,t))d\xi = \int_\Sigma (\vec{j}_{dep}\cdot\vec{n} - \vec{j}_{abl}\cdot\vec{n})dS \tag{11}$$

where Σ can be parametrized by

$$\begin{aligned} x &= \nu(\xi, t)\cos u, \\ y &= \nu(\xi, t)\sin u, , \\ z &= \xi, \end{aligned} \qquad \begin{aligned} 0 \le u \le 2\pi, \\ z \le \xi \le z + dz, \end{aligned} \tag{12}$$

and

$$\vec{n} = \frac{\cos u\mathbf{i} + \sin u\mathbf{j} - \nu_\xi(\xi, t)\mathbf{k}}{\sqrt{1 + \nu_\xi^2(\xi, t)}}. \tag{13}$$

Supposing enough regularity for the function ν equation (11) becomes[a]

$$\rho\frac{\partial \nu}{\partial t} = \frac{D\beta}{\psi}\frac{\partial T}{\partial r}(\nu, z) + \frac{A\eta Q}{\pi\epsilon_m\psi\nu^3}. \tag{14}$$

Recalling (3) we may integrate the above equation with $\nu(z, 0) = R$ getting

$$\sigma_d = R - \nu = \frac{D\beta t}{\psi\rho R}\left[\frac{hR(T_o - T_e)}{k}\exp\left\{-\frac{2\pi hR}{\rho c Q}z\right\} - \frac{A\eta Q}{\pi\epsilon_m D\beta R^2}\right]_+ \cdot$$

$$\cdot H(T_{cloud} - T_w), \tag{15}$$

where $[..]_+$ is the positive part, T_w is temperature at the wall and H is the Heaviside function (which guarantees that deposition is effective only after temperature at the wall has fallen below T_{cloud}). The quantity in square brackets must be nonnegative since deposition occurs only when molecular diffusion is stronger than ablation.

4. The deposition segment

It is clear from (15) that in order to have deposition the term $[..]_+$ has to be greater than 0 and $T_w < T_{cloud}$. We may prove that deposition occur only in the interval $[z_f, z_e]$, where

$$z_f = \frac{Q\rho c}{2\pi hR}\ln\left[\frac{T_o - T_e}{T_{cloud} - T_e}\right], \tag{16}$$

$$z_e = \frac{Q\rho c}{2\pi hR}\ln\left(\frac{(T_o - T_e)\pi\epsilon_m hD\beta R^3}{kA\eta Q}\right), \tag{17}$$

where z_f is the coordinate beyond which deposition may start (obtained imposing $T_w = T_{cloud}$) and z_e is the coordinate beyond which ablation

[a]In (14) the term containing T_z has been suppressed. Indeed such a quantity is much smaller than the others appearing in the equation and can be safely neglected.

is stronger than deposition (the quantity in the square brackets in (15) is negative). Of course the model makes sense only if $z_f < z_e$. We may distinguish three cases:

C1 $z_e > L$. Deposition takes place in the segment $[z_f, L]$.
C2 $z_f < z_e < L$. Deposition takes place in the segment $[z_f, z_e]$.
C3 $z_e < z_f$. Deposition never starts.

In this analysis we have not considered the possible desaturation of the oil. This will be the subject of section 7.

5. Ageing

The deposit is formed by wax and oil. The gradual release of oil from the deposit produces an increase of ψ. This phenomenon is known as ageing (or consolidation). Thus far we have assumed that the solid fraction ψ is constant. Now we want to analyze the case when the deposit evolves according to some kinetics. The choice of the latter has to be done according to experimental evidence. For the sake of simplicity here we propose the following simple model

$$\frac{\partial \psi}{\partial t} = \frac{1}{t_a}(1 - \psi), \tag{18}$$

where t_a is a characteristic consolidation time and where ψ depends only on time. Equation (18) means that the solid fraction growth rate is proportional to the oil fraction in the deposit. Such an equation can be integrated to get

$$\psi = 1 - (1 - \psi_o) \exp\left(-\frac{t}{t_a}\right), \tag{19}$$

where ψ_o is the initial deposit wax fraction.

6. The total mass of deposit

Now we are able to write the explicit formula to calculate the total amount of deposit formed in a given time interval $[0, t]$. Assuming to be in cases C1 or C2, the total mass of deposit (with oil inclusion) at a certain time t will be given by

$$M_{tot} = \int_{z_f}^{z_e} \rho\pi(R^2 - \nu^2(z,t))dz. \tag{20}$$

We have

$$M_{tot} = \frac{R^2}{T_{cloud}} \int_{z_f}^{z_e} \rho\pi \left[\frac{(T_o - T_e)}{\mu} \exp\left\{ -\frac{2\pi\alpha}{\mu Q} z \right\} - \frac{A\eta Q}{\pi\varepsilon_m D\beta R^2} \right].$$

yielding

$$M_{tot} = 2t_a D\beta \ln \left\{ \frac{1}{\psi_o} \left[\exp\left(\frac{t}{t_a} \right) - 1 \right] + 1 \right\}$$

$$\cdot \left[\frac{A\eta Q}{\pi\varepsilon_m D\beta R^2} (z_f - z_e) + (T_o - T_e) \frac{Q\rho c}{2\pi k} \right.$$

$$\left. \cdot \left(\exp\left\{ -\frac{2\pi h R z_f}{\rho c Q} \right\} - \exp\left\{ -\frac{2\pi h R z_e}{\rho c Q} \right\} \right) \right]. \tag{21}$$

7. Desaturation

Denoting with $G(z,t)$ the concentration of segregated wax in the bulk we write the balance

$$\left\{ \frac{\partial G}{\partial t} + V \frac{\partial G}{\partial z} \right\} \pi\nu^2 = 2\pi\nu \left\{ D\beta \frac{\partial T}{\partial r} - D\beta \frac{\partial T}{\partial z} \frac{\partial \nu}{\partial z} + \frac{A\eta Q}{\pi\varepsilon_m \nu^3} \right\}, \tag{22}$$

where we recall $Q = V\pi\nu^2$. If we look for the steady state solution $\hat{G}(z)$ and recall that the term containing $T_z\nu_z$ can be neglected we get

$$\hat{G}(z) = G_o + \frac{2\pi}{Q} \int_{z_f}^{z} \left\{ \nu D\beta \frac{\partial T}{\partial r} + \frac{A\eta Q}{\pi\varepsilon_m \nu^2} \right\} dz', \tag{23}$$

where G_o denotes the value of G at $z = z_f$. Recalling (3), the assumption $\nu \approx R$, and integrating (23) we get

$$\hat{G}(z) = G_o + \left[\frac{2A\eta}{\varepsilon_m R^2} (z - z_f) \right.$$

$$\left. + \frac{D\beta(T_o - T_e)}{\alpha} \left(\exp\{ -\frac{2\pi\alpha z}{\mu Q} \} - \exp\{ -\frac{2\pi\alpha z_f}{\mu Q} \} \right) \right], \tag{24}$$

which represents the concentration of the segregated phase for $z > z_f$. Desaturation may be achieved at a distance z_{des} such that $\hat{G}(z_{des}) = 0$, that is

$$G_o + \left[\frac{2A\eta}{\varepsilon_m R^2} (z_{des} - z_f) \cdot \right.$$

$$+\frac{D\beta(T_o - T_e)}{\alpha}\left(\exp\{-\frac{2\pi\alpha z_{des}}{\mu Q}\} - \exp\{-\frac{2\pi\alpha z_f}{\mu Q}\}\right)\Bigg] = 0. \qquad (25)$$

Of course the interesting case is when $z_{des} < z_e$. For $z > z_{des}$ deposition continues with the same rate as long as wax concentration $c(z,t)$ in the oil stays above the value of saturation concentration corresponding to the wall temperature. Thus, when $C_s(T_w(z)) < C_s(T_c(z))$, we write

$$\left\{\frac{\partial c}{\partial t} + V\frac{\partial c}{\partial z}\right\}\pi\nu^2 = 2\pi\nu\left\{D\beta\frac{\partial T}{\partial r} - D\beta\frac{\partial T}{\partial z}\frac{\partial \nu}{\partial z} + \frac{A\eta Q}{\pi\varepsilon_m\nu^3}\right\}, \qquad (26)$$

We remark that the solid wax removed by ablation dissolves immediately, slowing down the rate of decrease of c.

Referring once more to the steady state $\hat{c}$ we have

$$\hat{c}(z) = C_s(T_c(z_{des})) + \frac{2\pi}{Q}\int_{z_{des}}^{z}\left\{\nu D\beta\frac{\partial T}{\partial r} + \frac{A\eta Q}{\pi\varepsilon_m\nu^2}\right\}dz', \qquad (27)$$

The oil will be completely depleted of wax at z_s given by

$$\hat{c}(z_{des}) = C_s(T_w(z_s)), \qquad (28)$$

i.e. there is no concentration gradient of dissolved wax and consequently no deposition.

References

1. S. Correra, A. Fasano, L. Fusi, D. Merino-Garcia, *Calculating deposit formation in the pipelining of waxy crude oils*, to appear on Meccanica.
2. Kays W.M., Crawford M.E., *Convective Heat and Mass Transfer*, McGraw Hill (1980).

INVERSE MODELING IN GEOPHYSICAL APPLICATIONS

G. CURRENTI*, R. NAPOLI, D. CARBONE, C. DEL NEGRO, G. GANCI

Istituto Nazionale di Geofisica e Vulcanologia, Sezione di Catania, Catania, Italy

**E-mail: currenti@ct.ingv.it*

maglab.ct.ingv.it

The interpretation of the potential field data is an useful tool that allows for both investigating the subsurface structures and providing a quantitative evaluation of the geophysical process preceding and accompanying period of volcanic unrest. Potential field inversion problem are required to combine forward models with appropriate optimization algorithms and automatically find the best set of parameters that well matches the available observations. Indeed, investigations on the mathematical equations to be inverted, have revealed that these models are ill-posed and highly non-linear. Numerical methods for modeling potential field observations are proposed and applied on real dataset.

Keywords: Geophysical inversion; Potential Field; Mount Etna.

1. Introducton

The inversion problem in potential field modeling suffers from the ambiguity and instability of its solutions. The ambiguity arises from the inherent property of potential fields for which different combinations of parameters may lead to similar anomalies. Moreover, potential fields inversion is notoriously unstable. Because of ambiguity and instability of solutions, the inversion problem can be secured by narrowing the set of all possible solutions to a predefined solution class that allow for a unique and stable solution. The priori definition of the geometry (simplified bodies: spherical, rectangular, prismatic) of searched source and the priori recognition of the involved physical mechanism allows reducing the number of likely solutions considerably.[1] Potential field inversion method can be classified into two main categories depending on the type of the unknown parameters to be retrieved. Firstly, there are some methods looking for magnetization or density contrast values with fixed geometry sources. In this case linear inversion technique can be successfully applied. Secondly, there are those methods that look for geometrical determinations of the source. In this latter case the inversion problem is highly non-linear and robust nonlinear inversion techniques are needed. The first category of interpretation models is widely used in providing physical property distributions of subsurface geological structures, while the second one is generally applied in modeling potential field sources in volcanic areas. In the following two inversion approach are described to deal with linear and non-linear inversion problems and are applied on two case studies to appraise the goodness of the proposed methods.

2. 3D linear inversion of potential field data

Potential field inversion are aimed to provide a model which reconstructs, as well as possible, subsurface geological structures having a density/magnetization contrast with their surroundings. For the sake of simplicity, we consider the magnetic inverse problem. However, similar formulation can be applied to the case of gravity inversion method. The computational domain V which is supposed to surround the magnetic source is discretized using a finite number of $m = N_x * N_y * N_z$ rectangular prisms whose magnetization J_j is uniform inside each prism.[1] Using a 3D discrete numerical approach, the total anomaly field at $i - th$ observation point is computed by:

$$T_i = \sum_{j=1}^{n} J_i a_{ij} \tag{1}$$

where elements a_{ij} quantify the contribution to the total magnetic field anomaly T at $i - th$ observation point due to the magnetization of the $j - th$ prism.[2] Therefore, the inverse problem can be formulated as the solution of a system of n linear equations as:

$$Ax = T \tag{2}$$

where x is the m vector of unknown magnetization values of the prisms, T is the n vector of observed magnetic data, and A is a matrix with elements a_{ij} . The analytical expression of the a_{ij} term for a prismatic body was devised by Rao and Babu.[3] Based on discretization, the number of prisms is usually larger than the number of observation points, thus the linear inverse problem in Eq. (2) turns out to be unavoidably underdetermined. In such a case, the linear system leads to a solution with $m - n$ degree of freedom. A further difficulty in solving the system in Eq. (2) is due to the inherent non-uniqueness of the potential field: there are an infinite number of inverse models that can explain the same observed magnetic anomaly within error limits. This highlights that the magnetic inverse problem is ill-posed. That calls for some regularization techniques. In such a case, it is necessary to impose further constraints taking into account a priori knowledge about the solution when available. The idea of reducing the class of possible solution to some set on which the solution is stable lies on the fundamental concept of introducing a regularizing operator. The inverse problem can be re-formulated as an optimization problem aimed at finding the unknown magnetization values x that minimize a functional ϕ composed of a data misfit ϕ_d and a smoothing functional ϕ_m:

$$\phi = \phi_d + \phi_m = \frac{1}{2}[(Ax - T_{obs}^T)(Ax - T_{obs}) + \lambda(x - x_0)^T W^T W(x - x_0)] \quad (3)$$

where W is a weighting matrix and λ is the regularization parameter. Because of the lack of depth resolution, W is aimed to counterbalance the contribution of deeper prisms with respect to shallower ones. To ensure that the solution is geologically reasonable it is advisable to prescribe realistic bounds on the magnetization values on the basis of rock samples or available information about the local geology. The minimization of the quadratic functional in Eq. (3) subjected to bound constraints can be solved by using a Quadratic Programming (QP) algorithm based on active set strategy:[4]

$$min\phi = min[\frac{1}{2}x^T Qx - d^T x], L <= x <= U \quad (4)$$

where $Q = A^T A + \lambda W^T W$ and $d = A^T T + \lambda W^T W x_0$, and L and U are the vectors of lower and upper bounds. The quadratic formulation of the problem is solved iteratively by generating a sequence of feasible solutions that converge toward the optimal solution. The iteration is stopped when no relative improvements in the functional are achieved.

The inversion procedure described above was applied to analyze the anomalies detected by a ground magnetic survey of the Ustica island covering an area of about 9 km^2. The total-intensity anomaly field, obtained after data reduction process, shows the presence of a W-E striking magnetic anomaly in the middle of the island and other two intense anomalies, which seem to continue offshore, in the south-western and the north-eastern sides (Fig. 1).

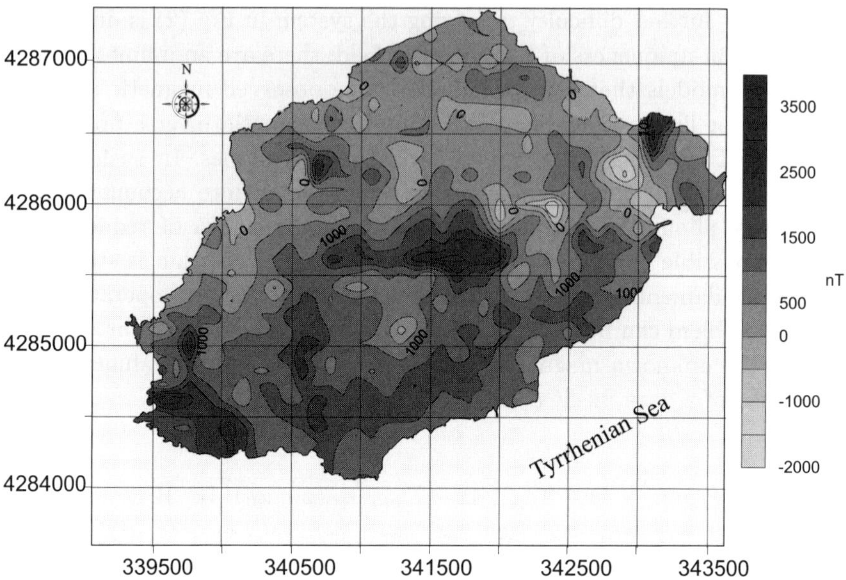

Fig. 1. Map of the total intensity magnetic field after reduction process.

In order to allow the maximum flexibility for the model to represent geologically realistic structures the island was represented as a crustal block, 4x3 km^2 in area and about 1.2 km in thickness, and was discretized into a set of rectangular prisms (0.125 x 0.125 x 0.15 km^3 in size). The anomaly field was inverted assuming that the average direction of total magnetization is close to Earth's present-day field direction (approximate inclination of 54° N, declination of 1° W). Moreover taking advantage of magnetizations measured on volcanic rocks of the island, the inversion of the data set was performed constraining the range of the magnetization value for the assumed sources from 0 to 10 A/m. The iterations were continued until the functional does not show significant improvements. The residual field, the difference between observed field and calcualte field is shown in Fig. 2. Only in the N-E and S-W area of the island two small residual anomalies remain probably due to a low resolution of the volume discretization in these zones.

The 3D model resulting from the constrained inversion of data set reveals the presence of three main magnetic trends, N-S, E-W and NE-SW, which are in good agreement with the surface geology, and are coincident with the main regional structural lineaments (Fig. 3).

In particular, the N-S and E-W trends, which are of more recent origin,

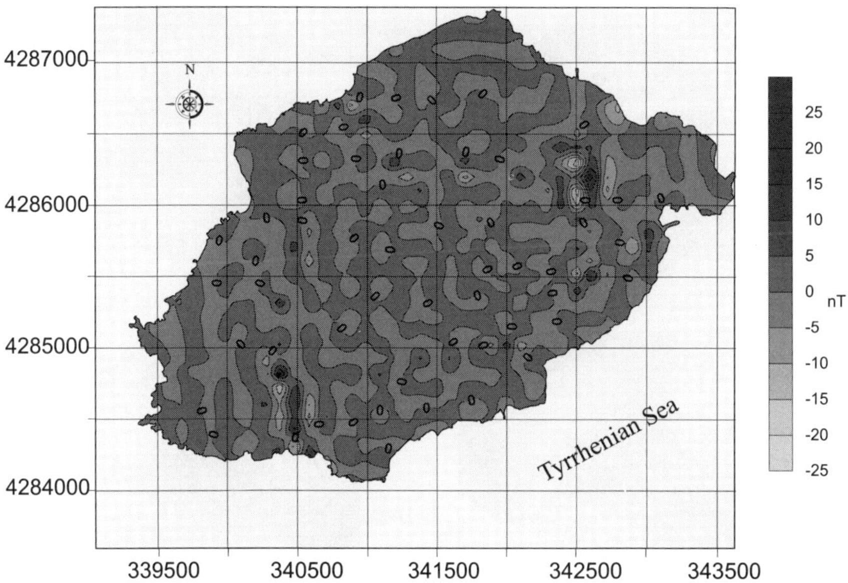

Fig. 2. Residual, total-field magnetic anomaly map produced by subtracting the computed field from observed field.

prevail in the shallow part of the model, while the NE-SW is relevant below 0.4 km b.s.l. At this depth the model reveals a magnetized body in the central area, which may be interpreted as the preferential area for magma storage and ascent and which supplied the feeding systems of the main subaerial volcanic centres of the island, Mt. Guardia dei Turchi and Mt. Costa del Fallo. Two other magnetized volumes were identified and ascribed to the small submarine/subaerial eruptive centres of the western island and to the younger cone of Capo Falconiera, respectively. These findings highlight how the regional tectonics has strongly influenced the structural and magmatic evolution of the Ustica volcanic complex producing preferential ways for magma ascents. Moreover,considering that the NE-SW trend prevails only in the deeper part of the model and is replaced at the shallow depths by the N-S and E-W orientations, it is possible to assume that a change of tectonic style and/or its orientation took place in the past[5] .

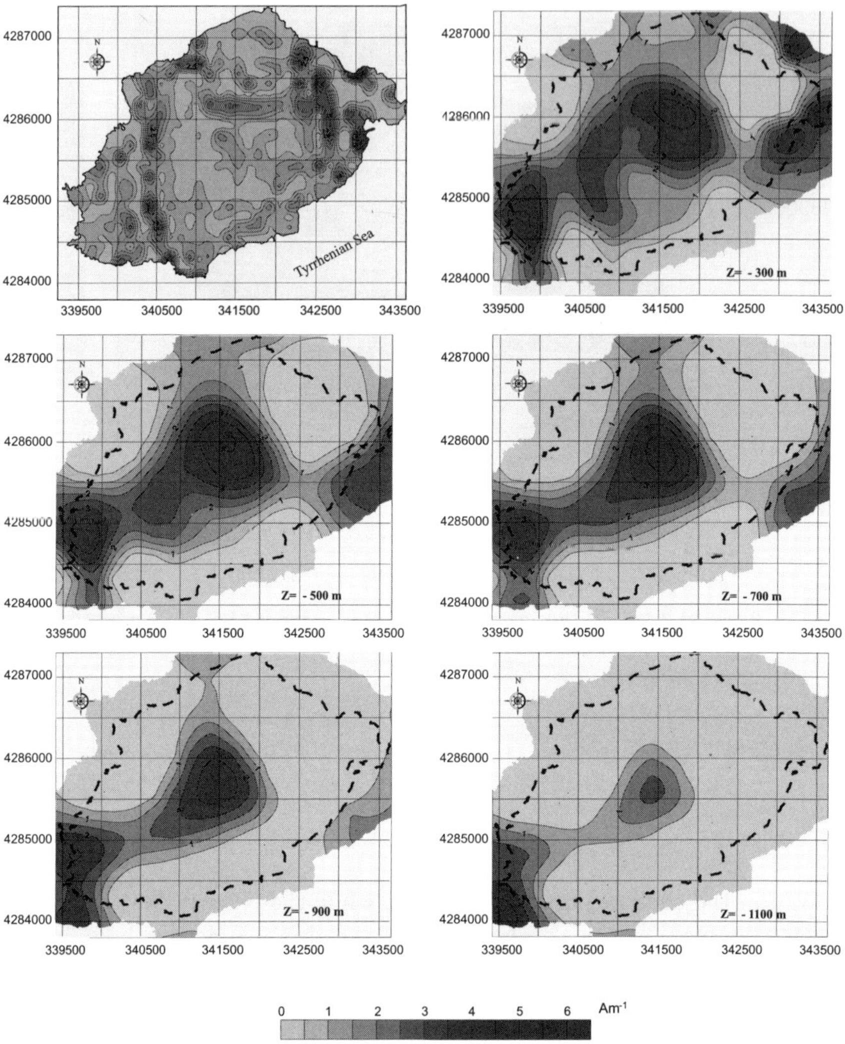

Fig. 3. The 3 D magnetization model of the Ustica volcanic complex. Horizontal sections of the uppermost part (above sea level) of the model and from 300m depth until 1100m depth.

3. Non-linear inversion of gravity anomaly by genetic algorithm technique

Attempts to model potential fields expected to preceed and accompany volcanic eruptions often involve a great deal of effort due to the complexity of the considered problem. The inversion problem deals with the identification of the parameters of a likely volcanic source that leads observable changes in potential field data recorded in volcanic areas. Indeed, investigations on the analytical models describing the involved geophysical processes, have revealed that the models are highly non-linear and, usually, characterized by several parameters. When non-linear models are involved, the inverse problem becomes difficult to solve through local optimization methods. Hence, elaborated inversion algorithms have to be implemented to efficiently identify the source parameters. We have investigated the use of Genetic Algorithms[6] (GAs) which perform a broad search over the parameter space using a random process with the aim of minimizing an objective function that quantifies the misfit between model values and observations. The GAs inversion strategy was applied on a gravity anomaly that grew up progressively in 5 months before the 2001 flank eruption of Mt. Etna along a East-West profile of stations on the southern slope of the volcano.[7] Between January and July 2001, the amplitude of the gravity change reached 80 μGal, while the wavelength of the anomaly was of the order of 15 km (Fig. 4).

Elevation changes observed through GPS measurements during a period spanning the 5-months gravity decrease, remained within 4-6 cm all over the volcano and within 2-4 cm in the zone covered by the microgravity profile. Since February 2001, an increase in the seismicity was observed,[8] with many earthquake locations clustered within a volume at a depth of about 4 km bsl and focal mechanisms pointing towards a prevailing tensile component. Therefore, we review both gravity and elevation changes by the Okubo model[9] which mathematically describes the effect of uniformly distributed tensile cracks on gravity and ground deformation field in an elastic homogeneous half-space medium.[10] The gravity solution consists of four contributions: (i) the free-air effect proportional to the uplift, (ii) the Bouguer change caused by the upheaved portion of the ground, (iii) the gravitational attraction of the crack-filling matter and (iv) the gravity field due to the redistribution of mass associated with the elastic displacements. This model implies inversion for nine parameters, $m = (A, X, Y, Z_1, Z_2, L, W, U, \delta\rho)$, whose description is reported in Table 1. To make the GA converge towards a solution which (i) best fits the observed data and (ii) has a good chance to be realistic from the volcanological point of view, we suitably restrict the parameter space to be investigated. The ranges of variability for the model parameters to be found are set on the grounds of the available geophysical and geological evidence. For the inversion procedure adopted in

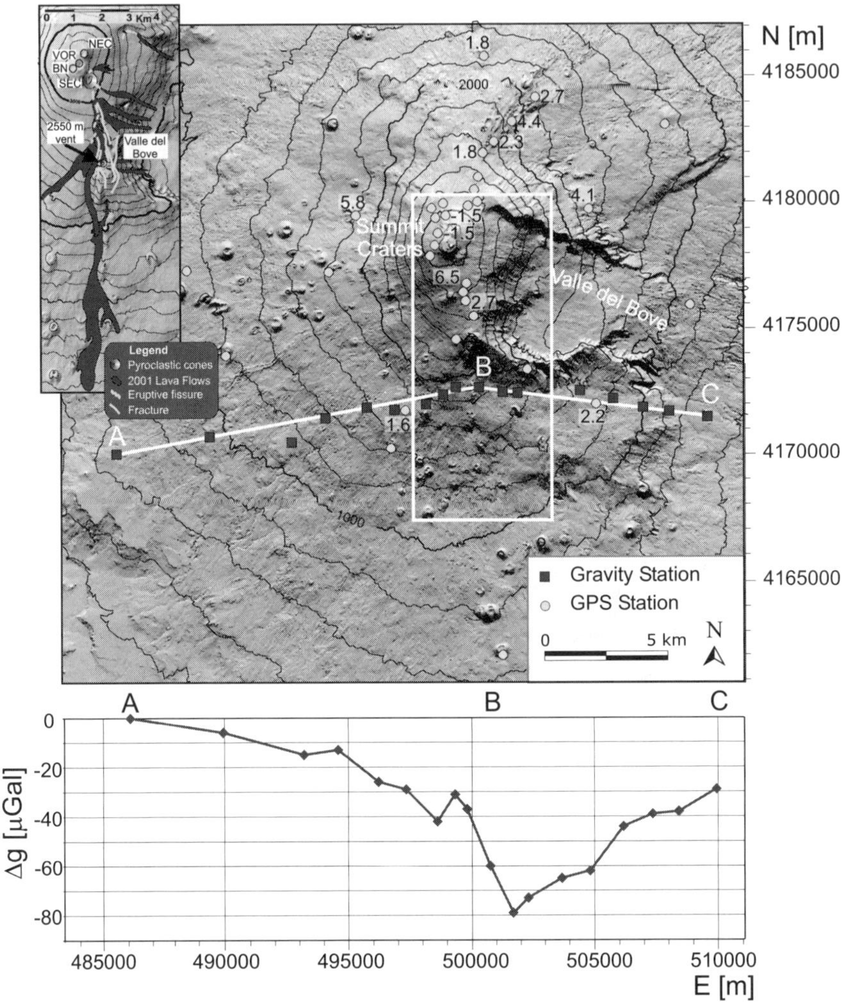

Fig. 4. Sketch map of Mt. Etna. Top: the gravity stations of the microgravity network along the East-West Profile (ABC) and the GPS stations measured in July 2000 and June 2001. The elevation changes higher than 1.5 cm, observed during the same interval, are also reported. The inset on the left shows the 2001 lava flow. Bottom: gravity change observed between January and July 2001 along the East-West Profile.

the present study, we have set the objective function equal to the χ^2 value that accounts for the measurements error defined by the standard deviation σ, as:

$$\chi^2 = \sum_k \frac{(M_k - C_k)^2}{\sigma_k^2} \tag{5}$$

where M_k are the measured data, C_k are the computed gravity variations and k is the number of available measurements. The parameters of the best model found through the GA are reported in Table 1, together with the edge values of each search range.

Table 1. The best model parameters found by the GA.

Parameter	Minimum	Maximum	Best value
Z_1 - Depth of the top (m b.s.l.)	0	2500	409
L - Length (m)	1000	5000	5000
H - Height (m)	3000	7000	7000
W - Thickness (m)	500	2000	500
A - Azimuth (from the North)	-45	0	-30
X - Northing Coordinate(m)	4170000	4180000	4173940
Y - Easting Coordnate(m)	495000	505000	503499
$\delta\rho$ - Density contrast (kg/m^3)	-2400	-2700	-2500
U - Extension (m)	0	2	2

The fit between observed and calculated gravity changes is very good (Fig. 5), with a residual of 7.43 μGal, well within the error on temporal gravity differences along the East-West profile (10 μGal at the 95% confidence interval). Some of the best values found by the GA correspond to one of the edge value of the corresponding search range. This is in principle not ideal since it states for the GA to have been unsuitably confined during its search procedure. A sensitivity analysis of the parameters whose best value coincides with an edge of the chosen range is accomplished. Sensitivity tests are computed for the L-H and W-U couples of parameters, while the other remaining parameters are held fixed at their best values. The tests highlighted that while U is a highly sensitive parameter, W appears not to be so much sensitive. This observation reflects the assumptions behind the model: U defines the volume increase (volume of new voids, if the density of the filling material is set to zero) predicted by the model, while W only changes the volume within which the new fractures are uniformly distributed. In principle, changes of W do not influence the net effect of the model-body, if W remains much smaller than the depth of the mass center of the body itself. As for the L and H, they are sensitive parameters and the obtained best values appears to be a good compromise between the needs of (a) not rising the value of these parameters towards an implausible extent and (b) assessing a satisfactory fit. In conclusion, the sensitivity tests confirm that the GA accomplished its task satisfactorily. Results show that, although it is possible to explain the observed gravity changes by means of the proposed analytical formulation, calculated elevation changes are significantly higher than observed. This finding could imply that another mechanism, allowing a significant density decrease without significant deformation, coupled the tensile mechanism due to the formation of new cracks, increasing the negative gravity effect while leaving the displacements at the surface unchanged. The only mechanism allowing a density decrease at depth without surface displacements is the loss of mass from a magma reservoir. To keep the maximum elevation change within 2-4 cm, a value of extension U less than 1 m should be assumed.

Using this value as an upper bound for the extension of the fracture zone, only about 50% of the observed gravity decrease can be explained using the Okubo model. Under the assumption that the new-forming tensile cracks and the loss of mass come from the same inferred source, it results that an about 10^{11} kg mass should be lost to contribute the missing 50% of the observed gravity decrease. The estimated mass corresponds to a minimum magma volume of $35*10^6$ m^3. It is worth to note that, during late 2000 and the first months of 2001, an almost continuous activity, with lava emission and Strombolian explosions from the summit Southeast Crater (SE), was observed.[11] The volume of the products emitted between January 2001 and the start of the main flank eruption is estimated to range between 13 and $20*10^6$ m^3. On the basis of our calculation, the volume emitted from the SE crater is of the same order of magnitude as the volume lost from the estimated gravity source. The hypothesis that the volume emitted from the summit SE crater, during the January-June 2001 period, was supplied by the same source in which tensile cracks formed needs to be further investigated to understand (i) which phenomenon could have triggered the mass transfer from the deep source volume to the surface and (ii) how this phenomenon relates to the formation of tensile fractures.

4. Conclusion

We illustrate two procedures dealing with linear and non-linear inversion of potential field data. Although the inversion problem suffers from the ambiguity and instability of its solutions, numerical methods allow for narrowing the set of all possible solutions and providing a unique and stable solution. As for large-scale linear inverse problem, a 3D inversion of magnetic field

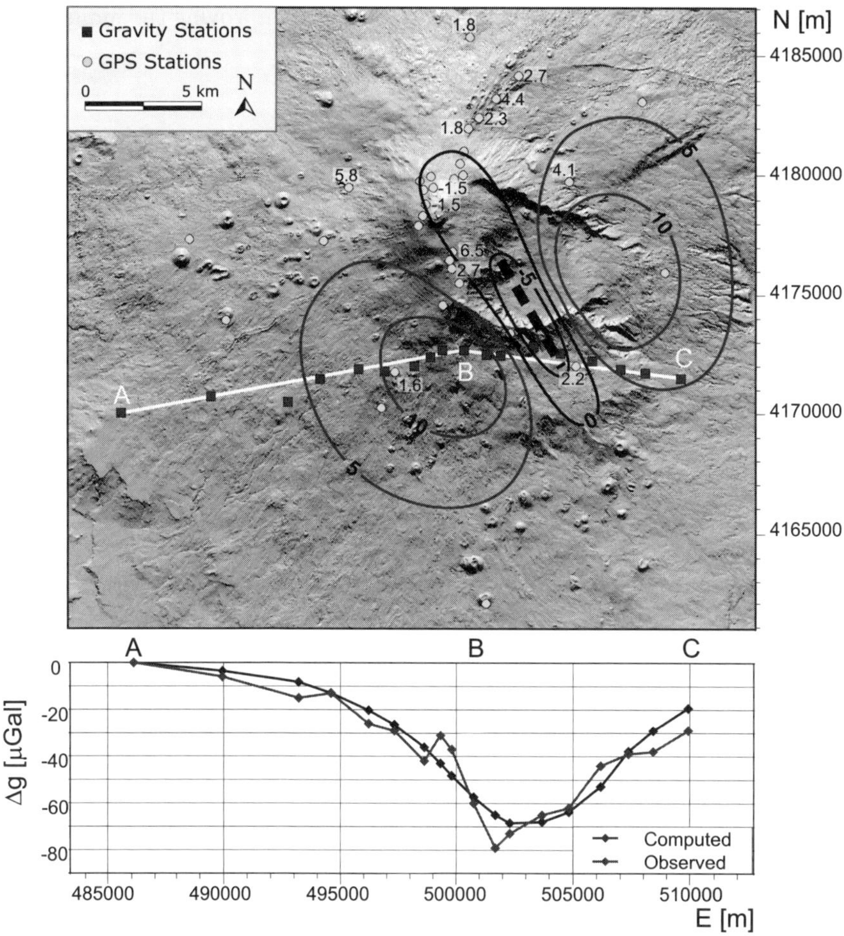

Fig. 5. Results based on the best model. Top: contour map (at 5 cm intervals) of computed elevation changes; surface projection of the source (dashed line); observed (July 2000 - June 2001) elevation changes higher than 1.5 cm. Bottom: observed (January - July 2001) and computed gravity anomalies.

data was performed by means of QP algorithm to produce a magnetization model that provides useful information of the subsurface geological structure of Ustica volcanic complex. As for non-linear potential field inversion, a GA techinique was proposed to infer the parameters of a volcanic source that justifies the growth of a negative gravity anomaly preceding the 2001 Etna eruption. Our findings demonstrate that the identification and interpretation of potential field data can be a useful instrument both for detecting subsurface geological structure model and improving the monitoring of active volcanoes.

References

1. R. J. Blakely, *Potential Theory in Gravity and Magnetic Applications* (Cambridge University Press, New York, 1995).
2. M. Fedi and A. Rapolla, *Geophysics* **64**, 452 (1999).
3. D. B. Rao and N. R. Babu, *Geophysics* **56**, 1729 (1991).
4. P. Gill, W. Murray, D. Ponceleon and M. Saunders, *Solving reduced KKT systems in barrier methods for linear and quadratic programming*, Technical Report SOL 917, Stanford University (Stanford, CA, 1991).
5. R. Napoli, G. Currenti and C. Del Negro, *Bull. Volc.* **in print** (2006).
6. G. Currenti, C. Del Negro and G. Nunnari, *Geophys. J. Int.* **162**, 1 (2005).
7. G. Carbone, D.and Budetta and F. Greco, *J. Geoph. Res.* **108**, 2556 (2003).
8. A. Bonaccorso, S. DAmico, M. Mattia and D. Patane, *Pure. Appl. Geophys.* **161**, 1469 (2004).
9. S. Okubo and H. Watanabe, *Geophys. Res. Lett.* **16**, 445 (1989).
10. D. Carbone, G. Currenti and C. Del Negro, *Bull. Volc.* **in print** (2006).
11. N. C. Lautze, A. J. L. Harris, J. E. Bailey, M. Ripepe, S. Calvari, J. Dehn, S. K. Rowland and K. Evans-Jones, *J. Volcan. Geotherm. Res.* **137**, 231 (2004).

PROTEOMIC MULTIPLE SEQUENCE ALIGNMENTS: REFINEMENT USING AN IMMUNOLOGICAL LOCAL SEARCH

V. CUTELLO, G. NICOSIA, M. PAVONE and I. PRIZZI
Department of Mathematics and Computer Science
University of Catania
Viale A. Doria 6, 95125 Catania, Italy
E-mail: {vctl, nicosia, mpavone, prizzi}@dmi.unict.it

An immunological algorithm has been designed and implemented to optimize multiple proteomic sequence alignments initially generated by CLUSTALW and by a random method. The experimental results show that the refinement algorithm has improved proteomic sequence alignments, producing 52 distinct alignments on average for each instance.

Keywords: Bio-informatics, multiple sequence alignments, protein sequences, immunological algorithms, evolutionary algorithms.

1. Introduction

The most effective method to discover structural or functional similarities among proteins, is to compare multiple proteins of various *phylogenetic* distances. The Multiple Sequence Alignment (MSA) of proteins plays a central role in molecular biology, as it can reveal the constraints imposed by structure and function on the evolution of whole protein families [1]. MSA has been used for building phylogenetic trees, for the identification of conserved motifs, to find diagnostic patterns families, and for predicting secondary and tertiary structures of RNA and protein sequences [2]. In order to be able to align a set of bio-sequences, a reliable objective function for the measurement of an alignment in terms of its biological plausibility through an analytical or computational function is needed. The alignment quality is often the limiting factor in the analysis of biological sequences — defining an appropriate and efficient objective function can remove this limitation, and this is still an active research field [3]. A simple objective function used for the purpose is the *weighted sums-of-pairs* (SP) with affine gap penalties [4]. Here each sequence receives a weight which is proportional to the amount of independent information it contains [5] and the cost of the multiple alignment is equal to the sum of the cost of all the weighted pair-wise substitutions. Since the knowledge about the structure of the search space for MSA is not enough to guide an effective search towards the best solution, several *evolutionary algorithms* (EA) have been developed to solve such a problem and, in general, computational biology problems. Evolutionary computation is applied to problems where heuristics are not available, or where the size of the search space precludes an exhaustive search for the optimal solution.

In this research work we tackle MSA instances using an Immunological Algorithm (IA), inspired by the human Clonal Selection Principle, called *IMSA*. *IMSA* incorporates specific perturbation operators for MSA of amino-acid sequences, and the obtained results show that the designed IA is comparable to state-of-art MSA algorithms. It is important to highlight that *IMSA* is able to produce several optimal or sub-optimal alignments, comparable to those obtained by other approaches. This is the crucial feature of the EAs, in general, and of the algorithm, IMSA, used in this research work in particular.

In section 2 the objective function used in the experiments is presented. In section 3 the immunological algorithm is introduced and section 4 contains a discussion of the obtained results and our conclusions.

2. Proteomics Multiple Sequence Alignments

One of the most important and popular computational sequence analysis problem is to determine if two, or more, biological sequences have common sub-sequences. However, to check the similarities between two or more sequences, there are two primary issues that need to be faced: the choice of an objective function that assesses the biological alignment quality and the design of an effective algorithm to optimize the given objective function. The alignment quality is often the limiting factor in biological analyses of amino-acid sequences; defining a proper objective function is a crucial task.

The alignment score of a pair of sequences is computed as the sum of substitution matrix scores, derived by a probabilistic model. For instance, for two amino-acids, aa_i and aa_j, we need a measure of the probability that they have a common ancestor, or that one aa is the result of one or several mutations of the other: let M be an $(\ell \times \ell)$ scoring matrix, where ℓ is the cardinality of the alphabet Σ; for any two characters a and b of the alphabet Σ, we have the following properties: $M(a,b) = M(b,a)$, for all $a, b \in \Sigma$;

$M(a, -) = GEP$, where GEP is a fixed gap penalty; $M(-, -) = 0$. The symbol dash '-' represents gaps. In general a gap of length h has a penalty score of $h \times GEP$, where $GEP < 0$ is the fixed gap (extension) penalty. This is called the *linear gap penalty function*. From a biological point of view a more appropriate penalty score is the *affine gap penalty function*, (AGPS): given an aligned sequence $\hat{S}_i$, the first gap receives a *gap opening penalty*, $GOP < GEP < 0$, which is stronger than penalty for gap extending spaces. Hence, a gap of length h has a cost of $GOP + (h-1)GEP$. The most common scoring matrices are the PAM and BLOSUM series. These scoring matrices have been inspired by observed mutations in nature. In order to minimise redundant information, each sequence usually receives a weight which is proportional to the amount of independent information it contains. This kind of information can be derived from a phylogenetic tree for the sequences.

The classical objective function used to measure the biological alignment quality is the *weighted sums-of-pairs* with affine gap penalties [4]: each sequence receives a weight proportional to the amount of independent information that it contains [5] and the cost of the multiple alignment is equal to the sum of the costs of all the weighted pairwise substitutions:

$$\max_{\hat{S}} \left(\sum_{i=1}^{n-1} \sum_{j=i+1}^{n} WSS(\hat{S}_i, \hat{S}_j) + \sum_{i=1}^{n} AGPS(\hat{S}_i) \right). \tag{1}$$

Sequence weights are determined by constructing a guide tree from known sequences.

For multiple protein sequence alignment, the weighted sum-of-pairs with affine gap penalties is a popular objective function included in many MSA packages. The problem of finding the multiple alignment is a NP-hard problem [6].

3. *IMSA*, an Immunological Algorithm

In this paper we present an immunological algorithm, IMSA, to tackle the multiple sequence alignment problem. It incorporates two different strategies to create the initial population, as well as new hypermutation operators, specific operators for solving MSA, which insert or remove gaps in the sequences. Gap columns which have been matched are moved to the end of the sequence. The remaining elements (amino acids in this work) and existing gaps are shifted into the freed space.

IMSA considers antigens (Ags) and B cells. The Ag is a given MSA instance, and B cells a set of alignments, that have solved (or approximated) the initial problem. In tackling the MSA Ags and B cells are represented by a sequence matrix. In particular, let $\Sigma = \{A, R, N, D, C, E, Q, G, H, I, L, K, M, F, P, S, T, W, Y, V\}$ be the twenty amino acid alphabet, and let $S = \{S_1, S_2, \ldots, S_n\}$ be the set of $n \geq 2$ sequences with length $\{\ell_1, \ell_2, \ldots, \ell_n\}$, such that $S_i \in \Sigma^*$. Then an Ag is represented by a matrix of n rows and $max\{\ell_1, \ldots, \ell_n\}$ columns, whereas each B cell is represented by an $(n \times \ell)$ matrix, with $\ell = (\frac{3}{2} \cdot max\{\ell_1, \ldots, \ell_n\})$. By using such a representation *IMSA* was able to develop more *compact alignments*.

To create the initial population of d candidate alignments, we used two different strategies. The first strategy is based on the use of random *"offsets"* to shift the initial sequences and it is called *random_initialization*. Such a model works by choosing randomly an offset in the range $[0, (\ell - \ell_i)]$ with a uniform distribution, and then the sequence S_i is shifted from an offset position towards the right side of the row i, of the current B cell. The second way of initializing the population is to seed the initial population with CLUSTALW and *CLUSTALW-seeding*. *IMSA* uses both strategies together: the second strategy creates a percentage of initial alignments using CLUSTALW and the remaining initial alignments are determined by the creation of random offsets. We propose the two models together to avoid the algorithm getting trapped in local optima. All the results shown in this paper were obtained using a combination of the two introduced strategies (80% of the B cell population is introduced by CLUSTALW seeding and the remaining 20% is introduced by random_initialization using the random offsets).

IMSA also incorporates the classical *static cloning operator*, which clones each B cell *dup* times producing an intermediate population $P^{(clo)}$ of $N_c = d \times dup$ B cells, where d is the population size.

Two specific hypermutation operators are used for the multiple sequence alignments, which insert or remove gaps in the sequences. Such operators are the *GAP operator* and the *BlockShuffling operator*. Both of them act on the cloned B cells ($P^{(clo)}$) and generate two new populations, $P^{(gap)}$ and $P^{(block)}$, respectively. The *GAP operator* consists of two procedures: one inserts adjacent sequences of gaps (*InsGap*) and the other removes them (*RemGap*). Initially, the GAP operator chooses what procedure to apply by using a random uniform distribution, i.e. it is randomly decided if a number of adjacent gaps is to be inserted into the sequences or removed. Then a number k, in the range $[1, \theta]$, of (adjacent) gaps is randomly chosen, where θ represents a percentage of the alignments length (ℓ). The results shown

```
IMSA (d, dup, τ_B, T_max)
t ← 0;
FFE ← 0;
N_c ← d × dup;
P^(t) ← Initialize_Population(d);
Strip_Gaps(P^(t));
Evaluate(P^(t));
FFE ← FFE + d;
while (FFE < T_max)do
    P^(clo) ← Cloning (P^(t), dup);
    P^(gap) ← Gap_operators (P^(clo));
    Strip_Gaps(P^(gap));
    Evaluate(P^(gap));
    FFE ← FFE + N_c;
    P^(block) ← BlockShuffling_operators (P^(clo));
    Compute_Weights();
    Normalize_Weights();
    Strip_Gaps(P^(block));
    Evaluate(P^(block));
    FFE ← FFE + N_c;
    (P_a^(t), P_a^(gap), P_a^(block)) = Elitist-Aging(P^(t), P^(gap), P^(block), τ_B);
    P^(t+1) ← (μ + λ)-Selection(P_a^(t), P_a^(gap), P_a^(block));
    t ← t + 1;
end_while
```

in section 4, were obtained setting $\theta = 2\%\ell$. INSGAP can be summarised in the following steps: split the n sequences in z groups; from experimental results, $z = 2$ is the better setting for the performances of *IMSA*. Hence, we rephrase this step as follows: randomly choose a value $m \in [1, n[$, and split the n sequences into two groups: from sequence 1 to m, and from $(m+1)$ to n; randomly choose two integer values x and y, in such a way that k adjacent gaps are inserted beginning from column x for the first group, and from column y for the second. Choose randomly a subsequence shift direction D, either left or right. Finally, it inserts the k adjacent gaps in the relative positions for each sequence, and it shifts the subsequence to the D direction. During the shifting phase, it is possible to miss $n \geq 0$ bits with value 1; in this case, InsGap will select n bits with value 0, different from the k gaps inserted, and they will be flipped to 1, rebuilding the correct sequence.

REMGAP, simply, removes k adjacent gaps, and moves the subsequences towards a randomly chosen direction, either left or right.

The *BlockShuffling* operator is based on the block definition and it moves aligned blocks to the left or to the right: a block is selected in each alignment starting from a random point in a sequence. *IMSA* includes three different approaches the operator: *BlockMove* moves whole blocks either to the left or to the right; *BlockSplitHor* divides the blocks in two levels, upper and lower, and shifts only one level, chosen randomly; *BlockSplitVer*, randomly chooses a column in the block, it divides the block into two sides (left and right), and shifts only one side, randomly chosen as well.

After the two hypermutation operators are used, *IMSA* moves matched gap columns to the right end side of the matrix, with the STRIP_GAPS($P^{(*)}$) function. This function is always applied before the fitness function is evaluated.

The aging operator, used by the algorithm eliminates old B cells in the populations $P^{(t)}$, $P^{(gap)}$ and $P^{(block)}$, thus maintaining high diversity in order to avoid premature convergence. The maximum number of generations a B cell can remain in the population is determined by the parameter τ_B : when a B cell reaches $\tau_B + 1$ generations, it is erased from the current population, even if it is a good candidate solution. The only exception is made for the best B cell present in the current population; we call this model *Elitist Aging*.

Hence, a new population $P^{(t+1)}$ of d B cells is obtained by selecting the best "survivors" to the aging operator, using the $(\mu + \lambda)$-selection operator (with $\mu = d$ and $\lambda = 2N_c$). Such operator reduces an offspring B cell population of size $\lambda \geq \mu$ to a new parent population of size μ; it guarantees monotonicity in the evolution dynamics.

Table 1 shows the pseudo-code of the described *IMSA* algorithm, where *Evaluate(P)* computes the sum-of-pairs objective function of each B cell in the population P (i.e. the proposed alignment quality, using the equation 1), and the functions COMPUTE_WEIGHTS() and NORMALIZE_WEIGHTS() compute and normalize the weights of the sequences using a rooted tree, which is used for the evaluation of the objective function.

Finally, T_{max} is the maximum number of fitness function evaluations, that we used as termination criterion.

4. Results and Conclusions

To evaluate the biological alignment quality produced by *IMSA*, we tested it using the classical benchmark BAliBASE version 1.0 and version 2.0. BAliBASE (Benchmark Alignment dataBASE) is a database developed to evaluate and compare multiple alignment programs containing high quality (manually refined) multiple sequence alignments. BAliBASE is divided into two versions: the first version contains 141 reference alignments and is divided into five hierarchical reference sets containing twelve representative alignments. Moreover, for each alignment the *core blocks* are defined. They are the regions which can be reliably aligned and they represent 58% of residues in the alignments. The remaining 42% are in ambiguous regions which cannot be reliably aligned. Reference 1 contains alignments of equidistant sequences with similar length, reference 2 contains alignments of a family (closely related sequences with > 25% identity) and 3 "orphan" sequences with < 20% identity, reference 3 consists of up to four families with < 25% identity between any two sequences from different families and references 4 and 5 contain sequences with large N/C-terminal extensions or internal insertions. For an extensive explanation of all references please refer to [3]. In the second version, BAliBASE v.2.0 [20], all alignments present in the first version have been manually verified and it includes three new reference sets: repeats, circular permutations and trans-membrane proteins. It consists of 167 reference alignments with more than 2100 sequences. The three new references contain 26 protein families with 12 distinct repeat types, 8 trans-membrane families and 5 families with inverted domains.

One interesting and favourable feature of *IMSA* is its ability to produce several optimal or sub-optimal alignments. In this way, *IMSA* gives biologists more tools to better understand and study the evolution process of proteins. In figure 1 we show two different alignments produced by *IMSA* for the BAliBASE instances *1ad2*, in the left plot, and *1aym3*, in the right one, on Reference 1. In particular, the left plot shows two different alignments with the same SP and CS scores. In the right plot two alignments are showed with different SP and CS scores. In both figures the different alignment sub-sequence obtained by *IMSA* are highlighted in grey colour. Figures 2 and 3 show different alignments produced by *IMSA* on instances of Reference 3 (*1uky*) and Reference 5 (*1qpg*), as well.

The obtained results showed in the next tables were obtained using a robust experimental protocol : $d = 10, dup = 1, \tau_B = 33, T_{max} = 2\times10^5$ and 50 independent runs. Moreover, we used the following substitution matrices: BLOSUM45 for Ref1v1 and Ref 3, with $GOP = 14, GEP = 2$; BLOSUM62 for Ref1v2, Ref 2, Ref 4 and Ref 5, with $GOP = 11, GEP = 1$; BLOSUM80 for Ref1v3, with $GOP = 10, GEP = 1$. Table 4 shows the average SP score obtained by the described alignment tools on every instance set of BAliBASE v.1.0. As it can be seen in the table, *IMSA* performs well on the Reference 2 and Reference 3 sets. The values obtained aid to raise the overall score, which is higher compared to the results published by the Bioinformatic platform of Strasbourg. In table 3 we show the ability of *IMSA* to improve the best initial alignment produced by CLUSTALW on BAliBase v.1.0 benchmark. One can see as *IMSA* represents an effective refinement methodology of the initial alignment; specifically in the overall

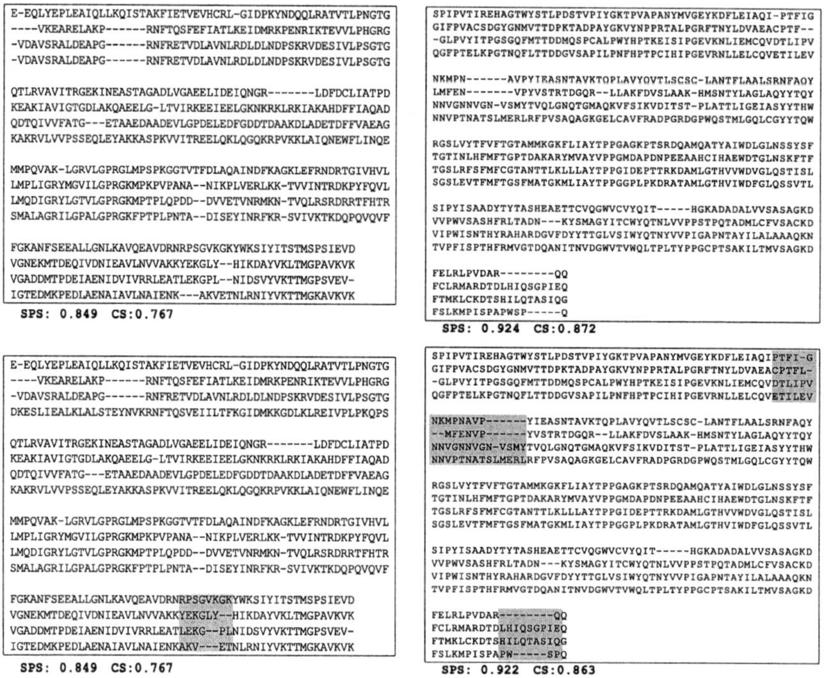

Fig. 1. Optimal and sub-optimal alignments produced by *IMSA* for the *1ad2* (left) and *1aym3* (right) instances of Reference 1 (V2). We show the SP and CS scores, and it is highlight in grey colour the different between the two alignments.

Table 2. SP values given by several methods on the BAliBase v.1.0 benchmark.

Aligner	Ref 1 (82)	Ref 2 (23)	Ref 3 (12)	Ref 4 (12)	Ref 5 (12)	Overall (141)
IMSA	80.7	**88.6**	**77.4**	70.2	82.0	**79.7**
DIALIGN [15]	77.7	38.4	28.8	**85.2**	**83.6**	62.7
CLUSTALX [9]	85.3	58.3	40.8	36.0	70.6	58.2
PILEUP8 [8]	82.2	42.8	33.3	59.1	63.8	56.2
ML_PIMA [14]	80.1	37.1	34.0	70.4	57.2	55.7
PRRP [16]	**86.6**	54.0	48.7	13.4	70.0	54.5
SAGA [18]	70.3	58.6	46.2	28.8	64.1	53.6
SB_PIMA [14]	81.1	37.9	24.4	72.6	50.7	53.3
MULTALIGN [7]	82.3	51.6	27.6	29.2	62.7	50.6

it produced an average SP score 10.5 times better than the initial one.

In table 4 we show the average SP and CS values obtained by the tools

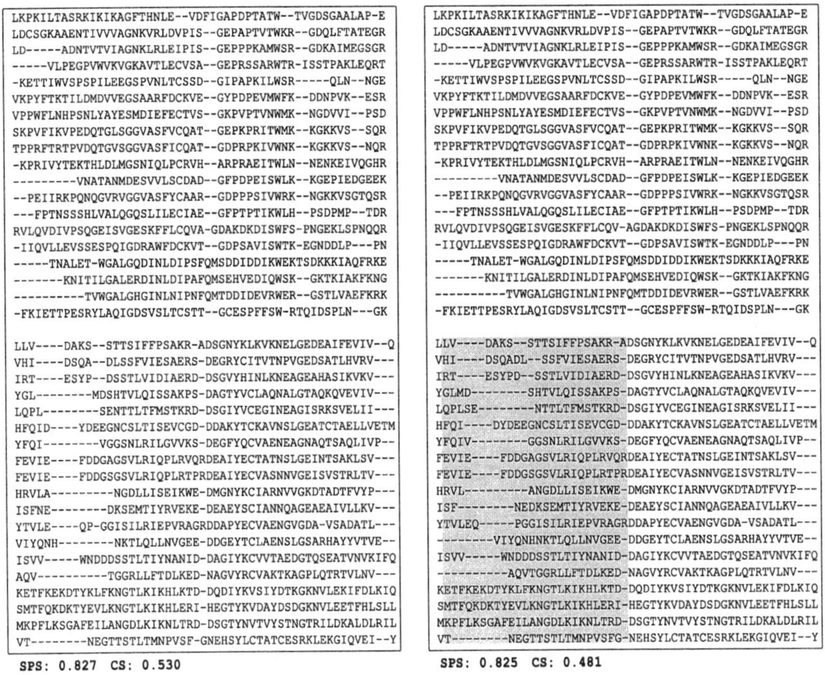

Fig. 2. Optimal and sub-optimal alignments produced by *IMSA* for the *1uky* instance for Reference 3 (medium) We show the SP and CS scores, and it is highlight in grey colour the different between the two alignments.

Table 3. Performance of *IMSA* with respect the initial population $P^{(t=0)}$ produced by CLUSTALW and *CLUSTALW-seeding*, on BAliBase v.1.0 benchmark.

Aligner	Ref 1 (82)	Ref 2 (23)	Ref 3 (12)	Ref 4 (12)	Ref 5 (12)	Overall (141)
CLUSTALW-seeding	77.1	63.1	63.7	65.7	78.4	69.2
IMSA	80.7	88.6	77.4	70.2	82.0	79.7
Improvement	+3.6	+25.5	+13.7	+4.5	+3.6	+10.5

on every group of instances belonging to the BAliBASE v.2.0 database. The Column Score is defined as the number of correctly aligned columns present in the generated alignments, divided by the total number of aligned columns in the core blocks of the reference alignment.

The values used in table 4 are drawn from data reported in [12]. *IMSA*

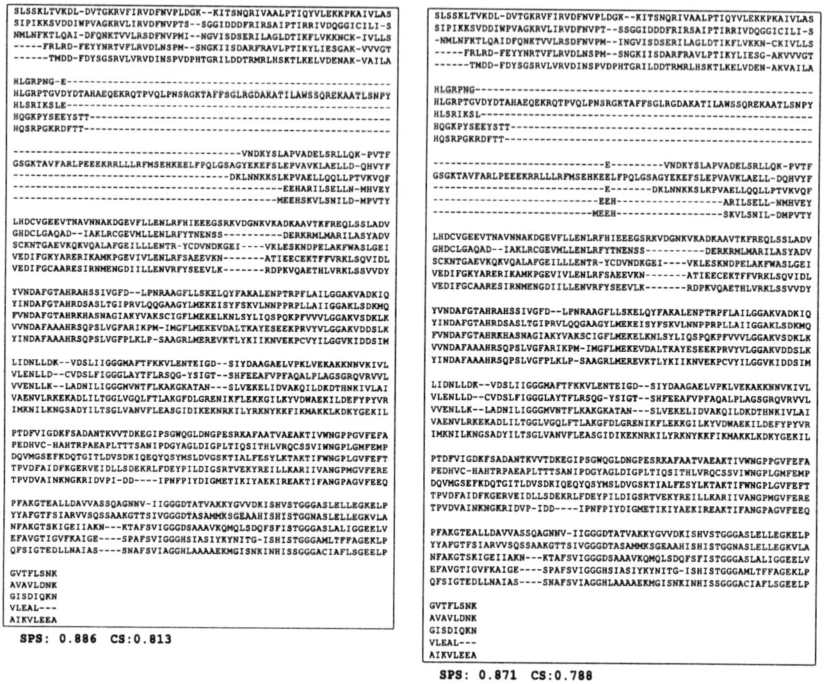

Fig. 3. Optimal and sub-optimal alignments for the *1qpg* instance of Reference 5.

obtains comparable values of SP score on Ref 1, Ref2 and Ref 5 — despite the fact that the value obtained on Reference 3 is the fourth best value. This table also shows that future efforts should focus on improving the CS metric. In the last column of table 4, we show the average number of improved alignments (*NIA*) with respect the initial population produced by *CLUSTALW-seeding*, as described in section 3.

Finally, table 5 shows the average SP and CS values obtained by *IMSA* and *CLUSTALW-seeding* to create the initial population $P^{(t=0)}$, on each reference belonging to the BAliBASE v.2.0 database

We have designed an Immunological Algorithm, called *IMSA*, to tackle the multiple sequence alignment problem. Such algorithm presents two different strategies to create the initial population, and specific mutation operators. To measure the alignment quality we have tested the proposed algorithm on the classical benchmark BAliBASE version 1.0 and version 2.0. A favourable feature of *IMSA* is the ability of generating more than a single sub-optimal alignment, for every MSA instance. This behaviour

Table 4. Alignment accuracies given by several methods on the the BAliBASE v.2.0 benchmark for multiple sequence alignment [12]. The last column shows the average number of improved alignments (*NIA*) produced by *IMSA*.

Aligner	Ref 1 (82)		Ref 2 (23)		Ref 3 (12)		Ref 4 (12)		Ref 5 (12)		Overall (141)		
	SP	CS	SP	CS	SP	CS	SP	CS	SP	CS	SP	CS	*NIA*
SPEM [12]	**90.8**	83.9	93.4	57.3	81.4	56.9	**97.4**	**90.8**	97.4	**92.3**	**91.5**	78.6	1
MUSCLE [17]	90.3	**84.7**	64.4	60.9	82.2	61.9	91.8	74.8	**98.1**	92.1	91.0	**78.7**	1
PROBCONS [13]	90.0	83.9	**94.0**	**62.6**	**82.3**	**63.1**	90.9	73.6	**98.1**	91.7	90.8	78.4	1
T-COFFEE [11]	86.8	80.0	93.9	58.5	76.7	54.8	92.1	76.8	94.6	86.1	88.2	74.6	1
PRALINE [19]	90.4	83.9	**94.0**	61.0	76.4	55.8	79.9	53.9	81.8	68.6	88.2	73.9	1
CLUSTALW [10]	85.8	78.3	93.3	59.3	72.3	48.1	83.4	62.3	85.8	63.4	85.7	70.0	1
IMSA	83.4	65.3	92.1	41.3	78.6	36.2	73.0	31.9	83.6	56.9	81.4	46.3	**52**

Table 5. Performance of *IMSA* with respect the initial population $P^{(t=0)}$ produced by *CLUSTALW-seeding*, on BAliBase v.2.0 benchmark.

Aligner	Ref 1 (82)		Ref 2 (23)		Ref 3 (12)		Ref 4 (12)		Ref 5 (12)		Overall (141)	
	SP	CS	SP	CS	SP	CS	SP	CS	SP	CS	SP	CS
CLUSTALW-seeding	77.1	64.9	85.5	40.7	68.3	34.9	64.1	29.9	73.8	51.4	73.7	44.3
IMSA	83.4	65.3	92.1	41.3	78.6	36.2	73.0	31.9	83.6	56.9	81.4	46.3
Improvement	+6.3	+0.4	+6.6	+0.6	+10.3	+1.3	+8.9	+2.0	+9.8	+5.5	+7.7	+2

is due to the stochastic nature of the algorithm and populations evolved during the convergence process. The alignment process is not affected by the presence of distant sequences, and this can be considered an other advantage of *IMSA*. From experimental results, the scoring function used by *IMSA* produces high SP values and low CS scores, therefore future work will be focus on the improvement of the CS score values using the T-Coffee scoring function.

References

1. I. Eidhammer, I. Jonassen and W. R. Taylor, *Protein Bioinformatics* (Wiley, Chichester, West Sussex, UK, 2004).
2. R. Durbin, A. K. S. Eddy, A. Krogh and G. Mitchison, *Biological sequence analysis* (Cambridge University Press, Cambridge, UK, 2004).

3. J. D. Thompson, F. Plewniak, R. Ripp, J. C. Thierry and O. Poch, Towards a Reliable Objective Function for Multiple Sequence Alignments, *J. on Mol. Biol.* **301**, 937 (2001).
4. S. F. Altschul and D. J. Lipman, Trees stars and multiple biological sequence alignment, *SIAM J. on App. Maths.* **49**, 197 (1989).
5. S. F. Altschul, R. J. Carroll and D. J. Lipman, Weights for data related by a tree, *J. on Mol. Biol.* **207**, 647 (1989).
6. P. Bonizzoni and G. D. Vedova, The Complexity of Multiple Sequence Alignment with SP-score that is a Metric, *Theor. Computer Science* **259**, 63 (2001).
7. F. Corpet, Multiple sequence alignment with hierarchical clustering, *Nuc. Acids Research* **16**, 10881 (1988).
8. Genetics Computer Group, Wisconsin package v.8 (1993), `http://www.gcg.com`.
9. J. D. Thompson, T. J. Gibson, F. Plewniak, F. Jeanmougin and D. G. Higgins, The ClustalX windows interface: flexible strategies for multiple sequence alignment aided by quality analysis tools, *Nuc. Acids Research* **24**, 4876 (1997).
10. J. D. Thompson, D. G. Higgins and T. J. Gibson, CLUSTAL W: improving the sensitivity of progressive multiple sequence alignment through sequence weighting, position-specific gap penalties and weight matrix choice, *Nuc. Acids Research* **22**, 4673 (1994).
11. C. Notredame, D. G. Higgins and J. Heringa, T-Coffee: a novel method for fast and accurate Multiple Sequence Alignment, *J. on Mol. Biol.* **302**, 205 (2000).
12. H. Zhou and Y. Zhou, SPEM: Improving multiple sequence alignment with sequence profiles and predicted secondary structures, *Bioinformatics* **21**, 3615 (2005).
13. C. B. Do, M. S. P. Mahabhashyam, M. Brudno and S. Batzoglou, ProbCons: Probabilistic consistency-based multiple sequence alignment, *Genome Research* **15**, 330 (2005).
14. R. F. Smith and T. F. Smith, Pattern-induced multi-sequence alignment (PIMA) algorithm employing secondary structure-dependent gap penalties for use in comparative protein modelling, *Prot. Engineering* **5**, 35 (1992).
15. B. Morgenstern, K. Frech, A. Dress and T. Werner, DIALIGN: Finding local similarities by multiple sequence alignment, *Bioinformatics* **14**, 290 (1998).
16. O. Gotoh, Further improvement in methods of group-to-group sequence alignment with generalized profile operations, *Bioinformatics* **10**, 379 (1994).
17. R. C. Edgar, MUSCLE: multiple sequence alignment with high accuracy and high throughput, *Nuc. Acids Research* **32**, 1792 (2004).
18. C. Notredame and D. G. Higgins, SAGA: sequence alignment by genetic algorithm, *Nuc. Acids Research* **24**, 1515 (1996).
19. V. A. Simossis and J. Heringa, PRALINE: a multiple sequence alignment toolbox that integrates homology-extended and secondary structure information, *Nuc. Acids Research* **33**, 289 (2005).
20. A. Bahr, J. D. Thompson, J. C. Thierry and O. Poch, BAliBASE (Benchmark Alignment dataBASE): Enhancements for Repeats, Transmembrane Sequences and Circular Permuations, *Nuc. Acids Research* **29**, 232 (2001).

CONVERGENCE TO SELF-SIMILARITY IN AN ADDITION MODEL WITH POWER-LIKE TIME-DEPENDENT INPUT OF MONOMERS

F. P. DA COSTA

Universidade Aberta, DCET
Rua Fernão Lopes 9, 2ºDto, P-1000-132 Lisboa, Portugal
and
Universidade Técnica de Lisboa, Instituto Superior Técnico, CAMGSD
Av. Rovisco Pais 1, P-1049-001 Lisboa, Portugal

JOÃO T. PINTO

Universidade Técnica de Lisboa, Instituto Superior Técnico,
Departamento de Matemática and CAMGSD
Av. Rovisco Pais 1, P-1049-001 Lisboa, Portugal

RAFAEL SASPORTES

Universidade Aberta, DCET
Rua Fernão Lopes 9, 2ºDto, P-1000-132 Lisboa, Portugal

In this note we extend the results published in Ref. 1 to a coagulation system with Becker-Döring type interactions and time-dependent input of monomers $J_1(t)$ of power-like type: $J_1(t)/(\alpha t^{\omega}) \to 1$ as $t \to \infty$, with $\alpha > 0$ and $\omega > -\frac{1}{2}$. The general framework of the proof follows Ref. 1 but a different strategy is needed at a number of points.

Keywords: Dynamics of non-autonomous ODEs, Coagulation equations, Long-time behaviour, Self-similar behaviour.

1. Introduction

There has been a recent upsurge of mathematical work in the field of coagulation-type equations (see, for instance, References 2 and 3, and references therein) of which a sizable portion has been dedicated to dynamical questions, particularly those of convergence to self-similar behaviour (see the references above and also Ref. 4). In the current note we consider this type of problems in the context of the following addition model (i.e., a coagulation equation with Becker-Döring type interactions) with time-dependent input of monomers, $J_1(t)$. Calling a cluster with j identical particles a j-*cluster*, and a single particle a 1-cluster, or a monomer, the kinetic scheme associated with the above mentioned process of cluster growth is

$$j\text{-cluster} + 1\text{-cluster} \longrightarrow (j+1)\text{-cluster}, \qquad j \geq 1.$$

We assume the existence of a monomer source, possibly time-dependent. Assuming also that the mass action law of chemical kinetics is valid, and denoting by $c_j = c_j(t)$ the concentration of j-clusters at time t, the following system of ordinary differential equations describes the time evolution of the concentrations c_j :

$$\begin{cases} \dot{c}_1 = J_1(t) - c_1^2 - c_1 \displaystyle\sum_{j=1}^{\infty} c_j \\ \dot{c}_j = c_1 c_{j-1} - c_1 c_j, \quad j \geq 2. \end{cases} \tag{1}$$

Systems of this type are used, for instance, as mean-field models of submonolayer growth in epitaxial deposition (cf., e.g., Ref. 5).

This work is a follow-up study to Ref. 1 and the reader should consult that paper for a more comprehensive explanation of the background and motivation involved. In Ref. 1 monomer input terms $J_1(t) = \alpha t^{\omega}$, with $\alpha > 0$ and $\omega > -\frac{1}{2}$, were considered. This is clearly a rather restrictive class, and here we provide an extension of those results to monomer input of the type $J_1(t) = \alpha t^{\omega}(1+\varepsilon(t))$ where $\varepsilon(\cdot)$ is a continuous function satisfying $\varepsilon(t) \to 0$ as $t \to +\infty$. Most, but not all, of the results in Ref. 1 follow with just this general assumption on ε. This is true, in particular, of those concerning convergence to a self-similar profile. Those about the rate of convergence of the bulk quantity $\sum_{j=1}^{\infty} c_j$ will need an extra assumption on the decay rate of the perturbation $\varepsilon(t)$.

2. General Approach and Statement of Results

Our main goal is to study the long-time behaviour and approach to self-similarity of solutions to Equation (1). The general approach, used also in Refs. 6 and 1 is to consider the auxiliary variable $c_0(t) = \sum_{j=1}^{\infty} c_j(t)$, representing the total amount, at time t, of clusters of every possible size j. This quantity satisfies the differential equation $\dot{c}_0 = J_1(t) - c_0 c_1$, and so

(1) can be written as

$$\begin{cases} \dot{c}_0 = J_1(t) - c_0c_1, \\ \dot{c}_1 = J_1(t) - c_0c_1 - c_1^2, \\ \dot{c}_j = c_1c_{j-1} - c_1c_j, \quad j \geq 2. \end{cases} \tag{2}$$

From this we conclude firstly that the dynamic behaviour of the monomer concentration $c_1(t)$ is determined only by the bidimensional (c_0, c_1)-system,

$$\begin{cases} \dot{c}_0 = J_1(t) - c_0c_1 \\ \dot{c}_1 = J_1(t) - c_0c_1 - c_1^2, \end{cases} \tag{3}$$

and, secondly, after introducing the new time scale $t \longmapsto \varsigma(t) := \varsigma_0 + \int_{t_0}^t c_1(s)ds$, the dynamics of the j-cluster becomes determined by the (infinite dimensional) lower triangular linear equations

$$\widetilde{c}_j{}' = \widetilde{c}_{j-1} - \widetilde{c}_j, \qquad j \geq 2, \tag{4}$$

where $\widetilde{c}_j(\varsigma) := c_j(t(\varsigma))$ and $(\cdot)' = \frac{d}{d\varsigma}$, which can be explicitly solved in terms of $\widetilde{c}_1$ by a repeated application of the variation of constants formula:

$$\widetilde{c}_j(\varsigma) = e^{-\varsigma} \sum_{k=2}^{j} \frac{\varsigma^{j-k}}{(j-k)!} c_k(0) + \frac{1}{(j-2)!} \int_0^\varsigma \widetilde{c}_1(\varsigma - s) s^{j-2} e^{-s} ds. \tag{5}$$

We shall consider the following general assumption for the input of monomers $J_1(t)$:

(H1) $J_1(t) = (1 + \varepsilon(t))\alpha t^\omega$, where $\alpha > 0$, $\omega > -\frac{1}{2}$, and $\varepsilon(t)$ is a continuous function satisfying $\varepsilon(t) \to 0$ as $t \to +\infty$.

The case where $\varepsilon(t) \equiv 0$ was considered in Ref. 1 and the particular case where, additionally to that, we had time independent input $\omega = 0$ was first studied in Ref. 6. Both these studies contributed to the rigorous justification of formal results in Ref. 7. The case of polynomial-like monomer input term considered in this note has never been investigated before, not even at a formal level.

The main idea behind our approach is that, since $\varepsilon(t) \to 0$ as $t \to +\infty$, the presence of this perturbation term should not be felt at large times, and so the results in Ref. 1 should remain valid. In particular, we should expect the following to hold true:

Theorem 2.1. *Assume (H1) holds, and let* (c_0, c_1) *be any solution of Eq. (3). Then, as* $t \to +\infty$, *we have*

(i) $\left(\frac{1+2\omega}{3\alpha^2}\right)^{\frac{1}{3}} t^{-\frac{1+2\omega}{3}} c_0(t) \longrightarrow 1,$

(ii) $\left(\frac{3}{(1+2\omega)\alpha}\right)^{\frac{1}{3}} t^{\frac{1-\omega}{3}} c_1(t) \longrightarrow 1.$

Using the result in Theorem 2.1 (ii) and translating it to the phase variable $\widetilde{c}_1$ in the time scale ς we can use it in Eq. (5) to obtain information on the long time behaviour of the j−cluster concentration $\widetilde{c}_j(\varsigma)$, in particular concluding that the behaviour in Theorem 2.1 (ii) holds true with $c_1(t)$ substituted by $c_j(t)$, for every $j \geq 1$. Furthermore, and more interesting from a dynamical point of view, the information about $\widetilde{c}_1$ allows us to use Eq. (5) in order to prove that solutions to (1) converge to the similarity profiles in Ref. 1 as $j, t \to +\infty$. In fact, defining

$$Q(\omega) := \left(\frac{3}{(1+2\omega)\alpha}\right)^{\frac{1}{2+\omega}} \left(\frac{2+\omega}{3}\right)^{r} \qquad \text{where } r := \frac{1-\omega}{2+\omega},$$

the following holds true:

Theorem 2.2. *Assume (H1). Let (c_j) be any solution of (1) with initial data $c_j(0) \in \ell_1$. Let $\varsigma(t)$ and $\widetilde{c}_j(\varsigma)$ be as above. Then*

(i) $\lim_{\substack{j,\varsigma\to+\infty \\ \eta=j/\varsigma \text{ fixed} \\ \eta\neq 1}} Q(\omega)\varsigma^r \widetilde{c}_j(\varsigma) = \begin{cases} (1-\eta)^{-r} & \text{if } \eta < 1 \\ 0 & \text{if } \eta > 1, \end{cases}$

(ii) *furthermore, if $c_j(0) = 0$ for $j \geq 2$,*

$$\lim_{\substack{j,\varsigma\to+\infty \\ \xi=\frac{j-\varsigma}{\sqrt{\varsigma}} \text{ fixed} \\ \xi\in\mathbb{R}}} \left(\frac{\pi}{2}\right)^{\frac{1}{2}} Q(\omega)\, \varsigma^{\frac{r}{2}}\, \widetilde{c}_j(\varsigma) = e^{-\frac{1}{2}\xi^2} \int_0^{+\infty} y^{1-2r} e^{-\xi y^2 - \frac{1}{2}y^4}\, dy.$$

Remark 2.1. The extra assumption on the initial data in Theorem 2.2 (ii) is most likely unnecessary but we are presently unable to overcome it (see Section 6.2 of Ref. 6 for an explanation of the technical problem involved).

In Fig. 1 and 2 we present plots of the similarity limits given in Theorem 2.2 for several values of ω. Note that the profiles in Fig. 2 provide a kind of inner expansion of the jump discontinuities occurring in the profiles in Fig. 1 when $\omega \leq 1$.

The proof of Theorem 2.2 is in no way dependent on the function $\varepsilon(t)$, since it is based on the analysis of Eq. (5) which does not involve any information concerning the input of monomers other than that carried over by the $\widetilde{c}_1(\varsigma)$. We direct the interested reader to Refs. 1 and 6 for the proof.

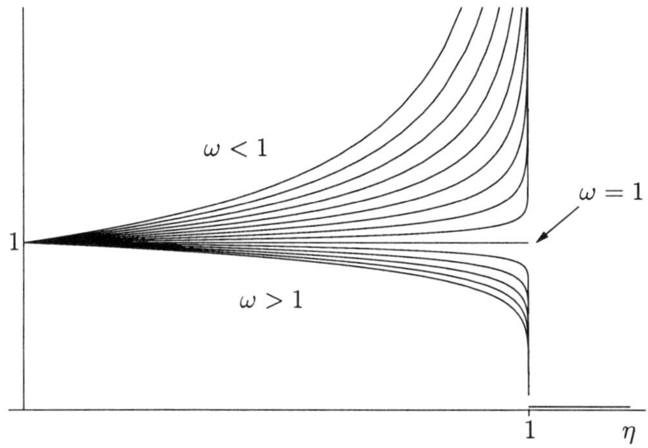

Fig. 1. Graphs of the similarity limits in Theorem 2.2 (i) for values of ω below and above 1 in steps of 0.1

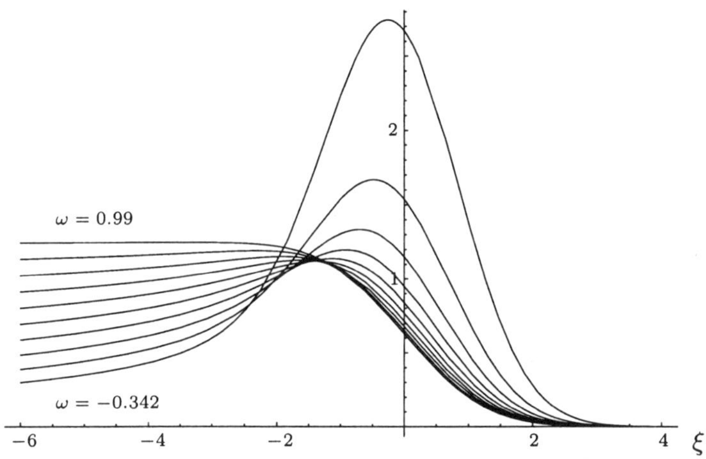

Fig. 2. Graphs of the similarity limits in Theorem 2.2 (ii) for values of ω from −0.342 to 0.99 in steps of 0.148.

In the remaining of this note we concentrate on the proof of Theorem 2.1, and specially in those parts of the proof that differ from the proof of the corresponding result in Ref. 1.

3. On the Proof of Theorem 2.1

The general plan of the proof of Theorem 2.1 is analogous to that of the corresponding theorem in Ref. 1: defining a new time scale $t \mapsto \tau$ so that $\frac{d\tau}{dt} = \left(\frac{3\alpha^2}{1+2\omega}\right)^{1/3} t^{\frac{1+2\omega}{3}}$, letting

$$\begin{cases} x(\tau) := \left(\frac{3}{(1+2\omega)\alpha}\right)^{\frac{1}{3}} t(\tau)^{\frac{1-\omega}{3}} c_1(t(\tau)) \\ y(\tau) := \left(\frac{1+2\omega}{3\alpha^2}\right)^{\frac{1}{3}} t(\tau)^{-\frac{1+2\omega}{3}} c_0(t(\tau)), \end{cases}$$

and using (H1), system (3) becomes

$$\begin{cases} x' = (1+\varepsilon(\tau) - xy) - A\tau^{-\frac{1}{2}}x^2 + B\tau^{-1}x \\ y' = (1+\varepsilon(\tau) - xy)\cdot A\tau^{-\frac{1}{2}} - A^2\tau^{-1}y\,, \end{cases} \tag{6}$$

where $A := \left(\frac{1+2\omega}{4+2\omega}\right)^{\frac{1}{2}}$, $B := \frac{1-\omega}{4+2\omega}$, and, in order not to overload the notation we keep denoting by $\varepsilon(\cdot)$ the function $\varepsilon(t(\cdot))$. To prove Theorem 2.1 it is sufficient to prove that all non-negative solutions (x, y) to system (6) satisfy $(x(\tau), y(\tau)) \to (1,1)$ as $\tau \to +\infty$, and in order to obtain this result we may start by proving the positivity and relative boundedness of the solution vector (meaning that y [resp. x] is bounded iff x [resp. y] is bounded away from zero); these are proved exactly as in Lemmas 1 and 2 of Ref. 1.

Next we need to prove the boundedness of the orbit $(x(\tau), y(\tau))$ when $\tau \to +\infty$. This is where things start to look a bit different from what happened in Ref. 1. By (H1) we need only to consider times large enough so that $|\varepsilon(\tau)| < \epsilon$ (where ϵ can be chosen arbitrarily small) if $\tau > T_\epsilon$ say. Observe that from the y−equation in system (6) we have $y' < 0$ if $(x, y) \in \Omega_\epsilon^- := \{1 + \epsilon - xy < 0\}$ and so $y(\tau)$ can only escape to $+\infty$ if the orbit ultimately remains in $\Omega_\epsilon^+ := \mathbb{R}^{2+} \setminus \Omega_\epsilon^-$. But to show that this cannot occur an argument similar to the one used in step 1 of the proof of Lemma 3 in Ref. 1 is enough, with the natural changes of the geometrical setting provided by Fig. 3. Without loss of generality we can assume the initial point P_{τ_0} is in Ω_ϵ^+ with $P_{\tau_0} > \max\{1, \tau_{L,\epsilon}\}$, where $\tau_{L,\epsilon} := \left(2\left(\frac{L^2}{1+\epsilon} - \frac{1+\epsilon}{L}\right)\right)^{-2}$, and $L > \sqrt[3]{1+\epsilon}$ is the ordinate of P_{τ_0}.

Using the differential equations (6) we easily conclude that, in the region of Fig. 3 where $y > \frac{L^2}{1+\epsilon}x$ and $x < \frac{1+\epsilon}{L}$ the following holds true:

$$y' < \left(x' + \left(\frac{1+\epsilon}{L} - \frac{L^2}{1+\epsilon} - BA^{-1}\tau^{-\frac{1}{2}}\right)xA\tau^{-\frac{1}{2}}\right)A\tau^{-\frac{1}{2}} < x'A\tau^{-\frac{1}{2}} < x',$$

where the last two inequalities hold true for all $\tau > \tau_0 > \max\{1, \tau_{L,\epsilon}\}$. But this is the same inequality as in Ref. 1 and from this uniform bound on

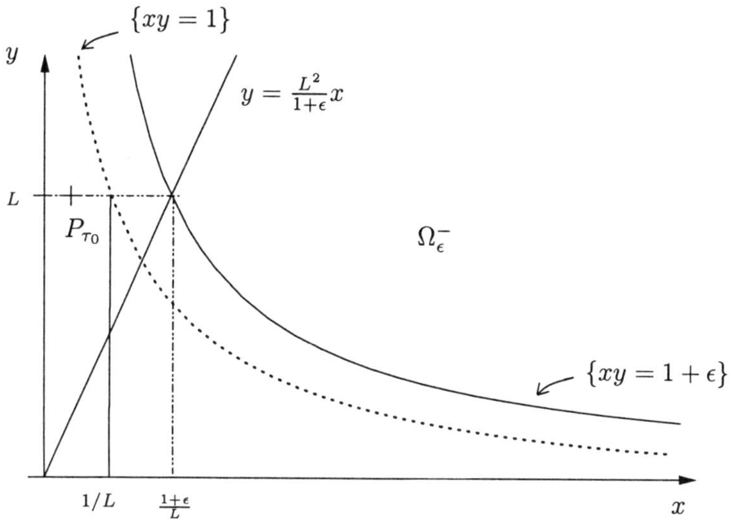

Fig. 3. Geometric setting for the analysis of the possibility of an orbit starting at $P_{\tau 0}$ to escape to infinity due to $y(\tau) \to +\infty$ as $\tau \to +\infty$.

the slope $\frac{dy}{dx}$ of the orbit, and the behaviour of y' in Ω_ϵ^- , we immediately conclude $y(\tau)$ cannot diverge to $+\infty$ as $\tau \to +\infty$. The analysis of what happens for large x , and in particular the impossibility for $x(\tau)$ to become unbounded as $\tau \to +\infty$, requires in the present case an altogether different approach to that used in Ref. 1. There we used a method based on the behaviour of the orbit on the level sets of an auxiliary function approaching the (fixed) hyperbola $\{xy = 1\}$. This essentially meant a change to a kind of moving reference system. Now this method is not likely to work in this case without unnatural (and unreasonable) restrictions on the perturbation $\varepsilon(\tau)$. Hence, we now use a novel approach. Let us suppose an orbit of (6) satisfies $x(\tau_n) \to +\infty$, for some sequence τ_n such that $\tau_n \to +\infty$. Then this orbit must eventually leave $\Omega^-_{\varepsilon(\tau)} := \{1 + \varepsilon(\tau) < xy\}$ since otherwise we would have $x' = 1 + \varepsilon(\tau) - xy - A\tau^{-\frac{1}{2}}x^2 + B\tau^{-1}x < A\tau^{-\frac{1}{2}}x(-x + BA^{-1})\tau^{-\frac{1}{2}} <$ 0 for all sufficiently large τ, a contradiction. An easy argument, as the one in Ref. 1 shows that for all large times, the orbit cannot enter $\Omega^-_{\varepsilon(\tau)}$. Furthermore, for all large enough τ it easily follows from Eqs. (6) that

$$A\tau^{-\frac{1}{2}}x' - y' = -\left(x - \frac{y}{x} - \frac{B}{A}\tau^{-\frac{1}{2}}\right) A^2\tau^{-1}x < 0. \tag{7}$$

Now suppose $x' > 0$ for all sufficiently large τ. From (7) this entails $y' >$ $A\tau^{-\frac{1}{2}}x' > 0$, which leads to a contradiction since the orbit would eventually enter $\Omega^-_{\varepsilon(\tau)}$ for large enough τ. So, in order to have $x(\tau_n) \to +\infty$ as $\tau_n \to +\infty$, there must exist an infinite sequence of time intervals in which $x' < 0$, and in at least some subintervals we must have $y' < 0$, since otherwise the orbit would enter in $\Omega^-_{\varepsilon(\tau)}$ anyway, or it would remain bounded. The differential inequality (7) implies that the situation depicted in Fig. 4 is essentially the only possible one.

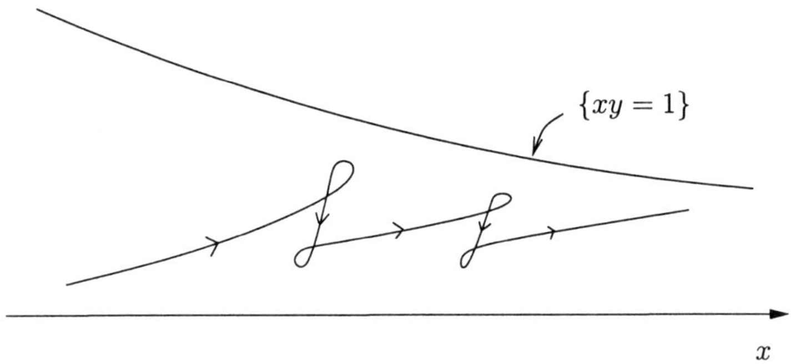

Fig. 4. The only possibility to have an orbit with $x(\tau) \to +\infty$ as $\tau \to +\infty$. Note that (7) implies that $y'(\tau) > 0$ whenever $x'(\tau) > 0$.

Observe that for the orbit to decrease in the y component it must first start decreasing in x, since (7) forces $A\tau^{-\frac{1}{2}}x' < y'$. This explains why the orbit should essentially be as shown in Fig. 4, with that type of self-crossings. But this is impossible. To see why, let us concentrate our attention in Fig. 5, where a portion of the orbit of Fig. 4 at the start of one of its $y-$descent sections is enlarged showing a self-crossing P that occurs at two instants, τ_1 and τ_2, with $\tau_1 < \tau_2$.

At τ_1 we have $x'(\tau_1) > 0$ and (7) gives $\frac{dy}{dx}(\tau_1) > A\tau_1^{-\frac{1}{2}}$, and at τ_2 we have $x'(\tau_2) < 0$, which implies that (7) now means $\frac{dy}{dx}(\tau_2) < A\tau_2^{-\frac{1}{2}}$. But since $\tau_2 > \tau_1$, we conclude that $\frac{dy}{dx}(\tau_2) < A\tau_2^{-\frac{1}{2}} < A\tau_1^{-\frac{1}{2}} < \frac{dy}{dx}(\tau_1)$ in contradiction with what we concluded above should happen at a self-crossing point. This concludes the proof that (x, y) must remain bounded (and bounded away from zero).

With this information we can now start to identify the orbit's $\omega-$limit, by showing that the $\omega-$limit set of any orbit is contained in the hyperbola $\{xy = 1\}$. In order to obtain this result we consider an auxiliary function $h(\tau) := x(\tau)y(\tau)$; as was done in Ref. 1, if (x, y) satisfies (7) then $h(\tau)$

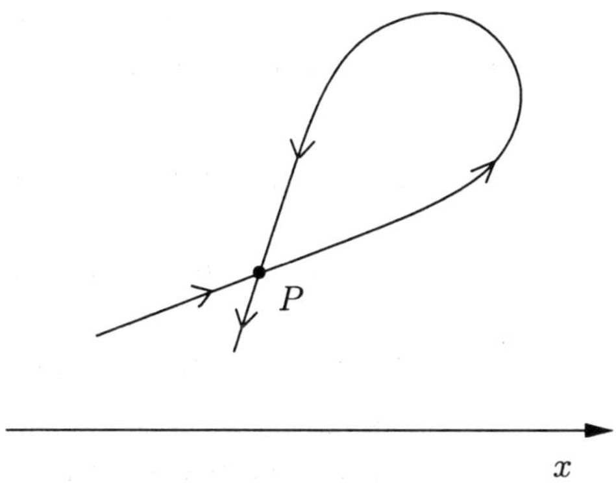

Fig. 5. Situation described in the text for a self-intersecting point of an orbit having $x(\tau) \to +\infty$ as $\tau \to +\infty$.

satisfies the following linear (in h) differential equation:

$$h'(\tau) = (1+\varepsilon(\tau))\left(y + A\tau^{-\frac{1}{2}}x\right) - \alpha(\tau)h(\tau), \tag{8}$$

where

$$\alpha(\tau) := y(\tau) + 2A\tau^{-\frac{1}{2}}x(\tau) + (A^2 - B)\tau^{-1}. \tag{9}$$

Using the variation of constants formula we can write the solution to (8) as

$$h(\tau) = h(\tau_0)e^{-\int_{\tau_0}^{\tau}\alpha(s)ds} + \int_{\tau_0}^{\tau}(1+\varepsilon(s))\left(y(s) + \frac{Ax(s)}{s^{\frac{1}{2}}}\right)e^{-\int_s^{\tau}\alpha(\theta)d\theta}ds. \tag{10}$$

We need to show that $h \to 1$. Using Eq. (10), we show that the first term goes to zero and that the second one goes to 1, as $\tau \to +\infty$. Since y is bounded away from zero, there exists a constant $L_y > 0$ such that $y(\tau) \geq L_y$, and so

$$\begin{aligned} e^{-\int_{\tau_0}^{\tau}\alpha(s)ds} &= \left(\frac{\tau_0}{\tau}\right)^{A^2-B} e^{-\int_{\tau_0}^{\tau}\left(y(s)+2As^{-1/2}x(s)\right)ds} \\ &\leq \left(\frac{\tau_0}{\tau}\right)^{A^2-B} e^{-\int_{\tau_0}^{\tau}y(s)ds} \leq \left(\frac{\tau_0}{\tau}\right)^{A^2-B} e^{-L_y(\tau-\tau_0)}, \end{aligned}$$

showing that the first term in Eq. (10) goes to zero (exponentially) as $\tau \to +\infty$.

Next we show that the second term in Eq. (10) goes to 1 as $\tau \to +\infty$. Using again the fact that $|\varepsilon(\tau)| < \epsilon$ if $\tau > T_\epsilon$, we can estimate the second term in Eq. (10) by splitting the integral as $\int_{\tau_0}^{\tau} = \int_{\tau_0}^{T_\epsilon} + \int_{T_\epsilon}^{\tau}$. The integral over $[\tau_0, T_\epsilon]$ converge to zero as $\tau \to \infty$, since we are integrating an exponentially decaying function on a fixed compact interval. On $[T_\epsilon, \tau]$, we can write

$$(1-\epsilon)\int_{T_\epsilon}^{\tau} g(s,\tau)ds \le \int_{T_\epsilon}^{\tau}(1+\varepsilon(\tau))g(s,\tau)ds \le (1+\epsilon)\int_{T_\epsilon}^{\tau} g(s,\tau)ds.$$

where $g(s,\tau) := \left(y(s) + As^{-\frac{1}{2}}x(s)\right)e^{-\int_s^\tau \alpha(\theta)d\theta}$. Using the results from the proof of Lemma 4 in Ref. 1 we have $\int_{T_\epsilon}^{\tau} g(s,\tau)ds \to 1$ as $\tau \to +\infty$ and this completes the proof of the convergence of h.

Inspired by the approach in Ref. 1, the next steps to locate the ω−limit set of the orbits will be, first to prove that, for each orbit, its ω−limit set, (which, by the previous result about h, is an arc of the hyperbola $\{xy = 1\}$) contains the point $(1,1)$, and second to conclude that that arc degenerates into the single point $(1,1)$. Resorting, as in Ref. 1, to the auxiliary function $b(\tau) := y(\tau) - A\tau^{-\frac{1}{2}}x(\tau)$ we discover that its dynamics is governed by the differential equation

$$b' = (-b + x^2)A^2\tau^{-1},$$

which is exactly the same as in Ref. 1, i.e., the dynamic behaviour of b is independent of $\varepsilon(\tau)$ and so all the results proved when $\varepsilon(\tau) \equiv 0$ in Lemmas 5 and 6 of Ref. 1 remain valid for the present case. Hence the two steps referred to above can be completed and the proof of Theorem 2.1 is achieved.

4. Remarks on the Rate of Convergence

A question that naturally comes to mind at this point is to ask at what rate do solutions converge to their final states. Whatever the mathematical sense we may give this question, its elucidation is bound to include the long time behaviour of the quantity $J_1(t) - c_0(t)c_1(t)$ (if we are thinking in terms of solutions to (3)) or of $1 + \varepsilon(\tau) - x(\tau)y(\tau)$ (if we are considering solutions to (6).)

In the case $\varepsilon(t) \equiv 0$ treated in Ref. 1 the assumptions in Theorem 2.1 are sufficient to prove that

$$\frac{1 - x(\tau)y(\tau)}{A\tau^{-\frac{1}{2}}} \longrightarrow 1 \quad \text{as} \quad \tau \to +\infty, \tag{11}$$

and hence to have, as $t \to +\infty$,

$$\left(\frac{3}{(1+2\omega)\alpha}\right)^{\frac{2}{3}} t^{2\frac{1-\omega}{3}}(\alpha t^\omega - c_0(t)c_1(t)) \longrightarrow 1. \tag{12}$$

In order to obtain similar results when $\varepsilon(t)$ is not identically zero we need further assumptions on this function besides those stated in (H1), namely we need to assume something about its decay rate to zero. The following hypothesis is sufficient to ensure (12) holds also in this case.

(H2) The function $\varepsilon(t)$ is continuously differentiable and, as $t \to +\infty$, satisfies $\varepsilon(t) = o(t^{-\frac{2+\omega}{3}})$, and $\dot{\varepsilon}(t) = o(t^{-\frac{1-\omega}{3}})$.

Remark 4.1. Observe that, by the definition of the time scale τ given in the beginning of Section 3, we easily conclude that $\tau^{-\frac{1}{2}} = \mathcal{O}(t^{-\frac{2+\omega}{3}})$, from which it follows that we must have $\varepsilon(\tau) = o(\tau^{-\frac{1}{2}})$ and also $\varepsilon'(\tau) = \dot{\varepsilon}\frac{dt}{d\tau} = o(\tau^{-\frac{1}{2}})$.

The way the proof of (11) proceeds is analogous to the proof of Lemmas 4 and 7 in Ref. 1: in the present case the expression for $(1 - h(\tau))/\tau^{-1/2}$ has the additional additive contribution coming from the perturbation term $\varepsilon(\tau)$, namely

$$-\tau^{\frac{1}{2}} \int_{\tau_0}^{\tau} \varepsilon(s) \left(y(s) + As^{-\frac{1}{2}} x(s) \right) e^{-\int_s^\tau \alpha(\theta)d\theta} ds, \tag{13}$$

with $\alpha(\cdot)$ the function defined in (9). The analysis of (13) proceeds as follows: first write it as

$$-\tau^{\frac{1}{2}} \int_{\tau_0}^{\tau} \varepsilon(s)\alpha(s) e^{-\int_s^\tau \alpha(\theta)d\theta} ds + \tau^{\frac{1}{2}} \int_{\tau_0}^{\tau} \varepsilon(s) \left(\frac{Ax(s)}{s^{1/2}} + \frac{A^2 - B}{s} \right) e^{-\int_s^\tau \alpha(\theta)d\theta} ds, \tag{14}$$

then observe that the first integral in (14) can be written as follows

$$\begin{aligned} \int_{\tau_0}^{\tau} \varepsilon(s)\alpha(s) e^{-\int_s^\tau \alpha(\theta)d\theta} ds &= e^{-\int_{\tau_0}^\tau \alpha(s)ds} \int_{\tau_0}^{\tau} \varepsilon(s) \frac{d}{ds} \left(e^{\int_{\tau_0}^s \alpha(\theta)d\theta} \right) ds \\ &= \varepsilon(\tau) - \varepsilon(\tau_0) e^{-\int_{\tau_0}^\tau \alpha(s)ds} - \int_{\tau_0}^{\tau} \varepsilon'(s) e^{\int_{\tau_0}^s \alpha(\theta)d\theta} ds \end{aligned} \tag{15}$$

where the last equality was obtained using integration by parts. Now, the estimates developed in the proof of Lemma 4 of Ref. 1 together with the assumption (H2), in the version presented in Remark 4.1, allow us to control the second integral in (14) and the right-hand side of (15), thus completing the proof of (11), and thus also the corresponding version of (12) that now reads as

$$\left(\frac{3}{(1+2\omega)\alpha} \right)^{\frac{2}{3}} t^{2\frac{1-\omega}{3}} \left((1+\varepsilon(t))\alpha t^\omega - c_0(t)c_1(t) \right) \xrightarrow[t \to +\infty]{} 1.$$

5. Final Remarks

In this note we considered the addition model Eq. (1) with power-like time-dependent input of monomers (i.e., addition of monomers at a rate satisfying **(H1)**). We proved that, considering the long time and cluster size limit with either $\eta := j/\varsigma$ or $\xi := (j-\varsigma)/\sqrt{\varsigma}$ constant (where ς is an appropriate rescaling of the original time variable), the solutions approach universal profiles, independently of the initial data. This type of convergence, referred to in the literature as self-similar behaviour, has been the subject of a number of recent studies in the context of Smoluchowski coagulation systems (cf. Refs. 2, 3, 4). The present study extends our knowledge of such a behaviour in Beker-Döring like coagulation equations, by complementing recent results, obtained for constant (Ref. 6) and for particular time-dependent (Ref. 1) inputs of monomers, to a much larger class of time-independent inputs. Since our results include the cases where the input rate of monomers can have a rather erratical behaviour (albeith continuous and asymptotically approaching a power αt^{ω}, with $\omega > -\frac{1}{2}$), it provides a further indication of the robustness of the convergence to self-similar behaviour in this class of systems.

References

1. F. P. da Costa and R. Sasportes, Dynamics of a non-autonomous ODE system occurring in coagulation theory, J. Dynam. Differential Equations, published on-line: 3 January 2007, DOI: 10.1007/s10884-006-9067-5.
2. F. P. da Costa, M. Grinfeld, W. Lamb and J. A. D. Wattis (eds.), Coagulation-Fragmentation Processes, *Physica D* **222**, pp. 1–166 (2006) (special issue).
3. Ph. Laurençot and S. Mischler, On Coalescence Equations and Related Models, in *Modelling and Computational Methods for Kinetic Equations*, Eds. P. Degond, L. Pareschi, G. Russo (Birkhäuser, Boston, 2004) pp. 321–356.
4. F. Leyvraz, Scaling theory and exactly solved models in the kinetics of irreversible aggregation, *Phys. Rep.* **383**, 95 (2003).
5. J. G. Amar, M. N. Popescu, F. Family, Rate-equation approach to island capture zones and size distributions in epitaxial growth, *Phys. Rev. Letters* **86**, 3092 (2001).
6. F. P. da Costa, H. van Roessel and J. A. D. Wattis, Long-time behaviour and self-similarity in a coagulation equation with input of monomers, *Markov Processes Relat. Fields* **12**, 367 (2006).
7. J. A. D. Wattis, Similarity solutions of a Becker-Döring system with time-dependent monomer input, *J. Phys. A: Math. Gen.* **37**, 7823 (2004).

LIVING SHELL-LIKE STRUCTURES

A. DI CARLO* and V. VARANO†

SMFM@DiS, Università "Roma Tre",
Via Corrado Segre 4/6, I-00146 Roma, Italy
** E-mail: adicarlo@mac.com*
† E-mail: v.varano@uniroma3.it

V. SANSALONE

CNRS UMR 7052 B2OA, Laboratoire de Mécanique Physique,
Faculté de Sciences et Technologie, Université Paris XII–Val-de-Marne,
61, avenue du Général de Gaulle, 94010 Créteil CEDEX, France
E-mail: vittorio.sansalone@univ-paris12.fr

A. TATONE

DISAT, Università degli Studi dell'Aquila,
Facoltà di Ingegneria, I-67040 Monteluco di Roio (AQ), Italy
Email: tatone@ing.univaq.it
http://ing.univaq.it/~tatone

A decade ago, the so-called Kröner-Lee decomposition—primarily introduced to discern between elastic and (visco-)plastic strains—was given a broader scope and a deeper interpretation than the original ones, as describing the interplay between the actual and the relaxed configuration of each body element. The main intended application was to growth mechanics of soft living tissues. In 2002, a novel (tensorial) balance law governing the time evolution of the relaxed configuration was devised, and endowed with a proper constitutive theory, thus establishing the foundations of a dynamical theory of material remodelling.

Material remodelling does not describe explicitly the chemistry or whatever else is acting behind the changes in material structure. However, it does account explicitly for the power expended by the biochemical control system, which is of the essence for modelling the mechanics of living tissue. Material remodelling discriminates active from passive remodelling, while treating both on the same footing. Thus it provides mechanistic models of living materials without conceiving of them as inert materials engineered with magic constitutive recipes. The present study develops a toy model of saccular aneurysms, focussing on the two-way coupling between growth and stress.

Keywords: Material remodelling; Growth mechanics; Growing spherical shells; Soft tissue; Saccular aneurysms.

1. Introduction

Soft shell-like structures are ubiquitous in living organisms, ranging from organelles and cell membranes to lymph and blood vessels, the alimentary canal and respiratory ducts, the urinary tract, and the uterus. The passive mechanical response of these structures—a key feature of their physiological and pathological functioning—is highly diversified and rather subtle. However, a much more elusive issue is their ability to grow and remodel, in a way which is both biochemically controlled and strongly coupled with the prevailing mechanical conditions. While the characterization of the passive mechanical response of soft tissue is progressing at a reasonably fast pace nowadays, we find that growth mechanics is definitely the weakest link in the modelling chain. For this reason, we focus on the two-way coupling between growth and stress, which we study using the apparatus of the theory of material remodelling, set forth in Ref. 1 and further developed, expounded and applied in Refs. 2–6.

In Sec. 2 we introduce the model of a pressurized vessel which may undergo large deformations—both *passive* (visco-elastic) and *active* (accretive)—while keeping a spherically symmetric shape. Since we regard this as a drastically simplified model of saccular aneurysms, a short introductory section on real aneurysms is in order. Section 1.1 draws mostly from Refs. 7–10.

1.1. *Saccular aneurysms*

According to Yonekura,[9] saccular aneurysms can be classified into four types (see Fig. 1 (top)):

(1) the aneurysm ruptures within a time span as short as several days to several months after formation;
(2) the aneurysm builds up slowly for a few years after formation and ruptures in this process;
(3) the aneurysm keeps growing slowly for many years without rupturing;
(4) the aneurysm grows up to a certain size (probably under 5 mm in diameter) and thereafter remains unchanged.

Fig. 1 (bottom) reproduces the cartoon where Humphrey[7] has summarized the somewhat unpredictable evolution—either ill-fated or well-behaved—of a saccular aneurysm.

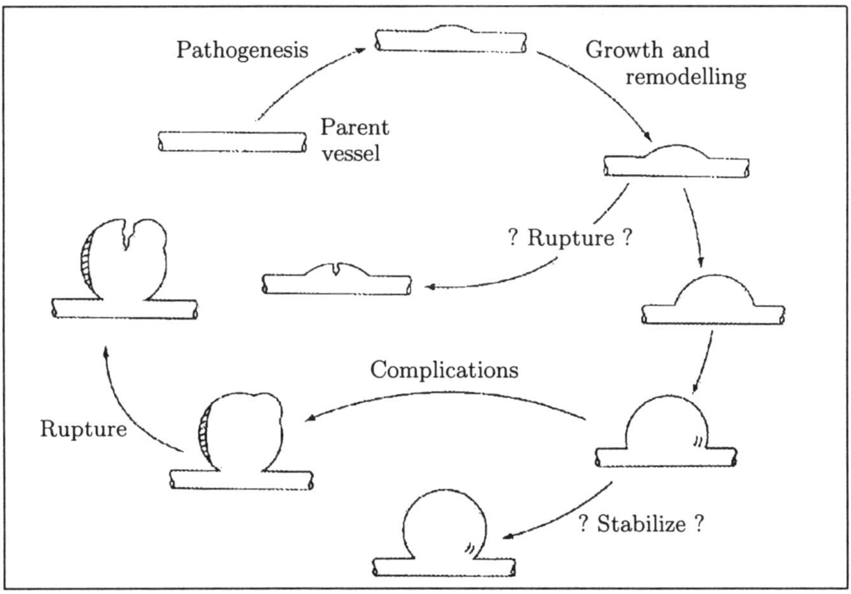

Fig. 1. Evolution paths of saccular aneurysms: (top) Process of growth and rupture: each row pictures one type of development (see text); each column corresponds to an aneurysm's lifetime: days to months for the 2nd, years for the 4th, decades for the 5th (schematics reproduced from Ref. 9); (bottom) Cartoon reproduced from Ref. 7.

Histological analyses provide limited information on the underlying mechanobiological processes. Here is an excerpt from Frösen *et al.*:[8]

> The cellular mechanisms of degeneration and repair preceding rupture of the saccular cerebral artery aneurysm wall need to be elucidated for rational design of growth factor or drug-releasing endovascular devices. [...] Before rupture, the wall of saccular cerebral artery aneurysms undergoes morphological changes associated with remodelling of the aneurysm wall. Some of these changes, like SMC [smooth muscle cell] proliferation and macrophage infiltration, likely reflect ongoing repair attempts that could be enhanced with pharmacological therapy. [...] The morphological changes that result from the MH [myointimal hyperplasia] and matrix destruction are collectively referred to as remodelling of the vascular wall. Although MH is an adaptation mechanism of arteries to hemodynamic stress, in SAH [subarachnoid hemorrhage] patients, for undefined reasons, vascular wall remodelling [is] insufficient to prevent SCAA [saccular cerebral artery aneurysm] rupture.

To sum up, wall remodelling is generally believed[10] to be stress driven. When the arterial wall is unduly stressed, some repair mechanisms get triggered. Their working, however, is still poorly understood.

2. Mathematical Model

In order to concentrate on growth mechanics, we strive to minimize all accessory difficulties, by tailoring an exceedingly simplified model of a saccular aneurysm. Our toy model consists in a highly deformable three-dimensional pressure vessel, constrained in such a way as to undergo only spherical symmetrical motions. Such a strong hypothesis curtails all technical difficulties related to finite kinematics and the allied dynamical issues; tensor algebra and analysis get elementary—though nontrivial, because of curvature and topology—, and a transparent treatment in components is made available by the exceptional existence of natural coordinates, provided by a spherical coordinate system. These features allow us to paraphrase the theory of material remodelling in terms perhaps more digestible than those in Refs. 1–6. However, the reader should be aware that simplicity is not synonymous with clarity, since in a highly simplified setting distinct general concepts may easily collapse into a single quantity and become confused. Warnings will be issued lest the naïve reader be caught in the most treacherous traps.

2.1. *Geometry & kinematics*

To *a priori* satisfy the above mentioned symmetry constraint, we conceive of a *paragon shape* $\mathscr{D}$ of the vessel $\mathscr{B}$ consisting in the (open) difference of two balls centred at $\boldsymbol{x}_\circ \in \mathscr{E}$, the three-dimensional Euclidean ambient space. Let ξ_-, ξ_+ be the radii of the two balls, with $\xi_- < \xi_+$. From now on, we shall identify each body-point $\boldsymbol{b}$ in $\mathscr{B}$ and on its boundary $\partial\mathscr{B}$ with the place $\boldsymbol{\kappa}(\boldsymbol{b})$ it has in the assumed *paragon configuration* $\boldsymbol{\kappa} : \overline{\mathscr{B}} \leftrightarrow \overline{\mathscr{D}}$. In turn, each place $\boldsymbol{x} \in \mathscr{D}$ will be identified with the triple of its spherical coordinates $(\widehat{\xi}(\boldsymbol{x}), \widehat{\vartheta}(\boldsymbol{x}), \widehat{\varphi}(\boldsymbol{x}))$, where $\widehat{\xi}(\boldsymbol{x}) = \|\boldsymbol{x} - \boldsymbol{x}_\circ\|$ is the radius of $\boldsymbol{x}$ and $\widehat{\vartheta}(\boldsymbol{x}), \widehat{\varphi}(\boldsymbol{x})$ are coordinates of its projection on the unit sphere. (Since all fields of interest will depend only on radius and time, there is no need to detail $\widehat{\vartheta}$ and $\widehat{\varphi}$.)

All (*gross*) *placement* of $\mathscr{B}$ will be described through the corresponding *transplacement*

$$\begin{aligned} \mathsf{p} : \mathscr{D} &\rightarrow \mathscr{E} \\ \boldsymbol{x} &\mapsto \boldsymbol{x}_\circ + \rho\,(\widehat{\xi}(\boldsymbol{x}))\, \boldsymbol{e}_{\mathsf{r}}(\widehat{\vartheta}(\boldsymbol{x}), \widehat{\varphi}(\boldsymbol{x}))\,, \end{aligned} \tag{1}$$

where $\boldsymbol{e}_{\mathsf{r}}(\vartheta, \varphi)$ is the outward unit normal to the sphere at (ϑ, φ). Therefore, the (smooth) placements of $\mathscr{B}$ compatible with the symmetry constraint are ultimately parameterized by the set of (smooth) real-valued, monotonically increasing maps

$$\rho : [\,\xi_-, \xi_+] \rightarrow \mathbb{R}\,, \tag{2}$$

which provide the *actual radius* $\rho(\xi)$ as a function of the *paragon radius* ξ. Henceforth, we will abridge notations by assuming that, whenever a place $\boldsymbol{x} \in \mathscr{D}$ is intended unambiguously, the triple $(\xi, \vartheta, \varphi)$ stands for $(\widehat{\xi}(\boldsymbol{x}), \widehat{\vartheta}(\boldsymbol{x}), \widehat{\varphi}(\boldsymbol{x}))$.

All spherically symmetric vector fields $\boldsymbol{v} : \mathscr{D} \rightarrow \mathsf{V}\mathscr{E}$ (with $\mathsf{V}\mathscr{E}$ the translation space of $\mathscr{E}$) admit the following parameterization, in terms of a scalar field $v : [\,\xi_-, \xi_+] \rightarrow \mathbb{R}$, which provides the *radial* component of $\boldsymbol{v}$ (its only strict component):

$$\boldsymbol{v}(\boldsymbol{x}) = v(\xi)\, \boldsymbol{e}_{\mathsf{r}}(\vartheta, \varphi)\,. \tag{3}$$

Similarly, spherically symmetric tensor fields $\mathbf{L} : \mathscr{D} \rightarrow \mathsf{V}\mathscr{E} \otimes \mathsf{V}\mathscr{E}$ are linear combinations of the two fields of *orthogonal projectors*

$$\mathbf{P}_{\mathsf{r}}(\boldsymbol{x}) := \boldsymbol{e}_{\mathsf{r}}(\vartheta, \varphi) \otimes \boldsymbol{e}_{\mathsf{r}}(\vartheta, \varphi)\,, \qquad \mathbf{P}_{\mathsf{h}}(\boldsymbol{x}) := \mathbf{I} - \mathbf{P}_{\mathsf{r}}(\boldsymbol{x}) \tag{4}$$

which depend only on (ϑ, φ), weighted with scalar fields that depend only on ξ, representing the *radial* and *hoop* components of $\mathbf{L}$, respectively:

$$\mathbf{L}(\boldsymbol{x}) = \mathrm{L}_{\mathsf{r}}(\xi)\, \mathbf{P}_{\mathsf{r}}(\vartheta, \varphi) + \mathrm{L}_{\mathsf{h}}(\xi)\, \mathbf{P}_{\mathsf{h}}(\vartheta, \varphi)\,. \tag{5}$$

In particular, the gradient of the transplacement (1) reads:

$$\nabla \mathsf{p}|_{\boldsymbol{x}} = \rho'(\xi)\,\mathbf{P}_{\mathsf{r}}(\vartheta,\varphi) + \frac{\rho(\xi)}{\xi}\,\mathbf{P}_{\mathsf{h}}(\vartheta,\varphi)\,, \tag{6}$$

where ρ' denotes the derivative of the radius-to-radius map (2). Of course, both components of $\nabla\mathsf{p}$ depend on the *single* scalar field ρ.

In order to distinguish growth from passive deformation, we postulate that, at each time $\tau \in \mathscr{T}$ (the *time line*, identified with the real line), there exists a dynamically distinguished tensor field $\mathbb{P}(\tau)$—smoothly depending on time—which we call *prototypal transplant* or, briefly, *prototype.* The assignment of a gross placement and a prototype to each time defines a *refined motion* $(\mathsf{p},\mathbb{P})$. The idea to refine the gross motion in this way dates back to Kröner[11] and Lee,[12] who introduced the notion of an "intermediate" configuration in the sixties, to distinguish between elastic and visco-plastic strains. Much later Rodriguez, Hoger and McCulloch imported that notion into biomechanics, reinterpreting it as the "zero-stress reference state" of a growing body element, to quote verbatim from their 1994 paper.[13] Since there is no reason why the tensor field $\mathbb{P}(\tau)$ should be the gradient of any (gross) placement,[a] it has *two* independent component:

$$\mathbb{P}(\boldsymbol{x},\tau) = \alpha_{\mathsf{r}}(\xi,\tau)\,\mathbf{P}_{\mathsf{r}}(\vartheta,\varphi) + \alpha_{\mathsf{h}}(\xi,\tau)\,\mathbf{P}_{\mathsf{h}}(\vartheta,\varphi)\,. \tag{7}$$

The *warp* $\mathbf{F}$, defined by the *Kröner-Lee decomposition*

$$\mathbf{F} := (\nabla\mathsf{p})\,\mathbb{P}^{-1} = \lambda_{\mathsf{r}}\,\mathbf{P}_{\mathsf{r}} + \lambda_{\mathsf{h}}\mathbf{P}_{\mathsf{h}}\,, \tag{8}$$

gauges how the actual transplant of body elements, characterized by $\nabla\mathsf{p}$, differs from the prototypal transplant $\mathbb{P}$. Since all spherically symmetric tensor fields are symmetric-valued (orthogonal projectors are symmetric), $\mathbf{F}$ coincides with the *stretch* $\mathbf{U}$, and its radial and hoop components are the fields of *principal stretches.* From Eqs. (6)–(8), one readily obtains:

$$\lambda_{\mathsf{r}}(\xi,\tau) = \frac{\rho'(\xi,\tau)}{\alpha_{\mathsf{r}}(\xi,\tau)}\,, \qquad \lambda_{\mathsf{h}}(\xi,\tau) = \frac{\rho(\xi,\tau)}{\xi\alpha_{\mathsf{h}}(\xi,\tau)}\,. \tag{9}$$

The *velocity* realized along the refined motion $(\mathsf{p},\mathbb{P})$ is, by definition, the pair consisting of the *gross velocity* $\dot{\mathsf{p}}$ and the *growth velocity* $\dot{\mathbb{P}}\,\mathbb{P}^{-1}$:

$$\begin{aligned} \dot{\mathsf{p}}(\boldsymbol{x},\tau) &= \dot{\rho}(\xi,\tau)\,\mathbf{e}_{\mathsf{r}}(\vartheta,\varphi)\,, \\ \dot{\mathbb{P}}\,\mathbb{P}^{-1}(\boldsymbol{x},\tau) &= \frac{\dot{\alpha}_{\mathsf{r}}(\xi,\tau)}{\alpha_{\mathsf{r}}(\xi,\tau)}\,\mathbf{P}_{\mathsf{r}}(\vartheta,\varphi) + \frac{\dot{\alpha}_{\mathsf{h}}(\xi,\tau)}{\alpha_{\mathsf{h}}(\xi,\tau)}\,\mathbf{P}_{\mathsf{h}}(\vartheta,\varphi)\,, \end{aligned} \tag{10}$$

[a]Beware that spherical symmetry blots out the distinction between *local* and *global* obstructions to compatibility.

where a superposed dot denotes time differentiation. The linear space of instantaneous *test velocities* $\mathfrak{T}$, comprising all smooth fields $\boldsymbol{x} \mapsto (\boldsymbol{v}, \mathbb{V})$, with $\boldsymbol{v}$ vector-valued and $\mathbb{V}$ tensor-valued, will play a central role in Sec. 2.2.

2.2. *Dynamics: brute and accretive forces; balance principle*

The basic balance structure of a mechanical theory is encoded in the way in which forces expend *working* on a general test velocity. Because of the compound structure of test velocities, force splits here additively into a *brute force*, dual to $\boldsymbol{v}$, and an *accretive force*, dual to $\mathbb{V}$. To be specific, we postulate that the working expended on $(\boldsymbol{v}, \mathbb{V})$ is given by

$$\int_{\mathscr{D}} \left(\mathbb{A}^{\mathsf{i}} \cdot \mathbb{V} - \mathbf{S} \cdot \nabla \boldsymbol{v} \right) + \int_{\mathscr{D}} \mathbb{A}^{\mathsf{o}} \cdot \mathbb{V} + \int_{\partial \mathscr{D}} \boldsymbol{t}_{\partial \mathscr{D}} \cdot \boldsymbol{v}\,, \tag{11}$$

where the integrals are taken with respect to the bulk volume and surface area of body elements in their paragon configuration—to be called *paragon volume* and *paragon area*, for short. The distinction between the *inner* working, given by the first bulk integral in Eq. (11), and the *outer* working, given by the remaining sum, is *not* germane to balance and was brought forward to this section just to save space. It will be discussed in Sec. 2.4. The *inner* and *outer accretive couples* per unit paragon volume $\mathbb{A}^{\mathsf{i}}$, $\mathbb{A}^{\mathsf{o}}$ and the (*brute*) *Piola stress* $\mathbf{S}$—also a specific couple—take values in $\mathsf{V}\mathscr{E} \otimes \mathsf{V}\mathscr{E}$; the (*brute*) *boundary-force* per unit paragon area $\boldsymbol{t}_{\partial \mathscr{D}}$ take values in $\mathsf{V}\mathscr{E}$. Because of spherical symmetry, Eq. (11) boils down to the one-dimensional representation:

$$\int_{\xi_-}^{\xi_+} \left(\mathrm{A_r V_r} + 2\,\mathrm{A_h V_h} - \mathrm{S_r}\, v' - 2\,\mathrm{S_h}\, v/\xi \right) 4\,\pi\,\xi^2 d\xi \; + \; \left(4\,\pi\,\xi^2\, t\, v \right)\Big|_{\xi_\mp}, \tag{12}$$

with the obvious meaning of the components $\mathrm{S_r}, \mathrm{S_h}$ of $\mathbf{S}$ and t of $\boldsymbol{t}_{\partial \mathscr{D}}$, and making use of the position:

$$\mathbb{A} := \mathbb{A}^{\mathsf{i}} + \mathbb{A}^{\mathsf{o}} = \mathrm{A_r}\,\mathbf{P_r} + \mathrm{A_h}\,\mathbf{P_h}\,. \tag{13}$$

Balance laws are provided by the *balance principle* stating that, at each time, the working expended on any test velocity should be zero. Via standard localization arguments, this yields the local statements of balance:

$$\left.\begin{aligned} 2\big(\mathrm{S_r}(\xi) - \mathrm{S_h}(\xi) \big) + \xi\, \mathrm{S_r'}(\xi) &= 0 \\ \mathrm{A_r}(\xi) = \mathrm{A_h}(\xi) &= 0 \end{aligned}\right\} \quad (\xi_- < \xi < \xi_+), \tag{14}$$
$$\mp \mathrm{S_r}(\xi_\mp) = t_\mp\,.$$

2.3. *Energetics*

To parametrize the state of the body, an additional energetic descriptor is needed. We postulate the existence of a real-valued *frcc cnergy* measure, such that the energy available to any part $\mathscr{P}$ of $\mathscr{D}$ is given by

$$\Psi(\mathscr{P}) = \int_{\mathscr{P}} J\psi\,, \tag{15}$$

where the density ψ is the free energy per unit *prototypal* volume and

$$J := \det(\mathbb{P}) = \alpha_{\mathsf{r}}\,\alpha_{\mathsf{h}}^2 > 0\,, \tag{16}$$

so that $J\psi$ is the free energy per unit *paragon* volume, the integral in Eq. (15) being taken with respect to the paragon volume. Within the present symmetry-restricted theory, only spherically symmetric subsets of $\mathscr{D}$ are to be considered as body-parts.

2.4. *Constitutive issues*

The *inner* force represents the interactions among the degrees of freedom resolved by the theory, *i.e.*, described by the refined motion $(\mathsf{p}, \mathbb{P})$; the *outer force*, on the contrary, represents the interactions between these d.o.f.'s and those whose evolution is *not* described by $(\mathsf{p}, \mathbb{P})$. In the present theory of the biomechanics of growth, the outer accretive couple $\mathbb{A}^o$ plays a primary role, representing the mechanical effects of the biochemical control system, finely distributed in the bulk of $\mathscr{B}$. Ignoring the chemical d.o.f.'s—as we do—does not allow us to neglect their feedback on mechanics.

The constitutive theory of inner forces rests on two main pillars, altogether independent of balance: the principle of material indifference to change in observer, and the dissipation principle. In the present context, the first of these principles is idle, since only the trivial action of the group of changes in observer is compatible with spherical symmetry.

2.4.1. *Dissipation principle*

We call *power expended* along a refined motion at any given time the opposite of the working expended by the inner force constitutively related to that motion on the velocity realized along the motion at the given time. Hence, the power expended measures the working done by a putative outer force balanced with the constitutively determined inner force. The *dissipation principle* we enforce requires that the *power dissipated*—defined as the difference between the power expended along a refined motion and the time derivative of the free energy along that motion—should be non-negative, for all body-parts, at all times. This localizes into:

$$\mathbf{S}\cdot(\nabla\dot{\mathsf{p}}) - \mathbb{A}^{\mathsf{i}}\cdot(\dot{\mathbb{P}}\,\mathbb{P}^{-1}) - (J\,\psi)^{\cdot} \geq 0\,. \tag{17}$$

2.4.2. *Free energy and inner force*

We posit that the value of the free energy $\psi(\boldsymbol{x},\tau)$ depends solely on the value of the warp $\mathbf{F}(\boldsymbol{x},\tau)$: there exists a map ϕ such that

$$\psi(\boldsymbol{x},\tau) = \phi\,(\,\lambda_{\mathsf{r}}(\xi,\tau), \lambda_{\mathsf{h}}(\xi,\tau); \xi\,)\;. \tag{18}$$

The requirement that inequality (17) be satisfied along all refined motions is fulfilled if and only if for each ξ (which will be dropped from now on) the constitutive mappings for the (brute) stress $\mathbf{S}$ and the inner accretive couple $\mathbb{A}^{\mathsf{i}}$ satisfy the following equalities:[b]

$$\begin{aligned} \mathrm{S}_{\mathsf{r}} &= J\phi_{,\mathsf{r}}/\alpha_{\mathsf{r}} + \overset{+}{\mathrm{S}}_{\mathsf{r}}\,, & \mathrm{S}_{\mathsf{h}} &= J\phi_{,\mathsf{h}}/\alpha_{\mathsf{h}} + \overset{+}{\mathrm{S}}_{\mathsf{h}}\,, \\ \mathrm{A}^{\mathsf{i}}_{\mathsf{r}} &= J\,[\,\mathrm{S}_{\mathsf{r}}\,\alpha_{\mathsf{r}}\,\lambda_{\mathsf{r}}/J - \phi\,] + \overset{+}{\mathrm{A}}_{\mathsf{r}}\,, & \mathrm{A}^{\mathsf{i}}_{\mathsf{h}} &= J\,[\,\mathrm{S}_{\mathsf{h}}\,\alpha_{\mathsf{h}}\,\lambda_{\mathsf{h}}/J - \phi\,] + \overset{+}{\mathrm{A}}_{\mathsf{h}}\,, \end{aligned} \tag{19}$$

where the *extra-energetic* components $(\overset{+}{\mathrm{S}}_{\mathsf{r}}, \overset{+}{\mathrm{S}}_{\mathsf{h}})$ and $(\overset{+}{\mathrm{A}}_{\mathsf{r}}, \overset{+}{\mathrm{A}}_{\mathsf{h}})$ make the *reduced dissipation inequality* identically satisfied:

$$\overset{+}{\mathrm{S}}_{\mathsf{r}}\,\alpha_{\mathsf{r}}\,\dot{\lambda}_{\mathsf{r}} + 2\overset{+}{\mathrm{S}}_{\mathsf{h}}\,\alpha_{\mathsf{h}}\,\dot{\lambda}_{\mathsf{h}} - \overset{+}{\mathrm{A}}_{\mathsf{r}}\,\dot{\alpha}_{\mathsf{r}}/\alpha_{\mathsf{r}} - 2\overset{+}{\mathrm{A}}_{\mathsf{h}}\,\dot{\alpha}_{\mathsf{h}}/\alpha_{\mathsf{h}} \geq 0\,. \tag{20}$$

In Eqs. (19) $\phi_{,\mathsf{r}}$ and $\phi_{,\mathsf{h}}$ are shorthands for the derivatives of ϕ with respect to the radial and hoop stretches, λ_{r} and λ_{h}, respectively.

We regard all dissipative mechanisms extraneous to growth to be negligible, assuming the extra-energetic brute stress to be null: $\overset{+}{\mathrm{S}}_{\mathsf{r}} = \overset{+}{\mathrm{S}}_{\mathsf{h}} = 0$. Then, we make inequality (20) satisfied in the most facile—though scarcely warranted—way, letting each component of the extra-energetic accretive couple be simply proportional to the homonymous component of the growth velocity through a prescribed *negative* scalar factor:

$$\overset{+}{\mathrm{A}}_{\mathsf{r}} = -JD_{\mathsf{r}}\,\dot{\alpha}_{\mathsf{r}}/\alpha_{\mathsf{r}}\,, \qquad \overset{+}{\mathrm{A}}_{\mathsf{h}} = -JD_{\mathsf{h}}\,\dot{\alpha}_{\mathsf{h}}/\alpha_{\mathsf{h}}\,, \tag{21}$$

the radial and the hoop *reluctance to growth* (per unit *prototypal* volume) being positive: $D_{\mathsf{r}} > 0\,,\; D_{\mathsf{h}} > 0\,.$

---

[b]Notice that the two bracketed quantities in Eqs. (19) are just the radial and hoop components of the *Eshelby tensor* $\mathbb{E} := (J^{-1}\mathbf{S}\,\mathbb{P}^{\top})\mathbf{F}^{\top} - \phi\mathbf{I}$ in disguise.

2.4.3. *Characterizing the passive mechanical response of soft tissue: incompressible elasticity*

Soft tissue—as all of soft matter—may be considered *elastically* incompressible (beware: growth may well change volume!):

$$\det \mathbf{F} = \lambda_{\mathsf{r}}\, \lambda_{\mathsf{h}}^2 = 1 \quad \Longleftrightarrow \quad \lambda_{\mathsf{r}} = 1/\lambda_{\mathsf{h}}^2 \,. \tag{22}$$

The incompressibility constraint (22) is maintained by a *reactive* inner force, which is requested to expend null working on all *divergence-free* test velocity. The ensuing set of reactions is parameterized by a scalar field $\overset{\bowtie}{\pi}$:[c]

$$\overset{\bowtie}{\mathbf{S}} = J\overset{\bowtie}{\pi}\left(\frac{1}{\alpha_{\mathsf{r}}\,\lambda_{\mathsf{r}}}\,\mathbf{P}_{\mathsf{r}} + \frac{1}{\alpha_{\mathsf{h}}\,\lambda_{\mathsf{h}}}\,\mathbf{P}_{\mathsf{h}}\right), \qquad \overset{\bowtie}{\mathbb{A}} = J\overset{\bowtie}{\pi}\,\mathbf{I}\,. \tag{23}$$

The active component of the inner force stems from the free-energy density (18) restricted to the constraint manifold:

$$\widetilde{\phi} : \lambda \mapsto \phi\,(\,1/\lambda^2,\, \lambda\,)\,. \tag{24}$$

Finally, collecting the active and reactive components, we get:

$$\begin{aligned} \mathrm{S}_{\mathsf{r}} &= \frac{J}{\alpha_{\mathsf{r}}\,\lambda_{\mathsf{r}}}\left(\overset{\bowtie}{\pi} - (\lambda_{\mathsf{h}}/3)\,\widetilde{\phi}'\right), & \mathrm{S}_{\mathsf{h}} &= \frac{J}{\alpha_{\mathsf{h}}\,\lambda_{\mathsf{h}}}\left(\overset{\bowtie}{\pi} + (\lambda_{\mathsf{h}}/6)\,\widetilde{\phi}'\right), \\ \mathrm{A}_{\mathsf{r}}^{\mathsf{i}} &= J\left(\mathrm{T}_{\mathsf{r}} - \widetilde{\phi} - D_{\mathsf{r}}\,\dot{\alpha}_{\mathsf{r}}/\alpha_{\mathsf{r}}\right), & \mathrm{A}_{\mathsf{h}}^{\mathsf{i}} &= J\left(\mathrm{T}_{\mathsf{h}} - \widetilde{\phi} - D_{\mathsf{h}}\,\dot{\alpha}_{\mathsf{h}}/\alpha_{\mathsf{h}}\right), \end{aligned} \tag{25}$$

where $\mathrm{T}_{\mathsf{r}} = J^{-1}\mathrm{S}_{\mathsf{r}}\,\alpha_{\mathsf{r}}\,\lambda_{\mathsf{r}}$ and $\mathrm{T}_{\mathsf{h}} = J^{-1}\mathrm{S}_{\mathsf{h}}\,\alpha_{\mathsf{h}}\,\lambda_{\mathsf{h}}$ are the radial and hoop components of the *Cauchy stress* $\mathbf{T} = (J\,\det(\mathbf{F}))^{-1}\,\mathbf{S}\,\mathbb{P}^{\top}\mathbf{F}^{\top}$.

The constitutive function $\widetilde{\phi}$ may be reasonably specified as follows:[7]

$$\widetilde{\phi}(\lambda) = (c/\delta)\, \exp\!\big((\Gamma/2)\,(\lambda^2 - 1)^2\big)\,, \tag{26}$$

where the moduli c and Γ may be identified—at least in principle—by performing biaxial traction tests on membrane samples, whose *relaxed* thickness is δ. According to Kyriacou and Humphrey[14] and Haslach and Humphrey,[15] the best fit to the experimental findings of Scott *et al.*[16] on aneurysmal tissue is given by $c = 0.88\,\mathrm{N/m}$ and $\Gamma = 12.99$.

---

[c]The parameter $\overset{\bowtie}{\pi}$ is to be interpreted as (the opposite of) a *pressure*, since the reactive Cauchy stress $\overset{\bowtie}{\mathbf{T}} = (J\,\det(\mathbf{F}))^{-1}\,\overset{\bowtie}{\mathbf{S}}\,\mathbb{P}^{\top}\mathbf{F}^{\top}$ equals $\overset{\bowtie}{\pi}\,\mathbf{I}$.

2.4.4. *Characterizing the active mechanical response of living tissue: constitutive recipes for the outer accretive couple*

In the intended application, the brute boundary-force $\boldsymbol{t}_{\partial\mathscr{D}}$ represents essentially the intramural blood pressure. To a first approximation, it may be assigned a constant value (10 KPa).[d] The key assumption is the one concerning the outer accretive couple $\mathbb{A}^o$, whose constitutive prescription should hopefully short-circuit the complex—and ill-understood—sensing/actuating mechanobiological functions that control vascular wall remodelling.

We put forward a preliminary, crude proposal, along lines akin to those of Ref. 10. We posit a homeostatic *target value* $\mathrm{T}_{\mathsf{h}}^{\odot}$ of the hoop component of the Cauchy stress and prescribe the outer accretive couple $\mathbb{A}^o$ as follows:

$$\begin{aligned} \mathrm{A}_{\mathsf{r}}^{o} &= J\left(G_{\mathsf{r}}\,(\mathrm{T}_{\mathsf{h}} - \mathrm{T}_{\mathsf{h}}^{\odot}) - \mathrm{T}_{\mathsf{r}} + \widetilde{\phi}\right), \\ \mathrm{A}_{\mathsf{h}}^{o} &= J\left(G_{\mathsf{h}}\,(\mathrm{T}_{\mathsf{h}}^{\odot} - \mathrm{T}_{\mathsf{h}}) - \mathrm{T}_{\mathsf{h}} + \widetilde{\phi}\right), \end{aligned} \tag{27}$$

where $G_{\mathsf{r}}, G_{\mathsf{h}}$ are positive *control gains.* Under this hypothesis, the evolution law for the prototypal transplant $\mathbb{P}$ takes the form:

$$\begin{aligned} \dot{\alpha}_{\mathsf{r}}/\alpha_{\mathsf{r}} &= (G_{\mathsf{r}}/D_{\mathsf{r}})\,(\mathrm{T}_{\mathsf{h}} - \mathrm{T}_{\mathsf{h}}^{\odot}), \\ \dot{\alpha}_{\mathsf{h}}/\alpha_{\mathsf{h}} &= (G_{\mathsf{h}}/D_{\mathsf{h}})\,(\mathrm{T}_{\mathsf{h}}^{\odot} - \mathrm{T}_{\mathsf{h}}). \end{aligned} \tag{28}$$

Notice that $\dot{\alpha}_{\mathsf{r}} \gtrless 0$ while $\dot{\alpha}_{\mathsf{h}} \lesseqgtr 0$ when $\mathrm{T}_{\mathsf{h}} \gtrless \mathrm{T}_{\mathsf{h}}^{\odot}$.

3. Concluding Remarks

We are presently attempting to fathom the computational depths of this model, numerically elusive despite its modest complexity. We defer therefore the presentation of numerical results to a later moment. In the meantime, a modicum of self-criticism is in order. Of course, the extreme geometrical and kinematical limitations of the present model need to be removed, if we want to develop a versatile mechanical theory of growing shells. However, its weakest point is elsewhere. In our opinion, a major conceptual improvement would be in distinguishing between different remodelling mechanisms. In the case at hand, at least three such mechanisms come to mind: passive viscous slipping between cells and various components of the extracellular matrix; active recovery due to cell adhesion and contractility; cell proliferation and collagen production. We are presently striving to formalize them separately, in order to include them individually into our model.

[d]The brute bulk-force, playing a negligible role, has been neglected altogether.

References

1. A. DiCarlo and S. Quiligotti, *Mech. Res. Comm.* **29**, 449 (2002).
2. A. Di Carlo, Surface and bulk growth unified, in *Mechanics of Material Forces*, eds. P. Steinmann and G. A. Maugin (Springer, New York, NY, 2005), pp. 53–64.
3. P. Nardinocchi and L. Teresi, Stress driven remodeling of living tissues, in *Proceedings of the FEMLAB Conference 2005*, (Stockholm, Sweden, 2005).
4. A. DiCarlo, S. Naili, S. Quiligotti and L. Teresi, Modeling bone remodeling, in *Proceedings of the COMSOL Multiphysics Conference 2005 Paris*, eds. J.-M. Petit and J. Daluz), (Paris, France, 2005).
5. A. DiCarlo, S. Naili and S. Quiligotti, *C. R. Mécanique*, **334**, 651 (2006).
6. M. Tringelová, P. Nardinocchi, L. Teresi and A. DiCarlo, The cardiovascular system as a smart system, in *Topics on Mathematics for Smart Systems*, eds. V. Valente and B. Miara (World Scientific, Singapore, 2007).
7. J. D. Humphrey, *Cardiovascular Solid Mechanics: Cells, Tissues, and Organs* (Springer, New York, NY, 2001).
8. J. Frösen, A. Piippo, A. Paetau, M. Kangasniemi, M. Niemeä, H. Juha and J. Jääskeläinen, *Stroke* **35**, 2287 (2004), `http://stroke.ahajournals.org/cgi/content/full/35/10/2287`.
9. M. Yonekura, *Neurologia medico-chirurgica* **44**, 213 (2004), `http://www.jstage.jst.go.jp/article/nmc/44/4/44_213/_article`.
10. S. Baek, K. R. Rajagopal and J. D. Humphrey, *Journal of Elasticity* **80**, 13 (2005).
11. E. Kröner, *Arch. Rational Mech. Anal.* **4**, 273 (1960).
12. E. H. Lee, *J. Appl. Mechanics* **36**, 1 (1969).
13. E. K. Rodriguez, A. Hoger and A. D. McCulloch, *J. Biomechanics* **27**, 455 (1994).
14. S. K. Kyriacou and J. D. Humphrey, *J. Biomechanics* **29**, 1015 (1996).
15. H. W. Haslach and J. D. Humphrey, *Int. J. Non-Linear Mech.* **39**, 399 (2004).
16. S. Scott, G. G. Ferguson and M. R. Roach, *Can. J. Physiol. Pharmacol.* **50**, 328 (1972).

Acknowledgements

The work of one or more of the four authors was supported by different funding agencies: GNFM-INdAM (the Italian Group for Mathematical Physics), MIUR (the Italian Ministry of University and Research) through the Project "Mathematical Models for Materials Science" and others, the Fifth European Community Framework Programme through the Project HPRN-CT-2002-00284 ("Smart Systems"), IMA (Institute for Mathematics and its Applications, Minneapolis, MN), Université Paris 12 Val de Marne.

DYNAMICS OF MATERIALS WITH A DEFORMABILITY THRESHOLD

A. FARINA, A. FASANO, L. FUSI*

Università Degli Studi di Firenze,
Dipartimento di Matematica "U. Dini",
Viale Morgagni 67/a, 50134, Firenze.
**E-mail: fusi@math.unifi.it*

K. R. RAJAGOPAL

Texas A&M University
Department of Mechanical Engineering
College Station, Texas 77843, USA
E-mail: krajagopal@tamu.edu

Some biological tissues exhibit a sharp reduction of deformability beyond some stress threshold, below which they may be considered elastic. This behaviour can be seen as a limit situation in which the material becomes undeformable beyond some deformation threshold. As a model problem we have considered the motion of a layer of such a material in which one boundary is kept fixed while a tangential stress beyond threshold is applied to the other. The corresponding mathematical model is formulated as a hyperbolic free boundary problem in which at each time the interface is made of points that has reached the stretching threshold.

Keywords: Implicit constitutive theories; Free boundary problems; Wave equation.

1. Introduction

Most models describing biological tissues like tendons or ligaments are based on a constitutive relation formulated as a one-to-one mapping between stress and strain. In such a way the stress is derived differentiating a scalar potential function which is commonly referred to as elastic energy. The knowledge of such a mapping allows to predict the deformation of the material when it undergoes specific loads.

Considering the 1D case, the simplest constitutive equation used to describe the mechanical behavior of tendons and ligaments is a stress-strain relation in which the stress is an increasing function of the strain (for instance exponential, see [2]). The system can be thus described by classical hyperelasticity.

Such a behavior can be idealized considering a material that beyond a certain stress is no longer capable of deformation, for instance a material that up to a certain stress behaves elastically and then like a rigid body (see Fig. 1). In a sense the latter property can be interpreted as the "dual"

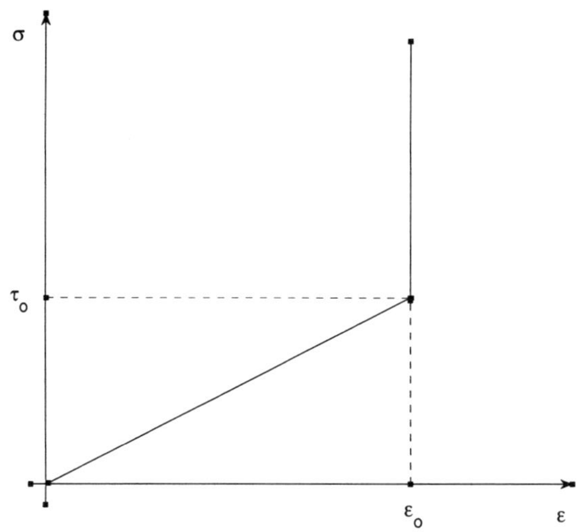

Fig. 1. Stress σ vs deformation gradient ϵ.

to the behavior of Bingham fluids [1] where shear is produced only beyond yield stress. Here any deformation is prevented beyond some threshold.

In such a case the system becomes dissipative and the stress exihibits a jump across the surface separating the deformable and fully stretched region. The power dissipated is proportional to the stress jump and the constitutive relation cannot be retrieved as the limit of non-dissipative models.

The problem we have studied is the one-dimensional case of a layer of material with one face kept fixed while a known shear is applied to the other. The layer is divided into two subdomains by an interface evolving with time. In one part the material is deformable (neo-Hookean elastic solid) and in the other is fully stretched. The constitutive equation is given in an implicit form [5] and the plot describing the stress-strain relation is a graph.

Although conceptually simple the mathematical model reveals quite complex. Because of this we have considered only some particular sets of initial and boundary data and we have not studied the model in its general form. The velocity of the interface separating the deformable and the fully stretched parts is greater than the speed of sound (supersonic free boundary) except for a very particular set of initial data (negative initial velocity of the elastic part with continuity of the stress across the interface).

Balance of energy has been carried out obtaining an explicit formula for the power dissipated that shows that dissipation can occur only if the jump of the stress is positive and if the deformable region is reducing. The details of this work are reported in [3].

2. The mathematical model

We consider a homogeneous slab of thickness h loaded on the top surface with a known shear stress $\widehat{\sigma}(t) > \tau_o$, where τ_o is the stress threshold. Let $x = X + f\,(y,t)$ be a pure shear motion, $f\,(y,t)$ being the unknown displacement. The elastic and fully stretched regions are divided by a sharp interface $\mathcal{S}$ denoted by $y = s(t)$, with $0 \leq s(t) \leq h$. In the deformable region $0 < y < s(t)$ the motion equation is

$$f_{tt} - c^2 f_{yy} = 0, \qquad (1)$$

where $c^2 = \mu/\rho$ is the speed of sound (μ being the elastic modulus and ρ being density). In the fully stretched part[a] the displacement is given by

$$f\,(y,t) = f\left(s^-,t\right) + \frac{\tau_o}{\mu}\,(y-s)\,, \qquad y \in [s,h]\,,$$

implying

$$v^{(s)}\,(t) = f_y\left(s^-,t\right)\dot{s} + f_t\left(s^-,t\right) - \frac{\tau_o}{\mu}\dot{s}, \qquad (2)$$

$$a^{(s)}\,(t) = f_{yy}\left(s^-,t\right)\dot{s}^2 + 2f_{yt}(s^-,t)\dot{s} + f_y(s^-,t)\ddot{s} + f_{tt}(s^-,t) - \frac{\tau_o}{\mu}\ddot{s}, \quad (3)$$

$v^s(t)$ and $a^s(t)$ being velocity and acceleration in the fully stretched region. In the deformable part velocity and acceleration are $f_t(y,t)$, $f_{tt}(y,t)$

[a]With $^-$, $^+$ we denote the limits from the deformable and fully stretched part respectively.

respectively. Equations (2), (3) yield[b]

$$[\![v]\!]_{\mathcal{S}} = -\frac{\tau_o - \sigma\left(s^-,t\right)}{\mu}\,\dot{s}\,, \tag{4}$$

$$[\![a]\!]_{\mathcal{S}} = -\frac{\tau_o - \sigma\left(s^-,t\right)}{\mu}\,\ddot{s} + \frac{\dot{s}}{\mu}\frac{d\sigma\left(s^-,t\right)}{dt} + \frac{\dot{s}}{\mu}\sigma_t\left(s^-,t\right), \tag{5}$$

where $\sigma(s^-,t) = \mu f_y(s^-,t)$ denotes the Cauchy stress at the interface. From conservation of linear momentum $\rho[\![v]\!]_{\mathcal{S}}\,\dot{s} = -[\![\sigma]\!]_{\mathcal{S}}$. Recalling (4) we get

$$\sigma\left(s^+,t\right) = \sigma\left(s^-,t\right) + \frac{\dot{s}^2}{c^2}\left(\tau_o - \sigma\left(s^-,t\right)\right). \tag{6}$$

The conservation of linear momentum in the fully stretched part is expressed by

$$\rho\, a^{(s)}(t)(h-s) = \widehat{\sigma} - \sigma\left(s^+,t\right), \tag{7}$$

Substitution of (3), (6) into (7) gives

$$\left[(h-s)\left(f_y - \frac{\tau_o}{\mu}\right)\ddot{s} + \left((h-s)\,f_{yy} + \frac{\tau_o}{\mu} - f_y\right)\dot{s}^2\right.$$

$$\left.+2\left(h-s\right)f_{yt}\dot{s} + \left(h-s\right)f_{tt} + c^2 f_y\right]_{y=s^-} = \frac{\widehat{\sigma}}{\rho}, \tag{8}$$

that is the evolution equation for the interface (free boundary) separating the fully stretched and elastic regions. Equation (8) must be coupled with the initial data

$$s\left(0\right) = s_o,\;\; 0 \le s_o \le h, \qquad \frac{\dot{s}^2\left(0\right)}{c^2}\left(\tau_o - \sigma\left(s_o^-,0\right)\right) = [\![\sigma]\!]_{\mathcal{S}}\big|_{t=0}\,. \tag{9}$$

[b] $[\![..]\!]_{\mathcal{S}}$ expresses the jump across the interface $\mathcal{S}$.

The free boundary problem in the elastic region is given by

$$\begin{cases} f_{tt} - c^2 f_{yy} = 0, \\ f(0,t) = 0, \\ f_y(s(t),t) = \dfrac{\sigma(s^-,t)}{\mu}, \quad \text{with } \sigma(s^-,t) \le \tau_o, \\ \left[(h-s)\left(f_y - \dfrac{\tau_o}{\mu}\right)\ddot{s} + \left((h-s) f_{yy} + \dfrac{\tau_o}{\mu} - f_y\right)\dot{s}^2 + \right. \\ \left. + 2(h-s) f_{yt}\dot{s} + (h-s) f_{tt} + c^2 f_y\right]_{y=s^-} = \dfrac{\widehat{\sigma}(t)}{\rho}, \\ f(y,0) = f_o(y), \\ f_t(y,0) = f_1(y), \\ s(0) = s_o, \\ \dfrac{\dot{s}^2(0)}{c^2}\left(\tau_o - \sigma(s_o^-,0)\right) = [\![\sigma]\!]_{\mathcal{S}}\big|_{t=0}, \end{cases} \tag{10}$$

where f_o, f_1 are the initial data.

In problem (10) one condition is missing since the system contains the extra unknown $\sigma(s^-,t)$. Depending on the initial and boundary data, the problem itself selects the additional information which is required to close the system. This depends on whether $[\![\sigma]\!]_{\mathcal{S}} = 0$, that is $\sigma(s^-,t) = \tau_o$, or $[\![\sigma]\!]_{\mathcal{S}} > 0$. When the stress is continuous across $\mathcal{S}$, namely $\sigma(s^-,t) \equiv \tau_o \equiv \sigma(s^+,t)$, $(10)_3$ and $(10)_4$ reduce to

$$f_y(s(t),t) = \frac{\tau_o}{\mu}, \tag{11}$$

$$\left[f_{ty}(s^-,t)\,\dot{s} + c^2 f_{yy}(s^-,t)\right](h-s) = \frac{\widehat{\sigma} - \tau_o}{\rho}, \tag{12}$$

respectively. On the other hand, if $[\![\sigma]\!]_{\mathcal{S}} > 0$ (which entails $|\dot{s}| \geqslant c$, see proposition below) $f(y,t)$ is essentially determined by the initial data via d'Alembert formula and equation $(10)_3$ can be used to evaluate $\sigma(s^-,t)$. We have the following

Proposition 2.1. *Let (f,s,σ) be a solution of problem (10) and $\widehat{\sigma}(t) > \tau_o$, for some $t \ge 0$, then:*

1. *If $|\dot{s}| < c$, then $[\![\sigma]\!]_{\mathcal{S}} = 0$ (there is no jump of the stress across the interface).*
2. *If $|\dot{s}| = c$, then $\sigma(s^+,t) = \tau_o$ (and thus $[\![\sigma]\!]_{\mathcal{S}} \ge 0$).*
3. *If $|\dot{s}| > c$, then either $\sigma(s^-,t) = \tau_o = \sigma(s^+,t)$ (i.e. $[\![\sigma]\!]_{\mathcal{S}} = 0$) or $\sigma(s^-,t) < \tau_o < \sigma(s^+,t)$.*
4. *If $[\![\sigma]\!]_{\mathcal{S}} = 0$, i.e. $\sigma(s^-,t) = \sigma(s^+,t) = \tau_o$, then relation (6) is identically satisfied irrespectively of $\dot{s}$.*

5. *If* $[\![\sigma]\!]_S > 0$*, then* $[\![v]\!]_S \neq 0$ *(if* $\dot{s} \neq 0$*),* $\tau_o > \sigma(s^-, t)$ *and* $|\dot{s}| \geqslant c$*.*

Proof. Equation (6) can be rewritten as

$$\sigma\left(s^+, t\right) - \tau_o = \left(\tau_o - \sigma\left(s^-, t\right)\right)\left(\frac{\dot{s}^2}{c^2} - 1\right), \tag{13}$$

from which 1.-5. follow. □

3. The case $[\![\sigma]\!]_S = 0$

Existence and uniqueness of a solution for this case can be proved exploiting the D'Alembert representation formula [4]. The corresponding situation is rather artificial from a physical point of view, since very peculiar conditions must be imposed on the data. Indeed, if the initial data satisfies some compatibility conditions and

$$W_1 \leq h f_o''(y) - \frac{h}{c} f_1'(y) \leq W_2, \qquad \forall\ y \in\ [-s_o, s_o], \tag{14}$$

with W_1, W_2 positive constants, we have $[\![\sigma]\!]_s = 0$. Condition (14) implies that the initial velocity of the elastic region is negative.

Proposition 3.1. *A time* θ *can be computed such that a unique solution* (f, s) *to problem (10) (with* $(10)_3$*,* $(10)_4$ *replaced by (11) and (12)), exists for* $t \in [0, \theta)$*, with the property* $-c < \dot{s} < 0$*.*

It can also be proved that, when $\widehat{\sigma} > \tau_o$ is constant and

$$h f_o''(y) - \frac{h}{c} f_1'(y) = const$$

there exists a solution with a stationary interface and the system comes to a stop (it becomes fully stretched everywhere) at time $t = 2s_o/c$.

4. The case $[\![\sigma]\!]_S \neq 0$

Let us now consider the case when $f_o(y) \equiv f_1(y) \equiv 0$ (i.e. the system is initially at rest) and the applied stress $\widehat{\sigma}$ increases in time, i.e. $\widehat{\sigma}'(t) > 0$ from $\widehat{\sigma}(0) = 0$, and, at some time $t_o < s_o/c$, $\widehat{\sigma}(t_o) = \tau_o$. The dynamics is now characterized by a jump of σ across $\mathcal{S}$, with the interface travelling faster than the speed of sound of the deformable medium. Of course now (14) is not fulfilled. We can prove the following

Proposition 4.1. *Let* (f, σ, s) *be a solution of problem (10) for* $t > t_o$*. Then* $s(t) < h - c(t - t_o)$*, i.e. we have a supersonic interface.*

Proof. We initially suppose that the free boundary is given by

$$\Sigma: \ y = h - c(t - t_o).$$

Exploiting d'Alembert representation formula formula we can prove that in the vicinity of Σ

$$f(y,t) = \frac{1}{\mu}\int_{h-ct}^{y} \widetilde{\sigma}(\xi + ct - h))d\xi, \tag{15}$$

where $\widetilde{\sigma}(y) = \widehat{\sigma}(c^{-1}y)$. Representation formula (15) yields $\mu f_{y_{|\Sigma}} = \tau_o$. This implies that $\sigma(s^-,t) = \tau_o$ and, from (4), (13), $[\![\sigma]\!]_{\mathcal{S}} = [\![v]\!]_{\mathcal{S}} = 0$. Therefore

$$\left.\frac{\partial f}{\partial t}\right|_{\Sigma} = v^{(s)}(t) = const, \quad \forall t > t_o \quad \Rightarrow \quad a^{(s)}(t) = \frac{dv^{(s)}(t)}{dt} = 0, \tag{16}$$

which is a contraddiction unless $\widehat{\sigma}(t) = \tau_o$ (see (7)). We then assume that $\mathcal{S}$ is located on the right of Σ, namely

$$s(t) > h - c(t - t_o), \ \ t \geq t_o.$$

and consider the domain $D_\delta = \{h - c(t - t_o) < y < s(t), \ t_o < t < t_o + \delta\}$, for some positive (and "small") δ. Next we consider the following Goursat problem for the unknown $w = f_y$

$$\begin{cases} w_{tt} - c^2 w_{yy} = 0, \ \ (y,t) \in D_\delta, \\ \\ w|_\Sigma = \dfrac{\tau_o}{\mu}, \qquad\qquad t_o \leq t < t_o + \delta, \\ \\ w|_{\mathcal{S}} = \dfrac{\tau_o}{\mu}, \qquad\qquad t_o \leq t < t_o + \delta, \end{cases}$$

whose unique solution is $w = f_y = \tau_o/\mu$, $\forall\ (y,t) \in D_\delta$. Recalling (3), we get

$$a^{(s)}(t) = 0, \quad \forall\, t \geq t_o.$$

Moreover $\sigma(s^-,t) = \tau_o$, and from (6), $\sigma(s^+,t) = \tau_o$ for those t such that $\dot{s}(t) > -c$ (which must exist according to our assumption). Thus

$$\underbrace{\rho a^{(s)}(t)(h - s)}_{=0} = \widehat{\sigma}(t) - \tau_o,$$

which is an evident contradiction. This same argument shows that in a neighborhood of $t = t_o$ there cannot be any point on Σ in which the free boundary has points in common with the characteristic and proceeds into the region lying above the characteristic itself. We conclude that $s(t) < h - c(t - t_o)$, $\forall\, t > t_o$, as stated in the proposition. $\square$

5. Some peculiar situations for the case $[\![\sigma]\!]_{\mathcal{S}} \neq 0$

The problem for a general applied stress $\widehat{\sigma}(t)$ is exceedingly complex so we have focused on some particular cases. When the applied load is $\widehat{\sigma}(t) = (\tau_o t)/t_o$ then

$$s(t) = h - \frac{3c}{2}(t - t_o).$$

When $\widehat{\sigma}(t) = (\tau_o t)/t_o$ in $[0, t_o]$ and $\widehat{\sigma}(t) = \tau_o$ for $t > t_o$ then $s(t)$ concides with the characteristic

$$s(t) = h - c(t - t_o).$$

When $\widehat{\sigma}$ is constant and greater than τ_o for all times $t \geq 0$ then

$$s(t) = h - \gamma c t,$$

where $\gamma = \sqrt{\widehat{\sigma}/\tau_o}$.

6. Energy considerations

As stated in the introduction, in general the system is dissipative. We show here how to calculate the power dissipated proving that it is proportional to the jump of the stress across $\mathcal{S}$ and that it can occur only if the fully stretched region is expanding (i.e. $\dot{s} < 0$). The total kinetic energy and the total elastic energy are given by

$$K = \int_0^h \frac{1}{2}\rho v^2 dy, \tag{17}$$

and

$$\Psi = \int_0^h \psi dy, \tag{18}$$

with ψ given by

$$\psi = \begin{cases} \dfrac{\mu}{2} f_y^2, & \text{if } 0 \leq |\sigma| < \tau_o, \quad \text{(deformable part)}, \\ \\ \psi_o = \dfrac{\tau_o^2}{2\mu}, & \text{if } |\sigma| \geq \tau_o, \quad \text{(fully stretched part)}. \end{cases} \tag{19}$$

Thus

$$\Psi = \int_0^s \frac{\mu}{2} f_y^2 dy + \int_s^h \frac{\tau_o^2}{2\mu} dy \quad \text{or} \quad \Psi = \int_0^s \frac{\sigma^2(y,t)}{2\mu} dy + \int_s^h \frac{\tau_o^2}{2\mu} dy. \tag{20}$$

The global energy balance for the system can be written as

$$\frac{d}{dt}(K+\Psi) = \mathcal{P}^{ext} - \mathcal{P}^{diss}, \tag{21}$$

where $\mathcal{P}^{ext}$ is the power exerted by external forces, namely

$$\mathcal{P}^{ext} = \widehat{\sigma}\, v\,(h,t) = \widehat{\sigma}\, v^{(s)}(t),$$

and $\mathcal{P}^{diss}$ denotes the global power dissipated. We have the following

Proposition 6.1. *Energy is dissipated if and only if $\dot{s} < 0$ and $\sigma\,(s^+,t) > \tau_o$. More precisely, the dissipated power is given by*

$$\mathcal{P}^{diss} = \frac{[\![v]\!]_{\mathcal{S}}}{2}\left(\sigma\left(s^+,t\right) - \tau_o\right). \tag{22}$$

Moreover $\mathcal{P}^{diss}$ is necessarily non-negative and $\mathcal{P}^{diss} > 0$ is compatible only with $\dot{s} < 0$ and $\sigma(s^+,t) > \tau_o$ (in other words a regression of the fully strained region cannot lead to dissipation and requires $\sigma(s^+,t) = \tau_o$).

Proof. We start considering the time derivative of K and Ψ. We have

$$\frac{dK}{dt} = \mathcal{P}^{ext} - \int_0^s \sigma v_y dy - \frac{\rho}{2}[\![v^2]\!]_{\mathcal{S}}\,\dot{s} - [\![\sigma\, v]\!]_{\mathcal{S}}\,. \tag{23}$$

$$\frac{d\Psi}{dt} = \int_0^s \sigma v_y dy - \frac{1}{2\mu}\left(\tau_o^2 - \sigma\left(s^-,t\right)^2\right)\dot{s}\,. \tag{24}$$

Substitution of (23), (24) into (21) yields

$$\mathcal{P}^{diss} = [\![\sigma\, v]\!]_{\mathcal{S}} + \dot{s}\left[\frac{1}{2\mu}\left(\tau_o^2 - \sigma\left(s^-,t\right)^2\right) + \frac{\rho}{2}[\![v^2]\!]_{\mathcal{S}}\right]. \tag{25}$$

The r.h.s of (25) can be manipulated to obtain

$$\mathcal{P}^{diss} = \frac{[\![v]\!]_{\mathcal{S}}}{2}\left(\sigma\left(s^+,t\right) - \tau_o\right), \tag{26}$$

which, recalling that $\rho[\![v]\!]_{\mathcal{S}}\,\dot{s} = -[\![\sigma]\!]_{\mathcal{S}}$, becomes

$$\mathcal{P}^{diss} = \frac{[\![\sigma]\!]_{\mathcal{S}}}{2\rho\,\dot{s}}\left(\tau_o - \sigma\left(s^+,t\right)\right). \tag{27}$$

Expression (27) proves that when there is no jump of the stress across $\mathcal{S}$ the system do not dissipate energy. The fact that $\mathcal{P}^{diss} > 0$ comes from the second law of thermodynamics. In particular, since $[\![\sigma]\!]_{\mathcal{S}}(\tau_o - \sigma(s^+,t)) \leq 0$, $\mathcal{P}^{diss} > 0$ entails $\dot{s} < 0$. □

7. Comparison with a piecewise linear elastic model

Here we want to prove that the model we have presented cannot be retrieved as the limit of non-dissipative models. Here we do not consider a general hyperelastic model but, for the sake of simplicity, we focus on the simple stress-strain relation depending on the parameter λ[c]

$$\sigma = \begin{cases} \mu\,\epsilon\,, & \text{if} \quad 0 \leq \epsilon \leq \dfrac{\tau_o}{\mu}, \\ \\ \mu\lambda^2\left(\epsilon - \dfrac{\tau_o}{\mu}\right) + \tau_o, & \text{if} \quad \epsilon > \dfrac{\tau_o}{\mu}, \end{cases} \qquad \text{with } \lambda^2 > 1, \tag{28}$$

where $\epsilon = f_y$. Model (28) is hyperelastic, since the stress can be derived by the following elastic energy (expressed in terms of σ)

$$\psi(\sigma) = \begin{cases} \dfrac{\sigma^2}{2\mu}, & \text{if} \quad 0 \leq \sigma \leq \tau_o, \\ \\ \dfrac{(\sigma - \tau_o)^2}{2\mu\lambda^2} + \dfrac{\tau_o}{\mu\lambda^2}(\sigma - \tau_o) + \dfrac{\tau_o^2}{2\mu}, & \text{if} \quad \sigma > \tau_o. \end{cases} \tag{29}$$

Differently from the fully stretched model we now have one more information for the presence of the elastic potential ψ, namely energy conservation. Indeed we have the following jump conditions

$$\text{kinematics}: \; f(s^-,t) = f(s^+,t)\,, \;\; \Rightarrow [\![v]\!]_S = -\dot{s}[\![f_y]\!]_S\,, \tag{30}$$

$$\text{dynamics}: \; \rho\dot{s}[\![v]\!]_S = -[\![\sigma]\!]_S\,, \tag{31}$$

$$\text{energy conservation}: \; \dot{s}\left(\frac{\rho}{2}[\![v^2]\!]_S + [\![\psi]\!]_S\right) = -[\![\sigma\, v]\!]_S\,. \tag{32}$$

Proposition 7.1. *Energy conservation implies that stress is continuous*

$$\sigma(s^+,t) = \sigma(s^-,t) = \tau_o, \quad \Leftrightarrow \quad [\![\sigma]\!]_S \equiv 0. \tag{33}$$

Proof. Using (30) and (31), we get from (32)

$$\frac{1}{2}[\![\sigma]\!]_S[\![f_y]\!]_S + [\![\psi]\!]_S = \sigma(s^+,t)[\![f_y]\!]_S,$$

or equivalently

$$[\![\psi]\!]_S = \left(\frac{\sigma(s^+,t) + \sigma(s^-,t)}{2}\right)[\![f_y]\!]_S. \tag{34}$$

---

[c] In practice we are considering an elastic material characterized by two elastic moduli: μ if $\sigma < \tau_o$ and $\mu\lambda^2$ if $\sigma > \tau_o$.

From (28) and (29)

$$[\![f_y]\!]_{\mathcal{S}} = \frac{\sigma(s^+,t)-\tau_o}{\mu\lambda^2} + \frac{\tau_o - \sigma(s^-,t)}{\mu}, \tag{35}$$

$$[\![\psi]\!]_{\mathcal{S}} = \frac{\sigma^2(s^+,t)-\tau_o^2}{2\mu\lambda^2} - \frac{\sigma^2(s^-,t)-\tau_o^2}{2\mu}. \tag{36}$$

Substituting (35) and (36) into (34) we get

$$(\tau_o - \sigma(s^-,t))(\sigma(s^+,t)-\tau_o)(1-\lambda^2) = 0. \tag{37}$$

Thus either $\sigma(s^-,t) = \tau_o$ or $\sigma(s^+,t) = \tau_o$. From (30) and (31)

$$\rho\dot{s}^2[\![f_y]\!]_{\mathcal{S}} = [\![\sigma]\!]_{\mathcal{S}},$$

that is

$$\dot{s}^2(\sigma(s^+,t)-\tau_o) + \lambda^2\dot{s}^2(\tau_o - \sigma(s^-,t)) = c^2\lambda^2(\sigma(s^+,t)-\sigma(s^-,t)). \tag{38}$$

If $\sigma(s^+,t) = \tau_o$, then, from (38),

$$\dot{s}^2(\tau_o - \sigma(s^-,t)) = c^2(\tau_o - \sigma(s^-,t)),$$

implying $\sigma(s^-,t) = \tau_o$ or $\dot{s}^2 = c^2$ (the latter meaning that $s(t) = h - c\,(t-t_o)$, i.e. the characteristic with slope $-c$). It is easy to check that $\dot{s}^2 = c^2 \iff \sigma(s^-,t) = \tau_o$. On the other hand, if $\sigma(s^-,t) = \tau_o$ (and consequently $\dot{s}^2 = c^2$), then, from (38)

$$(\sigma(s^+,t)-\tau_o) = \lambda^2(\sigma(s^+,t)-\tau_o)),$$

which yields $\sigma(s^+,t) = \tau_o$, since $\lambda^2 > 1$. Thus we conclude that in the "bi-elastic" model $s(t) = h - c\,(t-t_o)$ and (33) follows. □

For every $\lambda > 0$ model (28) is non-dissipative. One can show that the fully stretched model cannot be retrieved in the limit $\lambda \to \infty$, the reason lying essentially in the fact that the limit tends to preserve its hyperelastic character. Actually it is easy to check that equation (32) entails $\mathcal{P}^{diss} = 0$ for any λ (consistent with energy conservation).

Remark 7.1. The result just obtained can be extended to a more general non–linear stress–strain relationship $\sigma = \varphi_\lambda(\varepsilon) = \dfrac{d\psi_\lambda(\varepsilon)}{d\varepsilon}$.

References

1. R.B. Bird, W.E. Stewart, E.N. Lightfoot, *Transport phenomena*, Wiley, (1960).
2. Y.C. Fung, *Biomechanics: Mechanical properties of living tissue*, Springer-Verlag, (1993).
3. A. Farina, A. Fasano, L. Fusi, K.R. Rajagopal *Modelling of materials with a stretching threshold*, to be published on Mathematical Models and Methods in Applied Sciences.
4. L. Fusi, A. Farina, *An extension of the Bingham model to the case of an elastic core*, Adv. Math. Sci. Appl., **13**, 113-163, (2003).
5. K.R. Rajagopal, *On implicit constitutive theories*, Applications of Mathematics, **48**, 279-319, 2003.

ON THE STABILITY OF SEMI-LAGRANGIAN ADVECTION SCHEMES UNDER FINITE ELEMENT INTERPOLATIONS

R. FERRETTI* and G. PERRONE

Dipartimento di Matematica, Università di Roma Tre,
Roma, Italy
**E-mail: ferretti@mat.uniroma3.it*
www.mat.uniroma3.it

We investigate about the stability of Semi-Lagrangian schemes when finite-element type reconstructions are used. This choice leads to a scheme whose stability cannot be characterized by means of the classical Fourier analysis. In the paper, we propose a technique to estimate eigenvalues of the scheme in the case of a uniform mesh, and extend it to some situation of non-uniform spacing. The explicit computation is carried out for the $\mathbb{P}_2$ case.

Keywords: Semi-Lagrangian schemes; Finite-element interpolation; Stability.

1. Introduction

Large time-step schemes of Semi-Lagrangian type have been introduced in the present form by Wiin-Nielsen[6] for Numerical Weather Prediction problems, and in the the last two decades have been extended to high-order versions and to a number of different problems, including Computational Fluid Dynamics, conservation laws, first- and second-order Hamilton–Jacobi and Dynamic Programming equations (see,[5,3]).

The basic idea of Semi-Lagrangian (SL) schemes can be shown on the simple model problem of a one-dimensional, constant coefficient advection equation:

$$\begin{cases} u_t + au_x = 0, & \text{in } [0,1] \times [0,+\infty] \\ u(x,0) = u_0(x) & \text{in } [0,1] \\ u(0,t) = u(1,t) & \text{for } t \geq 0 \end{cases} \tag{1}$$

with periodic boundary conditions in $x = 0$ and $x = 1$. A semi–Lagrangian approximation for (1) reads

$$\begin{cases} u_j^{n+1} = I[U^n](x_j - a\Delta t) \\ u_j^0 = u_0(x_j) \end{cases} \tag{2}$$

where $U^n = (u_1^n, \ldots, u_q^n)^T$, $q = 1/h$ being the number of space nodes (with space step h), $I[U](x)$ is a numerical interpolation of the values u_j computed at the point x, and we assume that if $x_j - a\Delta t \notin [0,1]$, then it should be brought within the interval $[0,1]$ by 1–periodicity. Note that (2) mimics the construction of the exact solution by means of characteristics.

In principle, the reconstruction $I[U](x)$ needs not to be linear with respect to the values u_j, and in fact several implementations of SL schemes perform this step with nonlinear techniques, like monotone, Essentially Non–Oscillatory (ENO) or Weighted Essentially Non–Oscillatory (WENO) interpolations. In the sequel, however, we will assume that $I[U](x)$ is a Lagrange interpolation (hence, linear w. r. to U), which gives the form

$$\begin{cases} U^{n+1} = \Psi U^n \\ U^0 = (u_0(x_1), \ldots, u_0(x_q))^T. \end{cases} \tag{3}$$

with $\Psi = (\psi_{jk})$ a constant matrix. Two main ways can be used to construct a Lagrange interpolation $I[U]$, which result in different structures of the matrix Ψ. In the first case, the basis functions for the interpolation are fixed, so that

$$I[U](x) = \sum_{k=1}^{q} u_k \psi_k(x). \tag{4}$$

Typically, the basis functions ψ_k are of finite element type and the computational grid can be unstructured. However, even for the model problem (1) and with an evenly spaced 1–D grid, this choice does not lead in general to a banded matrix Ψ unless for $\mathbb{P}_1$ reconstructions.

In the second case, the basis functions depend on x: for example a cubic reconstruction is typically performed using two nodes left and two nodes right of x, so that if $x_l < x < x_{l+1}$, then

$$I[U](x) = \sum_{k=l-1}^{l+2} u_k L_{kl}(x), \tag{5}$$

where

$$L_{kl}(x) = \prod_{\substack{m = l-1 \\ m \neq k}}^{l+2} \frac{x - x_m}{x_k - x_m}. \tag{6}$$

In this case, it is easy to check by the definition that for the model problem (1), if the grid is evenly spaced then Ψ is a banded matrix.

Although numerical experiments show that SL schemes are unconditionally stable, the many efforts done to prove this fact in a rigorous way for reconstructions of degree higher then $\mathbb{P}_1$ have only led to partial results. Such results (see,[21]) are in the direction of Von Neumann stability, and therefore only applicable to reconstructions in a form like (5), (6). In this paper we propose a Von Neumann type of analysis to the finite element SL scheme, focusing on the characterization of the eigenvalues of the matrix Ψ. Filling the gap towards a complete stability analysis will be the object of a forthcoming study.

The outline of the paper is the following. In section 2 we will motivate the analysis with a counterexample. Section 3 shows the basic ideas and techniques on the case of $\mathbb{P}_1$ reconstructions. In sections 4 and 5 we extend this methods to arbitrary order finite elements and carry out the explicit computations on the $\mathbb{P}_2$ case. Finally, section 6 is devoted to the analysis of nonuniform space grids.

2. A counterexample

Let us take into consideration equation (1) with $a = -1$, $\Delta t = h/2$, so that the point $x_j - a\Delta t$ falls at the midpoint of the segment $[x_j, x_{j+1}]$. Let the total number q of nodes be even with $x_1 = 0$, $x_q = 1 - h$.

Assume first that the reconstruction I is performed by means of $\mathbb{P}_2$ finite elements, piecewise quadratic on each interval between two successive odd nodes. Carrying out the computations, it turns out that the matrix Ψ is given by

$$\Psi = \begin{pmatrix} 3/8 & 3/4 & -1/8 & 0 & 0 & 0 & \cdots & 0 \\ -1/8 & 3/4 & 3/8 & 0 & 0 & 0 & \cdots & 0 \\ 0 & 0 & 3/8 & 3/4 & -1/8 & 0 & \cdots & 0 \\ \cdots & \cdots & \cdots & \cdots & \cdots & \cdots & \cdots & \cdots \\ 0 & 0 & \cdots & 0 & -1/8 & 3/4 & 3/8 & 0 \\ -1/8 & 0 & \cdots & 0 & 0 & 0 & 3/8 & 3/4 \\ 3/8 & 0 & \cdots & 0 & 0 & 0 & -1/8 & 3/4 \end{pmatrix} \tag{7}$$

We can easily compute l^1 and l^∞ norms for the matrix as:

$$|\Psi|_1 := \max_k \sum_j |\psi_{jk}| = \max\left(\frac{3}{2}, 1\right) = \frac{3}{2} \tag{8}$$

$$|\Psi|_\infty := \max_j \sum_k |\psi_{jk}| = \frac{5}{4}. \tag{9}$$

As easily, we can find a vector U^n (e.g. a step function) for which neither of the two norms is conserved after an iteration.

This is a well–known fact, however. What is more surprising, is that despite the results of l^2 stability in the banded case, in this finite element case even the norm $|\Psi|_2$ exceeds the unity. This can be seen defining the matrix $B = (b_{ij}) = \Psi^T \Psi$ and recalling that

$$|\Psi|_2 := \sqrt{\sup_x \frac{x^T B x}{x^T x}}. \tag{10}$$

Now, if we compute the diagonal elements of B, it turns out that, for i even, $b_{ii} = 9/8$. Hence, using e_i instead of the maximizer in the norm definition, we obtain

$$\sup_x \frac{x^T B x}{x^T x} \geq \frac{e_i^T B e_i}{e_i^T e_i} = \frac{9}{8}, \tag{11}$$

so that we also have $|\Psi|_2 \geq \sqrt{9/8} \approx 1.061$ (more precisely, $|\Psi|_2 \approx 1.118$ computed by MATLAB with $q = 100$). Note also that such a bound on the norm of Ψ remains constant for $h \to 0$ under the relationship $\Delta t = h/2$, so that no chance exists to obtain estimates like $|\Psi|_2 \leq (1 + C\Delta t)$. On the other hand, if a quadratic reconstruction would be performed using (for example) always one node left and two nodes right of the point $x_j - a\Delta t$ then, according to the Von Neumann analysis in,[21] we would have $|\Psi|_2 \leq 1$.

Such considerations motivate an *ad hoc* stability analysis for the finite element SL scheme. What will be pursued here is again a Von Neumann type of analysis, verifying only a *necessary* stability condition, in the form

$$|\rho_k| \leq 1 \tag{12}$$

with ρ_k a generic eigenvalue of Ψ. However, since the matrix Ψ is no longer banded, we expect the standard tools of Fourier analysis to be inapplicable, and the analysis will focus on an alternative technique to characterize its eigenvalues. We will start with the case of a uniform grid, and give in the last section some results concerning the nonuniform case.

3. An example in detail: the $\mathbb{P}_1$ scheme

Although for the $\mathbb{P}_1$ scheme (which corresponds, for small Courant numbers, both to the Courant–Isaacson–Rees and to the Godunov scheme) it is very easy to prove stability via a monotonicity argument, we will nevertheless use this easier case to show the basic ideas and results. For self-similarity reasons, the scheme will be rewritten on the basis of two non-dimensional parameters, the number of subintervals N and the Courant number $\lambda = -a\Delta t/h$. We will also denote the matrix Ψ, whenever useful, as $\Psi(\lambda)$.

First, note that it is not restrictive to assume that the Courant number is in $[0,1]$. In fact, if $\lambda > 1$, we can look at $\Psi(\lambda)$ as the product of two matrices:

$$\Psi(\lambda) = Q_{[\lambda]}\Psi(\lambda - [\lambda]) \tag{13}$$

(here, $[\cdot]$ denotes as usual the integer part). The matrix $Q_{[\lambda]}$ is a permutation matrix which shifts the columns by $[\lambda]$ places. This matrix is orthonormal and has therefore the only effect to rotate the eigenvalues of $\Psi(\lambda - [\lambda])$ which in turn is constructed with a Courant number in $[0,1]$.

Now, for this scheme the matrix Ψ has the form:

$$\Psi = \begin{pmatrix} 1-\lambda & \lambda & 0 & \cdots & 0 & 0 & 0 \\ 0 & 1-\lambda & \lambda & \cdots & 0 & 0 & 0 \\ \cdots & \cdots & \cdots\cdots & \cdots & \cdots & \cdots \\ 0 & 0 & 0 & \cdots & 1-\lambda & \lambda & 0 \\ 0 & 0 & 0 & \cdots & 0 & 1-\lambda & \lambda \\ \lambda & 0 & 0 & \cdots & 0 & 0 & 1-\lambda \end{pmatrix}. \tag{14}$$

Assuming that N is the number of subintervals, the nodes are numbered as $x_1, \ldots, x_N$, and the periodicity condition identifies x_{N+1} with x_1. The condition which characterizes ρ as an eigenvalue for the eigenvector $v = (v_1, \ldots, v_N)^T$ reads

$$\rho v_j = (1-\lambda)v_j + \lambda v_{j+1}$$

that is,

$$v_{j+1} = \frac{\rho - (1-\lambda)}{\lambda} v_j.$$

Starting from $j = 1$ and iterating N times this procedure, we obtain

$$v_{N+1} = \left(\frac{\rho - (1-\lambda)}{\lambda}\right)^N v_1$$

and at last, taking into account the periodicity condition,

$$\left(\frac{\rho - (1-\lambda)}{\lambda}\right)^N = 1. \tag{15}$$

Condition (15) shows that

- There exists a transformation between $\mathbb{C}$ and $\mathbb{C}$ mapping the eigenvalues into the N-th roots of the unity, so that if ρ is an eigenvalue, the corresponding root of the unity is

$$z = \frac{\rho - (1 - \lambda)}{\lambda}$$

- there exists a curve (which is the inverse image of the unity circle of the complex plane) to which any eigenvalue ρ belongs, so that the curve itself depends only on λ and the position of the eigenvalues on the curve depends only on N.

Such assertions are well-known in the case of a banded matrix Ψ. We will show, however, that they can be suitably extended (especially the second one) to the block-banded case under consideration.

Neglecting the role of N in (15) we get the expression of the curve containing the eigenvalues. More precisely, we have the condition:

$$|\rho - (1 - \lambda)| = |\lambda| \tag{16}$$

that is, a circle centered at $1 - \lambda$ and with radius λ (note that this is precisely the boundary of the Gershgorin discs of the matrix Ψ). Therefore, it results that, for any N, the eigenvalues are contained in the unity disc. Fig. 1 shows the eigenvalues of Ψ for different values of $\lambda \in [0, 1]$.

4. A general eigenvalue analysis for the SL–FE scheme

Assume that the interval $[0, 1]$ is split into N subintervals, the interpolation degree being p on each subinterval, and $\lambda \in (0, 1)$. The total number of nodes is $q = Np$ and $h = 1/(Np)$ is the space step between nodes. Accordingly, the matrix Ψ has the structure

$$\Psi = \begin{pmatrix} A & B & 0 & \cdots & 0 & 0 & 0 \\ 0 & A & B & \cdots & 0 & 0 & 0 \\ \cdots & \cdots & \cdots & \cdots & \cdots & \cdots & \cdots \\ 0 & 0 & 0 & \cdots & A & B & 0 \\ 0 & 0 & 0 & \cdots & 0 & A & B \\ B & 0 & 0 & \cdots & 0 & 0 & A \end{pmatrix}.$$

with $A, B \in \mathbb{R}^{p \times p}$, and B of the form

$$B = \left(b \mid 0 \right),$$

b being a column vector of $\mathbb{R}^p$. More explicitly, setting $k = 1 + (j - 1)p$ (that is, denoting by x_k the left extremum of the j–th subinterval), we can

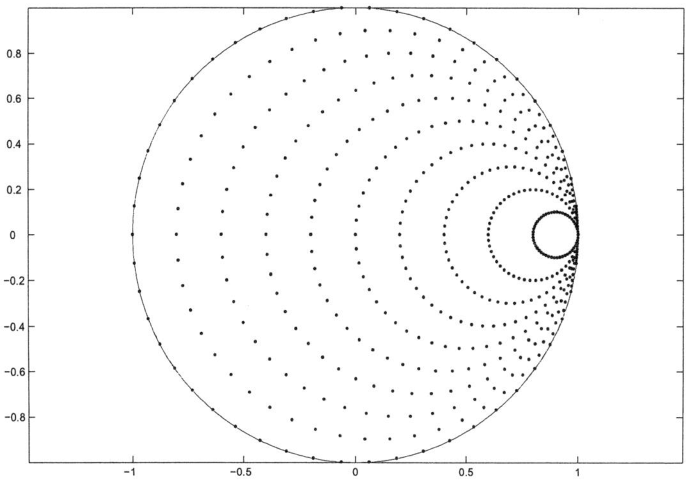

Fig. 1. Eigenvalues of the $\mathbb{P}_1$ scheme for $\lambda = 0.1, 0.2, \dots, 1$ and 50 nodes.

write the j–th block of the scheme as

$$\begin{cases} u_k^{n+1} = \phi_0(\lambda)u_k^n + \cdots + \phi_p(\lambda)u_{k+p}^n \\ \vdots \\ u_{k+p-1}^{n+1} = \phi_0(\lambda+p-1)u_k^n + \cdots + \phi_p(\lambda+p-1)u_{k+p}^n \end{cases} \tag{17}$$

where $\phi_0(\cdot), \dots, \phi_p(\cdot)$ are the basis functions expressed in a reference interval with unity step between nodes. Now, assuming ρ is an eigenvalue for the eigenvector v, we have from (17):

$$\begin{cases} \big[\phi_0(\lambda) - \rho\big]v_k + \cdots + \phi_p(\lambda)v_{k+p} = 0 \\ \vdots \\ \phi_0(\lambda+p-1)v_k + \cdots + \\ \quad + \big[\phi_{p-1}(\lambda+p-1) - \rho\big]v_{k+p-1} + \phi_p(\lambda+p-1)v_{k+p} = 0 \end{cases} \tag{18}$$

Since (18) has p equations and $p+1$ unknowns, it is possible, for example via Gauss elimination, to express v_{k+p} as a function of v_k in the form

$$v_{k+p} = f(\rho, \lambda)\, v_k \tag{19}$$

with $f(\cdot,\cdot)$ an iterated composition of rational functions. As it has been shown in the $\mathbb{P}_1$ example, repeating for N subintervals and imposing the periodicity condition we obtain:

$$f(\rho, \lambda)^N - 1, \tag{20}$$

that is, passing to the moduli,

$$\left|f(\rho, \lambda)\right|^N = 1. \tag{21}$$

In this case, the equation of the curve containing the eigenvalues is given in implicit form by

$$\left|f(\rho, \lambda)\right| = 1. \tag{22}$$

We show in Fig. 2 the eigenvalues of the $\mathbb{P}_3$ scheme for different values of λ and 51 nodes.

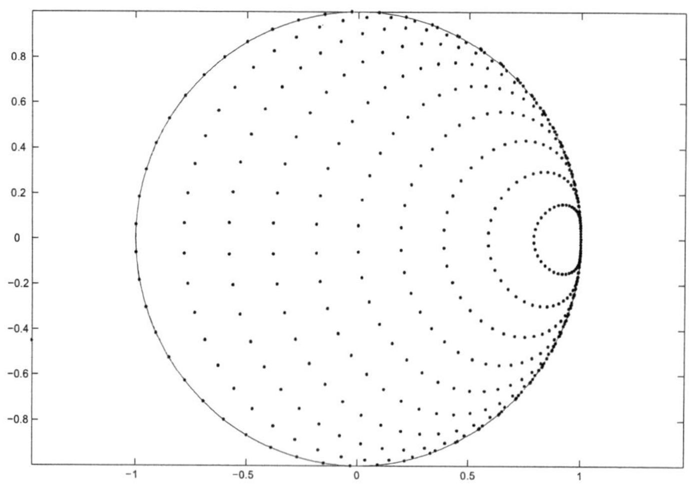

Fig. 2. Eigenvalues of the $\mathbb{P}_3$ scheme for $\lambda = 0.1,\ 0.2, \ldots,\ 1$ and 51 nodes.

5. The $\mathbb{P}_2$ case

We treat separately the $\mathbb{P}_2$ case, for which the basis functions in (17) have the form:

$$\begin{aligned}\phi_0(\lambda) &= \frac{1}{2}(\lambda-1)(\lambda-2)\\ \phi_1(\lambda) &= \lambda(2-\lambda)\\ \phi_2(\lambda) &= \frac{1}{2}\lambda(\lambda-1)\end{aligned}$$

and allow for an explicit computation. More precisely, we have the following

Theorem 5.1. *Assume that* $0<\lambda<1$. *Then the constant step,* $\mathbb{P}_2$ *scheme of the form* (17), *with* $p=2$ *and* ϕ_0, ϕ_1, ϕ_2 *given above, satisfies condition* (12).

Proof. We will refer to[4] for the detailed computations, while sketching here only the main steps. Carrying out the computations outlined in section 4, we obtain for (19) the explicit expression:

$$v_{k+2} = \left[\frac{\rho-\phi_0(\lambda)+\frac{\phi_1(\lambda)\phi_0(\lambda+1)}{\rho+\phi_1(\lambda+1)}}{\phi_2(\lambda)-\frac{\phi_1(\lambda)\phi_2(\lambda+1)}{\rho+\phi_1(\lambda+1)}}\right] v_k$$

which, via some algebra, gives the condition:

$$\left[\frac{2\rho^2+(\lambda+4)(\lambda-1)\rho+(\lambda-2)(\lambda-1)}{\lambda(\lambda-1)\rho+\lambda(\lambda+1)}\right]^N = 1.$$

Passing to the moduli, and writing $\rho = x+iy\in\mathbb{C}$, we obtain the equation

$$\begin{aligned}&y^4+\left[2x^2+(\lambda^2+3\lambda-4)x+(2\lambda^3-\lambda^2-3\lambda+2)\right]y^2+\\ &+\left[x^4+(\lambda^2+3\lambda-4)x^3+(2\lambda^3+\lambda^2-9\lambda+6)x^2+\right.\\ &\left.+(-5\lambda^2+9\lambda-4)x-2\lambda^3+3\lambda^2-3\lambda+1\right]=0.\end{aligned}$$

By working in the auxiliary variable y^2 we obtain, for $x\in[1-2\lambda,1]$, y^2 as the following function of x and λ:

$$y^2 = \frac{-\left[2x^2+(\lambda+4)(\lambda-1)x+(2\lambda^2+\lambda-2)(\lambda-1)\right]+\sqrt{\Delta(x,\lambda)}}{2}$$

with

$$\begin{aligned}\Delta(x,\lambda) &= \lambda^2(\lambda-1)(\lambda+7)x^2+2\lambda^2(\lambda-1)(2\lambda^2+7\lambda-7)x+\\ &\quad+\lambda^2(4\lambda^4-4\lambda^3-11\lambda^2+22\lambda-7).\end{aligned}$$

Finally, for $x \in [1-2\lambda, 1]$, it is easy to verify that $y^2 < 1 - x^2$, which amounts to prove that eigenvalues are in the unity disc of $\mathbb{C}$. □

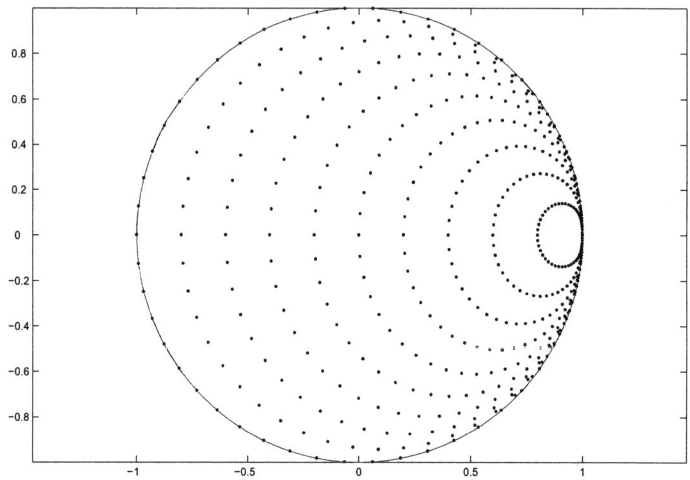

Fig. 3. Eigenvalues of the $\mathbb{P}_2$ scheme for $\lambda = 0.1, 0.2, \ldots, 1$ and 50 nodes.

6. Nonuniform space grids

In this situation, if $H_j = ph_j$ is the measure of the j–th subinterval, we can define *local* Courant numbers

$$\lambda_j = -\frac{a\Delta t}{h_j}.$$

We set again $k = 1 + (j-1)p$ with $j = 1, \ldots, N$, so that the l–th block of the scheme reads now

$$\begin{cases} u_k^{n+1} = \phi_0(\lambda_j)u_k^n + \cdots + \phi_p(\lambda_j)u_{k+p}^n \\ \vdots \\ u_{k+p-1}^{n+1} = \phi_0(\lambda_j + p - 1)u_k^n + \cdots + \phi_p(\lambda_j + p - 1)u_{k+p}^n. \end{cases} \tag{23}$$

Theorem 6.1. *Assume that, for any j, $0 < \lambda_{\min} \leq \lambda_j \leq \lambda_{\max} < 1$. Assume moreover that the constant step scheme* (17) *satisfies condition*

(12) *for any* $\lambda \in [\lambda_{\min}, \lambda_{\max}]$. *Then condition* (12) *is also satisfied by the nonconstant step scheme* (23).

Proof. Following the same procedure as above, condition (20) is now replaced by

$$\prod_{j=1}^{N} f(\rho, \lambda_j) = 1. \tag{24}$$

Keeping the analogy with the uniform mesh case, the condition which characterizes the curve containing the eigenvalues may be rewritten as

$$\prod_{j=1}^{N} |f(\rho, \lambda_j)| = 1 \tag{25}$$

which is the counterpart of condition (21). Now, all the term in the product satisfy the bounds

$$\min_j |f(\rho, \lambda_j)| \le |f(\rho, \lambda_j)| \le \max_j |f(\rho, \lambda_j)| \tag{26}$$

so that

$$\min_j |f(\rho, \lambda_j)|^N \le \prod_{j=1}^{N} |f(\rho, \lambda_j)| \le \max_j |f(\rho, \lambda_j)|^N \tag{27}$$

and we can conclude that if condition (12) is satisfied with a uniform grid for any $\lambda \in [\lambda_{\min}, \lambda_{\max}]$, then it is also satisfied in the nonuniform case. □

We show in Fig. 4 (in crosses) the eigenvalues of the $\mathbb{P}_2$ scheme, obtained with 25 subintervals of random measure and $0.014 \le \lambda_j \le 0.77$, compared with (in points) the eigenvalues associated to the extreme values of the local Courant numbers.

Conclusions

We have proposed a technique to characterize eigenvalues of SL schemes with finite element reconstructions, which do not allow in general to use Fourier analysis tools. The curve containing the eigenvalues may be easily expressed in implicit form; computing its explicit expression is feasible for the $\mathbb{P}_2$ scheme but seems to be too complex in general. In the case of a nonuniform mesh, we have proved a partial result which holds under small Courant numbers. Forthcoming studies will include the estimation of iterated powers of Ψ, as well as a more general setting of the nonuniform mesh case.

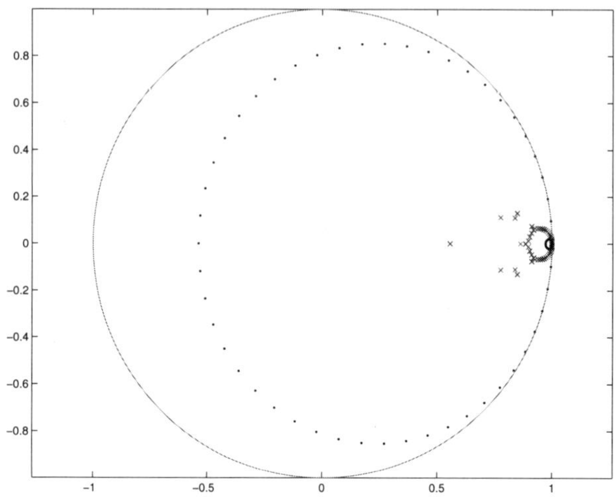

Fig. 4. Eigenvalues of the $\mathbb{P}_2$ scheme for $0.014 \leq \lambda_j \leq 0.77$ and 50 nodes.

References

1. N. Besse, M. Mehrenberger, *Convergence of classes of high-order semi-Lagrangian schemes for the Vlasov–Poisson system*, preprint, Université Louis Pasteur, Strasbourg, France (2006).
2. M. Falcone, R. Ferretti, *SIAM J. Num. Anal.* **35**, 909 (1998).
3. M. Falcone, R. Ferretti, *J. Comput. Phys.* **175**, 559 (2002).
4. G. Perrone, Sulla stabilità dei metodi semi-Lagrangiani di ordine alto, degree thesis, Universitá di Roma Tre (Roma, 2006).
5. A. Staniforth, J. Côté, *Mon. Wea. Rev.* **119**, 2206 (1991).
6. A. Wiin-Nielsen, *Tellus* **11**, 180 (1959).

A METHOD TO TEST BALLASTED HIGH-SPEED RAILWAY TRACKS

GIORDANO FRANCESCHINI AND ALBERTO GARINEI
Department of Industrial Engineering, University of Perugia
via G. Duranti 67, 06125 Perugia (Italy)
E-mail: giorda@unipg.it

The transit of high speed trains produces vibrations in the surrounding environment that can be harmful for the buildings next to the railway. To know, in progress of the railway construction, the entity of the vibrations transmitted, it is very useful to determine a force input that simulates the train passage, in order to apply it to fixed positions on the top of the embankment, to perform measures in situ. This paper concerns the study of this input in parametric form through a simple analytical model. We model the track as a beam on elastic Winkler ground subjected to an alternate moving load. Using the superposition principle, we analyse the load of every axle, comprehending its alternate components, and therefore the train's one. A comparison with models FEM points out the adequacy of the parametric model to the objectives. The force, at various speeds, is the input we can apply through simple actuators positioned on the top of the embankment.

1. Introduction

The results contained in this work are formulated to support experimental activities, which are currently in progress, on embankments of the "Quadruplicamento Ferroviario Veloce Milano-Bologna (Railway Quadruplicating Fast Milan-Bologna)" under construction, in order to preview the vertical vibrations transmitted from the trains to the receivers next to the railway. These experimental activities are oriented to the knowledge of the performances of the embankment to chek the correspondence to the detailed lists.

The study of the vibrations can be reduced to the analysis of three successive blocks:

block 1: input: action induced by the train on the railroad - output: action on the top of the embankment.

block 2: input: action on the top of the embankment - output: action inside the embankment and then the base of the same one.

block 3: input: action on the base of the embankment - output: action on the base of the receiver.

We take care over the analysis of the first two blocks, and over the study of the components of the vibrations "at low frequency" (less than $50hz$), that can interest the civil structures.

The aim is to give a methodology to determine in parametric form the load variable at varying time to be applied, with actuators, in fixed positions on the top of the embankment, in order to preview the behavior of the track and of the embankment that carry high-speed trains. This will allow to perform experimental activities during the construction of the railway.

To replace the effect of a moving load with a load variable at varying time, applied to a fixed point, involves approximations; the spread between the results shall be acceptable in the judgements of the planner and the test board.

The hypotheses to develop the analytical model are: the linearity of the phenomena, the elastic behavior of the beam (the track) and of the support (embankment and ground); therefore we don't consider dissipative effects. The axle force is described by the equation:

$$F(t) = F_0 + \sum_{i=0}^{h} F_i \cos(\Omega_i t + \phi_i). \tag{1}$$

The amplitudes and the frequencies can be measured. The data of tests carried out by the FIAT FERROVIARIA company on a ETR500 show that, beyond constant component, there are meaningful components relative to two or three frequencies under $10hz$ [1].

We analyse block 1, in paragraph 2, considering the railroad modelled as a rectilinear and homogenous elastic beam, of infinite length, supported on the embankment modelled as a continuous Winkler elastic support and subjected to a concentrated sinusoidal load moving with constant speed.

As it known [2], in this model the deflection of the track $y(x,t)$ is expressible in analytic form and this permit to easily evaluate the influence of characteristic parameters (speed or force frequency).

The deformation of the track, that is also the deformation of the top of the embankment, is the input for the second block and therefore is the reference for the construction of test signal to be applied to a fixed point.

To estimate how much the analytical result, with all the simplification introduced, could be meaningful for the study of the behavior of the displacements and stresses of the embankment, we use the results contained in [3]. In this report is described a detailed numerical analysis to validate the analytical model, using method FEM with ANSYS code. We compare the results for the system trails - sleepers - ballast - subballast - embankment submitted directly to the mobile sinusoidal action (FEM1 in the next) with those of a model FEM for the embankment subject to the reaction of the railroad and therefore to a force proportional (with the Winkler's constant) to the deformed calculated (FEM2 in the next). The paragraph 3 refers to an example of the validity and a synthesis of results more meaningful for our considerations.

With these results, in our opinion, the analytical results became widely acceptable as start point to determine the test signal that must have the same effects of the project signal. The procedure for the determination of test signal is described in paragraph 4.

The procedure, that it is always based on results and analytical procedures, allows to determine the temporal law of the fixed force to be applied. It is therefore possible to reproduce with an actuator the "fictitious" input signal for various speeds and frequencies, for the series of tests to carry out. The determination of the force law constitutes the preliminary step to the construction of the test device.

This paper is a synthesis of the original work [4], where there are all the procedures and calculations.

2. The deformed of the railroad

The deflection $y(x,t)$ of the infinite beam on elastic ground, under concentrated load $F(t) = F_0 \cos \Omega t$ moving with speed v, is descrbed by the equation:

$$EJ\frac{\partial^4 y}{\partial x^4} + \rho\frac{\partial^2 y}{\partial t^2} + ky = F_0 \cos \Omega t \cdot \delta(x - vt) \tag{2}$$

with null deflection and spin for $x \to \pm\infty$.We model the set rails-sleepers as an infinite beam: we consider the sleepers uniformly distributed with contribution to the only density. In the equation (2) EJ is the flexional rigidity of the rails, ρ is the density of the rails and the sleepers, k is the Winkler's foundation coefficient, $\delta(x - vt)$ is the Dirac delta.

The solution of (2) (here and after, "solution" is meaning "solution in weak sense") is [2] :

$$y(r,t) = \begin{cases} \frac{F_0}{2EJ}\{e^{b_1 r})[q_1 \cos(\Omega t + ar) - q_2 \sin(\Omega t + ar)] + \quad r < 0 \\ \quad e^{b_2 r}[w_1 \cos(\Omega t - ar) + w_2 \sin(\Omega t - ar)]\} \\ \\ \frac{F_0}{2EJ}\{e^{-b_1 r}[q_1 \cos(\Omega t + ar) + q_2 \sin(\Omega t + ar)] + \quad r > 0 \\ \quad e^{-b_2 r}[w_1 \cos(\Omega t - ar) - w_2 \sin(\Omega t - ar)]\} \end{cases} \tag{3}$$

where $r = x - vt$ and

$$q_1 = \frac{4a^2 - b_1^2 + b_2^2}{b_1\left[(4a^2 - b_1^2 + b_2^2)^2 + (4ab_1)^2\right]}, \quad q_2 = \frac{4a}{(4a^2 - b_1^2 + b_2^2)^2 + (4ab_1)^2]}$$

$$w_1 = \frac{4a^2 + b_1^2 - b_2^2}{b_2\left[(4a^2 + b_1^2 - b_2^2)^2 + (4ab_2)^2\right]}, \quad w_2 = \frac{4a}{[(4a^2 + b_1^2 - b_2^2)^2 + (4ab_2)^2]}.$$

The parameters a, b_1, b_2 $(b_1 > b_2)$ are the positive system solutions:

$$2a^2 - b_1^2 - b_2^2 = \tfrac{\rho v^2}{EJ} \; 2a(b_1^2 - b_2^2) = \tfrac{2\rho\Omega v}{EJ} \; (a^2 + b_1^1)(a^2 + b_2^2) = \tfrac{k - \rho\Omega^2}{EJ}. \tag{4}$$

We have the conditions: $k - \rho\Omega^2 > 0$ and $b_2^2 > 0$. Both are widely verified for parameter values that are meaningful in our case. The solution allows to analyze easily the effect due to the speed and the frequency variation. The (2) show that both the displacement of every point of the railway at varying time and the deflection of the various points of the railroad in a determined instant, are the sum of sinusoidal functions modulated with an exponential decay. The nonzero frequency of force implies that we have some points of the track not directly subjected to the load and that only the points subjected to the maximum load reach the maximum deformation:

$$y_{\max}^{(v,\Omega)} = \frac{F_0}{2EJ}(q_1 + w_1). \tag{5}$$

The solution of the static problem [5] is obtained for $v = \Omega = 0$; the maximum deflection is:

$$y_{\max}^{(0,0)} = \frac{F_0}{8EJ}\left(\frac{4EJ}{k}\right)^{3/4} \tag{6}$$

In the case of alternate load applied to a fixed point $v = 0, \Omega \neq 0$, we have resonance for $\Omega^2 = k/\rho$; in case of constant moving load $v \neq 0, \Omega = 0$, we have critical speed $v_{cr} = (4EJk/\rho^2)^{1/4}$.

We can describe the speed and/or pulsation effect on the maximum deformation defining the ratio:

$$Y(v, \Omega) \equiv \frac{y_{\max}}{y_{\max}^{(0,0)}} = 4\left(\frac{k}{4EJ}\right)^{3/4} \frac{b_1 + b_2}{b_1 b_2[(b_1 + b_2)^2 + 4a^2]}. \tag{7}$$

Fig. 1 shows (7), when the following data are assumed [7,8]:

rail UIC60: flexional rigidity $EJ_r = 6.38\ 10^6\ Nm^2$ and linear density $63Kg/m$; sleeper: length $1.80m$, width $0.22m$, height $0.20m$, volume $2500Kg/m^3$, and therefore mass $198Kg.$; distance centre sleepers: $0.8m$; foundation coefficient $k_r = 57.58\ 10^6 N/m^2$. The data for the equation (2) are therefore $EJ = 12.76\ 10^6 Nm^2$, $k = 115.2\ 10^6 N/m^2$, $\rho = 373Kg/m$.

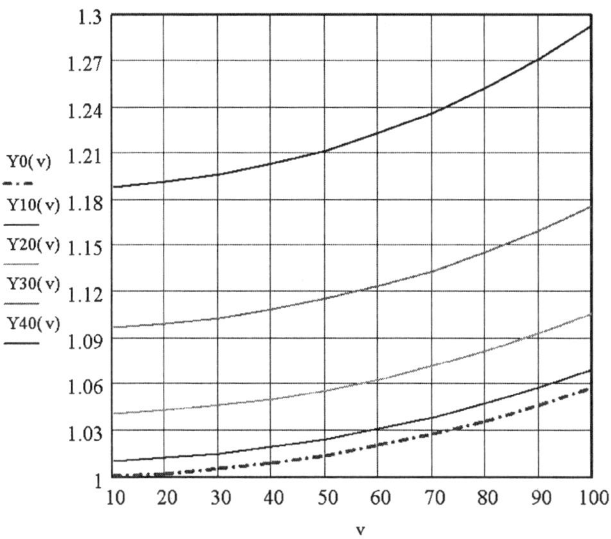

Fig. 1 - $Y\Omega(v)$ represents $Y(v, \Omega)$ for different values of Ω

The track deflectionis usefull to estimate its resistance in exercise conditions and to determine the deformation (or the force) on the ground surface or on the top of the embankment.

3. Validity of the model

To validate the model that describe the system under the action of alternate mobile loads, and in particular to determine the behavior of the embankment, we analyse the results of a series of F.E.M. simulations with code ANSYS [3]. With the two outlines described in the introduction, we define the behavior of the embankment and compare it to the various quotas, both in terms of displacement and of normal stress, at various train speed and load frequencies. For the simulation, beyond to the data indicated in the previous paragraph, we assume:

material	thickness $[m]$	elastic coeff. $[N/m^2]$	density $[Kg/m^3]$	coeff. Poisson
ballast	0.35	$2\,10^8$	2000	0.4
subballast	0.12	$7.5\,10^8$	2000	0.4
embankment	3.53	10^8	1900	0.35

and $v = 61.1m/s$, $\Omega = 12.56/s$ $F_0 = 2\,10^4 N$.

The relative diagrams referring to the displacements, in Fig. 2 and 3, show that, at different quota of embankment, the displacements calculated with FEM2 are greater than those calculated with FEM1.

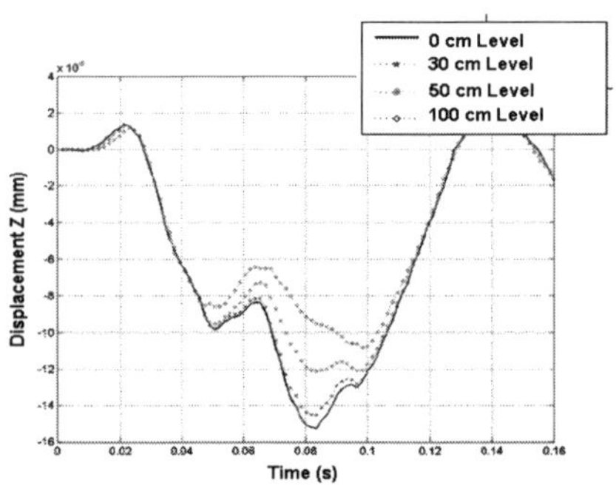

Fig. 2 - FEM1 model -dispacement

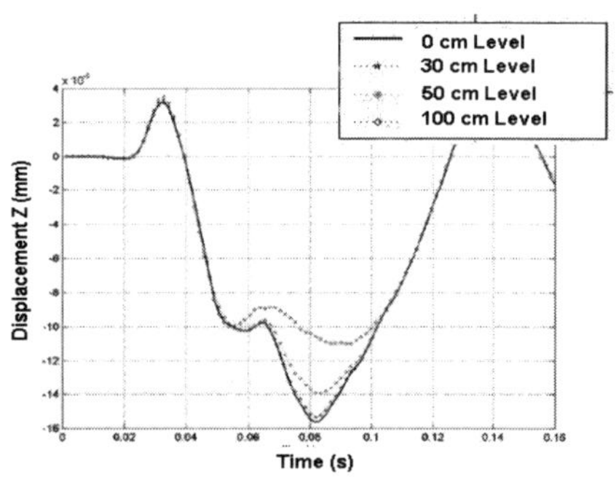

Fig. 3 - FEM2 model - displacement

The relative diagrams referring to the normal stress, in Fig. 4 and 5, show that on the top of the embankment, at the ballast quota, the strain calculated with FEM1 is greater than the one calculated with FEM2; on increasing the distance from the top, FEM2 decrease in monotonic way, while FEM1 is not monotone, as we can note from the peak at quota 0.4m.

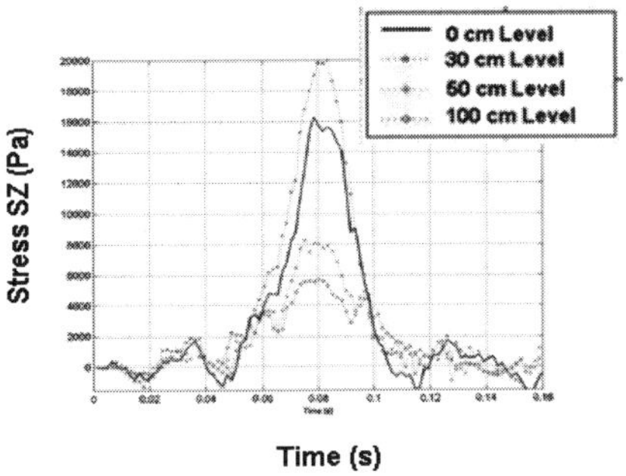

Fig. 4 - FEM1 model - stress

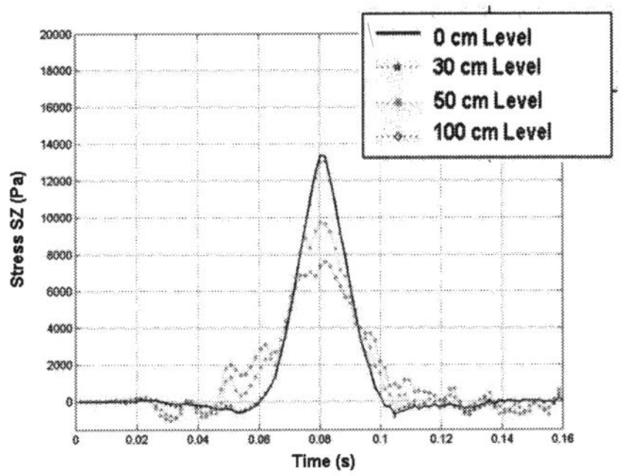

Fig. 5 - FEM2 model - stress

We explain the different behavior with the quota, analysing the different sleepers model. In FEM2 they are included "with continuity" in the beam, with single increasing effects of the constant value of the linear density. In FEM1, the sleepers are independent elements of greater rigidity than the ballast and subballst, and therefore they can provoke effects of localization, at least until to a certain quota.

We observe that the force assigned in FEM2 is obtained from the deformation of the railroad, calculated assuming continuous the elastic support; in FEM1 the load transmitted in concentrated way b the sleepers is distributed on the ballast and the subballast with a thickness of $0.5m$; we have a continuous distribution on the embankment from this quota.

Under the $0.5m$ the stresses calculated with FEM2 are greater than those from FEM1.

So under the $0.5m$ the results of FEM2 are worse (but comparable) than those of FEM1.

Therefore we can conclude that with this model we can determine useful results in order to set testing system and perform experimental activities during the construction of the embankment.

The diagrams at $0.50m$, with filters pass-low with range of $50Hz \sim 120Hz$, show that the sleepers give contributions beyond the $120Hz$.

4. The steady load test

According to the previous considerations, we can use the deformed shape (3) for the determination of a "parametric" test signal, that we can apply to a fixed position to predict, with experimental tests, both the track and the embankment behavior. The signal test, must supply, by experimental activities in progress, although in a limited but meaningful zone of the track-embankment system, information about thc displaccmcnts and the stresses that can describe the effective situation at full capacity.

The aim of this analysis is to determine the functional shape of a time dependent load , $F_f(t)$, applied by actuators to a fixed position on the top of the embankment, in order to reproduce, at least in the interesting zone, the effect of the moving load. The analytical expression allows to easily estimate the load for various parameters and therefore to predict different situations in relation to the different train models that will traverse the railway at various speeds.

The (3) shows how the displacement, induced by the moving load in every point of the railroad, varies at varying time, in particular in the points in which the maximum is reached, for example $x = 0$. The choice of the force shape is made through analitical considerations, because at varying time in every point of the beam, the displacement is expressed by exponential and sinusoidal and therefore by functions that, derived, remain of the same type. Then we can assume as force applied to a fixed point $(x = 0)$, the equation:

$$F_f(t) = ky(0,t) + \sum_{j=1}^{n} h_j \frac{d^j y(0,t)}{dt^j} \tag{8}$$

in which the only arbitrariness are given from the constants h_j, that can be determined in order to make similar the two deflections $y(x,t)$ and $y_f(x,t)$ at least around the maximum value. The values of h_j will depend on the choice of the characteristic parameters. In the extreme case $v = 0$,we have:

$$F_f(t) = F(t). \tag{9}$$

Considering (3), we can obtain the (9), assuming:

$$F_f(t) = ky(0,t) + h_2 \frac{d^2 y(0,t)}{dt^2} \tag{10}$$

with

$$h_2|_{v=0} = \Omega^{-2}[k - 2(4EJ)^{1/4}(k - \rho\Omega^2)^{3/4}].$$

The h_2 can assume values both positive and negative depending on the choices of the other characteristic parameters; for the chosen data, we have $h = -4.6\ 10^5 Kg/m$.

Also for $v \neq 0$ we can consider the shape (10) meaningful to reproduce the deformed shape next to the maximum value. Therefore it is enough to consider the only arbitrary parameter h_2 that we can determine with the correct condition; subsequently we will omit the index to make it easier.

The determination of the railroad deflection $y_f(x,t)$ under the load action $F_f(t)$ (10) concentrated in the point $x = 0$ is obtained by solving the equation:

$$EJ\frac{\partial^4 y_f}{\partial x^4} + \rho\frac{\partial^2 y_f}{\partial t^2} + ky_f = F_f(t)\delta(x) \tag{11}$$

with the same conditions for $x \to \pm\infty$ of the moving load case. The solution is:

$$y_f(x,t) = \begin{cases} \frac{F_0}{2EJ}\{e^{b_1vt}[q_{11}^{\pm}(x)\cos(\Omega - av)t + q_{22}^{\pm}(x)\sin(\Omega - av)t] + & t < 0, \\ +e^{b_2vt}[w_{11}^{\pm}(x)\cos(\Omega + av)t - w_{22}^{\pm}(x)\sin(\Omega + av)t]\} & \\ & \\ \frac{F_0}{2EJ}\{e^{-b_1vt}[q_{11}^{\pm}(x)\cos(\Omega - av)t - q_{22}^{\pm}(x)\sin(\Omega - av)t] + & t > 0, \\ +e^{-b_2vt}[w_{11}^{\pm}(x)\cos(\Omega + av)t + w_{22}^{\pm}(x)\sin(\Omega + av)t]\} & \end{cases} \tag{12}$$

The functions $q_{11}^{+}(x)$, $q_{22}^{+}(x)$, must be assumed for $x > 0$ and the functions $q_{11}^{-}(x), q_{22}^{-}(x)$ for $x < 0$. We have:

$$\begin{aligned} q_{11}^{\pm}(x) &= \mp\tfrac{\beta}{b_1C}(e^{\mp b_1x}\sin ax \pm e^{\mp ax}\cos b_1x)\mp \\ &\quad \mp\tfrac{\gamma}{b_1C}(-e^{\mp b_1x}\cos ax + e^{\mp ax}\sin b_1x) \\ q_{22}^{\pm}(x) &= \pm\tfrac{\gamma}{b_1C}(e^{\mp b_1x}\sin ax + e^{\mp ax}\sin b_1x) - \\ &\quad -\tfrac{\beta}{b_1C}(e^{\mp b_1x}\cos ax + e^{\mp ax}\cos b_1x). \end{aligned} \tag{13}$$

The functions $w_{11}^{+}(x)$, $w_{11}^{-}(x)$, $w_{22}^{+}(x)$, $w_{22}^{-}(x)$ are obtained from the correspondents $q_{11}^{+}(x)$, $q_{11}^{-}(x)$, $q_{22}^{+}(x)$, $q_{22}^{-}(x)$ replacing b_1 with b_2 in (13) and substituting b_1 and b_2 in the coefficients β, γ, C (the procedure and calculation details are in the Appendix of [4].

The solution introduces a symmetry of the displacements respect to x as consequence of the action $F_f(t)$, concentrated in the point $x = 0$, and a symmetry respect to the time as consequence of the time dependent load.

These symmetries do not make possible to approximate the deformed shape at varying time, under the moving load action, in the points far from the one of maximum deflection. It is however meaningful the approximation of the deformed shape at varying time for the points next to the loaded point (approximately $x = 0$), and in the interval next to the one of maximum load (approximately $t = 0$) as the following diagrams show. For the values of the previously chosen parameters, in Fig. 10 we have the two deformed $y(x, 0)$ and $y_f(x, 0)$ and in Fig. 11 we have the displacements $y(0, t)$ and $y_f(0, t)$.

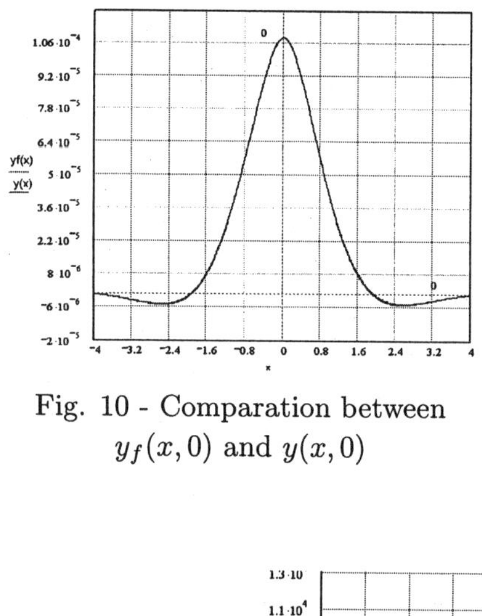

Fig. 10 - Comparation between $y_f(x, 0)$ and $y(x, 0)$

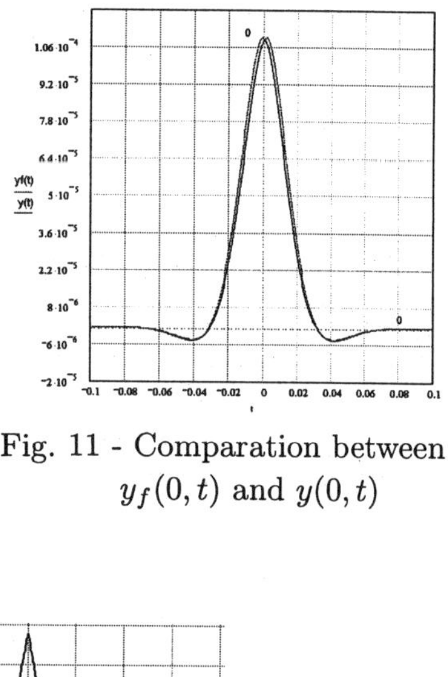

Fig. 11 - Comparation between $y_f(0, t)$ and $y(0, t)$

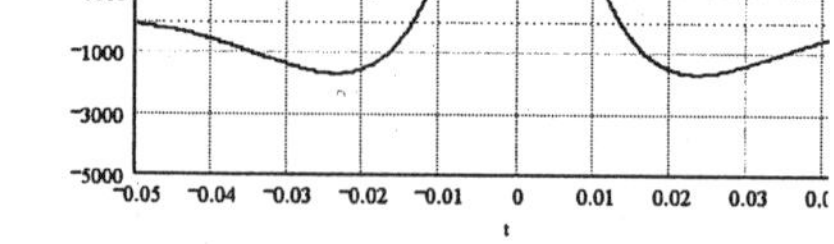

Fig. 12 - Force $F_f(t)$

The parameter is $h = -6.2\ 10^3 Kg/m$. The force $F_f(t)$ is represented in the diagram of Fig. 12. The different behavior in the neighbourhood of

$t = 0$ (10) is a consequence of the choice of the shape for $F_f(t)$ and can be removed considering further terms in (8). However the application of a filter $120H$.shows that the anomaly is due to components with frequencies outside the range of interest.

The assumption of a more general $F_f(t)$ allows to have a better approximation, having more parameters, but involves heavier calculations , that however can be developed with the same procedure. The choice (10) is already useful for the purposes the analysis was for and at the same time permits to set the calculation methodology for more elaborate situations.

5. Conclusions

The aim of the analysis undertaken was to develope a methodology in order to determine the law of the force to be applied to a fixed position on the top of the embankment, that could produce the deformed shape under a moving load. We developed a calculation for a moving sinusoidal force and we obtained the results depending on the parameters. The superimposition of the effects allowed to determine the law of the force corresponding to a moving force in the form (1). With the same hipothesis, we determined the signal corresponding to the loads applied to more axles. We observed also that the superimposition of the deformed depends on the type and on the speed of the train.

Once validated for the test the fixed signal, and defined the approximations acceptable, we planned the experimental activities consisting in a series of tests applying the force with an actuator and measuring the displacements and the stresses at different quotas of the embankment.

All this is essential in the study of the mechanical behaviour of the points of the embankment, in order to calculate the transfer functions directly, with an easier and faster methodology through which we can replace the traditional devices.

References

1. Garinei A. - *Simulazione su banco della corsa prova dei locomotori* - Tesi di laurea in Ingegneria Meccanica, Biblioteca di Ingegneria - Universita' di Perugia (2000)
2. Mathews P.M. - *Vibrations of a beam on elstic foundation* - ZAMM, 38, pp. 105-115 (1958)
3. Balli R. - *Prove sui rilevati* - Rapporto tecnico dell'Osservatorio Ambientale - Quadruplicamento Veloce Milano -Bologna, www. simibo.it (2003)

4. Franceschini G.Garinei A. - A method to test ballasted high-speed railway tracks - CD-ROM Proceedings of VII SIMAI Congress, Venezia (2004)
5. Belluzzi O. - *Scienza dele costruzioni* - Zanichelli Bologna (1941)
6. Fryba L. - *Vibration of solids and structures under moving loads* - Noordhoff International Publishing Groningen, The Netherlands (1972)
7. Esveld C. - *Modern railway track* - MRT Production (2001)
8. Natoni F. - *Rottura delle rotaie in esercizio* - Ingegneria ferroviaria, 7, CIFI - Roma (2002)

A SIMPLE VARIATIONAL DERIVATION OF SLENDER RODS THEORY

LORENZO FREDDI* ALESSANDRO LONDERO**

Dipartimento di Matematica e Informatica, Università di Udine, via delle Scienze 206, 33100 Udine, Italy
**E-mail: freddi@dimi.uniud.it **E-mail: londero@dimi.uniud.it*

ROBERTO PARONI

Dipartimento di Architettura e Pianificazione, Università di Sassari, Palazzo del Pou Salit, Piazza Duomo, 07041 Alghero, Italy,
E-mail: paroni@uniss.it

We present an asymptotic analysis of the three-dimensional problem for a thin linearly elastic cantilever $\Omega_\varepsilon = \varepsilon\omega \times (0, \ell)$ as ε goes to zero. By assuming ω simply connected and under suitable assumptions on the given loads, we show that the 3D problem converges in a variational sense to the classical dimensional models for extension, flexure and torsion of slender rods.

Keywords: slender rods; thin beams; linear elasticity; Γ-convergence.

1. Introduction

Structures with one or two dimensions much smaller than the remaining are very often encountered in engineering problems. The peculiar geometry of these thin structures suggests a lower, two or one, dimensional modelling. Classically, these lower dimensional models are based on some a-priori assumptions inspired by the smallness of certain dimensions. In the seventies new techniques which make circumvent the use of any a-priori assumption have been developed. The French school tuned a method based on a rigorous asymptotic expansion, while the Italian school followed the inspiration of E. De Giorgi[3] and adopted the use of Γ-convergence theory. Since then Γ-limits of energy functionals have been successfully applied to derive one or two-dimensional models of a variety of thin structures starting from linear as well as non linear three-dimensional elasticity.

In 1994, Anzellotti, Baldo and Percivale[1] derived variational models for linearly elastic homogeneous and isotropic rods and plates by using Γ-asymptotic developments (see also Percivale[10]). They deduce the mechanical behavior of the beam by calculating two different Γ-limits, one for the extensional problem and one for the flexural and torsional problems. The two Γ-limits are originated by different scalings of the energy functionals and correspond to terms of different order in the asymptotic development.

In this paper, by suitably scaling the axial component of the displacement in the three-dimensional energies and using a technique developed in Freddi, Morassi and Paroni,[4,5] we obtain, in an easier way, the extensional, flexural and torsional problems for a linearly elastic homogeneous and isotropic slender rod with only one Γ-limit.

Notation. Unless otherwise stated, we use the Einstein summation convention and we index vector and tensor components as follows: Greek indices α, β and γ take values in the set $\{1,2\}$ and Latin indices i, j, h in the set $\{1,2,3\}$. The component k of a vector v will be denoted either with $(v)_k$ or v_k and an analogous notation will be used to denote tensor components. $\mathcal{E}_{\alpha\beta}$ denotes the Ricci's symbol, that is $\mathcal{E}_{11} = \mathcal{E}_{22} = 0$, $\mathcal{E}_{12} = 1$ and $\mathcal{E}_{21} = -1$. $L^2(A;B)$ and $H^s(A;B)$ are the standard Lebesgue and Sobolev spaces of functions defined on A and taking values in B, with the usual norms $\|\cdot\|_{L^2(A;B)}$ and $\|\cdot\|_{H^s(A;B)}$, respectively. When $B = \mathbb{R}$ or when the right set B is clear from the context, we will simply write $L^2(A)$ or $H^s(A)$, sometimes even in the notation used for norms. Convergence in the norm will be denoted by $\rightarrow$ while weak convergence is denoted with $\rightharpoonup$.

With a little but harmless abuse of notation, we use to call "sequences" even those families indicized by a continuous parameter ε which, throughout the whole paper, will be assumed to belong to the interval $(0,1]$.

2. The 3-Dimensional problem

Let $\omega \subset \mathbb{R}^2$ be a simply connected, bounded, open set with a Lipschitz boundary. We consider a three-dimensional body $\Omega_\varepsilon \subset \mathbb{R}^3$, where $\Omega_\varepsilon := \omega_\varepsilon \times (0,\ell)$, $\omega_\varepsilon := \varepsilon\omega$, $\varepsilon \in (0,1]$ and $\ell > 0$. For any $x_3 \in (0,\ell)$ we further set $S_\varepsilon(x_3) := \omega_\varepsilon \times \{x_3\}$. Henceforth we shall refer to Ω_ε as the reference configuration of the body and denote by

$$Eu(x) := \mathrm{sym}(Du(x)) := \frac{Du(x) + Du^T(x)}{2},$$

the strain of $u : \Omega_\varepsilon \rightarrow \mathbb{R}^3$. The material is assumed to be homogeneous and isotropic, so that $\mathbb{C}A = 2\mu A + \lambda(\mathrm{tr}A)I$ for every symmetric matrix A. I

denotes the identity matrix of order 3. We assume $\mu > 0$ and $\lambda \geq 0$ so to have, for every symmetric tensor A,

$$\mathbb{C}A \cdot A \geq \mu |A|^2, \tag{1}$$

where $\cdot$ denotes the scalar product. Define

$$H^1_{\#}(\Omega_\varepsilon; \mathbb{R}^3) := \{u \in H^1(\Omega_\varepsilon; \mathbb{R}^3) : u = 0 \text{ on } S_\varepsilon(0)\}.$$

Due to the coercivity condition (1) and the strict convexity of the integrand, the energy functionals

$$\mathcal{F}_\varepsilon(u) := \frac{1}{2}\int_{\Omega_\varepsilon} \mathbb{C}Eu \cdot Eu \, dx - \int_{\Omega_\varepsilon} b^\varepsilon \cdot u \, dx$$

admit for every $\varepsilon > 0$ a unique minimizer among all competing displacements $u \in H^1_{\#}(\Omega_\varepsilon; \mathbb{R}^3)$.

3. The rescaled problem

To state our results it is convenient to stretch the domain Ω_ε along the transverse directions x_1 and x_2 in a way that the transformed domain does not depend on ε. Let us therefore set $\Omega := \Omega_1$, $S(x_3) := S_1(x_3)$ and let $p_\varepsilon : \Omega \to \Omega_\varepsilon$ be defined by $p_\varepsilon(y) = p_\varepsilon(y_1, y_2, y_3) = (\varepsilon y_1, \varepsilon y_2, y_3)$. Let us consider the following 3×3 matrix

$$H^\varepsilon v := (\frac{D_1 v}{\varepsilon}, \frac{D_2 v}{\varepsilon}, D_3 v),$$

where $D_i v$ denotes the column vector of the partial derivatives of v with respect to y_i. We will use moreover the following notation

$$E^\varepsilon v := \operatorname{sym}(H^\varepsilon v), \quad W^\varepsilon v := \operatorname{skw}(H^\varepsilon v)$$

and also denote by $Wv := W^1 v$ the skew symmetric part of the gradient. Let $H^1_{\#}(\Omega; \mathbb{R}^3) := \{v \in H^1(\Omega; \mathbb{R}^3) : v = 0 \text{ on } S(0)\}$ and consider the rescaled energy $F_\varepsilon : H^1_{\#}(\Omega; \mathbb{R}^3) \to \mathbb{R} \cup \{+\infty\}$ defined by

$$F_\varepsilon(v) := \frac{1}{\varepsilon^2}\mathcal{F}_\varepsilon(v \circ p_\varepsilon^{-1}) = I_\varepsilon(v) - \int_\Omega b^\varepsilon \circ p_\varepsilon \cdot v \, dy,$$

where $I_\varepsilon(v) := \frac{1}{2}\int_\Omega \mathbb{C}E^\varepsilon v \cdot E^\varepsilon v \, dy$. We further suppose the loads to have the following form

$$b_1^\varepsilon \circ p_\varepsilon(y) = \varepsilon^2 b_1(y) - \varepsilon \frac{m(y_3)}{I_O} y_2, \quad b_2^\varepsilon \circ p_\varepsilon(y) = \varepsilon^2 b_2(y) + \varepsilon \frac{m(y_3)}{I_O} y_1, \tag{2}$$
$$b_3^\varepsilon \circ p_\varepsilon(y) = \varepsilon b_3(y),$$

with $b = (b_1, b_2, b_3) \in L^2(\Omega; \mathbb{R}^3)$, $m \in L^2(0, \ell)$ and $I_O := \int_\omega (y_1^2 + y_2^2)\, dy_1\, dy_2$ is the polar moment of inertia of the section ω. With the loads given by (2), the energy $F_\varepsilon(v)$ can be rewritten as

$$F_\varepsilon(v) = I_\varepsilon(v) - \varepsilon^2 \int_\Omega b \cdot \big(v_1, v_2, \frac{v_3}{\varepsilon}\big)\, dy - \varepsilon^2 \int_0^\ell m\, \vartheta^\varepsilon(v)\, dy_3, \qquad (3)$$

where we have set

$$\vartheta^\varepsilon(v)(y_3) := \frac{1}{I_O} \int_\omega \Big(\frac{y_1}{\varepsilon} v_2(y_1, y_2, y_3) - \frac{y_2}{\varepsilon} v_1(y_1, y_2, y_3)\Big)\, dy_1 dy_2. \qquad (4)$$

We note that if $v \in L^2(\Omega; \mathbb{R}^3)$ then $\vartheta^\varepsilon(v) \in L^2(0, \ell)$. A similar statement holds if we replace L^2 with H^1.

4. Compactness lemmata

To prove the compactness of the displacements we need the following scaled Korn inequality.

Theorem 4.1. *There exists a positive constant K such that*

$$\int_\Omega \Big(\big|(u_1, u_2, \frac{u_3}{\varepsilon})\big|^2 + |H^\varepsilon u|^2\Big) dy \le \frac{K}{\varepsilon^2} \int_\Omega |E^\varepsilon u|^2\, dy,$$

for every $u \in H^1_\#(\Omega; \mathbb{R}^3)$ and every $\varepsilon \in (0, 1]$.

Proof. The inequality $\int_\Omega |H^\varepsilon u|^2\, dy \le (K/\varepsilon^2) \int_\Omega |E^\varepsilon u|^2\, dy$ simply follows by rescaling the Korn's inequality of Anzellotti, Baldo and Percivale,[1] Theorem A.1. To show that $\int_\Omega |(u_1, u_2, u_3/\varepsilon)|^2\, dy \le (K/\varepsilon^2) \int_\Omega |E^\varepsilon u|^2\, dy$, it suffices to set $v := (u_1, u_2, u_3/\varepsilon)$, notice that $|E^\varepsilon u| \ge \varepsilon |Ev|$ and apply the standard Korn inequality to v on Ω (see, for instance, Oleinik, Shamaev and Yosifian,[9] Theorem 2.7). □

Let $H_{BN}(\Omega; \mathbb{R}^3) := \{v \in H^1_\#(\Omega; \mathbb{R}^3) : (Ev)_{i\alpha} = 0 \text{ for } i = 1, 2, 3,\ \alpha = 1, 2\}$ be the space of Bernoulli-Navier displacements on Ω. It can be characterized also as follows (see Le Dret,[8] Section 4.1)

$$\begin{aligned} H_{BN}(\Omega; \mathbb{R}^3) = \{&v \in H^1_\#(\Omega; \mathbb{R}^3) : \exists\, \xi_\alpha \in H^2_\#(0, \ell), \exists\, \xi_3 \in H^1_\#(0, \ell) \\ &\text{such that } v_\alpha(y) = \xi_\alpha(y_3),\ v_3(y) = \xi_3(y_3) - y_\alpha \xi'_\alpha(y_3)\}. \end{aligned} \qquad (5)$$

In the remaining part of this section we assume u^ε to be a sequence of functions in $H^1_\#(\Omega; \mathbb{R}^3)$ such that

$$\|E^\varepsilon u^\varepsilon\|_{L^2(\Omega; \mathbb{R}^{3\times 3})} \le C\varepsilon, \qquad (6)$$

for some constant C and for every $\varepsilon \in (0,1]$.

Theorem 4.2. *Let (6) hold. Then, for any sequence of positive numbers ε_n converging to 0 there exist a subsequence (not relabelled) and a couple of functions $v \in H_{BN}(\Omega;\mathbb{R}^3)$ and $\vartheta \in L^2(\Omega)$ such that (as $n \to +\infty$)*

$$(u_1^{\varepsilon_n}, u_2^{\varepsilon_n}, \frac{u_3^{\varepsilon_n}}{\varepsilon_n}) \rightharpoonup v \textit{ in } H^1(\Omega;\mathbb{R}^3), \tag{7}$$

$$W^{\varepsilon_n}u^{\varepsilon_n} \rightharpoonup \begin{pmatrix} 0 & -\vartheta & D_3v_1 \\ \vartheta & 0 & D_3v_2 \\ -D_3v_1 & -D_3v_2 & 0 \end{pmatrix} \textit{ in } L^2(\Omega;\mathbb{R}^{3\times 3}). \tag{8}$$

Proof. It is convenient to set $v^\varepsilon := (u_1^\varepsilon, u_2^\varepsilon, u_3^\varepsilon/\varepsilon)$. Since $|E^\varepsilon u^\varepsilon| \geq \varepsilon|Ev^\varepsilon|$, by (6), Ev^ε is uniformly bounded in $L^2(\Omega;\mathbb{R}^{3\times3})$ and by Korn's inequality v^ε is uniformly bounded in $H^1(\Omega;\mathbb{R}^3)$. It then exists a $v \in H^1_\#(\Omega;\mathbb{R}^3)$ and a subsequence of ε_n such that $v^{\varepsilon_n} \rightharpoonup v$ in $H^1(\Omega;\mathbb{R}^3)$. Again, it is easy to check that $|(E^\varepsilon u^\varepsilon)_{i\alpha}| \geq |(Ev^\varepsilon)_{i\alpha}|$, thus, using (6) we deduce that $C\varepsilon \geq \|(Ev^\varepsilon)_{i\alpha}\|_{L^2(\Omega)}$ and consequently, as $n \to \infty$, $(Ev)_{i\alpha} = 0$ for $i = 1,2,3$ and $\alpha = 1,2$. Hence $v \in H_{BN}(\Omega;\mathbb{R}^3)$.

Using (6) and Theorem 4.1 we obtain that the sequence $H^{\varepsilon_n}u^{\varepsilon_n}$ is bounded in $L^2(\Omega;\mathbb{R}^{3\times3})$ so that, up to subsequences, it weakly converges in $L^2(\Omega;\mathbb{R}^{3\times3})$ to a matrix $H \in L^2(\Omega;\mathbb{R}^{3\times3})$. Since, from (6), $E^{\varepsilon_n}u^{\varepsilon_n} \to 0$ in $L^2(\Omega;\mathbb{R}^{3\times3})$, we have $W^{\varepsilon_n}u^{\varepsilon_n} \rightharpoonup H$ in $L^2(\Omega;\mathbb{R}^{3\times3})$. In particular, H is, almost everywhere, a skew-symmetric matrix. Since $(H^\varepsilon u^\varepsilon)_{13} = u^\varepsilon_{1,3} = v^\varepsilon_{1,3}$ and $(H^\varepsilon u^\varepsilon)_{23} = u^\varepsilon_{2,3} = v^\varepsilon_{2,3}$, we deduce that $(H)_{13} = v_{1,3}$ and $(H)_{23} = v_{2,3}$. We conclude the proof by denoting $(H)_{12} := -\vartheta$. □

Let $\wp$ denote the projection of $L^2(\omega;\mathbb{R}^2)$ on the subspace

$$\mathcal{R}_2 = \{r \in L^2(\omega;\mathbb{R}^2) : \exists\, \varphi \in \mathbb{R},\ c \in \mathbb{R}^2 : r_1(y) = -y_2\varphi + c_1,\ r_2(y) = y_1\varphi + c_2\}$$

of the infinitesimal rigid displacements on ω. It is easy to see that $\mathcal{R}_2$ is a closed subspace of $H^1(\omega;\mathbb{R}^2)$ (see also Freddi, Morassi and Paroni[4]). Moreover, if $w \in L^2(\omega;\mathbb{R}^2)$ we have that

$$(\wp w)_\alpha = \mathcal{E}_{\beta\alpha}y_\beta\Big(\frac{1}{I_O}\int_\omega \mathcal{E}_{\gamma\delta}y_\gamma w_\delta \, dy_1dy_2\Big) + \frac{1}{|\omega|}\int_\omega w_\alpha \, dy_1dy_2, \tag{9}$$

where $\mathcal{E}_{\alpha\beta}$ denote the Ricci's symbol. The two-dimensional Korn's inequality then writes as

$$\|w - \wp w\|^2_{H^1(\omega;\mathbb{R}^2)} \leq C\|Ew\|^2_{L^2(\Omega;\mathbb{R}^{2\times2})} \tag{10}$$

for all $w \in H^1(\omega;\mathbb{R}^2)$.

Lemma 4.1. *Under assumption* (6) *and the notation of Theorem 4.2 and of* (4) *we have*

1. $\|\vartheta^\varepsilon(u^\varepsilon) + (W^\varepsilon u^\varepsilon)_{12}\|_{L^2(\Omega)} \leq C\varepsilon$ *for every* $\varepsilon \in (0,1]$;
2. $\vartheta^\varepsilon(u^\varepsilon) \rightharpoonup \vartheta$ *in* $L^2(\Omega)$; *therefore* ϑ *does not depend on* y_1 *and* y_2;
3. $\vartheta \in H^1_\#(\Omega)$.

Proof. It is convenient to set $w^\varepsilon := (u_1^\varepsilon/\varepsilon, u_2^\varepsilon/\varepsilon, u_3^\varepsilon/\varepsilon^2)$. Then for almost $y_3 \in (0,\ell)$ and any $\varepsilon \in (0,1]$ we consider the projection of the first two components of $w^\varepsilon(\cdot, y_3)$. From (9) and recalling (4) we have

$$(\wp w^\varepsilon)_\alpha = \mathcal{E}_{\beta\alpha} y_\beta \vartheta^\varepsilon(u^\varepsilon) + \frac{1}{|\omega|}\int_\omega w_\alpha^\varepsilon\, dy_1 dy_2.$$

Since $(Ew^\varepsilon)_{11} = (E^\varepsilon u^\varepsilon)_{11}$, $(Ew^\varepsilon)_{12} = (E^\varepsilon u^\varepsilon)_{12}$ and $(Ew^\varepsilon)_{22} = (E^\varepsilon u^\varepsilon)_{22}$, we get $\|(Ew^\varepsilon)_{\alpha\beta}\|_{L^2(\Omega;\mathbb{R}^{2\times2})} = \|(E^\varepsilon u^\varepsilon)_{\alpha\beta}\|_{L^2(\Omega;\mathbb{R}^{2\times2})}$ for $\alpha,\beta = 1,2$. Then, integrating (10) on $(0,\ell)$ and taking into account (6), we deduce that

$$\int_0^\ell \|w^\varepsilon - \wp w^\varepsilon\|_{H^1(\omega;\mathbb{R}^2)} dy_3 \leq C\|E^\varepsilon u^\varepsilon\|_{L^2(\Omega;\mathbb{R}^{3\times3})} \leq C\varepsilon$$

and then $\|D_\alpha(w_\beta^\varepsilon - \wp w_\beta^\varepsilon)\|_{L^2(\Omega;\mathbb{R})} \to 0$ for $\alpha,\beta = 1,2$. Since $(W\wp w^\varepsilon)_{12} = -\vartheta^\varepsilon(u^\varepsilon)$ and $(Ww^\varepsilon)_{12} = (W^\varepsilon u^\varepsilon)_{12}$, we obtain, from the identity

$$\vartheta^\varepsilon(u^\varepsilon) = -(W\wp w^\varepsilon)_{12} = (W(w^\varepsilon - \wp w^\varepsilon))_{12} - (W^\varepsilon u^\varepsilon)_{12},$$

the first claim of the Lemma.

Using (8), for $\varepsilon \to 0$, we obtain that $\vartheta^\varepsilon(u^\varepsilon) \rightharpoonup \vartheta$ in $L^2(\Omega)$. From the fact that $\vartheta^\varepsilon(u^\varepsilon)$ does not depend on y_1 and y_2, the same holds for ϑ.

Setting $w^\varepsilon := (u_1^\varepsilon/\varepsilon, u_2^\varepsilon/\varepsilon, u_3^\varepsilon/\varepsilon^2)$, the proof of part *3* proceeds along the same lines of that of Lemma 4.6 of Freddi, Morassi and Paroni.[4] □

Lemma 4.2. *Under the same assumption and with the notation of Theorem 4.2 the following identities hold in* $L^2(\Omega)$

$$E_{33} = D_3 v_3, \tag{11}$$
$$-D_2 E_{13} + D_1 E_{23} = D_3 \vartheta, \tag{12}$$

where, up to subsequences, E_{33}, E_{13} *and* E_{23} *are, respectively, the limits of* $(E^\varepsilon u^\varepsilon)_{33}/\varepsilon$, $(E^\varepsilon u^\varepsilon)_{13}/\varepsilon$ *and* $(E^\varepsilon u^\varepsilon)_{23}/\varepsilon$ *in the weak convergence of* $L^2(\Omega)$.

Proof. To prove (11) it suffices to notice that $(E^\varepsilon u^\varepsilon)_{33}/\varepsilon = D_3(u_3^\varepsilon/\varepsilon)$ and apply (7). Let's prove (12). From (6) we deduce that, up to subsequences, $(E^\varepsilon u^\varepsilon)_{13}/\varepsilon \rightharpoonup E_{13}$ and $(E^\varepsilon u^\varepsilon)_{23}/\varepsilon \rightharpoonup E_{23}$ in $L^2(\Omega)$. To characterize $E_{13}, E_{23} \in L^2(\Omega)$ note that

$$D_3(W^\varepsilon u^\varepsilon)_{12} = D_2\Big(\frac{(E^\varepsilon u^\varepsilon)_{13}}{\varepsilon}\Big) - D_1\Big(\frac{(E^\varepsilon u^\varepsilon)_{23}}{\varepsilon}\Big),$$

in the sense of distributions. Hence for $\psi \in C_c^\infty(\Omega)$ we obtain

$$\int_\Omega (W^\varepsilon u^\varepsilon)_{12} D_3\psi\, dy = \int_\Omega \frac{(E^\varepsilon u^\varepsilon)_{13}}{\varepsilon} D_2\psi\, dy - \int_\Omega \frac{(E^\varepsilon u^\varepsilon)_{23}}{\varepsilon} D_1\psi\, dy.$$

Passing to the limit in the previous equality we find

$$\int_\Omega -\vartheta D_3\psi\, dy = \int_\Omega E_{13} D_2\psi\, dy - \int_\Omega E_{23} D_1\psi\, dy.$$

Thus $D_3\vartheta = -D_2 E_{13} + D_1 E_{23}$ in the sense of distributions, hence in $L^2(\Omega)$ since $\vartheta \in H^1_\#(\Omega)$. □

5. The limit energy

Let us consider the usual De Saint Venant - Kirchhoff energy density

$$f(A) = \frac{1}{2}\mathbb{C}A \cdot A = \mu|A|^2 + \frac{\lambda}{2}|\mathrm{tr}A|^2$$

and define $f_0(\alpha,\beta) := \min\{f(A) : A \in \mathrm{Sym},\ A_{13}^2 + A_{23}^2 = \alpha^2,\ A_{33} = \beta\}$.

A simple computation shows that

$$f_0(\alpha,\beta) := 2\mu\alpha^2 + \frac{E}{2}\beta^2, \tag{13}$$

where $E = \mu(2\mu + 3\lambda)/(\mu + \lambda)$ is the Young modulus.

Lemma 5.1. *Let u^ε be a sequence of functions in the space $H^1_\#(\Omega;\mathbb{R}^3)$. If $\sup_\varepsilon \left(F_\varepsilon(u^\varepsilon)/\varepsilon^2\right) < +\infty$, then (6) holds for some constant $C > 0$.*

Proof. Setting $v^\varepsilon := (u_1^\varepsilon, u_2^\varepsilon, u_3^\varepsilon/\varepsilon)$, the proof proceeds exactly along the same lines of that of Lemma 5.1 of Freddi, Morassi and Paroni.[4] □

The above Lemma 5.1 and Lemma 4.2 imply that the family of functionals $(1/\varepsilon^2)F_\varepsilon$ is coercive with respect to the weak convergence of the sequence $q_\varepsilon(u^\varepsilon) := \left(u_1^\varepsilon, u_2^\varepsilon, u_3^\varepsilon/\varepsilon, (W^\varepsilon u^\varepsilon)_{12}\right)$ in the space $H^1(\Omega;\mathbb{R}^3) \times L^2(\Omega;\mathbb{R})$, uniformly with respect to ε. Hence, for any sequence u^ε which is bounded in energy, that is $(1/\varepsilon^2)F_\varepsilon \le C$ for a suitable constant $C > 0$, and satisfies the boundary conditions, that is $u^\varepsilon = 0$ on $S(0)$, the corresponding sequence $q_\varepsilon(u^\varepsilon)$ is weakly relatively compact in $H^1(\Omega;\mathbb{R}^3) \times L^2(\Omega;\mathbb{R})$.

Now we introduce some auxiliary functions defined on ω. The so-called *Prandtl stress function* is defined as the unique solution ψ of

$$\begin{cases} \Delta\psi = -2 \\ \psi \in H_0^1(\omega). \end{cases} \tag{14}$$

Since ω is simply connected, then it remains defined, up to constants, the *torsion function* φ defined by

$$\begin{cases} D_1\psi = -D_2\varphi - y_1 \\ D_2\psi = D_1\varphi - y_2. \end{cases} \tag{15}$$

It's easy to see that

$$\begin{cases} \Delta\varphi = 0 \ \text{ in } \omega \\ D\varphi \cdot n = -y_1 n_2 + y_2 n_1 \ \text{ on } \partial\omega, \end{cases} \tag{16}$$

where $n = n(\sigma)$ is the normal unit vector to $\partial\omega$ at the point σ.

Theorem 5.1. *Let ψ be the Prandtl stress function defined above and let $F : H^1_\#(\Omega;\mathbb{R}^3) \times H^1_\#(\Omega;\mathbb{R}) \to \mathbb{R} \cup \{+\infty\}$ be defined by*

$$F(v,\vartheta) := \int_\Omega f_0\Big(\frac{|D\psi|\, D_3\vartheta}{2}, D_3 v_3\Big)\, dy - \int_\Omega b \cdot v\, dy - \int_0^\ell m\vartheta\, dy_3 \tag{17}$$

if $v \in H_{BN}(\Omega;\mathbb{R}^3)$, and $+\infty$ otherwise. As $\varepsilon \to 0$, the sequence of functionals $(1/\varepsilon^2)F_\varepsilon$ defined in (3) and (4) Γ-converges to the functional F, in the following sense:

(1) (liminf inequality) for every sequence of positive numbers ε_k converging to 0 and for every sequence $\{u^k\} \subset H^1_\#(\Omega;\mathbb{R}^3)$ such that

$$(u_1^k, u_2^k, \frac{u_3^k}{\varepsilon_k}) \rightharpoonup v \text{ in } H^1(\Omega;\mathbb{R}^3), \qquad (W^{\varepsilon_k} u^k)_{12} \rightharpoonup -\vartheta \text{ in } L^2(\Omega),$$

we have

$$\liminf_{k\to+\infty} \frac{F_{\varepsilon_k}(u^k)}{\varepsilon_k^2} \geq F(v,\vartheta);$$

(2) (recovery sequence) for every sequence of positive numbers ε_k converging to 0 and for every $(v,\vartheta) \in H^1_\#(\Omega;\mathbb{R}^3) \times H^1_\#(\Omega;\mathbb{R})$ there exists a sequence $\{u^k\} \subset H^1_\#(\Omega;\mathbb{R}^3)$ such that

$$(u_1^k, u_2^k, \frac{u_3^k}{\varepsilon_k}) \rightharpoonup v \text{ in } H^1(\Omega;\mathbb{R}^3), \qquad (W^{\varepsilon_k} u^k)_{12} \rightharpoonup -\vartheta \text{ in } L^2(\Omega),$$

and

$$\limsup_{k\to+\infty} \frac{F_{\varepsilon_k}(u^k)}{\varepsilon_k^2} \leq F(v,\vartheta).$$

Proof. Let us prove the liminf inequality. Without loss of generality we may suppose that

$$\liminf_{k\to+\infty} \frac{F_{\varepsilon_k}(u^k)}{\varepsilon_k^2} = \lim_{k\to+\infty} \frac{F_{\varepsilon_k}(u^k)}{\varepsilon_k^2} < +\infty.$$

Then Lemma 5.1 applies to the sequence $(1/\varepsilon_k^2)F_{\varepsilon_k}(u^k)$. Hence assumption (6) is fulfilled and the results of Section 4, namely Lemma 4.1 and Lemma 4.2, hold true.

Looking at the expressions (3) and (4) of the functional F_ε, and setting $L_\varepsilon := F_\varepsilon - I_\varepsilon$ the work done by loads, using Lemma 4.1 and the convergence assumptions on the sequence (u^k) we can see that

$$\frac{L_{\varepsilon_k}(u^k)}{\varepsilon_k^2} = \int_\Omega b\cdot(u_1^k, u_2^k, \frac{u_3^k}{\varepsilon_k})\,dy + \int_0^\ell m\vartheta^{\varepsilon_k}(u^k)\,dy_3 \to \int_\Omega b\cdot v\,dy + \int_0^\ell m\vartheta\,dy_3.$$

Thus we have only to prove that

$$\liminf_{k\to+\infty} \frac{I_{\varepsilon_k}(u^k)}{\varepsilon_k^2} \geq \int_\Omega f_0\Big(\frac{|D\psi|\,D_3\vartheta}{2}, D_3v_3\Big)\,dy. \tag{18}$$

By definition of f and f_0 and using (13), we observe that

$$\frac{1}{2}\mathbb{C}A\cdot A \geq 2\mu(A_{13}^2 + A_{23}^2) + \frac{E}{2}A_{33}^2.$$

Then we get

$$\frac{I_{\varepsilon_k}(u^k)}{\varepsilon_k^2} \geq 2\mu \int_\Omega \Big(\frac{(E^{\varepsilon_k}u^k)_{13}^2}{\varepsilon_k^2} + \frac{(E^{\varepsilon_k}u^k)_{23}^2}{\varepsilon_k^2}\Big)\,dy + \frac{E}{2}\int_\Omega \frac{|(E^{\varepsilon_k}u^k)_{33}|^2}{\varepsilon_k^2}\,dy.$$

Using Lemma 4.1 and Lemma 4.2 then we have

$$\liminf_{k\to+\infty} \frac{I_{\varepsilon_k}(u^k)}{\varepsilon_k^2} \geq 2\mu\int_\Omega (E_{13}^2 + E_{23}^2)\,dy + \frac{E}{2}\int_\Omega |D_3v_3|^2\,dy. \tag{19}$$

From equation (12), i.e. $-D_2E_{13} + D_1E_{23} = D_3\vartheta$, which we can rewrite as

$$D_2\big(E_{13} + \frac{y_2}{2}D_3\vartheta\big) = D_1\big(E_{23} - \frac{y_1}{2}D_3\vartheta\big) \;\text{ in } \mathcal{D}'(\Omega),$$

and the weak version of Poincaré's Lemma (see Girault and Raviart,[7] Theorem 2.9) we can find a function $\widehat{\varphi} \in L^2((0,\ell); H_m^1(\omega))$ such that

$$\begin{cases} E_{13} = D_1\widehat{\varphi} - \dfrac{y_2}{2}D_3\vartheta \\ E_{23} = D_2\widehat{\varphi} + \dfrac{y_1}{2}D_3\vartheta, \end{cases}$$

where $H_m^1(\omega) := \{v \in H^1(\omega) : \fint_\omega v = 0\}$. Thus

$$\int_\Omega (E_{13}^2 + E_{23}^2)\,dy \geq \inf_{\widehat{\varphi}} \int_\Omega |D_1\widehat{\varphi} - \frac{y_2}{2}D_3\vartheta|^2 + |D_2\widehat{\varphi} + \frac{y_1}{2}D_3\vartheta|^2\,dy, \tag{20}$$

where the infimum is taken over all functions $\widehat{\varphi}$ in $L^2((0,\ell);H^1_m(\omega))$. Furthermore, we now show that the infimum is achieved and we characterize a minimizer $\widehat{\varphi}$. First, by using Green's identities and the fact that ϑ depends only on y_3, we have

$$\int_\Omega (|D_1\widehat{\varphi} - \frac{y_2}{2}D_3\vartheta|^2 + |D_2\widehat{\varphi} + \frac{y_1}{2}D_3\vartheta|^2)\,dy =$$
$$= \int_\Omega \left[|D_\alpha\widehat{\varphi}|^2 + \frac{1}{4}(y_2^2+y_1^2)|D_3\vartheta|^2 + \mathrm{div}(-y_2\widehat{\varphi}, y_1\widehat{\varphi})D_3\vartheta\right] dy$$
$$= \int_\Omega |D_\alpha\widehat{\varphi}|^2\,dy + \frac{I_O}{4}\int_0^\ell |D_3\vartheta|^2\,dy_3 + \int_0^\ell D_3\vartheta \int_{\partial\omega}(-y_2n_1+y_1n_2)\widehat{\varphi}\,ds\,dy_3$$

where D_α denotes the gradient with respect to y_1, y_2 and $n=(n_1,n_2)$ is the normal unit vector to $\partial\omega$. Let us define

$$E(\widehat{\varphi}) := \int_\Omega |D_\alpha\widehat{\varphi}|^2\,dy + \int_0^\ell D_3\vartheta \int_{\partial\omega}(-y_2n_1+y_1n_2)\widehat{\varphi}\,ds\,dy_3.$$

The existence of a minimizer of $E(\widehat{\varphi})$ in the Hilbert space $L^2((0,\ell);H^1_m(\omega))$ follows from a standard application of the direct method of the Calculus of Variations. Let now $\widehat{\varphi}$ be a minimizer. Then it follows, by taking appropriate variations, that

$$\begin{cases} \mathrm{div}D_\alpha\widehat{\varphi} = \Delta\widehat{\varphi} = 0 \quad \text{in } \omega \\ D_\alpha\widehat{\varphi}\cdot n = \dfrac{D_3\vartheta}{2}(-y_1n_2+y_2n_1) \quad \text{on } \partial\omega \end{cases}$$

for almost every $y_3 \in (0,\ell)$. We note that $D_\alpha\widehat{\varphi}$ depends linearly by $D_3\vartheta$ on $\partial\omega$. If $\varphi \in H^1_m(\omega)$ is the solution of (16) having zero mean value, then

$$\widehat{\varphi} = \frac{1}{2}\varphi D_3\vartheta. \tag{21}$$

By putting together (15), (19), (20) and (21) we obtain the liminf inequality

$$\liminf_{k\to+\infty} \frac{I_{\varepsilon_k}(u^k)}{\varepsilon_k^2} \geq \int_\Omega (\frac{E}{2}|D_3v_3|^2 + \frac{\mu}{2}|D\psi|^2|D_3\vartheta|^2)\,dy$$

that is (18).

Let us now find a recovery sequence. Let $F(v,\vartheta) < +\infty$, otherwise there is nothing to prove. Then $v \in H_{BN}(\Omega;\mathbb{R}^3)$ and $\vartheta \in H^1_\#(\Omega;\mathbb{R})$.

We first assume further that v and ϑ are smooth and equal to zero near by $y_3=0$. By (5) there exists ξ smooth and equal to zero near by $y_3=0$ such that $v_\alpha(y) = \xi_\alpha(y_3)$, and $v_3(y) = \xi_3(y_3) - y_\alpha\xi'_\alpha(y_3)$. Let $u^{0,\varepsilon}$ be the sequence defined by

$$u_1^{0,\varepsilon} = \xi_1 - \varepsilon y_2 \vartheta + \varepsilon^2 \frac{\nu}{2}\big(- y_2^2 \xi_1'' + y_1^2 \xi_1'' + 2y_1 y_2 \xi_2''\big) - \varepsilon^2 \nu y_1 \xi_3'$$
$$u_2^{0,\varepsilon} = \xi_2 + \varepsilon y_1 \vartheta + \varepsilon^2 \frac{\nu}{2}\big(- y_1^2 \xi_2'' + y_2^2 \xi_2'' + 2y_1 y_2 \xi_1''\big) - \varepsilon^2 \nu y_2 \xi_3'$$
$$u_3^{0,\varepsilon} = \varepsilon(\xi_3 - y_1 \xi_1' - y_2 \xi_2') + \varepsilon^2 \varphi D_3 \vartheta$$

where $\nu = \lambda/2(\lambda+\mu)$ is the Poisson's coefficient, and φ is the torsion function with zero mean value. We have that $u^{0,\varepsilon}$ is equal to zero in $y_3 = 0$ and it is easily checked that, as $\varepsilon \to 0$,

$$\frac{(E^\varepsilon u^{0,\varepsilon})_{11}}{\varepsilon} \to -\nu D_3 v_3, \qquad \frac{(E^\varepsilon u^{0,\varepsilon})_{22}}{\varepsilon} \to -\nu D_3 v_3$$
$$\frac{(E^\varepsilon u^{0,\varepsilon})_{33}}{\varepsilon} \to D_3 v_3, \qquad \frac{(E^\varepsilon u^{0,\varepsilon})_{12}}{\varepsilon} \to 0$$
$$\frac{(E^\varepsilon u^{0,\varepsilon})_{13}}{\varepsilon} \to \frac{1}{2}(D_1\varphi - y_2) D_3\vartheta, \quad \frac{(E^\varepsilon u^{0,\varepsilon})_{23}}{\varepsilon} \to \frac{1}{2}(D_2\varphi + y_1) D_3\vartheta,$$

and $(W^\varepsilon u^{0,\varepsilon})_{12} \to -\vartheta$, in $L^2(\Omega)$. Therefore, performing computations, we obtain that

$$\frac{I_\varepsilon(u^{0,\varepsilon})}{\varepsilon^2} \to \int_\Omega \big(\frac{E}{2}|D_3 v_3|^2 + \frac{\mu}{2}|D\psi|^2 |D_3\vartheta|^2\big)\, dy.$$

It is also easy to check that the following estimates are satisfied

$$\Big|\frac{1}{\varepsilon^2} F_\varepsilon(u^{0,\varepsilon}) - F(v,\vartheta)\Big| \le \varepsilon C(v,\vartheta),$$
$$\|(W^\varepsilon u^{0,\varepsilon})_{12} + \vartheta\|_{L^2(\Omega)} \le \varepsilon C(v,\vartheta),$$
$$\Big\| \Big(u_1^{0,\varepsilon}, u_2^{0,\varepsilon}, \frac{u_3^{0,\varepsilon}}{\varepsilon}\Big) - v \Big\|_{H^1(\Omega)} \le \varepsilon C(v,\vartheta),$$

where $C(v,\vartheta)$ depends only on v and ϑ. Hence, in this case, (u^{0,ε_k}) is a recovery sequence. In the general case, i.e. $v \in H_{BN}(\Omega;\mathbb{R}^3)$ and $\vartheta \in H^1_\#(\Omega;\mathbb{R})$, a standard diagonal argument concludes the proof. □

6. Convergence of minima and minimizers

For every $\varepsilon \in (0,1]$ let us denote by $\tilde{u}^\varepsilon$ the solution of the following minimization problem

$$\min\big\{F_\varepsilon(u) : u \in H^1(\Omega;\mathbb{R}^3),\ u = 0 \text{ on } S(0)\big\}.$$

The existence of the solution can be proved by the direct method of the Calculus of Variations and the uniqueness follows by the strict convexity of the functional F_ε.

Corollary 6.1. *The following minimization problem for the* Γ*-limit functional* F *defined in* (17)

$$\min\left\{F(v,\vartheta) : v \in H_{BN}(\Omega;\mathbb{R}^3),\ \vartheta \in H^1(0,\ell),\ v = 0 \text{ on } S(0),\ \vartheta(0) = 0\right\}$$

admits a unique solution $(\tilde{v},\tilde{\vartheta})$. *Moreover, as* $\varepsilon \to 0$,

1. $(\tilde{u}_1^\varepsilon, \tilde{u}_2^\varepsilon, \tilde{u}_3^\varepsilon/\varepsilon) \rightharpoonup \tilde{v}$ *in* $H^1(\Omega;\mathbb{R}^3)$;
2. $(W^\varepsilon \tilde{u}^\varepsilon)_{12} \rightharpoonup -\tilde{\vartheta}$ *in* $L^2(\Omega)$;
3. $(1/\varepsilon^2)F_\varepsilon(\tilde{u}^\varepsilon)$ *converges to* $F(\tilde{v},\tilde{\vartheta})$.

Proof. Property *3* and the weak convergence in *1* and *2* follow from the Γ-convergence Theorem 5.1, the uniform coercivity of the sequence $(1/\varepsilon^2)F_\varepsilon$ and the variational property of Γ-convergence (see for instance Dal Maso[2] or Freddi and Paroni,[6] Proposition 3.4). □

References

1. G. Anzellotti, S. Baldo and D. Percivale, *Dimension reduction in variational problems, asymptotic development in* Γ*-Convergence and thin structures in elasticity*, Asymptot. Anal. **9** (1994), 61–100.
2. G. Dal Maso, *An Introduction to* Γ*-Convergence*, Birkhuser (1993).
3. E. De Giorgi and T. Franzoni, *Su un tipo di convergenza variazionale*, Atti Accad. Naz. Lincei Rend. Cl. Sci. Fis. Mat. Natur. (8) **58** (1975), 842–849.
4. L. Freddi, A. Morassi and R. Paroni, *Thin-walled beams: the case of the rectangular cross-section*, J. Elasticity **76** (2005), 45–66.
5. L. Freddi, A. Morassi and R. Paroni, *Thin-walled beams: a derivation of Vlassov theory via* Γ*-convergence*, To appear on J. Elasticity.
6. L. Freddi and R. Paroni, *The energy density of martensitic thin films via dimension reduction*, Interfaces Free Bound. **6** (2004), 439–459.
7. V. Girault and P. A. Raviart, *Finite Element Methods for Navier-Stokes Equations*, Springer-Verlag (1986).
8. H. Le Dret, *Problemes Variationnels dans les Multi-domaines. Modlisation des Jonctions et Applications*, Masson (1991).
9. O. A. Oleinik, A. S. Shamaev and G. A. Yosifian, *Mathematical Problems in Elasticity and Homogenization*, North-Holland (1992).
10. D. Percivale, *Thin elastic beams: the variational approach to St. Venant's problem*, Asymptot. Anal. **20** (1999), 39–59.

KINETIC MODELS FOR NANOFLUIDICS

ALDO FREZZOTTI

Dipartimento di Matematica del Politecnico di Milano
Piazza Leonardo da Vinci 32 - 20133 Milano, Italy

One-dimensional flows of a simple liquid through nanochannels are studied by numerical solution of Enskog-Vlasov kinetic equation which provides and approximate but accurate description of a fluid whose molecules interact through Sutherland potential. The accuracy of the results is assessed by comparisons with molecular dynamics simulations. The deviation from hydrodynamic behavior is studied as a function of relevant flow parameters. Finally, it is shown that the model allows a natural extension capable of describing fluid-wall interaction by the same formalism.

Keywords: Microfluidics; Nanofluidics; Kinetic theory; Monte Carlo simulations; Molecular dynamics

1. Introduction

Nanofluidics is a relatively young and interdisciplinary science whose aim is the study of fluid flows around or within nanosized structures.[1,2] The growing number of applications requires the parallel development of theoretical tools capable of providing a more general description of fluid flows at length scales comparable with molecular sizes. As a matter of fact, several studies[3] have shown that hydrodynamic equations (HE) fail to give a correct description of fluid flows in nano-channels. The reasons for HE failure can be found in the strongly non-local structure of some fundamental fluid properties (stress tensor, heat flux vector) which appears at the nanoscale and cannot be easily approximated by local expressions.[4] In many situations, the largest deviations from hydrodynamic behavior are observed in the vicinity of solid boundaries, whereas the fluid bulk can often be accurately described by HE. Therefore, it is tempting to try extending the validity of HE by developing slip boundary conditions, following the methods of kinetic theory of dilute gases.[5] However, the task appears considerably more complex in the case of dense gases or liquids[6] and slip coefficients have been often obtained from molecular dynamics (MD) simulations.[7] Although MD techniques provide an extremely powerful tool in nanofluidics studies, it is clear that an intermediate level of fluid description is necessary to bridge the gap between pure MD numerical experiments and hydrodynamics. Kinetic theory of dense fluids[8] provides a number of theoretical methods of various sophistication and complexity which may be applied to obtain a generalized hydrodynamic approach[6] in the form of slip boundary conditions or non-local constitutive relationships.[9] The kinetic approach can also be convenient form the numerical point of view, since kinetic equations can be efficiently solved by particle schemes which are computationally less demanding that MD.[10] The formulation of kinetic equations for dense fluids in which the molecular mean free path is of the order of the molecular size, is still an open problem, the main difficulty being the correct modeling of molecular correlations necessary to obtain the closure relationships to truncate the BBGKY hierarchy.[8] The phenomenological theory proposed by Enskog[11] to describe a dense hard sphere fluid has been later improved,[12,13] however the generalization has been obtained at the expense of tractability of the resulting equation. The research work described in the present paper aims at studying a simple extension of Enskog kinetic equation which describes a fluid whose molecules interact through Sutherland's potential which combines hard sphere interaction with a soft attractive potential tail. As shown in Refs. 14,15, the adoption of simplifying assumptions on pair correlations leads to a closed kinetic equation for the one-particle distributions function. The resulting equation is often referred to as Enskog-Vlasov (EV) equation since it differs from the original Enskog equation because of an additional term which describes a self-consistent force field generated by the attractive potential tail. Kinetic equations of the EV type have been applied in several studies of equilibrium and non-equilibrium structure of non-uniform dense fluids,[16–18] however previous investigations are limited to the study of equilibrium density profiles and/or to obtaining hydrodynamic equations. In this paper, EV equation is solved numerically by a particle scheme[10,17] without introducing additional assumptions beside those intrinsic in the equation itself. The numerical method is applied to study simple flows in nanochannels of simple geometry. Furthermore, it is shown that the kinetic model can be extended to incorporate the interaction of the fluid with solid walls within the same formalism.

2. The mathematical model

Following Refs. 14,15, we consider a fluid composed by spherical and identical molecules of mass m_1, interacting by Sutherland potential

$$\phi^{(11)}(\rho) = \begin{cases} +\infty & \rho < \sigma_1 \\ -\overline{\phi}^{(11)} \left(\frac{\rho}{\sigma_1}\right)^{-\gamma^{(11)}} & \rho \geq \sigma_1 \end{cases} \tag{1}$$

which results from the superposition of a hard sphere potential and a soft potential tail depending on the distance ρ between the centers of two interacting molecules. The hard sphere diameter is σ_1, whereas $\overline{\phi}^{(11)}$ and $\gamma^{(11)}$ are two positive constants which are related to the value of the right limit of $\phi^{(11)}(\rho)$ at $\rho = \sigma_1$ and to the range of the soft interaction, respectively. It is possible to obtain the following *exact* kinetic equation for the one-particle distribution function

$$\frac{\partial f_1}{\partial t} + \boldsymbol{v} \circ \nabla_{\boldsymbol{r}} f_1 = -\nabla_{\boldsymbol{v}} \circ \left[\int_{\rho > \sigma_1} \hat{\boldsymbol{k}} \frac{d\phi^{(11)}}{d\rho} f_2(\boldsymbol{r}, \boldsymbol{v}, \boldsymbol{r}_1, \boldsymbol{v}_1 | t) \, d\boldsymbol{r}_1 \, d\boldsymbol{v}_1 \right] +$$
$$\sigma_1^2 \int \left[f_2(\boldsymbol{r}, \boldsymbol{v}^*, \boldsymbol{r} + \sigma\hat{\boldsymbol{k}}, \boldsymbol{v}_1^* | t) - f_2(\boldsymbol{r}, \boldsymbol{v}, \boldsymbol{r} - \sigma\hat{\boldsymbol{k}}, \boldsymbol{v}_1 | t) \right] (\boldsymbol{v}_r \circ \hat{\boldsymbol{k}})^+ \, d\boldsymbol{v}_1 \, d^2\hat{\boldsymbol{k}} \tag{2}$$

In Eq. (2) $f_1(\boldsymbol{r}, \boldsymbol{v}|t)$ denotes the one-particle distribution function of molecular velocity $\boldsymbol{v}$ at spatial location $\boldsymbol{r}$ at time t, whereas $f_2(\boldsymbol{r}, \boldsymbol{v}, \boldsymbol{r}_1, \boldsymbol{v}_1|t)$ is the pair distribution function. The first integral at *r.h.s* of Eq. (2) represents the soft tail contribution to the rate of change of f_1, being $\rho = \|\boldsymbol{r}_1 - \boldsymbol{r}\|$ and $\hat{\boldsymbol{k}}$ the unit vector $\frac{\boldsymbol{r}-\boldsymbol{r}_1}{\rho}$. The contribution of hard collisions is given by the second integral where $\boldsymbol{v}^*$ and $\boldsymbol{v}_1^*$ are the post-collisional velocity vectors of two colliding molecules, $\boldsymbol{v}_r$ is the relative velocity $\boldsymbol{v}_1 - \boldsymbol{v}$. The integral over $\hat{\boldsymbol{k}}$ is limited to the hemisphere where the condition $\boldsymbol{v}_r \circ \hat{\boldsymbol{k}} > 0$ holds.
Eq.(2) is exact but of little use, since it also involves the pair distribution function $f_2(\boldsymbol{r}, \boldsymbol{v}, \boldsymbol{r}_1, \boldsymbol{v}_1|t)$. A closed equation for the one-particle distribution function is obtained by the following two approximations:

(a) In the hard sphere collision integral in Eq.(2), it is assumed that
$$f_2(\boldsymbol{r}, \boldsymbol{v}, \boldsymbol{r} - \sigma_1\hat{\boldsymbol{k}}, \boldsymbol{v}_1|t) = \chi^{(11)}(\boldsymbol{r}, \boldsymbol{r} - \sigma_1\hat{\boldsymbol{k}}| \{n\}) f_1(\boldsymbol{r}, \boldsymbol{v}|t) f_1(\boldsymbol{r} - \sigma_1\hat{\boldsymbol{k}}, \boldsymbol{v}_1|t) \tag{3}$$
being $\chi^{(11)}(\boldsymbol{r}, \boldsymbol{r} - \sigma_1\hat{\boldsymbol{k}}| \{n\})$ the contact value of the pair correlation function in a hard sphere fluid.

(b) Pair correlations are completely neglected in the soft potential contribution in Eq.(2). Accordingly, it is assumed that
$$f_2(\boldsymbol{r}, \boldsymbol{v}, \boldsymbol{r}_1, \boldsymbol{v}_1|t) = f_1(\boldsymbol{r}, \boldsymbol{v}|t) f_1(\boldsymbol{r}_1, \boldsymbol{v}_1|t) \tag{4}$$

Taking into account Eqs.(3,4) and dropping the subscript, Eq.(2) takes the form

$$\frac{\partial f}{\partial t} + \boldsymbol{v} \circ \nabla_{\boldsymbol{r}} f + \frac{\boldsymbol{F}^{(11)}(\boldsymbol{r}|t)}{m} \circ \nabla_{\boldsymbol{v}} f = C^{(11)}(f, f) \tag{5}$$

$$\boldsymbol{F}^{(11)}(\boldsymbol{r}|t) = \int_{\|\boldsymbol{r}_1 - \boldsymbol{r}\| > \sigma_1} \frac{d\phi^{(11)}}{d\rho} \frac{\boldsymbol{r}_1 - \boldsymbol{r}}{\|\boldsymbol{r}_1 - \boldsymbol{r}\|} n(\boldsymbol{r}_1|t)\, d\boldsymbol{r}_1 \tag{6}$$

$$C^{(11)}(f,f) = \sigma_1^2 \int \Big\{ \chi^{(11)}(\boldsymbol{r}, \boldsymbol{r} + \sigma_1 \hat{\boldsymbol{k}}|\,\{n\}) f(\boldsymbol{r} + \sigma_1 \hat{\boldsymbol{k}}, \boldsymbol{v}_1^*|t) f(\boldsymbol{r}, \boldsymbol{v}^*|t) - \chi^{(11)}(\boldsymbol{r}, \boldsymbol{r} - \sigma_1 \hat{\boldsymbol{k}}|\,\{n\}) f(\boldsymbol{r} - \sigma \hat{\boldsymbol{k}}, \boldsymbol{v}_1|t) f(\boldsymbol{r}, \boldsymbol{v}|t) \Big\} (\boldsymbol{v}_r \circ \hat{\boldsymbol{k}})^+ d\boldsymbol{v}_1 d^2\hat{\boldsymbol{k}} \tag{7}$$

Eq.(5), also named Enskog-Vlasov kinetic equation, describes a hard sphere fluid under the action of the self-consistent force field (see Eq.(6)) generated by the soft attractive tail. In the Standard Enskog Theory (SET) $\chi^{11}(\boldsymbol{r}, \boldsymbol{r} - \sigma_1 \hat{\boldsymbol{k}}|\,\{n\})$ is approximated by using the value of the pair correlation function in a fluid in *uniform equilibrium* with density $n(\frac{\boldsymbol{r}+\boldsymbol{r}_1}{2}|t)$. An approximate, but accurate expression for $\chi_{SET}(n)$ can be obtained from the equation of state of the hard sphere fluid proposed by Carnahan and Starling,[19] as

$$\chi_{SET}(n) = \frac{1}{nb}\left(\frac{p^{hs}}{n\kappa T} - 1\right) = \frac{1}{2}\frac{2-\eta}{(1-\eta)^3}, \qquad b = \frac{2\pi\sigma_1^3}{3}, \qquad \eta = \frac{\pi\sigma_1^3 n}{6} \tag{8}$$

SET theoretical properties have been considerably improved in Revised Enskog Theory (RET)[12] where $\chi^{(11)}$ is the contact value of the pair correlation function in a fluid in *non-uniform equilibrium.* The use of RET formulation is more difficult since $\chi^{(11)}(\boldsymbol{r}, \boldsymbol{r} - \sigma_1 \hat{\boldsymbol{k}}|\,\{n\})$ is a functional of the density field $n(\boldsymbol{r}|t)$. Although an expression for $\chi(\boldsymbol{r}, \boldsymbol{r} - \sigma \hat{\boldsymbol{k}}|\,\{n\})$ can be obtained as a formal cluster expansion in the density, in practical applications simpler approximations are recommended. Following Ref.20, in the present work the pair correlation function at contact has been computed as

$$\chi^{(11)}(\boldsymbol{r}, \boldsymbol{r} - \sigma_1 \hat{\boldsymbol{k}}|\,\{n\}) = \chi_{SET}[\overline{n}(\boldsymbol{r} - \sigma \frac{\hat{\boldsymbol{k}}}{2})] \tag{9}$$

$$\overline{n}(\boldsymbol{r}|t) = \frac{3}{4\pi\sigma_1^3} \int n(\boldsymbol{r}_1|t) w(\boldsymbol{r}, \boldsymbol{r}_1)\, d\boldsymbol{r}_1 \tag{10}$$

$$w(\boldsymbol{r}, \boldsymbol{r}_1) = \begin{cases} 1 & \|\boldsymbol{r}_1 - \boldsymbol{r}\| < \sigma_1 \\ 0 & \|\boldsymbol{r}_1 - \boldsymbol{r}\| > \sigma_1 \end{cases} \tag{11}$$

According to the three expressions above, a functional form for $\chi^{(11)}$ is obtained from the simpler χ_{SET} by replacing the actual value of the density at the contact point of two colliding spheres with the value of the density field *averaged* over a spherical volume of radius σ_1. Similar approximations are used in density functional theories of non-uniform fluids and considerably improve the results of SET.

2.1. *Fluid-wall interaction and boundary conditions*

When dealing with problems in which the fluid interacts with solid walls, it is necessary to formulate appropriate boundary conditions for Eq.(5). The simplest approach to fluid-wall interaction modeling is obtained by assuming that the wall is replaced by a smooth and impenetrable surface which act on $f(\boldsymbol{r}, \boldsymbol{v}|t)$ according to the following expression:[5]

$$(\boldsymbol{v} \circ \boldsymbol{n}) f(\boldsymbol{r}, \boldsymbol{v}|t) = \int_{\boldsymbol{v}_1 \circ \boldsymbol{n} < 0} K_w(\boldsymbol{v}_1 \to \boldsymbol{v}) |\boldsymbol{v}_1 \circ \boldsymbol{n}| f(\boldsymbol{r}, \boldsymbol{v}_1|t)\, d\boldsymbol{v}_1 \tag{12}$$

In Eq.(12), $\boldsymbol{n}$ is a unit vector normal to the wall surface and directed towards the fluid region whereas $K_w(\boldsymbol{v}_1 \to \boldsymbol{v})$ is a scattering kernel which gives the probability that a molecule impinging on the wall with velocity between $\boldsymbol{v}_1$ and $\boldsymbol{v}_1 + d\boldsymbol{v}_1$ is instantaneously re-emitted at the same location with velocity between $\boldsymbol{v}$ and $\boldsymbol{v}+d\boldsymbol{v}$. The rigorous determination of $K_w(\boldsymbol{v}_1 \to \boldsymbol{v})$ requires solving the complex dynamics of a fluid molecule in interaction with wall molecules, therefore the scattering kernel expression is derived from phenomenological models.[5] Maxwell's model is the most widely used choice for $K_w(\boldsymbol{v}_1 \to \boldsymbol{v})$ which takes the form:

$$K_w(\boldsymbol{v}_1 \to \boldsymbol{v}) = \alpha(\boldsymbol{v} \circ \boldsymbol{n}) \frac{1}{2\pi(RT_w)^2} \exp\left(-\frac{v^2}{2RT_w}\right) + \\ (1-\alpha)\delta(\boldsymbol{v} - \boldsymbol{v}_1 + 2(\boldsymbol{v}_1 \circ \boldsymbol{n})\boldsymbol{n}), \qquad 0 \le \alpha \le 1 \tag{13}$$

In Eq.(13) T_w is the wall temperature, whereas α is a coefficient which sets the probabilities of diffuse and specular reemission. Although the adoption of boundary conditions in the form given in Eqs. (12,13) is possible in modeling dense fluid flows, it is worth stressing that their derivation is based on the assumption that the time scale of fluid-wall interaction is much shorter than the time scale of fluid-fluid interaction. Such an assumption is certainly justified in dilute gas dynamics, but it represents an oversimplification in studying dense fluids flows which require a more detailed modeling of fluid-wall interaction.[7] As a compromise between the completely phenomenological approach outlined above and a more realistic (but computationally expensive) approach based on MD simulations,[7] a kinetic model of fluid wall interaction can be formulated. For simplicity, we consider the one-dimensional flow of a liquid in a channel bounded by two infinite planar parallel walls. The walls separation is denoted by $2L_z$ and the motion of the fluid is observed in a Cartesian reference frame whose x and y axis are parallel to the walls, whereas the coordinate z spans the gap between the walls. The origin O of the reference frame is located at distance L_z from the walls.

It is assumed that the walls are composed by spherical molecules having a diameter σ_2, mass m_2 and number density $n_2(z)$ given by the following expression:

$$n_2(z) = \begin{cases} n_w & |z| > L_z \\ 0 & |z| \leq L_z \end{cases} \tag{14}$$

n_w being the constant value of the wall density. Although no cxplicit assumption is made about the interaction among wall molecules, it is assumed that each wall is in a state of equilibrium described by the velocity distribution functions

$$f_w(z, \boldsymbol{v}) = \begin{cases} \frac{n_w}{(2\pi R_2 T_L)} \exp\left[-\frac{(\boldsymbol{v}-\boldsymbol{u}_L)^2}{2R_2 T_L}\right] & z \leq -L_z \\ \frac{n_w}{(2\pi R_2 T_R)} \exp\left[-\frac{(\boldsymbol{v}+\boldsymbol{u}_R)^2}{2R_2 T_R}\right] & z \geq L_z \end{cases} \tag{15}$$

In the above expressions, T_L and T_R denote the temperature of the left and right wall, respectively. The walls are allowed to a have velocities $\boldsymbol{u}_L$ and $\boldsymbol{u}_R$, parallel to the wall themselves. Accordingly, it is assumed that $\boldsymbol{u}_L = -\boldsymbol{u}_R = U_w \hat{\boldsymbol{x}}$, being $\hat{\boldsymbol{x}}$ a unit vector parallel to the x axis. The gas constant R_2 is defined as $\frac{k_B}{m_2}$, where k_B is the Boltzmann constant.

In complete analogy with the treatment of fluid molecules interaction, it is assumed that wall molecules interact with fluid molecules through the following Sutherland potential

$$\phi_{12}(\rho) = \begin{cases} +\infty & \rho < \sigma_{12} \\ -\overline{\phi}^{(12)} \left(\frac{\rho}{\sigma_{12}}\right)^{-\gamma^{(12)}} & \rho \geq \sigma_{12} \end{cases} \tag{16}$$

where the hard sphere radius is now defined as $\sigma_{12} = \frac{\sigma_1+\sigma_2}{2}$. The microscopic description of the fluid motion can be strongly simplified if one assumes that the superposition of long range tails of the interaction potential $\phi^{(12)}(\rho)$ only produces an average steady force field. Fluctuations due to the random motion of wall and fluid molecules are taken into account only in the short range hard sphere potential, as described below. It can be easily shown that the force field $F_z^{(12)}(z)$ generated by the soft tails of wall molecules is given by the following expression:

$$F_z^{(12)}(z) = 2\pi\overline{\phi}^{(12)} \left[\sigma_{12}^{\gamma^{(12)}} \int_{|z-z'|>\sigma_{12}} \frac{(z'-z)n_2(z')}{|z-z'|^{\gamma^{(12)}}}\, dz' + \int_{|z-z'|\leq\sigma_{12}} (z'-z)n_2(z')\, dz' \right] \tag{17}$$

The short range interaction of fluid and wall molecules can be described by a term having the same structure of the collision integral $C^{(11)}(f,f)$:

$$C^{(12)}(f_w,f) = \sigma_{12}^2 \int \Big\{ \chi^{(12)}(z, z+\sigma_{12}k_z|\{n_2\}) f_w(z+\sigma_{12}k_z, \boldsymbol{v}_1^*) f(z,\boldsymbol{v}^*|t) - \chi^{(12)}(z, z-\sigma_{12}k_z|\{n_2\}) f_w(z-\sigma_{12}k_z, \boldsymbol{v}_1|t) f(z,\boldsymbol{v}|t) \Big\} (\boldsymbol{v}_r \circ \hat{\boldsymbol{k}})^+ d\boldsymbol{v}_1 d^2\hat{\boldsymbol{k}} \tag{18}$$

The collision term in Eq.(18) is a linear functional of f since f_w is given and it is not modified by collisions. The pair correlation function $\chi^{(12)}$ can be approximated by the expression given in Eq. (8) in which n_2 is set equal to the wall number density n_w.

The final form of the kinetic equation for f is obtained by adding the field $F_z^{(12)}(z)$ and the fluid-wall collision term to Eq. (5)

$$\frac{\partial f}{\partial t} + \xi_z \frac{\partial f}{\partial z} + \frac{F_z(z|t)}{m_1}\frac{\partial f}{\partial \xi_z} = C^{(11)}(f,f) + C^{(12)}(f_w,f) \tag{19}$$

$$F_z(z|t) = F_z^{(12)}(z) + F_z^{(11)}(z|t) \tag{20}$$

$$F_z^{(11)}(z|t) = 2\pi\overline{\phi}^{(11)} \left[\sigma_1^{\gamma^{(11)}} \int_{|z-z'|>\sigma_1} \frac{(z'-z)n_1(z'|t)}{|z-z'|^{\gamma^{(11)}}}\, dz' + \int_{|z-z'|\leq\sigma_1} (z'-z)n_1(z'|t)\, dz' \right] \tag{21}$$

3. Numerical results

Kinetic equations in the form described above can be efficiently solved numerically by the particles schemes described in Refs. 10,17. The numerical method has been applied to study a simple Couette flow as a model problem and obtain the dependence of its properties from the relevant flow parameters. In a first series of computations, the soft tail of the interaction potential $\phi^{(11)}$ has been suppressed and Maxwell model has been adopted to model fluid-wall interaction. A number of solutions have been obtained varying the wall separation $2L_z$ and the averaged fluid density $n_0 = \frac{1}{2L_z}\int_{-L_z}^{L_z} n(z)\, dz$.

The nominal shear rate $\beta = \frac{U_w}{L_z}$ has been fixed to a small value to obtain an almost isothermal flow. A parallel series of MD simulations have also been performed to obtain the "exact" behavior of a dense hard sphere fluid and assess the accuracy of kinetic theory predictions in the simplest situation. The same boundary conditions given in Eqs. (12,13) have been used in MD simulations in which the motion of 10^4 hard spheres has been computed from the exact collision dynamics. Figure 1 shows normalized density profiles in a hard sphere liquid for two different values of the average reduced density $\eta_0 = \frac{\pi\sigma_1^3}{6}n_0$. The channel width is $11\sigma_1$, but the region accessible to the centers of the spherical molecules is only $10\sigma_1$ wide along z. It

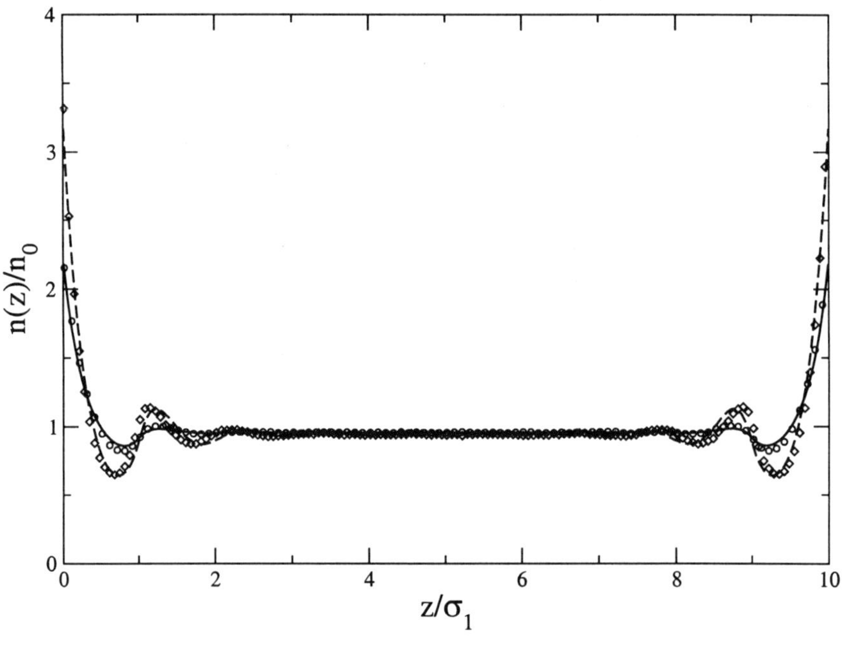

Fig. 1. Density profiles in the Couette flow of a dense hard sphere gas between rigid walls with full accommodation [$\alpha = 1$, in Eq. (13)]. Solid line: EV density profile $\eta_0 = 0.2$, $\beta = Uw/L_z = 0.0597$; dashed line: EV density profile $\eta_0 = 0.3$, $\beta = Uw/L_z = 0.1262$; circles: MD density profile $\eta_0 = 0.2$, $\beta = Uw/L_z = 0.0597$; diamonds: MD density profile $\eta_0 = 0.3$, $\beta = Uw/L_z = 0.1262$.

is interesting to observe that density is not constant since collisions push molecules toward the walls. Density oscillations indicate partial ordering of molecular layers which becomes more evident for higher values of n_0. The comparison of EV results with companion MD simulations shows that in the density range considered here, the agreement is rather good. Velocity profiles are shown in Figure (2). In spite of the strong density variations in the vicinity of the walls, velocity profiles exhibit a more regular shape and show little deviation from an overall linear behavior which would be found in a hydrodynamic treatment of the problem. However, the average slope of $u_x(z)$ profiles is different from the nominal slope β since the fluid velocity at locations $\pm(L_z - \sigma_1/2)$ is not equal the velocity of the walls. The deviation from hydrodynamic behavior of Couette flow in nanosized

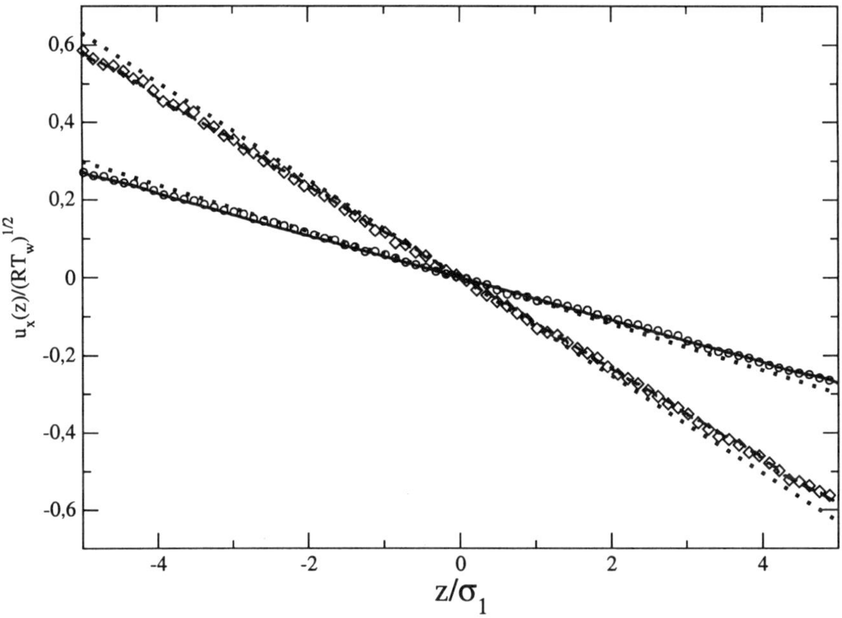

Fig. 2. Velocity profiles in the Couette flow of a dense hard sphere gas between rigid walls with full accommodation [$\alpha = 1$, in Eq. (13)]. Solid line: EV density profile $\eta_0 = 0.2$, $\beta = Uw/L_z = 0.0597$; dashed line: EV density profile $\eta_0 = 0.3$, $\beta = Uw/L_z = 0.1262$; circles: MD density profile $\eta_0 = 0.2$, $\beta = Uw/L_z = 0.0597$; diamonds: MD density profile $\eta_0 = 0.3$, $\beta = Uw/L_z = 0.1262$.

channels is best appreciated by considering the behavior of the P_{xz} component of the stress tensor which, both in the hydrodynamic and kinetic treatment, is constant across the channel and depends only on external flow parameters. Figure (3) shows the numerical value of P_{xz} obtained by solving EV equation for different Couette flows in which the nominal shear rate β and average density n_0 were kept fixed while varying the channel width $2L_z$. The results obtained from kinetic theory have been compared with the values of P_{xz} obtained from the viscosity of a dense hard sphere gas $\mu(n_0, T_w)^8$ using both the nominal shear rate β and an effective value of the shear rate, $\overline{\beta}$. The latter has been obtained by fitting the EV velocity profile with a straight line, hence it takes into account velocity slip at the walls. The results clearly show that hydrodynamic predictions based on

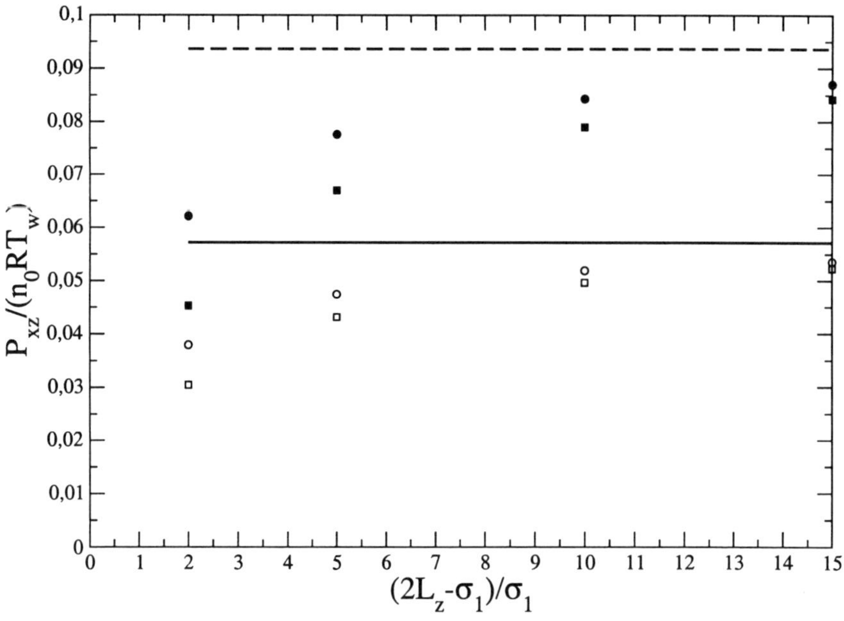

Fig. 3. P_{xz} vs.channel width in the Couette flow of a dense hard sphere gas between rigid walls with full accommodation [$\alpha = 1$, in Eq. (13)]. Squares: EV prediction, $\eta_0 = 0.2$, $\beta =$; circles: hydrodynamics predictions with effective shear rate $\overline{\beta}$, $\eta_0 = 0.2$, $\beta =$; solid line: hydrodynamics predictions with nominal shear rate β, $\eta_0 = 0.2$, $\beta =$; filled squares: EV prediction, $\eta_0 = 0.25$, $\beta =$; filled circles: hydrodynamics predictions with effective shear rate $\overline{\beta}$, $\eta_0 = 0.25$, $\beta =$; dashed line: hydrodynamics predictions with nominal shear rate β, $\eta_0 = 0.25$, $\beta =$.

nominal shear rate are not accurate, on the other hand including slip effects through the effective shear rate considerably reduces the distance between kinetic theory and hydrodynamic values of P_{xz}.

A more realistic description of fluid-wall interaction can be obtained by replacing the rigid wall, which scatters molecules according to Eq. (13),

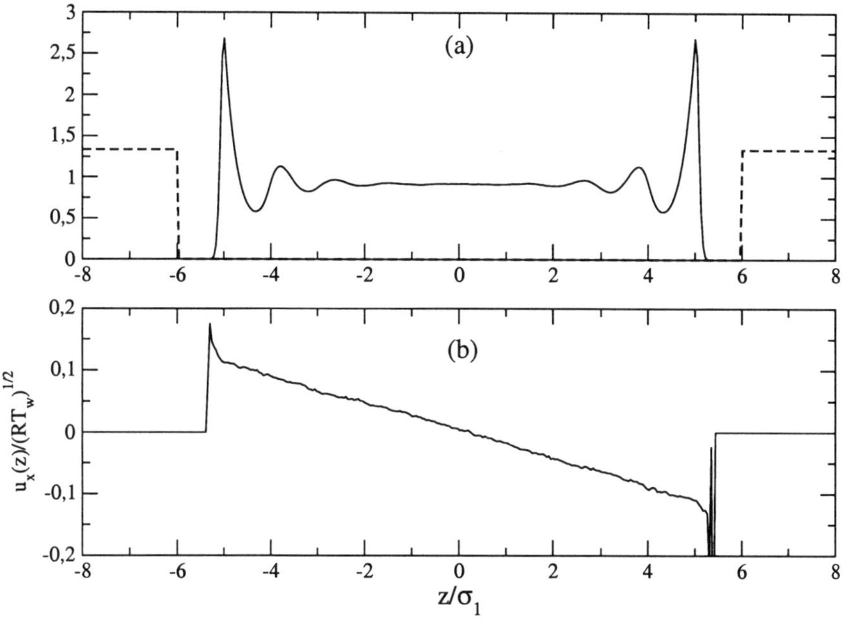

Fig. 4. Couette flow of a dense hard sphere fluid with fluid-wall interaction described by Eqs.(19,18). $\overline{\phi}^{(11)} = \overline{\phi}^{(12)} = 0$, $\sigma_1 = \sigma_2$, $m_2/m_1 = 4.875$, $U_w = 0.2\sqrt{RT_w}$, $\eta_0 = 0.3$. Graph (a)- Solid line: liquid density $n(z)/n_0$; dotted line: $n_2(z)\sigma_2^3$. Graph (b) - normalized velocity profile $u_x(z)/\sqrt{RT_w}$

with the kinetic model which includes the collision term $C^{(12)}(f_w, f)$. Although a complete account of the model properties cannot be given here, it is worth mentioning that numerical experiments show that the scattering patterns of molecules from a "solid" surface described by Eq. (18) compare well with experimental data[5] which exhibit a rather marked deviation from Maxwell's model. Moreover, the model predicts different accommodation coefficients for energy and momentum as well as the correct behavior of the energy accommodation coefficient when the mass ratio m_2/m_1 is changed. Figure 4 shows the results of a Couette flow simulation in a hard sphere liquid ($\phi^{(11)} = 0$) having an average reduced density η_0 equal to 0.3 and flowing between two walls described by Eq. (18). The wall separation is $12\sigma 1$, the wall velocities are $\pm 0.2\sqrt{RT_w}$ and the mass ratio m_2/m_1 is set equal to 4.875. The finite extent of fluid-wall interaction causes the fluid to "feel" walls when molecule centers are at distance σ_{12} from the nominal wall "surface". However, the walls are no longer impenetrable and the liquid confinement observed in Figure 4(a) is a result of the collisions with wall molecules. An adsorbed layer of liquid molecules can be clearly observed in the regions where the density falls rapidly to zero in the vicinity of the walls and the velocity profile suffers an abrupt deviation from the almost linear behavior shown in liquid bulk. The velocity slip effect is stronger than in the simulations described above, indicating that the effective tangential momentum accommodation coefficient is not unity

References

1. G. E. Karniadakis, N. Aluru and A. Beskok, *Microflows and nanoflows : fundamentals and simulation* (Springer, New York, 2005).
2. J. G. T. Eijkel and A. van den Berg, *Microfluid Nanofluid* **1**, 249 (2005).
3. K. P. Travis, B. D. Todd and D. J. Evans, *Phys. Rev. E* **55**, 4288 (1996).
4. J. Zhang, B. D. Todd and K. P. Travis, *J. Chem. Phys.* **121**, 10778 (2004).
5. C. Cercignani, *The Boltzmann Equation and Its Applications* (Springer-Verlag, Berlin, 1988).
6. L. A. Pozhar and K. E. Gubbins, *J. Chem. Phys.* **99**, 8970 (1993).
7. M. Cieplak, J. Koplik and J. R. Banavar, *Physical Review Letters* **86**, 803 (2001).
8. P. Resibois and M. DeLeener, *Classical kinetic theory of fluids* (J. Wiley & Sons, New York, 1977).
9. D. J. Evans and G. P. Morriss, *Statistical Mechanics of Nonequilibrium Liquids* (Academic Press, London, 1990).
10. A. Frezzotti, *Phys. Fluids* **9**, 1329 (1997).
11. D. Enskog, *Kungl. Svenska Vet.-Ak. Handl.* **63**, 3 (1921).
12. H. van Beijeren and M. H. Ernst, *Physica* **68**, 437 (1973).
13. L. A. Pozhar and K. E. Gubbins, *J.Chem. Phys.* **94**, 1367 (1991).
14. M. Grmela, *J. Stat. Phys.* **3**, 347 (1971).
15. J. Karkheck and G. Stell, *J. Chem. Phys.* **75**, 1475 (1981).
16. H. T. Davis, *J. Chem. Phys.* **86**, 1474 (1986).
17. A. Frezzotti, L. Gibelli and S. Lorenzani, *Phys. Fluids* **17**, 012102 (2005).
18. Z. Guo, T. S. Zhao and Y. Shi, *Phys. Fluids* **18**, 067107 (2006).
19. N. Carnahan and K. Starling, *J. Chem. Phys* **51**, 635 (1969).
20. J. Fischer and M. Methfessel, *Phys. Rev. A* **22**, 2836 (1980).

THERMODYNAMICS OF PIEZOELECTRIC MEDIA WITH DISLOCATIONS

D. GERMANÓ
University of Messina, Department of Mathematics
Salita Sperone 31, 98166 Messina, Italy
e-mail: dgermano@dipmat.unime.it

L. RESTUCCIA
University of Messina, Department of Mathematics
Salita Sperone 31, 98166 Messina, Italy
e-mail: lrest@dipmat.unime.it

In this paper we apply a geometrical theory of thermodynamics with vectorial and tensorial internal variables to a non-conventional model of piezoelectric crystals with dislocations, where a dislocation core tensor *à* la Maruszewski, its gradient and its flux describe the internal structure of these media. Using the concepts of process and transformation we derive the relevant structure of the entropy 1-form with the explicit existence conditions of the entropy function.

1. Introduction

In [1] and [2] in the framework of the extended irreversible thermodynamics a non conventional thermodynamical model for piezoelectric crystals defective by dislocation lines was given introducing as internal variables in the state space a second order dislocation core tensor *à* la Maruszewski [3], its gradient and its flux. The complete set of the laws of state and of the constitutive relations and the entropy production were established for isotropic and anisotropic media. The models for piezoelectrics with dislocations may have relevance in many fundamentals technological sectors: in systems using S.A.W.(Surface Acoustic Waves) for nondestructive testing; in Radar Technology; in the use of thin films to produce very high frequency vibrations; in ultrasonics. The structure of the dislocation lines resembles a network of infinitesimally capillary tubes in a elastic solid. Then, these defects acquired during a process of fabrication can self propagate because of changed conditions and favorable surrounding conditions, provoking a premature fracture. In this paper, taking into account the results obtained in [1] and [2], we construct a geometric model for the thermodynamics of piezoelectric crystals with dislocations, following [4], [5] and [6] - [10], and using the concepts of *process* and *transformation* we derive the relevant structure of the entropy 1-form with the explicit existence conditions of the entropy function. In [11], [12] and [13] geometrical models for perfect piezoelectric media without defects, for polarizable media with internal variables and for deformable dielectrics with a non-Euclidean structure were derived in the same geometrized framework.

2. The dislocation core tensor model

Now, we recall the model developed by Maruszewski in [3] in order to study in a simpler way piezoelectric media, in which dislocation lines, having a structure of capillary channels, disturb the periodicity of the crystal lattice [3], [14], [15] (see Fig.1). The interatomic distances are not conserved in

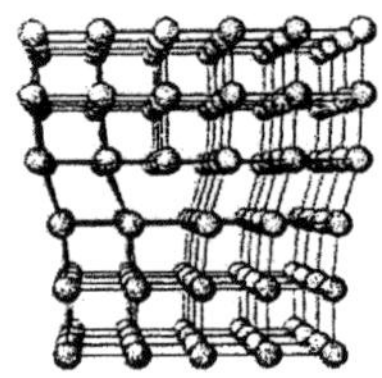

Figure 1. An edge dislocation structure (after[14]).

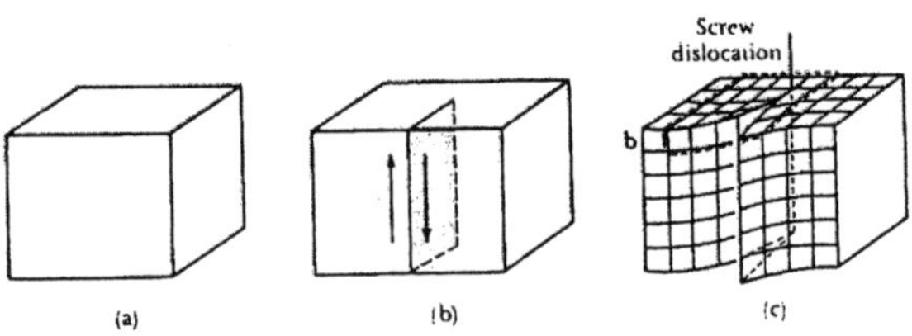

Figure 2. The perfect crystal (a) is cut and shared one atom spacing (b) and (c). The line along which shearing occurs is a screw dislocation. A *Burgers vector* **b** is required to close a loop of equal atom spacings around the screw dislocation (after[16]).

the direct neighborhood of the dislocation line in comparison to distances in the remaining part of the lattice. The diameter of the core is comparable with the lattice parameter and its shape depends on the kind of dislocation (see for example Fig. 2, [16]) and on the way in which the core is created. Their existence should not be omitted in the analysis of kinetic processes as diffusion of mass or charges, transport of heat, recombination of charge carriers, etc. There are many possibilities to describe the dislocation lines

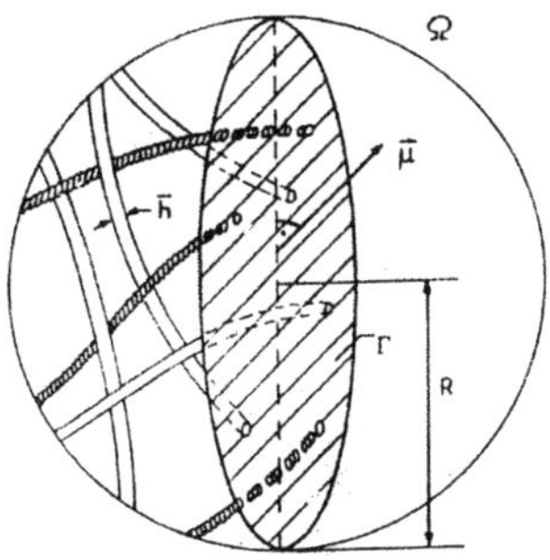

Figure 3. Characteristics of the pore-core structure ($\bar{h} \ll R$) (after [3])

distribution. In this paper we use a *dislocation core tensor* created in [3] by Maruszewski, based on the fact that the dislocation core lines resemble a network of infinitesimally thin tubes and have the same geometrical structure of porous channels. Thus, the definition and the introduction of the dislocation core tensor is based on Kubik's ideas concerning a very interesting geometrical model of a porous body [17] filled by fluid (see Fig. 3). In such a medium Kubik introduces a so-called structural permeability tensor, responsible for the structure of a system of thin capillary channels, in the following way

$$\bar{v}(\mathbf{x})_i = r_{ij}(\mathbf{x}, \boldsymbol{\mu}) \overset{*}{v}_j (\mathbf{x}, \boldsymbol{\mu}), \qquad where \tag{1}$$

$$\bar{\mathbf{v}}(\mathbf{x}) = \frac{1}{\Omega} \int_{\Omega^p} \mathbf{v}(\boldsymbol{\xi}) d\Omega, \;\; \boldsymbol{\xi} \in \Omega^p, \quad \overset{*}{\mathbf{v}}(\mathbf{x}, \boldsymbol{\mu}) = \frac{1}{\Gamma^p} \int_{\Gamma} \mathbf{v}(\boldsymbol{\xi}) d\Gamma, \;\; \boldsymbol{\xi} \in \Gamma^p \tag{2}$$

are the bulk-volume average of the fluid velocity and the corresponding pore-area average of the fluid velocity. In $(2)_1$ Ω is a representative elementary sphere volume of a porous skeleton filled with fluid, large enough to provide a representation of all the statistical properties of the pore space Ω^p, being $\Omega = \Omega^s + \Omega^p$, with Ω^s the solid space. In $(2)_2$ Γ is the section of central sphere with normal vector $\boldsymbol{\mu}$ and Γ^p is the pore area of Γ. Equation

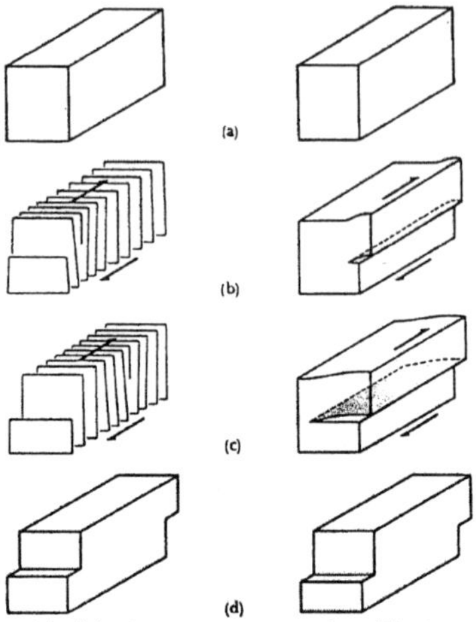

Figure 4. A shear force acting on a dislocation introduced into a perfect crystal (a) causes the dislocation to move through the crystal until a step is created (d). The crystal is now deformed (after [16])

(1) gives a linear mapping between the bulk volume average fluid velocity and the local velocity of fluid particles passing through the pore area Γ^p. Now, following Maruszewski, basing on previous definitions, for any flux q_i of some physical field transported trough a cobweb of lines we postulate that

$$\bar{q}(\mathbf{x})_i = r_{ij}(\mathbf{x}, \boldsymbol{\mu}) \overset{*}{q}_j (\mathbf{x}, \boldsymbol{\mu}), \quad \textit{where} \quad r_{ij}(\mathbf{x}, \boldsymbol{\mu}) = \Gamma a_{ij}(\mathbf{x}, \boldsymbol{\mu}). \qquad (3)$$

In equation (3) the tensor r_{ij} expresses a structure of dislocation cores and a_{ij} is a new tensor called *dislocation core tensor* that refers r_{ij} to the surface Γ . It expresses the core structure and deals with the anisotropy of the crystal (dependence on $\mathbf{x}$ and $\boldsymbol{\mu}$). Its unit is m^{-2}. Moreover, the components of a_{ij} form a kind of continuous representation of the number of dislocations which cross the surface Γ. Investigations show that during many physical processes occurring in a defective crystal the temporal evolution of the dislocation field is of great importance (a_{ij} is also dependent on time). In Fig. 4 there are two examples of the motion of dislocation lines.

3. Fundamental laws

We use the standard Cartesian tensor notation in rectangular coordinate system. We refer the motion of our material system to a current configuration $\mathcal{K}_t$. Now, we recall the model presented in [1] and [2] in the framework of the extended rational thermodynamics with internal variables, where a defective piezoelectric crystal is considered in which the following fields interact with each other: the thermal field described by the temperature ϑ, its gradient $\theta_{,i}$ and the heat flux q_i; the electromagnetic field described by the electromotive intensity $\mathcal{E}_i$ (the electric field referred to an element of matter at time t, i.e. the so called comoving frame $\mathcal{K}_c$) and the magnetic induction B_i per unit volume; the dislocation field described by the dislocation core tensor a_{ij}, its gradient $a_{ij,k}$ and the dislocation flux $\mathcal{V}_{ijk}$; the elastic field described by the total stress tensor T_{ij} and the strain tensor ε_{ij} defined as $\varepsilon_{ij} = \frac{1}{2}(u_{i,j} + u_{j,i})$, where u_i are the components of the displacement vector. Thus, the independent variables are represented by the set

$$C = \{\varepsilon_{ij}, \mathcal{E}_i, B_i, \theta, a_{ij}, \mathcal{V}_{ijk}, q_i, \theta_{,i}, a_{ij,k}\}\,. \tag{4}$$

This specific choice shows that the relaxation properties of the thermal field, the dislocation field are taken into account. However, we ignore the corresponding effect for the mechanical properties so that T_{ij} is not in the set of equation (4) (the viscoelastic properties of the material are excluded). All the processes occurring in the considered body are governed by three groups of laws. The first one deals with *the electromagnetic field* governed by Maxwell's equations in the following form [18]

$$\in_{ijk} E_{k,j} + \frac{\partial B_i}{\partial t} = 0, \quad D_{i,i} = 0, \quad \in_{ijk} H_{k,j} - \frac{\partial D_i}{\partial t} = 0, \quad B_{i,i} = 0, \tag{5}$$

where there are neither currents nor electric charge, the magnetic field, the electric field and the magnetization per unit volume are given by

$$\mathbf{H} = \frac{1}{\mu_0}\mathbf{B}, \qquad \mathbf{E} = \frac{1}{\varepsilon_0}(\mathbf{D} - \mathbf{P}), \qquad \mathbf{M} = 0, \tag{6}$$

ε_0, μ_0 denote the permittivity and permeability of vacuum and $\mathbf{D}$ and $\mathbf{P}$ are the electric displacement and the electric polarization per unit volume, respectively. The second group deals with the following classical balance equations:

the continuity equation

$$\rho + \rho v_{i,i} = 0, \tag{7}$$

where ρ denotes the mass density, v_i is the velocity of the body point and a over imposed dot denotes the material derivative;

the momentum balance

$$\rho\dot{v}_i - T_{ji,j} - \in_{ijk} \overset{\Delta}{P}_j B_k - P_j \mathcal{E}_{i,j} - f_i = 0, \qquad where \tag{8}$$

$$\overset{\Delta}{P}_j = \dot{P}_i + P_i v_{k,k} - P_k v_{i,k}, \qquad \mathcal{E}_i = E_i + \in_{ijk} v_j B_k, \tag{9}$$

and f_i is a body force;

the moment of momentum balance

$$\in_{ijk} T_{jk} + c_i = 0, \tag{10}$$

where c_i is a couple per unit volume. In [1] it was demonstrated that c_i vanishes and then T_{ij} is a symmetric tensor;

the internal energy balance

$$\rho\dot{U} - T_{ji}v_{i,j} - \rho\mathcal{E}_i\dot{\mathcal{P}}_i + q_{i,i} - \rho r = 0, \tag{11}$$

where U is the internal energy density, $P_i = \rho\mathcal{P}_i$, r is the heat source distribution and $v_{i,j} = L_{ij}$, where L_{ij} is the rate deformation given by $\mathbf{L} = \dot{\mathbf{F}}\mathbf{F}^{-1}$, with $\mathbf{F}$ the rate of deformation (which is assumed to be invertible with inverse $\mathbf{F}^{-1}$). Mass forces and heat source distribution will be neglected in the following.

The third group of laws deals with *the rate properties of the dislocation core tensor, the dislocation flux and the heat flux*, respectively

$$\overset{*}{a}_{ij} + \mathcal{V}_{ijk,k} - A_{ij}(C) = 0, \quad \overset{*}{\mathcal{V}}_{ijk} - V_{ijk}(C) = 0, \quad \overset{*}{q}_i - Q_i(C) = 0, \tag{12}$$

where $\overset{*}{a}_{ij} = \dot{a}_{ij} - \Omega_{ik}a_{kj} - \Omega_{jk}a_{ik}$; $\overset{*}{\mathcal{V}}_{ijk} = \dot{\mathcal{V}}_{ijk} - \Omega_{il}\mathcal{V}_{ljk} - \Omega_{jl}\mathcal{V}_{ilk} - \Omega_{kl}\mathcal{V}_{ijl}$, $\overset{*}{q}_i = \dot{q}_i - \Omega_{ij}q_j$, and A_{ij} is the source-like term which may, among others, deals with the annihilation of dislocations of apposite signs, V_{ij} is the source term for the dislocation flux and the superimposed asterisk denotes the Zaremba-Jaumann derivative (see [19]-[21] for the form of these equations). All processes considered here should be admissible from the thermodynamical point of view. It means that they should not contradict the second law

$$\rho\dot{S} + J_{S\,k,k} - \frac{\rho r}{\theta} \geq 0,$$

where S denotes the entropy per unit mass and $\mathbf{J}_S$ is the entropy flux associated with the fields of the set C given by

$$\mathbf{J}_S = \frac{1}{\theta}\mathbf{q} + \mathbf{k}, \tag{13}$$

with $\mathbf{k}$ an additional term called *extra entropy flux density*.
In [1] and [2] all the following constitutive functions $\mathbf{Z} = \tilde{\mathbf{Z}}(C)$,

$$\mathbf{Z} = \{T_{ij}, P_i, c_i, U, A_{ij}, V_{ijk}, Q_i, S, \Phi_i, \pi_{ij}\} \tag{14}$$

(with π_{ij} an opportune affinity), were obtained by analyzing the entropy inequality by Liu's theorem (see [22]) and using Smith theorem (see [23]) with the help of the isotropic polynomial representations of the proper constitutive functions, satisfying the objectivity principle ([20], [21]).

4. A geometric model for the thermodynamics of piezoelectrics defective by dislocations

Now, taking into account the results obtained in [1] and [2] we construct a geometric model for the thermodynamics of piezoelectrics defective by dislocations. Now, introducing the concepts of *process* and *transformation* we derive the expressions for the existence of an entropy function and the entropy 1-form. We consider a material element and we define the state space at time t as the set B_t of all state variables which "fit" the configuration of the element at time t. The leading idea consists in first introducing time as a variable entering explicitly the state functions and to treat it on equal footing as the other state variables. B_t is assumed to have the structure of a finite dimensional manifold. The "total state space" is the disjoint union

$$\mathcal{B} = \bigcup_t \{t\} \times B_t \tag{15}$$

with a given natural structure of fibre bundle over $\mathbb{R}$ where time flows ([4], [5]). We call it the thermodynamic bundle. If the instantaneous state space B_t does not vary in time (i.e. there is an abstract space B such that $B_t \simeq B$ for all instant of time t), then $\mathcal{B}$ has the topology of the Cartesian product $\mathcal{B} \simeq \mathbb{R} \times B$. Moreover, we consider an abstract space of *processes* ([4] - [6], [9] and [10]), i.e. a set Π of functions

$$P_t^i : [0, t] \to \mathcal{G}, \tag{16}$$

where $[0, t]$ is any time interval, the space $\mathcal{G}$ being a suitable target space, i a label ranging in an unspecified index set for all allowed processes and $t \in \mathbb{R}$ the so called *duration* of the process. For the given state space B we suppose that the set Π is such that the following hold:

- $\exists D : \Pi \to \mathcal{P}(B)$ is the set of all subsets of B; D is the *domain function* and $D_t^i \equiv D(P_t^i)$ is called the domain of the i-th process (of duration t);

- $\exists R : \Pi \to \mathcal{P}(B)$; R is the *range function* and $R_t^i \equiv R(P_t^i)$ is called the range of the i-th process (of duration t);
- considering the restrictions

$$P_t^i = P_t^i \mid_{[0,\tau]} \qquad (\tau \le t) \tag{17}$$

new processes are obtained ('restricted processes') and they satisfy the following condition:

$$\forall \tau < t \qquad D(P_t^i) \subseteq D(P_\tau^i). \tag{18}$$

Incidentally, this implies that $\bigcap_{\tau=0}^{t} D(P_\tau^i) = D(P_t^i)$, where t is the maximal duration.

A continuous function then is defined

$$\chi : \mathbb{R} \times \Pi \to C^0(B, B) \tag{19}$$

$$(t, P_t^i) \to \rho_t^i,$$

with $\rho_t^i : \sigma_0 \in D_t^i \subseteq B \to \rho_t^i(\sigma_0) = \sigma_t \in R_t^i \subseteq B$, so that for any instant of time t and for any process $P_t^i \in \Pi$ a continuous mapping, ρ_t^i, called *transformation* (induced by the process) is generated, which gives point by point a correspondence between the initial states σ_0 and the final state σ_t. Now, we introduce a function of time

$$\lambda_{\sigma_0}^i(\tau) = \begin{cases} \sigma_0 & if \ \ \tau = 0 \\ \rho_t^i(\sigma_0) & if \ \tau \in]0, t] \end{cases}$$

such that the transformation for the medium is given by

$$\delta : \mathbb{R} \longrightarrow \mathbb{R} \times B$$

$$\tau \longrightarrow \delta(\tau) = (\tau, \lambda_{\sigma_0}^i(\tau)).$$

With these positions the transformation is interpreted as a curve δ in the union of all state spaces such that it intersects the instantaneous state space just once. We assume that the behavior of the material element is described by the state variables $(\mathbf{F}, \mathrm{U}, \mathbf{D}, \mathbf{B}, \mathbf{a}, \boldsymbol{\mathcal{V}}, \mathbf{q}, \nabla\theta, \nabla\mathbf{a})$ (see equ. (4)) and the state space $B = \mathrm{Lin}(\mathfrak{V}) \oplus \mathbb{R} \oplus \mathfrak{V} \oplus \mathfrak{V} \oplus \mathcal{W}_1 \oplus \mathcal{W}_2 \oplus \mathfrak{V} \oplus \mathfrak{V} \oplus \mathrm{Lin}(\mathcal{W}_1)$, where $\mathcal{W}_1$ and $\mathcal{W}_2$ are vector spaces accounting for the internal variables a_{ij} and $\mathcal{V}_{ijk}$. The process $\mathbf{P}_t$ is defined by the following functions $P_t^i(\tau) \;=\; [\mathbf{L}(\tau), \mathrm{h}(\tau), \boldsymbol{\mathcal{H}}(\tau), \boldsymbol{\Xi}(\tau), \boldsymbol{\Upsilon}(\tau), \mathbf{V}(\tau), \mathbf{Q}(\tau), \gamma(\tau), \boldsymbol{\Gamma}(\tau)]$, with $\mathrm{h}(t) = \frac{1}{\rho}(\mathcal{E} \cdot \dot{\mathbf{P}}\rho - \mathcal{E} \cdot \mathbf{P}\dot{\rho}) - \nabla \cdot \mathbf{q}$, $\boldsymbol{\Upsilon}(t) = -\nabla \cdot \boldsymbol{\mathcal{V}} + \mathbf{A}(\mathbf{z})$ (see equ.s (20), (5)-(12)). The space $\mathcal{G}$ is given by $\mathcal{G} = \mathrm{Lin}(\mathfrak{V}) \oplus \mathbb{R} \oplus \mathfrak{V} \oplus \mathfrak{V} \oplus \mathcal{W}_1 \oplus \mathcal{W}_2 \oplus \mathfrak{V} \oplus \mathfrak{V} \oplus \mathrm{Lin}(\mathcal{W}_1)$. Following the usual method (see [4] and [9]) we assume that the transformation induced by the process, ρ_t^i, is governed by the following dynamical system

$$\begin{cases} \dot{\mathbf{F}} = \mathbf{L}(\tau)\mathbf{F}(\tau), \\ \rho\dot{\mathrm{U}} = \mathbf{T}(\delta)\cdot\mathbf{L}(\tau) + \mathrm{h}(\tau), \\ \dot{\mathbf{D}} = \boldsymbol{\mathcal{H}}(\tau), \\ \dot{\mathbf{B}} = \boldsymbol{\Xi}(\tau), \\ \dot{\mathbf{a}} = \boldsymbol{\Upsilon}(\tau), \\ \dot{\boldsymbol{\mathcal{V}}} = \mathbf{V}(\tau), \\ \dot{\mathbf{q}} = \mathbf{Q}(\tau), \\ \dot{\nabla\theta} = \gamma(\tau), \\ \dot{\nabla\mathbf{a}} = \boldsymbol{\Gamma}(\tau), \end{cases} \tag{20}$$

where we have considered the material time derivative of the variables $\mathbf{a}$, $\boldsymbol{\mathcal{V}}$ and $\mathbf{q}$ (see [20]-[21] for this choice in constructing the geometrical model). Moreover the following constitutive functions are defined (see (14))

$\theta : \mathbb{R}\times B \longrightarrow \mathbb{R}^{++}$, $\mathbf{T} : \mathbb{R}\times B \longrightarrow Sym(\mathfrak{V})$, $\mathbf{P} : \mathbb{R}\times B \longrightarrow \mathfrak{V}$,
$\mathbf{A} : \mathbb{R}\times B \longrightarrow \boldsymbol{\mathcal{W}}_1$, $\mathbf{V} : \mathbb{R}\times B \longrightarrow \boldsymbol{\mathcal{W}}_2$, $\mathbf{Q} : \mathbb{R}\times B \longrightarrow \mathfrak{V}$.

The set $(B, \Pi, \theta, \mathbf{T}, \mathbf{P}, \mathbf{A}, \mathbf{V}, \mathbf{Q})$ defines the simple material element under consideration (see [9]). The dynamical system (20) determines a linear morphism G defined on fibre bundle of process in the following way
$G : TB \longrightarrow TB$, having the following form

$$G : (\mathbf{F}, U, \mathbf{D}, \mathbf{B}, \mathbf{a}, \boldsymbol{\mathcal{V}}, \mathbf{q}, \nabla\theta, \nabla\mathbf{a}, \mathbf{L}, \mathrm{h}, \boldsymbol{\mathcal{H}}, \boldsymbol{\Xi}, \boldsymbol{\Upsilon}, \mathbf{V}, \mathbf{Q}, \gamma, \boldsymbol{\Gamma}) \\ \rightarrow (\mathbf{F}, U, \mathbf{D}, \mathbf{B}, \mathbf{a}, \boldsymbol{\mathcal{V}}, \mathbf{q}, \nabla\theta, \nabla\mathbf{a}, \dot{\mathbf{F}}, \dot{U}, \dot{\mathbf{D}}, \dot{\mathbf{B}}, \dot{\mathbf{a}}, \dot{\boldsymbol{\mathcal{V}}}, \dot{\mathbf{q}}, \dot{\nabla\theta}, \dot{\nabla\mathbf{a}}),$$

which in matrix form is expressed by:

$$\left(\mathbf{F}, U, \mathbf{D}, \mathbf{B}, \mathbf{a}, \boldsymbol{\mathcal{V}}, \mathbf{q}, \nabla\theta, \nabla\mathbf{a}, \dot{\mathbf{F}}, \dot{U}, \dot{\mathbf{D}}, \dot{\mathbf{B}}, \dot{\mathbf{a}}, \dot{\boldsymbol{\mathcal{V}}}, \dot{\mathbf{q}}, \dot{\nabla\theta}, \dot{\nabla\mathbf{a}}\right)^T =$$

$$= \begin{pmatrix} \mathbf{I} & \mathbf{0} \\ \mathbf{0} & \mathbf{A} \end{pmatrix} (\mathbf{F}, U, \mathbf{D}, \mathbf{B}, \mathbf{a}, \boldsymbol{\mathcal{V}}, \mathbf{q}, \nabla\theta, \nabla\mathbf{a}, \mathbf{L}, \mathrm{h}, \boldsymbol{\mathcal{H}}, \boldsymbol{\Xi}, \boldsymbol{\Upsilon}, \mathbf{V}, \mathbf{Q}, \gamma, \boldsymbol{\Gamma})^T \text{ with}$$

$$\mathbf{A} = \begin{pmatrix} \mathbf{F} & 0 & 0 & 0 & 0 & 0 & 0 & 0 & 0 \\ \frac{\mathbf{T}}{\rho} & \frac{1}{\rho} & 0 & 0 & 0 & 0 & 0 & 0 & 0 \\ 0 & 0 & 1 & 0 & 0 & 0 & 0 & 0 & 0 \\ 0 & 0 & 0 & 1 & 0 & 0 & 0 & 0 & 0 \\ 0 & 0 & 0 & 0 & 1 & 0 & 0 & 0 & 0 \\ 0 & 0 & 0 & 0 & 0 & 1 & 0 & 0 & 0 \\ 0 & 0 & 0 & 0 & 0 & 0 & 1 & 0 & 0 \\ 0 & 0 & 0 & 0 & 0 & 0 & 0 & 1 & 0 \\ 0 & 0 & 0 & 0 & 0 & 0 & 0 & 0 & 1 \end{pmatrix}. \tag{21}$$

By using this system of differential equations, following standard procedures (see [4] and [9]) in this geometrical structure we are able to introduce an "entropy function" in the following way:

$$s(t) = -\int_0^t \frac{1}{\rho} \nabla \cdot \mathbf{J}_S d\tau, \tag{22}$$

where $\mathbf{J}_S$ is defined according to equation (13). Then we get

$$s = \int_0^t -\frac{1}{\rho\theta} \nabla \cdot \mathbf{q} d\tau + \int_0^t \frac{1}{\rho\theta^2} \mathbf{q} \cdot \nabla\theta d\tau - \int_0^t \frac{1}{\rho} \nabla \cdot \mathbf{k} d\tau. \tag{23}$$

Using the internal energy balance we obtain the following expression for $\nabla \cdot \mathbf{q}$:

$$\nabla \cdot \mathbf{q} = -\rho\dot{U} + \mathbf{T} \cdot (\dot{\mathbf{F}}\mathbf{F}^{-1}) - \frac{\dot{\rho}}{\rho} \boldsymbol{\mathcal{E}} \cdot \mathbf{P} + \boldsymbol{\mathcal{E}} \cdot \dot{\mathbf{P}}, \tag{24}$$

so that the final expression for the entropy function is calculated as a path integral along the path δ between the initial and the actual states σ_0 and σ_t into the space $\mathbb{R} \times B$ of all the thermodinamic variables together with the independent time variable:

$$s(\rho_t, \sigma_0, t) = \int_\delta \Omega = \int_0^t -\tfrac{1}{\rho\theta}(\mathbf{T}\mathbf{F}^{-T}) \cdot d\mathbf{F} - \tfrac{1}{\rho\theta}(\mathbf{E} + \mathbf{v} \wedge \mathbf{B}) \cdot d\mathbf{D} + \tfrac{1}{\theta} dU +$$
$$\left[\tfrac{1}{\rho\theta^2} \mathbf{q} \cdot \nabla\theta + \tfrac{1}{\rho^2\theta} \dot{\rho}(\mathbf{E} + \mathbf{v} \wedge \mathbf{B}) \cdot \mathbf{P} + \tfrac{\varepsilon_0}{2\rho\theta} \dot{\mathbf{E}}^2 + \tfrac{\varepsilon_0}{\rho\theta}(\mathbf{v} \wedge \mathbf{B}) \cdot \dot{\mathbf{E}} - \tfrac{1}{\rho} \nabla \cdot \mathbf{k}\right] d\tau, \tag{25}$$

where we have used the relation $\mathbf{T} \cdot (\dot{\mathbf{F}}\mathbf{F}^{-1}) = (\mathbf{T}\mathbf{F}^{-T}) \cdot \dot{\mathbf{F}}$, with $\mathbf{F}^{-T} = (\mathbf{F}^{-1})^T$ (T denoting matrix transposition) and equations $(6)_2$ and $(9)_2$. For the explicit expressions of $\mathbf{T}$, $\mathbf{P}$, $\mathbf{k}$ and $\mathbf{q}$ see [1] and [2]. In (25) the entropy function defines a 1-form Ω in $\mathbb{R} \times B$ called the *entropy 1-form.* In components the entropy 1-form Ω becomes:

$$\Omega = \omega_\mu dq^\mu + \omega_0 dt = \omega_A dq^A, \quad (A = 1, 2, ..., 10) \tag{26}$$

where $q^A = (\mathbf{F}, \mathbf{D}, U, \mathbf{B}, \mathbf{a}, \boldsymbol{\mathcal{V}}, \mathbf{q}, \nabla\theta, \nabla\mathbf{a}, t)$ and

$$\omega_A \equiv \left(\left(-\frac{1}{\rho\theta}\mathbf{T}\mathbf{F}^{-T}\right), \left[-\frac{1}{\rho\theta}(\mathbf{E} + \mathbf{v} \wedge \mathbf{B})\right], \left(\frac{1}{\theta}\right), 0, 0, 0, 0, 0, 0,\right.$$

$$\left.\left[\frac{1}{\rho\theta^2} \mathbf{q} \cdot \nabla\theta + \frac{1}{\rho^2\theta} \dot{\rho}(\mathbf{E} + \mathbf{v} \wedge \mathbf{B}) \cdot \mathbf{P} + \frac{1}{\rho\theta} \epsilon_0 (\mathbf{E} + \mathbf{v} \wedge \mathbf{B}) \cdot \dot{\mathbf{E}} - \frac{1}{\rho} \nabla \cdot \mathbf{k}\right]\right).$$

Thus, by differentiation a 2-form is obtained:

$$d\Omega = \frac{1}{2} A_{\mu\lambda} dq^\mu \wedge dq^\lambda + E_\lambda dt \wedge dq^\lambda, \tag{27}$$

with $\quad A_{\mu\lambda} = \partial_\mu \omega_\lambda - \partial_\lambda \omega_\mu \quad$ *and* $\quad E_\lambda = \partial_0 \omega_\lambda - \partial_\lambda \omega_0.$

Applying the closure conditions for the 1-form, the necessary conditions for the existence of the entropy function during the processes under consideration are:

$$\partial_U\left(-\tfrac{1}{\rho\theta}\mathbf{T}\mathbf{F}^{-T}\right)=\partial_{\mathbf{F}}\left(\tfrac{1}{\theta}\right),\quad \partial_{\mathbf{D}}\left(-\tfrac{1}{\rho\theta}\mathbf{T}\mathbf{F}^{-T}\right)=\partial_{\mathbf{F}}\left[-\tfrac{1}{\rho\theta}(\mathbf{E}+\mathbf{v}\wedge\mathbf{B})\right],$$
$$\partial_{\mathbf{D}}\left(\tfrac{1}{\theta}\right)=\partial_U\left[-\tfrac{1}{\rho\theta}(\mathbf{E}+\mathbf{v}\wedge\mathbf{B})\right],$$
$$\tfrac{\partial\omega_A}{\partial q^B}=0\qquad(A=1,2,3,10;\ B=4,5,...,10),$$
$$\partial_t\left(-\frac{1}{\rho\theta}\mathbf{T}\mathbf{F}^{-T}\right)=\partial_{\mathbf{F}}\left[\frac{1}{\rho\theta^2}\mathbf{q}\cdot\nabla\theta+\frac{1}{\rho^2\theta}\dot{\rho}(\mathbf{E}+\mathbf{v}\wedge\mathbf{B})\cdot\mathbf{P}+\frac{1}{\rho\theta}\epsilon_0(\mathbf{E}+\mathbf{v}\wedge\mathbf{B})\cdot\dot{\mathbf{E}}-\frac{1}{\rho}\nabla\cdot\mathbf{k}\right],$$

$$\partial_t\left(\frac{1}{\theta}\right)=\partial_U\left[\frac{1}{\rho\theta^2}\mathbf{q}\cdot\nabla\theta+\frac{1}{\rho^2\theta}\dot{\rho}(\mathbf{E}+\mathbf{v}\wedge\mathbf{B})\cdot\mathbf{P}+\frac{1}{\rho\theta}\epsilon_0(\mathbf{E}+\mathbf{v}\wedge\mathbf{B})\cdot\dot{\mathbf{E}}-\frac{1}{\rho}\nabla\cdot\mathbf{k}\right],$$
$$\partial_t\left[-\frac{1}{\rho\theta}(\mathbf{E}+\mathbf{v}\wedge\mathbf{B})\right]=\partial_{\mathbf{D}}\left[\frac{1}{\rho\theta^2}\mathbf{q}\cdot\nabla\theta+\frac{1}{\rho^2\theta}\dot{\rho}(\mathbf{E}+\mathbf{v}\wedge\mathbf{B})\cdot\mathbf{P}+\frac{1}{\rho\theta}\epsilon_0(\mathbf{E}+\mathbf{v}\wedge\mathbf{B})\cdot\dot{\mathbf{E}}-\frac{1}{\rho}\nabla\cdot\mathbf{k}\right].$$

These relations give the necessary conditions characterizing a sort of "irrotationality" of the entropy 1-form during the transformation analyzed. If the entropy of 1-form in eq. (25) is closed and its coefficients are regular, this form is exact and the existence of an upper-potential satisfying relation $S(\sigma_t)-S(\sigma_0)\geq s$ is ensured (see [9] and [10]). Starting from the entropy 1-form it is possible to investigate and to introduce an extended thermodynamical phase space in a suitable way (see [24]).

References

1. S. Giambò , B. Maruszewski, L. Restuccia, On a nonconventional thermodynamical model of a defective piezoelectric crystal, *Journal of Technical Physics* (Polish Academy of Sciences, Warszawa), **43**(2), 155 (2002).
2. L. Restuccia, B.T.Maruszewski, Thermomechanics of piezoelectrics defective by dislocations, Proceedings of the 7th Conference, Venice, Italy, 20-24 September, 2004,in: Applied and Industrial Mathematics in Italy, Series on Advances in Mathematics for Applied Sciences, Vol. 69, edited by M. Primicerio, R. Spigler, V.Valente, World Scientific , 475 (2005).
3. B.Maruszewski, On a dislocation core tensor, *Phys. stat. sol. (b)* **168**, 59 (1991).
4. M. Dolfin, M. Francaviglia and P. Rogolino, A geometric perspective on irreversible thermodynamics with internal variables, *J. Non-Equilib. Thermodyn.* **23**, 250 (1998).
5. M. Dolfin, M. Francaviglia and P. Rogolino, A geometric model on the thermodynamics of simple materials, *Periodica Polytechnica Serie Mech. Eng.* **43**, 29 (1999).
6. W. Noll, A mathematical theory of the mechanical behaviour of continuous media, *Arch. Rat. Mech. Anal.*, **2**, 197 (1958).
7. B. D. Coleman and M. E. Gurtin, Thermodynamics with internal state variables, *J. Chem. Phys.*, **47**, 597(1967).
8. W. Noll, A new mathematical theory of simple materials, *Arch. Rat. Mech. Anal.*, **48** (1972).

9. B. D. Coleman, D. R. Owen, A mathematical foundation for thermodynamics, *Arch. Rat. Mech. Anal.* **54**, 1 (1974).
10. D.R. Owen, A first course in the mathematical foundations of thermodynamics, Springer-Verlag, New York (1984).
11. D. Germanó, L. Restuccia, A geometrical model of piezoelectric media, to be published.
12. M.Francaviglia, L.Restuccia, P.Rogolino, Entropy production in polarizable bodies with internal variables, *Journal of Non-Equilibrium Thermodynamics*, **29**, 1 (2004).
13. M. Dolfin, M. Francaviglia, L. Restuccia, Thermodynamics of deformable dielectrics with a non-Euclidean structure as internal variable, Technische Mechanik ,**24**, 137 (2004).
14. C. Kittel , Introduction to Solid State Physics, John Wiley and Sons, 3rd Ed., New York (1966).
15. F. R. N. Nabarro, Theory of Crystal Dislocations, Clarendon Press, Oxford (1967).
16. D. Hull, Introduction to Dislocation, Pergamon Press, London, Oxford(1975)
17. J. Kubik, A macroscopic description of geometrical pore structure of porous solids, *Int. J. Engng. Sci.***24**, 971 (1986).
18. G. A. Maugin, Continuum Mechanics of Electromagnetic Solids, North-Holland, Amsterdam (1988).
19. C. Truesdell, R.A. Toupin, The classical field theories , Handbuch der Physik, **III**/1, S. Flugge, Springer Verlag, Berlin (1960).
20. W.Muschik, L.Restuccia, Changing the observer and moving materials in continuum physics: objectivity and frame-indifference. Technische Mechanik, **22**, 2, 152(2002).
21. H. Hermann, W. Muschik, G. Ruckner, L. Restuccia, Constitutive mappings and the non-objective part of material frame indifference, Proc.s of The International Symposium on Trends in Continuum Physics TRECOP'04, Poznan, Poland, 17-19 November, 2004, Publishing House of Poznan University of Technology,Poland, Eds. B. T. Maruszewski, W. Muschik, A. Radowicz, 128 (2004).
22. I. S. Liu, The Method of Lagrange multipliers for exploitation of the entropy principle, *Arch. Rat. Mech. Anal.*, **46**, 131 (1972).
23. G. F. Smith, On isotropic functions of symmetric tensors,skew-symmetric tensors and vectors, *Int. J. Engng.Sci.*, **9**, 899 (1971).
24. S.Preston, J.Vargo, Indefinite metric of R. Mrugala and the geometry of thermodynamical phase space, International Conference and Summerschool, Proc.s THERMOCON '05, Thermal Theories of Continua: Survey and Developments, September 25-30, Messina, Italy, 2005, Atti Accademia Peloritana dei Pericolanti di Messina, 2007.

ESTIMATING THE DIFFUSION PART OF THE COVARIATION BETWEEN TWO VOLATILITY MODELS WITH JUMPS OF LÉVY TYPE

F. GOBBI

Dipartimento di Matematica per le Decisioni, Università degli Studi di Firenze,
E-mail: fgobbi@ds.unifi.it

C. MANCINI

Dipartimento di Matematica per le Decisioni, Università degli Studi di Firenze,
E-mail: cecilia.mancini@dmd.unifi.it

In this paper we consider two processes driven by diffusions and jumps. We consider both finite activity and infinite activity jump components. Given discrete observations we disentangle the covariation between the two diffusion parts from the co-jumps.
A commonly used approach to estimate the diffusion covariation part is to take the sum of the cross products of the two processes increments; however this estimator can be highly biased in the presence of jump components, since it approaches the quadratic covariation containing also the co-jumps. Our estimator is based on a threshold principle allowing to isolate the jumps. As a consequence we find an estimator which is consistent. In the case of finite activity jump components the estimator is also asymptotically Gaussian. We assess the performance of our estimator for finite samples on four different simulated models.

Keywords: co-jumps, diffusion correlation coefficient, finite activity jumps, infinite activity Lévy jumps, threshold estimator.

1. Introduction

We consider two state variables evolving as follows

$$dX_t^{(1)} = a_t^{(1)} dt + \sigma_t^{(1)} dW_t^{(1)} + dJ_t^{(1)},$$

$$dX_t^{(2)} = a_t^{(2)} dt + \sigma_t^{(2)} dW_t^{(2)} + dJ_t^{(2)},$$

for $t \in [0,T]$, T fixed, where $W_t^{(2)} = \rho_t W_t^{(1)} + \sqrt{1-\rho_t^2} W_t^{(3)}$; $W^{(1)} = (W_t^{(1)})_{t\in[0,T]}$ and $W^{(3)} = (W_t^{(3)})_{t\in[0,T]}$ are independent Wiener processes.

$J^{(1)}$ and $J^{(2)}$ are possibly correlated pure jump processes. We are interested in the identification both of the covariation $\int_0^T \rho_t\sigma_t^{(1)}\sigma_t^{(2)}dt$ between the two diffusion parts and of the *co-jumps* $\Delta J_t^{(1)}\Delta J_t^{(2)}$, the simultaneous jumps of $X^{(1)}$ and $X^{(2)}$. Given discrete equally spaced observations $X_{t_j}^{(1)}, X_{t_j}^{(2)}$, $j = 1..n$, in the interval $[0,T]$ (with $t_j = j\frac{T}{n}$), a commonly used approach to estimate $\int_0^T \rho_t\sigma_t^{(1)}\sigma_t^{(2)}dt$ is to take the sum of cross products $\sum_{j=1}^n (X_{t_j}^{(1)} - X_{t_{j-1}}^{(1)})(X_{t_j}^{(2)} - X_{t_{j-1}}^{(2)})$; however, this estimate can be highly biased when the processes $X^{(q)}$ contain jumps; in fact, such a sum approaches the global quadratic covariation $[X^{(1)}, X^{(2)}]_T = \int_0^T \rho_t\sigma_t^{(1)}\sigma_t^{(2)}dt + \sum_{0\le t\le T}\Delta J_t^{(1)}\Delta J_t^{(2)}$ containing also the co-jumps. It is crucial to single out the time intervals where the jumps have not occurred. Our estimator is based on a threshold criterion (Mancini, [19]) allowing to isolate the jump part. In particular, we asymptotically identify whether each process has jumped or not in a given time interval $]t_{j-1}, t_j]$, depending on whether the increment $|X_{t_j} - X_{t_{j-1}}|$ is too big with respect to a proper function (*threshold*) of the length $h \doteq t_j - t_{j-1}$ of the time interval. We derive an asymptotically unbiased estimator of the continuous part of the covariation process as well as of the co-jumps. More precisely, the following threshold estimator

$$\tilde{v}_{1,1}^{(n)}(X^{(1)}, X^{(2)})_T = \sum_{j=1}^n \Delta_j X^{(1)} 1_{\{(\Delta_j X^{(1)})^2 \le r(h)\}} \Delta_j X^{(2)} 1_{\{(\Delta_j X^{(2)})^2 \le r(h)\}},$$

is a truncated version of the realized quadratic covariation and it is shown to be consistent to $\int_0^T \rho_t\sigma_t^{(1)}\sigma_t^{(2)}dt$, as the number n of observations tends to infinity. We use results of Barndorff-Nielsen and Shephard [4] and Barndorff-Nielsen et al. [7] who analyzed a related problem in absence of jumps. Moreover, in the case where each $J^{(q)}$ is a *finite activity* jump process (i.e. only a finite number of jumps can occur, along each path, in each finite time interval) we show that our estimator is asymptotically Gaussian and converges with speed $\sqrt{h}$. We consider deterministic equally spaced observation times t_j, however our results hold even when the observations are not equally spaced ([19]).
The threshold criterion originated in Mancini ([17]) to separate the diffusion and the jump parts within a univariate parametric Poisson-Gaussian model. The criterion was shown to work even in nonparametric frameworks in [18], [19] and [13].

In the literature of non parametric inference for stochastic processes driven by diffusions plus jumps, some other approaches have been proposed to separate the diffusion part and the jump part given discrete observations.

However each one of such approaches have been applied to univariate processes, so to our knowledge no one dealt with the problem of disentangling the diffusion correlations from the co-jumps.
Berman ([8]) introduces a *power variation* estimator of the sum of given powers of the jumps in diffusion processes plus jumps. This is developed also in [12], [15], [5], [20] and [13]. Barndorff-Nielsen and Shephard define and develop (for instance in [5]) *multipower variation* estimators of integrals $\int_0^T (\sigma_t^{(1)})^p dt$ of the diffusion coefficient, for powers $p \in]0,2[$. These are developed also in [20] and [13]. Bandi and Nguyen ([3]) give pointwise estimators of the diffusion coefficients $a^{(1)}(x)$, $\sigma^{(1)}(x)$, assuming $a_t^{(1)} \equiv a^{(1)}(X_t), \sigma_t^{(1)} \equiv \sigma^{(1)}(X_t)$, and of infinitesimal moments of the jump process $J^{(1)}$, assumed to have finite activity of jump. Such estimators are based on the Nadaraya-Watson kernel method.
We adopt the threshold method here since it is a more effective way to identify (asymtoptically) each interval $]t_{j-1}, t_j]$ where J has jumped. In fact in the univariate parametric case the threshold estimator of $\int_0^T (\sigma_s^{(1)})^2 ds$ is more efficient (in the Cramer-Rao inequality lower bound sense) than the multipower variations estimators.

Applications of the theory we present here is of strong interest in finance, in particular in financial econometrics (see e.g. [2]) and in the framework of portfolio risk management ([10]).

2. The framework

Given a filtered probability space $(\Omega, \mathcal{F}, (\mathcal{F}_t)_{t\in[0,T]}, P)$, let $X^{(1)} = (X_t^{(1)})_{t\in[0,T]}$ and $X^{(2)} = (X_t^{(2)})_{t\in[0,T]}$ be two real processes defined by

$$\begin{aligned} X_t^{(1)} &= \int_0^t a_s^{(1)} ds + \int_0^t \sigma_s^{(1)} dW_s^{(1)} + dJ_s^{(1)}, \quad t \in [0,T], \\ X_t^{(2)} &= \int_0^t a_s^{(2)} ds + \int_0^t \sigma_s^{(2)} dW_s^{(2)} + dJ_s^{(2)}, \quad t \in [0,T], \end{aligned} \tag{1}$$

where

- **Assumption A1**. $W^{(1)} = (W_t^{(1)})_{t\in[0,T]}$ and $W^{(2)} = (W_t^{(2)})_{t\in[0,T]}$ are two correlated Wiener processes, with $\rho_t = Corr(W_t^{(1)}, W_t^{(2)})$, $t \in [0,T]$; we can write

  $$W_t^{(2)} = \rho_t W_t^{(1)} + \sqrt{1-\rho_t^2}\, W_t^{(3)},$$

  where $W^{(1)}$ and $W^{(3)}$ are independent Wiener processes.

- **Assumption A2**. The diffusion stochastic coefficients $\sigma^{(q)} = (\sigma_t^{(q)})_{t\in[0,T]}$, $a^{(q)} = (a_t^{(q)})_{t\in[0,T]}$, $q = 1,2$, and $\rho = (\rho_t)_{t\in[0,T]}$ are adapted cdlg; $a^{(q)}$ are locally bounded.

- **Assumption A3**. For $q = 1,2$

$$J^{(q)} = J_1^{(q)} + \tilde{J}_2^{(q)},$$

where $J_1^{(q)}$ are finite activity jump processes

$$J_{1t}^{(q)} = \int_0^t \gamma_s^{(q)} dN_s^{(q)} = \sum_{k=1}^{N_t^{(q)}} \gamma_{\tau_k^{(q)}}, \quad q = 1,2,$$

where $N^{(q)} = (N_t^{(q)})_{t\in[0,T]}$ are counting processes with $E[N_T^{(q)}] < \infty$; $\{\tau_k^{(q)},\ k = 1,...,N_T^{(q)}\}$ denote the instants of jump of $J_1^{(q)}$ and $\gamma_{\tau_k^{(q)}}$ denote the sizes of the jumps occurred at $\tau_k^{(q)}$. We assume

$$P(\gamma_{\tau_k^{(q)}} = 0) = 0, \quad \forall\ k = 1,...,N_T^{(q)},\ q = 1,2. \tag{2}$$

Denote, for each $q = 1,2$, $\underline{\gamma}^{(q)} = \min_{k=1,\ldots,N_T^{(q)}} |\gamma_{\tau_k^{(q)}}|$. By condition (2), a.s. we will have $\underline{\gamma}^{(q)} > 0$.

- **Assumption A4**. $\tilde{J}_2^{(q)}$ are infinite activity Lévy pure jump processes of small jumps,

$$\tilde{J}_{2t}^{(q)} = \int_0^t \int_{|x|\leq 1} x\ \tilde{\mu}^{(q)}(dx, ds), \tag{3}$$

where $\mu^{(q)}$ is the Poisson random measure of the jumps of $\tilde{J}_2^{(q)}$, $\tilde{\mu}^{(q)}(dx, ds) = \mu^{(q)}(dx, ds) - \nu^{(q)}(dx)ds$ is its compensated measure, where $\nu^{(q)}$ is the Lévy measure of $\tilde{J}_2^{(q)}$ (see [10]).

Each $\nu^{(q)}$ has the property that $\nu^{(q)}(\mathbb{R} - \{0\}) = \infty$, which characterizes the fact that the path of $\tilde{J}_2^{(q)}$ jumps infinitely many times on each compact time interval. $\tilde{J}_2^{(q)}$ is a compensated sum of jumps, each of which is bounded in absolute value by 1, so that substantially $J_1^{(q)}$ accounts for the "big" (bigger in absolute value than $\underline{\gamma}^{(q)}$) and rare jumps of $X^{(q)}$, while $\tilde{J}_2^{(q)}$ accounts for the very frequent and small jumps.

Remark 2.1. Note that if the finite activity jump terms $J_1^{(q)}$ are Lévy processes, then they are of compound Poisson type: $N^{(q)}$ are simple Poisson

processes with constant intensities $\lambda^{(q)}$ and for each q the random variables $\gamma_{\tau_k^{(q)}}$ are i.i.d, for $k = 1, ..., N_T^{(q)}$, and satisfy condition (2).
If $J^{(q)}$ is a pure jump Lévy process, it is always possible to decompose it as

$$J^{(q)} = J_1^{(q)} + \tilde{J}_2^{(q)},$$

([10]) where J_1 is a compound Poisson process accounting for the jumps bigger in absolute value than 1, and $\tilde{J}_2$ is like as in (3).

Let, for each n, $\pi_n^{[0,T]} = \{0 = t_{0,n} < t_{1.n} < \cdots < t_{n,n} = T\}$ be a partition of $[0, T]$. In this paper we assume equally spaced subdivisions, i.e. $h_n := t_{j,n} - t_{j-1,n} = \frac{T}{n}$ for every $n = 1, 2, ...$ Hence $h_n \to 0$ as $n \to \infty$. Let $\Delta_{j,n}X$ be the increment $X_{t_{j,n}} - X_{t_{j-1,n}}$. To simplify notations we will write h in place of h_n and $\Delta_j X$ in place of $\Delta_{j,n}X$.

- **Assumption A5**. We choose a deterministic function, $h \mapsto r(h)$, satisfying the following properties

$$\lim_{h\to 0} r(h) = 0, \quad \lim_{h\to 0} \frac{h log \frac{1}{h}}{r(h)} = 0.$$

We denote $r(h)$ by r_h. Denote also, for each $q = 1, 2$,

$$D_t^{(q)} = \int_0^t a_s^{(q)} ds + \int_0^t \sigma_s^{(q)} dW_s^{(q)},$$

the diffusion part of $X^{(q)}$.

3. Finite activity jumps: consistency and central limit theorem

In this section we assume that each $\tilde{J}_2^{(q)} \equiv 0$. The main tool for the construction of our estimators is the following

Theorem 3.1. *(Mancini, 2005) If $\tilde{J}_2^{(q)} \equiv 0$, under the assumptions **A1-A3**, and choosing r_h like as in **A5**, we have that a.s. for sufficiently small h*

$$1_{\{(\Delta_j X^{(q)})^2 \le r_h\}} = 1_{\{\Delta_j N^{(q)} = 0\}}, \quad j = 1, 2, ..., n, \quad q = 1, 2.$$

Now we construct our threshold estimators.

Definition 3.1. We define for $r, l \in \mathbb{N}$

$$v_{r,l}^{(n)}(X^{(1)}, X^{(2)})_T = h^{1-\frac{r+l}{2}} \sum_{j=1}^{n} (\Delta_j X^{(1)})^r (\Delta_j X^{(2)})^l,$$

$$w^{(n)}(X^{(1)}, X^{(2)})_T = h^{-1} \sum_{j=1}^{n-1} \prod_{i=0}^{1} \Delta_{j+i} X^{(1)} \prod_{i=0}^{1} \Delta_{j+i} X^{(2)}.$$

and their analogous *threshold* versions

$$\tilde{v}_{r,l}^{(n)}(X^{(1)}, X^{(2)})_T = h^{1-\frac{r+l}{2}} \sum_{j=1}^{n} (\Delta_j X^{(1)})^r 1_{\{(\Delta_j X^{(1)})^2 \leq r_h\}} (\Delta_j X^{(2)})^l 1_{\{(\Delta_j X^{(2)})^2 \leq r_h\}},$$

$$\tilde{w}^{(n)}(X^{(1)}, X^{(2)})_T = h^{-1} \sum_{j=1}^{n-1} \prod_{i=0}^{1} \Delta_{j+i} X^{(1)} 1_{\{(\Delta_{j+i} X^{(1)})^2 \leq r_h\}} \prod_{i=0}^{1} \Delta_{j+i} X^{(2)} 1_{\{(\Delta_{j+i} X^{(2)})^2 \leq r_h\}}.$$

$v_{r,l}^{(n)}(X^{(1)}, X^{(2)})_T$ and $w^{(n)}(X^{(1)}, X^{(2)})_T$ are first used in [4] to estimate the covariation in the case of diffusion processes. $\tilde{v}_{r,l}^{(n)}(X^{(1)}, X^{(2)})_T$ and $\tilde{w}^{(n)}(X^{(1)}, X^{(2)})_T$ are modified versions for the case of jump diffusion processes where, by remark 3.1, we exclude from the sums the terms containing some jumps.

Remark 3.1. (Consistency) As a consequence of theorem 3.2 below we will have that in fact $\tilde{v}_{1,1}^{(n)}(X^{(1)}, X^{(2)})_T$ is a consistent estimator of $\int_0^T \rho_t \sigma_t^{(1)} \sigma_t^{(2)} dt$ in the presence of finite activity jumps within the processes $X^{(q)}$.

For any semimartingale Z, let us denote

$$\Delta Z_s = Z_s - Z_{s-},$$

the size of the jump of Z at time s.

Remark 3.2. Clearly we will have an estimate of the sum of the co-jumps simply subtracting the diffusion covariation from the quadratic covariation estimators:

$$\sum_{j=1}^{n} \Delta_j X^{(1)} \Delta_j X^{(2)} - \tilde{v}_{1,1}^{(n)}(X^{(1)}, X^{(2)})_T \xrightarrow{P} \sum_{0 \leq s \leq T} \Delta J_s^{(1)} \Delta J_s^{(2)},$$

as $n \to \infty$. Therefore an estimate of each $\Delta J_s^{(1)} \Delta J_s^{(2)}$ will be obtained using

$$\Delta_j X^{(1)} \Delta_j X^{(2)} - \Delta_j X^{(1)} 1_{\{(\Delta_j X^{(1)})^2 \leq r_h\}} \Delta_j X^{(2)} 1_{\{(\Delta_j X^{(2)})^2 \leq r_h\}} =$$

$$\Delta_j X^{(1)} 1_{\{(\Delta_j X^{(1)})^2 > r_h\}} \Delta_j X^{(2)} 1_{\{(\Delta_j X^{(2)})^2 > r_h\}}$$

with j such that $s \in]t_{j-1}, t_j]$.

In view of the practical application of our estimator we are now interested in the speed of convergence of $\tilde{v}_{1,1}^{(n)}(X^{(1)}, X^{(2)})_T$. The main result of this section is to show that a central limit theorem holds with speed $\sqrt{h}$. We need some preliminary results.

Proposition 3.1. *If* $\tilde{J}_2^{(q)} \equiv 0$, *under the assumptions* ***A1-A3***, *and choosing* r_h *like as in* ***A5***, *we have*

$$\tilde{v}_{2,2}^{(n)}(X^{(1)}, X^{(2)})_T - \tilde{w}^{(n)}(X^{(1)}, X^{(2)})_T \xrightarrow{P} \int_0^T (1 + \rho_t^2)(\sigma_t^{(1)})^2 (\sigma_t^{(2)})^2 dt.$$

•

We now are ready to present our central limit theorem relative to the threshold estimator $\tilde{v}_{1,1}^{(n)}(X^{(1)}, X^{(2)})_T$. We will use the multivariate consequence of theorem 1 in Barndorff-Nielsen and Shephard ([6]).

Theorem 3.2. *If* $\tilde{J}_2^{(q)} \equiv 0$, *under the assumptions* ***A1-A3***, *and choosing* r_h *like as in* ***A5***, *we have*

$$\frac{\tilde{v}_{1,1}^{(n)}(X^{(1)}, X^{(2)})_T - \int_0^T \rho_t \sigma_t^{(1)} \sigma_t^{(2)} dt}{\sqrt{h}\sqrt{\tilde{v}_{2,2}^{(n)}(X^{(1)}, X^{(2)})_T - \tilde{w}^{(n)}(X^{(1)}, X^{(2)})_T}} \xrightarrow{d} Z,$$

where Z *has law* $N(0,1)$.

•

4. Infinite activity jumps: consistency

In this section we deal with the case in which the jump components have possibly infinite activity: $\tilde{J}_2^{(q)}$ can be non zero. However the infinite activity of jump can be mild, like as e.g. for the Variance Gamma process ([10]), or wild, like as e.g. for a pure jump α-stable process with α around 2. A way to measure the wildness of activity of a Lévy process is to check the integrability of powers of $|x|$ with respect to its Lévy measure. $\int_{|x|\le 1} x^2 \nu(dx) < \infty$ for any Lévy process having Lévy measure ν. For smaller powers of $|x|$ in general the integral can be infinite. The smaller is the power $\delta > 0$ for which $\int_{|x|\le 1} x^\delta \nu(dx) < \infty$ the milder is the activity of jump. It is defined the following Blumenthal-Gatoor index

Definition 4.1. (Blumenthal-Gatoor index) Let J be a pure jump Lévy process with Lévy measure ν. The Blumenthal-Gatoor index of J is the real number $\alpha \in [0, 2[$ defined by

$$\alpha := \inf\{\delta > 0, \int_{|x|\le 1} |x|^\delta \nu(dx) < \infty\}.$$

A (finite activity) compound Poisson process and the (infinite activity) Variance Gamma process have $\alpha = 0$. A β-stable process has Blumenthal-Gatoor index equal to β. The Normal Inverse Gaussian process and the Generalized Hyperbolic Lévy motion have $\alpha = 1$. If J has Blumenthal-Gatoor index $\alpha < 1$ then it has finite variation, whereas if $\alpha > 1$ it has infinite variation. The CGMY process ([9]) has $\alpha = Y$.
In the following we will need an additional assumption

- **Assumption A6**. We assume that each $\nu^{(q)}$ has density $f^{(q)}$ such that for $x \to 0$

$$f^{(q)}(x) \approx \frac{L^{(q)}(x)}{|x|^{1+\alpha_q}},$$

  for some function $L^{(q)}$ having finite nonzero limit as $|x| \to 0$, where α_q is the Blumenthal-Gatoor index of $J^{(q)}$.

The models most used in practice in the financial literature, e.g. the Variance Gamma process, the Normal Inverse Gaussian process, the α-stable processes, the CGMY models, the Generalized Hyperbolic Lévy motion, satisfy **A6**.

Under assumption **A6**, we can control the asymptotic behaviour (for $\varepsilon \to 0$) of integrals of kind $\int_{|x|\le\varepsilon} |x|^k \nu^{(q)}(dx)$, $k = 1, 2$, which we will need in the following. In fact, we have, as $\varepsilon \to 0$,

$$\int_{|x|\le\varepsilon} x^2 \nu^{(q)}(dx) = O(\varepsilon^{2-\alpha}),$$

$$\int_{\varepsilon<|x|\le 1} x \nu^{(q)}(dx) = O(1 - \varepsilon^{1-\alpha}).$$

To prove the consistency of our threshold estimator we need some notations. Recall that

$$D_t^{(q)} = \int_0^t a_s^{(q)} ds + \int_0^t \sigma_s^{(q)} dW_s^{(q)},$$

and denote

$$Y_t^{(q)} = D_t^{(q)} + J_{1t}^{(q)},$$
$$\eta_q^2(\varepsilon) = \int_{|x|\le\varepsilon} x^2 \nu^{(q)}(dx).$$

So $X_t^{(q)} = Y_t^{(q)} + \tilde{J}_{2t}^{(q)}$, $q = 1, 2$. Note that since $\tilde{J}_{2t}^{(q)} = \int_0^t \int_{|x|\le 1} x\, \tilde{\mu}^{(q)}(dx, ds)$ we have

$$E\tilde{J}_{2t}^{(q)} = 0,$$
$$Var(\tilde{J}_{2t}^{(q)}) = t \int_{|x|\le 1} x^2 \nu^{(q)}(dx) = t\eta_q^2(1).$$

Theorem 4.1. *(Consistency) Let $(X_t^{(1)})_{t\in[0,T]}$ and $(X_t^{(2)})_{t\in[0,T]}$ two processes of the form (1). Assume **A1-A6** are satisfied. Then*

$$\tilde{v}_{1,1}^{(n)}(X^{(1)}, X^{(2)})_T \xrightarrow{P} \int_0^T \rho_t \sigma_t^{(1)} \sigma_t^{(2)} dt,$$

as $n \to \infty$.

•

Remark 4.1. (Estimate of the co-jumps). In this framework of infinite activity jumps, note that

$$1_{\{(\Delta_j X^{(q)})^2 \le r_h\}} = 1_{\{\Delta_j N^{(q)}=0, |\Delta \tilde{J}_2^{(q)}| \le 2\sqrt{r_h}\}},$$

so that this time

$$\Delta_j X^{(1)} 1_{\{(\Delta_j X^{(1)})^2 > r_h\}} \Delta_j X^{(2)} 1_{\{(\Delta_j X^{(2)})^2 > r_h\}}$$

contains both the finite activity co-jumps and the small infinite activity co-jumps. However the contribution of the latter terms decreases as $n \to \infty$.

References

1 AÏT-SAHALIA, Y. (2004). Disentangling volatility from jumps, *Journal of Financial Economics*, 74, 487-528.

2 ANDERSEN, T.G., BOLLERSLEV, T., DIEBOLD, F.X. (2005): *Roughing it up: Including jump components in the measurement, modeling and forecasting of return volatility.* Manuscript, Northwestern University, Duke University and University of Pennsylvania.

3 BANDI, F.M., NGUYEN, T.H. (2003): On the function estimation of jump diffusion models. *Journal of Econometrics, 116*, 1, 293-328.

4 BARNDORFF-NIELSEN, O.E., SHEPHARD, N. (2004): Econometric analysis of realized covariation: high frequency based covariance, regression and correlation in financial economics. *Econometrica*, (2004), *72*, 885-925.

5 BARNDORFF-NIELSEN, O.E., SHEPHARD, N. (2004b): Power and bipower variation with stochastic volatility and jumps, (with discussion). *Journal of Financial Econometrics*, *2*, 1-48.

6 BARNDORFF-NIELSEN, O.E., SHEPHARD, N. (2006): Variation, jumps and high frequency data in financial econometrics, in *Advanced in Economics and Econometrics. Theory and Applications*, Ninth World Congress eds Richard Blundell, Persson Torsten, Whitney K Newey, Econometric Society Monographs, Cambridge U niversity Press

7 BARNDORFF-NIELSEN, O.E., GRAVENSEN, S.E., JACOD, J., PODOLSKIJ, M. and SHEPHARD, N. (2005). A central limit theorem for realised power and bipower variation of continuous semimartingales. Technical report, 2004. to appear in *From Stochastic Analysis to Mathematical Finance*, Festschrift for Albert Shiryaev.

8 BERMAN, S.M. (1965) Sign-invariant random variables and stochastic processes with sign invariant increments. *Trans. Amer. Math. Soc*, 119, 216-243.

9 CARR, P., GEMAN, H., MADAN, D., YOR, M. (2002): The finite structure of asset returns: An empirical investigation. *Journal of Business*, 75.

10 CONT, R., TANKOV, P. (2004): *Financial Modelling with jump processes.* Chapman and Hall-CRC.

11 GOBBI, F. (2006): *Estimating the diffusion part of the covariation between two stochastic volatility models with Lévy jumps.* Ph.D thesis, Department of Statistics, University of Florence.

12 HUDSON, W. N., MASON, J. D. (1976) Variational sums for additive processes. *Proc. Amer. Math. Soc*, 55, 395- 399.

13 JACOD, J. (2006): *Asymptotic properties of realized power variations and associated functions of semimartingales.* working paper,

14 KARATZAS, I., SHREVE, S.E. (1999): *Brownian motion and stochastic calculus*, Springer

15 LEPINGLE, D. (1976) La variation d'ordre p des semi-martingales. *Z. Wahrscheinlichkeitstheorie und Verw. Gebiete*, 36, 295-316.

16 MADAN, D.B. (2001): Purely discontinuous asset price processes, *Advances in Mathematical Finance*, Eds. J. Cvitanic, E. Jouini and M. Musiela, Cambridge University Press

17 MANCINI, C., (2001). Disentangling the jumps of the diffusion in a geometric jumping Brownian motion, *Giornale dell'Istituto Italiano degli Attuari*, Volume LXIV, Roma, 19-47

18 MANCINI, C. (2004): Estimation of the parameters of jump of a general

Poisson diffusion model. *Scandinavian Actuarial Journal*, *1*, 42–52.
19 MANCINI, C. (2005): *Estimating the integrated volatility in stochastic volatility models with Lévy type jumps.* Working paper, Dipartimento di Matematica per le Decisioni, Universit di Firenze.
20 WOERNER, J. (2006): Power and Multipower variation: inference for high frequency data, in *Stochastic Finance*, eds A.N. Shiryaev, M. do Rosário Grossinho, P. Oliviera, M. Esquivel, Springer, 343-364.

MODELING HORIZONTAL COASTAL FLOWS: ASSESSING THE ROLE OF VISCOUS CONTRIBUTIONS

G. GROSSO

DIAm, Università di Genova, via Montallegro 1,
Genova, 16145, Italy
E-mail: ggrosso@diam.unige.it

M. BROCCHINI* and A. PIATTELLA

Istituto di Idraulica e Infrastrutture Viarie, Università Politecnica delle Marche,via
Brecce Bianche 12,
Ancona, 60131, Italy
**E-mail: m.brocchini@univpm.it, URL: www.diam.unige.it/brocchin*

Description of horizontal mixing of shallow coastal flows induced by large-scale horizontal eddies is given on the basis of numerical solutions of the Non-linear Shallow Water Equations and sensitivity of the results on viscous-type dissipation is analyzed. Such analysis is the first step to determine a suitable framework for Horizontal Large Eddy Simulation computations of coastal flows performed by means of depth-averaged equations. Statistics of mixing are used to assess the value of the chosen viscous closure.

Keywords: NSWE; Coastal flow mixing; Viscous dissipations; HLES.

1. The Problem

Nearshore dynamics are deeply influenced by large-scale coherent vortical structures (macrovortices) similar to those of two-dimensional turbulence. They are often analyzed in terms of depth-averaged properties like in the case of the classic Nonlinear Shallow Water Equations (NSWE). We want to understand enstrophy and energy evolution of these macrovortices, to define a suitable parametrization for numerical schemes and, in the end, to produce a reliable Horizontal Large Eddy Simulation (HLES) model.

The shedding of macrovortices is mainly due to a spatially-nonuniform breaking of the incoming waves. This may be induced by various reasons, but the major cause of breaking unevenness is due to topography, which is often characterized by longshore natural or manmade, isolated or almost continue, features. Macrovortices can have important morphodynamic effects and deeply influence beach erosion. Moreover their presence is also fundamental for nearshore mixing dynamics, which is important for water quality evaluation and management. Typically, the evolution of passive tracers is predicted by means of depth-averaged convection–diffusion equations where the eddy diffusivity (K) depends on the turbulence structure.

Our analysis is based on use of the following NSWE equations:

$$d_{,t} + (ud)_{,x} + (vd)_{,y} = 0, \tag{1}$$

$$u_{,t} + uu_{,x} + vu_{,y} + gd_{,x} = gh_{,x} + F_x - B_x, \tag{2}$$

$$v_{,t} + uv_{,x} + vv_{,y} + gd_{,y} = gh_{,y} + F_y - B_y, \tag{3}$$

where u and v are the longshore and onshore velocity components and d is the total water depth given by the sum of the still water level h and the surface elevation η (see also figure 1). B is the bottom friction term, which is always accounted for, while F is the dissipative body force, due to breaking and turbulence, which can either be explicitly or implicitly accounted for. In the former case a viscous approach is used, while in the latter a pseudoinviscid one is considered. The aim of the present work is to produce a reliable HLES model (long-term goal), to define a suitable parametrization and numerical scheme (intermediate-term goal) and to understand enstrophy and energy evolution through statistics of lagrangian mixing (short-term goal). The present paper focuses on the short-term goal.

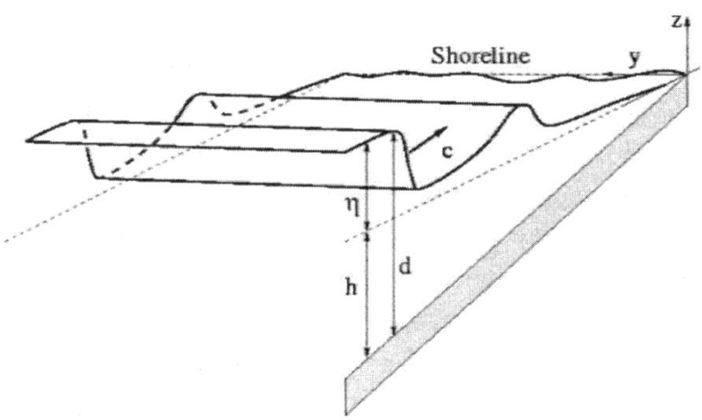

Fig. 1. The NSWE framework.

In the pseudoinviscid NSWE framework, in the absence of shock-type solutions, no generation of either vorticity ω or potential vorticity Ω is possible, i.e.

$$\frac{D\omega}{Dt} = -\omega\nabla_H \cdot \boldsymbol{v} = -\frac{\omega}{d}\frac{Dd}{Dt}, \qquad \frac{D\Omega}{Dt} = 0, \tag{4}$$

where

$$\nabla_H = \left(\frac{\partial}{\partial x}, \frac{\partial}{\partial y}\right), \qquad \omega = v_{,x} - u_{,y}, \qquad \Omega = \frac{\omega}{d}. \tag{5}$$

In particular ω can only be transported or locally intensified or reduced if the total depth d increases or decreases while following a water column which represents a coherent body of water of constant volume; following the same water column, the potential vorticity Ω is conserved. Moreover bottom friction acts at scales of few wavelengths or waveperiods.

However if shocks are present in the domain, jump conditions hold across the discontinuity and introduce a generation mechanism of vorticity or potential vorticity not accounted for in the absence of shocks. In the pseudoinviscid approach, breaking waves are modelled as shocks and the above generation mechanism can be applied to various nearshore flow conditions. More details on the generation mechanism of vorticity because of topographically-induced breaking can be found in Ref. 1.

2. Mixing in quasi-2D turbulence

In a shallow turbulent flow, characterized by large-scale coherent structures, the evolution of tracers and the flow dynamics are so intimately connected that the knowledge of the former may give a predictive key for the latter. Analysis of experimental data of coastal mixing has revealed that typical regimes of 2D turbulence also characterize the flow induced by submerged breakwaters. In particular, enstrophy cascading is seen to dominate the flow induced by one single structure, while rip current shearing dominates the flow due to arrays of breakwaters.

Although the primary and typical means for describing the turbulence behaviour is the spectral energy density $E(k)$, the properties of scalar mixing, like the particle separation D, depend so strictly on the flow that are most often used to indirectly characterize the flow itself. This is, typically, achieved by measuring D and extrapolating the $E(k)$-regime or measuring $E(k)$ and extrapolating the D-regime.

Flow regimes of turbulent mixing are, usually, separated with use of suitable time scales, like the Lagrangian decorrelation time T_L, which can be defined as the time necessary for each particle to loose memory of its initial velocity. Hence, on the basis of absolute statistics, we can define two different regimes: the ballistic regime, in which the absolute dispersion typically increases quadratically in time while the absolute diffusivity undergoes a linear growth ($\langle X^2 \rangle \propto t^2$ and $K^{(1)} \propto t$ for $t \ll T_L$), and the brownian regime, in which the absolute dispersion increases linearly in time, determining a constant diffusivity ($\langle X^2 \rangle \propto t$ and $K^{(1)} \approx$ cost. for $t \gg T_L$). Ultimately particle separations reach the scale of the energy containing eddies and the individual particle velocities become uncorrelated.

A more restrictive definition of the small-time regime is required when using relative statistics and we introduce the time $T_p \ll T_L$ for which, soon after deployment, pairs of particles have lost memory of their initial separation D_0. Since T_p depends on the particle initial separation, it cannot be uniquely defined and it makes more sense to think in terms of a large-space limit defined as the size of the energy containing eddies. Hence, for small separations, a behaviour similar to the ballistic regime is observed, $\langle D^2 \rangle \propto t^2$ and $K^{(2)} \propto (\epsilon D_0)^{2/3} t$ for $t < T_P$, where $\langle D^2 \rangle$ is the relative dispersion and $K^{(2)}$ is the relative diffusivity; for separations larger than L_L, instead, the particles are uncorrelated and the relative dispersion becomes absolute with a brownian regime for which the relative diffusivity is constant and approximately equal to twice the absolute diffusivity ($K^{(2)} \approx 2K^{(1)}$ for $L \geq L_L$ and $t \geq T_L$). In the intermediate regime $T_P < t < T_L$, various behaviours can be observed in dependence of the evolution modalities of large-scale coherent vortices: in the case of a dominance of enstrophy cascade, $\langle D^2 \rangle \propto \exp(\beta^{1/3} t)$ and $K^{(2)} \propto \beta^{1/3} \langle D^2 \rangle$; viceversa, in the presence of a background shear, $K^{(2)} \propto D^{4/3}$. More details on application of the mentioned statistics to shallow flows characterized by uneven topography can be found in Ref. 4.

There, details can also be found on some large-scale laboratory experiments which provide useful benchmarks on nearshore mixing. During the experiments, two distinct configurations were analyzed representative of prototype conditions scaled down through a Froude similarity at a 1:30 length: the 'single breakwater configuration' (SBC), in which a submerged structure is far from any other structure, and the 'rip current configuration' (RCC), in which the submerged breakwaters are separated by narrow gaps. The list of input waves can be found in Ref. 2. Floaters, i.e. 10-25 wooden spheres with diameter of 25mm-42mm, were released around the breakwaters and their meandering tracked with a fixed videocamera. The results pertaining to all statistically-equivalent cases were very similar, repeatability of their salient features being robust.

3. The HLES model

The long term goal of the present work is to produce a reliable HLES model. In a Large Eddy Simulation model, to the field defined in the continuous space, filtered fields (i.e. large scale fields) are associated with the aid of a filter such as a gaussian filter or a box filter or a sharp cutoff filter. The fluctuations at the scales smaller than the mesh grid dimension are referred to as subgrid-scale fields. The problem of the subgrid scale modelling is to express the subgrid scale terms as functions of the large scale field. One of the most used methods to solve this problem has been introduced by the Smagorinsky model: in it the whole unknown subgrid scales are modelled with the aid of an eddy viscosity exactly as was done for the Reynolds stresses in non-homogeneus turbulence. If the flow is quasi-two-dimensional in the large scales and three dimensional in the small ones, a Horizontal Large Eddy Simulation model can be used. In it the subgrid scales modelling to be developed has to take into account the two-dimensional dynamics of the large scales (i.e. conservation of kinetic energy and enstrophy) and also possible interactions with small scale three-dimensional turbulence.

In the present work, numerical computations with a viscous version of the NSWE are used to assess the role of k_c (cut-off wavenumber, inverse of the mesh grid) and of the eddy viscosity ν_T. Theoretical studies allow for inspection of both generation and dissipation of enstrophy. Equations (1), (2) and (3) are the classical formulation of the viscous NSWE, where the expressions of the viscous terms are

$$F_x = \frac{(dT_{xx}),_x + (dT_{xy}),_y}{d}, \qquad F_y = \frac{(dT_{xy}),_x + (dT_{yy}),_y}{d}, \tag{6}$$

with

$$T_{xx} = 2\nu_T u,_x \,, \qquad T_{xy} = T_{yx} = \nu_T(u,_y + v,_x), \qquad T_{yy} = 2\nu_T v,_y \,. \tag{7}$$

3.1. *Numerical Implementation*

To investigate the generation and the evolution of macrovortices at submerged breakwaters, numerical solutions of the NSWE have been used. The bathymetry used reproduces the one obtained during the laboratory experiments when the steady-state was reached. The dimension of the mesh grid is of 0.05 m in the x-direction and of 0.075 m in the y-direction. The flow conditions are those of the E test of Ref. 2.

A shock capturing, Godunov type solver is used (see Ref. 3), which is based on a second-order accurate Weighted Averaged Flux Method (see

Ref. 3). Intercell fluxes are evaluated by means of an exact Riemann solver which allows for a very accurate description of the solution at a limited computational cost. The solver uses an operator splitting technique, which allows to solve separately the homogeneous set of equations and the set of equations accounting for the source terms, such that the initial value problem is replaced by two subsequent initial value problems (e.g. Ref. 5). We focused mainly on the second step of the splitting procedure, i.e. working on the routine which solves the problem

$$u_{,t} = gh_{,x} + F_x - B_x, \tag{8}$$

$$v_{,t} = gh_{,y} + F_y - B_y. \tag{9}$$

3.1.1. *The constant eddy viscosity model (CEM)*

We first implemented a simple algorithm in which the mass and momentum diffusion terms are represented by the following expressions:

$$F_x = \nu_T \nabla^2 u, \qquad F_y = \nu_T \nabla^2 v \tag{10}$$

where ν_T =const. so that equations (8) and (9) become

$$\frac{\partial u}{\partial t} = 2\nu_T \frac{\partial^2 u}{\partial x^2} + \nu_T \frac{\partial^2 v}{\partial y^2} + \nu_T \frac{\partial^2 v}{\partial xy} + 2\nu_T \frac{\partial u}{\partial x}\frac{\partial d}{\partial x} + \nu_T \frac{\partial d}{\partial y}\left(\frac{\partial v}{\partial x} + \frac{\partial u}{\partial y}\right) + g\frac{\partial h}{\partial x} - B_x \tag{11}$$

and

$$\frac{\partial v}{\partial t} = 2\nu_T \frac{\partial^2 v}{\partial y^2} + \nu_T \frac{\partial^2 v}{\partial x^2} + \nu_T \frac{\partial^2 u}{\partial xy} + 2\nu_T \frac{\partial v}{\partial y}\frac{\partial d}{\partial y} + \nu_T \frac{\partial d}{\partial x}\left(\frac{\partial u}{\partial y} + \frac{\partial v}{\partial x}\right) + g\frac{\partial h}{\partial y} - B_y, \tag{12}$$

which we solve by means of a simple explicit method based on centered finite differences. The following derivatives are then used:

$$u_{,t} = \frac{\partial u}{\partial t} = \frac{u_{i,j}^{n+1} - u_{i,j}^n}{\Delta t}, \quad u_{,x} = \frac{\partial u}{\partial x} = \frac{u_{i+1,j}^n - u_{i-1,j}^n}{2\Delta x}, \quad u_{,y} = \frac{\partial u}{\partial y} = \frac{u_{i,j+1}^n - u_{i,j-1}^n}{2\Delta y},$$

$$u_{,xx} = \frac{\partial^2 u}{\partial x^2} = \frac{u_{i+1,j}^n - 2u_{i,j}^n + u_{i-1,j}^n}{(\Delta x)^2}, \quad u_{,yy} = \frac{\partial^2 u}{\partial y^2} = \frac{u_{i,j+1}^n - 2u_{i,j}^n + u_{i,j-1}^n}{(\Delta y)^2},$$

$$u_{,xy} = \frac{\partial^2 u}{\partial x \partial y} = \frac{1}{2\Delta x}\left(\frac{u_{i+1,j+1}^n - u_{i+1,j-1}^n}{2\Delta y} - \frac{u_{i-1,j+1}^n - u_{i-1,j-1}^n}{2\Delta y}\right). \tag{13}$$

We can express the d, h and v derivatives analogously; hence equations (11) and (12) can be written as follows:

$$\begin{aligned} u_{i,j}^{n+1} &= u_{i,j}^n + \nu_T \Delta t[2u_{,xx} + u_{,yy} + v_{,xy} + 2u_{,x}\, d_{,x} + d_{,y}\,(v_{,x} + u_{,y})] \\ &\quad - \Delta t B_x + \Delta t g h_{,x} \\ v_{i,j}^{n+1} &= v_{i,j}^n + \nu_T \Delta t[2v_{,yy} + v_{,xx} + u_{,xy} + 2v_{,y}\, d_{,y} + d_{,x}\,(u_{,y} + v_{,x})] \\ &\quad - \Delta t B_y + \Delta t g h_{,y}\,. \end{aligned} \tag{14}$$

3.1.2. *The variable eddy viscosity model (VEM)*
A second, more complex, algorithm, has been implemented in which the viscous terms are expressed in their complete formulation of equations (6), typical of the classical NSWE viscous approach, and the eddy viscosity is

$$\nu_T = \alpha u_* d \text{ with } u_* = \alpha_1 \sqrt{gd}, \text{ that is } \nu_T = \alpha_2 d\sqrt{gd} = \alpha_2 g^{1/2} d^{3/2}. \quad (15)$$

By substituting equation (15) in equations (7), using equations (6) in equations (8) and (9) and expanding derivatives, we get:

$$\begin{aligned}\frac{\partial u}{\partial t} &= 2\nu_T \frac{\partial^2 u}{\partial x^2} + \nu_T \frac{\partial^2 v}{\partial y^2} + \nu_T \frac{\partial^2 v}{\partial xy} + 2\nu_T \frac{\partial u}{\partial x}\frac{\partial d}{\partial x} + \nu_T \frac{\partial d}{\partial y}\left(\frac{\partial v}{\partial x} + \frac{\partial u}{\partial y}\right) \\ &+ \nu_T d\left[3\frac{\partial d}{\partial x}\frac{\partial u}{\partial x} + \frac{3}{2}\frac{\partial d}{\partial y}\frac{\partial u}{\partial y} + \frac{\partial d}{\partial y}\frac{\partial v}{\partial x}\right] + g\frac{\partial h}{\partial x} - B_x \qquad (16)\end{aligned}$$

$$\begin{aligned}\frac{\partial v}{\partial t} &= 2\nu_T \frac{\partial^2 v}{\partial y^2} + \nu_T \frac{\partial^2 v}{\partial x^2} + \nu_T \frac{\partial^2 u}{\partial xy} + 2\nu_T \frac{\partial v}{\partial y}\frac{\partial d}{\partial y} + \nu_T \frac{\partial d}{\partial x}\left(\frac{\partial u}{\partial y} + \frac{\partial v}{\partial x}\right) \\ &+ \nu_T d\left(\frac{3}{2}\frac{\partial d}{\partial x}\frac{\partial u}{\partial y} + \frac{3}{2}\frac{\partial d}{\partial x}\frac{\partial v}{\partial x} + 3\frac{\partial d}{\partial y}\frac{\partial v}{\partial y}\right) + g\frac{\partial h}{\partial y} - B_y. \qquad (17)\end{aligned}$$

To solve equations (16) and (17), we implemented a simple second-order accurate algorithm which uses centered differences:

$$\begin{aligned}u_{i,j}^{n+1} &= u_{i,j}^{n} + \nu_T \Delta t[2u_{,xx} + u_{,yy} + v_{,xy} + 2u_{,x}\, d_{,x} + d_{,y}\,(v_{,x} + u_{,y})] \\ &+ \nu_T d\Delta t\left[3d_{,x}\, u_{,x} + \frac{3}{2}d_{,y}\, u_{,y} + \frac{3}{2}d_{,y}\, v_{,x}\right] - \Delta t B_x + \Delta t g h_{,x} \qquad (18)\end{aligned}$$

$$\begin{aligned}v_{i,j}^{n+1} &= v_{i,j}^{n} + \nu_T \Delta t[2v_{,yy} + v_{,xx} + u_{,xy} + 2v_{,y}\, d_{,y} + d,x(u_{,y} + v_{,x})] \\ &+ \nu_T d\Delta t\left[\frac{3}{2}d_{,x}\, u_{,y} + \frac{3}{2}d_{,x}\, v_{,x} + 3d_{,y}\, v_{,y}\right] - \Delta t B_y + \Delta t g h_{,y}\,, \qquad (19)\end{aligned}$$

Mixed derivatives are discretized over 8 nodes.

4. The Numerical Solutions vs. The Benchmark Flow

Figures 2 and 3 show the results obtained through the laboratory experiments in two representative cases (SBC and RCC), in terms of absolute and relative statistics. For small times, the absolute statistics computed from the available data usually well match those of a 'ballistic regime', also called 'microturbulence regime', characterized by a dispersion only caused by the microturbulence, the macroturbulence due to waves and large scale eddies being absent. The absolute dispersion shows a typical quadratic growth in time for both the SBC and the RCC. One second 'intermediate' regime, defined as the 'growth regime' is observed for larger times, the duration of which is function both of the wave forcing and of the topographic features. In this regime the macroturbulence dominates. The growth regime is characterized by a t^3-power law. A final 'equilibrium regime' is found, in which transport and decay of the main large-scale features balance to give a sort of mean steady state. In this regime the statistics behave like in the case of a 'brownian regime', thus we find an almost linear growth of the absolute dispersion. The asymptotic value of the absolute dispersion for the cases of SBC is greater than that of the RCC. Note also that the absolute diffusivity is smaller for short waves than that of long waves and the absolute diffusivity is also directly proportional to the offshore wave height. The asymptotic regimes are well characterized by absolute statistics.

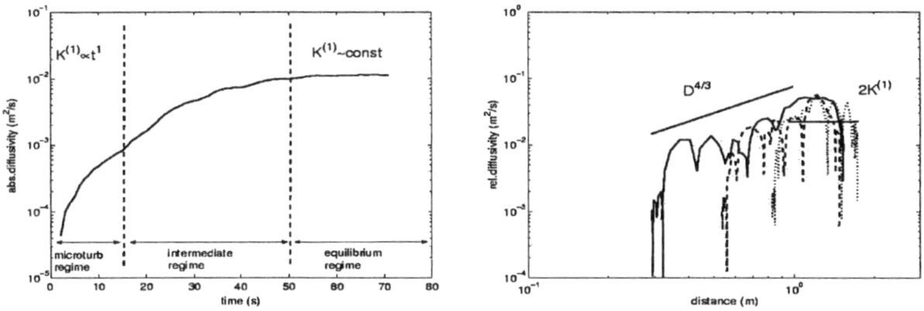

Fig. 2. Experimental absolute diffusivity (left panel) and relative diffusivity (right panel) in the RCC (Test E of Ref. 2).

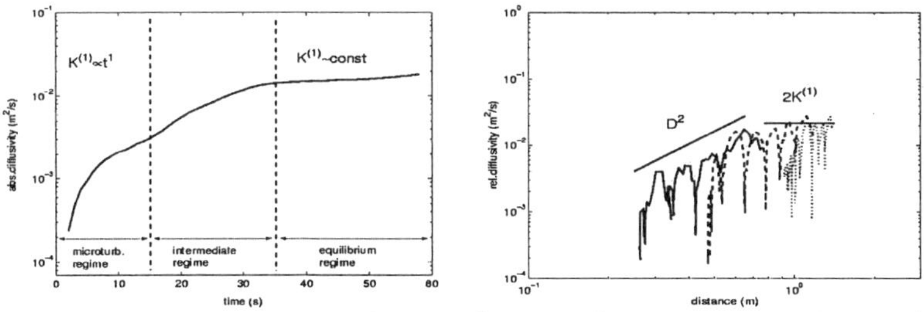

Fig. 3. Experimental absolute diffusivity (left panel) and relative diffusivity (right panel) in the SBC (Test E of Ref. 2).

Three initial separations are used to evaluate the relative dispersion: the smallest and the largest values are used to separate, respectively, the small- and large-scale turbulence. Different dispersion and diffusivity behaviours are found for the intermediate regime of the two configurations. While for the RCC, both relative and absolute statistics are those typical of a shear-dominated flow, in the case of SBC evidence has been found of an enstrophy cascade. In agreement with the results of the relative dispersion, for the scales of the intermediate regime, the diffusivities exhibit different power laws. In the case of the SBC, $K_2 \propto D^2$, which seems to suggest an enstrophy cascade. On the contrary, in the case of the RCC, $K_2 \propto D^{4/3}$, the horizontal shear of the rip currents causing the $D^{4/3}$ dependence. The mixing induced by the SBC is significantly more anisotropic than that induced by the RCC and, while the latter is dominated by a strong offshore and a weak longshore transport, the former is characterized by a strong longshore and a weak crosshore transport.

Comparison of the numerical solutions with the benchmark flow has been made in terms of absolute and relative diffusivity. To reproduce at best experimental statistics, the initial distribution of the floaters has been chosen to coincide with that used in the experiments. Although this choice, which is characterized by a relatively small number of particles (10 to 25), provides noisier statistics than those obtained with a random distribution of more numerous particles, we prefer the use of experimental tracers to better compare with experimental statistics.

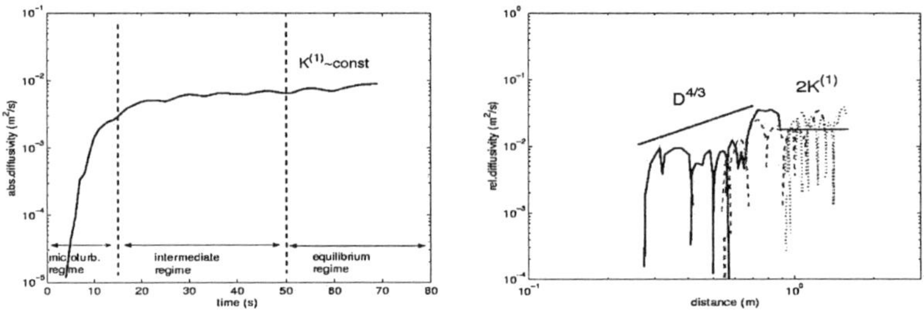

Fig. 4. Numerical absolute diffusivity (left panel) and relative diffusivity (right panel) for the RCC (Test E of Ref. 2) obtained by using the CEM ($\nu_T = 5 \times 10^{-3} \mathrm{m}^2/\mathrm{s}$).

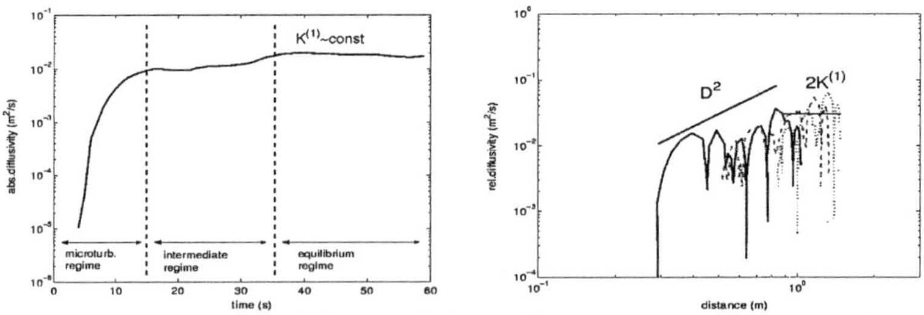

Fig. 5. Numerical absolute diffusivity (left panel) and relative diffusivity (right panel) for the SBC (Test E of Ref. 2) obtained by using the CEM ($\nu_T = 10^{-3} \mathrm{m}^2/\mathrm{s}$).

In figure 4, numerical estimates for the absolute and relative diffusivity for the RCC are shown, obtained by using the CEM. Comparison with figure 2, shows that the three regimes are fairly well reproduced. For the RCC, the best fit between experimental and numerical values of K_1 is obtained when

$\nu_T = 5 \times 10^{-3}\mathrm{m}^2/\mathrm{s}$: K_1 appears to be overestimated for $\nu_T < 10^{-3}\mathrm{m}^2/\mathrm{s}$ and underestimated for $\nu_T = 10^{-2}\mathrm{m}^2/\mathrm{s}$, which, thus, represents an upper bound for the viscosity. Also for the relative statistics, we can state that the numerical results fairly well reproduce the experimental trends: the statistics of the RCC are typical of a shear-dominated flow, while those of the SBC exhibit a different behaviour, which seems to suggest an enstrophy cascade. Also for the relative statistics of the RCC, K_2 appears to be overestimated for $\nu_T < 10^{-3}\mathrm{m}^2/\mathrm{s}$, well match the experimental data for $\nu_T = 5 \times 10^{-3}\mathrm{m}^2/\mathrm{s}$ and underestimated for $\nu_T = 10^{-2}\mathrm{m}^2/\mathrm{s}$. So the best fit between experimental and numerical values of the relative diffusivity and dispersion is obtained for $\nu_T = 5 \times 10^{-3}\mathrm{m}^2/\mathrm{s}$.

In figure 5, numerical estimates for the absolute and relative diffusivity of the SBC are shown, as obtained by means of the CEM. By comparing with figure 2, we find that the best fit between experimental and numerical values of K_1 is obtained when $\nu_T = 10^{-3}\mathrm{m}^2/\mathrm{s}$, but the experimental absolute diffusivity and the numerical one show a rather different trend. However since the figures axes are different, the asymptotic values do not differ that much. For both configurations the absolute diffusivity predicted by the numerical computations decreases with increasing ν_T (this trend is predicted also using random floaters). Computations made with $\nu_T = 10^{-3}\mathrm{m}^2/\mathrm{s}$ overestimate the experimental value of K_2, while those made with $\nu_T = 5 \times 10^{-3}\mathrm{m}^2/\mathrm{s}$ well reproduce the trend of the experimental relative diffusivity, still overestimating it. An increase in ν_T leads to a better estimation of K_2, but also to a worsening of the quadratic D-power law for the relative diffusivity.

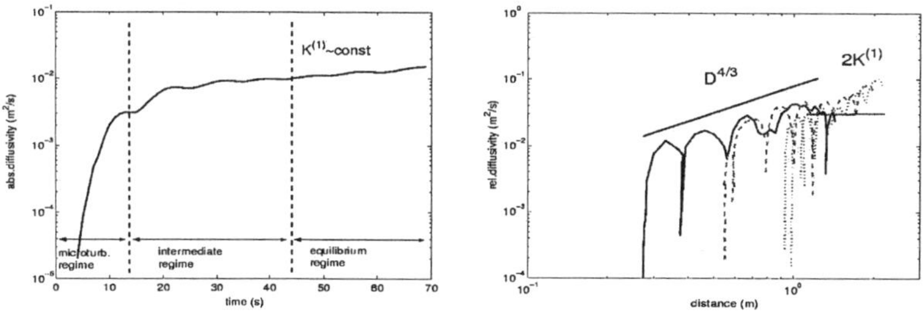

Fig. 6. Experimental absolute diffusivity (left panel) and relative diffusivity (right panel) for the RCC (Test E of Ref. 2) obtained by using the VEM with $\alpha_2 = 0.022$.

Figures 6 and 7 show numerical estimates of the absolute and relative diffusivity for the RCC, obtained with the VEM of equation (15). Different values of α_2 have been tested; in figures 6 and 7 we show the most significative ones. For a value of α_2 such that $\nu_T = 5 \times 10^{-3}\mathrm{m}^2/\mathrm{s}$ at the breakwater

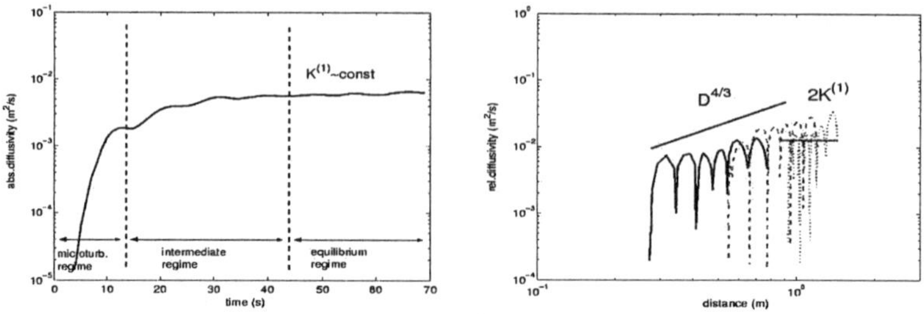

Fig. 7. Experimental absolute diffusivity (left panel) and relative diffusivity (right panel) for the RCC (Test E of Ref. 2) obtained by using the VEM with $\alpha_2 = 0.038$.

seaward foot (i.e. $\alpha_2 = 0.022$), a good overall fit is found between experimental and numerical values of K_1. Instead for a value of α_2 such that $\nu_T = 5 \times 10^{-3} \mathrm{m}^2/\mathrm{s}$ at the seaward boundary of the computational domain (i.e. $\alpha_2 = 0.038$), K_1 cannot be well reproduced. The relative statistics, instead, suggest a different behaviour. In the case of the RCC for $\alpha_2 = 0.022$, a good fit is found between experimental and numerical values of K_2 and the typical shear flow power law of $D^{4/3}$ is well reproduced. For $\alpha_2 = 0.038$, K_2 is underestimated and the decay laws worsen too.

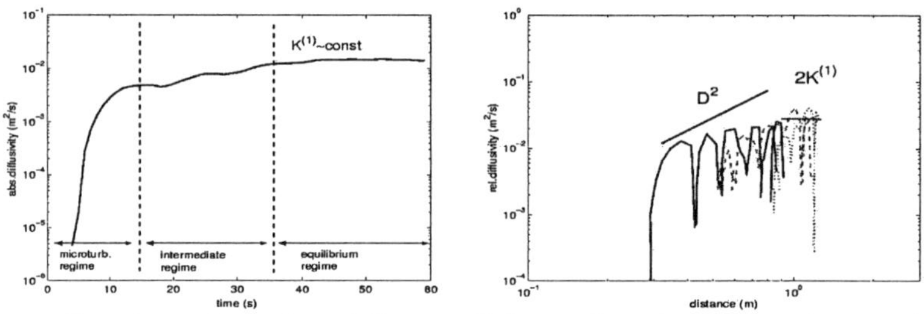

Fig. 8. Experimental absolute diffusivity (left panel) and relative diffusivity (right panel) for the SBC (Test E of Ref. 2) obtained by using the VEM with $\alpha_2 = 0.022$.

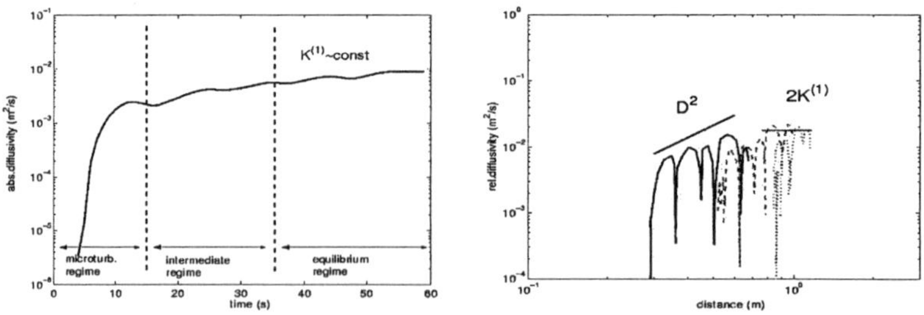

Fig. 9. Experimental absolute diffusivity (left panel) and relative diffusivity (right panel) for the SBC (Test E of Ref. 2) obtained by using the VEM with $\alpha_2 = 0.038$.

Figures 8 and 9 show numerical estimates of the absolute and relative diffusivity for the SBC, obtained with the VEM (see equation (15)). Again different values of α_2 have been tested. For $\alpha_2 = 0.022$, K_1 is just slightly underestimated by the computations while the decay law is well reproduced. For $\alpha_2 = 0.038$, K_1 is even more underestimated and the decay law worsens too. For $\alpha_2 = 0.022$, K_2 is well estimated by the numerical results, but the decay law does not reproduce well the typical enstrophy cascade power law. For $\alpha_2 = 0.038$, the estimate of K_2 is almost the same, but the decay law improves. In summary, from both absolute and relative statistics, we find that the value of α_2 which reproduces the best choice for the eddy viscosity obtained by the CEM (using the breakwater offshore water depth), gives the best fit for almost all parameters (usually that means $\alpha_2 = 0.022$, which gives $\nu_T = 5 \times 10^{-3}\mathrm{m}^2/\mathrm{s}$ at the breakwater seaward foot).

5. Conclusions

The main findings of our study can be summarized as follows:

- a description of macrovortex-induced mixing of nearshore flows has been given on the basis of numerical solutions of viscous NSWE;
- sensitivity of the results to various viscous-type closures has been studied on the basis of statistics of mixing, rather than through velocity spectra;
- both constant and variable (equation (15)) eddy viscosity closures have been used. The best overall agreement between experimental and numerical statistics is found for $\alpha_2 = 0.022$ which represents the physical condition of $\nu_T = 5 \times 10^{-3}\mathrm{m}^2/\mathrm{s}$ at the breakwater seaward foot.

This first, important, step towards a suitable NSWE-HLES framework, is being complemented by ongoing studies of the role of frequency dispersion and various parameterizations of the SGS.

References

1. M. Brocchini, A.B. Kennedy, L. Soldini & A. Mancinelli *Topographically-controlled, breaking wave-induced macrovortices. Part 1. Widely separated breakwaters.* (J. Fluid Mech. **507**, pp. 289-307, 2004).
2. A. Piattella, M. Brocchini & A. Mancinelli *Topographically-controlled, breaking wave-induced macrovortices. Part 3. The mixing features.* (J. Fluid Mech. **559**, pp. 81-106, 2006).
3. M. Brocchini, R. Bernetti, A. Mancinelli & G. Albertini *An efficient solver for nearshore flows based on the WAF method.* (Coast. Engng. **43**, pp. 105-129, 2001).
4. A. Kennedy, M. Brocchini, L. Soldini, & E. Gutierrez *Topographically-controlled, breaking wave-induced macrovortices. Part 2. Changing geometries.* (J Fluid. Mech. **559**, pp. 57-80, 2006).
5. E.F. Toro *Shock-capturing methods for free-surface shallow flows.* (J. Wiley and Sons, 2001).

ELECTRONIC TRANSPORT CALCULATION OF FINITE SINGLE-WALLED CARBON NANOTUBE SYSTEMS IN THE TWO-TERMINAL GEOMETRY

A. LA MAGNA* and I. DERETZIS

Istituto per la Microelettronica e Microsistemi
Stradale Primosole 50, Catania 95121, Italy
**E-mail: antonino.lamagna@imm.cnr.it*
www.imm.cnr.it

We have used the non-equilibrium Green's function formalism combined with a Tight-Binding and an Extended Hückel Hamiltonian for the transport calculation of finite single-walled carbon nanotube systems attached to Au(111) metallic contacts. Findings have revealed the limits of the Tight-Binding approach regarding its use in the Carbon Nanotube context. Moreover, we have thoroughly investigated the channel-blocking effects provoked by interference from the metallic contacts as well as the changes in the transmission function caused by variations in the metal-tube interface geometry.

Keywords: Carbon Nanotubes; Transport Calculation; Non-equilibrium Green's functional formalism; Tight-Binding; Extended Hückel.

1. Introduction

As current trends in the microelectronics industry move towards the fabrication of innovative devices in the molecular level, the need for a realistic modeling of various nanoscale systems becomes crucial. Carbon Nanotubes (CNTs) have been already classified as one of the most promising structures in the context of molecular electronics, as numerous implementations of CNT-based Field-Effect Transistors[1] demonstrate. This popularity emerges from the particular mechanical and electrical properties of CNTs, such as their high mechanical strength, their elasticity, but mostly, their diameter and folding angle dependent conducting behaviour. It has been both theoretically[2] and experimentally[3] shown that changes in the CNT helicity reflect on the way that carbon nanotubes conduct current. On the strict basis of the latter CNTs can behave as metals, semimetals or semiconductors.

Recent theoretical approaches for the transport calculation of CNT-based systems have varied both in terms of the geometrical reconstruction as well as in terms of the computational methodology. Electronic behaviour studies of infinite CNTs,[4] although enlightening for some aspects of ballistic molecular conduction, did not account for the fundamental aspect of scattering by the contacts. On the other hand, finite CNT system approximations (usually of a metal-CNT-metal junction formation) have used both semiempirical[5] (Tight-Binding, Extended Hückel) and *ab initio*[6] methods, each one having its *pros and cons.* Tight-Binding (TB) techniques have permitted large-scale simulations whereas they failed in describing realistically the chemical bond between interface carbon and metal atoms, as well as CNT wall curvature effects. On the contrary, *ab initio* estimations have introduced a high level of accuracy in the description of quantum effects, but have been computationally too heavy for calculations that involve more than some decades of atoms. Finally, Extended Hückel (EH) methods introduced a higher level of complexity with regard to other semiempirical approaches, yielding results close to first-principles data.[7]

In our approach we use the non-equiliblium Green's function formalism[8,9] with two semiempirical Hamiltonians (the π-electron TB one as well as the all-valence-electron EH one) for the transport calculation of various finite single-walled CNT systems attached to gold metallic contacts. The objectives of this analysis fall within two main categories. Firstly, we confront the results obtained by the two Hamiltonians and recognise the limits of the TB approach with regard to its more sophisticated EH counterpart. Secondly, using thereon the EH, we study various contact-tube interfacial configurations (contact-CNT distance and symmetry) that could influence the conduction mechanism.

2. Methodology

We use the non-equilibrium Green's function formalism for the quantum calculation of transport for various finite size CNTs embedded between two semi-infinite metallic contacts in a two-terminal geometry. Our approach is based on the single particle retarded Green's function matrix

$$G = [ES - H - \Sigma_L - \Sigma_R]^{-1} \tag{1}$$

where E is the scalar energy, H is the 'device' Hamiltonian in an appropriate basis set, S is the overlap matrix in that basis set and $\Sigma_{L,R}$ is the self energy which includes the effect of scattering due to the left (L) and right (R) contacts. The overlap matrix S represents the inter-relationship between the basis set of functions used to write the Hamiltonian matrix. If the basis functions result orthogonal and normalized (e.g. tight-binding case), S is the identity matrix I, elsewhere (e.g. extended Hückel case) the matrix elements need to be calculated. Self energy $\Sigma_{L,R}$ terms can be expressed as

$$\Sigma = \tau g_s \tau^\dagger \tag{2}$$

where g_s is the surface Green function specific to the contact type and τ is the Hamiltonian relative to the mutual interaction between the CNT and the contact.[10] g_s has been calculated using the decomposition of the semi-infinite contact in periodically repeated slices and applying the proper lattice symmetry of the contact material.[11] A Landauer-type expression can be used for the current calculation in case of coherent transport:

$$I = \frac{2e}{h} \int_{-\infty}^{+\infty} dET(E)[f(E,\mu_L) - f(E,\mu_R)] \tag{3}$$

where $T(E) = Tr[\Gamma_L G \Gamma_R G^\dagger]$ is the transmission as a function of energy and $\Gamma_{L,R} = i[\Sigma_{L,R} - \Sigma^\dagger_{L,R}]$. In equation (3), $f(E,\mu_{L,R})$ represents the Fermi-Dirac distribution of electrons in the contact at chemical potential $\mu_{L,R}$.

We use two semiempirical Hamiltonian matrixes, the Extended Hückel and the Tight Binding one. Single particle EH Hamiltonian elements are given by:

$$H_{i,j} = \begin{cases} -V_i & \text{if } i = j \\ \frac{S_{i,j}}{2}(V_i + V_j) & \text{if } i \neq j \end{cases} \tag{4}$$

where the indices i, j run over all valence orbitals (approximated with Slater type orbital functions), while overlap matrix $S_{i,j}$ and the diagonal elements V_i can be calculated using the appropriate parameterisation.[9,12] On the other hand in the TB model the basis functions are the π orbitals centred at each atom. The single particle Hamiltonian elements are:

$$H_{i,j} = \begin{cases} \epsilon_0 & \text{if } i = j \\ -t & \text{if } i,j \text{ are next neighbours} \\ 0 & \text{otherwise} \end{cases} \tag{5}$$

Here indices i, j run over all atoms and the constants ϵ_0 and t are 0 and 2.66eV respectively for inner carbon atoms of the CNT.[13] In order to attain a reliable comparison of the results obtained by the two methods, the hopping integral t_{Au-Au} between two gold atoms in the TB approximation has been set as so in order to produce the same surface Density of States at the Fermi level with the EH model. Moreover, we have appositely fitted[14] the t_{C-Au} hopping parameter and have concluded that a $t_{C-Au} = 0.45 \times t_{Au-Au}$ parameterisation produces similar I-V spectra for both hamiltonian models within a large range of the potential across the $V = 0$ value.

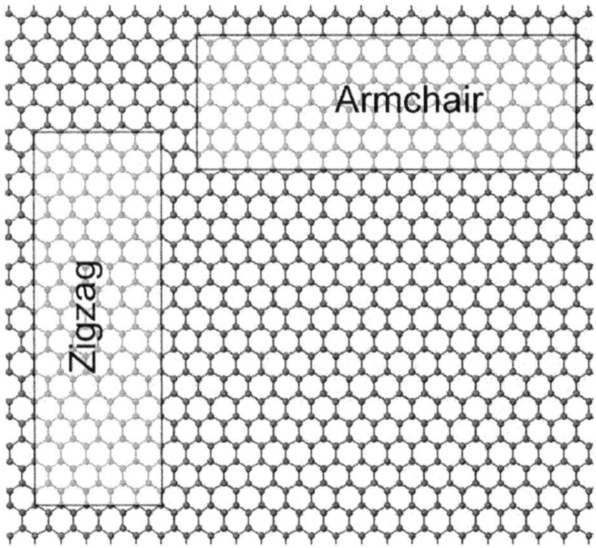

Fig. 1. Carbon nanotubes can be 'generated' by rolling up a graphene sheet. Depending on the direction that this sheet is rolled up, armchair (with basis vector indices (n, n)), zigzag (with basis vector indices $(n, 0)$) and chiral nanotubes (with (n,m) indices and $n \neq m \neq 0$) can be formed.

3. TB and EH transmission functions for armchair and zigzag CNTs

We have performed a number of simulations in order to calculate transmission functions for various metal-CNT-metal systems, using Au(111) as a reference contact metal. We have applied a 'stick and ball' geometrical configuration between the pad and the CNT with each interface carbon atom lying within three gold atoms, although not always maintaining equal distances from them. At this point we have fixed the contact-tube distance at 1 Å.[5] Figures 2 and 3 plot transmission functions of various armchair and zigzag CNTs (see figure 1) using the EH and the TB Hamiltonians. Transmission curves for the respective infinite (perfectly contacted) CNTs are also shown as dashed lines. Infinite conductance has a step-like behaviour, where each step corresponds to the number of spin degenerate bands available (each having a conductance equal to the conductance quantum $G_0 = \frac{2e^2}{h}$). A lack of bands that cross the system's Fermi energy (E_F) denotes a conductance gap typical of semiconducting CNTs, evidenced by a zero transmission value in the two figures. Contrary, tubes with transmission values other than zero have a metallic character. The superiority of the EH model over the TB one can be quickly identified by the fact that it is the only one to predict a secondary conductance gap around E_F for the (9,0) tube. Such minor gap is typical of semimetallic zigzag nanotubes and can be evidenced only when the band structure is calculated by means of theoretical approaches beyond the zone-folding approximation (e.g. using *ab initio* Hamiltonians).[15] When it comes to the finite tubes, the infinite conduction plateaux is replaced by peaks that represent the discrete set of energy states that substitute the continuous bands.

Results obtained for small radius CNTs show that the discrepancy between the two methods is noteworthy. In fact, the (4,0) CNT that has a semiconducting character with TB, behaves as a metal for the EH. In this case the deviation observed lies in the fact that for thin CNTs $\sigma - \pi$ rehybridisation effects take place due to wall curvature. EH is the only method capable of predicting them since it involves all valence orbitals in the estimation of the Hamiltonian matrix. Likewise, for the metallic (2,2) CNT where TB shows a $2\times G_0$ plateaux around E_F (typical of metallic nanotubes), EH shows levels relative to $1\times G_0$ and $3\times G_0$. Therefore, a first important consideration that can be stated is that the TB approximation is inadequate when it comes to transport calculation of small radius nanotubes (less than ~3.5Å). On the other hand, our findings for bigger radius CNTs show that given a well-fitted parameterisation the two models seem to better agree, although one has to take into account that there always exist small-scale disparities that are important for currents of the order of μA.

The significance of the metallic contact on the conduction mechanism can be evidenced by the differences between finite and infinite transmission spectra. The peaks of the finite spectra for the (6,6) CNT do not reach

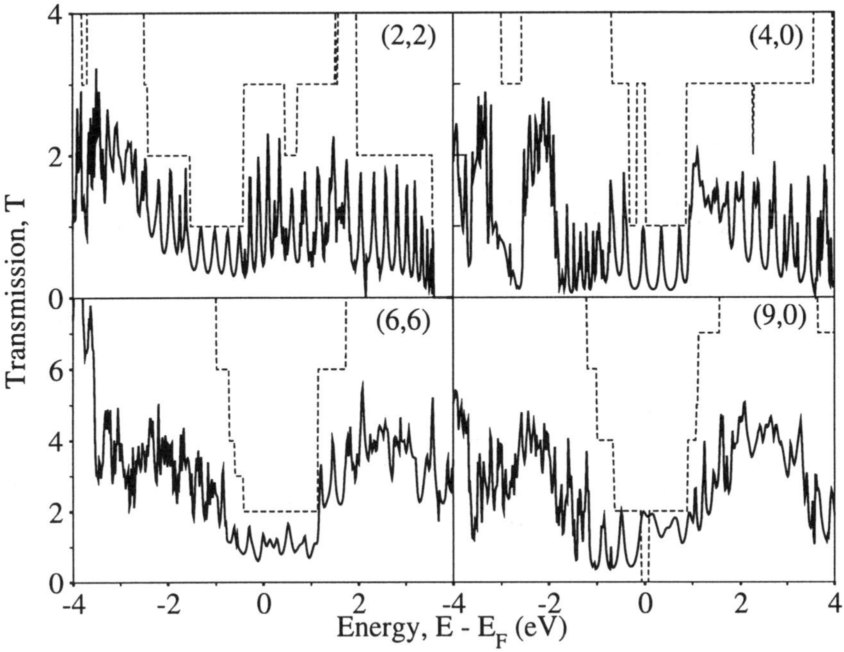

Fig. 2. Transmission as a function of energy for a 20 unit-cell (2,2), a 10 unit-cell (4,0), a 14 unit-cell (6,6) and an 8 unit-cell (9,0) CNT, calculated with the EH model. Transmission lines of infinite systems are also shown as dashed lines. Fermi energy of the respective infinite CNTs is set to zero.

the infinite conductance plateau near E_F by neither of the two models. In such cases the effect of contact interference between conduction channels can be fundamental for the overall conductance of the system. On the other hand the (9,0) tube with the same type of contact does not show analogous behaviours and peaks reach the infinite conductance plateau. Equivalently, (2,2) and (4,0) CNTs demonstrate peaks that arrive at the $1\times G_0$ level near E_F, but never at the $3\times G_0$ one. Since interference is purely a contact effect, a direct conclusion could be that the chemical nature of the electrode is not the only factor which rules contact transparency, since by fixing the latter while changing the CNT type, complex contact interference behaviours can be observed.

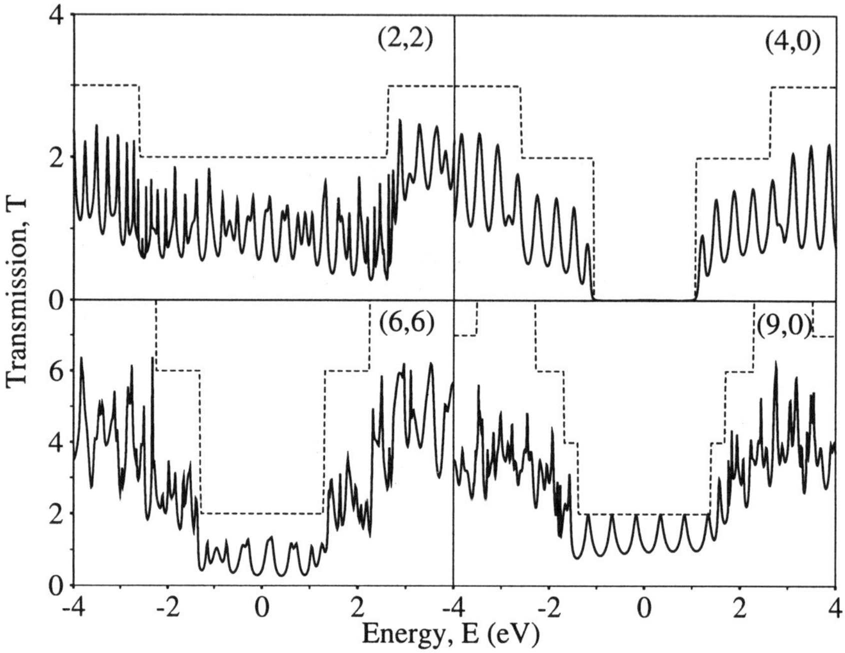

Fig. 3. Transmission as a function of energy for a 20 unit-cell (2,2), a 10 unit-cell (4,0), a 14 unit-cell (6,6) and an 8 unit-cell (9,0) CNT, calculated with the TB model. Transmission lines of infinite systems are also shown as dashed lines.

4. Influence of the contact-CNT interface geometry on transport

Interfacial reconstruction effects are of a well-known importance when it comes to CNT transport. As such, we have decided to study the influence of electrode positioning with respect to the CNT on the systems conducting behaviour. Rather than focusing on an optimised 2-terminal structure, we have investigated a wide range of metal-CNT geometries of various distances and symmetries that can be considered closer to the experimental aspect of the problem. Since the TB Hamiltonian is quantitatively inadequate for the description of the contact-CNT chemical bonding character, we have preferred the more accurate EH method for the completion of these simulations. We considered two 8-unit cell CNTs (a (9,0) and a (10,0)), which we 'sandwiched' between the two gold pads, using at first the same symmetrical configuration between C and Au interface atoms as above. We initially focused on the role of distance between the metallic electrodes and the CNTs, calculating the transmission functions of two diverse snapshots of our systems that varied this distance from 1 to 2 Å. Simulation outcomes (figure 4) reveal an important influence of the positioning factor for both types of tubes. The resulted transmission curve for the (9,0) system implies a low conductance system for a 2-Å contact-tube distance (shorter and thinner peaks around E_F), while when bringing the pad closer a respective increase in the current flow is observed with peaks arriving at $2\times G_0$. In the case of the (10,0) CNT the transmission spectra becomes even more remarkable since for a 2-Å distance peaks appear inside the semiconducting gap. Such phenomenon is typical of unformed chemical bonds between interface atoms at the dangling ends of the CNT. In fact, when contacts are brought closer such peaks disappear and the semiconducting character of this tube can be evidenced by the conductance gap around E_F. These theoretical results can have a strong laboratory impact since they imply that Au electrode probes that a more distant than 1.5-2 Å from the CNTs ends cannot be considered transparent.

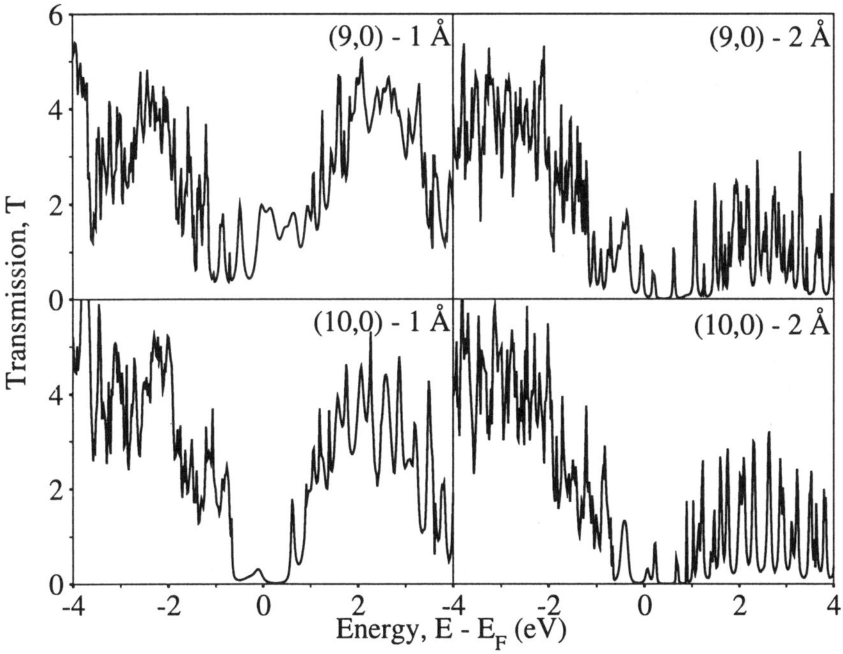

Fig. 4. Transmission as a function of energy for a (9,0) and a (10,0) system (8 unit-cells long) with a varying contact distance of 1 Å and 2 Å.

Apart from the distance factor, a realistic representation of the interfacial geometry between the contacts and the CNT is important in understanding the factors that influence CNT transport. For such reason we have performed simulations for the same types of tubes as before but randomly varying this time the symmetry between the electrode and the tube, provoking an arbitrary twist and move of the device with respect to the contacts. The obtained results displayed slight changes in the transmission spectra whereas the overall behaviour of the systems remained unaltered. It can therefore be stated that a contact-tube symmetry effect is present, whereas this plays a marginal role. Finally, it is important to mention that same procedure simulations have been also performed for other types of CNTs with similar findings.

5. Discussion

In this paper we have studied the electronic transport properties of various finite carbon nanotube systems using the non-equilibrium Green's function formalism with a TB and an EH Hamiltonian. Our analysis has demonstrated that the TB approach fails in describing adequately the conduction mechanism when it comes to small radius CNTs (where valence orbital re-hybridisation effects take place), or when an accurate description of the carbon-metal bond is needed. It can be still used though as an auxiliary computational tool for large scale calculations of bigger radius nanotubes, given an appropriate parameterisation. Using the EH Hamiltonian thereon we have attempted to give an insight on the effect that the contact-CNT interfacial reconstruction has on transport, demonstrating that the contact-nanotube distance can be considered crucial, whereas that of contact-CNT symmetry does not influence the conduction mechanism in a significant manner. Taking therefore into account recent CNT-based FET device architectures,[1] we can state that we can expect quite homogeneous conductance properties among equivalent (same metal, same size, same helicity) devices. However, a strong distance variance should arise when a contact is represented by a metallized atomic force microscope tip used as an electrical nanoprobe in conductance measurements. In this case, our results indicate that a careful analysis should be performed in order to understand the experimental results.

Acknowledgments

The authors would like to thank P. Alippi for the useful discussions and V. Privitera and S. Scalese, who indicated the experimental relevance of the problem in study.

References

1. Joachim C, Gimzewski J K, Aviram A, *Nature* **408**, 541 (2000); Martel R, Schmidt T, Shea R, Hertel T and Avouris Ph, *Appl. Phys. Lett.* **73**, 2447 (1998); A. Bachtold, P. Hadley, T. Nakanishi, C. Dekker, *Science* **294**, 1317 (2001).
2. Mintmire J W, Dunlap B I and White C T, *Phys. Rev. Lett.* **68**, 631 (1992); Hamada N, Sawada S and Oshiyama A, *Phys. Rev. Lett.* **68**, 1579 (1992); Saito R, Fujita M, Dresselhaus G and Dresselhaus M S, *Phys. Rev. B* **46**, 1804 (1992); Saito R, Fujita M, Dresselhaus G and Dresselhaus M S, *Appl. Phys. Lett.* **60**, 2204 (1992).
3. Odom T W, Huang J, Kim P and Lieber C M, *Nature* **391**, 624 (1998); Wildöer J W G, Venema L C, Rinzler A G, Smalley R E and Dekker C, *Nature* **391**, 5962 (1998).
4. Chico L, Benedict L X, Louie S G and Cohen M L, *Phys. Rev. B* **54**, 2600 (1996).
5. Rochefort A, Avouris P, Lesage F and Salahub D R, *Phys. Rev. B* **60**, 13824 (1999).
6. Palacios J J, Pérez-Jiménez A J, Louis E, SanFabián E and Vergés J A, *Phys. Rev. Lett.* **90**, 106801 (2003).
7. Rochefort A, Salahub D R and Avouris P, *J. Phys. Chem. B* **103**, 6416 (1999).
8. Datta S, Non-Equilibrium Green's Function (NEGF) Formalism: An elementary Introduction, *Proceedings of the International Electron Devices Meeting (IEDM)*, IEEE Press (2002).
9. Zahid F, Paulsson M and Datta S, Electrical conduction through molecules: Advanced Semiconductors and Organic Nano-Techniques, ed. H Morkoc, New York: Academic (2003).
10. A detailed derivation of the expression which connects the self-energy due to the contact to the contact surface Green function is reported by S. Datta in: Electronic Transport in Mesoscopic Systems, H. Ahmed, M. Pepper, A. Broers (Eds.), Cambridge University Press (1995).
11. Lopez-Sancho M, Lopez-Sancho J and Rubio J, *J. Phys. F. Met. Phys.* **14**, 1205 (1984); Sanvito S, Lambert C J, Jefferson J H, Bratkovsky A M, *Phys. Rev. B* **59**, 11936 (1999).
12. Mulliken R S, Rieke C A, Orloff D and Orloff H, *J. Chem. Phys.* **17**, 1248 (1949).
13. Krompiewski S, *J. Magn. Magn. Mater.* **272-276**, 1645 (2004).
14. Deretzis I, La Magna A, *Nanotechnology* **17**, 5063 (2006).
15. Reich S, Thomsen C and Maultzsch J, Carbon Nanotubes: Basic Concepts and Physical Properties (Weinheim: WileyVCH) (2004).

MULTILEVEL GRADIENT METHOD WITH BÉZIER PARAMETRISATION FOR AERODYNAMIC SHAPE OPTIMISATION

M. MARTINELLI* and F. BEUX**

Scuola Normale Superiore di Pisa, Piazza dei Cavalieri 7, 56126 Pisa, Italy
*E-mails: *max.martinelli@sns.it and **fbeux@sns.it*

A multilevel approach based on a set of embedded parametrisations is proposed in the context of gradient methods for optimum shape design in aerodynamics. This method extends an existing multilevel gradient-based formulation to another type of control subspaces, giving, in this way, another preconditioner for the gradient. In particular, Bézier control points associated with the property of degree elevation are involved instead of shape grid-points coordinates and polynomial interpolation. The behaviour of the new formulation is illustrated on different 2D inverse problems for inviscid flows.

Keywords: Optimum shape design; gradient methods; multilevel strategies.

1. Introduction

Since optimum shape design problems in aerodynamics are characterised by the high computational cost of the objective function evaluation, the improvement of optimisation algorithm efficiency appears as an important task. In the present work, we are interested in gradient-like methods in which the computation of a discrete gradient is considered, i.e. the differentiation is performed on the fully discretised governing equations. In this context, a natural choice of parametrisation is to consider the coordinates of the grid-points localised on the shape which should be optimised. Indeed, this kind of parametrisation allows a direct correlation with the explicit representation of the shape in the discrete cost functional. Nevertheless, in this case, non-smooth profiles can appear during the shape optimisation process, and moreover, the large number of variables involved has also a negative effect on the convergence rate. In [1], a multilevel strategy, in which the minimisation is done alternatively on different subsets of control parameters according to multigrid-like cycles, has been defined in order to reduce these drawbacks. More particularly, using shape grid-point coordinates as design variables, a hierarchical parametrisation was defined considering different subsets of parameters extracted from the complete parameterisation, which can be prolongated to the higher level by linear mapping. This approach acts as a smoother and, on another hand, makes the convergence rate of the gradient-based method low dependent of the number of control parameters. The concept of a multilevel approach for aerodynamic shape optimisation has also been proposed in [2], but, considering a set of Bézier control points as control variables, and gradient-free methods as optimisation algorithm.

In the present study, a new multilevel strategy based on the use of Bézier control points, but in the context of a gradient-based method, is proposed. Indeed, the present algorithm can be understood as a multilevel strategy as defined in [1], in which a particular prolongation operator is applied. Nevertheless, it is grounded on some basic concepts yet proposed in [2], as, for instance, the degree elevation property of the Bézier curves, and, can be also interpreted as a multilevel strategy in which the control parameters are Bézier control points instead of shape grid-points.

In the first section, the multilevel formulation based on a change of control space, as initially introduced in [1], is recalled as well as extended to the case of an affine prolongation operator. The second section is dedicated to the new multilevel approach which considers a subset of Bézier control points as parametrisation on the sub-levels. In particular, an affine prolongation operator and a way to construct a family of consistent sub-parametrisations are defined. Furthermore, the algorithm is also reinterpreted as a descent method for Bézier control points as control variables is proposed. Finally, numerical experiments are presented for different 2D inverse problems for inviscid flows in order to evaluate the behaviour of the present approach.

2. Gradient-based method and multilevel approach

2.1. *Space change as preconditioning for gradient methods*

Let us consider the following optimisation problem:

$$\text{Find } \bar{\gamma} \in U \text{ such that } j(\bar{\gamma}) = \min_{\gamma \in U} j(\gamma) \tag{1}$$

in which $j : U \to I\!R$ is a differentiable functional and U a Hilbert space. Moreover, instead of a direct minimisation of j in U, it can be also envisaged a minimisation of j in the subset $f(V) \subset U$, in which V is a second Hilbert space and f an application from V to U. It can be formulated, equivalently, as the minimisation of $\mathcal{J} = j \circ f$ in V, i.e.:

$$\text{Find } \bar{\alpha} \in V \text{ such that } \mathcal{J}(\bar{\alpha}) = \min_{\alpha \in V} [j \circ f](\alpha) \tag{2}$$

Furthermore, let us consider the particular case in which f is an affine application from V to U, and thus, it exists $b \in U$ and $P \in \mathcal{L}(V, U)$, i.e being a linear continuous application from V to U, such that $f : \alpha \to P\alpha + b$. Then, since the derivative of f verifies $f' = P$, the Fréchet derivative of $\mathcal{J}$ at $\alpha \in V$ can be expressed as follows:

$$\forall h \in V \quad \mathcal{J}'(\alpha)(h) = [j \circ f]'(\alpha)(h) = j'(\gamma)(f'(\alpha)(h)) = j'(\gamma)(Ph) \tag{3}$$

in which γ is defined by $\gamma = f(\alpha)$.

From eq. (3) the following relation is obtained in terms of gradient:

$$\forall h \in V \quad \langle \text{grad}_V \mathcal{J}(\alpha), h \rangle_V = \langle \text{grad}_U j(\gamma), Ph \rangle_U = \langle P^* \text{grad}_U j(\gamma), h \rangle_V \tag{4}$$

where $P^* \in \mathcal{L}(U, V)$ is the adjoint of P and $\langle ., . \rangle_U$ and $\langle ., . \rangle_V$ are the inner products associated to U and V respectively.

Then, thanks to the previous relation (3) between the two gradients, solving the minimisation problem (2) through a gradient descent method corresponds to the following iterative algorithm:

$$\alpha_0 \in V \text{ given,} \qquad \text{for } r \geq 0 \quad \alpha_{r+1} = \alpha_r - \omega_r P^* \text{grad}_U\, j(f(\alpha_r))$$

Nevertheless, applying the operator f permits to go back to space U, and thus, to obtain:

$$f(\alpha_{r+1}) = P\alpha_{r+1} + b = f(\alpha_r) - \omega_r PP^* \text{grad}_U\, j(f(\alpha_r))$$

Thus, considering as initial solution $\gamma_0 = f(\alpha_0)$, the following iterative algorithm is finally defined in U:

$$\text{for } r \geq 0 \quad \gamma_{r+1} = \gamma_r - \omega_r PP^* \text{grad}_U\, j(\gamma_r) \tag{5}$$

Furthermore, the resulting algorithm is a weak descent method in U, i.e. for any $\omega_r > 0$ small enough, we have $j(\gamma_{r+1}) \leq j(\gamma_r)$. Indeed,

$$j(\gamma_r + p_r) = j(\gamma_r) + \langle j'(\gamma_r), p_r \rangle_U + o(||p_r||_U)$$

in which p_r is the descent direction, i.e. $p_r = PP^* \text{grad}_U\, j(\gamma_r)$. Thus for $\omega_r > 0$ small enough, the following inequality is verified:

$$j(\gamma_{r+1}) = j(\gamma_r) - \omega_r ||P^* \text{grad}_U\, j(\gamma_r)||_V + o(|\omega_r|) \leq j(\gamma_r).$$

For a discrete problem in finite dimension, the two Hilbert spaces can be, typically, $U = \mathbb{R}^p$ and $V = \mathbb{R}^q$, p, q being integers. Then, the linear operator P and its adjoint P^* are associated to a matrix $M \in \mathbb{R}^{p \times q}$ and its transpose respectively, and thus, the algorithm can be rewritten as:

$$\text{for } r \geq 0 \quad \gamma_{r+1} = \gamma_r - \omega_r \, MM^T g_r \tag{6}$$

with $g_r = \operatorname{grad} j(\gamma_r) \in \mathbb{R}^p$. Note that, by construction, MM^T is a square symmetrical positive semidefinite matrix. Furthermore, since the minimisation in V appears uniquely through the matrix MM^T while the gradient computation is only done in U, this algorithm can be also interpreted as a preconditioned gradient method.

2.2. *A hierarchical shape grid-point parametrisation*

An optimisation algorithm based on space change as presented in Sec. 2.1 has been initially proposed in [1] for the case of a linear operator, i.e. for $f = P$. In this study, the ordinates of the grid-points, localised on the shape which should be optimised, have been chosen as control variables γ. Then, a set of points, extracted from the complete set of shape grid-points, is considered as sub-parametrisation α while a Hermitian cubic interpolation is defined as prolongation operator f.

Moreover, instead of considering a single space V, the cost functional is minimised alternatively on different control subspaces of decreasing dimension. More precisely, a family of embedded sub-parametrisations is considered, in which for each increase of level the number of points is doubled. At a particular level, the prolongation operator is defined by $P^{(l)} = P^L_{L-1} \circ \cdots \circ P^{l+1}_l$ where L corresponds to the finest level, i.e. to the complete parametrisation, while P^{i+1}_i is the cubic interpolation used for the prolongation from level i to the next one. In practice, at each optimisation iteration r corresponds a particular level l, and following (5), minimising on this coarse level l corresponds to replace the gradient g_r by the descent direction $p^{(l)}_r = P^{(l)}(P^{(l)})^* g_r$. The choice of the particular subspace is determinate by a strategy of level changes similar to multilevel/multigrid strategies used for the resolution of partial differential equations (as, for instance, V-cycles).

Note that the multilevel approach elaborated in [1] is strongly linked to the particular type of parametrisation used. Nevertheless, the formulation described in Sec. 2.1 is rather more general since, providing that the prolongation operator be affine, any type of design variables and sub-parametrisations can be proposed. In particular, it should be interesting to define a multilevel approach associated to a polynomial representation of the shape. Indeed, the kind of parametrisation is often used allowing a more compact description with only few control parameters. In this context, the Bézier curves, in which a shape is characterised by control points, act as a basic tool.

3. Multilevel method associated to Bézier parametrisation

3.1. *Sub-parametrisations through Bézier control points*

A Bézier curve of degree n can be defined according to its parametrisation, i.e.:

$$S(t) = (x(t), y(t))^T = \sum_{q=0}^{n} B_q^n(t) S_q \quad \text{with} \quad t \in [0,1] \tag{7}$$

where $S_q = (\mathrm{x}_q, \mathrm{y}_q)^T$ is the q-th Bézier control point while $B_q^n(t)$ corresponds to the q-th Bernstein polynomial of degree n.
Thus, for each set of parameters $(t_k)_{k=0,m}$ with $t_0 = 0 < t_1 < \cdots < t_m = 1$ corresponds a set of points on the Bézier curve depending of the particular choice of Bézier control points $(S_q)_{k=0,n}$. As done in [1], let us consider as control variables the ordinates of the shape grid points whereas, the abscissas $x_0^\Gamma, \cdots, x_m^\Gamma$ are defined by the knowledge of the initial mesh and frozen during the optimisation process.

Note that, since the extremities of the shape to be optimised are fixed, and due to the property of endpoints interpolation of the Bézier curve, y_0 and y_n are given parameters defined by

$$\mathrm{y}_0 = y(0) = y_0^\Gamma \text{ and } \mathrm{y}_n = y(1) = y_m^\Gamma$$

while the $(m-1)$ other shape ordinates are obtained through eq. (9) by:

$$y_k^\Gamma = y(t_k) = \sum_{q=1}^{n-1} B_q^n(t_k)\mathrm{y}_q + \underbrace{B_0^n(t_k)\mathrm{y}_0 + B_n^n(t_k)\mathrm{y}_n}_{b_k} \tag{8}$$

Thus, the control variables $\gamma = (y_1^\Gamma, \cdots, y_{m-1}^\Gamma)^T$ are related to $\alpha = (\mathrm{y}_1, \cdots, \mathrm{y}_{n-1})^T$ by the following affine operator:

$$\begin{aligned} f : \mathbb{R}^{n-1} &\longrightarrow \mathbb{R}^{m-1} \\ \alpha &\longmapsto \gamma = M\alpha + b \quad b \in \mathbb{R}^{m-1}, M \in \mathbb{R}^{m-1 \times n-1} \end{aligned}$$

where

$$\begin{cases} M_{kq} = B_q^n(t_k), & \text{for } q = 1, n-1 \text{ and } k = 1, m-1 \\ b_k = (1-t_k)^n y_m^\Gamma + (t_k)^n y_0^\Gamma, & \text{for } k = 1, m-1 \end{cases}$$

Then, according to the formulation defined in Sec. 2.1, it can be envisaged to define a strategy similar to the original multilevel approach [1], where instead of a subset of boundary grid-points, each sub-parametrisation is a set of Bézier control points. Indeed, let us consider the following descent direction at iteration r and level l:

$$d_r^{(l)} = M^{(l)}(M^{(l)})^T g_r \quad \text{with} \quad M_{ij}^{(l)} = B_j^{n_l}(t_i^{(l)}). \tag{9}$$

Note that the algorithm is formally identical to the one proposed in [1]; indeed, the two algorithms only differ by the particular choice $M^{(l)}$, i.e. the choice of the specific subspace used in the multilevel process. Nevertheless, it is not so obvious to describe the different levels since the sub-parametrisation definition occurs, here, in a less natural way. Indeed, the abscissas of shape grid-points $(\{x_k^\Gamma\}_{k=0,m})$ being given, at each level, $X^{(l)} = (x_0^{(l)}, \cdots, x_{n_l}^{(l)})^T$ with $n_l > n_{l-1}$ and $T^{(l)} = (t_0^{(l)}, \cdots, t_m^{(l)})^T$ should be defined consistently with (7), i.e.:

$$x_k^\Gamma = x(t_k^{(l)}) = \sum_{q=0}^{n_l} B_q^{n_l}(t_k^{(l)})\mathrm{x}_q^{(l)} \quad \text{for} \quad k = 0, \cdots, m \tag{10}$$

Note that the definition of the different sub-parametrisations is not dependent of the particular optimisation iteration r, and thus, should be done as a preprocessing.

3.2. *Sub-parametrisation based on degree elevation*

In order to define an adequate parametrisation at level l, the property of degree elevation, which allows to increase the degree and the number of control points of a Bézier curve, is used here. Note that the good features of the degree-elevation property of Bézier curves have been pointed out in [2] and, already, employed to construct an embedded parametrisation.

Given a Bézier curve of degree s associated to the $s+1$ control points $S_q = (\mathrm{x}_q, \mathrm{y}_q)^T$, the same geometrical curve can be also understood as a Bézier curve of degree $s+1$ considering a new set of $s+2$ control points $\bar{S}_q = (\bar{\mathrm{x}}_q, \bar{\mathrm{y}}_q)^T$ obtained from S_q as follows:

$$\bar{S}_q = \frac{q}{s+1} S_{q-1} + \left(1 - \frac{q}{s+1}\right) S_q \quad \text{for } q = 0, \cdots, s+1 \tag{11}$$

An interesting feature is that the distribution of the parameters t over the Bézier curve does not change by degree elevation, and thus:

$$\sum_{q=0}^{s} B_q^s(t)\,\mathrm{x}_q = \sum_{q=0}^{s+1} B_q^{s+1}(t)\,\bar{\mathrm{x}}_q \qquad \forall t \in [0,1] \tag{12}$$

Consequently, if X, a set of abscissas is consistent with T, a set of parameters t (i.e., the relation (10) is verified) then the consistency with respect to T is preserved by applying the degree elevation algorithm on X. Thus, let us suppose that the parametrisation on the coarsest level has been yet defined with $X^{(0)}$ and $T^{(0)}$ consistent, i.e. with (10) verified for $l = 0$. Then, keeping the parameters t_k unchanged on all the levels, i.e. $T^{(l)} = T^{(0)}$ for all $l > 0$, one can construct $X^{(l)}$ for $l > 0$ by applying successively (11) until to obtain $n_l + 1$ abscissas. More precisely, $X^{(l)}$ is obtained from $X^{(l-1)}$ by applying $n_l - n_{l-1}$ times the degree elevation algorithm, and thus, we obtain a family of embedded sub-parametrisations with a progressive increase of the number of control points.

Note that, the conditions at the endpoints $\mathrm{x}_0^{(l)} = x_0^\Gamma$ and $\mathrm{x}_n^{(l)} = x_m^\Gamma$ are automatically verified at each level if these relations occur for $l = 0$. Furthermore, additional geometrical constraints are often imposed on the shape: for instance, a vertical tangent at the origin is a standard constraint at the leading edge for airfoil profiles. With a Bézier curve, the derivatives at the endpoints can be easily managed, and in particular, a vertical tangent at the origin can be enforced by the condition $\mathrm{x}_0 = \mathrm{x}_1$. It is easy to see that the degree elevation also preserves this condition.

3.3. *Construction of a consistent coarsest sub-level*

The simplest way to construct an initial set of abscissas for Bézier control points is to consider the set reduced to the endpoints, i.e. $\tilde{X}_0 = (x_0^\Gamma, x_m^\Gamma)$. In this case, eq. (10) with $n_l = 1$ gives the following expression for the parameters t_k:

$$t_k = \frac{x_k^\Gamma - x_0^\Gamma}{x_m^\Gamma - x_0^\Gamma} \quad \text{for } k = 0, \cdots, m.$$

On another hand, any $\tilde{X}_l$ obtained by applying successively the degree-elevation process starting from $\tilde{X}_0$ yields an uniform distribution of points on the interval $[x_0^\Gamma, x_m^\Gamma]$. Furthermore, if, as for the case of the nozzle inverse problem presented in Sec. 4, the $(x_k^\Gamma)_{k=0,m}$ are uniformly distributed, it directly follows that the parameters t_k are also uniformly distributed. Thus, a consistent coarsest level can be easily obtained by simply choosing an uniform distribution for both $X^{(0)}$ and $T^{(0)}$.

Let us, now, consider the case in which a vertical tangent should be imposed at the origin. In this case, $\tilde{X}_0 = (x_0^\Gamma, x_0^\Gamma, x_m^\Gamma)^T$ gives the simplest initial set of abscissas for Bézier control points. $\tilde{X}_0$ is associated to the following distribution of the parameters t_k:

$$t_k = \sqrt{\frac{x_k^\Gamma - x_0^\Gamma}{x_m^\Gamma - x_0^\Gamma}} \quad \text{for } k = 0, \cdots, m. \tag{13}$$

A consistent coarsest level is then obtained considering $T^{(0)}$ defined by (13) and $X^{(0)}$ being any $\tilde{X}_l$ obtained from $\tilde{X}_0$ by applying successively the degree-elevation process.

3.4. *Reinterpretation of the Bézier-based algorithm*

In the present formulation, even if each sub-parametrisation corresponds to the ordinates of a set of Bézier control points, these control points do not explicitly appear in the definition of the descent direction since only the parameters $(t_k)_{k=0,m}$ are directly involved in (9). Thus, the knowledge of the particular position of the Bézier control points is not required in the practical algorithm implementation. Nevertheless, an explicitly dependence on these control points can be, also, reintroduced in the formulation.

Let us consider a particular level l, with $0 \leq l < L$, L corresponding to the finest level of Bézier control points.

$$\text{for } k = 1, m-1 \qquad y_k^\Gamma = \sum_{q=1}^{n_l - 1} B_q^{n_l}(t_k)\, \mathrm{y}_q^{(l)} + b_k^{(l)} = \sum_{q=1}^{n_L - 1} B_q^{n_L}(t_k) \mathrm{y}_q^{(L)} + b_k^{(L)}$$

The last equality is due to the fact that $(\mathrm{y}_q^{(L)})_{q=0,n_L}$ is obtained from $(\mathrm{y}_q^{(l)})_{q=0,n_l}$ by a degree-elevation process, i.e. using iteratively the relation (12) for the ordinates. Thus, the prolongation operator at level l, i.e. $f^{(l)} : \alpha \longmapsto \gamma = M^{(l)}\alpha + b^{(l)}$, verifies

$$f^{(l)} = f^{(L)} \circ d_l^L$$

in which $f^{(L)} : \beta \longmapsto \gamma = M^{(L)}\beta + b^{(L)}$ is the operator relating the finest level of Bézier control points to the shape grid-points while d_l^L corresponds to the prolongation operator from level l to level L.

Note that the degree-elevation process is linear, and thus, excluding the endpoints, the application d_s, which furnishes s internal control points from $s-1$ ones, is affine. More precisely, d_s is defined by:

$$\begin{array}{rl} d_s : I\!\!R^{s-1} & \longrightarrow I\!\!R^s \\ \alpha & \longmapsto D_s\alpha + \lambda_s \end{array} \quad \text{with} \quad \begin{cases} \lambda_s = \dfrac{1}{s+1}(y_0^\Gamma, 0, \cdots, 0, y_m^\Gamma)^T \\ \\ (D_s)_{ij} = \dfrac{1}{s+1}\left((s+1-i)\delta_{ij} + i\delta_{(i-1)j}\right) \end{cases}$$

Consequently, the prolongation operator from level l to the finest level L of Bézier control points can be expressed as $d_l^L = d_{n_L-1} \circ \cdots \circ d_{n_l}$, and thus, is an affine application associated to the matrix $\mathcal{D}_l^L = D_{n_L-1} \cdots D_{n_l+1} D_{n_l}$. In conclusion, at a particular level l, one can directly relate the current set of Bézier control points with the shape grid-points through $f^{(l)}$, or alternatively, apply successively the operators d_l^L and $f^{(L)}$. The two ways are equivalent from a theoretical view point, even if they can differ in the implementation. Thus, the following algorithm can be used instead of eqs. (6) and (9):

$$\gamma_{r+1} = \gamma_r - \omega_r \, M^{(L)} \mathcal{D}_l^L (\mathcal{D}_l^L)^T (M^{(L)})^T \, g_r \tag{14}$$

In (14), even if the degree elevation explicitly appears through $\mathcal{D}_l^L$, the control variables are still the ordinates of the shape grid-points. Nevertheless, since the application d_l^L is affine it can be also envisaged to directly use the ordinates of the Bézier control points on the finest level as control variables keeping the same sub-parametrisations in a multilevel approach. Note that, in this case, there is a strong analogy with the approach proposed in [2] since, in the both algorithms, the degree-elevation property is directly used to prolongate from a coarse level to the complete set of design variables. The new functional to be minimised becomes, here, $\mathcal{J}^{(L)} = j \circ f^{(L)}$ while d_l^L corresponds to the prolongation operator from level l to the complete parametrisation. The following iterative algorithm is then obtained:

$$\beta_{r+1} = \beta_r - \omega_r \, \mathcal{D}_l^L (\mathcal{D}_l^L)^T \, \mathcal{G}_r$$

where $\beta = \alpha^{(L)} = (y_1, \cdots, y_{n_L-1})^T$ and $\mathcal{G}_r$ is the gradient of $\mathcal{J}^{(L)}$ with respect to β at iteration r, which can be expressed as follows using (4):

$$\mathcal{G}_r = \text{grad}_\beta \mathcal{J}^{(L)}(\beta_r) = \left(P^{(L)}\right)^* \text{grad}_\gamma j(\gamma_r) = (M^{(L)})^T \, g_r \tag{15}$$

As soon as β_{r+1} is obtained, the shape grid-points can be also updated by $\gamma_{r+1} = f^{(L)}(\beta_{r+1})$. Then, since $f^{(L)}$ is affine, the following expression is achieved for γ_{r+1}:

$$\gamma_{r+1} = \gamma_r - \omega_r \, M^{(L)} \mathcal{D}_l^L (\mathcal{D}_l^L)^T \, \mathcal{G}_r \tag{16}$$

Furthermore, from the relation (15) between the two gradients $\mathcal{G}_r$ and g_r, it appears that the algorithm defined by (16) coincides with the one defined in (14). Thus, the approach described in Sec. 3.1 to 3.3 can be also interpreted as a multilevel gradient-based method in which Bézier control points are taken as control variables.

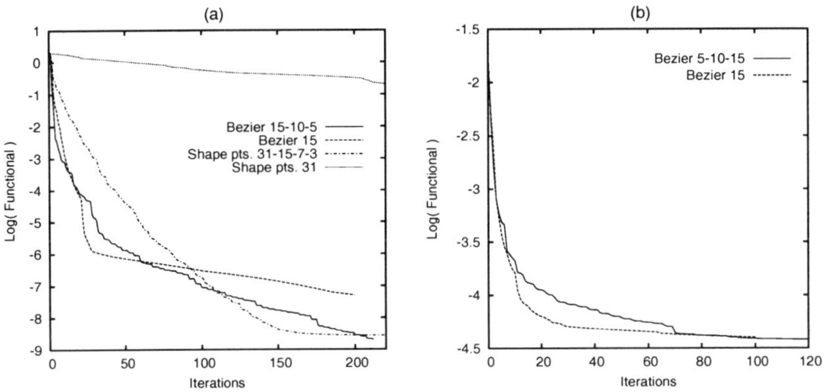

Fig. 1. (a) Test-case 1: Comparison between shape grid-points and Bézier parametrisations for 1-level and V-cycle multilevel approaches; (b) Test-case 2: Comparison between 1-level and V-cycle multilevel approaches for the Bézier-based parametrisation.

4. Numerical experiments on 2D inverse problems

4.1. *Test-case 1: a 2D nozzle inverse problem*

The first test-case, already used for the multilevel approach associated to shape grid-point coordinates parametrisation [1], is a 2D convergent-divergent nozzle inverse problem for inviscid subsonic flows (the flow is modelled, here, by the Euler equations). Here, the particular inverse problem is characterised by an initial constant-section nozzle and a target sine shape. For the considered mesh of 1900 nodes, 63 shape grid-points are available. Since, the abscissas of the shape grid-points points are uniformly distributed, an uniform distribution is also taken for the abscissas of the Bézier control points at each level (see Sec. 3.3). Fig. 1a shows the convergence behaviour obtained with different strategies. A reference run has been obtained considering the original parametrisation based on shape grid-point and a one-level strategy with the finest sub-level (31 parameters). Consistently with [1], a large convergence improvement is obtained considering a V-cycle multilevel approach. Concerning the new set of sub-parametrisation based on Bézier control points, even the one-level approach with 15 parameters gives interesting results. Indeed, at least for the decrement of the first 6 orders of magnitude of the cost functional, a better behaviour of this one-level strategy is observed with respect to the multilevel approach associated to the classical sub-parametrisation. The corresponding multilevel strategy yields ulterior improvements in the final part of the convergence history. Nevertheless, interpreted as an optimisation with respect to the

Bézier control points, this multilevelling appears less impressive. Indeed, it seems that the principal gain is due to the use of a Bézier-based parametrisation more than the change of control sub-spaces.

4.2. *Test-case 2: an airfoil inverse problem*

In this second test-case, starting from a symmetric NACA0012 airfoil, the cambered RAE2822 should be rebuilt. The initial flow conditions are characterised by a far-field Mach number of 0.734 and an angle of incidence of 2.79^o. Furthermore, as in the previous test-case, only inviscid flows are considered. A mixed unstructured/structured mesh of 3282 nodes has been generated on a circular computational domain centred on the airfoil. The definition of the Bézier-based sub-parametrisations have been done, here, imposing a vertical tangent at the leading edge, and thus, following the process proposed in Sec. 3.3 associated to eq. (13).

The direct use of the parametrisation based on shape grid-points seems more difficult for this optimisation problem, in particular, near the training edge where non physical profiles appear. This problem can be solved by imposing geometrical constraints, considering only coarser sub-levels in order to increase the smoothing or/and modifying the criterion for the choice of the descent step ω_r. Nevertheless, we choose, here, to consider only the Bézier-based sub-parametrisations. Fig. 1b shows the convergence behaviour for a one-level strategy using 15 Bézier control points as well as a V-cycle multilevel strategy on three sub-levels (5, 10 and 15 parameters). It can be observed a lower decrease of the cost functional with respect to the previous test-case, and on another hand, the multilevel algorithm does not improve the convergence rate. The first behaviour can be explained by the fact that the target profile, i.e. the RAE2822 airfoil, is not very accurately approximated by Bézier curve of moderate degree. Indeed, for the present choice of $X^{(L)}$ and $T^{(L)}$ and 15 Bézier control points, the distance between the Bézier curvefit (obtained by a least squares approximation on the ordinates of the Bézier coefficients) and the airfoil shape is of about 10^{-8} for the NACA0012 shape while it is only about $5 10^{-4}$ for the RAE2822 profile. Furthermore, the Bézier control polygon is quite oscillatory for the approximate RAE2822 profile contrary to the previous case of nozzle shapes. Thus, the sub-parametrisation choice is, perhaps, less adequate for this second test-case, since, as pointed out in [3], in which a procedure of parametrisation adaptation is also proposed, the lack of regularity of the control polygon can have a negative effect on the convergence behaviour.

5. Conclusion

In the present study, the description of a multilevel gradient-based method for aerodynamic shape design is addressed. This approach starts from an existing formulation [1] based on an embedded parametrisation of shape grid-points and on interpolation operators. A new set of sub-parametrisations is proposed, in which a coarse level is described by using Bézier control points. More precisely, starting from a consistent coarsest level, the degree-elevation property of Bézier curves is applied to successively define the different finer levels. In this context, even if, in practice, the Bézier control points are not explicitly computed, the proposed algorithm can be also interpreted as a descent method for Bézier control points as control variables. Thus, it seems not so far from the approach proposed in [2], even if, in this case, the formulation is centred on gradient-free methods. The numerical experiments shows that the introduction of this new family of sub-parametrisations has suitable effects, if it is understood as an alternative gradient preconditioning for the optimisation with respect to the shape grid-points. Nevertheless, if the approach is interpreted as a multilevel approach considering the ordinates of the Bézier control points as control variables, the multilevelling efficiency has not been clearly demonstrated. Note that the optimisation procedure is subordinated to the particular distribution of the abscissas of the Bézier control points on the different sub-levels. The choice, done here, is very simple and low expensive (in particular, it is free of any resolution of least squares problems), but on another hand, it seems that it can be a limitation for the efficiency of the proposed approach. Note that, if the interest of this new formulation could be better established, since the Bézier curves act as a basic tool for polynomial shape representation, one can also envisage to extend the formulation to more complex curves as B-splines, and also, to 3D case through, for instance, tensorial Bézier parametrisation.

References

1. F. Beux and A. Dervieux, A hierarchical approach for shape optimization, *Engineering Computations*, **11**, 25-48, 1994.
2. J.-A. Désidéri, Hierarchical optimum-shape algorithms using embedded Bézier parameterizations, *Numerical Methods for Scientific Computing, Variational Problems and Applications*, Y. Kuznetsov *et al* eds., CIMNE, Barcelona, 2003.
3. B. Abou El Majd, J.-A. Désidéri and A. Janka, Nested and self-adaptive Bézier parameterization for shape optimization, *International Conference on Control, Partial Differential Equations and Scientific Computing*, Beijing, China, 13-16 Sept, 2004.

NONLINEAR EXACT CLOSURE FOR THE HYDRODYNAMICAL MODEL OF SEMICONDUCTORS BASED ON THE MAXIMUM ENTROPY PRINCIPLE

G. MASCALI

Department of Mathematics, University of Calabria and INFN-Gruppo c.Cosenza, Cosenza, 87036, Italy
**E-mail: g.mascali@unical.it*

V. ROMANO

Department of Mathematics and Informatics, University of Catania Catania, 95125, Italy
E-mail: romano@dmi.unict.it

An exact closure is obtained for the 8-moment model of semiconductors based on the maximum entropy principle.

Keywords: Semiconductors; hydrodynamical models; maximum entropy principle.

1. Kinetic model

Semiconductors are characterized by a sizable energy gap between the valence and the conduction bands. The energy band structure of crystals can be obtained at the cost of intensive numerical calculations (and also semi-phenomenologically) by means of the quantum theory of solids.[1] The electrons, which mainly contribute to the charge transport, are those with energy near the lowest conduction band minima, each neighborhood being called valley. In silicon, which is the material we will deal with in this paper, there are six equivalent ellipsoidal valleys along the main crystallographic directions Δ at about 85 % from the center of the first Brillouin zone, near the X points, which, for this reason, are termed as X-valleys.

In the derivation of macroscopic models, usually, the energy in each valley is represented by analytical approximations. Among these, the most common one is the Kane dispersion relation, which describes the energy $\mathcal{E}_A$

of the A-valley, measured from the bottom of the valley $\overline{\mathcal{E}_A}$, as

$$\mathcal{E}_A(k_A)\left[1+\alpha_A\mathcal{E}_A(k_A)\right]=\frac{\hbar^2 k_A^2}{2m_A^*},\quad \mathbf{k}_A\in\mathbb{R}^3. \tag{1}$$

$\mathbf{k}_A$ is the electron wave vector in the A-valley and k_A its modulus, m_A^* is the effective electron mass in the A-valley and $\hbar$ the reduced Planck constant. α_A is the non parabolicity parameter. In the sequel, in order to simplify the notation, the valley index is omitted.

The electron velocity $\mathbf{v}(\mathbf{k})$ depends on the energy $\mathcal{E}$ by the quantum relation

$$\mathbf{v}(\mathbf{k})=\frac{1}{\hbar}\nabla_{\mathbf{k}}\mathcal{E}.$$

Explicitly, we get in the Kane approximation of the dispersion relation

$$v^i=\frac{\hbar k^i}{m^*\left[1+2\alpha\mathcal{E}(k)\right]}. \tag{2}$$

In the semiclassical kinetic approach the charge transport in semiconductors is described by the *Boltzmann equation*[a]

$$\frac{\partial f}{\partial t}+v^i(\mathbf{k})\frac{\partial f}{\partial x^i}-\frac{eE^i}{\hbar}\frac{\partial f}{\partial k^i}=\mathcal{C}[f], \tag{3}$$

where $f(\mathbf{x},\mathbf{k},t)$ is the one electron distribution function and e the absolute value of the electron charge. In a multivalley description one has to consider a transport equation for each valley.

The electric field $\mathbf{E}$ is calculated by solving the Poisson equation for the electric potential ϕ

$$E_i=-\frac{\partial\phi}{\partial x_i}, \tag{4}$$

$$\nabla\cdot(\epsilon\nabla\phi)=-e(N_+-N_--n), \tag{5}$$

N_+ and N_- being the donor and acceptor density respectively (which depend only on the position) and n the electron number density

$$n=\int f d\mathbf{k}.$$

$\mathcal{C}[f]$ represents the effects due to scattering of electrons with phonons, impurities and with other electrons. After a collision the electron can remain in the same valley (intravalley scattering) or be drawn into another valley (intervalley scattering).

[a]Hereafter summation over repeated indices is understood

Under the assumption that the electron gas is dilute, the collision operator can be assumed in the linear form

$$\mathcal{C}[f] = \int [P(\mathbf{k}', \mathbf{k}) f(\mathbf{k}') - P(\mathbf{k}, \mathbf{k}') f(\mathbf{k})] \, d\mathbf{k}. \tag{6}$$

For the sake of brevity, we will consider only electron-phonon scatterings which can be summarized as follows:

- scattering with intravalley acoustic phonons (approximately elastic);
- electron-phonon intervalley inelastic scatterings, for which there are six contributions: the three g_1, g_2, g_3 and the three f_1, f_2, f_3 optical and acoustical intervalley scatterings[2]

m_e	electron rest mass	9.109510^{-28} g
m^*	effective electron mass	$0.32\ m_e$
T_L	lattice temperature	300 K
ρ	density	2330 g/cm^3
v_s	longitudinal sound speed	$9.18\ 10^5$ cm/sec
Ξ_d	acoustic-phonon deformation potential	9 eV
α	non parabolicity factor	0.5 eV^{-1}
ϵ_r	relative dielectric constant	11.7
ϵ_0	vacuum dieletric constant	1.24×10^{-22} C/(eV cm)

α	Z_f	$\hbar\omega(meV)$	(D_t K) (10^8 eV/cm)
g_1	1	12	0.5
g_2	1	18.5	0.8
g_3	1	61.2	11
f_1	4	19	0.3
f_2	4	47.4	2
f_3	4	59	2

In the elastic case

$$P^{(ac)}(\mathbf{k}, \mathbf{k}') = K_{ac}\, \delta(\mathcal{E} - \mathcal{E}'), \tag{7}$$

while for the inelastic scatterings

$$P^{\alpha}(\mathbf{k}, \mathbf{k}') = K_{\alpha} \left[(N_B^{(\alpha)} + 1)\delta(\mathcal{E}' - \mathcal{E} + \hbar\omega_{\alpha}) + N_B^{(\alpha)}\delta(\mathcal{E}' - \mathcal{E} - \hbar\omega_{\alpha}) \right], \tag{8}$$

where $\alpha = g_1, g_2, g_3, f_1, f_2, f_3$, $N_B^{(\alpha)}$ is the phonon equilibrium distribution obeying the Bose-Einstein statistics and $\hbar\omega_\alpha$ the phonon energy.

The parameters that appear in the scattering rates can be expressed in terms of physical quantities characteristic of the considered material

$$K_{ac} = \frac{k_B T_L \Xi_d^2}{4\pi^2 \hbar \rho v_s^2}, \quad K_\alpha = \frac{Z_{f\,\alpha}(D_t K)_\alpha^2}{8\pi^2 \rho \omega_\alpha},$$

where k_B is the Boltzmann constant, T_L the lattice temperature, Ξ_d the deformation potential of acoustic phonons, ρ the mass density of the semiconductor, v_s the sound velocity of the longitudinal acoustic mode, $(D_t K)_\alpha$ the deformation potential relative to the interaction with the α intervalley phonon and $Z_{f\,\alpha}$ the number of final equivalent valleys for the considered intervalley scattering.

2. Moment equations

Macroscopic models are obtained by taking the moments of the Boltzmann transport equation. In principle, all the hierarchy of the moment equations should be retained, but for practical purposes it is necessary to truncate it at a suitable order N. Such a truncation introduces two main problems due to the fact that the number of unknowns exceeds that of the equations:

i) the closure for higher order fluxes;

ii) the closure for the production terms.

As in gasdynamics, multiplying eq. (3) by a sufficiently regular function $\psi(\mathbf{k})$ and integrating with respect to $\mathbf{k}$, one gets the generic *moment equation*

$$\frac{\partial M_\psi}{\partial t} + \int \psi(\mathbf{k}) v^i(\mathbf{k}) \frac{\partial f}{\partial x^i} d\mathbf{k} + \frac{e}{\hbar} E^j \int f \frac{\partial \psi(\mathbf{k})}{\partial k^j} d\mathbf{k} = \int \psi(\mathbf{k}) \mathcal{C}[f] d\mathbf{k}, \quad (9)$$

with

$$M_\psi = \int \psi(\mathbf{k}) f d\mathbf{k},$$

the moment relative to the weight function ψ.

Various models employ different expressions of $\psi(\mathbf{k})$ and numbers of moments.

3. The maximum entropy principle

The maximum entropy principle (hereafter MEP) leads to a systematic way for obtaining constitutive relations on the basis of information theory (see[3–8] for a review).

According to the MEP if a given number of moments M_A, $A = 1, \ldots, N$, are known, the distribution function which can be used to evaluate the unknown moments of f, corresponds to the extremal, f_{ME}, of the entropy functional under the constraints that it exactly yields the known moments M_A

$$\int \psi_A f_{ME} d\mathbf{k} = M_A, \quad A = 1, \ldots, N. \tag{10}$$

f_{ME} is the least biased distribution, which can be used to estimate f, when only a finite number of moments of this latter are known.

Since the electrons interact with the phonons which describe the thermal vibrations of the ions placed at the points of the crystal lattice, in principle we should deal with a two component system (electrons and phonons). However, if one considers the phonon gas as a thermal bath at constant temperature T_L, only the electron component of the entropy must be maximized. Moreover, by considering the electron gas as sufficiently dilute, one can take the expression of the entropy obtained as limiting case of that arising in the Fermi statistics, that is

$$s = -k_B \int (f \log f - f) \, d\mathbf{k}. \tag{11}$$

If we introduce the Lagrangian multipliers Λ_A, the problem of maximizing s under the constraints (10) is equivalent to maximizing

$$\tilde{s} = \Lambda_A \left(M_A - \int \psi_A f d\mathbf{k} \right) - s,$$

the Legendre transform of s, without constraints, so that the equation

$$\delta \tilde{s} = 0$$

has to be solved. This gives

$$\left[\log f + \frac{\Lambda_A \psi^A}{k_B} \right] \delta f = 0.$$

Since the latter relation must hold for arbitrary δf, it follows that

$$f_{ME} = \exp \left[-\frac{1}{k_B} \Lambda_A \psi^A \right]. \tag{12}$$

In order to get the dependence of the Λ_A's on the M_A's, one has to invert the constraints (10). Then by taking the moments of f_{ME} and $\mathcal{C}[f_{ME}]$, one finds the closure relations for the fluxes and the production terms appearing in the balance equations. On account of the analytical difficulties this, in general, can be achieved only with a numerical procedure. However, apart from the computational problems, the balance equations are now a closed set of partial differential equations and with standard considerations in extended thermodynamics[4] it is easy to show that they form a quasilinear hyperbolic system.

When the Kane dispersion relation is used, the solvability of the maximum entropy problem has been proved in.[9]

4. The 8-moments model

Let us consider the balance equations for the density, the velocity, the energy and the energy flux, which correspond to the kinetic variables $1, \mathbf{v}, \mathcal{E}, \mathcal{E}\mathbf{v}$

$$\frac{\partial n}{\partial t} + \frac{\partial (nV^i)}{\partial x^i} = 0, \tag{13}$$

$$\frac{\partial (nV^i)}{\partial t} + \frac{\partial (nU^{ij})}{\partial x^j} + neE_j\, H^{ij} = nC_{V^i}, \tag{14}$$

$$\frac{\partial (nW)}{\partial t} + \frac{\partial (nS^j)}{\partial x^j} + neV_k E^k = nC_W, \tag{15}$$

$$\frac{\partial (nS^i)}{\partial t} + \frac{\partial (nF^{ij})}{\partial x^j} + neE_j G^{ij} = nC_{S^i}. \tag{16}$$

The macroscopic quantities involved in the balance equations are related to the one particle distribution function of electrons $f(\mathbf{x}, \mathbf{k}, t)$ by the following definitions

$$n = \int_{\mathbb{R}^3} f d\mathbf{k} \quad \text{is the electron density,}$$

$$V^i = \frac{1}{n}\int_{\mathbb{R}^3} f v^i d\mathbf{k} \quad \text{is the average electron velocity,}$$

$$W = \frac{1}{n}\int_{\mathbb{R}^3} \mathcal{E}(k) f d\mathbf{k} \quad \text{is the average electron energy,}$$

$$S^i = \frac{1}{n}\int_{\mathbb{R}^3} f v^i \mathcal{E}(k) d\mathbf{k} \quad \text{is the energy flux,}$$

$$U^{ij} = \frac{1}{n}\int_{\mathbb{R}^3} f v^i v^j d\mathbf{k} \quad \text{is the velocity flux,}$$

$$H^{ij} = \frac{1}{n}\int_{\mathbb{R}^3} \frac{1}{\hbar} f \frac{\partial}{\partial k_j}(v_i) d\mathbf{k},$$

$$F^{ij} = \frac{1}{n}\int_{\mathbb{R}^3} f v^i v^j \mathcal{E}(k) d\mathbf{k} \quad \text{is the flux of the energy flux,}$$

$$G^{ij} = \frac{1}{n}\int_{\mathbb{R}^3} \frac{1}{\hbar} f \frac{\partial}{\partial k_j}(\mathcal{E} v_i) d\mathbf{k},$$

$$C_{V^i} = \frac{1}{n}\int_{\mathbb{R}^3} \mathcal{C}[f]\hbar v^i d\mathbf{k} \quad \text{is the velocity production,}$$
$$C_W = \frac{1}{n}\int_{\mathbb{R}^3} \mathcal{C}[f]\mathcal{E}(k) d\mathbf{k} \quad \text{is the energy production,}$$
$$C_{S^i} = \frac{1}{n}\int_{\mathbb{R}^3} \mathcal{C}[f] v^i \mathcal{E}(k) d\mathbf{k} \quad \text{is the the energy flux production.}$$

These moment equations do not constitute a set of closed relations because of the fluxes and production terms. Therefore constitutive assumptions must be prescribed.

If we assume as fundamental variables n, V^i, W and S^i, which have a direct physical interpretation, the closure problem consists of expressing U^{ij}, H^{ij}, F^{ij}, G^{ij} and the moments of the collision term C_{V^i}, C_W and C_{S^i} as functions of n, V^i, W and S^i.

If we use the MEP to get the closure relations, we have to face with the problem of inverting the constraints (10) with $\psi_A = 1, \mathbf{v}, \mathcal{E}, \mathcal{E}\mathbf{v}$.

This problem has been overcame in[10,11] upon the ansatz of small anisotropy for f_{ME} since Monte Carlo simulations for electron transport in Si show that the anisotropy of f is small even far from equilibrium.

Here we will show that it is possible to invert the constraints (10) in an exact way assuming what follows

BASIC ASSUMPTION:

$$\mathbf{V} \quad \text{and} \quad \mathbf{S} \quad \text{are collinear.} \tag{17}$$

REMARK. The previous assumption is valid in the one dimensional case. In general it is not true. However, apart from the specific interest in semiconductor mathematical modeling, getting exact closure relations is itself of great interest in thermodynamical theories of non equilibrium and in particular gives relevant insights into the influence of the non linear terms.

5. Closure relations

The constraints (10) in the case under consideration explicitly read

$$n = \int_{\mathbb{R}^3} f_{ME} d\mathbf{k}, \quad V^i = \frac{1}{n}\int_{\mathbb{R}^3} f_{ME} v^i d\mathbf{k}, \tag{18}$$
$$W = \frac{1}{n}\int_{\mathbb{R}^3} \mathcal{E}(k) f_{ME} d\mathbf{k}, \quad S^i = \frac{1}{n}\int_{\mathbb{R}^3} f_{ME} v^i \mathcal{E}(k) d\mathbf{k}, \tag{19}$$

where

$$f_{ME} = \exp\left[-\left(\frac{\lambda}{k_B} + \lambda^W \mathcal{E} + \boldsymbol{\lambda}^V \cdot \mathbf{v}(\mathbf{k}) + \boldsymbol{\lambda}^S \cdot \mathbf{v}(\mathbf{k})\,\mathcal{E}\right)\right], \tag{20}$$

λ, λ^W, $\boldsymbol{\lambda}^V$ and $\boldsymbol{\lambda}^S$ being the Lagrangian multipliers relative to the density, energy, velocity and energy-flux respectively.

Thanks to the assumption (17),

$$\boldsymbol{\lambda}^V \cdot \mathbf{v}(\mathbf{k}) + \boldsymbol{\lambda}^S \cdot \mathbf{v}(\mathbf{k})\,\mathcal{E} = \left(|\boldsymbol{\lambda}^V| + |\boldsymbol{\lambda}^S|\,\mathcal{E}\right)|\boldsymbol{v}|\cos\vartheta,$$

ϑ being the angle between $\boldsymbol{V}$ and $\boldsymbol{v}$.

By expressing the elementary volume $d\mathbf{k}$ as $g(\mathcal{E})\,d\Omega$, where $g(\mathcal{E}) = \sqrt{2}(m^*)^{3/2}\,\hbar^{-3}\sqrt{\mathcal{E}(1+\alpha\mathcal{E})}(1+2\alpha\mathcal{E})$ is related to the density of states and $d\Omega$ is the element of solid angle, the constraints, after some algebra, become

$$n = \frac{2\pi(2m^*)^{3/2}}{\hbar^3} e^{-\frac{\lambda}{k_B}} d_0, \tag{21}$$

$$V = \frac{1}{d_0}\sqrt{\frac{2}{m^*}} \int_0^\infty \mathcal{E}(1+\alpha\mathcal{E}) e^{-\lambda^W \mathcal{E}} \left[\frac{\sinh A(\mathcal{E})}{A(\mathcal{E})^2} - \frac{\cosh A(\mathcal{E})}{A(\mathcal{E})}\right] d\mathcal{E}, \tag{22}$$

$$W = \frac{1}{d_0}\int_0^\infty \mathcal{E} e^{-\lambda^W \mathcal{E}} \frac{\sinh A(\mathcal{E})}{A(\mathcal{E})} \sqrt{\mathcal{E}(1+\alpha\mathcal{E})}(1+2\alpha\mathcal{E})\,d\mathcal{E}, \tag{23}$$

$$S = \frac{1}{d_0}\sqrt{\frac{2}{m^*}} \int_0^\infty \mathcal{E}^2(1+\alpha\mathcal{E}) e^{-\lambda^W \mathcal{E}} \left[\frac{\sinh A(\mathcal{E})}{A(\mathcal{E})^2} - \frac{\cosh A(\mathcal{E})}{A(\mathcal{E})}\right] d\mathcal{E}, \tag{24}$$

with

$$d_0 = \int_0^\infty e^{-\lambda^W \mathcal{E}} \frac{\sinh A(\mathcal{E})}{A(\mathcal{E})} \sqrt{\mathcal{E}(1+\alpha\mathcal{E})}(1+2\alpha\mathcal{E})\,d\mathcal{E},$$

where $A(\mathcal{E}) = (|\boldsymbol{\lambda}^V| + |\boldsymbol{\lambda}^S|\,\mathcal{E})|\boldsymbol{v}|$ and V and S are the relevant component of $\mathbf{V}$ and $\mathbf{S}$ in the chosen frame (of course here time and position are frozen).

The previous relations define the fields n, V, W and S in terms of the Lagrangian multipliers apart from an integration with respect to $\mathcal{E}$ which can be efficiently performed with Gauss-Laguerre quadrature formulas.

Inserting f_{ME} in the definition of fluxes one has

$$m^*U = \frac{2}{d_0}\int_0^\infty e^{-\lambda^W \mathcal{E}} \frac{[\mathcal{E}(1+\alpha\mathcal{E})]^{3/2}}{1+2\alpha\mathcal{E}} B(\mathcal{E})d\,\mathcal{E}, \tag{25}$$

$$m^*F = \frac{2}{d_0}\int_0^\infty \mathcal{E}\; e^{-\lambda^W \mathcal{E}} \frac{[\mathcal{E}(1+\alpha\mathcal{E})]^{3/2}}{1+2\alpha\mathcal{E}} B(\mathcal{E})d\,\mathcal{E}, \tag{26}$$

$$m^*H = \frac{1}{d_0}\int_0^\infty e^{-\lambda^W \mathcal{E}} \left[\sqrt{\mathcal{E}(1+\alpha\mathcal{E})}\frac{\sinh A(\mathcal{E})}{A(\mathcal{E})} - 4\alpha\frac{[\mathcal{E}(1+\alpha\mathcal{E})]^{3/2}}{(1+2\alpha\mathcal{E})^2} B(\mathcal{E})\right] d\,\mathcal{E}, \tag{27}$$

$$m^*G = \frac{1}{2\alpha}(1 - m^*H)\,. \tag{28}$$

with

$$B(\mathcal{E}) = \left[\frac{\sinh A(\mathcal{E})}{A(\mathcal{E})} - \frac{2}{A(\mathcal{E})^2}\left(\cosh A(\mathcal{E}) - \frac{\sinh A(\mathcal{E})}{A(\mathcal{E})}\right)\right].$$

Here U, F, G and H are the relevant components (that is those in the direction of $|\boldsymbol{V}|$) of the tensors U_{ij}, F_{ij}, G_{ij} and H_{ij}, respectively, .

Similarly inserting f_{ME} in the definition of the production terms, one has for the significant components in the case of elastic phonon scattering

$$C_V = -\frac{8\pi K_{ac} m^*}{\hbar^3\, d_0} \int_0^\infty e^{-\lambda^W \mathcal{E}} [\mathcal{E}(1+\alpha\mathcal{E})]^{3/2} (1+2\alpha\mathcal{E})\, B_1(\mathcal{E})\, d\,\mathcal{E}, \quad (29)$$

$$C_W = 0, \quad (30)$$

$$C_S = -\frac{8\pi K_{ac} m^*}{\hbar^3\, d_0} \int_0^\infty \mathcal{E}\, e^{-\lambda^W \mathcal{E}} [\mathcal{E}(1+\alpha\mathcal{E})]^{3/2} (1+2\alpha\mathcal{E})\, B_1(\mathcal{E})\, d\,\mathcal{E}, (31)$$

with

$$B_1(\mathcal{E}) = \left(\frac{\sinh A(\mathcal{E})}{A(\mathcal{E})^2} - \frac{\cosh A(\mathcal{E})}{A(\mathcal{E})} \right),$$

and in the case of inelastic phonon scattering

$$\begin{aligned} C_V = & -\frac{8\pi K_\alpha m^*}{\hbar^3 d_0} \int_0^\infty \Big[e^{-\lambda^W \mathcal{E}}\, \mathcal{E}(1+\alpha\mathcal{E}) N_B^{(\alpha)} \mathcal{N}_+(\mathcal{E}) B_1(\mathcal{E}) \\ & +(N_B^{(\alpha)}+1)\mathcal{E}_+(1+\alpha\mathcal{E}_+) e^{-\lambda^W \mathcal{E}_+} \mathcal{N}(\mathcal{E}) B_1(\mathcal{E}_+) \Big]\, d\,\mathcal{E}, \end{aligned} \quad (32)$$

$$\begin{aligned} C_W = & \frac{8\pi K_\alpha (m^*)^{3/2}}{\sqrt{2}\hbar^3 d_0} \hbar\omega_\alpha \int_0^\infty e^{-\lambda^W \mathcal{E}} \mathcal{N}_+(\mathcal{E}) \mathcal{N}(\mathcal{E}) \\ & \left[N_B^{(\alpha)} \frac{\sinh A(\mathcal{E})}{A(\mathcal{E})} - (N_B^{(\alpha)}+1) \frac{\sinh A(\mathcal{E}_+)}{A(\mathcal{E}_+)} \right] d\,\mathcal{E}, \end{aligned} \quad (33)$$

$$\begin{aligned} C_S = & -\frac{8\pi K_\alpha m^*}{\hbar^3 d_0} \int_0^\infty \Big[e^{-\lambda^W \mathcal{E}}\, \mathcal{E}^2(1+\alpha\mathcal{E}) N_B^{(\alpha)} \mathcal{N}_+(\mathcal{E}) B_1(\mathcal{E}) \\ & +(N_B^{(\alpha)}+1)\mathcal{E}_+^2(1+\alpha\mathcal{E}_+) e^{-\lambda^W \mathcal{E}_+} \mathcal{N}(\mathcal{E}) B_1(\mathcal{E}_+) \Big]\, d\,\mathcal{E}, \end{aligned} \quad (34)$$

with

$$\mathcal{N}(\mathcal{E}) = \sqrt{\mathcal{E}(1+\alpha\mathcal{E})}(1+2\alpha\mathcal{E}), \quad \mathcal{E}_+ = \mathcal{E} + \hbar\omega_\alpha, \quad \mathcal{N}_+(\mathcal{E}) = \mathcal{N}(\mathcal{E}_+).$$

6. Comparison of linear and nonlinear closure in the one dimensional bulk case

In the one dimensional homogenous problem the density equals the constant doping, while the balance equations of velocity, energy and energy-flux lead to the following system of ODEs

$$\frac{d}{dt}V = -eEH + C_V, \tag{35}$$

$$\frac{d}{dt}W = -eEV + C_W, \tag{36}$$

$$\frac{d}{dt}S = -eEG + C_S, \tag{37}$$

where the electric field E enters as a parameter. Once all the variables have been expressed in terms of the Lagrangian multipliers, the balance equations (35) -(37) can be rewritten as

$$\frac{d}{dt}\begin{pmatrix} \lambda^V \\ \lambda^W \\ \lambda^S \end{pmatrix} = J_0^{-1}\begin{pmatrix} -eEH + C_V \\ -eEV + C_W \\ -eEG + C_S \end{pmatrix}, \tag{38}$$

with J_0 Jacobian matrix

$$\frac{\partial(V, W, S)}{\partial(\lambda^V, \lambda^W, \lambda^S)}.$$

As initial conditions we consider the equilibrium state

$$V(0) = 0, \quad W(0) = W_0 = \frac{3}{2}k_B T_L, \quad S(0) = 0.$$

We recall that T_L is considered as constant.

In terms of Lagrangian multipliers the previous conditions read

$$\lambda^V = 0, \quad \lambda^W = \frac{1}{k_B T_L}, \quad \lambda^S = 0.$$

For the evaluation of the integrals the Gauss-Laguerre formulas with weights e^{-x} and $\sqrt{x}\, e^{-x}$ have been adopted. A Runge-Kutta method has been used for the numerical integration of the evolution equations.

In fig. 1 we compare the drift velocity and average energy in bulk silicon versus the electric field, obtained by using respectively the approximated closure based on the small anisotropy ansatz (AM) and the exact closure presented in this paper (EM). As can be seen the results are remarkably different at high fields, moreover if they are compared with the Monte Carlo ones shown in K. Tomizawa[12](p. 100, figure 3.11) together with the experimental data, it is possible to conclude that the results with the exact closure are considerably better. This shows that the anisotropy effects and the nonlinearity play an important role.

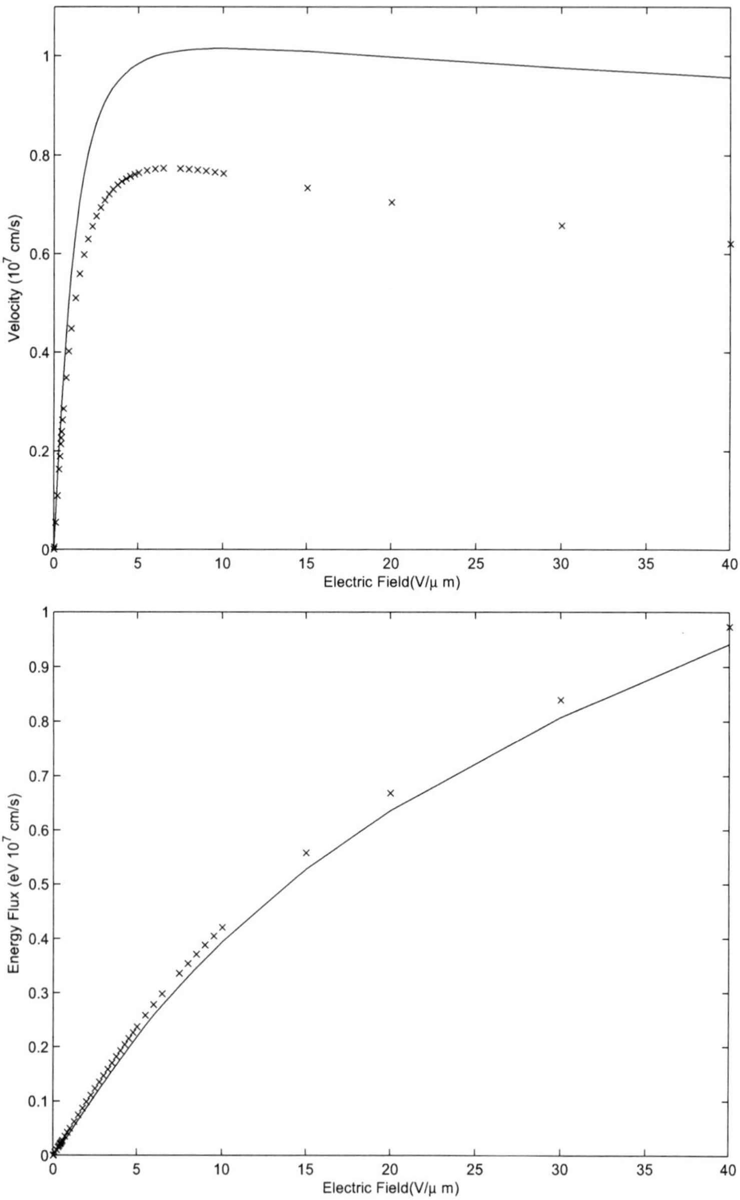

Fig. 1. Velocity and Energy vs Electric Field (bulk silicon), continuous line: the AM model, crosses: the EM model.

References

1. N. W. Ashcroft and N. D. Mermin, *Solid State Physics*, (Sounders College Publishing International Edition, Philadelphia, 1976).
2. C. Canali, C. Jacoboni, F. Nava, G. Ottaviani, A. Alberigi-Quaranta, *Phys. Rev B* **12** 2265 (1975).
3. E. T. Jaynes, *Phys. Rev B* **106** 620 (1957).
4. I. Müller and T. Ruggeri, *Rational Extended Thermodynamics*, (Springer-Verlag, Berlin, 1998).
5. D. Jou, J. Casas-Vazquez and G. Lebon, *Extended irreversible thermodynamics*, (Springer-Verlag, Berlin, 1993).
6. C. D. Levermore, *J. Stat. Phys* **83** 331 (1996).
7. N. Wu, *The maximum entropy method*, (Springer-Verlag, Berlin, 1997).
8. A.M. Anile, G. Mascali and V. Romano, in *Mathematical Problems in Semiconductor Physics, Lecture Notes in Mathematics 1832*, (Springer, Berlin, 2003).
9. M. Junk and V. Romano, *Cont. Mech. Thermodyn.* **17** 247 (2005).
10. A. M. Anile and V. Romano , *Cont. Mech. Thermodyn.* **11** 307 (1999).
11. V. Romano, *Cont. Mech. Thermodyn.* **12** 31 (2000).
12. K. Tomizawa, *Numerical simulation of sub micron semiconductor devices*, (Artech House, Boston 1993).

A THERMODYNAMICAL MODEL OF INHOMOGENEOUS SUPERFLUID TURBULENCE

M.S. MONGIOVI'

Dipartimento di Metodi e Modelli Matematici Università di Palermo, Viale delle Scienze, 90128 Palermo, Italy
E-mail:mongiovi@unipa.it

D. JOU

Departament de Física, Universitat Autònoma de Barcelona, 08193 Bellaterra, Catalonia, Spain
E-mail: david.jou@uab.es

In this paper we perform a thermodynamical derivation of a nonlinear hydrodynamical model of inhomogeneous superfluid turbulence. The theory chooses as fundamental fields the density, the velocity, the energy density, the heat flux and the averaged vortex line length per unit volume. The restrictions on the constitutive quantities are derived from the entropy principle, using the Liu method of Lagrange multipliers. The mathematical and physical consequences deduced by the theory are analyzed both in the linear and in the nonlinear regime. Field equations are written and the wave propagation is studied with the aim to describe the mutual interactions between the second sound and the vortex tangle.

Keywords: superfluid turbulence; nonequilibrium thermodynamics.

1. Introduction

Due to its quantum nature, the behavior of superfluid helium II is very different from that of ordinary fluids: it has an extremely low viscosity and temperature waves (second sound) propagate in it. An example of non classical behavior is heat transfer in counterflow experiments, characterized by no matter flow but only heat transport. Consider a channel with a heater at a closed end and open to the helium bath at the other end. Using an ordinary classical fluid, such as helium I, a temperature gradient can be measured along the channel, indicating a finite thermal conductivity. If helium II is used, and the heat flux inside the channel is lower than a critical value q_c, the temperature gradient is so small that it cannot be measured, so indicating that the liquid has an extremely high thermal conductivity. If the heat flux exceeds the critical value q_c, one observes an extra attenuation of second sound, which grows with the square of the heat flux. This phenomenon is known as "quantum turbulence" or "superfluid turbulence".[1,2] The damping force, known as "mutual friction", finds its origin in the interaction between the flow of excitations and a chaotic tangle of quantized vortices of equal circulation κ.

In many situations, the vortex tangle is assumed to be isotropic and may be described by introducing a scalar quantity L, the average vortex line length per unit volume (briefly called vortex line density). The evolution equation for L in counterflow superfluid turbulence has been formulated by Vinen.[3] Neglecting the influence of the walls, such an equation can be written as:[4]

$$\frac{dL}{dt} = AqL^{3/2} - BL^2, \tag{1}$$

with A and B coefficients dependent on the temperature.

This equation assumes homogeneous turbulence, i.e. that the value of L is the same everywhere in the system. However, homogeneity may be expected if the average distance between the vortex filaments is much smaller than the size of the system, but it will be not so for dilute vortex tangles.

The aim of this paper is to describe the coupling between the heat flux and the inhomogeneities in the vortex line density, both in the linear and in the nonlinear regimes. In fact, nonlinear phenomena are important in the study of superfluid turbulence, because the vortices can be formed when nonlinear second sound and shock waves are propagated. Therefore we formulate, using Extended Thermodynamics,[5,6] a nonlinear hydrodynamical model of quantum turbulence, in which the role of inhomogeneities is explicitly taken into account. This is important because second sound provides the standard methods of measuring the vortex line density L, and the dynamical mutual interplay between second sound and vortex lines may modify the standard results.

We will choose as fundamental fields the density ρ, the velocity $\mathbf{v}$, the internal energy density E, the heat flux $\mathbf{q}$, and the averaged vortex line density L. The relations which constrain the constitutive quantities are deduced from the entropy principle, using the Liu method of Lagrange multipliers.[7] The mathematical and physical consequences deduced by the theory are analyzed both in the linear and in the nonlinear regime. The vortex diffusion and the propagation of second sound and its interaction with vortex waves will be also considered.

2. Balance Equations and Constitutive Theory

Extended Thermodynamics (E.T.) offers a natural framework for the macroscopic description of liquid helium II. Indeed, in analogy with heat transport problem, using E.T., the relative motion of the excitations is well described by the dynamics of the heat flux. In Ref. 8, E.T. was applied to formulate a non-standard one-fluid model of liquid helium II, for laminar flows. Successively, a first thermodynamic study of turbulent flows was made in Ref. 9, where the presence of the vortex tangle was modelled through a constitutive relation. Here, we want to build up an hydrodynamical model of superfluid turbulence, which can describe also inhomogeneities and nonlinear phenomena.

2.1. *Balance Equations*

We consider, as a starting point the formulation of E.T. which uses the method of Lagrange multipliers accounting for the restrictions set by the balance equations[6,7]. For the fields ρ, $\rho\mathbf{v}$, $\rho\epsilon+\frac{1}{2}\rho v^2$, $\mathbf{q}$ and L general balance equations are chosen, which in terms of non convective quantities can be written:

$$\dot{\rho} + \rho\nabla\cdot\mathbf{v} = 0, \tag{2}$$

$$\rho\dot{\mathbf{v}} + \nabla\cdot\mathbf{J}^{\mathbf{v}} = 0, \tag{3}$$

$$\dot{E} + E\nabla\cdot\mathbf{v} + \nabla\cdot\mathbf{q} + \mathbf{J}^{\mathbf{v}}\cdot\nabla\mathbf{v} = 0, \tag{4}$$

$$\dot{\mathbf{q}} + \mathbf{q}\nabla\cdot\mathbf{v} + \nabla\cdot\mathbf{J}^{\mathbf{q}} = \sigma^{\mathbf{q}}, \tag{5}$$

$$\dot{L} + L\nabla\cdot\mathbf{v} + \nabla\cdot\mathbf{J}^{L} = \sigma^{L}. \tag{6}$$

Here $E = \rho\epsilon$ is the specific energy per unit volume, $\mathbf{J}^{\mathbf{v}}$ the stress tensor, $\mathbf{J}^{\mathbf{q}}$ the intrinsic part of the flux of the heat flux and $\mathbf{J}^{L}$ the flux of vortex lines; $\sigma^{\mathbf{q}}$ and σ^{L} are terms describing the net production of heat flux and vortices. In this system an upper dot denotes the material time derivative.

2.2. *Constitutive Theory*

Constitutive equations for the fluxes $\mathbf{J}^{\mathbf{v}}$, $\mathbf{J}^{\mathbf{q}}$ and $\mathbf{J}^{L}$ and the productions $\sigma^{\mathbf{q}}$ and σ^{L} are necessary to close the set of equations (2)-(6). To describe nonlinear phenomena, we choose for the fluxes the following general constitutive equations:

$$\mathbf{J}^{\mathbf{v}} = p_0(\rho, E, L, q^2)\mathbf{U} + a(\rho, E, q^2, L)\mathbf{q}\mathbf{q}, \tag{7}$$

$$\mathbf{J}^{\mathbf{q}} = \beta_0(\rho, E, L, q^2)\mathbf{U} + \gamma(\rho, E, q^2, L)\mathbf{q}\mathbf{q}, \tag{8}$$

$$\mathbf{J}^{L} = \nu_0(\rho, E, L, q^2)\mathbf{q}. \tag{9}$$

For the production term in the equation of the heat flux we will take the simple expression $\sigma_{\mathbf{q}} = -KL\mathbf{q}$ $(K > 0)$,[4] and for the one in the equation for the line density L we will choose Vinen's production and destruction terms (1).

Further restrictions on the constitutive relations are obtained imposing the validity of the entropy principle, applying the Liu procedure.[7] This method requires the existence of a scalar function $S = S\left(\rho, E, q^2, L\right)$ and a vector function $\mathbf{J}^S = \phi\left(\rho, E, q^2, L\right)\mathbf{q}$ of the fundamental fields, namely the entropy density and the entropy flux density respectively, such that the following inequality:

$$\begin{aligned}
&\dot{S} + S\nabla\cdot\mathbf{v} + \nabla\cdot\mathbf{J}^S - \Lambda^\rho\left[\dot{\rho} + \rho\nabla\cdot\mathbf{v}\right] - \Lambda^{\mathbf{v}}\cdot\left[\rho\dot{\mathbf{v}} + \nabla\cdot\mathbf{J}^{\mathbf{v}}\right]\\
&\quad - \Lambda^E\left[\dot{E} + E\nabla\cdot\mathbf{v} + \nabla\cdot\mathbf{q} + \mathbf{J}^{\mathbf{v}}\cdot\nabla\mathbf{v}\right]\\
&\quad - \Lambda^{\mathbf{q}}\cdot\left[\dot{\mathbf{q}} + \mathbf{q}\nabla\cdot\mathbf{v} + \nabla\cdot\mathbf{J}^{\mathbf{q}} - \sigma^{\mathbf{q}}\right]\\
&\quad - \Lambda^L\left[\dot{L} + L\nabla\cdot\mathbf{v} + \nabla\cdot\mathbf{J}^L - \sigma^L\right] \geq 0,
\end{aligned} \tag{10}$$

is satisfied for arbitrary fields ρ, $\mathbf{v}$, E, $\mathbf{q}$ and L. This inequality expresses the restrictions coming from the second law of thermodynamics. The quantities $\Lambda^\rho = \Lambda^\rho(\rho, E, L, q^2)$, $\Lambda^{\mathbf{v}} = \Lambda^v(\rho, E, L, q^2)\mathbf{q}$, $\Lambda^E = \Lambda^E(\rho, E, L, q^2)$, $\Lambda^{\mathbf{q}} = \lambda^q(\rho, E, L, q^2)\mathbf{q}$ and $\Lambda^L = \Lambda^L(\rho, E, L, q^2)$ are Lagrange multipliers, which are also objective functions of $\rho, E, \mathbf{q}$ and L.

The constitutive theory is obtained substituting (7)-(9) in (10) and imposing that the coefficients of all derivatives must vanish. After some lengthy calculations, we obtain $\Lambda_v = 0$, $a = 0$ and

$$dS = \Lambda_\rho d\rho + \Lambda_E dE + \lambda_q q_i dq_i + \Lambda_L dL, \tag{11}$$

$$-\rho\Lambda_\rho - \Lambda_E(E + p) - \lambda_q q^2 - \Lambda_L L = 0, \tag{12}$$

$$\phi = \Lambda_E + \lambda_q \gamma q^2 + \Lambda_L \nu, \tag{13}$$

$$d\phi = \lambda_q\left(d\beta + (1/2)\gamma dq^2 + q^2 d\gamma\right) + \Lambda_L d\nu. \tag{14}$$

It remains the following residual inequality for the entropy production:

$$\sigma^S = \mathbf{\Lambda}^{\mathbf{q}}\cdot\sigma^{\mathbf{q}} + \Lambda^L\sigma^L \geq 0. \tag{15}$$

In the following Sections the coefficients introduced in (7)-(9) will be examined in depth, and related to specific situations specially suitable to stress their physical meaning.

2.3. *Physical Interpretation of the Constitutive Relations*

In order to single out the physical meaning and relevance of the constitutive quantities and of the Lagrange multipliers, we analyze now in detail the relations obtained in the previous section. Denoting with Υ any of the scalar quantities S, ϕ, p, β, γ, ν, Λ_ρ, Λ_E, λ_q, Λ_L we put $\Upsilon_0(\rho, E, L) = \Upsilon(\rho, E, 0, L)$, obtaining, for the first-order coefficients, the following relations:

$$dS_0 = \Lambda_0^\rho d\rho + \Lambda_0^E dE + \Lambda_0^L dL, \tag{16}$$

$$S_0 - \rho\Lambda_0^\rho - \Lambda_0^E(E + p_0) - \Lambda_0^L L = 0, \tag{17}$$

$$\phi_0 = \Lambda_0^E + \Lambda_0^L \nu_0, \tag{18}$$

$$d\phi_0 = \lambda_0^q d\beta_0 + \Lambda_0^L d\nu_0. \tag{19}$$

We first introduce a "generalized temperature" as the reciprocal of the first-order part of the Lagrange multiplier of the energy:

$$\Lambda_0^E = \left[\frac{\partial S_0}{\partial E}\right]_{\rho,L} = \frac{1}{T}. \tag{20}$$

In the laminar regime (when $L = 0$), Λ_0^E reduces to the absolute temperature of thermostatics. In the presence of a vortex tangle the "generalized temperature" depends also on the line density L.

If we write now equations (16) and (17) in the following way:

$$dE = TdS_0 - T\Lambda_0^\rho d\rho - T\Lambda_0^L dL, \tag{21}$$

$$-T\Lambda_0^\rho = \frac{E}{\rho} - T\frac{S_0}{\rho} + \frac{p_0 + LT\Lambda_0^L}{\rho}, \tag{22}$$

we can define the "mass chemical potential" μ_0^ρ in turbulent superfluid as:

$$\mu_0^\rho = -T\Lambda_0^\rho = -T\left[\frac{\partial S_0}{\partial \rho}\right]_{E,L}, \tag{23}$$

and the "chemical potential of vortex lines" μ_0^L as:

$$\mu_0^L = -T\Lambda_0^L = -T\left[\frac{\partial S_0}{\partial L}\right]_{\rho,L}. \tag{24}$$

Indeed, in absence of vortices ($L = 0$), equation (21) is just Gibbs equation of thermostatics and the quantity (22) is the equilibrium chemical potential. The presence of vortices modifies the energy and the chemical potentials. For the chemical potential of vortex lines we will take the expression:[4]

$$\mu_0^L = \epsilon_V \ln\left(L/L^*\right), \tag{25}$$

where ϵ_V is the energy per unit length of the vortex lines[1,4], which depends essentially on T. In (25) L^* is a reference vortex line density, defined as the average length $< l >$ of the vortex loops composing the tangle, divided by the volume of the system, namely $L^* \equiv < l > /V$.

Consider now equations (18) and (19) which concern the expressions of the fluxes. Using definitions (20) and (24), we get:

$$\lambda_0^q d\beta_0 = d\left(\frac{1}{T}\right) - \nu_0 d\left(\frac{\mu_0^L}{T}\right). \tag{26}$$

From this equation, recalling that μ_0^L depends only on T and L, we obtain $\partial\beta_0/\partial\rho = 0$, and we put

$$\zeta_0 = \frac{\partial\beta_0}{\partial T} = -\frac{1}{T^2\lambda_0^q}\left[1 + \nu_0 T^2 \frac{\partial}{\partial T}\left(\frac{\mu_0^L}{T}\right)\right], \tag{27}$$

$$\chi_0 = \frac{\partial\beta_0}{\partial L} = -\frac{\nu_0}{T\lambda_0^q}\frac{\partial\mu_0^L}{\partial L} = -\frac{\nu_0}{T\lambda_0^q}\frac{\epsilon_V}{L}. \tag{28}$$

In Ref. 4 we have shown that it results $\lambda_0^q < 0$, $\zeta_0 \geq 0$, $\nu_0 \leq 0$ and $\chi_0 \leq 0$.

2.4. *The Constitutive Relations far from Equilibrium*

We analyze now the complete mathematical expressions far from equilibrium of the Lagrange multipliers, in order to shed some light on their physical meaning. Anyway, this is still an open topic for research, and we will only outline the general ideas.

First, we introduce the following quantity:

$$\theta = \frac{1}{\Lambda_E(\rho, E, L, q^2)}, \tag{29}$$

which, near equilibrium ($L = 0$, $\mathbf{q} = 0$) can be identified with the local equilibrium absolute temperature T. In the following we will choose as fundamental field the quantity θ, instead of the internal energy density E. In accord with Ref. 5, we will call θ "non-equilibrium temperature", a topic which is receiving much attention in current non-equilibrium Thermodynamics.[10]

We have shown that at equilibrium the quantities $-\Lambda_\rho/\Lambda_E$ and $-\Lambda_L/\Lambda_E$ can be interpreted as the equilibrium mass chemical potential (eq. (23)) and the equilibrium vortex line density chemical potential (eq. (24)). We define, consequently, as nonequilibrium chemical potentials the quantities:

$$\mu = -\frac{\Lambda_\rho}{\Lambda_E}, \qquad \text{and} \qquad \mu_L = -\frac{\Lambda_L}{\Lambda_E}. \tag{30}$$

Using these quantities, relations (11) and (12) can be written:

$$\theta dS = dE - \mu d\rho - \mu_L dL + \theta\lambda_q \mathbf{q}\cdot d\mathbf{q}, \tag{31}$$

$$\theta S = E + p - \rho\mu_\rho - L\mu_L + \theta\lambda_q q^2. \tag{32}$$

In particular, denoting with ϵ the specific energy and with s the nonequilibrium specific entropy, we see that the non equilibrium chemical potentials μ_ρ and μ_L must satisfy the relation:

$$\mu_\rho + \frac{L}{\rho}\mu_L = \epsilon - \theta s + \frac{p}{\rho} + \frac{\theta}{\rho}\lambda_q q^2. \tag{33}$$

In the absence of vortices ($\mu_L = 0$), this equation furnishes a generalized expression for the mass chemical potential in the laminar flow of superfluids (see also Ref. 11).

The theory developed here furnishes also the complete nonequilibrium expression of the entropy flux $\mathbf{J}^s$. Indeed, we can write:

$$\mathbf{J}^s = \left(\frac{1}{\theta} + \nu\Lambda_L + \gamma\lambda_q q^2\right)\mathbf{q} = \frac{1}{\theta}\left(\mathbf{q} - \mu_L \mathbf{J}^L + \theta\gamma\lambda_q q^2 \mathbf{q}\right). \tag{34}$$

This equation shows that, in a nonlinear theory of Superfluid Turbulence, the entropy flux is different from the product of the reciprocal nonequilibrium temperature and the heat flux, but it contains additional terms depending on the flux of heat flux and on the flux of line density.

This concludes, for the moment, our analysis of the thermodynamic restrictions on the coefficients of the constitutive relations, Further consequences of the nonlinear constitutive theory will be presented in a following paper. In the next section, we will explore two simple but physically relevant situations where the terms related to vortex density inhomogeneities play an especially explicit role.

3. Fields Equations

Substituting the constitutive equations (7)-(9) and the restriction $a = 0$ in system (2)-(6), the following system of field equations is obtained:

$$\dot{\rho} + \rho\nabla\cdot\mathbf{v} = 0, \tag{35}$$

$$\rho\dot{\mathbf{v}} + \nabla p = 0 \tag{36}$$

$$\rho\dot{\epsilon} + \nabla\cdot\mathbf{q} + p\nabla\cdot\mathbf{v} = 0, \tag{37}$$

$$\dot{\mathbf{q}} + \mathbf{q}\nabla\cdot\mathbf{v} + \nabla\beta + \nabla\cdot(\gamma\mathbf{q}\mathbf{q}) = \sigma^{\mathbf{q}}, \tag{38}$$

$$\dot{L} + L\nabla\cdot\mathbf{v} + \nabla\cdot(\nu\mathbf{q}) = \sigma^L, \tag{39}$$

Observe that in these equations there are the unknown quantities p, ϵ, β, γ and ν and the productions $\sigma^{\mathbf{q}}$ and σ^L. Concerning the fluxes, one observes that the five quantities p, ϵ, β, γ and ν cannot be totally independent, because they must satisfy relations (11)-(14).

In the following we pay a special attention to the two last equations, which contain the new effects on which we are focusing our attention.

3.1. *The drift velocity of the tangle*

In a hydrodynamical model of turbulent superfluids, the line density L acquires field properties: it depends on the coordinates, it has a drift velocity $\mathbf{v}_L$, and its rate of change must obey a balance equation of the general form:

$$\partial_t L + \nabla \cdot (L\mathbf{v}^L) = \sigma^L, \tag{40}$$

with $\mathbf{v}^L$ the drift velocity of the tangle and ∂_t stands for $\partial/\partial t$. If we now observe that the equation (39) can be written:

$$\partial_t L + \nabla \cdot (L\mathbf{v} + \nu\mathbf{q}) = \sigma^L, \tag{41}$$

comparing (40) and (41), we conclude that the drift velocity of the tangle, with respect to the container, is given by:

$$\mathbf{v}^L = \mathbf{v} + \frac{\nu}{L}\mathbf{q}. \tag{42}$$

Note that the velocity $\mathbf{v}^L$ does not coincide with the microscopic velocity of the vortex line element, but represents an averaged macroscopic velocity of this quantity. It is to make attention to the fact that often in the literature the microscopic velocity $\dot{\mathbf{s}}$ is denoted with $\mathbf{v}_L$.

Another possibility is to interpret $\nu\mathbf{q} = \mathbf{J}^L$ as the diffusion flux of vortices. Note that, in this model, if $\mathbf{q} = 0$, $\mathbf{J}^L$ also is zero.

3.2. *Vortex diffusion*

Here we will apply the general set of equations derived up to here to describe vortex diffusion. We will study this phenomenon neglecting non linear terms in the heat flux. In this case, equations (38) and (39) simplify as:

$$\dot{\mathbf{q}} + \zeta_0 \nabla T + \chi_0 \nabla L = \sigma^{\mathbf{q}} = -KL\mathbf{q}, \tag{43}$$

$$\dot{L} + L\nabla \cdot \mathbf{v} + \nu_0 \nabla \cdot \mathbf{q} = \sigma^L = -BL^2 + AqL^{3/2}, \tag{44}$$

with ζ_0 and χ_0 defined by (27)-(28).

Assume, for the sake of simplicity, that $T = constant$ and that $\mathbf{q}$ varies very slowly, in such a way that $\dot{\mathbf{q}}$ may be neglected. We find from (43) that $\chi_0 \nabla L = -KL\mathbf{q}$. Then, we may write

$$\mathbf{q} = -\frac{\chi_0}{KL}\nabla L. \tag{45}$$

Introducing this expression in equation (44), we find:

$$\dot{L} + L\nabla \cdot \mathbf{v} - \frac{\nu_0 \chi_0}{KL}\Delta L + \frac{\nu_0 \chi_0}{KL^2}(\nabla L)^2 = \sigma^L. \tag{46}$$

Then, we have for L a reaction-diffusion equation, which generalizes the usual Vinen's equation (1) to inhomogeneous situations. The diffusivity coefficient is found to be

$$D = \frac{\nu_0 \chi_0}{KL}. \tag{47}$$

Since $\lambda_0^q < 0$ and $K > 0$, it turns out that $D > 0$, as it is expected. Thus, the vortices will diffuse from regions of higher L to those of lower L.

If $\mathbf{v}$ vanishes, or if its divergence vanishes, equation (46), neglecting also the term in $(\nabla L)^2$, yields:

$$\dot{L} = -BL^2 + AqL^{3/2} + D\Delta L. \tag{48}$$

Equation (48) indicates two temporal scales for the evolution of L: one of them is due to the production-destruction term ($\tau_{decay} \approx [BL - AqL^{1/2}]^{-1}$) and another one to the diffusion: $\tau_{diff} \approx \frac{X^2}{D}$, where X is the size of the system. For large values of L, τ_{decay} will be much shorter and the production-destruction dynamics will dominate over diffusion; for small L, instead, diffusion processes may be dominant. This may be also understood from a microscopic perspective because the mean free path of vortex motion is of the order of intervortex spacing, of the order of $L^{-1/2}$, and therefore it increases for low values of L.

A more general situation for the vortex diffusion flux is to keep the temperature gradient in (43). In this more general case, $\mathbf{q}$ is not more parallel to ∇L but results

$$\mathbf{q} = -\frac{\chi_0}{KL}\nabla L - \frac{\zeta_0}{KL}\nabla T \tag{49}$$

in which case, it would become

$$\mathbf{J}^L = \nu_0 \mathbf{q} = -D\nabla L - D\frac{\zeta_0}{\chi_0}\nabla T. \tag{50}$$

The second term in (50) plays a role analogous to thermal diffusion -or Soret effect- in usual diffusion of particles. This kind of situations have not been studied enough in the context of vortex tangles, but they would arise in a natural way when trying to understand the behavior of quantum turbulence in the presence of a temperature gradient.

3.3. *Wave propagation in counterflow vortex tangles*

We briefly recall here the results on the propagation of second sound harmonic plane waves obtained in Ref. 4. Experiments show that in this case the velocity $\mathbf{v}$ is zero, and only the fields T, $\mathbf{q}$ and L are involved. The equations for these fields, under these hypotheses, expressing the energy in terms of T and L, are simply:

$$\rho c_V \dot{T} + \rho \epsilon_L \dot{L} + \nabla \cdot \mathbf{q} = 0, \tag{51}$$

$$\dot{\mathbf{q}} + \zeta_0 \nabla T + \chi_0 \nabla L = -KL\mathbf{q} \tag{52}$$

$$\dot{L} + \nu_0 \nabla \cdot \mathbf{q} = -BL^2 + AqL^{3/2}, \tag{53}$$

where $c_V = \partial\epsilon/\partial T$ is the specific heat at constant volume and $\epsilon_L = \partial\epsilon/\partial L \simeq \epsilon_V$.

These equations are enough for the discussion of the physical effects of the coupling of second-sound and the distorsion of the vortex tangle (represented by the inhomogeneities in L), which must be taken into account in an analysis of the vortex tangle by means of second sound. In fact, some of the previous hydrodynamical analyses of turbulent superfluids had this problem as one of their main motivations.[12]

A stationary solution of system (43)-(45) is given by

$$\mathbf{q} = \mathbf{q}_0 = (q_{10}, 0, 0), \quad L = L_0 = \frac{A^2}{B^2}[q_{10}]^2, \quad T = T_0(\mathbf{x}) = T^* - \frac{KL_0 q_{10}}{\zeta_0} x_1, \tag{54}$$

with $q_{10} > 0$.

Let $\mathbf{n}$ the unit vector in the direction of wave propagation. We assume heat flux parallel to x axis. It is seen that when the temperature wave is propagated along x axis, one obtains the following dispersion relation[4]

$$\begin{aligned} \omega^2 = {} & k^2\left[V_2^2(1-\rho\epsilon_L\nu_0)+\nu_0\chi_0\right] + N_1N_2 - i\omega(N_1+N_2) + i\frac{k^2}{\omega}V_2^2N_2 \\ & - ik\left[(\chi_0 + V_2^2\rho\epsilon_L)N_4 - \nu_0 N_3\right] \end{aligned} \tag{55}$$

while, when the wave is orthogonal to the heat flux $\mathbf{q}$, one obtains[4]

$$\begin{aligned} \omega^2 = {} & k^2\left[V_2^2(1-\rho\epsilon_L\nu_0)+\nu_0\chi_0\right] + N_1N_2 - i\omega(N_1+N_2) \\ & + i\frac{k^2}{\omega}V_2^2\left(N_2 - \frac{N_3N_4}{\omega + iN_1}\right), \end{aligned} \tag{56}$$

where $k = k_r + ik_s$ is the complex wave number and ω the real frequency, and where we have put $N_1 = KL_0$, $N_2 = 2BL_0 - (3/2)AL_0^{1/2}q_{10}$, $N_3 = Kq_{10}$ and $N_4 = Aq_{10}L_0^{3/2}$. In both equations (55) and (56), we have denoted with V_2 the quantity:

$$V_2^2 = \frac{\zeta_0}{\rho c_V}, \tag{57}$$

which, in the absence of vortices, coincides with the usual velocity of the second sound,[8,11]. As shown in Ref. 4, in the presence of vortices, this coefficient includes a positive contribution proportional to L, which shows that the speed of the waves increases when L increases.

We compare the result (56) with the result obtained in Ref. 9, where we supposed L a fixed quantity, and the term ν_0 was assumed to vanish, eliminating in this way the effects of the oscillations of $\mathbf{q}$ on the vortex line density L of the tangle. In that work, the dispersion relation for the second sound was:

$$\omega^2 = V_2^2 k^2 - i\omega K L_0. \tag{58}$$

Comparison of (56) with (58) shows that the distortion of the vortex tangle under the action of the heat wave, and its corresponding back reaction on the latter, implies remarkable changes in the velocity and the attenuation of the second sound, the latter effect depending on the relative direction between $\mathbf{q}_0$ and $\mathbf{n}$. Thus, if one uses (58) instead of (56) one obtains erroneous values for the average vortex line density L_0 and the friction coefficient, leading to an incorrect interpretation of the physical results.

4. Conclusions

The study of quantum turbulence in superfluids often assumes homogeneity of the vortex tangle line density L. In several situations this homogeneity will not hold, and the vortex lines will diffuse from the most concentrated to the less concentrated zones. For instance, vortex lines could be produced near the walls and migrate by diffusion to the bulk of the container until a homogeneous situation is reached. Furthermore, if vortex lines are flexible, they will be bent and their density will be compressed and rarefied by second-sound waves and this will produce an inhomogeneity in L, which, on its turn, will influence the propagation of second sound. This may be relevant in the interpretation of the experimental results on the speed and the attenuation of the second sound in terms of the average vortex line density of the system.

To incorporate these effects has been the main motivation of this paper. We have not limited ourselves to add a more general evolution equation for L, but we have tried to insure thermodynamical consistency of the mutual coupling of this equation and the equations considered previously for the other fields. We have worked in a macroscopic thermodynamic framework, which yields the several consequences of incorporating the additional terms to the evolution equations for the heat flux and the vortex line density. The thermodynamic consequences are shown as restrictions on the coefficients of the new terms. We have analyzed both in the linear and in the nonlinear regime the mathematical and physical consequences deduced by the theory. We have obtained in this way field equations for the relevant quantities, which have been applied to describe vortex diffusion and the propagation of harmonic plane waves. The study of nonlinear phenomena, as the propagation of non linear second sound and shock waves will be made in a following paper.

Acknowledgments

We acknowledge the support of the Acción Integrada España-Italia (Grant S2800082F HI2004-0316 of the Spanish Ministry of Science and Technology and grant IT2253 of the Italian MIUR). DJ acknowledges the financial support from the Dirección General de Investigación of the Spanish Ministry of Education under grant FIS 2006-12296-C02-01 and of the Direcció General de Recerca of the Generalitat of Catalonia, under grant 2005 SGR-00087. MSM and MS acknowledge the financial support from MIUR under grant "PRIN 2005 17439-003" and by "Fondi 60%" of the University of Palermo.

References

1. R.J. Donnelly, *Quantized vortices in helium II* (Cambridge University Press, Cambridge, U.K., 1991).
2. *Quantized Vortex Dynamics and Superfluid Turbulence*, edited by C.F. Barenghi, R.J. Donnelly and W.F. Vinen (Springer, Berlin, 2001).
3. W.F. Vinen, "Mutual friction in a heat current in liquid helium II. III. Theory of the mutual friction" Proc. Roy. Soc. London **A240**, 493-515. (1957).
4. M.S. Mongiovì and D. Jou , "A thermodynamical derivation of a hydrodynamical model of inhomogeneous superfluid turbulence". Phys. Rev. B, **75**, 024507 (2006).
5. D. Jou, J. Casas-Vázquez and G. Lebon, *Extended Irreversible Thermodynamics* (Springer-Verlag, Berlin Heidelberg, 2001).
6. I. Müller and T. Ruggeri, *Rational Extended Thermodynamics* (Springer-Verlag, New York, 1998).
7. I. Liu, "Method of Lagrange multipliers for exploitation of the entropy principle" Arch. Rat. Mech. Anal. **46**, 131-148 (1972).
8. M.S. Mongiovì, "Extended Irreversible Thermodynamics of Liquid Helium II" Phys. Rev. B **48**, 6276 (1993).
9. D. Jou, G. Lebon and M.S. Mongiovì, "Second sound, superfluid turbulence and intermittent effects in liquid helium II", Phys. Rev. B **66**, 224509(2002).
10. J. Casas-Vázquez and D. Jou,"Temperature in non-equilibrium states: a review of open problems and current proposals" Rep. Prog. Phys. **66**, 1937 (2003).
11. M.S. Mongiovì, "Non-linear Extended Thermodynamics of a Non-Viscous in Fluid the Presence of Heat Flux", **25**, 31-47 (2000).
12. M. Tsubota, T. Araki and W.F. Vinen, "Diffusion in an inhomogeneous vortex tangle", Physica B **329-333**, 224-225 (2003).

CONVERGENCE OF FINITE ELEMENTS ADAPTED FOR WEAKER NORMS

PEDRO MORIN

Departamento de Matemática, Facultad de Ingeniería Química, Instituto de Matemática Aplicada del Litoral, Universidad Nacional del Litoral, CONICET
Güemes 3450, S3000GLN Santa Fe, Argentina
E-mail: pmorin@math.unl.edu.ar
math.unl.edu.ar/˜pmorin/

KUNIBERT G. SIEBERT

Institut für Mathematik, Universität Augsburg
Universitätsstraße 14, D-86159 Augsburg, Germany
E-mail: siebert@math.uni-augsburg.de
scicomp.math.uni-augsburg.de/Siebert/

ANDREAS VEESER

Dipartimento di Matematica, Università degli Studi di Milano
Via C. Saldini 50, I-20133 Milano, Italy
E-mail: veeser@mat.unimi.it
www.mat.unimi.it/users/veeser/

We consider finite elements that are adapted to a (semi)norm that is weaker than the one of the trial space. We establish convergence of the finite element solutions to the exact one under the following conditions: refinement relies on unique quasi-regular element subdivisions and generates locally quasi-uniform grids; the finite element spaces are conforming, nested, and satisfy the inf-sup condition; the error estimator is reliable and appropriately locally efficient; the indicator of a non-marked element is bounded by the estimator contribution associated with the marked elements, and each marked element is subdivided at least once. This abstract convergence result is illustrated by two examples.

Keywords: Adaptivity, conforming finite elements, convergence.

1. Introduction and Outline

Adaptivity has become a popular technique to increase the efficiency of finite element methods for boundary values problems. In practice, finite element grids are adapted to various error notions: the energy norm, other norms, or the output of certain functionals applied to the solution. However, the theoretical underpinning of the methods in terms of convergence and complexity results essentially restrict, up to now, to the most immediate cases of the energy norm and the norm of the trial space.[2,5,6,8–10,14]

This paper presents a basic convergence result for finite elements that are adapted to a (semi)norm that is possibly weaker than the one of the trial space. To this end, §2 gives general assumptions on the problem itself, the refinement framework, the finite element spaces, the approximate solution, the a posteriori error estimator, the marking strategy, and the step REFINE. They ensure the convergence of both error in the weaker (semi)norm and associated estimator. The proof is obtained by generalizing the convergence proof of Ref. 11 in a straight-forward manner.

In §3 we illustrate this convergence result by two examples: Lagrange elements for the Poisson problem that are adapted for the mean square error and Raviart-Thomas or Brezzi-Douglas-Marini elements in a mixed discretization that are adapted for the mean square error of the flux.

2. Abstract Convergence for Weak Norms

We first describe the problem class and adaptive algorithm and then present the convergence result.

2.1. *Problem Class and Error Notion*

We consider linear boundary value problems that can be reformulated in the following weak form: given a real Hilbert space $\mathbb{V}$ with norm $\|\cdot\|$, a continuous bilinear form $\mathcal{B}:\mathbb{V}\times\mathbb{V}\to\mathbb{R}$, and an element $f\in\mathbb{V}^*$ of the dual space of $\mathbb{V}$, find

$$u\in\mathbb{V}: \qquad \mathcal{B}(u,w)=\langle f,\,w\rangle \qquad \forall w\in\mathbb{V}. \tag{1}$$

We suppose that the so-called *inf-sup (or Babuška-Brezzi) condition* holds: there exists $\alpha>0$ such that

$$\inf_{\substack{v\in\mathbb{V}\\ \|v\|=1}}\ \sup_{\substack{w\in\mathbb{V}\\ \|w\|=1}} \mathcal{B}(v,w)\geq\alpha, \qquad \inf_{\substack{w\in\mathbb{V}\\ \|w\|=1}}\ \sup_{\substack{v\in\mathbb{V}\\ \|v\|=1}} \mathcal{B}(v,w)\geq\alpha. \tag{2a}$$

Concerning the error notion, we are interested in a seminorm $|\cdot|$ that is weaker than $\|\cdot\|$: there exists $C\geq 0$ such that

$$\forall v\in\mathbb{V} \quad |v|\leq C\|v\|. \tag{2b}$$

Below, we will introduce *'local features'* into (1) by making assumptions on a mesh-dependent counterpart of $|\cdot|$ and its interplay with $\mathcal{B}$. To this end, we suppose that $\mathbb{V}$ is a subspace of $L_p(\Omega;\mathbb{R}^m)$, where $p \in (1,\infty)$, $m \in \mathbb{N}$, and Ω is the underlying domain in $\mathbb{R}^d$, $d \geq 2$, that can be meshed. In what follows, we suppress the dependence on the data Ω, f, and $\mathcal{B}$.

2.2. *Adaptive Algorithm*

The adaptive algorithm for approximating u in (1) is an iteration of the following main steps:

$$
\begin{aligned}
&(1)\ u_k := \mathsf{SOLVE}(\mathbb{V}(\mathcal{G}_k)).\\
&(2)\ \{\mathcal{E}_k(E)\}_{E\in\mathcal{G}_k} := \mathsf{ESTIMATE}(u_k, \mathcal{G}_k).\\
&(3)\ \mathcal{M}_k := \mathsf{MARK}(\{\mathcal{E}_k(E)\}_{E\in\mathcal{G}_k}, \mathcal{G}_k).\\
&(4)\ \mathcal{G}_{k+1} := \mathsf{REFINE}(\mathcal{G}_k, \mathcal{M}_k), \text{ increment } k.
\end{aligned}
\tag{3}
$$

In practice, a stopping test is used after step (2) for terminating the iteration; here we shall ignore it for notational convenience. The realization of these steps requires the following objects and modules:

Initial Grid and Framework for Refinement. An initial grid $\mathcal{G}_0$ of the domain Ω and a refinement procedure REFINE. The refinement procedure has two input arguments: a grid $\mathcal{G}$ and a subset $\mathcal{M} \subset \mathcal{G}$. All elements $E \in \mathcal{M}$ must be 'refined'. The input grid $\mathcal{G}$ can be the initial grid $\mathcal{G}_0$ or the output of a previous application of REFINE. A grid $\mathcal{G}'$ is called refinement of $\mathcal{G}$ whenever $\mathcal{G}'$ can be produced from $\mathcal{G}$ by a finite number of applications of REFINE. Initial grid and refinement procedure thus generate the set

$$\mathbb{G} := \{\mathcal{G} \mid \mathcal{G} \text{ is a refinement of } \mathcal{G}_0\}.$$

We shall write '$\preccurlyeq$' for '$\leq C$' where C may depend on data of (1), the class $\mathbb{G}$, and the modules ESTIMATE, MARK below, but not on a particular grid or the iteration number. Similarly, we say that some object is 'fixed' if it has the same dependencies.

We suppose that REFINE relies on *unique quasi-regular element subdivisions.* More precisely, there exist constants $q_1, q_2 \in (0,1)$ such that, irrespective of the grid $\mathcal{G}$, any element $E \in \mathcal{G}$ can be subdivided into $n(E) \geq 2$ subelements $E'_1, \dots, E'_{n(E)}$ such that

$$E = E'_1 \cup \cdots \cup E'_{n(E)}, \qquad |E| = |E'_1| + \cdots + |E'_{n(E)}|, \tag{4a}$$

$$q_1|E| \leq |E'_i| \leq q_2|E|, \qquad i = 1,\dots,n(E), \tag{4b}$$

where $|E|$ stands for the d-dimensional Lebesgue measure of E.

These unique element subdivisions generate a 'master forest' $\mathcal{F}$ of infinite trees, where each node corresponds to an element, its direct successors to its subelements, and the roots to the elements of the initial grid $\mathcal{G}_0$. A subforest $\hat{\mathcal{F}} \subset \mathcal{F}$ is called finite if it has a finite number of nodes. Any finite tree may have interior nodes, i.e. nodes with successors, and does have leaf nodes, i.e. nodes without any successor.

Any subdivision $\mathcal{S}$ of the domain Ω that is subordinated to $\mathcal{G}_0$ is uniquely associated with a finite subforest $\mathcal{F}(\mathcal{S})$ of $\mathcal{F}$, where the leaf nodes are the elements of the subdivision. Given $n \in \mathbb{N}$ and a subset $\hat{\mathcal{S}}$ of such subdivision $\mathcal{S}$, we denote by $\mathcal{F}_n(\mathcal{S}, \hat{\mathcal{S}})$ the subforest of $\mathcal{F}$ that consists of $\mathcal{F}(\mathcal{S})$ and all successors of elements in $\hat{\mathcal{S}}$ up to generation n.

We suppose that the class $\mathbb{G}$ is a subclass of the subdivisions of Ω subordinated to $\mathcal{G}_0$ and is *locally quasi-uniform* in that

$$\sup_{\mathcal{G}\in\mathbb{G}} \max_{E\in\mathcal{G}} \#N_{\mathcal{G}}(E) \preccurlyeq 1, \qquad \sup_{\mathcal{G}\in\mathbb{G}} \max_{E'\in N_{\mathcal{G}}(E)} \frac{|E|}{|E'|} \preccurlyeq 1, \tag{4c}$$

where $N_{\mathcal{G}}(E) := \{E' \in \mathcal{G} \mid E' \cap E \neq \emptyset\}$ denotes the set of neighbors of E in $\mathcal{G}$. The grids in $\mathbb{G}$ may have additional properties like conformity.

Finite Element Spaces and Mesh-Dependent Norms. We suppose that the finite element spaces $\mathbb{V}(\mathcal{G})$, $\mathcal{G} \in \mathbb{G}$, are conforming, nested, and satisfy a discrete inf-sup condition: for any $\mathcal{G}, \mathcal{G}' \in \mathbb{G}$, there hold

$$\mathbb{V}(\mathcal{G}) \subset \mathbb{V} \tag{5a}$$

$$\mathcal{G}' \text{ is a refinement of } \mathcal{G} \implies \mathbb{V}(\mathcal{G}) \subset \mathbb{V}(\mathcal{G}') \tag{5b}$$

$$\inf_{\substack{v\in\mathbb{V}(\mathcal{G})\\ \|v\|=1}} \sup_{\substack{w\in\mathbb{V}(\mathcal{G})\\ \|w\|=1}} \mathcal{B}(v,w) \geq \beta \tag{5c}$$

with some fixed $\beta > 0$.

Moreover, we suppose that, for each grid $\mathcal{G} \in \mathbb{G}$, there is a pair $|\cdot|_{\mathcal{G}}$, $|\!|\!|\cdot|\!|\!|_{\mathcal{G}}$ of possibly *mesh-dependent seminorms* that is associated with the weak seminorm $|\cdot|$ of the error notion and has the following properties:

- $|\cdot|_{\mathcal{G}} = |\cdot|_{\mathcal{G};\Omega}$ is a seminorm on $\mathbb{V}$ that is close to $|\cdot|$, p-subadditive with respect to the domain, and absolutely continuous with respect to the Lebesgue measure in the following sense: for all $v \in \mathbb{V}$,

$$|v| \preccurlyeq |v|_{\mathcal{G}} \preccurlyeq \|v\|, \tag{5d}$$

$$\sum_{i=1}^{n} |v|^p_{\mathcal{G};\omega_i} \preccurlyeq |v|^p_{\mathcal{G};\Omega}, \tag{5e}$$

$$|v|_{\mathcal{G}_k;\Omega_k} \to 0, \tag{5f}$$

where $\{\omega_i\}_{i=1}^n$ are disjoint subdomains of Ω, each one being a union of elements of $\mathcal{G}$, and $\{\Omega_k\}_k$ is a sequence of subdomains such that $|\Omega_k| \to 0$ and each Ω_k is a union of elements of a grid $\mathcal{G}_k$.

- $|\!|\!|\cdot|\!|\!|_{\mathcal{G}}$ is a seminorm on a subspace $\tilde{\mathbb{V}}(\mathcal{G})$ of $\mathbb{V}(\mathcal{G})$;
- the bilinear form is continuous with respect to the pair $|\cdot|_{\mathcal{G}}$, $|\!|\!|\cdot|\!|\!|_{\mathcal{G}}$ in a local sense: there is a constant $C_{\mathcal{B}} \geq 0$ such that, if ω is a union of elements of $\mathcal{G}$, then we have for any $v \in \mathbb{V}$ and any $w \in \tilde{\mathbb{V}}(\mathcal{G})$

$$w = 0 \text{ in } \Omega \setminus \omega \qquad \Longrightarrow \qquad \mathcal{B}(v, w) \leq C_{\mathcal{B}} |v|_{\mathcal{G};\omega} \, |\!|\!|w|\!|\!|_{\mathcal{G}} . \tag{5g}$$

The role of these mesh-dependent seminorms will become clear from the example in §3.1.

SOLVE. We suppose that the output $u_{\mathcal{G}} := \mathsf{SOLVE}\big(\mathbb{V}(\mathcal{G})\big)$ is the *Galerkin approximation* of u in $\mathbb{V}(\mathcal{G})$:

$$u_{\mathcal{G}} \in \mathbb{V}(\mathcal{G}) : \qquad \mathcal{B}(u_{\mathcal{G}}, w) = \langle f, w \rangle \qquad \forall w \in \mathbb{V}(\mathcal{G}). \tag{6}$$

Thanks to (5a) and (5c), the solution of (6) exists and is unique.

ESTIMATE. We suppose that $\{\mathcal{E}_{\mathcal{G}}(E)\}_{E\in\mathcal{G}} := \mathsf{ESTIMATE}(u_{\mathcal{G}}, \mathcal{G})$ has the following two properties for any grid $\mathcal{G} \in \mathbb{G}$: First, there holds the following *global upper bound* for the error in $|\cdot|$ of the Galerkin approximation $u_{\mathcal{G}}$:

$$|u_{\mathcal{G}} - u| \preccurlyeq \mathcal{E}_{\mathcal{G}}, \tag{7a}$$

where, given a subset $\hat{\mathcal{G}} \subset \mathcal{G}$, we define $\mathcal{E}_{\mathcal{G}}(\hat{\mathcal{G}}) := \big(\sum_{E\in\hat{\mathcal{G}}} \mathcal{E}_{\mathcal{G}}^p(E)\big)^{1/p}$ and set $\mathcal{E}_{\mathcal{G}} := \mathcal{E}_{\mathcal{G}}(\mathcal{G})$ and $\mathcal{E}_{\mathcal{G}}(\emptyset) := 0$.

Secondly, a *fixed finite subdivision depth implies a local lower bound with respect to a mesh-dependent dual seminorm of the residual.* More precisely, there is a fixed $n \in \mathbb{N}$ such that, for any element $E \in \mathcal{G}$ and any finer grid $\mathcal{G}' \in \mathbb{G}$ with $\mathcal{F}(\mathcal{G}') \supset \mathcal{F}_n\big(\mathcal{G}, N_{\mathcal{G}}(E)\big)$, there holds

$$\mathcal{E}_{\mathcal{G}}(E) \preccurlyeq \sup\Big\{\langle \mathcal{R}_{\mathcal{G}}, w\rangle \mid w \in \tilde{\mathbb{V}}\big(\mathcal{G}'; \omega_{\mathcal{G}}(E)\big), \, |\!|\!|w|\!|\!|_{\mathcal{G}'} \leq 1\Big\} + \operatorname{osc}_{\mathcal{G}}(E), \tag{7b}$$

where the oscillation indicator satisfies

$$\operatorname{osc}_{\mathcal{G}}(E) \preccurlyeq m(|E|)\Big(|u_{\mathcal{G}}|_{\mathcal{G};\omega_{\mathcal{G}}(E)} + \|D\|_{\mathbb{D}(\omega_{\mathcal{G}}(E))}\Big). \tag{7c}$$

Hereafter

- $\mathcal{R}_{\mathcal{G}} \in \mathbb{V}^*$ is the residual defined by

$$\langle \mathcal{R}_{\mathcal{G}}, w\rangle := \mathcal{B}(u_{\mathcal{G}}, w) - \langle f, w\rangle, \qquad \forall w \in \mathbb{V}; \tag{8}$$

- $\omega_{\mathcal{G}}(E) \subset \Omega$ is the patch (union) of elements in $N_{\mathcal{G}}(E)$;

- $\tilde{\mathbb{V}}(\mathcal{G}';\omega_\mathcal{G}(E))$ is the space of 'local test functions' given by

$$\tilde{\mathbb{V}}(\mathcal{G}';\omega_\mathcal{G}(E)) := \{w \in \tilde{\mathbb{V}}(\mathcal{G}') \mid w = 0 \text{ in } \Omega \setminus \omega_\mathcal{G}(E)\};$$

- $m : [0,\infty) \to [0,\infty)$ is a fixed, continuous, and nondecreasing function with $m(0) = 0$;
- $\mathbb{D}$ is another space with a norm that is p-subadditive and absolutely continuous with respect to the Lebesgue measure in the sense of (5e), (5f), and $D \in \mathbb{D}$ is given by the data of (1).

The global upper bound (7a) ensures that the error indicators do not overview any source of error. Inequality (7b) is the main step in proving a local lower error bound by Verfürth's constructive argument:[15] indeed, if one inserts (1) into (8) and recalls (5g), then (7b) readily yields the local lower error bound

$$\mathcal{E}_\mathcal{G}(E) \preccurlyeq |u_\mathcal{G} - u|_{\mathcal{G};\omega_\mathcal{G}(E)} + \operatorname{osc}_\mathcal{G}(E). \tag{9}$$

Thus, (7b) ensures, up to (7c) and the difference between $|\cdot|$ and $|\cdot|_\mathcal{G}$, the sharpness of the upper bound (7a) in a local sense. The presence of the oscillation indicator (7c) is discussed in Remark 4.7 of Ref. 11.

MARK. We suppose that the output $\mathcal{M} := \mathsf{MARK}(\{\mathcal{E}_\mathcal{G}(E)\}_{E\in\mathcal{G}}, \mathcal{G})$ of marked elements has the property

$$\forall E \in \mathcal{G} \setminus \mathcal{M} \qquad \mathcal{E}_\mathcal{G}(E) \le \mathcal{E}_\mathcal{G}(\mathcal{M}). \tag{10}$$

We suppose (10) only for convenience; in §5 of Ref. 11 we consider a weaker condition that is sufficient and essentially necessary for convergence.

REFINE. We suppose that the output grid $\mathcal{G}' := \mathsf{REFINE}(\mathcal{G}, \mathcal{M})$ satisfies the *minimal requirement*

$$\mathcal{F}(\mathcal{G}') \supset \mathcal{F}_1(\mathcal{G}, \mathcal{M}), \tag{11}$$

that is, each marked element of the input grid is subdivided at least once in the output grid. Additional elements in $\mathcal{G} \setminus \mathcal{M}$ may be refined in order to fulfill (4c) or to ensure that the output grid is in the class $\mathbb{G}$.

2.3. *Convergence*

We now state the main result of this paper. The difference to Theorem 2.1 in Ref. 11 is that here the grids are adapted to the error in the seminorm $|\cdot|$, which is weaker than the one of the trial space.

Theorem 2.1 (Abstract Convergence for Weak Norms). *Let u be the exact solution of* (1)*, suppose that there holds* (2)*, and that $\{u_k\}_k$ is the sequence of approximate solutions generated by iteration* (3)*.*

If the refinement framework, the finite element spaces, the modules SOLVE, ESTIMATE, MARK, and REFINE satisfy, respectively, (4), (5), (6), (7), (10)*, and* (11)*, then both error and estimator decrease to* 0*, that is*

$$|u_k - u| \to 0 \quad \text{and} \quad \mathcal{E}_k \to 0 \quad \text{as} \quad k \to \infty.$$

Proof. In view of Lemma 4.2 of Ref. 11, $\{u_k\}_k$ converges to some $u_\infty \in \mathbb{V}$ and it remains to show that $u_\infty = u$. To this end, proceed as in §4.2 of Ref. 11 with the following modifications: use (5g) instead of a 'local' continuity of $\mathcal{B}$ in terms of $\|\cdot\|$, sum p-powers of local (semi)norms instead of squares, and then exploit (5d) or (5f). □

3. Two applications

The following two applications focus on the error notion, which is really weaker that the norm of the trial space; further examples are in §3 of Ref. 11. In what follows, $h_{\mathcal{G}}$ stands for the meshsize function associated with $\mathcal{G} \in \mathbb{G}$.

3.1. *Mean Square Error in Poisson's Problem*

We apply iteration (3) to generate finite element solutions to the Poisson problem that adaptively approach the exact solution in the L_2-error.

Problem. Let $\Omega \subset \mathbb{R}^d$, $d \in \{2, 3\}$, be a bounded, polyhedral, and convex domain, set

$$\mathbb{V} = H_0^1(\Omega), \ \|\cdot\| = \|\nabla \cdot\|_{L_2(\Omega)}, \quad |\cdot| = \|\cdot\|_{L_2(\Omega)},$$

$$\mathcal{B}(v, w) = \int_\Omega \nabla v \cdot \nabla w, \quad v, w \in \mathbb{V},$$

and suppose $f \in L_2(\Omega)$. It is well known that there hold $f \in \mathbb{V}^*$ and (2).

Refinement framework. Let $\mathcal{G}_0$ be a suitable conforming triangulation of Ω into d-simplices and let $\mathbb{G}$ be the class of all triangulations that can be generated from $\mathcal{G}_0$ by iterative or recursive bisection; e.g. see Ref. 12. Then (4) is fulfilled with $n(E) = 2$, and $q_1 = q_2 = \frac{1}{2}$; the hidden constants in (4c) depend on $\mathcal{G}_0$. Moreover, $\mathbb{G}$ is a shape-regular family of triangulations.

Finite element spaces and mesh-dependent norms. For any $\mathcal{G} \in \mathbb{G}$, we choose Lagrange elements of any fixed order ℓ,

$$\mathbb{V}(\mathcal{G}) := \mathbb{LE}_\ell(\mathcal{G}) \cap H_0^1(\Omega) := \{v \in H_0^1(\Omega) \mid \forall E \in \mathcal{G} \ v_{|E} \in \mathbb{P}_\ell(E)\},$$

which is contained in $\mathbb{V}$. Since coercivity and continuity are handed down to a restriction of $\mathcal{B}$ and spaces of piecewise polynomials are nested on nested grids, (5a)-(5c) are valid with $\beta = 1$.

Moreover, we define the mesh-dependent norms as follows: given any $v \in \mathbb{V}$ and any union ω of elements of $\mathcal{G}$, we set

$$|v|_{\mathcal{G};\omega} = \left(\sum_{E \subset \omega} \|v\|^2_{L_2(E)} + \|h_{\mathcal{G}}^{1/2} v\|^2_{L_2(\partial E)} \right)^{1/2} \tag{12a}$$

and, for any $w \in \tilde{\mathbb{V}}(\mathcal{G}) = \mathbb{V}(\mathcal{G})$,

$$\|\!|w|\!\|_{\mathcal{G}} = \left(\sum_{E \in \mathcal{G}} \|D^2 w\|^2_{L_2(E)} + \|h_{\mathcal{G}}^{-1/2} \partial_n w\|^2_{L_2(\partial E)} \right)^{1/2}. \tag{12b}$$

Then $|\cdot|_{\mathcal{G};\Omega}$ is a norm on $\mathbb{V}$ and, in view of the scaled trace theorem $\|\cdot\|_{L_2(\partial E)} \preccurlyeq \|h_{\mathcal{G}}^{-1/2} \cdot\|_{L_2(E)} + \|h_{\mathcal{G}}^{1/2} \nabla \cdot\|_{L_2(E)}$ and the Poincaré inequality, (5d), (5e), and (5f) are valid. Moreover, $\|\!|\cdot|\!\|_{\mathcal{G}}$ is a norm on $\mathbb{V}(\mathcal{G})$ and (5g) is readily verified after an element-wise integration by parts.

Approximate solution and estimator. We suppose that SOLVE outputs the Galerkin approximation given by (6). Given such Galerkin solution $u_{\mathcal{G}}$ on a grid $\mathcal{G}$, the output of ESTIMATE is the standard residual estimator $\{\mathcal{E}_{\mathcal{G}}(E)\}_{E \in \mathcal{G}}$ for the $L_2(\Omega)$-error given by

$$\mathcal{E}^2_{\mathcal{G}}(E) := \|h_{\mathcal{G}}^{3/2} [\![\partial_n u_{\mathcal{G}}]\!] \|^2_{L_2(\partial E \setminus \partial\Omega)} + \|h^2_{\mathcal{G}}(f + \Delta u_{\mathcal{G}})\|^2_{L_2(E)}, \quad E \in \mathcal{G},$$

where $[\![\partial_n u_{\mathcal{G}}]\!]$ stands for the jump of the normal derivative of $u_{\mathcal{G}}$ across interelement sides. This estimator fulfills (7) with

$$n = \begin{cases} 3 & \text{if } d = 2, \\ 6 & \text{if } d = 3, \end{cases} \qquad \operatorname{osc}_{\mathcal{G}}(E) = \|h_{\mathcal{G}}(f - \bar{f}_{\mathcal{G}})\|_{L_2(\omega_{\mathcal{G}}(E))},$$

$$m(s) = s^{1/d}, \ s \in [0, \infty), \quad \mathbb{D} = L_2(\Omega), \quad D = f,$$

where $\bar{f}_{\mathcal{G}}$ is the L_2-projection of f on the space of possibly discontinuous piecewise polynomials of degree $\leq \ell - 1$; indeed, for (7a) see Prop. 3.8 in Ref. 15 and for (7b) see §6 of Ref. 10 but use (5g) with (12).

Marking strategy and refinement rule. Take any marking strategy ensuring that the biggest indicator is marked and require only that each marked simplex is bisected at least once. Then (10) and (11) are valid.

Under the above assumptions, Theorem 2.1 ensures that

$$\|u_k - u\|_{L_2(\Omega)} \to 0 \quad \text{and} \quad \mathcal{E}_k \to 0 \quad \text{as} \quad k \to \infty.$$

To our best knowledge, this is the first convergence result for the Poisson problem where the adaptation is not directed by an energy norm estimator.

3.2. *Mean Square Error of the Flux in Mixed Discretizations*

For mixed discretizations of the Poisson problem, we consider iteration (3) with an estimator for the approximation error in the flux.

Problem. Let Ω be a bounded, connected, polyhedral Lipschitz domain in $\mathbb{R}^2$. The mixed formulation of Poisson's problem and the error notion are given by

$$\mathbb{V} = \mathbf{V} \times \mathbb{Q} \quad \text{with} \quad \mathbf{V} = H(\operatorname{div};\Omega),\ \mathbb{Q} = L_2(\Omega)$$
$$\|v\|_{\mathbb{V}}^2 = \|\boldsymbol{v}\|_{L_2(\Omega;\mathbb{R}^2)}^2 + \|\operatorname{div}\boldsymbol{v}\|_{L_2(\Omega)}^2 + \|q\|_{L_2(\Omega)}^2, \quad |v| = \|\boldsymbol{v}\|_{L_2(\Omega;\mathbb{R}^2)}$$
$$\mathcal{B}(v,w) = \int_\Omega \boldsymbol{v}\cdot\boldsymbol{w} - \int_\Omega q\operatorname{div}\boldsymbol{w} + \int_\Omega \operatorname{div}\boldsymbol{v}\, r,$$

for $v = [\boldsymbol{v}, q]$, $w = [\boldsymbol{w}, r] \in \mathbb{V}$. Suppose $f \in L_2(\Omega)$, which is identified with $(\mathbf{0}, f) \in \mathbb{V}^*$. Then (2) is valid; see Example 1.2 in §II.1.2 of Ref. 3.

Refinement framework, finite element spaces, and seminorms. We use the same refinement framework as in §3.1 for $d = 2$ and choose Raviart-Thomas or Brezzi-Douglas-Marini elements of order ℓ or $\ell+1$ for the flux variable and piecewise polynomials of degree $\leq \ell$ for the scalar variable: given a triangulation $\mathcal{G} \in \mathbb{G}$, we set

$$\mathbb{V}(\mathcal{G}) = \mathbf{V}(\mathcal{G}) \times \mathbb{Q}(\mathcal{G}) \quad \text{with} \quad \mathbf{V}(\mathcal{G}) = \mathbb{RT}_\ell(\mathcal{G}) \text{ or } \mathbb{BDM}_\ell(\mathcal{G}),$$

where

$$\mathbb{Q}(\mathcal{G}) := \left\{q \in L_2(\Omega) \mid \forall E \in \mathcal{G}\ q_{|E} \in \mathbb{P}_\ell(E)\right\},$$
$$\mathbb{RT}_\ell(\mathcal{G}) := \left\{\boldsymbol{w} \in H(\operatorname{div};\Omega) \mid \forall E \in \mathcal{G}\ \boldsymbol{w}_{|E} \in \left(\mathbb{P}_\ell(E;\mathbb{R}^2) + \boldsymbol{x}\,\mathbb{P}_\ell(E)\right)\right\},$$
$$\mathbb{BDM}_\ell(\mathcal{G}) := \left\{\boldsymbol{w} \in H(\operatorname{div};\Omega) \mid \forall E \in \mathcal{G}\ \boldsymbol{w}_{|E} \in \mathbb{P}_{\ell+1}(E;\mathbb{R}^2)\right\}.$$

In both cases, the inclusion $\operatorname{div}\mathbf{V}(\mathcal{G}) \subset \mathbb{Q}(\mathcal{G})$ and (5a)-(5c) hold; see Prop. 1.1 in §IV.1.2 of Ref. 3.

Moreover, we let $|\cdot|_{\mathcal{G}} = |\cdot|$, which does not depend on $\mathcal{G}$. Then (5d)-(5f) are valid; the seminorm $|\!|\!|\cdot|\!|\!|_{\mathcal{G}}$ will be chosen below.

Approximate solution and estimator. Let SOLVE output the Galerkin solution of (6) and, writing $u_{\mathcal{G}} = [\boldsymbol{u}_{\mathcal{G}}, p_{\mathcal{G}}]$, we suppose that ESTIMATE outputs $\{\mathcal{E}_{\mathcal{G}}(E)\}_{E\in\mathcal{G}}$ given by

$$\mathcal{E}_{\mathcal{G}}^2(E) = \|h_{\mathcal{G}} \operatorname{rot}\boldsymbol{u}_{\mathcal{G}}\|_{L_2(E)}^2 + \|h_{\mathcal{G}}^{1/2}\, [\![\boldsymbol{u}_{\mathcal{G}}\cdot\boldsymbol{t}]\!]\,\|_{L_2(\partial E\cap\Omega)}^2 + \|h_{\mathcal{G}}(\bar{f}_{\mathcal{G}} - f)\|_{L_2(E)}^2,$$

where $\operatorname{rot} \boldsymbol{v} = \partial_{x_2} v_1 - \partial_{x_1} v_2$, $\bar{f}_{\mathcal{G}}$ stands for the $L_2(\Omega)$-orthogonal projection of f onto $\mathbb{Q}(\mathcal{G})$, and on any inter-element side, $\boldsymbol{t}$ stands for a fixed unit tangent vector. We shall prove that this estimator satisfies (7) with

$$n = 4, \quad m(s) = s^{1/d}, \ s \in [0, \infty), \quad \operatorname{osc}_{\mathcal{G}}(E) = \|h_{\mathcal{G}}(\bar{f}_{\mathcal{G}} - f)\|_{L_2(\omega_G(E))},$$
$$\mathbb{D} = L_2(\Omega), \text{ and } D = f.$$

Before embarking on the proper proof of the a posteriori bounds, we recall the orthogonal Helmholtz-decomposition (Theorem III.3.2 in Ref. 7):

$$L_2(\Omega; \mathbb{R}^2) = \nabla H^1(\Omega)/\mathbb{R} \oplus \mathbf{curl}\, H^1_{\partial\Omega}(\Omega), \tag{13}$$

where $H^1_{\partial\Omega}(\Omega)$ denotes the space of all $H^1(\Omega)$-functions that are constant on each connected component of $\partial\Omega$ and $\mathbf{curl}\,\phi = \left[-\partial_{x_2}\phi, \partial_{x_1}\phi\right]^T$, which has rot as adjoint operator. Note that $\phi \in H^1_{\partial\Omega}(\Omega)$ implies $\mathbf{curl}\,\phi \cdot \boldsymbol{n} = 0$ on $\partial\Omega$. The decomposition (13) appears in the relationship of error $\boldsymbol{u}_{\mathcal{G}} - \boldsymbol{u}$ and residual $\mathcal{R}$: If $w = [\mathbf{0}, -\psi] \in \mathbb{V}$ with $\psi \in \nabla H^1(\Omega)/\mathbb{R}$ normalized such that $\int_{\partial\Omega} (\boldsymbol{u}_{\mathcal{G}} - \boldsymbol{u}) \cdot \boldsymbol{n}\psi = 0$, then

$$\int_\Omega (\boldsymbol{u}_{\mathcal{G}} - \boldsymbol{u}) \cdot \nabla\psi = \mathcal{B}(u_{\mathcal{G}} - u, w) = \langle \mathcal{R}_{\mathcal{G}}, w\rangle = \int_\Omega (\bar{f}_{\mathcal{G}} - f)\psi \tag{14}$$

thanks to integration by parts, (1), and (6). Moreover, if $w = [\mathbf{curl}\,\phi, 0] \in \mathbb{V}$ with $\phi \in H^1_{\partial\Omega}(\Omega)$, then

$$\int_\Omega (\boldsymbol{u}_{\mathcal{G}} - \boldsymbol{u}) \cdot (\mathbf{curl}\,\phi) = \mathcal{B}(u_{\mathcal{G}} - u, w) = \langle \mathcal{R}_{\mathcal{G}}, w\rangle = \int_\Omega \boldsymbol{u}_{\mathcal{G}} \cdot \mathbf{curl}\,\phi \tag{15}$$

because $\mathbf{curl}\,\phi$ is divergence-free and again thanks to (1).

The proof of the upper bound (7a) can be established by exploiting both the relationship (14) for the gradient part and (15) for the **curl**-part of the error; proceed similarly to the proof of Thm. 3.1 in Ref. 1 and notice that, for $\phi \in H^1_{\partial\Omega}(\Omega)$, the interpolation operator in Ref. 13 allows to choose an approximation $\phi_{\mathcal{G}}$ that equals ϕ on $\partial\Omega$ and, thus, the estimator does not contain contributions on $\partial\Omega$.

We now derive the discrete local lower bound (7b), which appears to be new. Since $\|h_{\mathcal{G}}(\bar{f}_{\mathcal{G}} - f)\|_{L_2(E)}$ appears in the oscillation indicator, we only have to deal with terms that are related to (15). This suggests to construct discrete functions $\mathbf{curl}\,\phi \in \mathbf{V}(\mathcal{G}')$ for a suitable refinement $\mathcal{G}'$ of $\mathcal{G}$. To this end, we employ the Lagrange elements $\mathbb{LE}_{\ell+1}(\mathcal{G}')$ of §3.1. Given a subdomain $\omega \subset \Omega$, set $\mathbb{LE}_{\ell+1}(\mathcal{G}'; \omega) := \mathbb{LE}_{\ell+1}(\mathcal{G}') \cap H^1_0(\omega)$. Since continuity of $\phi \in \mathbb{LE}_{\ell+1}(\mathcal{G}')$ across interelement edges entails continuity of $\mathbf{curl}\,\phi \cdot \boldsymbol{n}$ across those edges and $\mathbf{curl}\,\phi$ is element-wise a polynomial of degree $\leq \ell$, we have $\mathbf{curl}\,\mathbb{LE}_{\ell+1}(\mathcal{G}';\omega) \subset \mathbf{V}(\mathcal{G}';\omega)$ for any union ω of elements in $\mathcal{G}$. This motivates to use $\tilde{\mathbb{V}}(\mathcal{G}') = [\mathbf{curl}\,\mathbb{LE}_{\ell+1}(\mathcal{G}'), 0]$ and we choose $|\!|\!|\cdot|\!|\!|_{\mathcal{G}} = |\cdot|$, which does not depend on $\mathcal{G}$. Thanks to (15) we obtain (5g).

To bound $\|h_{\mathcal{G}} \operatorname{rot} \boldsymbol{u}_{\mathcal{G}}\|_{L_2(E)}$ for a given element $E \in \mathcal{G}$, we now use a variant of Verfürth's constructive argument. We subdivide E by 3 bisections, thus creating a node inside E. Let λ_E be the continuous piecewise affine hat function associated with that node. Testing (15) with $v = [\mathbf{curl}\,\phi, 0]$ where $\phi = \lambda_E \operatorname{rot} \boldsymbol{u}_{\mathcal{G}} \in \mathbb{LE}_{\ell+1}(\mathcal{G}'; E)$ and standard scaling arguments then yield the desired bound. To proceed similarly for the remaining jump indicators, we need the following technical lemma.

Lemma 3.1. *Let S be an interval, decomposed into four subintervals $\mathcal{G}_S = \{S_1, \ldots, S_4\}$ of same size, and let P_S be the $L_2(S)$-orthogonal projection onto $\mathbb{LE}_{\ell+1}(\mathcal{G}_S) \cap H_0^1(S)$. Then $\|J\|^2_{L_2(S)} \preccurlyeq \int_S J P_S J$ for all $J \in \mathbb{P}_{\ell+1}(S)$.*

Proof. Thanks to a standard scaling argument, we only have to prove the claim for the interval $S = (0,1)$, decomposed by the points $\frac{1}{4}$, $\frac{1}{2}$, and $\frac{3}{4}$.
[1] We first show that, for any $J \in \mathbb{P}_{\ell+1}(S) \setminus \{0\}$, there exists a $\phi \in \mathbb{B} := \mathbb{LE}_{\ell+1}(\mathcal{G}_S) \cap H_0^1(S)$ with $\int_S J\phi \neq 0$. Suppose this is not the case, i.e. there is a $J \in \mathbb{P}_{\ell+1}(S) \setminus \{0\}$ such that $\int_S J\phi = 0$ for all $\phi \in \mathbb{B}$.

Let ϕ_1 be the continuous piecewise affine hat function at $\frac{1}{4}$. In view of our assumption, we have

$$\forall q \in \mathbb{P}_\ell(S) \qquad \int_0^{\frac{1}{2}} Jq\phi_1 = \int_S Jq\phi_1 = 0.$$

Since $\phi_1 > 0$ on $(0, \frac{1}{2})$, the left hand side defines weighted scalar product on $L_2(0, \frac{1}{2})$. Hence J has $\ell+1$ roots in $(0, \frac{1}{2})$. The same argument shows that J has also $\ell+1$ roots in $(\frac{1}{2}, 1)$. Since $J \in \mathbb{P}_{\ell+1}(S)$ and has $2\ell+2 > \ell+1$ roots in $(0,1)$, it has to vanish, which is a contradiction.
[2] Thanks to step 1, the $L_2(S)$-orthogonal projection P_S verifies $P_S J \neq 0$ for all $J \in \mathbb{P}_{\ell+1} \setminus \{0\}$. Consequently, the continuity of P_S gives

$$\min_{\|J\|_{L_2(S)}=1} \int_S J P_S J = \min_{\|J\|_{L_2(S)}=1} \|P_S J\|^2_{L_2(S)} = \alpha > 0,$$

which directly implies $\|J\|^2_{L_2(S)} \leq \alpha^{-1} \int_S J P_S J$ for all $J \in \mathbb{P}_{\ell+1}(S)$. □

To bound $\|h_{\mathcal{G}}^{1/2} [\![\boldsymbol{u}_{\mathcal{G}} \cdot \boldsymbol{t}]\!]\|_S$ for a given interelement side $S = E \cap E'$, we bisect E and E' four times, entailing a subdivision of S into four subintervals of same size. Testing (15) with $v = [\mathbf{curl}\,\phi, 0]$ where ϕ is an extension of $P_S([\![\boldsymbol{u}_{\mathcal{G}} \cdot \boldsymbol{t}]\!]_S)$ to $\mathbb{LE}_{\ell+1}(\mathcal{G}'; E \cup E')$, Lemma 3.1, and standard arguments then conclude the proof of (7b).

Marking strategy and refinement rule. We make the same assumptions on marking strategy and refinement rule in §3.1.

Under the above assumptions, Theorem 2.1 ensures that

$$\|\boldsymbol{u}_k - \boldsymbol{u}\|_{L_2(\Omega)} \to 0 \quad \text{and} \quad \mathcal{E}_k \to 0 \quad \text{as} \quad k \to \infty.$$

This generalizes the convergence result of Ref. 4 to Raviart-Thomas and Brezzi-Douglas-Marini elements of any fixed order.

References

1. A. ALONSO, *Error estimators for a mixed method*, Numer. Math. 74 (1996), 385–395.
2. I. BABUŠKA, M. VOGELIUS, *Feedback and adaptive finite element solution of one-dimensional boundary value problems*, Numer. Math. 44 (1984), 75-102.
3. F. BREZZI AND M. FORTIN, *Mixed and Hybrid Finite Element Methods*, Springer Series in Computational Mathematics 15, Springer (1991).
4. C. CARSTENSEN, R. H. W. HOPPE, *Error Reduction and Convergence For An Adaptive Mixed Finite Element Method*, Math. Comp. 75 (2006), 1033–1042.
5. Z. CHEN, J. FENG, *An adaptive finite element algorithm with reliable and efficient error control for linear parabolic problems*, Math. Comp. 73 (2004), 1167–1193.
6. W. DÖRFLER, *A convergent adaptive algorithm for Poisson's equation*, SIAM J. Numer. Anal. 33 (1996), 1106–1124.
7. V. GIRAULT, P. A. RAVIART. *Finite Element Approximation of the Navier-Stokes Equations*, Springer-Verlag, New York (1986).
8. K. MEKCHAY, R.H. NOCHETTO, *Convergence of adaptive finite element methods for general second order linear elliptic PDE*, SIAM J. Numer. Anal. 43 (2005), 1803–1827.
9. P. MORIN, R.H. NOCHETTO, K.G. SIEBERT, *Data oscillation and convergence of adaptive FEM*, SIAM J. Numer. Anal. 38 (2000), 466–488.
10. P. MORIN, R.H. NOCHETTO, K.G. SIEBERT, *Convergence of adaptive finite element methods*, SIAM Review 44 (2003), 631–658.
11. P. MORIN, K.G. SIEBERT, A. VEESER, *A basic convergence result for conforming adaptive finite elements*, preprint no. 1/2007, Dipartimento di Matematica "F. Enriques", Via C. Saldini 50, 20133 Milano, Italy.
12. A. SCHMIDT, K.G. SIEBERT, *Design of Adaptive Finite Element Software. The Finite Element Toolbox ALBERTA*, Springer, 2005.
13. L. R. SCOTT AND S. ZHANG, *Finite element interpolation of nonsmooth functions satisfying boundary conditions.* Math. Comp. 54 (1990) 483–493.
14. R. STEVENSON, *Optimality of a standard adaptive finite element method*, Found. Comput. Math. Published online: 5 July 2006. DOI 10.1007/s10208-005-0183-0.
15. R. VERFÜRTH, *A review of a posteriori error estimation and adaptive mesh-refinement techniques*, Adv. Numer. Math., John Wiley, Chichester, UK, 1996.

ENO/WENO INTERPOLATION METHODS FOR ZOOMING OF DIGITAL IMAGES

R. M. PIDATELLA, F. STANCO, C. SANTAERA

Dipartimento di Matematica ed Informatica
University of Catania
Viale A. Doria 6 - 95125 Catania, Italy
E-mails: rosa@dmi.unict.it, fstanco@dmi.unict.it, carmelosantaera@tiscali.it

In this paper we address the problem of producing an enlarged picture from a given digital image. We propose a zooming technique based on ENO and WENO schemes that are high order accurate finite difference schemes designed for problems with piecewise smooth solutions containing discontinuities. ENO and WENO schemes have been quite successful in applications, especially for problems containing both shocks and complicated smooth solution structures, such as compressible turbulence simulations and aeroacoustics. The algorithm works both on monochromatic images and RGB color pictures. Our experiments show that the proposed method is better than classical simple zooming techniques (e.g. pixel replication, bilinear interpolation). Moreover our algorithm is competitive both for quality and efficiency with bicubic interpolation.

Keywords: Zooming; Interpolation, ENO, WENO.

1. Introduction

In this paper we address the problem of producing a zoomed picture from a given digital image. This problem arises frequently whenever a user wishes to zoom in to get a better view of a given picture. There are several issues to take into account about zooming: unavoidable smoothing effects, reconstruction of high frequency details without the introduction of artifacts and computational efficiency both in time and in memory requirements. Several good zooming techniques are nowadays well known.[5,7–10,14,19] A generic zooming algorithm takes as input an RGB picture and provides as output a picture of greater size preserving as much as possible the information content of the original image. For a large class of zooming techniques this is achieved by means of some kind of interpolation; replication, bilinear and bicubic are the most popular choices and they are routinely implemented in commercial digital image processing software. Unfortunately, these methods, while preserving the low frequencies content of the source image, are not equally able to enhance high frequencies in order to provide a picture whose visual sharpness matches the quality of the original image. The methods proposed in this paper take into account information about discontinuities or sharp luminance variations while increasing the input picture. The research of new heuristic strategies able to outperform classical image processing techniques is nowadays the key-point to produce digital consumer engine (e.g. Digital Still Camera, 3G Mobile Phone, etc.) with advanced imaging application.[2] The key idea of our algorithm lies at the approximation level, where a nonlinear adaptive procedure is used to automatically choose the locally smoothest stencil, hence avoiding crossing discontinuities in the interpolation procedure as much as possible.

Our experiments show that the proposed method is better, in subjective quality than other well known methods such as pixel replication, bilinear interpolation, LAZA (locally adaptive zooming algorithm).[16] Moreover our algorithm is competitive both for quality and efficiency with bicubic interpolation, b-spline and other traditional techniques of zooming (e.g. Lanczos, s-spline). The technique proposed in this paper is simpler and much faster than fractal based zooming algorithms.[14,17]

The rest of the paper is organized as follows. Section 2 provides a description of ENO and WENO schemes. Section 3 provides a detailed description of the basic algorithm. Section 4 describes how to adapt the proposed algorithm to color images. Section 5 reports the results obtained together with a detailed discussion about related performance and weakness. Section 6 concludes the paper.

2. ENO and WENO schemes

In this section we provide a brief description of ENO and WENO schemes.

2.1. *ENO*

ENO is the first successful attempt to obtain a self similar uniformly high order accurate, yet essentially non-oscillatory, interpolation (i.e. the magnitude of the oscillations decays as $O(\Delta x^k)$ where k is the order of accuracy) for piecewise smooth functions.

In 1D we define cells, cell centers, and cell sizes by:

$$I_i = \left[x_{i-\frac{1}{2}}, x_{i+\frac{1}{2}}\right], \quad x_i \equiv \frac{1}{2}\left(x_{i-\frac{1}{2}} + x_{i+\frac{1}{2}}\right), \qquad (1)$$
$$\Delta x_i = x_{i-\frac{1}{2}} - x_{i+\frac{1}{2}}, \quad i = 1, 2, \ldots, N$$

We will face the following problem: given the cell averages of a function $v(x)$:

$$\overline{v}_i \equiv \frac{1}{\Delta x_i} \int_{x_{i-\frac{1}{2}}}^{x_{i+\frac{1}{2}}} v(\xi) d\xi \qquad (2)$$

find a polynomial of degree at most $k-1$, for each cell I_i, such that it is a $k-th$ order accurate approximation to the function $v(x)$ inside I_i:

$$p_i(x) = v(x) + O(\Delta x^k), \quad x \in I_i, \quad i = 1, \ldots, N \qquad (3)$$

In a given cell I_i, we choose a stencil $S(i)$ based on r cells to the left, s cells to the right, and I_i itself. If the order of accuracy has to be k, we have:

$$S(i) \equiv \{I_{i-r}, \ldots, I_{i+s}\} \qquad (4)$$

There is a unique polynomial $p(x)$ of $k-1$ degree whose cell average in $S(i)$ is that of $v(x)$:

$$\frac{1}{\Delta x_j} \int_{x_{j-\frac{1}{2}}}^{x_{j+\frac{1}{2}}} p(\xi) d\xi = \overline{v}_j \qquad (5)$$

Given the k cell averages $\overline{v}_{i-r}, \ldots, \overline{v}_{i-r+k-1}$ that are constants $c_{r,j}$ such that one can find a reconstructed value at the cell boundary $x_{i+\frac{1}{1}}$:

$$v_{i+\frac{1}{2}} = \sum_{j=0}^{k-1} c_{r,j} \overline{v}_{i-r+j} \qquad (6)$$

which is k-order accurate.[1]

To determine the local stencil, we develop a hierarchy that begins with one cell

$$S(i) = I_i = \left\{x_{i-\frac{1}{2}}, x_{i+\frac{1}{2}}\right\} \qquad (7)$$

and adds one cell at a time to the stencil from the two candidates on the left and on the right, based on the size of the two relevant Newton divided differences

$$v\left[x_{i-\frac{3}{2}}, x_{i-\frac{1}{2}}, x_{i+\frac{1}{2}}\right] \quad \text{and} \quad v\left[x_{i-\frac{1}{2}}, x_{i+\frac{1}{2}}, x_{i+\frac{3}{2}}\right] \qquad (8)$$

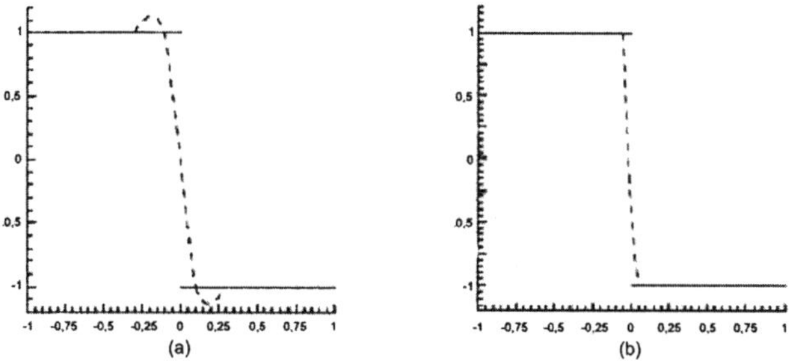

Fig. 1. (a) Fixed central stencil cubic interpolation; (b) ENO cubic interpolation for the step function. Solid: exact function; Dashed: interpolant piecewise cubic polynomials.

We choose the one with the least absolute value, achieving:

$$S(i) = \{I_{i-1}, I_i\} \equiv \left\{x_{i-\frac{3}{2}}, x_{i-\frac{1}{2}}, x_{i+\frac{1}{2}}\right\} \tag{9}$$

$$S(i) = \{I_i, I_{i+1}\} \equiv \left\{x_{i-\frac{1}{2}}, x_{i+\frac{1}{2}}, x_{i+\frac{3}{2}}\right\} \tag{10}$$

This procedure goes forward until we obtain the number of required points in the stencil.

Once we have defined the stencil

$$S(i) = \{I_{i-r}, \ldots, I_{i+s}\}, \quad k = r + s + 1 \tag{11}$$

one can compute the Eq. 6 where the constants $c_{r,j}$ depend on r, k and Δx^k.

We can find the constants $c_{r,j}$ from the Lagrange polynomials and the non-oscillatory behavior is provided by adapting the stencil. Fig. 1 shows the differences differences between the cubic interpolation and the ENO cubic interpolation.

2.2. *WENO*

Differently to ENO, in WENO schemes one uses a convex combination of all the stencils. Suppose the k candidate stencils

$$S_r(i) = \{x_{i-r}, \ldots, x_{i-r+k-1}\}, \quad r = 0, \ldots, k-1 \tag{12}$$

produce k different reconstructions to the value $v_{i+\frac{1}{2}}$:

$$v_{i+\frac{1}{2}}^{(r)} = \sum_{j=0}^{k-1} c_{r,j}\overline{v}_{i-r+j}. \tag{13}$$

Then we combine polynomials using weights ω_r

$$v_{i+\frac{1}{2}} = \sum_{r=0}^{k-1} \omega_r v_{i+\frac{1}{2}}^{(r)} \tag{14}$$

with $\sum_{r=0}^{k-1} \omega_r = 1$.

The weights ω_r are calculated using the following equations:

$$\omega_r = \frac{\alpha_r}{\sum_{s=0}^{k-1} \alpha_s}, \quad \alpha_r = \frac{d_r}{\beta_r}, \quad \beta_r = \sum_{l=1}^{k-1} \Delta x^{2l-1} \int_{x_{i-\frac{1}{2}}}^{x_{i+\frac{1}{2}}} \left(\frac{\partial^l p_r(x)}{\partial x^l}\right) dx \tag{15}$$

where β_r are the "*smoothness indicators*" and d_r is always positive and it satisfies the condition

$$\sum_{r=0}^{k-1} d_r = 1 \tag{16}$$

Many versions of WENO schemes have been developed in the last few years. These methods differ from WENO schemes in the dimension of the stencil or in the calculation of the "smoothness indicators". Some of these schemes are: OWENO (Optimized-WENO);[3] CWENO (Central-WENO);[4] WENO5M (Fifth-Order Mapped WENO).[6]

3. The proposed algorithm

In this section, we give a detailed description of the proposed algorithm. If S is a gray level image whose size is $n \times n$, the zooming algorithm creates a new image Z of size $(2n-1) \times (2n-1)$. Initially, the image Z contains pixels with known values and pixels unknown. More precisely, as first step the algorithm expands the source image S into a regular grid Z. Fig. 2(a) shows schematically the mapping function $E : S \rightarrow Z$ that disposes the original pixels into the new image. The expansion follows the equation:

$$E(S(i,j)) = Z(2i-1, 2j-1), \quad i,j = 1,2,\ldots,n. \tag{17}$$

The mapping E leaves undefined the value of all the pixels in Z with at least one even coordinate (white dots in Fig. 2(a)).

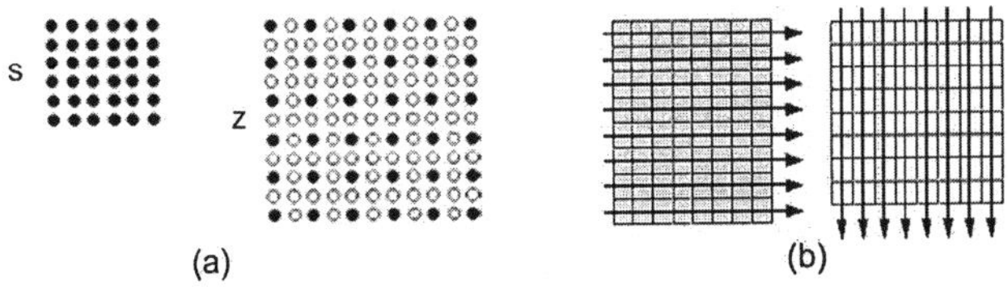

Fig. 2. (a) Zooming enlargement; (b) Scan order.

The second step of the algorithm is different if the approach is ENO or WENO based. It always works over the pixels still undefined. More precisely, first we interpolate the pixels in the odd rows and then all the others (Fig. 2(b)).

In the ENO Approach, if k is the degree of the method, the cell $I_{i,j}$ starts with a two point stencil

$$S_2(i,j) = \{x_{i,j-1}, x_{i,j+1}\} \tag{18}$$

For $l = 2, \ldots, k$ assuming

$$S_l(i,j) = \{x_{i,j+1}, \ldots, x_{i,j+2l-1}\} \tag{19}$$

is known, add one of the two neighboring points $x_{i,j-1}$ or $x_{i,j+2l+1}$ to the stencil, following the ENO procedure specified in section 2.1.

Once the stencil is found, we can calculate $v_{i,j}$ following the:

$$v_{i,j} = \sum_{l=0}^{k-1} c_{r,l} v_{i,j-(2r+1)+2l} \tag{20}$$

with r is the number of stencils at the left of I_i, and the coefficient $c_{r,l}$ are defined in.[1]

In the WENO approach, if the k candidate stencils are computed using ENO procedure in Eq. 19:

$$S_r(i,j) = \{x_{i,j-2r-1}, \ldots, x_{i,j-2r+2k-1}\}, \quad r = 0, \ldots, k-1 \tag{21}$$

we produce k different reconstructions to the value $v_{i,j}$

$$v_{i,j}^{(r)} = \sum_{l=0}^{k-1} c_{r,l} v_{i,j-(2r+1)+2l}, \quad r = 0, \ldots, k-1 \tag{22}$$

The constants d_r, and the *smoothness indicators* β_r, for all $r = 0, \ldots, k-1$ are described in Section 2.2.

The final reconstruction is

$$v_{i,j} = \sum_{r=0}^{k-1} \omega_r v_{i,j}^{(r)} \tag{23}$$

4. Zooming color pictures

The basic algorithm described above for gray scale pictures can be easily generalized to the case of RGB colored digital images. To change the space color we take advantage of the higher sensitivity of human visual system to luminance variations with respect to chrominance values. Hence it makes sense to allocate larger computational resources to zoom luminance values, while chrominance values may be processed with a simpler and more economical approach. This simple strategy is inspired by analogues techniques used by classical lossy image compression algorithms like JPEG and/or JPEG2000 vastly implemented in almost digital still camera engines. Accordingly we propose to operate as follows:

- Translate the original RGB picture into the YUV color model.
- Zoom the luminance values Y according with the basic algorithm described above.
- Zoom the U and V values using a simpler pixel replication algorithm.
- Back translates the zoomed YUV picture into an RGB image.

The results obtained with this basic approach are qualitatively comparable with the results obtained using bicubic interpolation over the three color channels. From the computational point of view, it is important to note how no significant difference in terms of timing response has been observed between the simple application of our approach to the three RGB planes and the approach described above (RGB-YUV conversion, Y zooming U, V replication, YUV-RGB conversion). Yet, in real applications (DSC, 3G Mobile phone,) the zooming process inside typical Image Generation Pipeline if present is realized just before compression: the YUV conversion is always performed as a crucial step to achieve visual lossless compression. In this case the color conversion itself does not introduce further computational costs.

5. Experimental results

The validation of a zooming algorithm requires the assessment of the visual quality of the zoomed pictures. Fig. 3 shows three examples of zooming pictures obtained with the proposed algorithm. Unfortunately this qualitative judgment involves qualitative evaluation of many zoomed pictures from a large pool of human observers and it is hard to be done in a subjective and precise way. For this reason several alternative quantitative measurements related to picture quality have been proposed and widely used in the literature. To validate our algorithm we have chosen both the approaches proposed in Refs.,[11,12,15] classical metrics and subjective tests. In particular we have used the cross-correlation and the PSNR (Peak Signal Noise to Ratio) between the original picture and the reconstructed picture to assess the quality of reconstruction. In our experimental contest we have first collected a test pool of 100 gray scale pictures. For each image I in this set we have first performed the following operations:

- reduction by decimation: a new picture I_d of half size of I is obtained taking only the pixels with both odd coordinates of the original picture;
- starting from I_d we have obtained the zoomed image;
- calculation of the following quantitative measurements between the original picture and the reconstructed picture: PSNR, cross-correlation coefficient and error threshold;
- calculation of the cpu-time;
- qualitative evaluation of the zoomed image.

The cross-correlation coefficient C between two pictures A and B is:

$$C = \left| \frac{\left(\sum_{k,l} A(k,l)B(k,l) - KLab\right)}{\sqrt{\left(\sum_{k,l} A^2(k,l) - KLa^2\right)\left(\sum_{k,l} B^2(k,l) - KLb^2\right)}} \right| \tag{24}$$

where a and b denote, respectively, the average value of picture A and B, K and L denote, respectively, width and length, in pixels, of images A and B. Notice that cross-correlation coefficients is between 0 and 1. The more the coefficient approaches 1, the better the reconstruction quality. The PSNR is calculated using the classical equation.

We have chosen simple replication, LAZA algorithm, bilinear and bicubic interpolation as comparing stones to assess the quality of our technique.

Fig. 3. (a) Original images; Examples of zoomed pictures with the Bryson-Levy (b) and Russo-Ferretti (c) based methods.

It is generally accepted that replication provides the worst quality-wise zooming algorithm while bicubic interpolation is considered one of the best options available. Table 1 and Table 2 show experimental results related, respectively, to ENO and WENO based algorithms of zooming.

Table 1. ENO results

	CPU Time	Cross-Correlation	PSNR
ENO2	5,23	0,9858	29,42
ENO3	15,72	0,9847	29,06
ENO4	34,38	0,9826	28,44
ENO5	71,58	0,9852	23,35
ENO3-Cross	15,18	0,9898	30,83
LAZA	12,75	0,9946	33,58
Replication	0,07	0,9854	29,32
Spline	0,93	0,9952	34,27
Bilinear	0,43	0,995	33,95
Bicubic	1,05	0,9953	34,36

The numerical results illustrated in these tables says that ENO schemes achieve not too good results if applied to the zooming $\times 2$ of a digital image due to the higher cpu-time, the quantitative measurements worse than bicubic and LAZA, and do not give good quality images. However, the WENO

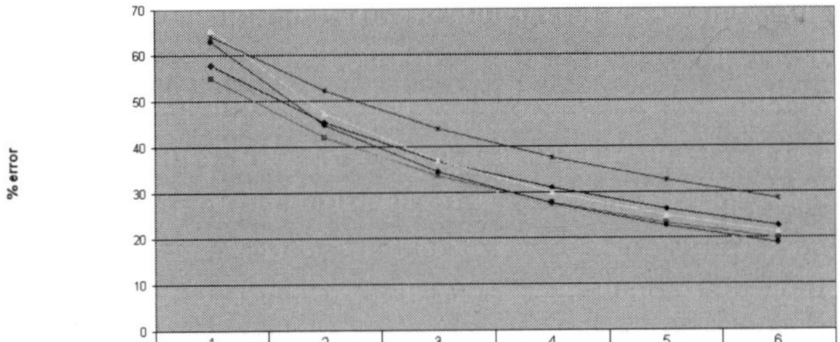

	1	2	3	4	5	6
WENO 3	57,84	45,34	36,78	30,97	26,33	22,69
WENO 5	54,93	42,09	33,76	27,85	23,30	19,80
WENO 7	65,45	47,37	36,99	30,04	24,90	21,31
LAZA	57,72	42,38	33,28	26,88	22,15	18,54
REPLCATION	64,03	52,29	43,95	37,65	32,61	28,54
BICUBIC	62,87	44,68	34,48	27,61	22,67	18,83

Threshold error

(a)

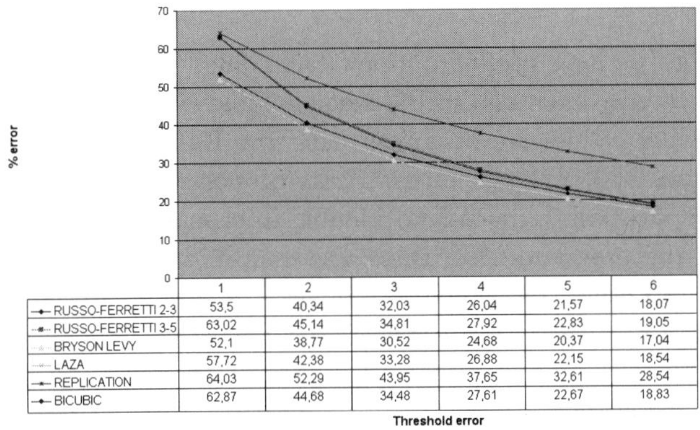

	1	2	3	4	5	6
RUSSO-FERRETTI 2-3	53,5	40,34	32,03	26,04	21,57	18,07
RUSSO-FERRETTI 3-5	63,02	45,14	34,81	27,92	22,83	19,05
BRYSON LEVY	52,1	38,77	30,52	24,68	20,37	17,04
LAZA	57,72	42,38	33,28	26,88	22,15	18,54
REPLICATION	64,03	52,29	43,95	37,65	32,61	28,54
BICUBIC	62,87	44,68	34,48	27,61	22,67	18,83

Threshold error

(b)

Fig. 4. Error values with different thresholds

schemes achieve very good results with small cpu-time, with quantitative measurements comparable to bicubic, and give good quality images.

The graphs in Fig. 4 show the average percentage of errors observed over the test pool as different tolerance threshold are considered. The WENO Russo-Ferretti[13] and Bryson-Levy[18] methods are always the best, since their values are always lower than the others.

Table 2. WENO results

	CPU Time	Cross-Correlation	PSNR
WENO3	1,08	0,9902	30,94
WENO5	1,5	0,9924	32,11
WENO7	1,86	0,9936	32,73
WENO7-2D	2,1	0,9946	33,47
OWENO1-2D	2,14	0,9947	33,58
Russo-Ferretti 2-3	1,56	0,9953	34,06
Russo-Ferretti 3-5	1,95	0,9957	34,34
Bryson Levy	1,46	0,9956	34,33
LAZA	12,99	0,995	33,58
Replication	0,05	0,9871	29,32
Spline	0,75	0,9955	34,27
Bilinear	0,42	0,9953	33,95
Bicubic	1,01	0,9957	34,36

6. Conclusions

In this paper we have proposed a new technique for zooming a digital picture, both in gray scale and in RGB colors. The experimental results show that satisfying results are obtained using the Russo-Ferretti and Bryson-Levy versions of WENO schemes. These proposed methods beat in quality pixel replication, bilinear and bicubic interpolation, LAZA algortihm. Moreover, the proposed algorithms are competitive both for quality and efficiency with other traditional techniques of zooming. Moreover, these algorithms preserve the image edges (major transition light-shade zones) in any direction without the introduction of artifacts and without to put out of focus the resulting images. The tests prove that these methods are very good to zooming medical pictures and images including writings. The proposed method, while of not greater complexity than bicubic interpolation provides qualitatively better results.

References

1. C.-W. Shu, Essentially Non-Oscillatory and Weighted Essentially Non-Oscillatory Schemes for Hyperbolic Conservation Laws, in Advanced numerical approximation of nonlinear hyperbolic equations, LNM1697, Ed. A. Quarteroni, pp.325-432, Spinger, 1997.
2. S. Battiato, M. Mancuso, An introduction to the digital still camera technology, ST Journal of System Research, Special Issue on Image Processing for Digital Still Camera, vol. 2, issue 2, 2001.
3. Z. J. Wang and R. F. Chen, Optimized Weighted Essentially Nonoscillatory Schemes for Linear Waves with Discontinuity, Journal of Computational Physics 174, pp. 381-404, 2001.

4. D. Levy, G. Puppo and G. Russo, Central WENO Schemes for Hyperbolic Systems of Conservation Laws, M2AN, Vol. 33, No 3, pp. 547-571, 1999.
5. D.F. Florencio, R.W. Schafer, Post-sampling aliasing control for images, Proceedings of International Conference on Acoustics, Speech and Signal Processing, Detroit, MI 2, pp. 893-896, 1995.
6. A. K. Henrick, T. D. Aslam, J. M. Powers, Mapped weighted essentially non-oscillatory schemes: Achieving optimal order near critical points, Journal of Computational Physics, vol. 207, issue 2, pp. 542-567, 2005.
7. H.S. Hou, H.C. Andrews, Cubic splines for image interpolation and digital filtering, IEEE Transactions on Acoustics, Speech, Signal Processing ASSP-26, issue 6, pp. 508-517, 1978.
8. A.K. Jain, Fundamentals of Digital Image Processing, Prenctice-Hall, Englewood Cliffs, NJ, 1989.
9. R.G. Keys, Cubic convolution interpolation for digital image processing, IEEE Transactions. on Acoustics, Speech, Signal Processing vol. 29, issue 6, pp. 1153-1160, 1981.
10. S.W. Lee, J.K. Paik, Image interpolation using adaptive fast B-spline filtering, Proceedings of International Conference on Acoustics, Speech, and Signal Processing vol. 5, pp. 177-179, 1993.
11. T.M. Lehmann, C. Gonner, K. Spitzer, Survey: interpolation methods in medical image processing, IEEE Transactions on Medical Imaging vol. 18, issue 11, 1999.
12. E. Maeland, On the comparison of interpolation methods, IEEE Transactions on Medical Imaging MI-7, pp. 213-217, 1988.
13. E. Carlini, R. Ferretti, G. Russo, A Weighted Essentially Non-oscillatory, large time-step schemes for Hamilton-Jacobi Equations, SIAM J. Sci. Comp., vol 27, pp. 1071-1901, 2005.
14. D.M. Monro, P.D. Wakefield, Zooming with implicit fractals, Proceedings of International Conference on Image Processing ICIP97 vol. 1, pp. 913-916, 1997.
15. J.A. Parker, R.V. Kenyon, D.E. Troxel, Comparison of interpolating methods for image resampling, IEEE Transactions on Medical Imaging MI-2, pp. 31-39, 1983.
16. S. Battiato, G. Gallo, F. Stanco, A locally adaptive zooming algorithm for digital images, Elsevier Image and. Vision Computing, vol. 20, issue 11, 2002.
17. E. Polidori, J.L. Dugelay, Zooming using iterated function systems, NATO ASI Conference on Fractal Image Encoding and Analysis, Trondheim, Norway,1995.
18. Steve Bryson, Doron Levy, High-Order Central Weno Schemes for multidimensional Hamilton-Jacobi equations, SIAM Journal on Numerical Analysis, vol. 41, pp.1339-1369, 2003.
19. P.V. Sankar, L.A. Ferrari, Simple algorithms and architecture for Bspline interpolation, IEEE Transactions on Pattern Analysis Machine Intelligence PAMI vol. 10, pp. 271-276, 1988.

FINITE ELEMENT DISCRETIZATIONS FOR THE DENSITY GRADIENT EQUATION

RENÉ PINNAU

Fachbereich Mathematik, Technische Universität Kaiserslautern

JORGE MAURICIO RUIZ V

Departamento de Matemáticas, Universidad Nacional de Colombia
Fachbereich Mathematik, Technische Universität Kaiserslautern

We consider finite element discretizations for the density gradient equation in quantum drift diffusion theory. Especially, we derive a finite element description of the nonlinear scheme presented in[3] and we compare it with an scheme based on linear interpolation. Numerical results are presented and the effect of vacuum boundary conditions is studied.

1. Introduction

For almost 50 years semiconductor design is based on the drift–diffusion (DD) model.[18] However due to the continuous miniaturization of the devices various quantum effects cannot be neglected, like quantum confinement or tunneling. To overcome this problem diverse extended models have been proposed which one can classify into full quantum models and quantum corrected models.[10] Especially, the Quantum Drift Diffussion (QDD) model[7] is a good candidate to be the successor of the DD model, since it adds quantum terms to the DD model in a compact and computationally efficient manner.[9,32] The QDD model was developed by M. G. Ancona and G. J. Iafrate in 1989 and the scaled unipolar stationary QDD model on the bounded domain $\Omega = (0,1)$ reads

$$-\varepsilon^2 \frac{\partial_{xx}\sqrt{n}}{\sqrt{n}} + \log(n) + V = F, \tag{1}$$

$$\partial_x(n\partial_x F) = 0, \tag{2}$$

$$-\lambda^2 \partial_{xx} V = n - C_{dop}. \tag{3}$$

The variables are the electron density n, the quantum quasi–Fermi potential F and the electrostatic potential V. The parameter ε is the scaled Planck constant, λ is the scaled Debye length and the function C_{dop} represents the concentration of fixed background ions. The equations (1)-(3) are subject to Dirichlet boundary conditions modeling the Ohmic contacts of the device

$$n = n_D, \quad V = V_D := V_{eq} + V_{ext} \quad F = F_D := F_{eq} + V_{ext} \quad \text{on } \partial\Omega, \tag{4}$$

where n_D, V_{eq} and F_{eq} are the equilibrium values of the charge concentrations, the potential and the quasi-Fermi level, respectively, and V_{ext} is the external applied voltage.

Several discretization schemes have been proposed for the solution of the coupled nonlinear system (1) - (3). These can be classified into linear and nonlinear schemes. Among the linear schemes are the piecewise linear finite element discretization developed in,[27] where also its stability properties are studied. In[3] a linear conservative scheme based on finite differences is presented.

However, due to the quantum effects that occur inside the device, the density might change by several orders of magnitude. Thus, such schemes require very fine grids in order to obtain reliable results, which implies a significant computational cost. To cope with such difficulties nonlinear schemes have to be used, like the finite difference nonlinear scheme,[3] which have been proved its efficiency in solving device examples involving quantum effects for coarse grids. Another line of research is presented in,[25] where the existence of a discrete solutions and error bounds are investigated.

This paper is organized as follows. In Section 2 we present two different finite element discretizations for the density gradient equation. Numerical results for different grid sizes and varying boundary conditions are discussed in Section 3. Finally, conclusions are given in Section 4.

2. The Finite Element Approaches

We consider here only the boundary value problem for the density gradient equation

$$-\varepsilon^2 \frac{\partial_{xx}\sqrt{n}}{\sqrt{n}} + \log(n) + V = 0, \tag{5}$$

$$n(0) = \alpha \qquad n(1) = \beta. \tag{6}$$

on the bounded domain $\Omega = (0,1)$ and for a given potential $V \in L^\infty(\Omega)$.

After multiplication with $\sqrt{n}$ and using an exponential transformation $n = e^{2u}$, in order to resolve better the large variations of the carrier density in the vicinity of inversion layers, we get the transformed problem in terms of the new unknown u:

$$-\varepsilon^2 \partial_{xx} e^u + e^u(2u + V) = 0, \tag{7}$$

$$u(0) = \frac{1}{2}\log(\alpha) \qquad u(1) = \frac{1}{2}\log(\beta). \tag{8}$$

The weak formulation reads now: Find $u \in u_D + H_0^1(\Omega)$ such that

$$\varepsilon^2 \int_\Omega e^u\, \partial_x u\, \partial_x \phi\, dx + \int_\Omega (2u+V)e^u\, \phi\, dx = 0, \quad \forall \phi \in H_0^1(\Omega), \tag{9}$$

where u_D is an $H^1(\Omega)$ of the boundary data.

In order to discretize (9), the interval $[0,1]$ is splitted into N subintervals $I_i = (x_{i-1}, x_i]$, $i = 1, ..., N$, with

$$0 = x_0 < x_1 < ... < x_{N-1} < x_N = 1.$$

We define the length of the intervals h_i and the maximal mesh spacing h by

$$h_i := x_i - x_{i-1}, \quad i = 1, ..., N, \quad \text{and} \quad h := \max_{i=1,\dots,N} h_i.$$

As a finite dimensional subspace of $H_0^1(\Omega)$ we use the space of piecewise linear functions

$$V_h := \{\phi \in H_0^1(\Omega) : \ \phi|_{I_i} \in \mathbb{P}_1, \ i = 0, ..., N\}$$

with basis $\{b_0, ..., b_N\}$ such that

$$b_i(x) = \begin{cases} \frac{x - x_i}{h_i} & \text{if } x \in I_i, \\ \frac{x_{i+1} - x}{h_i} & \text{if } x \in I_{i+1}, \\ 0 & \text{otherwise} . \end{cases}$$

The discretized problem may then be written in the form: Find $u_h \in u_D + V_h$ such that

$$\varepsilon^2 \int_\Omega e^{u_h} \, \partial_x u_h \, \partial_x b_i(x) \, dx \ + \int_\Omega (2u_h + V) e^{u_h} \, b_i(x) \, dx = 0, \tag{10}$$

for $i = 0, ..., N$.

To solve this nonlinear system of equations we need to evaluate first the integrals in (10). This can be done exactly or numerically.

Remark 2.1. The exact calculation of these integrals yields a highly nonlinear system. This encourages us to use quadrature rules instead.

Let us discuss two alternative ways to approximate the integrals in (10). The first approach leads us to the same nonlinear scheme proposed by Ancona in.[3] Hence, Ancona's discretization can be studied numerically and theoretically using finite element theory. The second method is based on linear interpolation of the second integrand and has the advantage of its simplicity in the treatment of the obtained nonlinear system.

2.1. *Finite Element derivation of Ancona's Scheme.*

To obtain a finite element version of Ancona's nonlinear scheme[3] we approximate both integrals in (10). For the first integral we use the midpoint rule and get

$$\begin{aligned} \varepsilon^2 \int_\Omega e^u \, \partial_x u \, \partial_x b_i(x) \, dx &= \varepsilon^2 \Big(\frac{u_i - u_{i-1}}{h_i^2}\Big) \int_{I_i} e^{u_h(x)} dx - \varepsilon^2 \Big(\frac{u_{i+1} - u_i}{h_{i+1}^2}\Big) \int_{I_{i+1}} e^{u_h(x)} dx \\ &\approx \varepsilon^2 \Big(\frac{u_i - u_{i-1}}{h_i}\Big) e^{u(x_{i-1/2})} - \varepsilon^2 \Big(\frac{u_{i+1} - u_i}{h_{i+1}}\Big) e^{u(x_{i+1/2})} \\ &= \varepsilon^2 \left(\frac{u_i - u_{i-1}}{h_i}\right) e^{(u_{i-1} \, + \, u_i)/2} \ - \ \varepsilon^2 \left(\frac{u_{i+1} - u_i}{h_{i+1}}\right) e^{(u_i \, + \, u_{i+1})/2} \end{aligned} \tag{11}$$

where $x_{i-1/2} = \dfrac{x_i + x_{i-1}}{2}$ for $i = 1, \dots, N$.

For the approximation of the second integral in (10) we proceed as follows

$$\begin{aligned}\int_\Omega (2u+V)e^u b_i\,dx &= \int_{x_0}^{x_{0+1/2}} e^u(2u+V)b_i\,dx + \sum_{j=1}^{N}\int_{x_{j-1/2}}^{x_{j+1/2}} e^u(2u+V)b_i\,dx \\ &+ \int_{x_{N+1/2}}^{x_{N+1}} e^u(2u+V)b_i\,dx \\ &= \int_{x_{(i-1)-1/2}}^{x_{(i-1)+1/2}} e^u(2u+V)b_i\,dx + \int_{x_{i-1/2}}^{x_{i+1/2}} e^u(2u+V)b_i\,dx \\ &+ \int_{x_{(i+1)-1/2}}^{x_{(i+1)+1/2}} e^u(2u+V)b_i\,dx, \quad (12) \\ &\approx \int_{x_{i-1/2}}^{x_{i+1/2}} e^u(2u+V)b_i\,dx, \quad (13) \\ &\approx (2u_i+V_i)\int_{x_{i-1/2}}^{x_{i+1/2}} e^u\,dx. \quad (14)\end{aligned}$$

The above approach considers the integrals between the midpoints of each subinterval I_i and use an open quadrature rule obtaining (13). Next since e^u never changes sign in $[x_{i-1/2}, x_{i+1/2}]$, we can apply the weighted mean-value theorem for integrals[8] to estimate (13) by (14).

Remark 2.1. From (13) it is observed that the terms $\int_{x_0}^{x_{0+1/2}} e^u(2u+V)b_i\,dx$ and $\int_{x_{N+1/2}}^{x_{N+1}} e^u(2u+V)b_i\,dx$, which contain the information given by the boundary condition are neglected. This is the cause for the sensitive behavior of the approximate solution near to the boundary (see also the discussion in[30]).

The integral (14) can be computed exactly which yields

$$\begin{aligned}0 &= \varepsilon^2\left(\frac{u_i-u_{i-1}}{h_i}\right)e^{(u_{i-1}+u_i)/2} - \varepsilon^2\left(\frac{u_{i+1}-u_i}{h_{i+1}}\right)e^{(u_i+u_{i+1})/2} \quad (15)\\ &+ (2u_i+V_i)\,e^{u_i}\left(\frac{h_i}{u_i-u_{i-1}}\left(1-e^{(u_{i-1}-u_i)/2}\right) + \frac{h_{i+1}}{u_{i+1}-u_i}\left(e^{(u_{i+1}-u_i)/2}-1\right)\right).\end{aligned}$$

Remark 2.2. If we introduce

$$e^{u(x)}|_{I_i} = A_i e^{\alpha_i x} \quad (16)$$

where

$$A_i := \exp\left(\frac{u_{i-1}\,x_i - u_i\,x_{i-1}}{h_i}\right) \quad \text{and} \quad \alpha_i := \frac{u_i-u_{i-1}}{h_i} \quad (17)$$

for $i = 1,\ldots,N$ and denote by $s_i = e^{u_i}$, then the discretization scheme (15) is equivalent to the nonlinear discretization scheme developed in.[3]

2.2. *The FE Scheme with Linear Interpolation*

The second alternative to approximate (10) is as follows. We replace (10) by:

Find $u_h \in u_D + V_h$ such that

$$\varepsilon^2 \int_\Omega e^{u_h}\, \partial_x u_h\, \partial_x b_i(x)\, dx + \int_\Omega (e^{u_h}\, (2u_h + V))^I\, b_i(x)\, dx = 0, \tag{18}$$

for $i = 1, ..., N$.

Now, we compute these integrals exactly and get

$$-\varepsilon^2 \frac{1}{h_i}\, e^{u_{i-1}} + \varepsilon^2 \Big(\frac{1}{h_i} + \frac{1}{h_{i+1}}\Big)\, e^{u_i} - \varepsilon^2 \frac{1}{h_{i+1}}\, e^{u_{i+1}} + \frac{1}{6} h_i\, e^{u_{i-1}}\, (2u_{i-1} + V_{i-1})$$
$$+ \frac{1}{3}(h_i + h_{i+1})\, e^{u_i}\, (2u_i + V_i) + \frac{1}{6} h_{i+1}\, e^{u_{i+1}}\, (2u_{i+1} + V_{i+1}) = 0, \tag{19}$$

for $i = 1, \ldots, N$.

Remark 2.3. Rewriting equation (19) in terms of the variable s_i we get

$$-\varepsilon^2 \frac{s_i}{h_i} + \varepsilon^2 \Big(\frac{1}{h_i} + \frac{1}{h_{i+1}}\Big)\, s_i - \varepsilon^2 \frac{s_i}{h_{i+1}} + \frac{1}{6} h_i\, s_{i-1}\, (2\log(s_{i-1}) + V_{i-1})$$
$$+ \frac{1}{3}(h_i + h_{i+1})\, s_i\, (2\log(s_i) + V_i) + \frac{1}{6} h_{i+1}\, s_{i+1}\, (2\log(s_{i+1}) + V_{i+1}) = 0.$$

We observe that our approach is closer to the linear conservative scheme proposed in.[3] The difference lies on the treatment of the second term.

3. Numerical results

To compare both discretization presented before, we study a MIS (Metal Insulator Semiconductor) diode in thermal equilibrium. The MIS diode consists of a uniformly doped piece of semiconductor coated with a thin layer of insulating material which carriers a metal contact called gate.

In particular, we study the behavior of the electron density in the semiconductor part of the device, which is described by the following boundary value problem for the electron density $n(x)$:

$$-\varepsilon^2 \Big(\frac{\partial_{xx}\sqrt{n}}{\sqrt{n}}\Big) + \log(n) + V = 0, \qquad \text{in } x \in (0,1) \tag{20}$$

$$n(0) = 0, \qquad n(1) = 1 \tag{21}$$

and the potential function V is given by

$$V(x) = x(\alpha x - \beta) e^{-\delta x};$$

where $\alpha = -47.57$, $\beta = 5.42$ and $\delta = 19.45$. This explicit form of $V(x)$ was obtained by performing an exponential fitting on precomputed data for the fully coupled problem.

Remark 3.1. It is clear that the discretization schemes presented here cannot fullfill the boundary condition at $x = 0$ due to their exponential character. For this reason we should take a very small value near to zero for the boundary value $n(0)$. For our numerical simulations purposes we have considered several values of $n(0)$ and used several uniform grids of variable sizes.

From Figures 1 to 3 we see that the finite element scheme with linear interpolation is very stable with respect to the imposed boundary conditions and the grid size. On the other hand, the nonlinear scheme presents yields a strong dependence on the choice of the boundary condition, which only improves for very small grid sizes, as Figure 3 shows. This sensitivity problem of the nonlinear scheme is also reported in.[30] Note, that there one also finds a variant of the nonlinear scheme, which does not have this drawback.

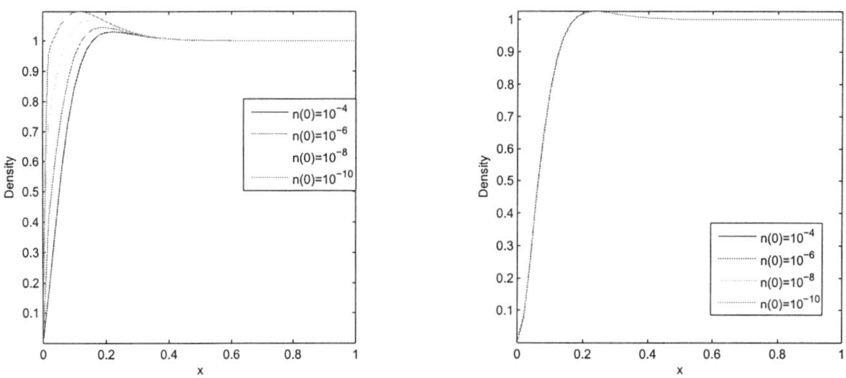

Fig. 1. Electron density obtained by using nonlinear (left) and finite element (right) scheme with $N = 50$ grid points and different boundary conditions at $x = 0$

4. Conclusions

We gave a finite element interpretation of Ancona's nonlinear scheme[3] and proposed a second approach based on an exponential transformation of variables and linear interpolation. Numerical tests indicate that the second approach is behaves for approximate vacuum boundary conditions better on coarser grids. In a forthcoming paper we use this finite element approach to prove consistency and convergence of both schemes.

References

1. R. A. Adams, *Sobolev Spaces*, 1st ed., Academic Press, New York, 1975.

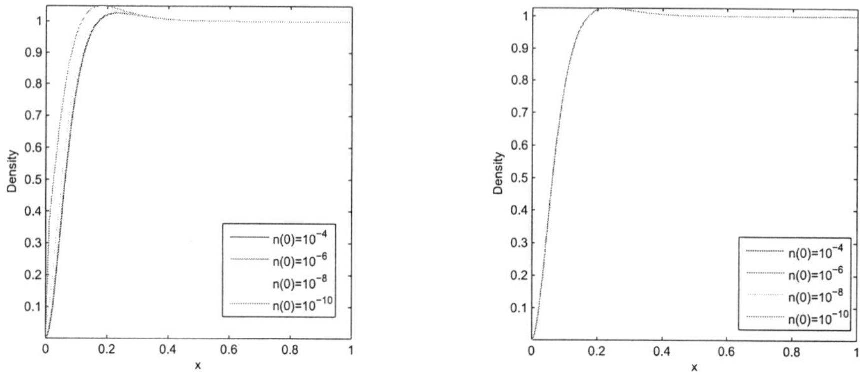

Fig. 2. Electron density obtained by using nonlinear (left) and finite element (right) schemes with $N = 375$ grid points and different boundary conditions at $x = 0$

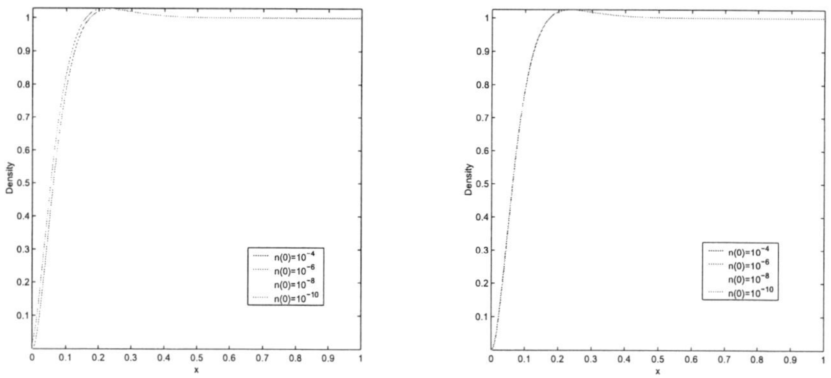

Fig. 3. Electron density obtained by using nonlinear (left) and finite element (right) schemes with $N = 1000$ grid points and different boundary conditions at $x = 0$

2. M. Ancona, *Equations of state for silicon inversion layers,* IEEE Trans. Elect. Devices, 47 (2000), pp. 1449-1456.
3. M. Ancona, *Finite-difference schemes for the density-gradient equations,* J. Comp. Elect., 1 (2002), pp. 435-443.
4. M. Ancona and H. Tiersten, *Macroscopic physics of the silicon inversion layer,* Phys. Rev., 35 (1987), pp. 7959-7965.
5. M. Ancona, D. Yergeau, Z. Yu, and B. Biegel, *On Ohmic boundary conditions for density gradient theory,* J. Comp. Elect., 1 (2002), pp. 103-107.
6. M. Ancona, Z. Yu, R. Dutton, P. Voorde, M. Cao, and D. Vook, *Density-gradient analysis of MOS tunneling*, IEEE Trans. Elect. Devices, 47 (2000), pp. 2310-2318.
7. M. G. Ancona and G. J. Iafrate, *Quantum correction of the equation of state of an electron gas in a semiconductor*, Phys. Rev. B, 39 (1989), pp. 9536-9540.

8. T. M. Apostol, *Mathematical Analysis*, Addison -Wessley, London, (1969).
9. B.A. Biegel, C. S. Rafferty, M.G. Ancona and Z. Yu. *Efficient Multi-Dimiensional Simulation of Quantum Effects in Advanced MOS Devices*, NAS Technical Report NAS-04-008 (2004).
10. B.A. Biegel, *Simulation of Ultra-Small Electronic Devices: The Classical-Quantum Transition Region*, NAS Technical Report 97-028, Oct. (1997).
11. F. Brezzi, I. Gasser, P.A MArkowich, and Ch. Schmeiser, *Thermal equilibrium states of the quantum hydrodynamic model for semiconductors*, Appl. Math. Lett., 8 (1995), pp. 47-52.
12. P. G. Ciarlet, *The Finite Element Method for Elliptic Problems*, 1st ed., North-Holland, Amsterdam, (1978).
13. H. K. Gummel, *A self consistent iterative scheme for one-dimensional steady state transistor calculations.* IEEE trans. Elec Dev., ED-11 (1964) pp. 455-465.
14. C. T. Kelly. *Iterative Methods for Linear and Nonlinear Equations,* SIAM, (1998).
15. D. Gilbarg and N. S. Trudinger, *Elliptic Partial Differential Equations of Second Order,* 1st ed., Springer-Verlag, Berlin, (1983).
16. A. Jüngel and R. Pinnau, *Global nonnegative solutions of a nonlinear fourth-order parabolic equation for quantum systems,* SIAM J. Math. Anal., 32 (2000), pp. 760-777.
17. A. Jüngel and R. Pinnau, *A positivity-preserving numerical scheme for a nonlinear fourth order parabolic system,* SIAM J. Numer. Anal., 39 (2001), pp. 385-406.
18. P. A. Markowich, *The Stationary Semiconductor Device Equations,* 1st ed., Springer-Verlag, Wien, (1986).
19. P. A. Markowich, C. A. Ringhofer, and C. Schmeiser, *Semiconductor Equations,* 1st ed., Springer-Verlag, Wien, (1990).
20. M. S. Mock, *Analysis of Mathematical Models of Semiconductor Devices,* Boole Press, Dublin, (1983).
21. R. Pinnau, *The linearized transient quantum drift diffusion model- stability of stationary states,* ZAMM. Z. Angew. Math. Mech., 80 (2000), pp. 327-344.
22. R. Pinnau, *A review on the quantum drift diffusion model,* Transport Theory Statist. Phys., 31 (2002), pp. 367-395.
23. R. Pinnau, *A Scharfetter-Gummel type discretization of the quantum drift diffusion model,* Proc. Appl. Math. Mech., 2 (2003), pp. 37-40.
24. R. Pinnau and A. Unterreiter, *The stationary current-voltage characteristics of the quantum drift-diffusion model,* SIAM J. Numer. Anal., 37 (1999), pp. 211-245.
25. R. Pinnau, *Uniform convergence of the exponentially fitted scheme for the quantum drift-diffusion model,* SIAM J. Numer. Anal. 42, No. 4, 1648-1668 (2004).
26. A.J. Quarteroni, A. Valli, *Numerical Approximation of Partial Differential Equations* Springer Verlarg, New york (1994)
27. C. Falco, E. Gatti, A.L. Lacaita and R. Sacco *Quantum-corrected drift-diffusion models for transport in semiconductor devices*, J. Comp. Phys., 204 (2), 533–561 (2005).
28. J. Stoer and R Bulrisch, *Introduction to numerical analysis* Springer Verlarg, New york (1993)
29. S.M Sze, *Physics of semiconductors Devices,* 2nd edition, Willey, New York (1981).
30. T. Tang, X. Wang, Y. Li *Discretization Scheme for the Density-Gradient Equation and Effect of Boundary Conditions,* J. Comp. Elect., 1 (2002), pp. 389-393.
31. A. Unterreiter, *The thermal equilibrium solution of a generic bipolar quantum hydrodynamic model,* Comm. Math. Phys., 188 (1997), pp. 69-88.
32. A. Wettstein, A. Schenk, and W. Fichtner, *Quantum device-simulation with the densitygradient model on unstructured grids,* IEEE Trans. Elect. Devices, 48 (2001), pp. 279-284.

33. E. Zeidler, Nonlinear Functional Analysis and Its Applications, 1st ed., Vol. II/A and II/B, Springer-Verlag, Berlin, (1990).

A NUMERICAL APPROACH TO THE DYNAMICS OF MAGNETOELASTIC MATERIALS

F. PISTELLA* and V. VALENTE

Istituto per le Applicazioni del Calcolo, CNR
Viale del Policlinico 137, Rome 00161, Italy
**E-mail: pistella@iac.rm.cnr.it*

We consider a system of nonlinear partial differential equations, which couples the magnetic and elastic processes, describing the dynamics of magnetostrictive materials. Finite difference schemes for the numerical study are proposed and numerical experiments on a specific test problem are carried out mainly to investigate on some evolutive phenomena such as the development and propagation of singular solutions.

Keywords: Coupled nonlinear systems; Magneto-elastic interactions; Finite difference approximation; Numerical simulation

1. Introduction

The paper deals with the numerical study of the evolution equations for a magnetoelastic material, that is a material which is capable of deformation and magnetization. We start from the differential equation, well known in the literature as the *Gilbert-Landau-Lifshitz* (refs. 4,6) equation in the form

$$\gamma^{-1}\mathbf{m}_t = -\,\mathbf{m} \times (\,\mathbf{H}_{\text{eff}} + \mathbf{m}_t) \tag{1}$$

The unknown $\mathbf{m}$, the magnetization vector, is a map from Ω (a bounded open set of $\mathbb{R}^d$, $d \geq 1$) to S^2 (the unit sphere of $\mathbb{R}^3$) and γ is a positive constant which represents the damping factor introduced to describe dissipative local phenomena. The magnetization distribution is well described by a free energy functional which we assume composed of three terms, namely the *exchange* energy E_{ex}, the *elastic* energy E_{el} and the *elastic-magnetic* energy E_{em}. Let $\mathbf{u}$ be the displacement vector, then the total free energy E for a deformable ferromagnet is given by

$$E(\mathbf{m},\mathbf{u}) \,=\, E_{\text{ex}}(\mathbf{m}) \,+\, E_{\text{em}}(\mathbf{m},\mathbf{u}) \,+\, E_{\text{el}}(\mathbf{u})$$

The effective field $\mathbf{H}_{\text{eff}}$ is obtained as the first variational derivative with respect to $\mathbf{m}$ of the total free energy $E(\mathbf{m}, \mathbf{u})$, that is formally

$$\mathbf{H}_{\text{eff}} = \partial_{\mathbf{m}} E(\mathbf{m}, \mathbf{u}) \tag{2}$$

We neglect other contributions to the free energy due, for example, to anisotropy and demagnetization because our interest is to analyze, in particular from the numerical point of view, the effect of the magneto-elastic interaction term on the phenomena connected with evolution processes. To the equation (1) we associate the evolution equation for the displacement $\mathbf{u}$, which we formally write as

$$\rho \mathbf{u}_{tt} = -\partial_{\mathbf{u}} E(\mathbf{m}, \mathbf{u}) \tag{3}$$

where ρ is a nonnegative parameter. In the next section we detail the three energetic terms and, after some simplifications, derive the dynamics equations which we use for studying the mutual effect of both elastic and magnetic processes. In particular we show the effects of the magnetic-elastic energy term and the kinetic energy term, modulated by the parameters λ and ρ respectively, in the appearance of singular solutions already pointed out for the only magnetic process both from numerical and theoretical point of views (see, for example, refs. 1,5,7,8). We propose, in section 3, two numerical schemes according to the choice $\rho > 0$ and $\rho = 0$. The numerical simulation on a specific test problem is shown and discussed in section 4.

2. The model

In the general three-dimensional theory (d=3), we assume $\Omega \subset \mathbb{R}^3$ is the volume of the ferromagnetic material at the time $t = 0$ and $\partial\Omega$ its boundary.

Let x_i, $i = 1, 2, 3$ be the position of a point $\mathbf{x}$ of Ω and denote by $u_i = u_i(\mathbf{x}, t)$, $i = 1, 2, 3$, the components of the displacement vector $\mathbf{u}$ and

$$\varepsilon_{kl}(\mathbf{u}) = \frac{1}{2}(u_{k,l} + u_{l,k}), \qquad k, l = 1, 2, 3$$

the deformation tensor where, as a common praxis, $u_{k,l}$ stands for $\dfrac{\partial u_k}{\partial x_l}$. Moreover we denote by $m_j = m_j(\mathbf{x}, t)$, $j = 1, 2, 3$ the components of the magnetization vector $\mathbf{m}$. In the sequel, where not specified, the Latin indices vary in the set {1,2,3} and the summation of repeated indices is assumed. We define

$$E_{\text{ex}}(\mathbf{m}) = \frac{1}{2} \int_{\Omega} a_{ij} m_{k,i} m_{k,j} \, d\Omega \tag{4}$$

where (a_{ij}) is a symmetric positive definite matrix which is supposed diagonal for most materials with all diagonal elements equal to a positive number a. The magneto-elastic energy for cubic crystals is assumed

$$E_{\text{em}}(\mathbf{m},\mathbf{u}) = \frac{1}{2}\int_{\Omega} \lambda_{ijkl} m_i m_j \varepsilon_{kl}(\mathbf{u}) d\Omega \tag{5}$$

where $\lambda_{ijkl} = \lambda_1\delta_{ijkl} + \lambda_2\delta_{ij}\delta_{kl} + \lambda_3(\delta_{ik}\delta_{jl} + \delta_{il}\delta_{jk})$ with $\delta_{ijkl} = 1$ if $i = j = k = l$ and $\delta_{ijkl} = 0$ otherwise. Moreover we introduce the elastic energy

$$E_{\text{el}}(\mathbf{u}) = \frac{1}{2}\int_{\Omega} \sigma_{klmn}\varepsilon_{kl}(\mathbf{u})\varepsilon_{mn}(\mathbf{u}) d\Omega \tag{6}$$

where σ_{klmn} is the elasticity tensor satisfying the symmetry property $\sigma_{klmn} = \sigma_{mnkl} = \sigma_{lkmn}$ and moreover the inequality

$$\sum(\sigma_{klmn}\varepsilon_{kl}\varepsilon_{mn}) \geq \beta \sum |\varepsilon_{kl}|^2$$

holds for some $\beta > 0$. We consider the tensor σ_{ijkl} in the form

$$\sigma_{ijkl} = \tau_1(\delta_{ijkl} + \delta_{ik}\delta_{jl} - \delta_{ij}\delta_{kl}) + \tau_2\delta_{ij}\delta_{kl}$$

and make further approximations. First of all we assume $\Omega \subset \mathbb{R}^2$ (i.e. we consider the case d=2) and neglect the components in plane of the displacement vector $\mathbf{u}$. Let λ and τ be two positive numbers, we assume $\mathbf{u} = (0, 0, w)$ and setting $\lambda_3 = \lambda, \tau_1 = \tau$, the functional E reduces to

$$E(t) = \frac{1}{2}\int_{\Omega} (a|\nabla \mathbf{m}|^2 + 2\lambda m_3(m_\alpha w_{,\alpha}) + \tau|\nabla w|^2) d\Omega \tag{7}$$

where the Greek indices vary in the set $\{1,2\}$.

The governing equations (1), (2), (3) become

$$\begin{cases} \gamma^{-1}\mathbf{m}_t - \mathbf{m} \times (a\Delta\mathbf{m} - \mathbf{m}_t - \lambda\mathbf{V}) = 0 \\ \\ \rho w_{tt} - \tau\Delta w - \lambda\,(m_\alpha m_3)_{,\alpha} = 0 \end{cases} \tag{8}$$

in $Q = \Omega \times (0, T]$ where $\mathbf{V} = [m_3 w_{,\alpha}, m_\alpha w_{,\alpha}]^T$.
The following initial and boundary conditions are assumed

$$w(\cdot, 0) = w^0, \;\; w_t(\cdot, 0) = w^1, \quad \mathbf{m}(\cdot, 0) = \mathbf{m}^0, \quad |\mathbf{m}^0| = 1 \quad \text{in } \Omega \tag{9}$$

$$w = 0, \qquad \frac{\partial \mathbf{m}}{\partial \boldsymbol{\nu}} = 0 \qquad \text{on} \quad \Sigma = \partial\Omega \times (0, T) \tag{10}$$

where $\boldsymbol{\nu}$ is the outer unit normal at the boundary $\partial\Omega$.

We introduce the functional $F(\mathbf{m}, w)$ defined as

$$F(\mathbf{m}, w) = \frac{a}{2}\int_{\Omega} |\nabla\mathbf{m}|^2 d\Omega + \frac{\tau}{4}\int_{\Omega} |\nabla w|^2 d\Omega + \frac{\rho}{2}\int_{\Omega} |w_t|^2 d\Omega \tag{11}$$

and put

$$F^0 = \frac{a}{2}\int_\Omega |\nabla \mathbf{m}^0|^2 d\Omega + \frac{\tau}{4}\int_\Omega |\nabla w^0|^2 d\Omega + \frac{\rho}{2}\int_\Omega |w^1|^2 d\Omega. \tag{12}$$

The following result has been proved (see refs. 9, 10)

Theorem 2.1 Given $w^0 \in H_0^1(\Omega;\mathbb{R})$, $w^1 \in L^2(\Omega;\mathbb{R})$ and $\mathbf{m}^0 \in (H^1(\Omega);\mathbb{R}^3)$ with $|\mathbf{m}^0| = 1$ a.e. in Ω, there exists a weak solution $(\mathbf{m}, w)$ to the problem (8), (9), (10) in the sense that:

- $\mathbf{m} \in H^1(Q;\mathbb{R}^3)$ with $|\mathbf{m}| = 1$ a.e. in Q, $w \in L^2(0,T;H_0^1(\Omega))$ and $w_t \in L^2(0,T;L^2(\Omega))$
- for each couple $(\mathbf{p}, g)$ such that $\mathbf{p} \in C^\infty(\bar{Q};\mathbb{R}^3)$ vanishing at $t = 0$ and $t = T$, and $g \in H^1(Q;\mathbb{R}) \cap C_0(Q;\mathbb{R})$ one has

$$\int_Q \left(\gamma^{-1}\mathbf{m}_t \cdot \mathbf{p} + a(\mathbf{m} \times \nabla \mathbf{m}) \cdot \nabla \mathbf{p} + (\mathbf{m} \times \mathbf{m}_t) \cdot \mathbf{p} + \lambda(\mathbf{m} \times \mathbf{V}) \cdot \mathbf{p}\right) dQ = 0$$

$$\int_Q \left(-\rho w_t g_t + \tau \nabla w \cdot \nabla g + \lambda\, m_3(m_\alpha g_{,\alpha})\right) dQ = 0$$

Moreover there exist two positive constants c_1 and c_2 such that if $\mathbf{m}$ and w are solutions of the problem (8), (9), (10) the following estimate holds

$$F(\mathbf{m}, w) \le c_1 F^0 + c_2. \tag{13}$$

We remark that from (13) one has in particular

$$\|\nabla \mathbf{m}\|_{L^2(\Omega)} \le \; const, \qquad \|\nabla w\|_{L^2(\Omega)} \le \; const$$

and hence, since $|\mathbf{m}| = 1$, lower and upper bounds for the total energy (7) can easily be found.

3. The Proposed Numerical Schemes

Let Ω a rectangle of $\mathbb{R}^2$ and i, j two indices such that $i = 0, 1, ..., I$ and $j = 0, 1, ..., J$. Consider a uniform partition of Ω in disjoint rectangles $\Omega_{i,j}$ with edges h_x, h_y and vertices $P_{i,j} = (x_i, y_j) = (ih_x, jh_y)$. Let $\Omega = \cup_{i,j}\Omega_{i,j}$ with $i = 0, 1, ..., I-1$ and $j = 0, 1, ..., J-1$.

We introduce a positive parameter δt and consider the uniform partition

$$t_n = n\delta t; \qquad n = 0, 1, ..., N; \qquad N\delta t = T$$

For any function $f(x, y, t)$ defined in the cylinder $\Omega \times [0, T]$ we introduce

$$f_{i,j}^n = f^n|_{P_{i,j}} = f(x_i, y_j, t_n)$$

and define the operators $\nabla^h = (\nabla^{h_x}, \nabla^{h_y})$, $\Delta^h = \Delta^{h_x} + \Delta^{h_y}$ by the centered finite difference approximations

$$\nabla^{h_x} f^n_{i,j} = \frac{f^n_{i+1,j} - f^n_{i-1,j}}{2h_x}, \qquad \Delta^{h_x} f^n_{i,j} = \frac{f^n_{i+1,j} - 2f^n_{i,j} + f^n_{i-1,j}}{h_x^2}$$

Analogous definition for $\nabla^{h_y}, \Delta^{h_y}$.

3.1. *The case $\rho > 0$*

We propose the following *explicit numerical scheme* for the problem (8), (9), (10)

- for $n = 1, 2, \quad i = 1, 2, ..., I-1$ and $j = 1, 2, ..., J-1$

$$\mathbf{m}^0_{i,j} = \mathbf{m}^0|_{P_{i,j}}, \quad w^0_{i,j} = w^0|_{P_{i,j}}, \quad w^1_{i,j} = w^0|_{P_{i,j}} + \delta t w^1|_{P_{i,j}} \tag{14}$$

$$\begin{aligned} &\gamma^{-1}\frac{\mathbf{m}^n_{i,j} - \mathbf{m}^{n-1}_{i,j}}{\delta t} = \frac{\mathbf{m}^n_{i,j} + \mathbf{m}^{n-1}_{i,j}}{2} \\ &\times \left(-\frac{\mathbf{m}^n_{i,j} - \mathbf{m}^{n-1}_{i,j}}{\delta t} + a\Delta^h \mathbf{m}^{n-1}_{i,j} - \lambda \mathbf{V}^{n-1,n-1}_{i,j}\right) \end{aligned} \tag{15}$$

$$\begin{aligned} &\rho\frac{w^2_{i,j} - 2w^1_{i,j} + w^0_{i,j}}{\delta t^2} = \tau\Delta^h w^1_{i,j} \\ &+\lambda\Big(\nabla^{h_x}((m_1)^0_{i,j}(m_3)^1_{i,j}) + \nabla^{h_y}((m_2)^0_{i,j}(m_3)^1_{i,j})\Big) \end{aligned} \tag{16}$$

- for $n = 2, 3, ..., N-1, \quad i = 1, 2, ..., I-1$ and $j = 1, 2, ..., J-1$

$$\begin{aligned} &\rho\frac{w^{n+1}_{i,j} - w^n_{i,j} - w^{n-1}_{i,j} + w^{n-2}_{i,j}}{2\delta t^2} = \frac{\tau}{2}\Delta^h(w^{n-1}_{i,j} + w^n_{i,j}) \\ &+\lambda\Big(\nabla^{h_x}((m_1)^{n-1}_{i,j}(m_3)^n_{i,j}) + \nabla^{h_y}((m_2)^{n-1}_{i,j}(m_3)^n_{i,j})\Big) \end{aligned} \tag{17}$$

$$\begin{aligned} &\gamma^{-1}\frac{\mathbf{m}^{n+1}_{i,j} - \mathbf{m}^{n-2}_{i,j}}{3\delta t} + \frac{\mathbf{m}^{n+1}_{i,j} + \mathbf{m}^{n-2}_{i,j}}{2} \times (\frac{\mathbf{m}^{n+1}_{i,j} - \mathbf{m}^n_{i,j} - \mathbf{m}^{n-1}_{i,j} + \mathbf{m}^{n-2}_{i,j}}{2\delta t}) \\ &= \frac{\mathbf{m}^{n+1}_{i,j} + \mathbf{m}^{n-2}_{i,j}}{2} \times (-\frac{\mathbf{m}^{n+1}_{i,j} - \mathbf{m}^{n-2}_{i,j}}{3\delta t} + \frac{a}{2}\Delta^h(\mathbf{m}^{n-1}_{i,j} + \mathbf{m}^n_{i,j}) - \lambda\mathbf{V}^{n-1,n}_{i,j}) \end{aligned} \tag{18}$$

where

$$\mathbf{V}_{i,j}^{n-1,k} = \begin{bmatrix} (m_3)_{i,j}^{n-1}\nabla^{h_x} w_{i,j}^k \\ (m_3)_{i,j}^{n-1}\nabla^{h_y} w_{i,j}^k \\ (m_1)_{i,j}^k \nabla^{h_x} w_{i,j}^{n-1} + (m_2)_{i,j}^k \nabla^{h_y} w_{i,j}^{n-1} \end{bmatrix}$$

with boundary conditions

$$\mathbf{m}_{1,j}^n - \mathbf{m}_{0,j}^n = \mathbf{m}_{I,j}^n - \mathbf{m}_{I-1,j}^n = \mathbf{m}_{i,1}^n - \mathbf{m}_{i,0}^n = \mathbf{m}_{i,J}^n - \mathbf{m}_{i,J-1}^n = 0 \quad (19)$$

$$w_{0,j}^n = w_{I,j}^n = 0, \quad w_{i,0}^n = w_{i,J}^n = 0 \quad (20)$$

The scheme (14)–(20) preserves the modulus. Indeed we have

Lemma 3.1 The numerical solution computed from (14)-(20) verifies

$$|\mathbf{m}_{i,j}^n| = |\mathbf{m}_{i,j}^0| = 1 \quad (21)$$

for each $i = 0, 1, ..., I, \quad j = 0, 1, ..., J, \quad n = 1, 2, ..., N.$

Proof. The equality (21) is easily proven taking the wedge product of (15) and (18) by $\mathbf{m}_{i,j}^{n-1} + \mathbf{m}_{i,j}^n$ and by $\mathbf{m}_{i,j}^{n+1} + \mathbf{m}_{i,j}^{n-2}$ respectively. □

Let us observe that the three-step approximations of the time derivatives introduced in (17), (18) are needed to achieve a suitable approximation of the magneto-elastic term (λ-term) and to afford the study of the stability of the previous scheme in the gradient norm. Let

$$||Gf^k||^2 = \frac{h_x h_y}{3} \sum_{i=1}^{I} \sum_{j=1}^{J} \sum_{s=-1}^{1} \left[\left(\frac{f_{i,j}^{k+s} - f_{i-1,j}^{k+s}}{h_x} \right)^2 + \left(\frac{f_{i,j}^{k+s} - f_{i,j-1}^{k+s}}{h_y} \right)^2 \right]$$

$$||Df^k||^2 = \frac{h_x h_y}{6} \left[\sum_{i=1}^{I-1} (S_{i,0}^k + S_{i,J}^k) + \sum_{j=1}^{J-1} (S_{0,j}^k + S_{I,j}^k) \right]$$

$$+ \frac{h_x h_y}{3} \sum_{i=1}^{I-1} \sum_{j=1}^{J-1} S_{i,j}^k + \frac{h_x h_y}{12} \left[S_{0,0}^k + S_{0,J}^k + S_{I,0}^k + S_{I,J}^k \right]$$

where

$$S_{i,j}^k = \left[\left(\frac{f_{i,j}^{k+1} - f_{i,j}^{k-1}}{\delta t} \right)^2 + \left(\frac{f_{i,j}^{k} - f_{i,j}^{k-1}}{\delta t} \right)^2 + \left(\frac{f_{i,j}^{k+1} - f_{i,j}^{k}}{\delta t} \right)^2 \right]$$

For the solution $\mathbf{m}_{i,j}^{n+1}, w_{i,j}^{n+1}$ computed according to the above scheme the following numerical stability result has been established (see ref. 2).

Theorem 3.1 Given $\rho, \tau, a, \lambda > 0$, for each choice of the step sizes h_x, h_y and δt which verify the condition

$$\delta t \leq \frac{1}{2} \min \left\{ \frac{h^2}{a}, h\sqrt{\frac{\rho}{\tau}} \right\}, \qquad h^{-2} = h_x^{-2} + h_y^{-2} \tag{22}$$

the following inequality holds

$$F_h^N \equiv \frac{a}{2}||G\mathbf{m}^{N-1}||^2 + \frac{\tau}{4}||Gw^{N-1}||^2 + \frac{\rho}{4}||Dw^{N-1}||^2 \leq \frac{a}{2}||G\mathbf{m}^1||^2 +$$

$$+\frac{3\tau}{4}||Gw^1||^2 + \frac{\rho}{2}||Dw^1||^2 + \frac{4\text{vol}(\Omega)\lambda^2}{\tau} + \frac{\delta t}{2}||D\mathbf{m}^1||^2 \leq 3F_h^1 + \text{const.}$$

We note that

Remark 3.1. For the numerical computations we need an explicit expression for $\mathbf{m}_{i,j}^{n+1}$ defined by (15),(18). We observe that these equations can be written in the general form

$$\mathbf{m}_{i,j}^{n+1} = C_* \mathbf{m}_{i,j}^{n+1} \times \mathbf{Q}_{i,j}^n + \tilde{\mathbf{Q}}_{i,j}^n = \mathbf{A}_{i,j}^n \mathbf{m}_{i,j}^{n+1} + \tilde{\mathbf{Q}}_{i,j}^n \tag{23}$$

where C_* is a real constant different from zero and $\mathbf{Q}_{i,j}^n$ and $\tilde{\mathbf{Q}}_{i,j}^n$ depend on the solution computed up to step n. Since at each point $P_{i,j}$ the determinant of the 3×3 matrix $\mathbf{I} - \mathbf{A}_{\mathbf{i,j}}^{\mathbf{n}}$ is different from zero and given by

$$det(\mathbf{I} - \mathbf{A}_{i,j}^n) = 1 + C_*^2 |\mathbf{Q}_{i,j}^n|^2$$

the required explicit formula easily follows.

3.2. *The case ρ=0*

Without loss of generality we can assume $a = \tau = 1$. We propose the following *implicit numerical scheme* for the problem (8), (9), (10) with ρ=0

- for $i = 1, ..., I-1$ and $j = 1, ..., J-1$

$$\Delta^h(w_{i,j}^0) + \lambda\Big(\nabla^{h_x}((m_1^0)_{i,j}(m_3^0)_{i,j}) + \nabla^{h_y}((m_2^0)_{i,j}(m_3^0)_{i,j})\Big) = 0 \tag{24}$$

- for $n = 1, 2, ..., N$, $\;i = 1, 2, ..., I-1$ and $j = 1, 2, ..., J-1$

$$\begin{cases} \gamma^{-1} \dfrac{\mathbf{m}_{i,j}^n - \mathbf{m}_{i,j}^{n-1}}{\delta t} = \dfrac{\mathbf{m}_{i,j}^n + \mathbf{m}_{i,j}^{n-1}}{2} \times \left(-\dfrac{\mathbf{m}_{i,j}^n - \mathbf{m}_{i,j}^{n-1}}{\delta t} + \Delta^h \mathbf{m}_{i,j}^{n-1} - \lambda \mathbf{V}_{i,j}^{n,n-1} \right) \\ \\ \Delta^h w_{i,j}^n + \lambda\Big(\nabla^{h_x}((m_1)_{i,j}^{n-1}(m_3)_{i,j}^n) + \nabla^{h_y}((m_2)_{i,j}^{n-1}(m_3)_{i,j}^n)\Big) = 0 \end{cases} \tag{25}$$

where, in this case

$$\mathbf{V}^{n,k}_{i,j} = \begin{bmatrix} (m_3)^n_{i,j} \nabla^{h_x} w^n_{i,j} \\ (m_3)^n_{i,j} \nabla^{h_y} w^n_{i,j} \\ (m_1)^k_{i,j} \nabla^{h_x} w^k_{i,j} + (m_2)^k_{i,j} \nabla^{h_y} w^k_{i,j} \end{bmatrix}$$

Moreover as boundary conditions (19),(20) are assumed.

The solvability of the problem (24),(25),(19), (20) follows from Lemma 3.1 and the following Theorem 3.2.

Theorem 3.2 Fixed $\mathbf{m}^{n-1}$, for δt small enough, there exists the solution $\mathbf{m}^n_{i,j}$ to (24),(25),(19),(20).

Proof. Fixed $\mathbf{m}^{n-1}$ with $n > 1$ we introduce the sequence of indices $\{n_l\}$ with $l = 0, 1, 2, ...$ and $n_0 = n - 1$ and the following iterative scheme

$$\begin{cases} \gamma^{-1} \dfrac{\mathbf{m}^{n_l+1}_{i,j} - \mathbf{m}^{n-1}_{i,j}}{\delta t} = \dfrac{\mathbf{m}^{n_l+1}_{i,j} + \mathbf{m}^{n-1}_{i,j}}{2} \times \left(- \dfrac{\mathbf{m}^{n_l+1}_{i,j} - \mathbf{m}^{n-1}_{i,j}}{\delta t} + \Delta^h \mathbf{m}^{n-1}_{i,j} - \lambda \mathbf{V}^{n_l, n-1}_{i,j} \right) \\ \Delta^h w^{n_l}_{i,j} + \lambda \Big(\nabla^{h_x} ((m_1)^{n-1}_{i,j} (m_3)^{n_l}_{i,j}) + \nabla^{h_y} ((m_2)^{n-1}_{i,j} (m_3)^{n_l}_{i,j}) \Big) = 0 \end{cases} \tag{26}$$

with

$$\mathbf{m}^{n_l}_{1,j} - \mathbf{m}^{n_l}_{0,j} = \mathbf{m}^{n_l}_{I,j} - \mathbf{m}^{n_l}_{I-1,j} = \mathbf{m}^{n_l}_{i,1} - \mathbf{m}^{n_l}_{i,0} = \mathbf{m}^{n_l}_{i,J} - \mathbf{m}^{n_l}_{i,J-1} = 0 \tag{27}$$

$$w^{n_l}_{0,j} = w^{n_l}_{I,j} = 0, \quad w^{n_l}_{i,0} = w^{n_l}_{i,J} = 0, \quad \forall i, j \tag{28}$$

The equation $(26)_1$ can be written in the form

$$\gamma^{-1} \mathbf{m}^{n_l+1}_{i,j} - \mathbf{m}^{n_l+1}_{i,j} \times \left(\mathbf{m}^{n-1}_{i,j} + \frac{\delta t}{2} \Delta^h \mathbf{m}^{n-1}_{i,j} - \frac{\delta t}{2} \lambda \mathbf{V}^{n_l,n-1}_{i,j} \right) =$$
$$\gamma^{-1} \mathbf{m}^{n-1}_{i,j} + \mathbf{m}^{n-1}_{i,j} \times \left(\mathbf{m}^{n-1}_{i,j} + \frac{\delta t}{2} \Delta^h \mathbf{m}^{n-1}_{i,j} - \frac{\delta t}{2} \lambda \mathbf{V}^{n_l,n-1}_{i,j} \right)$$

or equivalently in the form

$$\left(\gamma^{-1}\mathbf{I} - Y^{n_l}_{i,j}\right) \mathbf{m}^{n_l+1}_{i,j} = 2\gamma^{-1} \mathbf{m}^{n-1}_{i,j} - \left(\gamma^{-1}\mathbf{I} - Y^{n_l}_{i,j}\right) \mathbf{m}^{n-1}_{i,j}$$

where $\gamma^{-1}\mathbf{I} - Y^{n_l}_{i,j}$ is the 3×3 matrix

$$\left(\gamma^{-1}\mathbf{I} - Y^{n_l}_{i,j}\right) = \begin{pmatrix} \gamma^{-1} & -{y_3^{n_l}}_{i,j} & {y_2^{n_l}}_{i,j} \\ {y_3^{n_l}}_{i,j} & \gamma^{-1} & -{y_1^{n_l}}_{i,j} \\ -{y_2^{n_l}}_{i,j} & {y_1^{n_l}}_{i,j} & \gamma^{-1} \end{pmatrix}$$

and $y_1^{n_l}{}_{i,j}$, $y_2^{n_l}{}_{i,j}$, $y_3^{n_l}{}_{i,j}$ are the components of the vector $\mathbf{y}_{i,j}^{n_l}$ defined by

$$\mathbf{y}_{i,j}^{n_l} = \mathbf{m}_{i,j}^{n-1} + \frac{\delta t}{2}\Delta^h \mathbf{m}_{i,j}^{n-1} - \frac{\delta t}{2}\lambda \mathbf{V}_{i,j}^{n_l,n-1}$$

Since the determinant of the matrix $\left(\gamma^{-1}\mathbf{I} - Y_{i,j}^{n_l}\right)$ is positive, indeed we easily check that det $\left(\gamma^{-1}\mathbf{I} - Y_{i,j}^{n_l}\right) = \gamma^{-1}(\gamma^{-2} + |\mathbf{y}_{i,j}^{n_l}|^2)$, at each level l we can solve

$$\mathbf{m}_{i,j}^{n_{l+1}} = 2\gamma^{-1}\left(\gamma^{-1}\mathbf{I} - Y_{i,j}^{n_l}\right)^{-1}\mathbf{m}_{i,j}^{n-1} - \mathbf{m}_{i,j}^{n-1} \tag{29}$$

For the convergence of the internal iterative scheme (29), $(26)_2$, (27), (28) we take into account that at each level l the computed solution $\mathbf{m}_{i,j}^{n_{l+1}}$ lies on a one-ray sphere for any one-modulus initial datum.
The difference between two iteration gives

$$|\mathbf{m}_{i,j}^{n_{l+1}} - \mathbf{m}_{i,j}^{n_l}| = 2\gamma^{-1}\left|\left(\gamma^{-1}\mathbf{I} - Y_{i,j}^{n_l}\right)^{-1}\mathbf{m}_{i,j}^{n-1} - \left(\gamma^{-1}\mathbf{I} - Y_{i,j}^{n_{l-1}}\right)^{-1}\mathbf{m}_{i,j}^{n-1}\right|$$

Now we show that for δt small enough we obtain

$$\sup_{i,j}|\mathbf{m}_{i,j}^{n_{l+1}} - \mathbf{m}_{i,j}^{n_l}| \leq \mathcal{K}\sup_{i,j}|\mathbf{m}_{i,j}^{n_l} - \mathbf{m}_{i,j}^{n_{l-1}}|, \qquad \mathcal{K} < 1 \tag{30}$$

Indeed for each component of the vector $\mathbf{m}_{i,j}^{n_{l+1}} = (m_{k,i,j}^{n_{l+1}},\ k = 1,2,3)$ we can write from (29)

$$(m_{k,i,j}^{n_{l+1}} - m_{k,i,j}^{n_l}) = \mathfrak{F}_{k,i,j}(\{m_{k,i,j}^{n_l}\}) - \mathfrak{F}_{k,i,j}(\{m_{k,i,j}^{n_{l-1}}\})$$

where $\mathfrak{F}_{k,i,j}^{n_l}$ is a function depending, in general, on all the components of the vector $\mathbf{m}^{n_l}$, that is

$$\mathfrak{F}_{k,i,j}(\{m_{k,i,j}^{n_l}\}) = \mathfrak{F}_{k,i,j}(\{m_{k,i,j}^{n_l}\}_{k=1,2,3;i=1,\dots I;j=1,\dots,J})$$

Since

$$\frac{\partial \mathfrak{F}_{k,i,j}(\{m_{k,i,j}^{n_l}\})}{\partial m_{k,i,j}^{n_l}} = \lambda\,\delta t\,\mathfrak{G}_{k,i,j}(\{m_{k,i,j}^{n_l}\}$$

and thanks to the condition $|\mathbf{m}_{i,j}^{n_{l+1}}| = 1$ for each l, we get

$$\sup_{k,i,j}|\mathfrak{G}_{k,i,j}(\{m_{k,i,j}^{n_l}\}| \leq K$$

where K is a positive number. We obtain

$$|m_{k,i,j}^{n_{l+1}} - m_{k,i,j}^{n_l}| \leq \lambda\,\delta t\,K\sum_{k,i,j}|m_{k,i,j}^{n_l} - m_{k,i,j}^{n_{l-1}}|$$

From that we deduce

$$\sup_{k,i,j}|m_{k,i,j}^{n_{l+1}} - m_{k,i,j}^{n_l}| \leq \lambda\frac{\delta t}{h^2}K\sup_{k,i,j}|m_{k,i,j}^{n_l} - m_{k,i,j}^{n_{l-1}}|\sum_{k,i,j}h^2$$

and hence taking $\delta t \leq (3\lambda K vol(\Omega))^{-1}h^2$ we get (30). □

4. Numerical experiments

We apply the proposed schemes to a relevant test case concerning the development of singularities. We start from an analytical result in 2-D case given in ref. 3. In the domain $\Omega = [-r_o, r_o] \times [-r_o, r_o]$, $r_o = 1/2$ we consider the initial datum

$$\mathbf{m}^0 = \left[\frac{x}{r} sin(f(r)), \frac{y}{r} sin(f(r)), cos(f(r))\right] \tag{31}$$

where

$$f(r) = 8\pi r(r-1) \quad \text{for}\, r \le r_0, \quad f(r) = f(r_0) \text{ elsewhere}$$

Null initial data for the displacement w are chosen, i.e.

$$w^0 = 0, \qquad w^1 = 0 \tag{32}$$

We plot in the Figures 1–3 the behaviour in time of $||\nabla \mathbf{m}||^2_{L^\infty(\Omega)}$, $\frac{a}{2}||\nabla \mathbf{m}||^2_{L^2(\Omega)}$ and the total energy $E(t)$ (see (7)) computed according to the schemes proposed in the previous section 3.

In all the experiments we fix $\gamma^{-1} = 10^{-2}$, $\tau = 1$, $a = 1$ and the coupling parameter $\lambda = 30$, while the parameter ρ is varied to test its influence on the solution. Moreover in all tests the space and time steps, chosen according to (22), are $h_x = h_y = 4 \times 10^{-3}$ and $\delta t = 4 \times 10^{-6}$, respectively. We report

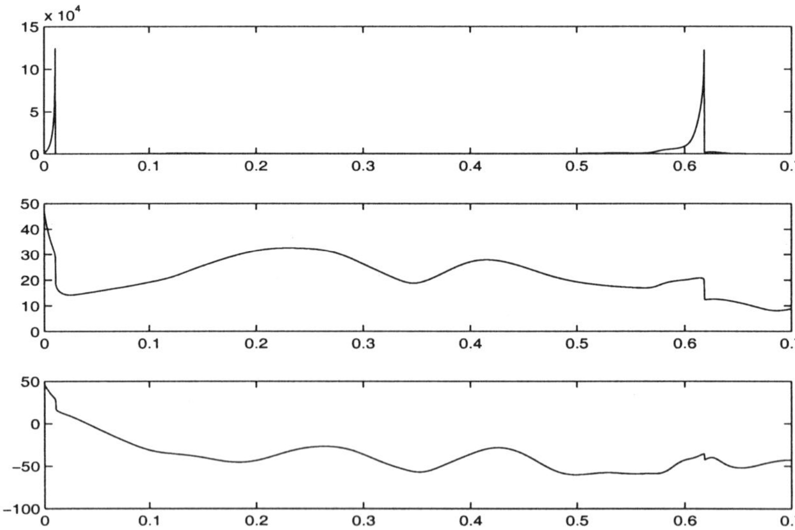

Fig. 1. From top to down, $||\nabla \mathbf{m}||^2_{L^\infty(\Omega)}$, $\frac{a}{2}||\nabla \mathbf{m}||^2_{L^2(\Omega)}$ and $E(\mathbf{m}, w)$ vs. time for $\lambda = 30$ and $\rho = 0.5$.

the results of the numerical simulation for a fixed value $\lambda = 30$ and different values of ρ. Putting $\rho = 0.5$ the first plot in Figure 1 reveals two blow-up in the time interval $[0, 0.7]$ and as well known each blow-up corresponds to a suddenly jump of the magnetic energy. Fixed $\lambda = 30$, two other tests

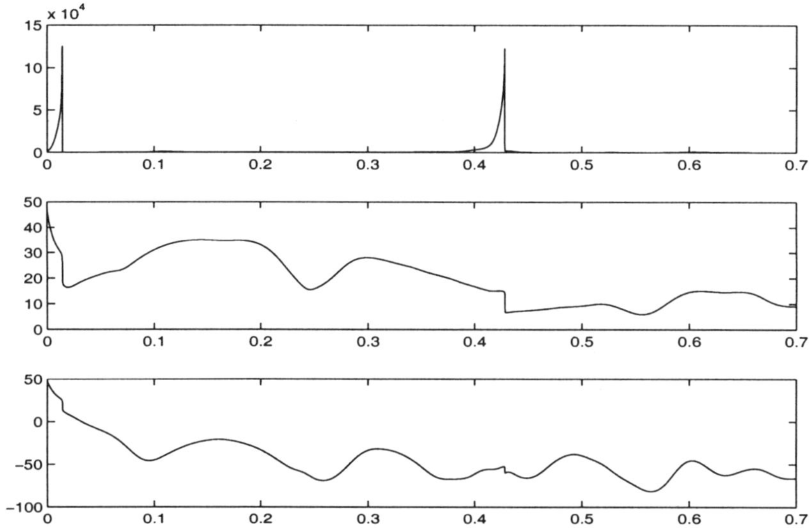

Fig. 2. From top to down, $||\nabla \mathbf{m}||^2_{L^\infty(\Omega)}$, $\frac{a}{2}||\nabla \mathbf{m}||^2_{L^2(\Omega)}$ and $E(\mathbf{m}, w)$ vs. time for $\lambda = 30$ and $\rho = 0.25$.

are carried out taking small values of ρ, that is $\rho = 0.25$ and $\rho = 10^{-4}$. In the first case we observe (Figure 2) that the time of the second blow-up decreases and it disappears in the last test reported in Figure 3 (continuous line); instead of it a step of uniformly distributed energy density occurs just after the single singularity (see Figure 3 at the bottom). An analogous qualitative behavior can be found applying the second scheme (broken line in Figure 3). The tests show that the first blow-up time remains unchanged for any value of ρ. The effect of ρ concerns only possible further singularities.

Acknowledgement

The work has been supported by the European Community under the contract HPRN-CT-2002-00284 (SMART-SYSTEMS).

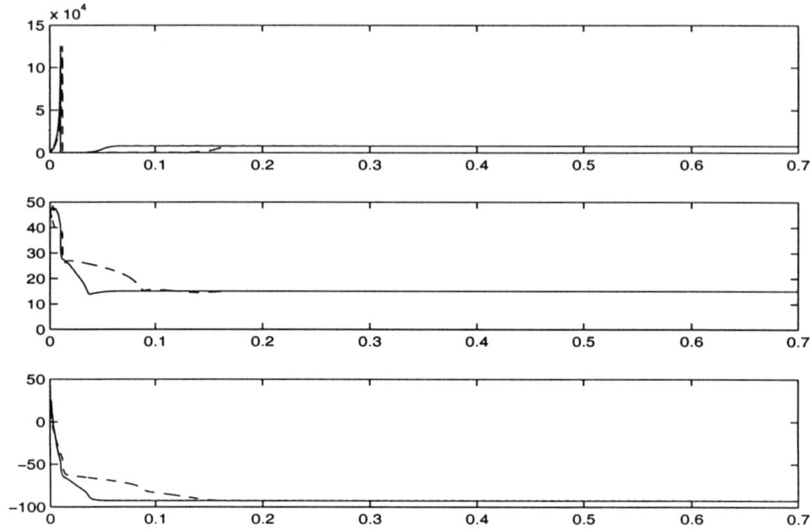

Fig. 3. From top to down, $||\nabla \mathbf{m}||^2_{L^\infty(\Omega)}$, $\frac{a}{2}||\nabla \mathbf{m}||^2_{L^2(\Omega)}$ and $E(\mathbf{m}, w)$ vs. time for $\lambda = 30$ and $\rho = 10^{-4}$(continuous line), $\rho = 0$ (broken line).

References

1. M. Bertsch, P. Podio-Guidugli, V. Valente, *On the dynamics of deformable ferromagnets I*. Ann. Mat. Pura Appl. (IV), CLXXIX (2001), 331–360.
2. M.M. Cerimele, F. Pistella, V. Valente, *Numerical Advances on an Evolutive Model for Magnetostrictive Materials.* To appear om Mathematics and Computers in Simulation.
3. K.–C. Chang, W.–Y. Ding, R. Ye, *Finite–Time Blow–up of the Heat Flow of Harmonic Maps from Surfaces.* J. Diff. Geom., 36 (1992), 507–515.
4. T.L. Gilbert, *A Lagrangian formulation of the gyromagnetic equation of the magnetization field.* Phys. Rev. 100 (1955) 1243.
5. B. Guo, M.C. Hong, *The Landau-Lifshitz equation of the ferromagnetic spin chain and harmonic maps.* Calc. Var. 1 (1993) 311- 334.
6. L. Landau, E. Lifshitz, *On the theory of the dispersion of magnetic permeability in ferromagnetic bodies.* Phys. Z. Sowjet. 8 (1935) 153.
7. F. Pistella, V. Valente, *Numerical stability for a discrete model in the dynamics of ferromagnetic bodies.* Numer. Meth. Part. Diff. Eq. 15 (1999) 544–557.
8. F. Pistella, V. Valente, *Numerical study of the appearance of singularities in ferromagnets.* Adv. in Math. Sci. Appl., 12 (2002), 803-816.
9. V. Valente, *An evolutive model for magnetostrictive interactions: existence of weak solutions.* SPIE-Proceedings on Smart Structures and Materials, Modeling, Signal Processing and Control, (2006).
10. V. Valente, G. Vergara Caffarelli, *On the dynamics of magnetoelastic interactions: existence of solutions and asymptotic behaviours.* Submitted

CONSTELLATIONS OF REPEATING SATELLITES FOR LOCAL TELECOMMUNICATION AND MONITORING SERVICES

MAURO PONTANI

Scuola di Ingegneria Aerospaziale, University of Rome "La Sapienza", via Eudossiana 16, 00184 Rome, Italy

Recently, telecommunication and navigation corporations have shown a growing interest in low and medium Earth orbit constellations due to several operational advantages, such as reduced power requirements and signal time delays with respect to geostationary platforms. An increased imaging resolution of Earth surface is also associated with the use of low orbits rather than high altitude orbits. This paper is focused on eccentric constellation configurations aimed at ensuring the continuous coverage of a target area, with reference to different operational requirements. This result is achieved by employing a correlation-based approach.

1. Introduction

In recent years, the relevant surge in designing satellite constellations in medium and low orbit is related to the growing interest by commercial telecommunication and navigation ventures. Basically, constellation build-up strategies must include, as first step, an accurate coverage analysis aimed at fixing an optimal set of parameters, such as the number of satellites and their spatial distribution, with reference to the required operational purposes.

Higher resolution imaging for Earth observation, reduced transmission power requirements and signal time delays are main advantages of deploying LEO and MEO satellite constellations with respect to geostationary platforms. On the other hand, the motion relative to Earth surface introduces several complexities: numerical methods for capturing the performance of constellation architecture are needed, since generalized analytical solutions are still elusive.

Common classes of constellations include Walker symmetric Rosette constellations [1-4], circular polar orbit constellations [5, 6], highly elliptical orbit constellations [7], geosynchronous orbit constellations, and polyhedral constellations [8].

A large set of parameters is involved in defining constellation configurations, so usually some basic assumptions are taken. This occurs, for instance, for Walker constellations, exhibiting a high degree of symmetry and suitability for coverage of large areas [1-4]. Yet, for special purposes, such as the continuous coverage of restricted areas, geometric hypotheses seem restrictive.

In a former paper [9] the author proposed a correlation-based approach for constellation design. All the satellites were assumed to be placed in repeating, equally inclined circular orbits with an identical altitude. As a matter of fact, a suitable way of predicting the constellation dynamics and performance along its whole operative life may be based on using repeating orbits, so that the analysis can be focused on the period of repetition only. A correlation-based approach was used in finding optimal configurations yielding the maximum duration of continuous coverage or the minimum revisit time, with reference to two distinct operational requirements.

This paper is concerned with constellation configurations aimed at ensuring the continuous coverage of a preselected target area. Satellite passes over the target area are translated into a binary time-dependent function, named visibility function, which identifies the time intervals of visibility of the satellite. The overlapping of visibility functions gives a basic idea about both gap intervals (when the target area is not in view by any satellite) and coverage intervals (when the area is visible from at least one satellite). A suitable way of increasing the continuous observation duration and simultaneously reducing the gap duration is based on minimizing the overlapping between two (or more) visibility functions. Such a condition is faced here by introducing a correlation function, which measures the overlapping of visibility functions. The correlation function depends on time delay between two consecutive satellite passes over the target. The search for minima of the correlation function allows the identification of the constellation configurations with minimum overlapping among satellite passes.

Two applicative examples show the effectiveness of the above strategy for constellation design, with reference to different requirements:

1. the design of a constellation of 4 satellites for a local telecommunication service
2. the design of a constellation of 8 satellites for local continuous monitoring

This paper is organized as follows: Section 2 introduces repeating ground track orbits and the basic hypotheses of the model employed in the constellation design. Orbit parameters for maximum visibility of the target area from a single satellite are derived in Section 3, with reference to two distinct applicative situations. Section 4 is concerned with the strategy for constellation distribution. The so-called "visibility function" is introduced and the correlation function is described as a mathematical tool to find the constellation configurations which meet the desired requirements. Section 5 presents the optimal constellations for the two mentioned applicative situations. Finally, the Appendix deals with the derivation of orbital parameters once time delays are known.

2. Repeating Ground Track Orbits

The constellation is assumed to be composed of several satellites placed in equally inclined eccentric orbits with identical semi-major axis (SMA), a, and eccentricity, e. Orbits are considered such that atmospheric drag, solar radiation pressure, and third body effects can be regarded as negligible for the purpose of constellation design. However, J_2 and J_3 perturbations are relevant and their effects can be critical for station keeping of LEO and MEO constellations. Specifically, the rotation of the apsidal line of an orbit is an undesirable effect caused by J_2 and J_3 perturbations. This phenomenon can be avoided by selecting orbital planes with inclination i equal to the critical value, $i_{cr} = \arccos\left(1/\sqrt{5}\right) = 63.43$ deg. Moreover, this value for inclination makes the perturbation J_3 ineffective and will be assumed from now on.

Generally speaking, J_2 perturbation only affects the right ascension of the ascending node (RAAN), Ω, the argument of perigee, ω, and the mean anomaly, M:

$$\dot{\Omega}(a,e,i) = -\frac{3}{2} J_2 \frac{R_E^2}{a^{\frac{7}{2}}} \sqrt{\mu_E} \left(1-e^2\right)^{-2} \cos i \tag{1}$$

$$\dot{\omega}(a,e,i) = \frac{3}{2} J_2 \frac{R_E^2}{a^{\frac{7}{2}}} \sqrt{\mu_E} \left(1-e^2\right)^{-2} \left(\frac{5}{2}\cos^2 i - \frac{1}{2}\right) \tag{2}$$

$$\dot{M}(a,e,i) = \frac{3}{2} J_2 \frac{R_E^2}{a^{\frac{7}{2}}} \sqrt{\mu_E} \left(1-e^2\right)^{-1.5} \left(1 - \frac{3}{2}\sin^2 i\right) + \sqrt{\frac{\mu_E}{a^3}} \tag{3}$$

where R_E and μ_E are Earth radius and planetary constant, respectively, and J_2 is Earth oblateness coefficient. The initial values of Ω, ω, and M are denoted with Ω_0, ω_0, M_0; $i = i_{cr}$ implies $\dot{\omega} \equiv 0$ so $\omega(t) = \omega_0 \; \forall t$.

All the satellites must have identical inclination, eccentricity, and SMA, in order to avoid differential actions by J_2 perturbation on each of them. This circumstance would cause a fast phase-displacement and a substantial alteration of the performance attainable by the constellation. These considerations justify the above assumption on identical SMA and eccentricity for all the satellites.

Repeating orbits are considered, since they allow prediction of the constellation performance during its whole operative life.

An orbit is said to be *repeating* when phased with Earth rotation, i.e. when the trajectory ground track is periodically repeated. This occurs if the satellite completes N_t orbits in m nodal days[a]:

$$mD_n(a,e) = N_t T_n(a,e) \quad \text{where} \quad T_n = \frac{2\pi}{\dot{\omega} + \dot{M}} \quad \text{and} \quad D_n = \frac{2\pi}{\omega_E - \dot{\Omega}} \tag{4}$$

In (4) T_n is the nodal orbital period[b], whereas D_n is the nodal day, both depending on orbit SMA, a, and eccentricity, e (the inclination i is set equal to i_{cr}); ω_E is the Earth rotation rate. In this paper repeating satellites completing 5 orbits in 1 nodal day will be considered; furthermore, a perigee altitude H_P equal to 700 km is assumed.

With the above hypotheses a and i are identified, since the following relationship holds:

$$r_P = a(1-e) \qquad \text{where} \qquad r_P = R_E + H_P \tag{5}$$

Hence, orbit SMA and eccentricity can be derived from (4) and (5): $a = 14407$ km and $e = 0.509$.

In addition, at the starting time the first satellite is assumed to be at perigee, so $M_{01} = 0$. Obviously, this assumption is not restrictive, since it is equivalent to the choice of the initial time, which is arbitrary.

For the first satellite, in Section 3 the remaining orbital parameters will be identified in such a way that the total duration of visibility is maximized during the period of repetition. All the remaining satellites are supposed to have the same motion relative to Earth surface, so that also their total duration of visibility is maximized. This property implies that all the satellites have the same ground track. Hence, the visible passes of each satellite over the target area are simply delayed with respect to the ones of the first satellite.

Due to J_2 perturbation, only the mean anomaly, M, and the RAAN, Ω, are time-dependent, so their starting values must be considered. Moreover, each satellite must have different values of initial RAAN and mean anomaly in order to have a ground track identical to the first. These values can be derived from time delays between each satellite and the first, as described in Appendix. The remaining four orbital parameters (i.e. a, e, i, and ω) are identical for all the satellites.

---

[a] The *nodal day* is the time required for the Earth to make a complete rotation with respect to the orbital plane

[b] The *nodal orbital period* is the time interval between two consecutive ascending node crossings

3. Argument of perigee and RAAN for maximum visibility of a target area

In the previous section some orbital parameters were derived for the first constellation satellite: a, e, i, and M_{01}. This section deals with the identification of the remaining parameters, Ω_{01} and ω.

First of all, the target area is delimited in longitude $(\lambda_l \le \lambda \le \lambda_u)$ and latitude $(\phi_l \le \phi \le \phi_u)$. Constellation design is related to operational requirements which refer to this area. In this paper, two basic operational requirements will be considered and the corresponding optimal constellation configurations will be presented:

A. a constellation of 4 satellites for a local telecommunication service
B. a constellation of 8 satellites for local continuous monitoring

Both configurations are required to ensure the continuous coverage of the preselected target area.

For the constellation A the following target area and minimum elevation angle, ε_{min}, are assumed:

A. $\lambda_l = 5$ deg, $\lambda_u = 20$ deg, $\phi_l = 35$ deg, $\phi_u = 50$ deg and $\varepsilon_{min} = 10$ deg

For the constellation B the following target area and minimum elevation angle, ε_{min}, are assumed:

B. $\lambda_l = 10$ deg, $\lambda_u = 15$ deg, $\phi_l = 41$ deg, $\phi_u = 43$ deg and $\varepsilon_{min} = 45$ deg

A numerical investigation over the full range of values for Ω_{01} and ω suggests their optimal choice, i.e. that maximizes the total duration of visibility of the target area. This investigation is based on the following considerations:

- the target area is in view if and only if the following four points are visible at a time: $P_1(\lambda_l, \phi_l)$, $P_2(\lambda_l, \phi_u)$, $P_3(\lambda_u, \phi_u)$, $P_4(\lambda_u, \phi_l)$
- for a given pair (Ω_{01}, ω), the time-dependent East, North and up coordinates $((x_k, y_k, z_k),\ k = 1, 2, 3, 4)$ in the local frame centered in P_k $(k = 1, 2, 3, 4)$ are calculated
- the satellite is in view from P_k if the following relationship holds:

$$\frac{z_k(t)}{\sqrt{[x_k(t)]^2 + [y_k(t)]^2}} \ge \tan \varepsilon_{min} \tag{6}$$

- each time interval is such that (6) hold for all the four points $\{P_k\}_{k=1,2,3,4}$. Hence, it can be identified by using this basic condition
- finally, the optimal values of Ω_{01} and ω are selected

With reference to the two mentioned applicative situations, the total duration of visibility (as a function of and Ω_{01} and ω) is shown in Fig. 1.

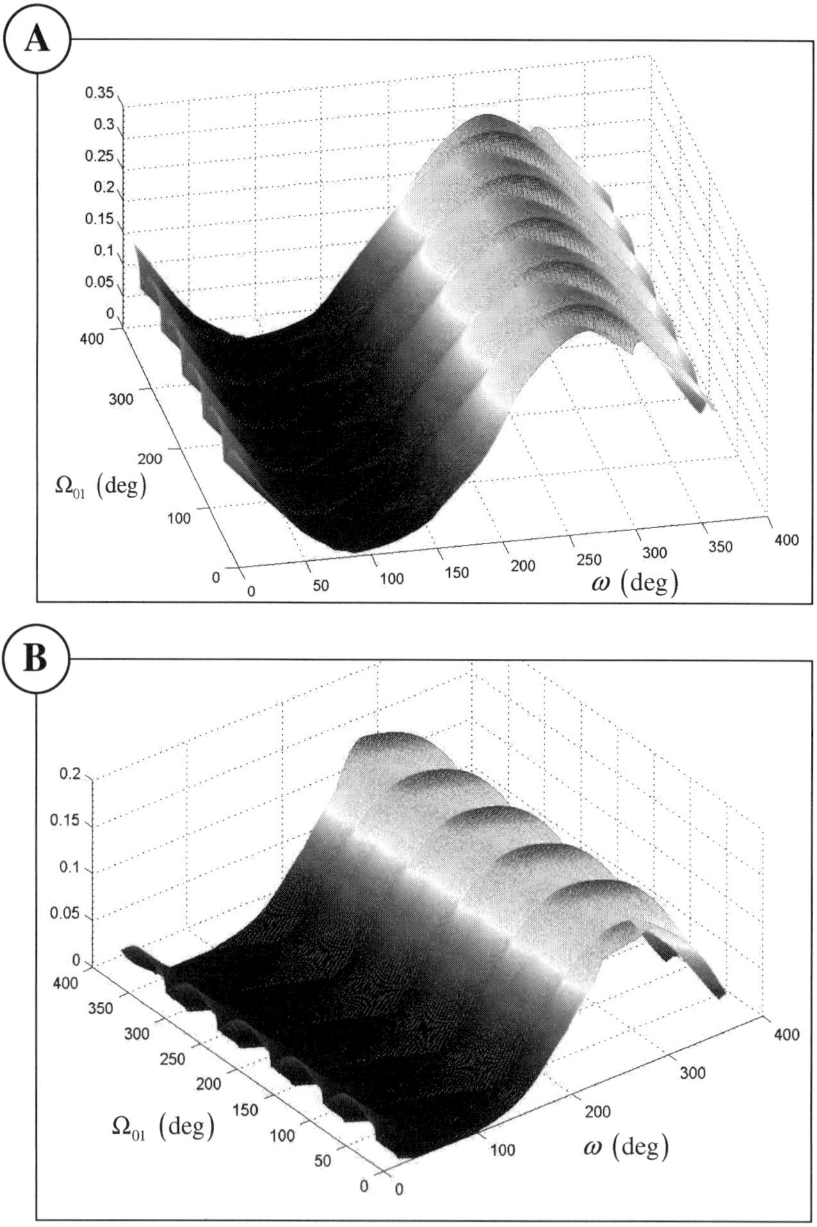

Figure 1. Total duration of visibility (as fraction of the period of repetition) depending on Ω_{01} and ω. An initial Greenwich sidereal time equal to 30 degrees is assumed

An inspection of the previous figure reveals that the total duration of visibility is a non smooth function of Ω_{01} and ω. Moreover, five equivalent maxima can be identified in both cases. Basically, this circumstance is linked to the total number of orbits in the single nodal day of repetition, which number is equal to 5 for the applicative situations at hand.

In conclusion, the following optimal values of Ω_{01} and ω are selected:

Constellation A: $\Omega_{01} = 167.8$ deg and $\omega = 271.4$ deg

Constellation B: $\Omega_{01} = 276.5$ deg and $\omega = 270.0$ deg

The corresponding visible ground tracks are illustrated in Fig. 2, which, in addition, shows target areas for both cases.

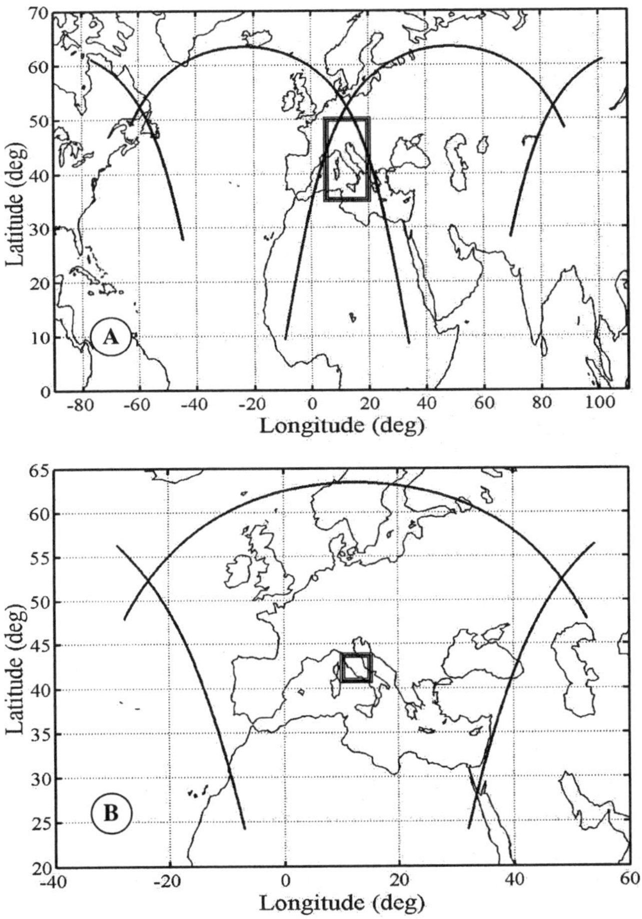

Figure 2. Visible ground track arcs and target areas for cases A and B

4. Strategy for constellation configuration

In this section the visibility function and a correlation integral are introduced. The latter will be the main tool for the constellation design.

For the first satellite an auxiliary time-dependent function, named visibility function, is defined as follows:

$$V_1(t) = \begin{cases} 0, & \text{if satellite 1 is not visible at instant t} \\ 1, & \text{if satellite 1 is visible at instant t} \end{cases} \qquad (0 \le t \le T) \quad (7)$$

T represents the time of repetition, equal to 1 nodal day in the applicative situations at hand.

Now, let n be the number of time intervals of visibility: $\left\{t_l^{(in)}\right\}_{l=1,\ldots,n}$ and $\left\{t_l^{(out)}\right\}_{l=1,\ldots,n}$ respectively identify the initial and terminal point of each of them. In addition, t_{0l} and T_l are defined by:

$$t_{0l} = 0.5\left(t_l^{(out)} + t_l^{(in)}\right) \qquad T_l = t_l^{(out)} - t_l^{(in)} \qquad (l = 1,\ldots,n) \quad (8)$$

The visibility function can be written as follows:

$$V_1(t) = \sum_{l=1}^{n} \mathrm{rect}_{T_l}(t - t_{0l}) \quad \text{where} \quad \mathrm{rect}_B(t) = \begin{cases} 0 & \text{if} \quad |t| > B/2 \\ 1 & \text{if} \quad |t| \le B/2 \end{cases} \qquad (B > 0) \quad (9)$$

Due to periodicity, the time intervals are repeated for $t > T$ so (9) can be formally extended to $V_1^P(t)$:

$$V_1^P(t) = \sum_{k=-\infty}^{\infty} V_1(t - kT) = \sum_{k=-\infty}^{\infty}\left[\sum_{l=1}^{n} \mathrm{rect}_{T_l}(t - kT - t_{0l})\right] \quad \text{with} \quad t \ge 0 \quad (10)$$

Now, let us suppose that at instant $t = \tau_{1,2}$ a second satellite crosses the equator (from South to North) at the same point relative to Earth surface. For this satellite the functions $V_2^P(t)$ and $V_2(t)$ are respectively $V_1^P(t)$ and $V_1(t)$ shifted forward by $\tau_{1,2}$:

$$V_2^P(t) = V_1^P(t - \tau_{1,2}) = \sum_{k=-\infty}^{\infty} V_2(t - kT) = \sum_{k=-\infty}^{\infty} V_1(t - \tau_{1,2} - kT), \qquad t \ge 0 \qquad (11)$$

The overlapping between two visibility functions can be measured through the following integral:

$$R_{1-2}(\tau_{1,2}) = \int_{-\infty}^{\infty} V_1^P(t) V_2(t)\, dt = \int_{-\infty}^{\infty} V_1^P(t) V_1(t - \tau_{1,2})\, dt \qquad (12)$$

Constellations of 2 satellites with minimum overlapping between visible passes of different satellites can be identified by imposing the minimization of

$R_{1-2}(\xi)$, which can be seen as a correlation function depending on $\xi = -\tau_{1,2}$. An analytic expression exists for $R_{1-2}(\tau_{1,2})$ [9]:

$$R_{1-2}(\tau_{1,2}) = \sum_{l=1}^{n}\left\{T_l \underset{T_l}{\text{tri}}(\tau_{1,2}) + T_l \underset{T_l}{\text{tri}}(\tau_{1,2} - T)\right\}\cdot$$
$$+ \sum_{\substack{\nu=1,\ldots n\\ \sigma=\nu+1,\ldots,n}} \left\{\underset{T_\nu,T_\sigma}{\text{trapz}}\left[\tau_{1,2} - (t_{0\nu} - t_{0\sigma})\right] + \underset{T_\nu,T_\sigma}{\text{trapz}}\left[\tau_{1,2} - (t_{0\sigma} - t_{0\nu})\right]\right\} \qquad (13)$$
$$+ \sum_{\substack{\nu=1,\ldots n\\ \sigma=\nu+1,\ldots,n}} \left\{\underset{T_\nu,T_\sigma}{\text{trapz}}\left[\tau_{1,2} - (T + t_{0\nu} - t_{0\sigma})\right] + \underset{T_\nu,T_\sigma}{\text{trapz}}\left[\tau_{1,2} - (T + t_{0\sigma} - t_{0\nu})\right]\right\}$$

where

$$\underset{B}{\text{tri}}(\xi) = \begin{cases} 1 - \dfrac{|\xi|}{B} & \text{if } \ 0 \le |\xi| \le B \\ 0 & \text{if } \ |\xi| \ge B \end{cases} \quad \text{and}$$
$$\underset{T_\nu,T_\sigma}{\text{trapz}}(\xi) = \frac{T_\nu + T_\sigma}{2} \underset{\frac{T_\nu + T_\sigma}{2}}{\text{tri}}(\xi) - \frac{|T_\nu - T_\sigma|}{2} \underset{\frac{|T_\nu - T_\sigma|}{2}}{\text{tri}}(\xi) \qquad (14)$$

Every single term in (13) vanishes for values of $\tau_{1,2}$ which can be easily identified through (14). These values, denoted with $\left\{\tau_{1,2}^{(h)}\right\}_{h=1,\ldots,s}$, correspond to local minima of $R_{1-2}(\tau_{1,2})$. Starting from each of them, the parameters M_{02} and Ω_{02} for the second satellite can be determined through the relationships included in Appendix. Hence, s constellations of 2 satellites can be generated: each of them is checked in order to verify if it is able to provide continuous coverage of the target area.

The extension to constellations of 4 and 8 satellites is straightforward.

5. Optimal constellation configurations

This section presents the optimal constellation configurations for the two mentioned applicative situations A and B. For both cases the two proposed constellations ensure the continuous coverage of the target area.

Figure 3 shows the 3-d views of the constellations, whereas Figure 4 reports the complete scheduling of visible passes over a single period of repetition for cases A and B.

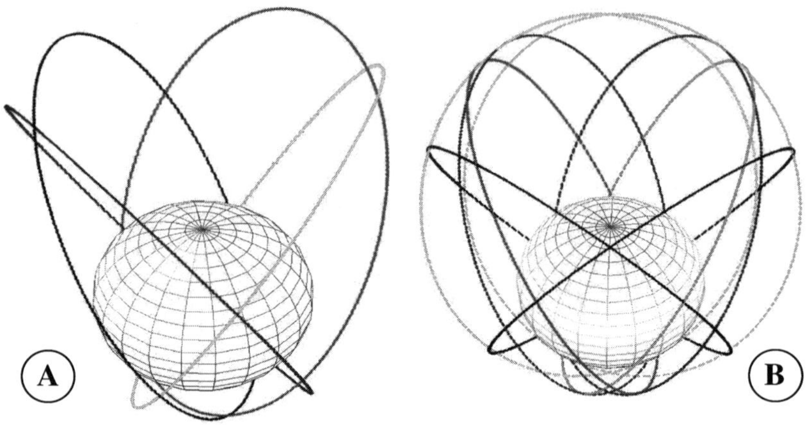

Figure 3. 3-d views of the optimal constellations for cases A and B

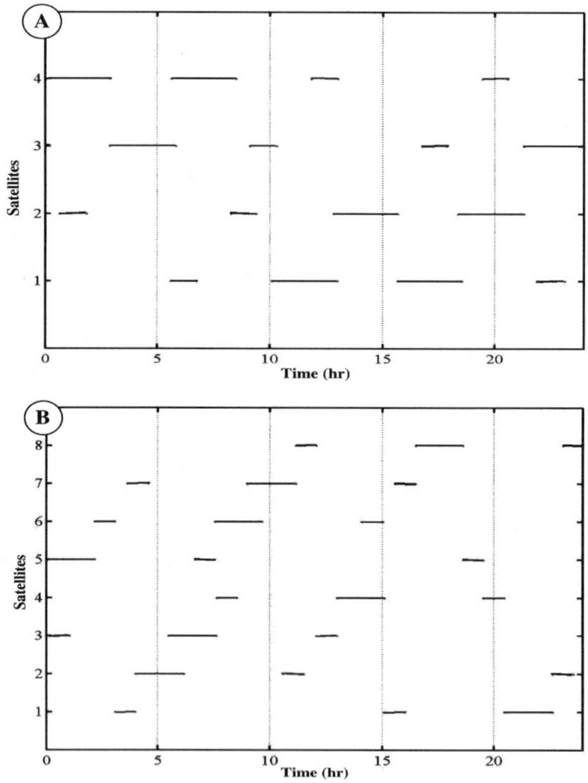

Figure 4. Scheduling of visible passes for optimal constellations (cases A and B)

6. Conclusions

This paper employs an alternative method for designing LEO and MEO eccentric satellite constellations for the local observation of a target area.

All the satellites are placed in equally inclined orbits with identical SMA, eccentricity and argument of perigee in order to avoid differential actions of J_2 perturbation on each of them. Then, visible passes are coded through a binary function, named visibility function. A correlation integral is introduced as an effective tool to find all time delays between consecutive visible passes over the target area. These time delays must minimize the correlation function. Hence, they correspond to constellation configurations with minimum overlapping between visible passes of distinct satellites. This appears a suitable way of ensuring the continuous coverage of the target area.

Two optimal constellation configurations are presented in Section 5, with reference to two applicative situations: a constellation of 4 satellites for a local telecommunication service and a constellation of 8 satellites for local monitoring.

With several revisions, this approach has been successfully applied also to the optimization of constellations of low Earth circular orbit satellites [9]. In this context, the annihilation of the correlation function was imposed. This condition allowed finding some optimal constellations with maximum continuous duration of visibility or minimum revisit time.

In conclusion, the method proposed seems quite efficient, since computational effort is limited. This is due to the existence of the analytical expression of the correlation function, which, moreover, allows finding solutions with great accuracy.

Appendix. Constellation orbital parameters

Constellation design depends on time delays, found by minimizing (13). This appendix deals with the derivation of the orbital parameters Ω_{02} and M_{02} for the second satellite of the constellation, once $\tau_{1,2}$ is known.

At the starting time, the first satellite is supposed to be at perigee, so $M_{01} = 0$. Hence, the second satellite, delayed by $\tau_{1,2}$ with respect to the first, at instant $t = \tau_{1,2}$ must be at perigee. It follows:

$$M_2\left(\tau_{1,2}\right) = 0 \quad \text{so} \quad M_{02} = M_{01} - \dot{M}\,\tau_{1,2} = -\dot{M}\,\tau_{1,2} \tag{15}$$

In addition, the satellites 1 and 2 must cross the equatorial plane (from South to North) at the same point relative to Earth surface, so the following relationship must hold:

$$\Omega_2\left(\tau_{1,2}\right) - \theta_g\left(\tau_{1,2}\right) = \Omega_{01} - \theta_{g0} \tag{16}$$

where $\theta_g(t) = \theta_{g0} + \omega_E t$ is the Greenwich sidereal time and θ_{g0} is its initial value. It follows:

$$\left(\Omega_{20} + \dot{\Omega}\tau_{1,2}\right) - \left(\theta_{g0} + \omega_E \tau_{1,2}\right) = \Omega_{01} - \theta_{g0} \quad \text{so} \quad \Omega_{20} = \Omega_{10} + \left(\omega_E - \dot{\Omega}\right)\tau_{1,2} \quad (17)$$

Orbital parameters for all the remaining satellites are determined in a similar way, making reference to their time delays with respect to the first satellite.

References

1. J. G. Walker, "Circular Orbit Patterns Providing Continuous Whole Earth Coverage", *Royal Aircraft Establishment, Tech. Rep. 70211*, Farnborough (UK), November 1970
2. J. G. Walker, "Some Circular Orbit Patterns Providing Continuous Whole Earth Coverage", *Journal of the British Interplanetary Society*, Vol. 24, 1971, pp. 369-384
3. J. G. Walker, "Continuous Whole Earth Coverage by Circular Orbit Satellite Patterns", *Royal Aircraft Establishment, Tech. Rep. 77044*, Farnborough (UK), March 1977
4. A. H. Ballard, "Rosette Constellations of Earth Satellites", *IEEE Transactions on Aerospace and Electronic Systems*, Vol. 16, No. 5, September 1980, pp. 656-673
5. M. H. Ullock, A. H. Schoen, "Optimum Polar Satellite Networks for Continuous Earth Coverage", *AIAA Journal*, Vol. 1, No. 1, January 1963, pp. 69-72
6. L. Rider, "Optimized Polar Orbit Constellations for Redundant Earth Coverage", *The Journal of the Astronautical Sciences*, Vol. 33, No. 2, April-June 1985, pp. 147-161
7. G. B. Palmerini, F. Graziani, "Polar Elliptic Orbit for Global Coverage Constellations", AAS/AIAA Astrodynamics Conference, Scottsdale, AZ, *AIAA Paper 94-3720*, August 1994, pp.120-129
8. J. E. Draim, "Three- and Four-Satellite Continuous-Coverage Constellations", *Journal of Guidance, Control, and Dynamics*, Vol. 8, No. 6, November-December 1985, pp. 725-730
9. M. Pontani, P. Teofilatto, "Satellite Constellations for Continuous and Early Warning Observation: a Correlation-Based Approach", *Journal of Guidance, Control, and Dynamics*, to be published

MONOMIAL ORDERS IN THE VAST WORLD OF MATHEMATICS

G. RESTUCCIA

Department of Mathematics, University of Messina,
C.da Papardo, salita Sperone, 31, 98166 Messina, (Italy)
E-mail: grest@dipmat.unime.it

The bi-sorted order on monomials in doubly multi-indexed variables is successfully employed to prove a conjecture on toric varieties associated to generalized graphs G_q, q an integer, $q > 1$, consisting of all walks of length q of a complete bipartite graph G.

Keywords: Monomial Algebras; Toric Ideals; Gröbner bases

Introduction

Algebraic models of phenomena arising from different fields make use of the polynomial ring P in a finite number of variables, with coefficients in a field K (of characteristic zero). The study of phenomena does not always require to consider a total order on the multiplicative set of monomials in P. However, in the last years,this is the more attractive direction. Moreover, the classic lexicographic order has been substituted by other orders, sometimes more powerful. Examples of total orders different from the lexicographic order can be found in recent papers of algebraic combinatorics and commutative computational algebra. Algebras of Veronese type $K[L]$ (subrings of polynomial rings) are present in a lot of models like graphs, generalized graphs, transportation problems, economy. Segre products of type Veronese algebras are employed to study bipartite graphs, generalized graphs, transportation problems and to study varieties of Veronese type and Segre varieties in algebraic geometry. The crucial points are the determination and the study of the toric ideal I of $K[L]$ that we would like to be generated from binomials of the lower possible degree (degree two). From the geometric point of view this means that our algebraic variety is an intersection of quadrics.The order on the monomials needs to determine a Gröbner basis G of I since the structure of G reflects strong properties

of $K[L]$ (the binomials of degree two of G produce very good properties of the algebra $K[L]$, like Koszul, strongly Koszul). The interest to have a Gröbner basis of low degree answers to the request to have S-couples of the same degree of equations, in the not linear case . The linear case is in fact completely known (theory of circuits). Our problems are the following:
1. Determination of the Gröbner basis of toric ideals.
2. Research of monomial orders that produce Gröbner bases of the minimum possible degree.
We will examine monomial orders in a polynomial ring with multi-indexed variables. They appear in the presentation of so called "subrings of mixed products". They come from ideals of mixed products studied in [4] and they can be associated to some special generalized graphs.
More precisely, let P and Q be two polynomial rings, in one set of variables (respectively $\underline{X} = (X_1, \ldots, X_n)$, $\underline{Y} = (Y_1, \ldots, Y_m)$), let I_r be the ideal of P generated by all square-free monomials $X_{i_1} \cdots X_{i_r}$, $1 \leq i_1 < i_2 < \cdots < i_r \leq n$ and J_k the ideal of Q generated by all square-free monomials $Y_{j_1} \cdots Y_{j_k}$, $1 \leq j_1 < j_2 < \cdots < j_k \leq m$. Given the integers k, r, s, t such that $k + r = s + t$, we consider the square-free monomial ideal of mixed products of the ring $R = K[X_1, \ldots, X_m; Y_1, \ldots, Y_n]$ given by $L = I_r J_k + I_s J_t$. If F is the minimal generating set of L consisting of square-free monomials, let $K[F]$ be the monomial mixed algebra, subring of $K[X_1, \ldots, X_m; Y_1, \ldots, Y_m]$. Let A be the polynomial ring over K with doubly multi-indexed variables such that $K[F] = A/I$. The monomial order we are going to utilize in A is the bi-sorted order. This order is neither lexicographic, nor reverselexicographic, but it is useful to succeed in our research. We refer to [3,chap.4], for the theory of the sorted order in one set of variables.
Project: The aim is to resolve a geometric conjecture for toric varieties.
Result: The conjecture has a positive answer for $K[F]$, where F is the minimal generating set of L,being L the generalized graph ideal of a complete bipartite graph. For F, given by L, the affine toric variety $X_{\mathcal{A}}$ defined by I is normal. Consequently, we deduce the normality of the semigroup $\mathbb{N}\mathcal{A}$, where $\mathcal{A}$ is the finite subset of Z^d, underlying the ring $K[F]$. Moreover, the semigroup $\mathcal{A}$ admits an unimodular triangulation [2, chap.13]. In fact in each case we determine a square-free quadratic Gröbner basis for I [3], that is sufficient for normality. The converse is not true. We have examples of projectively normal toric varieties without square-free initial toric ideal[3, Example 13.17].
For $\mathcal{A} = \{(0,0,0,0,1), (1,0,0,0,1), (0,1,0,0,1), (0,0,1,0,1), (1,1,1,1,1),$

$(1,1,2,2,1),(1,2,2,3,1),(1,2,3,4,1),(1,2,3,5,1)\}$, the 4- dimensional projectively normal toric variety $X_{\mathcal{A}}$ in $\mathbb{P}^8$ is such that no initial ideal of the toric ideal I is square-free. In the section 1 we recall class of mixed products ideals in a polynomial ring in two sets of variables. The connection with the generalized graph of a complete bipartite graph is explained. The conjectures 1 and 2 are enunciated. Section 2 is dedicated to the presentation of the bi-sorted order in two sets of variables and to the proof of the main theorem on the initial ideal of toric ideals. In the section 3 we resolve the conjecture 2 for normality of all classes of toric varieties associated to generalized graphs. Moreover, if the graph $G = G^{(1)} \cup G^{(2)}$ is not connected, with $G^{(1)}$ and $G^{(2)}$ complete graphs, we prove the conjectures 1 and 2 for the toric ring $K[F]$, $F = F_1 \cup F_2$, where F_1 is related to the generalized graph $G_s^{(1)}$ of walks of length s of $G^{(1)}$ and F_2 is related to the generalized graph $G_s^{(2)}$ of walks of length s of $G^{(2)}$.

1. Mixed products ideals

Consider two sets of variables $\underline{X} = (X_1, \dots, X_n)$, $\underline{Y} = (Y_1, \dots, Y_m)$, and the polynomial ring $R = K[\underline{X}, \underline{Y}] = K[X_1, \dots, X_n; Y_1, \dots, Y_m]$.

Definition 1.1. Let k, r, s, t integers, non negative, $k, r, s, t \geq 0$, such that $k + r = s + t$. We define a mixed products ideal the ideal of R

$$L = I_k J_r + I_s J_t,$$

when I_k is the ideal generated by all the square-free monomials of degree k in the variables $X_1, \dots, X_n$ and J_r is the ideal generated by all the square-free monomials of degree r in the variables $Y_1, \dots, Y_m$.

It is convenient to set $I_0 = J_0 = R$. By symmetry, we can see that there are essentially two cases:

(1) $L = I_k J_r + I_s J_t$,with $0 \leq k \leq s$
(2) $L = I_k J_r$, with $k \geq 1$ or $r \geq 1$.

In a joint work with Villarreal [4], we are able to give a complete classification of normal ideals of mixed products. More precisely, the only ideals that are normal are:

(a) $L = I_k J_r + I_{k+1} J_{r-1}$, $k \geq 0$, $r \geq 1$
(b) $L = I_k J_r$, $k \geq 1$, $r \geq 1$.
(c) $L = J_r + I_m J_t$ and $L = I_k + J_m I_t$

The ideals that are not normal are:

(1) $L = I_s + J_r$, $r \leq s \leq min\{m,n\}$, (for $s = min\{m,n\}$, L is normal)
(2) $L = J_r + I_s J_t$, $2 \leq s < n$
(3) $L = I_k J_r + I_s$, $2 \leq r < m$
(4) $L = I_k J_r + I_s J_t$, $s \geq k+2$, $k \geq 1$, $t \geq 1$.

Applications: If G is a complete bipartite graph, then the generalized graph ideal $I_q(G)$ is normal for $q \geq 2$. If G is a graph with n vertices $x_1, \ldots, x_n$, we have:

Definition 1.2. We call generalized graph ideal $I_q(G)$ the ideal of $K[X_1, \ldots, X_n]$ generated by the square free monomials $x_{i_1}, \ldots, x_{i_n}$ such that x_{i_j} is adjacent to $x_{i_{j+1}}$, for all $1 \leq j \leq q-1$

Now, let $X_1, \ldots, X_n$ and $Y_1, \ldots, Y_m$ be the vertex set of G. One may assume that the edges of G are precisely the pairs of the form $\{X_i, Y_j\}$. Therefore, we have:

$$I_q(G) = \begin{cases} I_t J_{t+1} + I_{t+1} J_t & \text{if} \quad q = 2t+1 \\ I_t J_t & \text{if} \quad q = 2t \end{cases}.$$

We will call G_q the new graph with edges ideal $I_q(G)$.

Example 1.1. $R = K[X_1, X_2; Y_1, Y_2, Y_3]$

$$I_3(G) = \{X_1Y_1X_2, X_1Y_2X_2, X_1Y_3X_2, Y_1X_1Y_2, Y_1X_1Y_3, Y_3X_1Y_1, Y_2X_1Y_3\} = I_2J_1 + I_1J_2$$

$$I_4(G) = \{X_1Y_1X_2Y_2, X_1Y_3X_2Y_1, X_2Y_3X_1Y_2, \ldots\} = I_2J_2.$$

Remark 1.1. We recall that an ideal $I \subset R$ is normal if $I^i = \bar{I}_i$, for every $i \geq 1$, $\bar{I}$ the integral closure of I in the integral domain R. Normality of ideals in the polynomial ring is a very important property [5]. The integral closure of monomial ideals has a simple geometric description, more visible in two variables [5]. As a consequence, if the graph G has the edge ideal I normal, we can have good informations on the walks of G and on the integral closure of the algebra $K[G]$ generated on K by the minimal set of monomials generators of I. In fact I normal implies $K[G]$ integrally closed as a subring of $K[\underline{X}] = K[X_1, ..., X_n]$, being $X_1, ..., X_n$ the vertices of G.

Other directions of application: Square-free Veronese subrings are directly involved in monomial subrings that can arise from mixed product ideals and in addition, Segre products, tensor products of square-free

Veronese subrings. Moreover, we have new classes of monomial subrings of general interest, from an algebraic and geometric point of view.

CONJECTURE 1: For any ideal L of mixed products, the monomial subring $K[F]$ is normal (L not necessarily normal).

Our monomial algebras will give examples where $K[F]$ is normal, but the ideal L and, consequently, the Rees Algebra $\mathcal{R}(L)$ is not normal [[5],chap. 7]. For instance, $L = I_s + J_s$, $2 \leq s \leq min\{m,n\}$ is not normal, but $K[F]$ is normal. We will use the following Criterion of normality: let $K[F]$ be a homogeneous monomial subring of the polynomial ring $K[\underline{X};\underline{Y}]$. If the toric ideal I of $K[F] = A/I$, A a polynomial ring over K, has a square-free initial ideal with respect to some term order $<$ on the monomial of A, then $K[F]$ is normal [[3], Prop.13.15].
The converse is not true. We have examples of projective normal toric varieties without square-free initial toric ideal (see Introduction).

CONJECTURE 2: For any ideal L of mixed products, the toric ideal of $K[F]$ has a square-free initial ideal for some term order on the monomials of A.

Our proposal is to prove the conjecture 2. As a consequence, the conjecture 1 will be true. For the proof we will use a special order on monomials in doubly multi-indexed variables, the *bi-sorted order*. This order is neither lexicographic, nor reverselexicographic, in general.
The sorted order on monomials in multi-indexed variables was introduced in ([3]).

2. Bi-sorted monomials in doubly multi-indexed variables

We fix two integers r and s and we consider the set

$$T_{n,m} = \{(i_1, \ldots, i_n; j_1, \ldots, j_m)\} \subset \mathbb{N}^n \oplus \mathbb{N}^m,$$

$i_1 + \cdots + i_n = r$, $j_1 + \cdots + j_m = s$, $0 \leq i_1 \leq r_1, \ldots, 0 \leq i_n \leq r_n$ and $0 \leq j_1 \leq s_1, \ldots, 0 \leq j_m \leq s_m$.

There is a natural bijection between the elements of $T_{n,m}$ and increasing bi-strings of length $r+s$ over the alphabet $A_1 = \{1, 2, \ldots, n\}$ from 1 until r and over the alphabet $A_2 = \{1, 2, \ldots, m\}$ from $r+1$ until s, having at most r_t occurrences of the letter t until r and at most s_t occurrences of the letter t from $r+1$ until s. Under this bijection the vector $(i_1, \ldots, i_n; j_1, \ldots, j_m) \in T_{n,m}$ is mapped to the increasing bi-string $u_1u_2\ldots u_r; v_1v_2\ldots v_s = \underbrace{11\ldots1}_{i_1-times}\underbrace{22\ldots2}_{i_2-times}\ldots\underbrace{nn\ldots n}_{i_n-times};\underbrace{11\ldots1}_{j_1-times}\underbrace{22\ldots2}_{j_2-times}\ldots\underbrace{mm\ldots m}_{j_m-times}$.
We write $X_{\underline{u};\underline{v}} = X_{u_1u_2\ldots u_r;v_1v_2\ldots v_s}$ for the corresponding variable in the

polynomial ring $K[\underline{X}]$. Let sort $(\cdot;\cdot)$ denote $*$ the operation which takes any bi-string over the alphabets A_1 and A_2 and sorts separately any string into increasing order.

Proposition 2.1. *The toric ideal defined by the set $T_{n,m}$ equals:*

$$I_{T_{n,m}} = (X_{\underline{u_1};\underline{v_1}} X_{\underline{u_2};\underline{v_2}} \cdots X_{\underline{u_\rho};\underline{v_\rho}} - X_{\underline{u'_1};\underline{v'_1}} X_{\underline{u'_2};\underline{v'_2}} \cdots X_{\underline{u'_\rho};\underline{v'_\rho}}) :$$

$$bi\text{-}sort(\underline{u}_1\underline{u}_2 \cdots \underline{u}_\rho; \underline{v}_1\underline{v}_2 \cdots \underline{v}_\rho) = bi\text{-}sort(\underline{u'}_1\underline{u'}_2 \cdots \underline{u'}_\rho; \underline{v}_1\underline{v'}_2 \cdots \underline{v'}_\rho).$$

Definition 2.1. A monomial $X_{\underline{u};\underline{v}} X_{\underline{u'};\underline{v'}} \cdots X_{\underline{u}^{(i)};\underline{v}^{(i)}} = X_{u_1u_2\cdots u_r;v_1v_2\cdots v_r} \cdot X_{u'_1u'_2\cdots u'_r;v'_1v'_2\cdots v'_r} \cdots X_{u_1{}^{(i)}u_2{}^{(i)}\cdots u_r{}^{(i)};v_1{}^{(i)}v_2{}^{(i)}\cdots v_r{}^{(i)}}$ of $K[\underline{X}]$ is said to be bi-sorted if

(1) $u_1 \leq u'_1 \leq \cdots \leq u_1{}^{(i)} \leq \cdots \leq u_r \leq u'_r \leq \cdots \leq u_r{}^{(i)}$

(2) $v_1 \leq v'_1 \leq \cdots \leq v_1{}^{(i)} \leq \cdots \leq v_r \leq v'_r \leq \cdots \leq v_r{}^{(i)}$

For monomials which are not bi-sorted we define the inversion numbers pair (a, b) as the numbers of inversions in the two strings 1) and 2) respectively. We define inversion in the string 1) or 2) a pair of indices (i, j) such that $i < j$ and $u_i > u_j$ in 1) and $v_i > v_j$ in the bi-string 2).

Proposition 2.2. *We have the following facts:*

(1) Every variable is bi-sorted
(2) Every power of variable is bi-sorted
(3) If a monomial is not bi-sorted, then it contains a quadratic factor which is not bi-sorted.
(4) For every binomial

$$f = X_{\underline{u_1};\underline{v_1}} X_{\underline{u_2};\underline{v_2}} \cdots X_{\underline{u_\rho};\underline{v_\rho}} - X_{\underline{u'_1};\underline{v'_1}} X_{\underline{u'_2};\underline{v'_2}} \cdots X_{\underline{u'_\rho};\underline{v'_\rho}} \in I_A$$

such that $bi - sort(\underline{u_1} \cdots \underline{u_\rho}; \underline{v_1} \cdots \underline{v_\rho}) = bi - sort(\underline{u'_1} \cdots \underline{u'_\rho}; \underline{v'_1} \cdots \underline{v'_\rho})$, there exists a binomial $X_{\underline{w_1};\underline{z_1}} \cdots X_{\underline{w_\rho};\underline{z_\rho}} - X_{\underline{u'_1};\underline{v'_1}} \cdots X_{\underline{u'_\rho};\underline{v'_\rho}} \in I_{T_{n,m}}$ such that the first monomial is bi-sorted.

Proof: We prove only 4). 1), 2), 3) are obvious. Consider

$$bi - sort(\underline{u}_1\underline{u}_2 \cdots \underline{u}_\rho; \underline{v}_1\underline{v}_2 \cdots \underline{v}_\rho) = \underline{w}_1\underline{w}_2 \cdots \underline{w}_\rho; \underline{z}_1\underline{z}_2 \cdots \underline{z}_\rho.$$

Then $g = X_{\underline{w}_1\underline{w}_2\cdots\underline{w}_\rho;\underline{z}_1\underline{z}_2\cdots\underline{z}_\rho} - X_{\underline{u}_1\underline{u}_2\cdots\underline{u}_\rho;\underline{v}_1\underline{v}_2\cdots\underline{v}_\rho} \in I_{T_{n,m}}$, and the first monomial is sorted. Finally, $g + f \in I_{T_{n,m}}$ and

$$g + f = X_{\underline{u}_1;\underline{z}_1} \cdots X_{\underline{w}_\rho;\underline{z}_\rho} - X_{\underline{u'}_1;\underline{v'}_1} \cdots X_{\underline{u'}_\rho;\underline{v'}_\rho}$$

Let $f \in K[\underline{X}]$ be a polynomial. We say that f is marked if the initial term in(f) of f is specified , where in(f) can be any term of f. Given a set $\mathcal{F}$ of marked polynomials, we define the reduction relation module F in the sense of the theory of Gröbner bases.

We say that $\mathcal{F}$ is coherently marked if there exists a term order $<$ on $K[\underline{X}]$ such that $in_<(f) = in(f)$ for all $f \in \mathcal{F}$. Then, if $\mathcal{F}$ is coherently marked, the reduction relation $\xrightarrow{\mathcal{F}}$ is Noetherian.

Theorem 2.1. *A finite set $\mathcal{F} \subset K[\underline{X}]$ of marked polynomials is coherently marked if and only if the reduction relation module $\mathcal{F}$ is noetherian, i.e., every sequence of reductions module $\mathcal{F}$ terminates.*

Theorem 2.2. *Let*

$$\mathcal{F} = \{\underline{X_{\underline{u}_1;\underline{v}_1} X_{\underline{u}_2;\underline{v}_2}} - X_{\underline{u'}_1;\underline{v'}_1} X_{\underline{u'}_2;\underline{v'}_2} : bi\text{-}sort(u_{11}u_{21}u_{12}u_{22}\cdots u_{1n}u_{2n};$$

$$v_{11}v_{21}v_{12}v_{22}\cdots v_{1m}v_{2m}) = u'_{11}u'_{21}u'_{12}u'_{22}\cdots u'_{1n}u'_{2n}; v'_{11}v'_{21}v'_{12}v'_{22}\cdots v'_{1m}v'_{2m}\}$$

be a set of marked binomials of $K[\underline{X}]$, with $u_1 = u_{11}\ldots u_{1n}$, $u_2 = u_{21}\ldots u_{2n}$, $v_1 = v_{11}\ldots v_{1m}$, $v_2 = v_{21}\ldots v_{2m}$, $u'_1 = u'_{11}\ldots u'_{1n}$, $u'_2 = u'_{21}\ldots u'_{2n}$, $v'_1 = v'_{11}\ldots v'_{1m}$, $v'_2 = v'_{21}\ldots v'_{2m}$. Then we have:

(1) Every monomial $m \in k[\underline{X}]$ is a normal form with respect to this reduction if and only if m is bi-sorted.
(2) If a monomial m_1 not bi-sorted reduces to another monomial m_2 using $\mathcal{F}$ at least one of the inversion numbers a and b of m_2 is strictly less than the corresponding inversion number of m_1.
(3) The reduction relation defined by $\mathcal{F}$ is Noetherian.

Proof:

(1) $\Rightarrow$) If the monomial is normal, it is not divisible for any initial term of the binomials, i.e. for any quadratic factor not bi-sorted. Then it is bi-sorted.
$\Leftarrow$) If the monomial is bi-sorted, it is not divisible for any initial term of the binomials. Then it is normal.
(2) If a monomial m_1 is not bi-sorted, then it contains a quadratic factor which is not bi-sorted. Then we can reduce m_1 to m_2 and we kill some not bi-sorted quadratic factors, hence the assertion.
(3) It is a consequence of the assertion 2).

Theorem 2.3. *There exists a term order $\prec$ on $K[\underline{X}]$ such that the bi-sorted monomials are precisely the $\prec$-standard monomials modulo the ideal*

I generated by the elements of $\mathcal{F}$. $\mathcal{F}$ is the reduced Gröbner basis of I and consequently $\mathrm{in}_{\prec} I$ is generated by square-free quadratic monomials.

Proof: Let $\mathcal{F}$ denote the above set of marked binomials. Consider the initial ideal $\mathrm{in}_{\prec} I$. We prove that every no bi-sorted monomial lies in $\mathrm{in}_{\prec} I$. Suppose that there exists a bi-sorted monomial m with respect to a set of variables (or to both the sets of variables) that lies in $\mathrm{in}_{\prec} I$. Then there exists a non-zero binomial $m_1 - m_2 \in I$ such that $m_2 \notin \mathrm{in}_{\prec} I$ (since m is the initial monomial). Then m_2 is bi-sorted and m_1 and m_2 are bi-sorted monomials which lie in the same residue class modulo I ($\overline{m_1} = \overline{m_2}$). It follows that $m_1 = m_2$, a contradiction. Then $\mathcal{F}$ is the reduced Gröbner basis of I with respect to $\prec$.

Example 2.1. $n = m = 3$, $r = s = 2$, $s_i = t_i = 1$

$K[\underline{X}] = K[X_{12;12}, X_{12;13}, X_{12;23}, X_{13;12}, X_{13;13}, X_{13;23}, X_{23;12}, X_{23;13}, X_{23;23}]$

$\mathcal{F} = \{X_{12;12}X_{23;23} - X_{23;12}X_{12;23}, X_{13;13}X_{23;23} - X_{23;13}X_{13;23}, \dots\}$
1223; 1223 = bi-sort(2132; 1223) and 1233; 1233 = bi-sort(2133; 1233)

Example 2.2. $n = 3$, $m = 4$, $r = 2$, $s = 3$ $s_i = t_i = 1$
$K[\underline{X}] = K[X_{12;123}, X_{12;124}, X_{12;134}, X_{12;234}, X_{13;234}, X_{13;123}, X_{23;123}, X_{23;124}, X_{23;234}, X_{23;134}]$, $\mathcal{F} = \{X_{12;123}X_{23;124} - X_{12;124}X_{23;123}, \dots\}$
1223; 112234 =bi-sort(1223; 112243).

3. Generalized graph ideals with normal associated subalgebra

In this number we present three families of mixed products ideals and we prove the conjecture 2 for the toric varieties associated to them.

Theorem 3.1. *Let $L = I_r J_s$ be a mixed products ideal, $r, s \geq 1$. Then we have:*

(1) The toric ideal I of $K[F]$ has an initial square-free monomial ideal generated in degree two;
(2) $K[F]$ is normal.

Proof:

(1) Let V_r be the set of minimal generators of I_r and V_s be the set of minimal generators of J_s, and consider the K-algebras $K[V_r], K[V_s]$. Consider the Segre product $K[V_r] * K[V_s] = K[F]$, where F is a set of

minimal generators of $K[V_r] * K[V_s]$ resulting by the multiplication of a generator of I_r for all generators of J_s.

$$K[V_r] \subset K[X_1, X_2, \dots X_n],\ K[V_s] \subset K[Y_1, Y_2, \dots Y_m]$$

$$K[V_r] * K[V_s] \subset K[X_1, X_2, \dots X_n; Y_1, Y_2, \dots Y_m].$$

Consider the semigroup $\mathbb{N}^n \oplus \mathbb{N}^m$. Any generator of $I_r J_s$ can be identified with an element of a subset $A \subset \mathbb{N}^n \oplus \mathbb{N}^m$. Then we can conclude by theorem 2.3.

(2) It follows from 1) or from the fact that the ideal $I_r J_s$is normal.

Remark 3.1. $K[F] = K[V_r] \otimes_S K[V_s]$ is the Segre product of the algebras $K[V_r]$ and $K[V_s]$, attached to the $G_r^{(1)}$ and $G_s^{(2)}$ with generalized graph ideals I_r and J_s of the complete graphs $G^{(1)}$ and $G^{(2)}$, and the resulting graph is the Segre product of $G_r^{(1)}$ and $G_s^{(2)}$. For $r = s$, $L = I_r J_r$ is the generalized graph ideal of the graph G_{2r}, G a complete bipartite graph G.

Theorem 3.2. *Let $L = I_s + J_s$, $s \leq min\ \{m, n\}$ be a mixed products ideal. Then we have:*

(1) The toric ideal I of $K[F]$ has a square-free initial ideal for some term order;
(2) $K[F]$ is normal.

Proof:

(1) Let V_s be the set of minimal generators of I_s and W_s be the set of minimal generators of J_s, consider the K-algebras $K[V_s]$ and $K[W_s]$.

$$K[V_s] \subset K[X_1, X_2, \dots X_n],\ K[W_s] \subset K[Y_1, Y_2, \dots Y_m].$$

Consider the subsets of $\mathbb{N}^n \oplus \mathbb{N}^m$:

$$A_1 = \{(i_1, \dots, i_n; 0, \dots, 0) \in \mathbb{N}^n \oplus \mathbb{N}^m\}$$

$$A_2 = \{(0, \dots, 0; j_1, \dots, j_m) \in \mathbb{N}^n \oplus \mathbb{N}^m\}$$

For A_1, $s = 0$, $i_1 + \cdots + i_n = r$. For A_2, $r = 0$, $j_1 + \cdots + j_m = s$. Consider $A_1 \cup A_2$. Let $\mathcal{F}_1$ and $\mathcal{F}_2$ be sets of marked binomials that arise from A_1 and A_2 respectively. We want to prove that $\mathcal{F}_1 \cup \mathcal{F}_2$ is the reduced Gröbner basis for the toric ideal I of $K[F] = K[V_s \cup W_s] = K[V_s] \otimes K[W_s]$, the tensor product of the $K-$algebras $K[V_s]$ and $K[W_s]$. Let $\prec_1$ be the sorted term order on the monomials in the multi-indexed variables $\underline{X}$ and let $\prec_2$ be the sorted term order on the monomials in the multi-indexed variables $\underline{Y}$. We introduce the term order $\prec$ on the monomials of $K[\underline{X};\underline{Y}]$. If m_1, m_2, n_1, n_2 are monomials, $m_1 \prec m_2 \Leftrightarrow m_1 \prec_1 m_2$, where $m_1, m_2 \in K[\underline{X}]$ and $n_1 \prec n_2 \Leftrightarrow n_1 \prec_2 n_2$, where $n_1, n_2 \in K[\underline{Y}]$ and $m_1 n_1 \prec m_2 n_2$ if $m_1 \prec_1 m_2$ or $m_1 = m_2$ and $n_1 \prec_2 n_2$. Then $\prec$ is a term order on all the monomials of $K[\underline{X};\underline{Y}]$. We want to prove that $\mathcal{F}_1 \cup \mathcal{F}_2$ is a reduced Gröbner basis for $A_1 \cup A_2$ with respect to term order $\prec$. The assertion follows by the structure of the toric ideal I.

(2) It follows by (1).

Remark 3.2. The ideal L is not normal, hence the conjecture 1 is true for $K[F]$. $K[F] = K[V_r \cup W_s] = K[V_r] \otimes K[W_s]$ is the tensor product of the algebras $K[V_r]$ and $K[W_s]$, attached to the complete graphs $G^{(1)}$ and $G^{(2)}$ with generalized graph ideals I_r and J_s, and the resulting graph is the tensor product of the generalized graphs $G_r^{(1)}$ and $G_s^{(2)}$.

Remark 3.3. So far we considered monomials $X_{\underline{u}_1;\underline{v}_1} \cdots X_{\underline{u}_\rho;\underline{v}_\rho}$ with the following properties:
1) the lengths of the strings of integers $\underline{u}_1, \ldots, \underline{u}_\rho$ are the same;
2) the lengths of the strings of integers $\underline{v}_1, \ldots, \underline{v}_\rho$ are the same.
In the same way for the bi-sorted order introduced for monomials as before, we can introduce a bi-sorted order on the monomials of type $X_{\underline{u}_1;\underline{v}_1} \cdots X_{\underline{u}_\rho;\underline{v}_\rho}$ with the following properties:
1′) the lengths of the strings of integers $\underline{u}_1, \ldots, \underline{u}_\rho$ are different;
2′) the lengths of the strings of integers $\underline{v}_1, \ldots, \underline{v}_\rho$ are different.

As a consequence all results on bi-sorted monomials will be still true, if we agree to not write the absent occurrences in the strings (the first or the second string have different lengths from the second or the first string). For instance, given the monomial $X_{1,23}X_{23,2}$, we write $123; 232$ instead of the bi-string $1 - 23; 232-$.

Theorem 3.3. *Let $L = I_k J_{k+1} + I_{k+1} J_k$, $k \geq 2$ a mixed products ideal. Then we have:*

(1) The toric ideal I of $K[F]$ has a square-free initial ideal for some term order.
(2) $K[F]$ is normal.

Proof:1) Let $I_{T_{k,k+1}}$ be the toric ideal defined by the set $T_{k,k+1}$ in the polynomial ring $K[\underline{U}]$ and $\mathcal{F}_1$ the correspondent reduced Gröbner basis of

$I_{T_{k,k+1}}$ with respect to the bi-sorted term order $\prec_1$ on the monomials in the variables $\underline{U}$. Let $I_{T_{k+1,k}}$ be the toric ideal defined by the set $T_{k+1,k}$ in the polynomial ring $K[\underline{V}]$ and $\mathcal{F}_2$ the correspondent reduced Gröbner basis of $I_{T_{k+1,k}}$ with respect the bi-sorted term order $\prec_2$ on the monomials in the variables $\underline{V}$. In the polynomial ring $K[\underline{U};\underline{V}]$ we consider the set of binomials: $G = \left\{U_{\underline{u}_1;\underline{v}_1}V_{\underline{u}_2;\underline{v}_2} - \underline{U_{\underline{u'}_1;\underline{v'}_1}V_{\underline{u'}_2;\underline{v'}_2}}; \underline{u}_1\underline{u}_2; \underline{v}_1\underline{v}_2 = \text{bi-sort}(\underline{u'}_1\underline{u'}_2; \underline{v'}_1\underline{v'}_2)\right\}$. Let $\prec$ be the order on the monomials of $k[\underline{U},\underline{V}]$, described in Theorem 3.2. We want to prove that the set $\{\mathcal{F}_1, \mathcal{F}_2, G\}$ is a reduced Gröbner basis for the toric ideal I and with respect to the order $\prec$. More precisely, we claim that the standard monomials with respect to $\prec$ are the monomials bi-sorted with respect to $\prec_1$ and with respect to $\prec_2$ and bi-sorted in the monomials UV. For the proof, we proceed in the same way that in Theorem 2.3.

2) By 1) or since L is normal.

Remark 3.4. The ideal $L = I_k J_{k+1} + I_{k+1} J_k$ is the generalized graph ideal of a complete bipartite graph G and the corresponding K-algebra $K[F]$ is related to the generalized graph G_{2k+1} of all walks of length $2k+1$ of G. It would be interesting to investigate other properties of this algebra and links with properties of G_{2k+1}.

Example 3.1. $L = I_2 J_1 + I_1 J_2$.

The generators of L are: $X_1X_2Y_1, X_1X_2Y_2, X_1X_2Y_3, X_1Y_1Y_2, X_2Y_1Y_2, X_1Y_1Y_3, X_2Y_1Y_3, X_1Y_2Y_3, X_2Y_2Y_3$. The corresponding variables are: $T_{12,1}, T_{12,2}, T_{12,3}, U_{1,12}, U_{2,12}, U_{1,13}, U_{2,13}, U_{1,23}, U_{2,23}$. The Gröbner basis for the toric ideal I_1 corresponding to $L_1 = I_2 J_1$ is empty, then we can assume on the variables $\underline{T}$ the order $T_{12,1} \succ T_{12,2} \succ T_{12,3}$.

The reduced Gröbner basis for the toric ideal I_2 corresponding to $L_2 = I_1 J_2$, is:

$g_1 = U_{1,12}U_{2,23} - \underline{U_{2,12}U_{1,23}}, g_2 = U_{1,13}U_{2,23} - \underline{U_{2,13}U_{1,23}}, g_3 = U_{1,12}U_{2,13} - \underline{U_{2,12}U_{1,13}}$, where the marked monomial is the initial not bi-sorted term of the binomial and the other monomial is bi-sorted. The other mixed relations are:

$$f_1 = T_{12,1}U_{1,23} - \underline{T_{12,3}U_{1,12}}; \quad f_2 = T_{12,1}U_{2,23} - \underline{T_{12,2}U_{2,13}}$$

$$f_3 = T_{12,1}U_{2,23} - \underline{T_{12,3}U_{2,12}}; \quad f_4 = T_{12,1}U_{1,23} - \underline{T_{12,2}U_{1,13}}$$

where the marked monomial is the initial term. The set $\{g_1, g_2, g_3, f_1, f_2, f_3, f_4\}$ is the reduced Gröbner basis for the toric ideal I. We have to calculate only the S-pairs:

$S(g_1, g_2) = -U_{2,12}g_2 + U_{2,13}g_1 = -U_{2,23}g_3$
$S(g_1, g_3) = -U_{1,23}g_3 + U_{1,13}g_1 = -U_{1,12}g_2$
$S(f_1, f_3) = -U_{1,12}f_3 + U_{2,12}f_1 = -T_{12,1}g_1$
$S(f_2, f_4) = -U_{2,13}f_4 + U_{1,13}f_2 = T_{12,1}g_2$
$S(f_2, g_2) = -T_{12,2}g_2 + U_{1,13}f_2 = -U_{2,23}f_4$
$S(f_3, g_1) = -T_{12,3}g_1 + U_{1,23}f_3 = -U_{2,23}f_1$
$S(f_3, g_3) = -T_{12,3}g_3 + U_{1,13}f_3 = U_{2,13}f_1 + T_{12,1}g_2$
$S(f_4, g_3) = -T_{12,2}g_3 + U_{2,12}f_4 = U_{1,12}f_2 - T_{12,1}g_1$
In any case, all S-pairs reduce to zero.

Example 3.2. $L = I_2J_1 + I_1J_2$. The new variables are:
$T_{12,1}, T_{13,1}, T_{23,1}, T_{12,2}, T_{13,2}, T_{23,2}, T_{12,3}, T_{13,3}, T_{23,3}$,
$U_{1,12}, U_{1,13}, U_{1,23}, U_{2,12}, U_{2,23}, U_{2,13}, U_{3,12}, U_{2,13}, U_{3,23}$.
The reduced Gröbner basis of I is $G_1 \cup G_2 \cup G_3$, where G_1 is the Gröbner basis for the toric ideal I_1 corresponding to $L_1 = I_2J_1$, G_2 is the Gröbner basis for the toric ideal I_2 corresponding to $L_2 = I_1J_2$, and $G_3 = \{T_{12,1}U_{2,23} - T_{12,2}U_{2,13}, T_{23,3}U_{2,13} - T_{23,1}U_{2,23}, \ldots\}$. Put:
$h_1 = T_{12,1}U_{2,23} - \underline{T_{12,2}U_{2,13}}$, $h_2 = T_{23,3}U_{2,13} - \underline{T_{23,1}U_{2,23}}$,
$h_3 = T_{13,2}U_{2,23} - \underline{T_{12,2}U_{3,23}}, \ldots$. For instance, we compute the S-pair:
$S(h_1, h_3) = -U_{2,13}h_3 + U_{3,23}h_1 = U_{2,23}h_4$, $h_4 = T_{12,1}U_{3,23} - \underline{T_{13,2}U_{2,13}}$
and $h_4 \in G_3$. So $S(h_1, h_3)$ reduces to zero by G_3. In any case all S-pairs reduce to zero by $G_1 \cup G_2 \cup G_3$.

Remark 3.5. No information we have about the order on monomials of the ring in the examples 3.1, 3.2. We don't need that order to find the Gröbner basis of the toric ideal I. Nevertheless, it would be interesting to know a new order that is compatible with the initial term of the marked binomials for simple examples as 3.1 and 3.2.

References

1. W. Fulton, *Introduction to Toric Varieties.* (Princeton University Press, 1993)
2. T. Oda, *Convex Bodies and Algebraic Geometry: an Introduction to the theory of Toric Varieties*, (Springer Verlag, New York,1998)
3. B. Sturmfels, *Gröbner Bases and Convex Polytopes*, (American Mathematical Society, 1996)
4. G.Restuccia and R.H. Villarreal, *On the normality of monomial ideals of mixed products*, Comunications in Algebra, **29(8)**, 3571 (2001)
5. R.H. Villarreal, *Monomial algebras*, Pure and Applied Mathematics, 2000.

GEOMETRIC MULTISCALE APPROACH BY OPTIMAL CONTROL FOR SHALLOW WATER EQUATIONS

FAUSTO SALERI*, EDIE MIGLIO*

*Laboratorio MOX, Dipartimento di Matematica, Politecnico di Milano
Via Bonardi 9, 20133, Milano, Italia
e-mail: {fausto.saleri,edie.miglio}@mate.polimi.it
web page: http://mox.polimi.it

In this paper we consider the coupling between one-dimensional and two-dimensional shallow water models. To this aim we introduce a domain decomposition technique with overlap to solve the coupled 1D-2D shallow water system. The proposed method is based on the introduction of suitable boundary control functions and generalizes, to the situation of heterogeneous differential problems in space, the method proposed in,[1] when the heterogeneity is confined at the differential level. We present some numerical results to assess the effectiveness of the proposed method.

Keywords: Optimal control; shallow water; heterogeneous domain decomposition.

1. Introduction

The heterogeneous domain decomposition method is a well established procedure to simulate phenomena that can be modeled by different partial differential equations in different regions of the computational domain (see, for an exahustive introduction,[2]). In the framework of the overlapping heterogeneous domain decomposition methods the so-called virtual control technique has been introduced in the last years to treat heterogeneous problems in an elegant mathematical context. The idea is based on the introduction of a control function on the subdomain interfaces which ensures that the solutions match on the overlapping region. In particular, starting from the papers,[34] and,[5] the method has been generalized to different situations such as pure advection and advection-diffusion equations,[1],[6] the plate blending problem[7] and fourth-order problems[8] ,[9].

In this paper we extend this approach to the case of the coupling of problems that are heterogeneous in space and in particular, we consider the coupling between one-dimensional and two-dimensional shallow water equations. This problem is motivated by the fact that in some physical situations, like harbours or rivers, one-dimensional hydrostatic models are satisfactory in a large part of the computational domain and only in some small regions, like bifurcations, a two-dimensional model is required. The problem is to find a suitable placement for the interfaces between 1D and 2D models: for this reason a coupling approach that admits in the overlapped regions the coexistence of 1D and 2D effects seems more robust that one based on a nonoverlapped decomposition (as proposed in[10] or, with a completely different approach, in[11]).

This paper provides a mathematical set up for the treatment of the coupling between 1D and 2D shallow water equations in overlapping subdomains and provides a first numerical assesment of the proposed method in an academic test cases.

2. The shallow water models and their discretization in time

In this section we introduce the shallow water models that we consider for the heterogeneous coupling.

2.1. *The 2D model*

We denote by Ω_2 an open limited regular set of $\mathbb{R}^2$ with boundary $\partial\Omega_2 = \Gamma_{in} \cup \Gamma_{out} \cup \Gamma_c$, $\mathbf{x} = (x_1, x_2)^T = (x, y)^T$, $\mathbf{x} \in \bar{\Omega}_2$ (see Figure 1). In Ω_2 the following *shallow water* system will be considered: for any $t > 0$, find $(\mathbf{u}, \xi)$ such that

$$\begin{cases} \dfrac{\partial \mathbf{u}}{\partial t} + g\nabla\xi + R(\mathbf{u})\mathbf{u} = \mathbf{f} \quad \text{in } \Omega_2, \\[2ex] \dfrac{\partial \xi}{\partial t} + \nabla \cdot (h\mathbf{u}) = 0 \quad \text{in } \Omega_2, \end{cases} \tag{1}$$

where $\mathbf{u} = (u, v)^T$ represents the average velocity of the fluid along the vertical direction, ξ is the position of the free surface (the elevation) with respect to an horizontal reference plane, $h = h_0 + \xi$ is the total depth, $-h_0$ is the position of the bottom with respect to the same reference level (see Figure 1, right), g is the gravity acceleration, $\mathbf{f} = (f_1, f_2)^T$ is a given source term and R is a positive function that models the friction effects on the bottom. We assume that $R(\mathbf{u}) = R_*|\mathbf{u}|/h$, where $R_* > 0$ is the friction coefficient, $|\mathbf{u}|^2 = (u^2 + v^2)$.

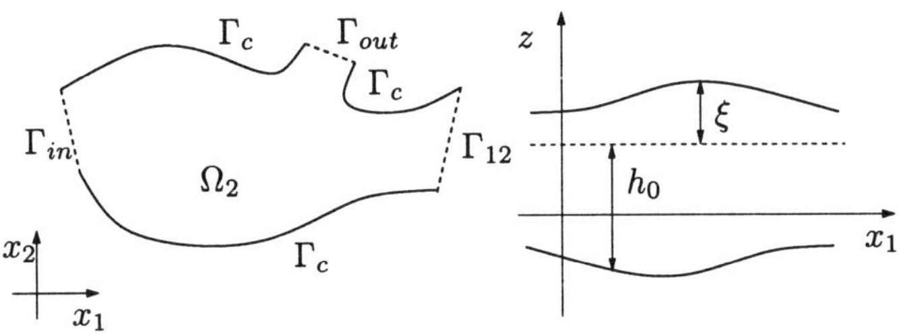

Fig. 1. Notations in the two-dimensional case

System (1) can be regarded as a model for an hydrostatic free surface fluid in which the convective (see Remark 2.1) and the diffusive terms in the momentum equation have been neglected.

From the mathematical view point (1) is a strictly hyperbolic system (see, e.g.,[12]) and should be completed with suitable boundary conditions (see Section 2.1.1) and the following initial conditions $\mathbf{u}(\mathbf{x},0) = \mathbf{u}_0(\mathbf{x})$, $\xi(\mathbf{x},0) = \xi_0(\mathbf{x})$ for any $\mathbf{x} \in \bar{\Omega}_2$, where $\mathbf{u}_0$ and ξ_0 are given functions. Since we are interested in the simulation of rivers or channels we can assume that the flow is sub-critical *i.e.* the Froude number $Fr = U/\sqrt{g\tilde{H}}$ is less than 1 (usually $0.1 \div 0.2$), where U is a characteristic speed of the fluid in the channel and $\tilde{H}$ is a typical vertical length scale.

In particular, we identify with Γ_{in} the part of the boundary $\partial\Omega_2$ that presents two incoming characteristics (the inlet), with Γ_{out} the part with one incoming characteristic (the outlet) and with Γ_c the closed part of the boundary (for instance, the coast) where we impose the slip condition $\mathbf{u}\cdot\mathbf{n} = 0$, $\mathbf{n} = (n_1, n_2)^T$ beeing the outward unit normal vector to $\partial\Omega_2$.

2.1.1. *Time-discretization*

To discretize in time system (1) we consider the following semi-implicit first order time advancing scheme: given $\mathbf{u}^{(0)} = (u_0, v_0)^T$ and $\xi^{(0)} = \xi_0$ for any $n \geq 0$ find

$$\begin{cases} \dfrac{u^{(n+1)} - u^{(n)}}{\Delta t} - lv^{(n+1)} + g\dfrac{\partial \xi^{(n+1)}}{\partial x_1} + R(\mathbf{u}^{(n)})u^{(n+1)} = f_1^{n+1}, \\ \dfrac{v^{(n+1)} - v^{(n)}}{\Delta t} + lu^{(n+1)} + g\dfrac{\partial \xi^{(n+1)}}{\partial x_2} + R(\mathbf{u}^{(n)})v^{(n+1)} = f_2^{n+1}, \\ \dfrac{\xi^{(n+1)} - \xi^{(n)}}{\Delta t} + \dfrac{\partial}{\partial x_1}\left(h^{(n)}u^{(n+1)}\right) + \dfrac{\partial}{\partial x_2}\left(h^{(n)}v^{(n+1)}\right) = 0, \end{cases} \qquad (2)$$

where $\Delta t > 0$, $f_i^{n+1} = f_i(t^{n+1})$, $i = 1, 2$, $t^n = n\Delta t$ and the dependence by $\mathbf{x}$ is understood. Setting

$$\alpha_1 = 1 + R(\mathbf{u}^{(n)})\Delta t, \qquad \beta_1 = l\Delta t,$$

$$g_1 = \Delta t f_1^{n+1} + u^{(n)} - g\Delta t \frac{\partial \xi^{(n+1)}}{\partial x_1}, \; g_2 = \Delta t f_2^{n+1} + v^{(n)} - g\Delta t \frac{\partial \xi^{(n+1)}}{\partial x_2},$$

we can deduce the following expression for the components of the velocity field

$$u^{(n+1)} = \frac{\alpha_1 g_1 + \beta_1 g_2}{\alpha_1^2 + \beta_1^2}, \; v^{(n+1)} = \frac{\alpha_1 g_2 - \beta_1 g_1}{\alpha_1^2 + \beta_1^2}. \tag{3}$$

Using these expressions on the continuity equation under suitable regularity assumptions on the unknowns, we obtain the following elliptic problem for the elevation $\xi^{(n+1)}$

$$-\nabla \cdot (A \nabla \xi^{(n+1)}) + \sigma \xi^{(n+1)} = \sigma \xi^{(n)} - \frac{\partial F_1}{\partial x_1} - \frac{\partial F_2}{\partial x_2}, \tag{4}$$

where $\sigma = 1/\Delta t$, $A = (a_{ij})$ with

$$a_{11} = \frac{g\alpha_1 h^{(n)} \Delta t}{\alpha_1^2 + \beta_1^2}, \; a_{12} = \frac{g\beta_1 h^{(n)} \Delta t}{\alpha_1^2 + \beta_1^2}, \; a_{21} = -a_{12}, \; a_{22} = a_{11},$$

and

$$F_1 = h^{(n)} \left[\alpha_1 (\Delta t f_1^{n+1} + u^{(n)}) + \beta_1 (\Delta t f_2^{n+1} + v^{(n)}) \right] / (\alpha_1^2 + \beta_1^2),$$

$$F_2 = h^{(n)} \left[\alpha_1 (\Delta t f_2^{n+1} + v^{(n)}) - \beta_1 (\Delta t f_1^{n+1} + u^{(n)}) \right] / (\alpha_1^2 + \beta_1^2).$$

Therefore, at each time-step the computation of $(\mathbf{u}^{(n+1)}, \xi^{(n+1)})$ requires the solution of the unsymmetrical elliptic problem (4) and the use of the relations (3) in order to reconstruct the velocity field. Problem (4) requires the imposition of a condition on $\xi^{(n+1)}$ on the whole boundary of Ω_2. Therefore, any boundary condition on the shallow water system (1) should be converted in a boundary condition for $\xi^{(n+1)}$. We have:

(1) on Γ_c we impose on the shallow water system the slip condition $\mathbf{u} \cdot \mathbf{n} = 0$. Therefore, from (3) we find that

$$0 = \mathbf{u}^{(n+1)} \cdot \mathbf{n} = \frac{1}{\alpha_1^2 + \beta_1^2} \left[(\alpha_1 g_2 - \beta_1 g_1) n_2 + (\alpha_1 g_1 + \beta_1 g_2) n_1 \right].$$

However,

$$\frac{h^{(n)}(\alpha_1 g_2 - \beta_1 g_1)}{\alpha_1^2 + \beta_1^2} = F_2 - a_{22} \frac{\partial \xi^{(n+1)}}{\partial x_2} - a_{21} \frac{\partial \xi^{(n+1)}}{\partial x_1},$$

$$\frac{h^{(n)}(\alpha_1 g_1 + \beta_1 g_2)}{\alpha_1^2 + \beta_1^2} = F_1 - a_{11} \frac{\partial \xi^{(n+1)}}{\partial x_1} - a_{12} \frac{\partial \xi^{(n+1)}}{\partial x_2},$$

thus the slip-condition is equivalent to the following non-homogeneous natural condition for the elliptic problem (4)

$$-A\nabla\xi^{(n+1)}\cdot\mathbf{n} = -\mathbf{F}\cdot\mathbf{n};$$

(2) on Γ_{out} we impose the Dirichlet boundary condition $\xi^{(n+1)} = \psi(t^{n+1})$ (where ψ is a given function) which is an essential condition for (4). In particular, we denote by Γ_{21} a subset of Γ_{out} where $\psi(t^{n+1}) = \lambda_2(t^{n+1})$. Note that in the following λ_2 will be an unknown as well as $\xi^{(n+1)}$;

(3) on Γ_{in} we impose a Dirichlet boundary condition for $\mathbf{u}^{(n+1)}$, that is $\mathbf{u}^{(n+1)} = \mathbf{u}_d(t^{n+1})$ for a given function $\mathbf{u}_d$. We project this condition along the outward normal direction to the boundary obtaining the following Neumann condition for (4)

$$-A\nabla\xi^{(n+1)}\cdot\mathbf{n} = -\mathbf{F}\cdot\mathbf{n} + h^{(n)}\mathbf{u}_d(t^{n+1})\cdot\mathbf{n}. \tag{5}$$

Remark 2.1 (convective term). *Usually the convective term* $(\mathbf{u}\cdot\nabla)\mathbf{u}$ *cannot be neglected in the momentum equation* $(1)_1$*. They can be easily included in the previous scheme if an explicit treatment (for instance using a Lagrangian approach see* e.g.[13]*) is considered.*

2.2. *The 1D model*

In a domain $\omega_1 \subset \mathbb{R}$ we consider the following one-dimensional problem (which is the one-dimensional counterpart of (1)): for any $t > 0$ find (v, η) such that

$$\begin{cases} \dfrac{\partial u}{\partial t} + g\dfrac{\partial \eta}{\partial x} + r(u)u = f \text{ in } \omega_1, \\ \dfrac{\partial \eta}{\partial t} + \dfrac{\partial}{\partial x}(ku) = 0 \qquad \text{in } \omega_1, \end{cases} \tag{6}$$

where u is the velocity field, η is the elevation, $k = \eta + k_0$ is the total depth of the water, $-k_0$ is the position of the bottom with respect an horizontal reference level, r is a positive function to model the friction effects (we assume $r(u) = r_*|u|/k$ with $r_* = const > 0$) and f is a given external force. System (6) describes the motion of a fluid in a channel with rectangular cross-section.

System (6) is strictly hyperbolic. As for the two-dimensional system we assume that the flow is subcritical and therefore one boundary condition is required at each end-point of ω_1. We denote the boundary of ω_1 by $\partial\omega_1 = \gamma_{in} \cup \gamma_{out}$. The boundary conditions for the system (6) are: $\eta = \eta_d(t)$ at γ_{out} and $\eta = \lambda_1(t)$ at γ_{in} for any $t > 0$, where $\eta_d(t)$ and $\lambda_1(t)$ are two given functions (in the sequel λ_1 will be unknown).

2.2.1. *Time-discretization*

For the time discretization of the system (6) we consider the same time-advancing scheme introduced in the previous section for the 2D system (1). Therefore, at each discrete level t^{n+1} we want to compute $(u^{(n+1)}, \eta^{(n+1)})$ such that

$$\begin{cases} \dfrac{u^{(n+1)} - u^{(n)}}{\Delta t} + g\dfrac{\partial \eta^{(n+1)}}{\partial x} + r(u^{(n)})u^{(n+1)} = f^{(n+1)}, \\ \dfrac{\eta^{(n+1)} - \eta^{(n)}}{\Delta t} + \dfrac{\partial (k^{(n)}u^{(n+1)})}{\partial x} = 0. \end{cases} \tag{7}$$

With similar computation to those of the one-dimensional case, we reduce the solution of problem (7) to the solution of the following elliptic problem

$$-\frac{\partial}{\partial x}\left(a\frac{\partial \eta^{(n+1)}}{\partial x}\right) + \gamma\eta^{(n+1)} = \gamma\eta^{(n)} - \frac{\partial F_1^{1D}}{\partial x}, \tag{8}$$

where $\gamma = 1/\Delta t$, $a = g\alpha_{1D}\Delta t\ k^{(n)}$, $\alpha_{1D} = 1/(1 + \Delta t\ r(u^{(n)}))$ and $F_1^{1D} = \alpha_{1D}k^{(n)}(u^{(n)} + \Delta t f^{(n+1)})$. At each time step we require that the following boundary conditions are satisfied: $\eta^{(n+1)} = \lambda_1\ (= \lambda_1(t_{n+1}))$ on γ_{in}, $\eta^{(n+1)} = \eta_d(t_{n+1})$ on γ_{out}, where η_d is a given function.

3. The heterogeneous problem and the optimal control approach

In this section we consider a physical situation in which it seems to be reasonable to consider an heterogeneous shallow water model in the computational domain $\Omega = \Omega_1 \cup \Omega_2$, one-dimensional in a subset $\omega_1 \subset \Omega_1$, two-dimensional in Ω_2. To analyze the problem of coupling of these two problems, we accept that in a subregion $\Omega_{12} = \Omega_1 \cap \Omega_2 \neq \emptyset$ the one and two-dimensional models coexist.

For the sake of simplicity, in the exposition we refer to a simple situation in which the computational domain is a straight rectangular channel of width L and constant cross-section. In this case Ω_2 is a rectangle and ω_1 is the straight line $(\bar{x}_1, \bar{x}_3) \times L/2$ (see Figure 2), that is the one-dimensional representation of the channel. Moreover, in the sequel we will identify γ_{in} with $\bar{x}_1$ and γ_{out} with $\bar{x}_3$.

To exchange informations between the 1D and the 2D domain one need to introduce an extension operator. In particular, given a one-dimensional quantity w defined in ω_1, we denote with

$$E_{2D}w = \{w(x_1, t),\ \forall (x_1, x_2) \in \Omega_{12}\} \tag{9}$$

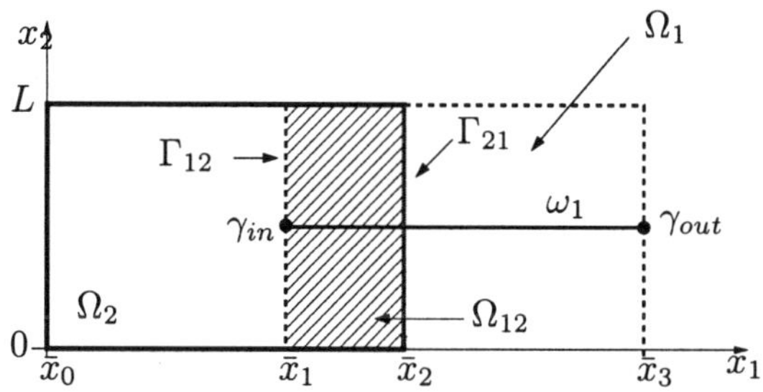

Fig. 2. The 2D-1D reference domain with the overlap region $\Omega_{12} = \Omega_1 \cap \Omega_2$ (shaded). In this case $\omega_1 = \{(\bar{x}_1, \bar{x}_3) \times L/2\}$, $\gamma_{in} = \{(\bar{x}_1, L/2)\}$ and $\gamma_{out} = \{(\bar{x}_3, L/2)\}$

the two-dimensional extension of w in the overlap region. Notice that, introducing the overlap region Ω_{12}, we assume implicitely that, if (u, η) is a solution of (8) then $\mathbf{u} = (E_{2D}u, 0)^T$ and $\xi = E_{2D}\eta$ are good approximations in Ω_{12} of the solution of the two dimensional shallow water system (1) (assuming $R_* = r_*$, $f_1 = E_{2D}f$, $f_2 = l\ E_{2D}u$).

To summarize, we want to solve the two-dimensional shallow water model (1) in Ω_2 and the one-dimensional one (6) in ω_1: using a boundary control approach. we solve this problem searching a "solution" in Ω_{12} that minimize in some sense the difference between the one-dimensional and the two-dimensional computed elevations. More in details, assuming that the solutions $(\mathbf{u}^{(n)}, \xi^{(n)})$ and $(u^{(n)}, \eta^{(n)})$ are available, then at time-step t^{n+1} compute

$$\begin{cases} -\nabla \cdot (A\nabla\xi^{(n+1)}) + \sigma\xi^{(n+1)} = \sigma\xi^{(n)} - \dfrac{\partial F_1}{\partial x_1} - \dfrac{\partial F_2}{\partial x_2} & \text{in } \Omega_2, \\ -A\nabla\xi^{(n+1)} \cdot \mathbf{n} = -\mathbf{F} \cdot \mathbf{n} & \text{on } \Gamma_c, \\ \xi^{(n+1)} = \psi(t^{n+1}) & \text{on } \Gamma_{in}, \\ -A\nabla\xi^{(n+1)} \cdot \mathbf{n} = -\mathbf{F} \cdot \mathbf{n} + h^{(n)}\mathbf{u}_d(t^{n+1}) \cdot \mathbf{n} & \text{on } \Gamma_{out}, \\ \xi^{(n+1)} = \lambda_2 & \text{on } \Gamma_{21}, \end{cases} \tag{10}$$

$$\begin{cases} -\dfrac{\partial}{\partial x}\left(a\dfrac{\partial\eta^{(n+1)}}{\partial x}\right) + \gamma\eta^{(n+1)} = \gamma\eta^{(n)} - \dfrac{\partial F_1^{1D}}{\partial x} & \text{in } \omega_1, \\ \eta^{(n+1)} = \eta_d(t^{n+1}) & \text{on } \gamma_{out}, \\ \eta^{(n+1)} = \lambda_1 & \text{on } \gamma_{in}, \end{cases} \tag{11}$$

where (λ_1, λ_2) are assumed now to be unknown functions (the *boundary control functions*). Notice that λ_1, that represents a one-dimensional elevation, is a constant function along Γ_{12}. To adjust the two-dimensional solution with the one-dimensional one we introduce the following minimization problem:

$$\inf_{\lambda_1,\lambda_2} J_\alpha(\lambda_1,\lambda_2), \tag{12}$$

where

$$J_\alpha(\lambda_1,\lambda_2) = \frac{\alpha}{2}\int_{\Gamma_{21}} \lambda_2^2 \, d\Gamma + \frac{\alpha}{2}\lambda_1^2 + J_0(\lambda_1,\lambda_2) \tag{13}$$

and

$$J_0(\lambda_1,\lambda_2) = \frac{1}{2}\int_{\Omega_{12}} (\xi^{(n+1)} - E_{2D}\eta^{(n+1)})^2 d\Omega.$$

The parameter α is a given nonnegative constant.

Note that if we are looking for a $\xi^{(n+1)}$ in the space $H^1(\Omega_2)$, we need to assume $\lambda_2 \in H^{1/2}(\Gamma_{21})$. Therefore, we denote by Λ the set of vectors $\boldsymbol{\lambda} = (\lambda_1,\lambda_2)$, where $\lambda_1 \in \mathbb{R}$ and λ_2 belongs in $H^{1/2}(\Gamma_{21})$ and the optimal control problem reads as:

find $\xi^{(n+1)}$, $\eta^{(n+1)}$ and $\boldsymbol{\lambda}$ such that (10)-(12) is satisfied.

In order to study the dependence of the solution on the boundary control $\boldsymbol{\lambda}$ we introduce some auxiliary problems.

Problem 1: find $\hat{\xi}^{(n+1)}$ such that

$$\begin{cases} -\nabla\cdot(A\nabla\hat{\xi}^{(n+1)}) + \sigma\hat{\xi}^{(n+1)} = \sigma\xi^{(n)} - \dfrac{\partial F_1}{\partial x_1} - \dfrac{\partial F_2}{\partial x_2} & \text{in } \Omega_2,\\ -A\nabla\hat{\xi}^{(n+1)}\cdot\mathbf{n} = -\mathbf{F}\cdot\mathbf{n} & \text{on } \Gamma_c,\\ \hat{\xi}^{(n+1)} = \psi(t^{n+1}) & \text{on } \Gamma_{in},\\ -A\nabla\hat{\xi}^{(n+1)}\cdot\mathbf{n} = -\mathbf{F}\cdot\mathbf{n} + h^{(n)}\mathbf{u}_d(t^{n+1})\cdot\mathbf{n} & \text{on } \Gamma_{out},\\ \hat{\xi}^{(n+1)} = 0 & \text{on } \Gamma_{21}. \end{cases} \tag{14}$$

Problem 2: find $\tilde{\xi}^{(n+1)}$ such that

$$\begin{cases} -\nabla\cdot(A\nabla\tilde{\xi}^{(n+1)}) + \sigma\tilde{\xi}^{(n+1)} = 0 \text{ in } \Omega_2,\\ -A\nabla\tilde{\xi}^{(n+1)}\cdot\mathbf{n} = 0 & \text{on } \Gamma_c\cup\Gamma_{out},\\ \tilde{\xi}^{(n+1)} = 0 & \text{on } \Gamma_{in},\\ \tilde{\xi}^{(n+1)} = \lambda_2 & \text{on } \Gamma_{21}. \end{cases} \tag{15}$$

Problem 3: find $\hat{\eta}^{(n+1)}$ such that

$$\begin{cases} -\dfrac{\partial}{\partial x}\left(a\dfrac{\partial\hat{\eta}^{(n+1)}}{\partial x}\right)+\gamma\hat{\eta}^{(n+1)}=\gamma\eta^{(n)}-\dfrac{\partial F_1^{1D}}{\partial x} & \text{in } \omega_1, \\ \hat{\eta}^{(n+1)}=\eta_d(t^{n+1}) & \text{on } \gamma_{out}, \\ \hat{\eta}^{(n+1)}=0 & \text{on } \gamma_{in}. \end{cases} \tag{16}$$

Problem 4: find $\tilde{\eta}^{(n+1)}$ such that

$$\begin{cases} -\dfrac{\partial}{\partial x}\left(a\dfrac{\partial\tilde{\eta}^{(n+1)}}{\partial x}\right)+\gamma\tilde{\eta}^{(n+1)}=0 \text{ in } \omega_1, \\ \tilde{\eta}^{(n+1)}=0 & \text{on } \gamma_{out}, \\ \tilde{\eta}^{(n+1)}=\lambda_1 & \text{on } \gamma_{in}. \end{cases} \tag{17}$$

Using the solutions of the problems (14)-(17) we reformulate the optimal control problem as follows: find $\tilde{\xi}^{(n+1)}$, $\tilde{\eta}^{(n+1)}$ and $\boldsymbol{\lambda}$ such that the equations (15), (17) are satisfied and

$$\inf_{\lambda_1,\lambda_2}\tilde{J}_\alpha(\lambda_1,\lambda_2), \tag{18}$$

where

$$\tilde{J}_\alpha(\lambda_1,\lambda_2)=\frac{\alpha}{2}\int\limits_{\Gamma_{21}}\lambda_2^2\,d\Gamma+\frac{\alpha}{2}\lambda_1^2+\tilde{J}_0(\lambda_1,\lambda_2)$$

with

$$\tilde{J}_0(\lambda_1,\lambda_2)=\frac{1}{2}\int\limits_{\Omega_{12}}(\tilde{\xi}^{(n+1)}-E_{2D}\tilde{\eta}^{(n+1)}+G^{(n+1)})^2d\Omega,$$
$$G^{(n+1)}=\hat{\xi}^{(n+1)}-E_{2D}\hat{\eta}^{(n+1)},$$

where the extension operator E_{2D} is defined in (9).

4. Numerical results

We present some numerical results in order to evaluate the proposed method in the special situation in which the one-dimensional channels are straight rectangular channels with constant width.

For the space discretization of the shallow water systems we have considered linear finite elements (see for instance[13]). To solve the optimality problem we use a conjugate gradient algorithm that requires at each iteration the solution of the primal and the dual problems.

The situation we are going to consider channel network of Figure 3.

The computational domain is decomposed into three parts: the two-dimensional one Ω_{2D}, the overlap regions Ω_i, $i = 1, \ldots, 6$, and the one-dimensional domain ω_i, $i = 1, \ldots, 4$. The domain is 800 meters lenght and the overlap regions are rectangles of lenght L, where L will be choose in order to modify the extension of the overlap region. We considere a quasi-uniform triangular grid of step $h = 5$ meters and a time-step $\Delta t = 120$ seconds.

On the vertical boundaries of the computational domain we impose an obscillatory behavior for the elevation in order to generate interacting waves expecially near the bifurcation, while in the other part of the boundary we impose $\mathbf{u} \cdot \mathbf{n} = 0$.

We consider two cases: $L = 10$ (which corresponds to have only three nodes on the overlap regions in the one-dimensional horizontal section) and $L = 20$. The number of the control unknowns is 23 and 25, respectively. The number of iterations required by the conjugate gradient method to obtain a residual less than 10^{-5} is approximately 43 in both cases (we consider an average in time of the number of iterations required at each time-step), while the computational time decreases as L decreases.

As expected, as the overlap decreases the ability of the heterogeneous approach to reproduce the correct two-dimensional solution decreases. Even so, for the minum overlap $L = 10$ the correspondence between the 1D-2D elevation with the full two-dimensional is satisfying as shown in Figures 4 where the elevation is represented as a function of the time at the control points $A = (207.9099, 2.4310)$, $B = (630.2859, 142.4734)$ and $C = (210.1170, 141.5462)$ (see Figure 3). A different situation occurs when we compare the horizontal velocities at the same points (Figure 5): with the minum overlap $L = 10$ a big difference between the 1D-2D and the full

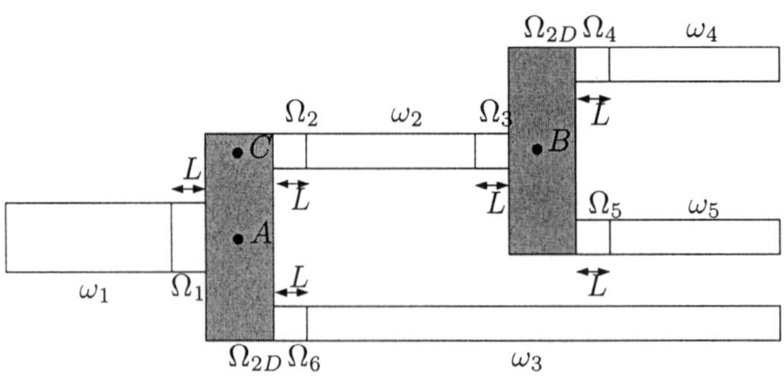

Fig. 3. The computational domain and the control points

2D velocities is present. Notice that in our approach only the elevation is controlled. Nevertheless, increasing the overlap the compairison is satisfying also for the velocity field (see the situation for $L = 20$ in Figure 6)

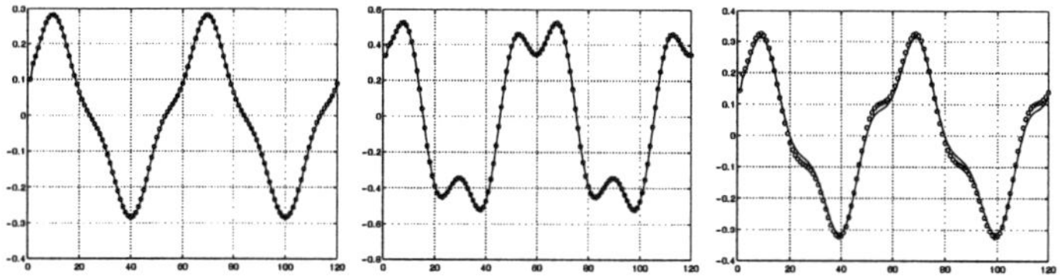

Fig. 4. A comparison between the computed elevation at the control points A, B and C (left, center and right) for the two-dimensional (circles) and the 1D-2D (solid line) model when $L = 10$

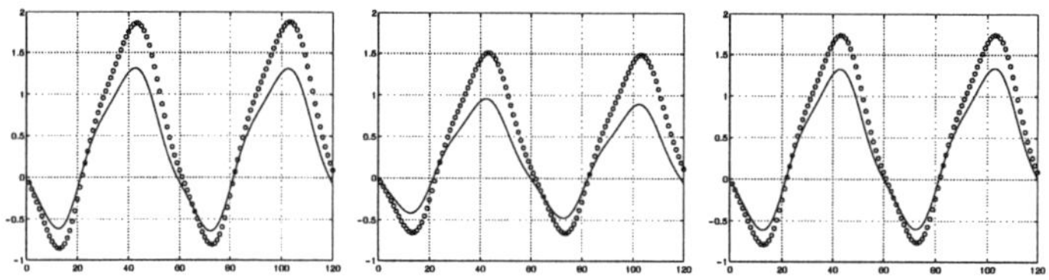

Fig. 5. A comparison between the horizontal component of the velocity at the control points A, B and C (left, center and right) for the two-dimensional (circles) and the 1D-2D (solid line) model when $L = 10$

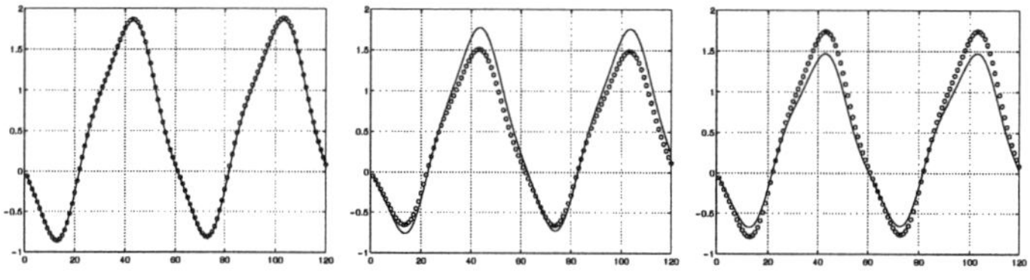

Fig. 6. A comparison between the horizontal component of the velocity at the control points A, B and C (left, center and right) for the two-dimensional (circles) and the 1D-2D (solid line) model when $L = 20$

5. Conclusions

In this paper we have introduced an optimal control approach for the coupling between 1D and 2D shallow water systems assuming an overlap region between the 1D and the 2D domain. Some preliminary numerical results shows the effectiveness of the proposed method.

References

1. V. Agoshkov, P. Gervasio and A. Quarteroni, *Russ. J. Numer Anal. Math. Modelling* **20**, 229 (2005).
2. A. Quarteroni and A. Valli, *Domain Decomposition Methods for Partial Differential Equations* (Oxford Science Publications, 1999).
3. J. Lions and O. Pironneau, *C. R. Acad. Sci. Paris Sèr. I Math. t.* **327**, 947 (1998).
4. J. Lions and O. Pironneau, *C. R. Acad. Sci. Paris Sèr. I Math. t.* **327**, 993 (1998).
5. J. Lions and O. Pironneau, *C. R. Acad. Sci. Paris Sèr. I Math. t.* **328**, 73 (1999).
6. P. Gervasio, J.-L. Lions and A. Quarteroni, *Numerische Mathematik* , 241 (2001).
7. P. Gervasio, *Comput. Methods Appl. Mech. Engrg.* **194**, 4321 (2005).
8. P. Gervasio, *Virtual Control for Fourth-Order Problems and for Heterogeneous Fourth-Order Second-Order Coupling*, in *Numerical Mathematics and Advanced Applications*, eds. F. Brezzi, A. Buffa, S. Corsaro and A. Murli (Springer, 2003), pp. 827–836.
9. P. Gervasio, J.-L. Lions and A. Quarteroni, Domain decomposition and virtual control for fourth order problems, in *Proceedings of 13th International Conference on Domain Decomposition Methods*, (Barcelona, Spain, 2002).
10. E. Miglio, S. Perotto and F. Saleri, *Nonlinear Analysis* **63**, 1885 (2005).
11. M. Amara, D. C. Papaghiuc and D. Trujillo, *Comput. Visual. Sci.* **6** (2004).
12. G. Whitham, *Linear and non linear waves* (Wiley-Interscience, Ney York, 1974).
13. V. Agoshkov, E. Ovchinnikov, A. Quarteroni and F. Saleri, *Mathematical Models and Methods in Applied Sciences* **4**, 533 (1994).

POLAR SITTER MISSION FOR CONTINUOUS OBSERVATION OF THE POLES

S. SGUBINI and S. PORFILI and C. CIRCI*

Scuola di Ingegneria Aerospaziale, Università di Roma La Sapienza
Via Eudossiana 16, 00184, Roma, Italy
**E-mail: silvano.sgubini@uniroma1.it; silvia.porfili@gmail.com; christian.circi@uniroma1.it*

The Polar Sitter mission provides a continuous observation of one of the Earth poles throughout the year. In order to perform this mission, equilibrium points of the circular restricted 3-body problem with radiation pressure have been chosen as candidate positions for the satellite. The tilt of the Earth spin axis with respect to the ecliptic plane prevents the sailcraft from keeping a constant equilibrium location over the poles, thus causing increasing values of the angle-of-sight between the satellite position and the North Pole. In this paper we show that, by studying the equilibrium points obtained with the use of solar sail propulsion, a minimisation of the angle-of-sight is possible, providing a significant improvement in the orbit geometry .

Keywords: Solar sailing; polar sitter mission; space engineering;

Introduction

Solar sails are very light structures made of high reflective materials which are capable of using solar radiation for the primary populsion system of a spacecraft. By producing a small but constant acceleration they can perform exotic orbits which would be unpractical with other propulsion systems.[1,2] Equilibrium positions which are solutions of the restricted 3-body problem of a Sun-Earth-sailcraft system can be exploited so as to observe high latitude regions of the Earth with a year long orbit around the Sun, as shown in Fig. 1. From such a position the weather, the magnetospheric conditions and other unique polar phenomena could be continuously monitored. In summer a satellite can be stationed directly over the North Pole, providing the best observing situation however, the tilt of the Earth spin axis with respect to the ecliptic plane doesn't allow the satellite to keep a permanent position on the polar axis during the year. This work aims at improving

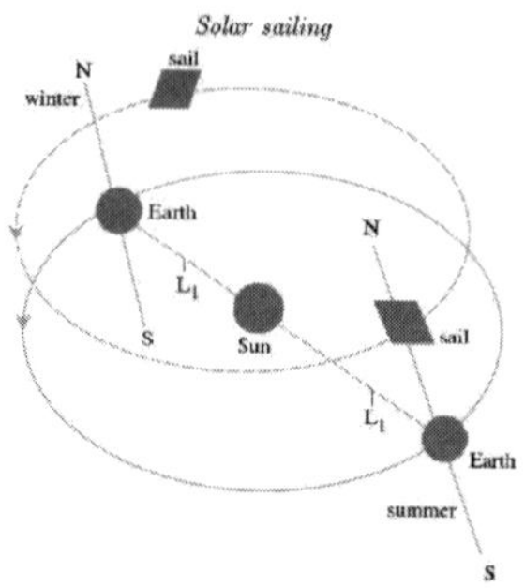

Fig. 1. Polar Sitter mission in a year

the satellite orbit in order to prolong the period of direct observation of the Earth Poles. Sections are divided as follows: in Sec. 1 two models for the photonic pressure are presented so as to define the dynamics of the problem. In Sec. 2 the restricted 3-body problem, modified by the presence of the solar pressure force, is introduced. In Sec. 3 the results for the two models are shown. In Sec. 4 an optimisation of the equilibrium orbit is given.

1. Solar radiation pressure force

The momentum transmitted to the sail by the Solar energy generates the so-called photonic pressure. Considering a perfectly reflecting solar sail, Fig. 2, the expression for the solar pressure force, in mass units, is equal to:

$$\vec{a} = \frac{\vec{f}_P}{m} = \left[\frac{2}{\sigma} \frac{W_E}{c} \left(\frac{R_E}{r_1} \right)^2 \cos^2 \alpha \right] \hat{n} \tag{1}$$

where R_E is the Sun-Earth distance, W_E is the mean solar energy flux at distance R_E from the Sun, $\vec{r}_1$ is the Sun-sail distance, c is the speed of light, $\hat{n}$ is the unit vector normal to the sail surface and the cone-angle α is the angle between $\hat{n}$ and $\vec{r}_1$. A very important parametrer which is commonly used to express the solar sail performance is the total solar sail mass per unit area or sail loading $\sigma = \frac{m_{sail}+m_{s/c}}{A_{sail}}$. In case the sail has a realistic reflectivity $\tilde{r} < 1$, Fig. 3, considering all other forms of reflection negligible, another expression for the solar pressure force is obtained:

$$\vec{a} = \frac{(1+\tilde{r})}{\sigma} \frac{W_E}{c} \left(\frac{R_E}{r_1} \right)^2 \cos^2 \alpha \, \hat{n} + \frac{(1-\tilde{r})}{\sigma} \frac{W_E}{c} \left(\frac{R_E}{r_1} \right)^2 \cos \alpha \sin \alpha \, \hat{t} \tag{2}$$

The force vector $\vec{f}_P$ is no longer normal to the sail surface because a part of the radiation is absorbed and thus not all the incident radiation is reflected.

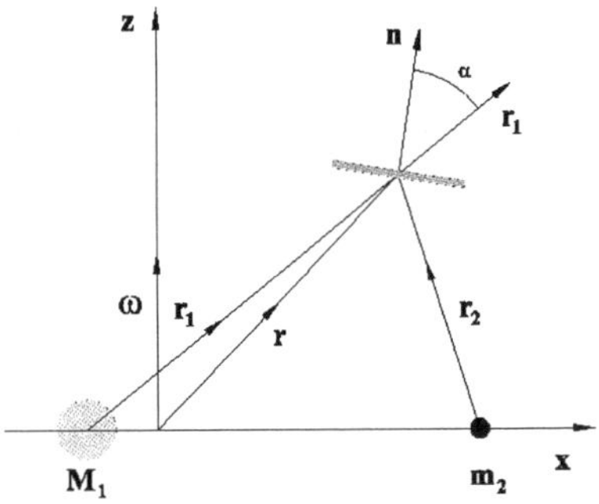

Fig. 2. Perfectly reflecting solar sail

Therefore, the resulting photonic pressure force is inclined of an angle ϕ in the direction of the vector $\vec{r}_1$ and is also reduced in magnitude with respect to the case of a perfectly reflecting solar sail.

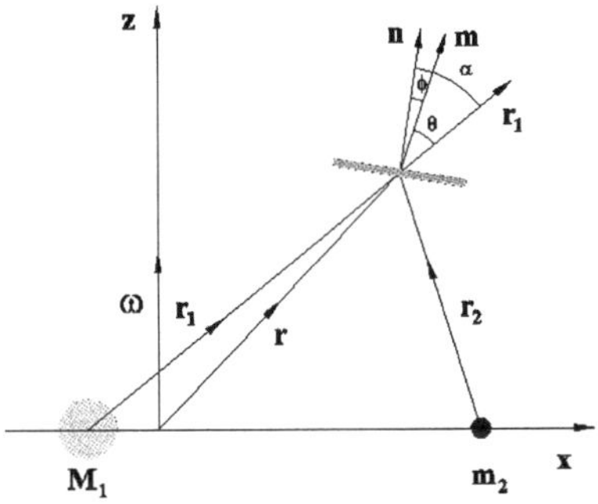

Fig. 3. Partially reflecting solar sail

2. Restricted 3-Body problem with solar pressure force

Solar sails can produce equilibrium solutions in the 3-body Sun-Earth-sail system which can be used so as to accomplish the Polar Sitter mission. The reference system adopted is a frame co-rotating with the primary bodies of mass m_1 (Sun) and m_2 (Earth). The frame is chosen so that it has the primary bodies placed along the x axis. The z axis is normal to the orbit plane and is in the direction of the angular speed vector ω. The y axis is always chosen so as to form a right-handed set of coordinate axes. The distance between the primaries and their total mass is set to unity, i. e.

$r_{12} = 1$ and $m_1 + m_2 = 1$. Having introduced the quantity $\mu = \frac{m_2}{m_1+m_2}$, it follows that $m_1 = 1 - \mu$ and $m_2 = \mu$. Similarly, the gravitational constant $G = 1$ so that we obtain that the co-rotating frame has a constant angular speed $\omega = 1$ and its period of rotation is 2π. The equation of motion for a solar sail in this reference frame is:

$$\ddot{\vec{r}} + 2\vec{\omega} \times \dot{\vec{r}} + \overrightarrow{\nabla U} = \vec{a} \tag{3}$$

where $\vec{r}$ is the position of the solar sail in the rotating frame and the potential U has the following expression:

$$U = -\left[\frac{1}{2}\left(x^2 + y^2\right) + \frac{1-\mu}{r_1} + \frac{\mu}{r_2}\right] \tag{4}$$

The acceleration due to the photonic pressure force is $\vec{a}$. Equilibrium solutions can be obtained when the first two terms of Eq. 3 vanish:

$$\overrightarrow{\nabla U} = \vec{a} \tag{5}$$

The regions of space where these new solutions exist are characterised by the condition:

$$\hat{r}_1 \cdot \overrightarrow{\nabla U} \geq 0 \tag{6}$$

which can be interpreted physically as the impossibility to point the solar radiation pressure acceleration vector directly towards the Sun. In order to study the equilibria the two reflection models are considered separately: firstly a perfectly reflecting solar sail is examined and then a more realistic sail with a partial reflectivity of order 0.9 is introduced.

Perfectly reflecting sail

In this case, the solar radiation pressure acceleration can be written in the following form:[3,4]

$$\vec{a} = \beta \frac{1-\mu}{r_1^2} \left(\hat{r}_1 \cdot \hat{n}\right)^2 \hat{n} \tag{7}$$

where the adimensional parameter β (lightness number) is the ratio of the solar radiation pressure acceleration at the distance of the Earth from the Sun, when the direction of the normal to the sail surface and the direction of the incident radiation are parallel (this value is commonly called characteristic acceleration), to the the gravitational acceleration due to the presence of the Sun, i. e.:

$$\beta = \frac{a_c}{a_g} = \frac{2}{\sigma} \frac{W_E}{c} \frac{R_E^2}{GM_1} = \frac{1.53}{\sigma} \tag{8}$$

It can be seen that there is an inverse relationship between the lightness number β and the sail loading σ. From now on we use the lightness number β, whenever possible, so as to work with dimensionless numbers. In this configuration the acceleration $\vec{a}$ has the same direction of unit vector $\hat{n}$, as shown in Fig. 2, and we can consider:

$$\hat{n} = \frac{\overrightarrow{\nabla U}}{\left|\overrightarrow{\nabla U}\right|} \qquad since \qquad |\hat{n}| = 1 \tag{9}$$

which can be used to express the sail attitude, defined with respect to the co-ordinate triad $(\hat{r}_1, \hat{\omega} \times \hat{r}_1, \hat{r}_1 \times (\hat{\omega} \times \hat{r}_1))$ centered on the sail, in terms of α and δ, Fig. 4.

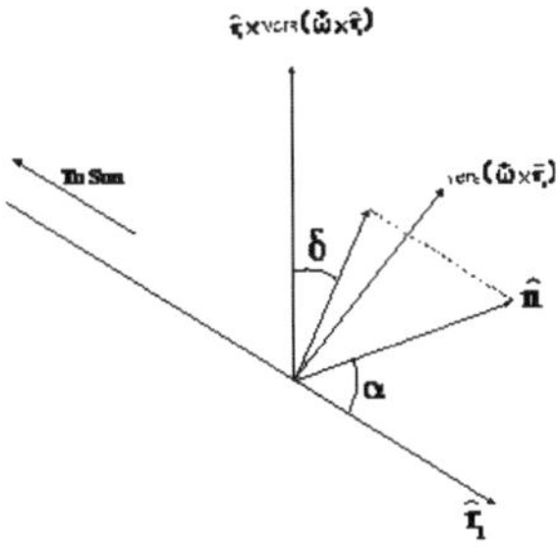

Fig. 4. Sail attitude in term of α and δ angles

$$\hat{n} = \cos\alpha\,\hat{r}_1 + \sin\alpha\sin\delta\;(\hat{\omega} \times \hat{r}_1) + \sin\alpha\cos\delta\;\hat{r}_1 \times (\hat{\omega} \times \hat{r}_1) \tag{10}$$

The cone-angle α can be optained as:

$$\tan\alpha = \frac{\left|\hat{r}_1 \times \overrightarrow{\nabla U}\right|}{\hat{r}_1 \cdot \overrightarrow{\nabla U}} \tag{11}$$

While, the clock-angle δ, which represents the angle between the projection of the unit vector $\hat{n}$ on the plane$(\hat{\omega} \times \hat{r}_1, \hat{r}_1 \times (\hat{\omega} \times \hat{r}_1))$ and the direction of the unit vector $(\hat{r}_1 \times (\hat{\omega} \times \hat{r}_1))$, can be written as:

$$\tan\delta = \frac{\left|[\hat{r}_1 \times (\hat{\omega} \times \hat{r}_1)] \times \left[\hat{r}_1 \times \left(\overrightarrow{\nabla U} \times \hat{r}_1\right)\right]\right|}{[\hat{r}_1 \times (\hat{\omega} \times \hat{r}_1)] \cdot \left[\hat{r}_1 \times \left(\overrightarrow{\nabla U} \times \hat{r}_1\right)\right]} \tag{12}$$

Using Eqs. 7 and 5 and taking a scalar product with unit vector $\hat{n}$, the lightness number β may be expressed as:

$$\beta = \frac{r_1^2}{1-\mu}\frac{\overrightarrow{\nabla U} \cdot \hat{n}}{(\hat{r}_1 \cdot \hat{n})^2} \tag{13}$$

With these positions, the set of five classical lagrange points is replaced by an infinite set of equilibrium solutions, generated by selecting the parameter β and the attitude angles α and δ in Eqs. 11, 12 and 13.

Partially reflecting sail

Considering the effect due to the sail absorption of a part of the photonic energy, the reflectivity of the solar sail is then inferior to unity ($\tilde{r} \leq 1$). The expression of the photonic pressure acceleration is in this case:[3]

$$\vec{a} = \frac{1}{2}\beta\,\frac{1-\mu}{r_1^2}\,(1+\tilde{r})\,(\hat{r}_1\cdot\hat{n})^2\,\hat{n} + \frac{1}{2}\beta\,\frac{1-\mu}{r_1^2}\,(1-\tilde{r})\,(\hat{r}_1\cdot\hat{n})\,\left(\hat{r}_1\cdot\hat{t}\right)\hat{t} \quad (14)$$

Now the photonic pressure force is not in the direction of the unit vector $\hat{n}$ but is inclined towards the direction of the incident radiation and the acceleration vector of the solar sail is no longer normal to the sail surface, but it will act in the direction of the unit vector $\hat{m}$. This situation brings to the presence of the angles ϕ (centre-line angle) and θ (cone-angle), as shown in Fig. 3. Again, for an equilibrium solution, the first two terms of the equation of motion (Eq. 3) must vanish, it follows that:

$$\overrightarrow{\nabla U} = a\,\hat{m} \quad (15)$$

The acceleration vector $\vec{a}$ is now in direction $\hat{m}$, therefore:

$$\hat{m} = \frac{\overrightarrow{\nabla U}}{\left|\overrightarrow{\nabla U}\right|} \qquad \textit{since} \qquad |\hat{m}| = 1 \quad (16)$$

The unit vector $\hat{m}$ can be defined by the cone-angle θ as:

$$\tan\theta = \frac{|\hat{r}_1 \times \overrightarrow{\nabla U}|}{\hat{r}_1\cdot\overrightarrow{\nabla U}} \quad (17)$$

Similarly, according to Eq. 13, the centre-line angle ϕ can be expressed as:

$$\tan\phi = \frac{1-\tilde{r}}{1+\tilde{r}}\tan\alpha \quad (18)$$

since $\alpha = \theta + \phi$. Combining Eqs. 15 and 14 and taking a scalar product with $\hat{n}$, assuming that $\hat{n}$ and $\hat{t}$ are normal, the lightness number β can be obtained as:

$$\beta = \frac{2r_1^2}{1-\mu}\,\frac{\overrightarrow{\nabla U}\cdot\hat{n}}{(1+\tilde{r})\,(\hat{r}_1\cdot\hat{n})^2} \quad (19)$$

Using Eqs. 11, 12, 17, 18 and 19 the sail attitude and the lightness number required for an equilibrium position of the problem can be found.

3. Equilibrium points

In Figs. 5 and 6 equilibrium solutions are shown in a region close to the Earth, in a plane (x, z) normal to the ecliptic plane which contains the primary bodies, while in Figs. 7 and 8 equilibrium surfaces are represented in a space close to the Earth. The curves S_1 and S_2 separate the zone of

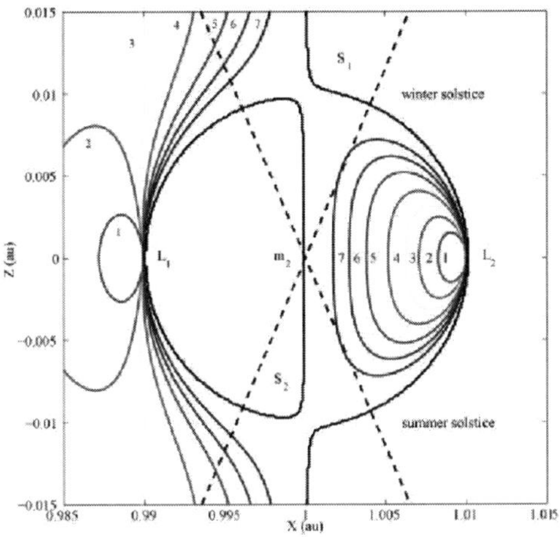

Fig. 5. Equilibrium solution contours in the $x - z$ plane (perfectly reflecting sail). Sail loading σ: [1] 76.500, [2] 38.250, [3] 25.500, [4] 15.300, [5] 7.650, [6] 3.825, [7] 1.530 (g/m^2)

existence of solutions from the zone without admissible solutions. The other curves are the equilibrium solutions at various values of the parameter β (or corresponding sail loading σ). Distances are in astronomical units (au). For a partially reflecting sail, it can be noted that close to the lagrange point L_1 solution curves are at greater distance from Earth compared to those for a perfectly reflecting sailcraft, whilst close to the lagrange point L_2, the zone of possible solutions is so highly reduced that there is no solution on the Earth polar axis. For this reason we focus our attention on solutions close to the lagrange point L_1.

4. Optimised performance of the Polar Sitter mission

At this point, a value of the sail loading and consequently of the Earth distance of the satellite has been selected. Figure 9 shows the distance of the satellite from Earth at summer solstice as a funcion of the sail loading. On x axis r_2 is the distance of the sailcraft from Earth, while on y axis is the corresponding σ. The chosen value for the sail loading σ is $10.2\frac{g}{m^2}$, which

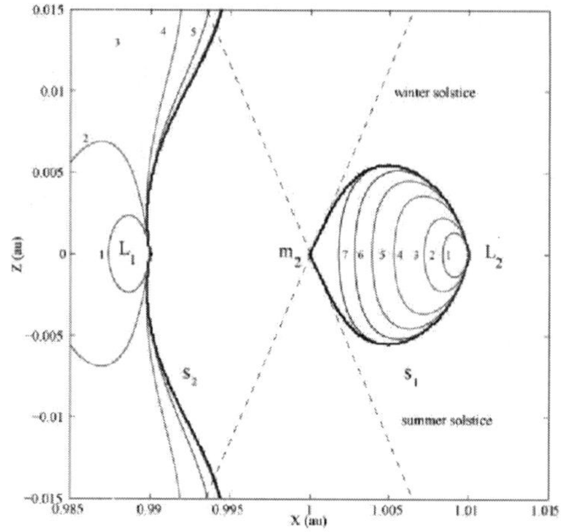

Fig. 6. Equilibrium solution contours in the $x-z$ plane (partially reflecting sail $\tilde{r} = 0.9$). Sail loading σ: [1] 76.500, [2] 38.250, [3] 25.500, [4] 15.300, [5] 7.650, [6] 3.825, [7] 1.530 (g/m^2)

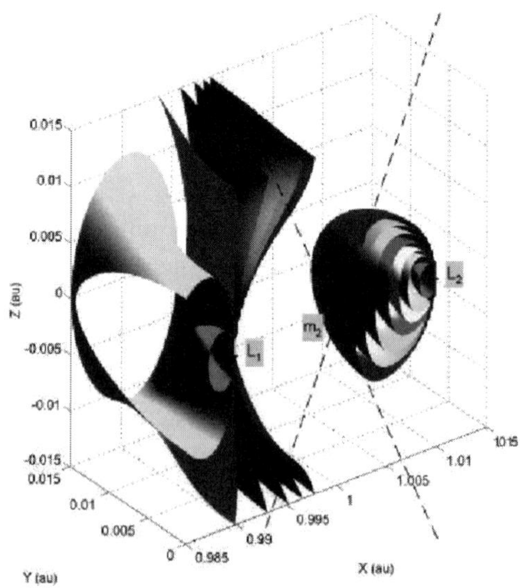

Fig. 7. Equilibrium solutions (perfectly reflecting sail). Sail loading σ: [1] 76.500, [2] 38.250, [3] 25.500, [4] 15.300, [5] 7.650, [6] 3.825, [7] 1.530 (g/m^2)

is obtained at a distance from the Earth of about $2.24 \times 10^6\ km$ at summer solstice. Figure 10 illustrates how equilibrium positions at summer solstice allow the sail to be directly above the North Pole, whilst the sailcraft is

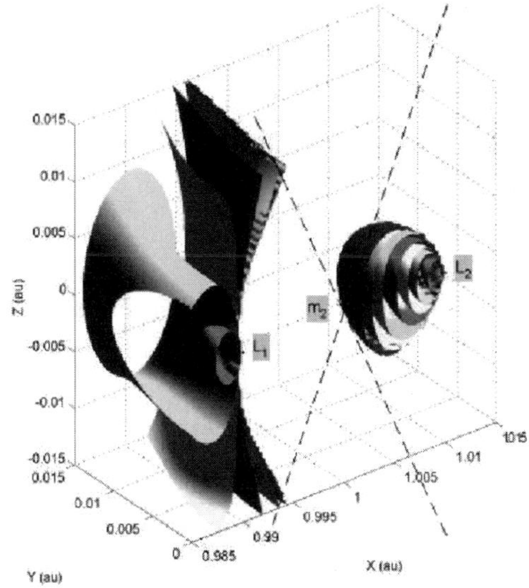

Fig. 8. Equilibrium solutions (partially reflecting sail). Sail loading σ: [1] 76.500, [2] 38.250, [3] 25.500, [4] 15.300, [5] 7.650, [6] 3.825, [7] 1.530 (g/m^2)

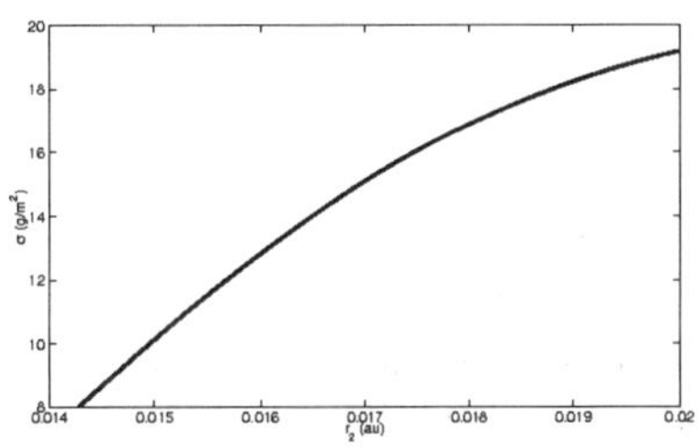

Fig. 9. Distance from the Earth at summer solstice of a sailcraft at various values of sail loading σ

in an opposite condition at winter solstice. It is then clear that at winter solstice, the angle-of-sight γ reaches the value $\gamma = 47°$,twice the inclination of the Earth spin axis with respect to the ecliptic. The aim of this work is to improve the visibility of the Earth polar regions for a perfectly reflecting sailcraft. The angle-of-sight γ is the parameter of interest. Once a chosen value for the sail loading σ has been selected, the angle γ depends on the solar sail attitude and therefore on the values of α and δ. The optimisation is expressed as: fixed $\sigma = \bar{\sigma} = 10.2 \frac{g}{m^2}$ determine α and δ so that $\gamma^* = min\ \gamma = min\ f(\bar{\sigma}, \alpha, \delta)$ with:

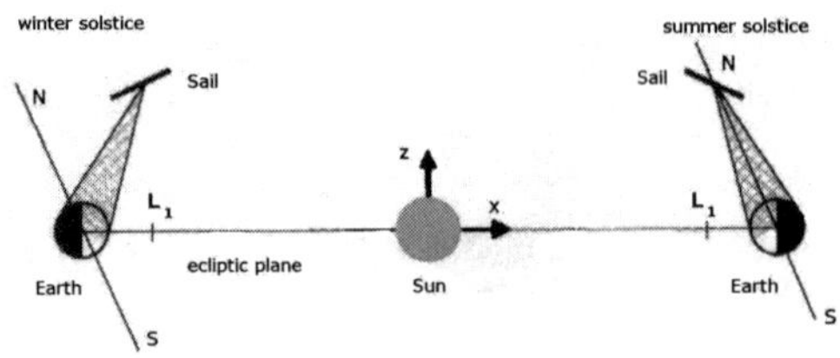

Fig. 10. Equilibrium posistions at summer solstice and at winter solstice

- 1. no limitation on the values of r_2
- 2. $r_2 \leq \bar{r}_2 = 0.0195\ au\ \left(2.9 \times 10^6\ km\right)$

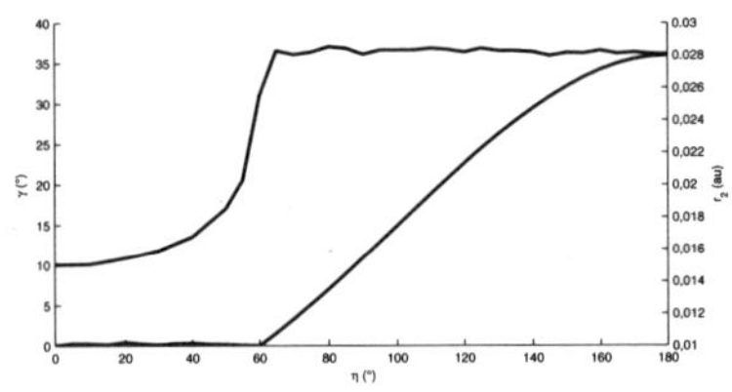

Fig. 11. Angle-of-sight γ and corresponding distance r_2 in half a year

The results obtained for the first case are shown in Fig. 11, where the distance of the sailcraft from Earth and the angle-of-sight γ are presented ($\eta = 0°$). Half a year is shown since the results behave symmetrically during the rest of the year. The maximum value of r_2 is $r_{2max} = 0.028\ au\ \left(4.2 \times 10^6\ km\right)$ and the angle-of-sight γ doesn't exceed the value $\gamma_{max} = 36°$. The second case $r_2 \leq \bar{r}_2 = 0.0195\ au\ \left(2.9 \times 10^6\ km\right)$ is shown

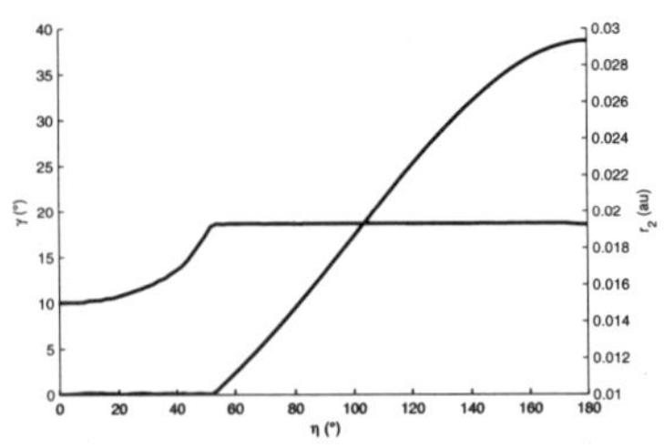

Fig. 12. Angle-of-sight γ and corresponding distance r_2 in half a year

in Fig. 12. The maximum distance from Earth reached by the satellite is $r_{2max} = 0.0195\ au\ \left(2.9 \times 10^6\ km\right)$ and the angle-of-sight γ doesn't exceed the value $\gamma_{max} = 39°$. In Figs. 13 and 14 the corresponding values for the

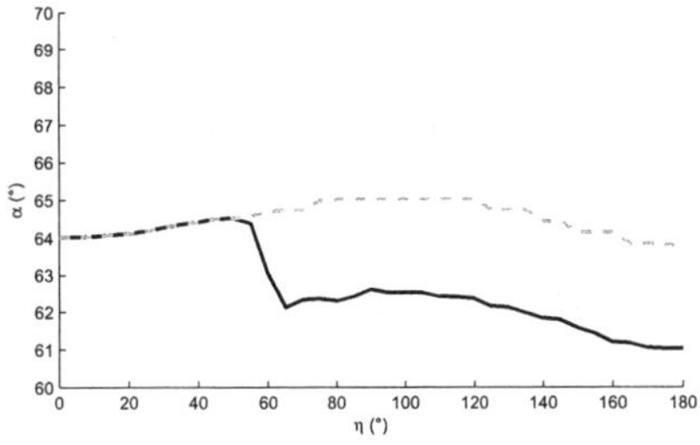

Fig. 13. Cone-angle α in half a year, first case (black) and second case (gray)

sail attitude angles are shown. The cone-angle α remains fixed around 60° in the first case and around 65° in the second one, whilst the clock-angle δ is almost negligible and always below 7° in the first case and below 8° in the second one.

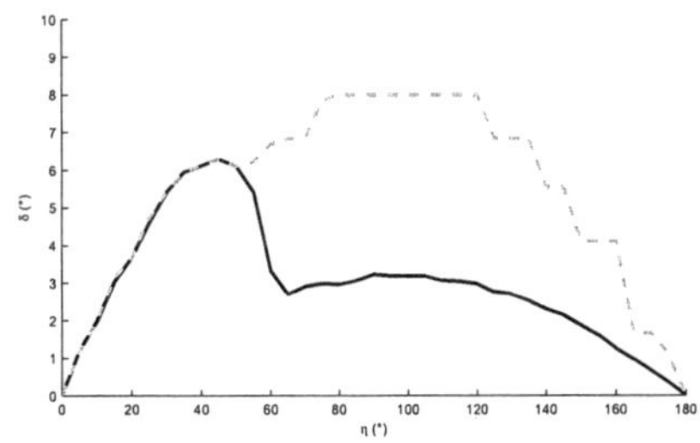

Fig. 14. Clock-angle δ in half a year, first case (black) and second case (gray)

Conclusions

The results obtained show that a Polar Sitter mission can observe the polar region with an angle-of-sight below 40° by setting determined values for the sail attitude angles for a prolonged period of time. In addition the observation can be optimised by changing conveniently the corresponding position of the solar sail on the equilibrium surface during the annual orbit, thus minimising the angle-of-sight γ. The results obtained in case there is a limitation on the maximum distance from the Earth of the satellite demonstrate that posing a constraint on the optimisation does not affect the value of the angle-of-sight. According to this consideration we reckon that further studies on the optimisation of the Polar Sitter mission can offer valuable improvements to the mission performance in terms of image resolution.

References

1. R. L. Forward, *J. Spacecraft Rockets* **28**, 606 (1991).
2. C. R. McInnes, *J. Guid. Control Dynam.* **22**, 185 (1999).
3. C. R. McInnes, *An examination of the constant polar orbit: discussion document dr-9809*, tech. rep., Dept. of Aerospace Engineering, Univ. of Glasgow, Scotland (1998).
4. C. R. McInnes, *Solar Sailing: Technology, Dynamics and Mission Applications* (Springer-Praxis, 1999).

PHASE EQUILIBRIA OF POLYDISPERSE HYDROCARBONS: MOMENT FREE ENERGY METHOD ANALYSIS

A. SPERANZA(*), F. DIPATTI(**) and A. TERENZI(***)

(*)(**) *Innovazione Industriale Tramite Trasferimento Tecnologico, I^2T^3 Onlus*
Dip. di Matematica Università di Firenze
V.le Morgagni 67/a Firenze, Italia
E-mail alessandro.speranza@i2t3.unifi.it
(***) *Snamprogetti S.p.A.*
Via Toniolo 1, Fano (PU), Italia

We analyze the phase equilibria of systems of polydisperse hydrocarbons by means of the recently introduced moment method. Hydrocarbons are modelled with the Soave-Redlick-Kwong and Peng-Robinson equations of states. Numerical results show no particular qualitative difference between the two equations of states. Furthermore, in general the moment method proves to be an excellent method for solving phase equilibria of polydisperse systems, showing excellent agreement with previous results and allowing a great improvement in generality of the numerical scheme and speed of computation.

1. Introduction

In this paper we analyze the phase behaviour of a mixture of hydrocarbons, by means of the moment method [1,2]. This method allows to reduce the number of degrees of freedom of the free energy, which normally depends on the concentration of each specie in the mixture, to a smaller number of moments of the density distribution which already appear in the excess part of the free energy. By doing this, one is able is reduce the number of equations needed to analyze the phase equilibria and, at the same time, by projecting the free energy onto the space generated by the moments only, to check for global and local stability of the phases [2].

The approximation made when introducing the moment free energy can be efficiently controlled and minimized by means of the adaptive method of choice of extra moments [3], which allows to reduce the deviation of the moment method solution from the exact solution, by simply retaining two extra moments, beyond the ones appearing in the excess free energy. This iterative method, which it can be proven, converges to the exact solution, as long as it converges at all, shows to represent an excellent compromise between approximation, which can easily be reduced to an error smaller than 0.01%, and computational speed. Furthermore, the resulting algorithm turns out to be stable and to be very little affected by the number of species in the mixture. In fact, as the number of unknowns is not increased by the increase of the number of species, the computation is hardly affected at all, with just a small influence on its global speed of computation, while no relevant error is introduced.

The numerical results agree very well with the results obtained with a widely diffused commercial program in different points of the phase diagram. The concentration of each component in the coexisting phases and the density of both phases are evaluated correctly. Clud point is detected exactly. Furthermore, the introduction of heavy species, up to n-C15, even present in very small amount, does not compromise either numerical results or computation.

2. Polydisperse hydrocarbons

In order to analyze the phase equilibria of hydrocarbons, we will refer to the two equations of state most widely used to describe them, *i.e.*, the Soave-Redlick-Kwong (SRK) [4] equation of state and the Peng-Robinson (PR) [5] equation of state. Both are cubic equations of state and thus are able to predict gas-liquid phase transitions. Although originally introduced for pure systems, as we will see, they are both easily extended to describe multicomponent,*i.e.*, polydisperse systems. As we will show, given the polydisperse form of two equations of state, one can easily obtain the Gibbs and Helmoltz free energies, by Legendre transforming, and therefore obtain the phase equilibrium equations that are to be solved, in order to fully analyze the phase behaviour of the system.

The SRK equation of state is generally written as

$$P = \frac{N\kappa_{\mathrm{B}}T}{V-b} - \frac{N^2\alpha(T)a_C}{V\,(V+b)} \tag{1}$$

where N is the total number of particles, V the total volume, κ_{B} the Boltzman constant and the parameters a, b and $\alpha(T)$ depend on the critical temperature and pressure, shape and size of the molucules etc., of the specific hydrocarbon.

The extension of the equation above to the case of polydisperse system is rather straightforward, if one introduces a set of parameters $a_{C,i}$, b_i, $\alpha_i(T)$

for each specie i and defines new global parameters B and D as follows

$$B = \sum_i N_i b_i \tag{2}$$

and

$$D(T) = \sum_{i,j} N_i N_j a_{ij}(T) \tag{3}$$

where

$$a_{ij}(T) = \sqrt{a_{C,i} a_{C,j} \alpha_i(T) \alpha_j(T)} \tag{4}$$

In this way, the polydisperse version of Eq. (1) is

$$P = \frac{N\kappa_B T}{V - B} - \frac{D}{V(V + B)} \tag{5}$$

where $N = \sum_i N_i$ is the total number of particles of the system.

The Helmoltz free energy can now be obtained simply by solving the equation

$$P = -\frac{\partial F}{\partial V}$$

and by introducing F_{id}, the ideal part of the free energy, *i.e.*, the free energy of a mixture of ideal gas

$$F_{\text{id}} = \kappa_B T \sum_i N_i \left(\ln \frac{N_i}{V} - 1 \right) \tag{6}$$

The free energy then turns out to be

$$F(\mathbf{n}, V, T) = F_{\text{id}} + N\kappa_B T \ln \frac{V}{V - B} - \frac{D}{B} \ln \frac{V + B}{V} \tag{7}$$

The above quantity is extensive, one can therefore define an intensive "free energy density" $f = F/Vol$. Introducing a density distribution $\rho(k) = N_k/V$ and multiplying by $\beta = 1/\kappa_B T$, the free energy density turns out to be

$$\beta f[\rho(k), T] = \beta f_{\text{id}} - \rho \ln(1 - \tilde{B}) - \frac{\tilde{D}}{\tilde{B}} \ln(1 + \tilde{B}) \tag{8}$$

where the ideal part is just

$$f_{\text{id}} = \kappa_B T \sum_k \rho(k) \left[\ln \rho(k) - 1 \right] \tag{9}$$

and we have defined two new parameters $\tilde{B}$ and $\tilde{D}$, by rescaling B and D with the volume

$$\tilde{B} = \frac{B}{V} = \sum_k b_k \rho(k) \tag{10}$$

$$\tilde{D} = \frac{\beta D}{V^2} = \sum_{k,j} \beta a_{k,j}(T) \rho(k) \rho(j) \tag{11}$$

The non-ideal part of the free energy in Eq. (8) is called excess free energy $\tilde{f}$ and contains the terms due to the interaction between the particles in a non-ideal gas.

The Gibbs free energy can now be obtained from the expressions above, simply by Legendre transforming F as $G(N, P, T) = \min_V \{F(N, V, T) + PV\}$. We now introduce a Gibbs free energy per particle $g = G/N$ and divide again the resulting function into the ideal part

$$\beta g_{\text{id}}[x(k), P, T] = \frac{\beta G_{\text{id}}}{N} = \sum_k x(k) \ln x(k) + \ln \beta P \tag{12}$$

and the excess part

$$\beta \tilde{g}[x(k), P, T] = \frac{\beta \tilde{G}}{N} = -\sum_k x(k) \left[\ln \beta PV + 1\right] + \frac{\beta \tilde{f}}{\rho} + \frac{\beta P}{\rho} \tag{13}$$

where the number fraction $x(k)$ of the specie k is just $x(k) = N_k/N = \rho(k)/\rho$, with $\rho = \sum_k \rho(k) = N/V$ the overall density.

From the above equations (12,13) we can now derive the phase equilibrium equations $\mu_k^a = \mu_k^b$ for the coexisting phases a and b, and each specie k as $\mu_k = \partial G/\partial N_k = \partial g/\partial x(k)$. For a system of M species, dividing in P phases, the phase equilibrium is therefore fully analyzed by solving a system of $(P-1)M$ equations, plus the M equations given by the conservation of the total number of particles, *i.e.*, $\sum_a x^a(k) = x^{(0)}(k)$, where $x^{(0)}(k)$ is the number density of the kth specie of the parent, in the MP unknowns $x^a(k)$. The values of P and T are set as external parameters.

As far as the Peng-Robinson equation of state is concerned, not much change is needed in the equations above. The PR equation is generally written as

$$P = \frac{N \kappa_B T}{V - B} - \frac{D}{V(V+B) + B(V-B)} \tag{14}$$

where B and D differ from the SRK case in the numerical coefficients of b_k and $a_{C,k}$. Once again, from the equation above one gets the excess part of

Helmoltz free energy as

$$\beta\tilde{f} = \rho \ln \frac{V}{V-B} + \frac{\sqrt{2}}{4}\frac{D}{BV} \ln \left(\frac{V+(1-\sqrt{2})B}{V+(1+\sqrt{2})B} \right) \tag{15}$$

Similarly, the excess part of the Gibbs free energy per particle turns out to be

$$\beta\tilde{g} = -\sum_k x(k)\left[\ln \beta PV + 1\right] + \frac{\beta\tilde{f}}{\rho} + \frac{\beta P}{\rho} \tag{16}$$

2.1. *Truncatable systems*

A polydisperse system is said to be *truncatable* when the excess part of its free energy, say the Helmoltz free energy, is a function a limited number of moments ρ_i of the density distribution $\rho(k)$

$$\rho_i = \sum_k w_i(k)\rho(k) \tag{17}$$

with given weight functions $w_i(k)$. In other words, the Helmoltz free energy of a truncatable system is

$$f[\rho(k), T] = \sum_k \rho(k)\left[\ln \rho(k) - 1\right] + \tilde{f}(\rho_i)$$

For a truncatable system one has

$$\beta\mu(k) = \frac{\partial(\beta f)}{\partial \rho(k)} = \ln \rho(k) + \sum_i w_i(k)\beta\tilde{\mu}_i$$

where the excess moment chemical potentials $\tilde{\mu}_i = \partial\tilde{f}/\rho_i$. By imposing the equality of the chemical potentials in all the coexisting phases, one gets that the density distribution of the coexisting phases must have the form

$$\rho^a(k) = R(k)\exp[-\beta\sum_i \tilde{\mu}_i^a w_i(k)] \tag{18}$$

By imposing the lever rule, *i.e.*, the conservation of the total number of particles per specie, $\sum_a v^a \rho^a(k) = \rho^{(0)}(k)$ ($\rho^{(0)}(k)$ is the density of the parent and $v^a = V^a/V$ is the volume occupied by the phase) one finds that the function $R(k)$ has the form

$$R(k) = \frac{\rho^{(0)}(k)}{\sum_a v^a \exp[-\beta\tilde{\mu}^a(k)]} = \frac{\rho^{(0)}(k)}{\sum_a v^a \exp[-\beta\sum_i \tilde{\mu}_i^a w_i(k)]} \tag{19}$$

Although formally solved through the two equations above, an actual numerical solution of the system is not easily found. Eq. (18) actually represents a set of, say, M (for M species of particles) self consistent all strongly coupled through the denominator of Eq. (19). The problem is that, although the excess free energy is a function just of the moments ρ_i, usually a much smaller number than the number of species, the ideal part of the free energy is still function of the whole density distribution $\rho(k)$. Ideally, one would like to reduce the problem to a smaller number of degrees of freedom, by expressing also the ideal part of the free energy as a function of the moments only. While this argument will be treated in the next section, here we will show that both the SRK and the PR equations of state are in fact truncatable.

In fact, it is rather easy to show that the equations of state generate two truncatable systems if one introduces two moments of the density distribution ρ_1 and ρ_2 as follows

$$\rho_1 = \tilde{B} = \sum_k b_k \rho(k) \tag{20}$$

$$\rho_2 = \sum_k d_k \rho(k) \tag{21}$$

where, from Eq. (4), $d_k = \sqrt{\beta a_{C,k} \alpha_k(T)}$. From the definition above and Eq. (3), we get that $\tilde{D} = \rho_2^2$. Plugging (20,21) into Eq. (8), we therefore get, for the SRK equation of states

$$\beta \tilde{f}(\rho, \rho_1, \rho_2) = -\rho \ln(1 - \rho_1) - \frac{\rho_2^2}{\rho_1} \ln(1 + \rho_1) \tag{22}$$

Note that in the above expression, the overall density ρ is itself a moment of the density distribution, with weight function $w_0(k) = 1$, as $\rho = \rho_0 = \sum_k \rho(k)$. In other words, the excess part of the free energy is fully described by the knowledge of only three moments of the density distribution ρ_0, ρ_1, ρ_2 and not on the whole distribution $\rho(k)$ itself.

For the PR equation of state the argument is again similar to the case of the SRK equation of states. Introducing again the three moments ρ_0, ρ_1 and ρ_2, we get that the excess part of the Helmoltz free energy is just

$$\beta \tilde{f}(\rho_0, \rho_1, \rho_2) = -\rho \ln(1 - \rho_1) - \frac{\sqrt{2}}{4} \frac{\rho_2^2}{\rho_1} \ln\left(\frac{1 + (1 - \sqrt{2})\rho_1}{1 + (1 + \sqrt{2})\rho_1}\right) \tag{23}$$

As far as the Gibbs free energy is concerned, it is easy to show [2] that the Gibbs free energy inherits its moment structure from the Helmoltz free energy. However, this time, one usually introduces normalized moments m_i of the number density distribution $x(k)$ defined as

$$m_i = \sum_k w_i(k) x(k) = \frac{\rho_i}{\rho_0} \tag{24}$$

Clearly this time $m_0 = \sum_k x(k) = 1$, thus, since $\tilde{f}$ depends on three moments, ρ_0, ρ_1, ρ_2, the Gibbs free energy depends itself on the overall density, which, however, is obtained from the equation of state, as a function of P, m_1 and m_2. In other words, the Gibbs free energy turns out to have one degree of freedom less than the Helmoltz free energy.

In any case, with the definitions above, one gets that the excess part of the Gibbs free energy for the SRK equation of states is just

$$\beta\tilde{g}(m_1, m_2) = \ln\frac{\rho_0}{\beta P} - 1 + \beta\frac{P}{\rho_0} - \ln(1-\rho_0 m_1) - \frac{m_2^2}{m_1}\ln(1+\rho_0 m_1) \quad (25)$$

Similarly, for the PR equation of state, we get

$$\beta\tilde{g}(m_1, m_2) = \ln\frac{\rho_0}{\beta P} - 1 + \beta\frac{P}{\rho_0} - \ln(1-\rho_0 m_1) + \frac{\sqrt{2}}{4}\frac{m_2^2}{m_1}\ln\left(\frac{1+(1-\sqrt{2})\rho_0 m_1}{1+(1+\sqrt{2})\rho_0 m_2}\right) \quad (26)$$

Note that in the two equations above we have omitted the dependence of $\tilde{g}$ on P and T.

3. The moment method

Truncatable systems allow to express the excess free energy as a function of a small, say M, number of moments only. However, as we saw in the previous section, the difficulty of solving the phase coexistence equations remains largely unaltered, as the ideal part of the free energy is still function of the full density (or number) distribution. Ideally, one would like to express the ideal free energy too, as a function of the moments only. This is in fact possible, by means of the moment method [1,2,6]. While the following description will refer mostly to the Helmoltz free energy, similar considerations can be made for the Gibbs free energy [2], with the introduction of the normalized moments and the number density distribution.

The moment method arises from the hypothesis, in fact verified in different works [1,3,6,7], that the excess free energy is mostly responsible for the phase behaviour of the whole system. This is in fact not surprising, as the ideal free energy is overall convex, and thus does not allow for phase separation. With this in mind, the moment free energy is constructed as follows. We subtract from the actual free energy a term $\rho(k)\ln\rho^{(0)}(k)$, where $\rho^{(0)}(k)$ is the density distribution of the parent. This term, as linear in the density $\rho(k)$, does not affect the phase behaviour, as it adds just a constant to the chemical potential $\mu(k) = \partial f/\partial\rho(k)$. The resulting function is then minimized with respect to $\rho(k)$ with the M moments appearing in the excess part as constraints (2 in the two previous cases). The minimum value of the resulting free energy is then found to be

$$f_{\mathrm{mom}}(\boldsymbol{\rho}) = \sum_{i}^{M} \lambda_i \rho_i - \rho_0 + \tilde{f}(\boldsymbol{\rho}) \tag{27}$$

where $\boldsymbol{\rho}$ is just a vector having the moments ρ_i as components and the λs are the M Lagrange multipliers. The minimum value of the free energy is reached for a density distribution from the family

$$\rho_{\mathrm{mom}}(k) = \rho^{(0)}(k) \exp\left(\sum_{i} \lambda_i w_i(k)\right) \tag{28}$$

From the moment free energy (27), one can also define the moment chemical potentials μ_i, as $\mu_i = \partial f_{\mathrm{mom}}/\partial \rho_i = \lambda_i + \tilde{\mu}_i$. The pressure is obtained from the Gibbs-Duhem relation as

$$P = \sum_{k} \mu(k)\rho(k) - f = \sum_{i} \mu_i \rho_i - f$$

It is easy to show that the above expression obtained from the moment free energy is in fact identical to the one obtained from the exact free energy [2]. Furthermore, it is easy to show that the moment free energy correctly detects the onset of phase coexistence. In other words, one can show [2,8] that any two phases coexist for the full system, if and only if, they coexist for the moment free energy, thus, $\mu^a(k) = \mu^b(k) \Leftrightarrow \mu_i^a = \mu_i^b$. Thus, at least up to the onset of the phase coexistence, the full solution in Eq. (18) actually belongs to the family in Eq. (28). Cloud point and shadow phases are then correctly detected by the moment free energy solution above. Furthermore, spinodals and critical/tricricital points are found exactly [2].

The enforcement of the lever rule only for the moments is in fact the only approximation we make in using the moment method, as this does not ensure the satisfaction of the complete levere rule, while, as mentioned earlier, equality of pressure and chemical potentials are ensured. However, as shown in details in Ref. [3,6,7], the approximation can be reduced efficiently by retaining extra moments and, in particular, by means of the adaptive method of choice of extra weight functions which allows to obtain a solution as close as wanted to the exact one, by retaining only two extra moments.

In order to give a more precise insight of the actual problem one has to solve in the case of the two equations of state mentioned earlier, let us sketch the resulting system of equations, obtained within the moment method approach. As mentioned, since we have P and T as external parameters, we move on to the Gibbs formalism. Thus we evaluate the gibbs free energy and from that, we calculate the moment chemical potentials as $\mu_i = \partial g_{\rm mom}/\partial m_i$, where m_i are the normalized moments. Let us now assume we have a Gas-Liquid demixing and let us call $\phi^a = N^a/N^{(0)}$ the fraction of particles in each phase (G or L). The phase coexistence is then fully solved by enforcing the equality of the moment chemical potentials and of the quantity $\Pi = g_{\rm mom} - \ln P - \sum_{i\neq 0} m_i\mu_i$, which is a sort of Legendre transform of the pressure [2], in all the coexisting phases. We must also enforce the conservation of the total number of particles, *i.e.*, $\sum_a N^a(k) = N^{(0)}(k)$. If we multiply by $w_i(k)$ on both sides and sum over k, we get, after rearranging, the lever rule for the normalized moments $\sum_a \phi^a m_i^a = m_i^{(0)}$, which is the condition we actually enforce. Thus, for the two equations of state, the system of equations we have to solve turns out to be

$$\begin{cases} \mu_1^{\rm G} &= \mu_1^{\rm L} \\ \mu_2^{\rm G} &= \mu_2^{\rm L} \\ \Pi^{\rm G} &= \Pi^{\rm L} \\ m_1^{(0)} &= \phi^{\rm G} m_1^{\rm G} + (1-\phi^{\rm G}) m_1^{\rm L} \\ m_2^{(0)} &= \phi^{\rm G} m_2^{\rm G} + (1-\phi^{\rm G}) m_2^{\rm L} \\ P &= P(\rho_0^{\rm G}, m_1^{\rm G}, m_2^{\rm G}) \\ P &= P(\rho_0^{\rm L}, m_1^{\rm L}, m_2^{\rm L}) \end{cases}$$

i.e., 7 equations in the 7 unknowns $\lambda_1^{\rm G}, \lambda_1^{\rm L}, \lambda_2^{\rm G}, \lambda_2^{\rm L}, \rho_0^{\rm G}, \rho_0^{\rm L}, \phi^{\rm G}$. The moment chemical potentials and the pressure are then for the SRK equation of state:

$$\begin{aligned} \beta\mu_1 &= \beta\lambda_1 + \frac{\rho_0}{1-\rho_0 m_1} - \frac{\rho_0 m_2^2}{m_1(1+\rho_0 m_1)} + \frac{m_2^2}{m_1^2}\ln(1+\rho_0 m_1) \\ \beta\mu_2 &= \beta\lambda_2 - 2\,\frac{m_2}{m_1}\ln(1+\rho_0 m_1) \\ \beta P &= \frac{\rho_0}{1-\rho_0 m_1} - \frac{\rho_0^2 m_2^2}{1+\rho_0 m_1} \end{aligned}$$

For the PR equation of state, the calculations are just slightly more complicated, as now we get:

$$\begin{aligned} \beta\mu_1 &= \beta\lambda_1 + \frac{\rho_0}{1-\rho_0 m_1} - \frac{\rho_0 m_2^2}{m_1(1+2\rho_0 m_1 - \rho_0^2 m_2^2)} - \frac{\sqrt{2}}{4}\frac{m_2^2}{m_1^2}\ln\left[\frac{1+(1-\sqrt{2})\rho_0 m_1}{1+(1+\sqrt{2})\rho_0 m_1}\right] \\ \beta\mu_2 &= \beta\lambda_2 + \frac{\sqrt{2}}{2}\frac{m_2}{m_1}\ln\left[\frac{1+(1-\sqrt{2})\rho_0 m_1}{1+(1+\sqrt{2})\rho_0 m_1}\right] \\ \beta P &= \frac{\rho_0}{1-\rho_0 m_1} - \frac{\rho_0^2 m_2^2}{1+2\rho_0 m_1 - \rho_0^2 m_1^2} \end{aligned}$$

4. Numerical results

As a way of example, in order to show the potential of the method described in the previous section, in this section we will show some numerical results obtained by applying it to a real case-study. We solve the phase equilibria for a mixture of 24 hydrocarbons, up to nC15, along a straight line crossing the (P, T) phase diagram. The calculations are done using the SRK equation of state, although no relevant difference is observed when using the PR equation of state. Our numerical results are compared with the results obtained with a commercial program (PVTsim), licensed to Snamprogetti s.p.a. (that provided the results). In Fig. 4, we plot the concentration in mole % of C1 (methane), in the two coexisting phases, against the temperature. Our result agrees very well with the points obtained with PVTsim.

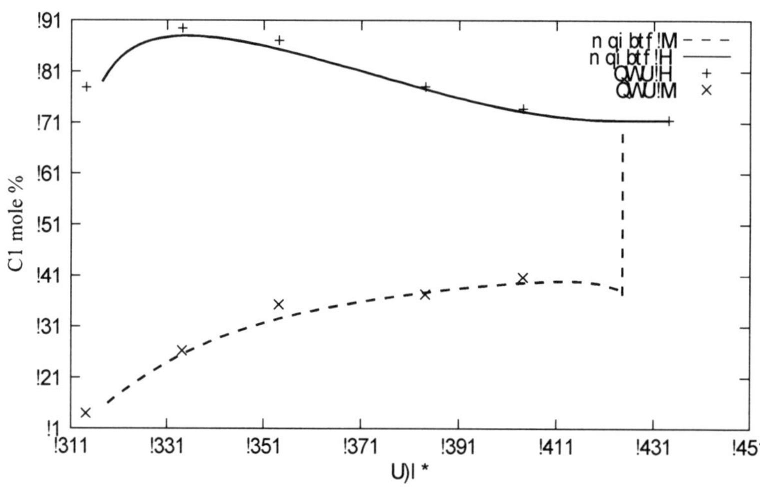

Figure 1. Concentration of methane against temperature in the two coexisting phasese (solid and dashed lines), compared to the results obtained with PVTsim (diamonds). The liquid phase appears correctly at lower temperature as the path across the phase diagram crosses the phase envelope. The cloud point is detected exactly by our method. Deep inside the coexistence region some small deviations appear, although it is not clear whether they are due to the moment method, or to the approximations introduced by PVTsim.

The concentration of both phases is evaluated correctly. Furthermore, the cloud point, *i.e.*, the point at which the liquid phase appears, is detected exactly on the phase envelope shown by PVTsim.

In Fig. 4 we show again the same case. Now we plot the concentration of n-C4 against the pressure. As the pressure-temperature path enters the coexistence region, liquid is found. This time the component represents less than 1% of the total composition of the gas, while it is about 10% of the total composition of the liquid. Again, even with a heavier hydrocarbon, our numerical results are in excellent agreement with the ones obtained with PVTsim.

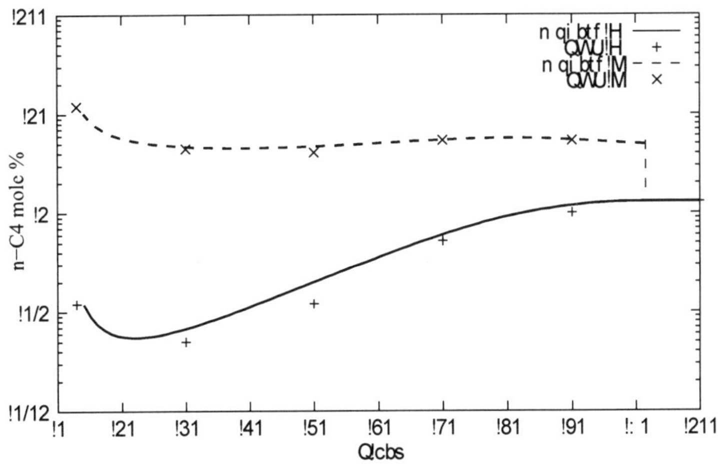

Figure 2. Concentration in mole % of n-C4 against pressure across the phase diagram. As pressure and temperature drop enough to enter the phase coexistence region, the liquid phase is found. The concentration the element in both phases are in excellent agreement with the ones obtained with PVTsim. Again, it is not clear whether the small deviations inside the coexistence region are due to our method, or rather to approximations and truncations introduced by PVTsim.

5. Conclusion

We have applied the moment method to the analysis of phase equilibria of mixture of hydrocarbons, using the SRK and PR equations of state. Our results show that not only the moment method is applicable as it correctly detects gas-liquid phase coexistence. Even with a large number of components in the mixture, our algorithm remains robust and numerical calculation fast. Furthermore, our numerical results agree quantitatively very well with the results obtained using a widely used commercial program (PVTsim). No further demixing, beyond the G-L coexistence is observed, however, it is not yet clear whether this depends on the equations of state used, or rather to the choice of the mixture of hydrocarbons. It may be possible to have the coexistence of more than two (G and L) phases, *e.g.*, more than one different liquid and/or gas phases, using a wider distribution of hydrocarbons.

As far as future developments are concerned, one could proceed to the analysis of a fully polydisperse case, *i.e.*, by introducing a continuous distribution of species. Clearly the two equations should first be extended to the continuous case, *e.g.*, by introducing a continuous dependence of the acentric factor on the size of particles. It is possible that further demixing appears using different distributions. The extension to the continuous case should not have any impact on our method. Clearly, the computation may be slightly slower, as in that case the moments have to be evaluated by integration rather than summation over a finite number of species; however, no formal or substantial adjustment is needed.

Finally, further applications of the moment method could be tried. For instance, one could try to apply it to metallurgy or other cases of industrial interest. Clearly, when solid phases appear, one would have to extend the analysis introducing spatial and/or orientational degrees of freedom. However, as previous results already show [3,6,7,9–11] this should not represent a limitation for the method.

Bibliography

1. P. Sollich and M. E. Cates, *Phys. Rev. Lett.* **80**, 1365 (1998).
2. P. Sollich, P. B. Warren and M. E. Cates, *Adv. Chem. Phys.* **116**, 265 (2001).
3. A. Speranza and P. Sollich, *J. Chem. Phys.* **117**, 5421 (2002).
4. G. Soave, *Chem. Eng. Sci.* **27**, 1197 (1972).
5. D. Y. Peng and D. B. Robinson, *Ind. Eng. Chem. Fundam.* **15**, 59 (1976).
6. N. Clarke, J. A. Cuesta, R. Sear, P. Sollich and A. Speranza, *J. Chem. Phys.* **113**, 5817 (2000).
7. A. Speranza and P. Sollich, Isotropic-nematic phase equilibria of polydisperse hard rods: The effect of fat tails in the length distribution, Submitted to *J. Chem. Phys.*, (2002).
8. A. Speranza, Effects of length polydispersity in colloidal liquid crystals, PhD thesis, King's College, University of London, (London, UK, 2002).
9. M. Fasolo and P. Sollich, *Physical Revew Letters* **91**, p. 068301 (2003).
10. M. Fasolo and P. Sollich, *Physical Review E* **70**, p. 041410 (2004).
11. M. Fasolo, P. Sollich and A. Speranza, *React. Funct. Polym.* **58**, 187 (2004).

OPTIMIZATION OF ELECTRONIC CIRCUITS

E.J.W. TER MATEN, T.G.A. HEIJMEN
NXP Semiconductors, Research, DMS - Physical Design Methods,
Hich Tech Campus 48, 5656 AE Eindhoven, The Netherlands
E-mail: {Jan.ter.Maten,Tino.Heijmen}@NXP.com

C. LIN and A. EL GUENNOUNI
Magma Design Automation,
TUE Campus, Den Dolech 2, Dommel Building Z-Wing 8, 5612 AZ Eindhoven, The Netherlands
E-mail: {Achie,Ahmed}@Magma-DA.com

Keywords: Global Optimization; Derivative Free; Robust Design; Augmented Lagrangian.

1. Introduction

Several types of parameters $\mathbf{p} = (\mathbf{x}, \mathbf{s}, \theta)$ influence the behaviour of electronic circuits and have to be taken into account when optimizing appropriate performance functions $f(\mathbf{p})$: design parameters $\mathbf{x}$, manufacturing process parameters $\mathbf{s}$, and operating parameters θ.

For optimizing one wants to minimise a performance function $f(\mathbf{p})$ while also several constraints have to be satisfied. The performance function $f(\mathbf{p})$ and the constraint functions $c(\mathbf{p})$ can be costly to evaluate and are subject to noise (for instance due to numerical integration effects). For both, the dependency on $\mathbf{p}$ can be highly nonlinear. In circuit simulation, sensitivities of $f(\mathbf{p})$ and $c(\mathbf{p})$ with respect to $\mathbf{p}$ are not always provided (several model libraries do not yet support the calculation of sensitivities). When the number of parameters increases adjoint sensitivity methods become of interest. For transient integration of linear circuits this is described in.[2] Recently, in[8] a more general procedure is described that also applies to nonlinear circuits and retains efficiency by exploiting (nonlinear) techniques from Model Order Reduction.

In this paper we will describe our in-house developed method for optimization and our experiences with it. We restrict ourselves to so-called derivative free methods, so we will not require sensitivities of $f(\mathbf{p})$ and $c(\mathbf{p})$ with respect to $\mathbf{p}$ from the circuit simulator. Also some new directions for further research will be described.

2. Constrained optimization by augmented Lagrangian

In this section we restrict our parameters $\mathbf{p}$ to the design variables $\mathbf{x}$, which can be geometrical quantities like transistor width W and length L. The designer can adjust them during the process of optimizing the performance of an actual design. The performance functions $f(\mathbf{x})$ typically consist of one or more circuit characteristics that the designer would like to maximize or minimize. Examples are: maximization of gain or bandwidth, and mimization of power dissipation or area. The constraints $c(\mathbf{p})$ are also related to circuit performance but these functions have an explicit target. Examples are: bandwidth $\geq$ 800MHz, and $49\% <$ duty cyle $< 51\%$. Formulating a proper set of performance functions and constraints is a non-trivial task. In practice, this requires trial-and-error and a fair amount of tuning. Obtaining the value of a performance function or constraint is done via one or more circuit analyses of the same or different types, e.g. DC, AC, Transient, Noise. Typically, the time required to perform an analysis is a limiting factor in the overall optimization approach.

The search for the optimal values of the optimization variables (OVs) $\mathbf{x}$ can be formulated as a nonlinear constrained optimization problem in n variables with m constraints,

$$\begin{aligned} &\text{minimize} \quad f(\mathbf{x}), \qquad \mathbf{x} = (x_1, x_2, \ldots, x_n), \\ &\text{subject to} \quad c_i(\mathbf{x}) \leq 0, \quad i = 1, \ldots, m, \\ &\qquad a_j \leq x_j \leq b_j,\ j = 1, \ldots, n, \end{aligned} \tag{1}$$

where x_i denotes the i-th OV. With $\mathbf{x}^\star$ we denote the point where the minimum occurs. The values of the objective function $f(\mathbf{x})$ and the constraining functions $c_i(\mathbf{x})$ are obtained from circuit simulation. The performance and stability of the optimization algorithm is affected by the *scaling* of the OVs, of $f(\mathbf{x})$ and of the $c_i(\mathbf{x})$.[6]

The Nelder-Mead algorithm can be used to minimize

$$\Phi(\mathbf{x}) = f(\mathbf{x}) + \mu \sum_{i=1}^{m} < c_i(\mathbf{x}) >^2 \tag{2}$$

where $< \alpha >= \max(\alpha, 0)$. If the minimum is taken at $\mathbf{x}_\mu$, one has $\lim_{\mu\to\infty} \mathbf{x}_\mu = \mathbf{x}^\star$, which means that the minimization of (2) becomes more and more ill-conditioned. Apart from that, the Nelder-Mead algorithm, which is simple to program, is not that easy to analyse. A similar conditioning problem happens when dealing with logarithmic barrier functions that provide an impassible barrier at the boundary of the feasible region.

Recently pattern search methods have been studied.[10] The most well-known method of this class is the method of Hooke-Jeeves.[7] The Nelder-Mead method

does not belong to that class (it can also beat pattern search methods). The pattern search methods are nice in the sense that they prove rather weak conditions under which the methods do converge, f.i. for the unconstrained problem $f \in C^1$.
By introducing slack variables $s_i \geq 0$, the augmented Lagrangian penalty function can be written as,[15]

$$\Phi_{\mathrm{ALAG,s}}(\mathbf{x}, \boldsymbol{\lambda}, \boldsymbol{\mu}, \mathrm{s}) = \underbrace{f(\mathbf{x}) - \sum_{i=1}^{m} \lambda_i [c_i(\mathbf{x}) + s_i]}_{\mathcal{L}(\mathbf{x},\boldsymbol{\lambda})} + \sum_{i=1}^{m} \mu_i [c_i(\mathbf{x}) + s_i]^2, \quad (3)$$

$$= f(\mathbf{x}) + \sum_{i=1}^{m} \{\mu_i [(c_i(\mathbf{x}) - \frac{\lambda_i}{2\mu_i}) + s_i]^2 - \frac{\lambda_i^2}{4\mu_i}\} \quad (4)$$

in which $\mathcal{L}$ is the standard Lagrangian. The parameters λ_i and μ_i are Lagrange multipliers and penalty factors, respectively. Pattern search methods for (3) converge when $f, c_i \in C^2$.[11] Minimization (4) over the slack variables s_i at optimal value $s_i^\star = \max\left[-c_i(\mathbf{x}) + \frac{\lambda_i}{2\mu_i}, 0\right]$ yields the simplified merit function that is used in,[6]

$$\Phi_{\mathrm{ALAG}}(\mathbf{x}, \boldsymbol{\lambda}, \boldsymbol{\mu}) = f(\mathbf{x}) - \sum_{i=1}^{m} \lambda_i \Psi(c_i(\mathbf{x}), \frac{\lambda_i}{2\mu_i}) + \sum_{i=1}^{m} \mu_i \Psi^2(c_i(\mathbf{x}), \frac{\lambda_i}{2\mu_i}) \quad (5)$$

$$= f(\mathbf{x}) + \sum_{i=1}^{m} \mu_i \tilde{\Psi}(c_i(\mathbf{x}), \frac{\lambda_i}{2\mu_i}), \quad (6)$$

where

$$\Psi(y, \xi) = \max\left[y, \xi\right], \quad \text{and} \quad \tilde{\Psi}(y, \xi) = \max\left[y - \xi, 0\right]^2 - \xi^2$$

The optimal point $(\mathbf{x}^\star, \boldsymbol{\lambda}^\star)$ becomes a stationary point of $\mathcal{L}$ and satisfies the Karush-Kuhn-Tucker (KKT) conditions,

$$\nabla_x \mathcal{L}(\mathbf{x}^\star, \boldsymbol{\lambda}^\star) = 0, \quad (7)$$
$$c_i(\mathbf{x}^\star) \;\leq\; 0, \quad \lambda_i^\star \;\leq\; 0, \quad \lambda_i^\star c_i(\mathbf{x}^\star) = 0, \quad i = 1, \ldots, m.$$

Note that in (6) $\forall \mu_i \neq 0$ $\Psi(c_i(\mathbf{x}^\star), \frac{\lambda_i^\star}{2\mu_i}) = 0$ and $\tilde{\Psi}(c_i(\mathbf{x}^\star), \frac{\lambda_i^\star}{2\mu_i}) = 0$. The KKT conditions imply that the projected gradient $\mathcal{P}(\nabla_x \Phi_{\mathrm{ALAG}}(\mathbf{x}^\star, \boldsymbol{\lambda}^\star, \boldsymbol{\mu})) = 0$, where

$$[\mathcal{P}(\nabla_x F(\mathbf{x}))]_j = \begin{cases} [\nabla_x F(\mathbf{x})]_j & \text{if } a_j < x_j < b_j \\ \min[\,[\nabla_x F(\mathbf{x})]_j, 0\,] & \text{if } a_j = x_j \\ \max[\,[\nabla_x F(\mathbf{x})]_j, 0\,] & \text{if } b_j = x_j \end{cases} \quad (8)$$

The reduction in dimension has been paid back by loss of smoothness due to the 'max' function. However, any algorithm can check for $c_i = \lambda_i/(2\mu_i)$ and in general this does not occur at a KKT point.[3] Thus, when locally enough started, pattern search methods will also converge for (6).

A conjecture is that under mild conditions there are $|\mu_i| < \infty$ such that $\Phi_{\text{ALAG}}(.,.,\boldsymbol{\mu})$ has a local minimum in $(\mathbf{x}^\star, \boldsymbol{\lambda}^\star)$. Also the location becomes independent of μ_i (when large enough). This happens for instance when $\forall i$ $c_i(\mathbf{x}) \equiv \tilde{c}_i(x_i)$ [a simple 1-D example is $f(x) = x^2$, $c(x) = x - 1$, for which $(\mathbf{x}^\star, \boldsymbol{\lambda}^\star) = (1, -2)$]. This indicates the better conditioning of the problem (6) when compared to (2).

For updating $\boldsymbol{\lambda}$ we observe the equalities [cf also Primal-Dual methods]:

$$\begin{aligned}\nabla_x \Phi_{\text{ALAG,s}}(\mathbf{x}, \boldsymbol{\lambda}, \boldsymbol{\mu}, \mathbf{s})\Big|_{\mathbf{s}=\mathbf{s}^\star} &= \nabla_x \Phi_{\text{ALAG}}(\mathbf{x}, \boldsymbol{\lambda}, \boldsymbol{\mu}) \\ &= \nabla_x \mathcal{L}(\mathbf{x}, \boldsymbol{\lambda}') \quad \text{for} \quad \lambda_i' = \lambda_i - \max[2\mu_i c_i(\mathbf{x}), \lambda_i].\end{aligned}$$

The basic method used to solve the optimization problem is the Method of Multipliers (Algorithm 2.1). In step 3 a trust region approach is applied on a response

Algorithm 2.1 Method of Multipliers[6]

1: Start: $k = 1, x_i = x_i^{(0)}, \lambda_i = \lambda_i^{(0)} = 0, \mu_i = \mu_i^{(0)}$
2: Set $\Phi(\mathbf{x}) = \Phi_{\text{RSM}}(\mathbf{x}, \boldsymbol{\lambda}^{(k-1)}, \boldsymbol{\mu}^{(k-1)}; \mathbf{x}^{(k-1)})$
3: Minimize: $\mathbf{x}^{(k)} = \text{argmin}_{\mathbf{x}} \Phi(\mathbf{x})$
4: If (update μ_i) $\mu_i^{(k)} = 10 * \mu_i^{(k-1)}$ (but $< \mu_{\text{max}}$!)
5: Else update $\lambda_i^{(k)} = \lambda_i^{(k-1)} - \max[2\mu_i^{(k-1)} c_i(\mathbf{x}^{(k-1)}), \lambda_i^{(k-1)}]$
6: Endif
7: Test for optimality
8: If not optimal set $k = k + 1$ and goto 2.

surface model around $\mathbf{x}^{(k-1)}$ to solve the minimization problem (for details see[6]). In step 4 we keep $\mu_i^{(k)} < \mu_{\text{max}}$. This is based on the basic observation made above: in the end we have trust that only the $\lambda_i^{(k)}$ should be able to do the work. Note that in step 5 linear convergence of the $\lambda_i^{(k)}$ can be improved by applying vector extrapolation techniques. The overall method is derivative free (no gradients of f and of c_i are required). Also the method is not sensitive to noise on f and of c_i. Convergence of a related (but simpler, because no grid is used) algorithm is discussed in,[15] assuming that $f, c_i \in C^2$ and that one assures $c_i(\mathbf{x}) \neq \lambda_i/(2\mu_i)$: error control is on $||\mathcal{P}(\nabla_x \Phi_{\text{ALAG}}(\mathbf{x}, \boldsymbol{\lambda}, \boldsymbol{\mu}))||$ and on $||\Psi(c_i(\mathbf{x}), \frac{\lambda_i}{2\mu_i})||$.

In our case[6] the whole approach is applied on a grid (i.e. all $\mathbf{x}^{(k)}$ are projected to the nearest grid point), that subsequently is refined, like in.[4] This prevents the algorithm from going too fast to a small scale (and then gets stuck in a local minimum). However, in practice equally appreciated, is that it also allows to store and re-use expensive parts of $f(\mathbf{x})$ and of $c_i(\mathbf{x})$ during the re-building of the response surface model in the optimization process, when $\boldsymbol{\lambda}$ or $\boldsymbol{\mu}$ are updated, and also when refining the grid.

Our trust region error control is described in.[6] Because we only have approximative gradients from the response surface model approximation, for final convergence, error control is based on

$$|f(\mathbf{x}^{(k)}) - f(\mathbf{x}^{(k-1)})| < \varepsilon_f(1 + |f(\mathbf{x}^{(k)})|), \tag{9}$$

$$||\mathbf{x}^{(k)} - \mathbf{x}^{(k-1)}|| < \varepsilon_x(1 + ||\mathbf{x}^{(k)}||) \tag{10}$$

The parameters $\varepsilon_f, \varepsilon_x$ are decreased when refining the grid. At initialization we apply a Uniform Design approach which is based on number theory. Actually this approach is problem independent. Here improvements can be expected by introducing additional techniques, like pattern search methods,[10] that add more information from the problem itself to improve initial starts for the above method. Also Kriging techniques using the DACE regression model[9] can be used in this phase, which additionally offer a more global optimization aspect, when also applied later in the process. The regression model includes effects due to correlation. Based on values $y_i = y(\mathbf{x}^{(i)}) = \mu + \epsilon(\mathbf{x}^{(i)})$, in which the $\epsilon(x_k^{(i)}) \approx \mathrm{N}(0, \sigma^2)$ are normally distributed and have correlation matrix $R = \mathrm{Corr}[\epsilon(\mathbf{x}^{(i)}), \epsilon(\mathbf{x}^{(j)})] = \exp[-\sum_k \theta_k |x_k^{(i)} - x_k^{(j)}|^{p_k}]$, an exact interpolant $\hat{Y}(\mathbf{x})$ can be derived, together with an estimation for the variance $s^2(\mathbf{x})$ [clearly $s^2(\mathbf{x}^{(i)}) = 0$]. The EGO algorithms[9,16] maximize the expected improvement for $Y(\approx \mathrm{N}(\hat{Y}, s^2)$. However, experience shows that only a small reduction in the number of overall function evaluations was obtained when including this approach in the start-up phase of our simulator. In this case all parameters were involved. Note that the method becomes expensive when the number of parameters increase. It may make sense to limit the Kriging approach to the parameters for which the augmented Lagrangian varies only moderately.

3. Discrete optimization

Considerable interest has been shown for discrete optimization methods since continuous optimization methods were well established in the late 1980s. In general, discrete optimization methods are aimed at solving the so-called mixed-discrete nonlinear programming problem, where the term mixed-discrete indicates that both discrete and continuous design variables are present.

Current discrete optimization methods can be classified as either deterministic or probabilistic.[1] Probabilistic methods have been applied to solve engineering optimization problems for a long time. The most well known probabilistic methods

for both continuous and discrete optimization are Genetic Algorithms[14] and Simulated Annealing.[17] One major advantage of probabilistic methods is their ability to directly deal with discrete design variables. However, these methods are extremely expensive (because of a large number of function evaluations), and thus impractical for the optimization of electronic circuits.
Various approaches have been developed to apply deterministic methods for discrete and mixed-discrete optimization problems.[5] The simplest and least expensive method for obtaining a discrete solution is by rounding the continuous solution to the value of the closest discrete values. However, this rounding process can easily result in stagnation and convergence problems since extra local minimas are introduced. Therefore, the applicability of the rounding methods may be limited.
The branch and bound method is probably the best known and most frequently used discrete optimization method. The method is based on the sequential analysis of a discrete tree for each variable, the computational cost grows exponentially with the number of design variables. Therefore, the method is also extremely expensive.
In our in-house developed method, a dynamic rounding approach was adopted. Design variables are rounded to the value of the closest discrete values at each evaluation step. The approach has the advantage that no extra local minimums are generated. Stagnation problems are avoided by the use of a component-wise trust region method when solving the quadratic problems generated by the algorithm. The component-wise trust region method allows us to select the space where the new design variable candidate should be stamped by forcing the trust radius of each discrete design variable to be larger than a desired value.

4. Toward Robust Design

The additional types of parameters, the manufacturing process parameters $\mathbf{s}$ and operating parameters $\boldsymbol{\theta}$, introduce additional requests.

- The first ones, $\mathbf{s}$, reflect statistical process fluctuations, like T_{ox} (oxide thickness), and V_{th0} (threshold voltage), substrate doping, characterized by their distribution functions.
- The operating parameters $\boldsymbol{\theta}$ include supply voltage V_{sup} and temperature T. Here additional constraints are found that require $c(\mathbf{x}, \boldsymbol{\theta}) \leq 0$ should hold for all $\boldsymbol{\theta}$ within some interval.

The *operating parameters* $\boldsymbol{\theta}$ imply simple additional inequality constraints that fit the already defined framework. The dependency with respect to $\boldsymbol{\theta}$ ususally is quite smoothly. An interesting test function to consider the effect of operating parameters is

$$c(x,\theta) = (\theta - 2)^2 \frac{x\ \sin(x)}{2}, \quad \text{with } x \in [0,6], \ \ \theta \in [0,5]. \tag{11}$$

When starting at $(x = 5, \theta = 0)$ a local maximum at $(x = 6, \theta = 0)$ is easily found. An even higher value of c may be preferred here as a more robust result with respect to θ when one likes to satisfy $c(x, \theta) \geq 0$. Here Kriging techniques were of help to improve the result. Note that this is a particular form of robust design. The problem can be simplified by restricting θ to a discrete set $\theta_i \in \Theta$. In an outer loop the bounds on the constraints can be adaptively updated. New values of θ_i can be selected based on the Kriging approach to determine next interpolation points.

Most *statistical parameters* s appear as transistor model parameters and cannot be modified by the designer if a fully qualified production process is assumed. However, production variation must be taken into account when designing a circuit. For robust design one can think to optimize a combination of f and of its (higher) derivatives (H^1-norm optimization, say), which requires the need to evaluate these derivatives (sensitivities). Also spreads of several σ's have to be taken into account. The occurence of process parameters s also makes the functions f and c_i stochastic. So one needs to consider the effect of the transformed probability density function while also correlations have to be taken into account. With each $s_j \approx N(\mu_j, \sigma_j)$, the common normal density function $\phi(\mathbf{s}) = C\exp[-\frac{1}{2}\beta^2(\mathbf{s})]$ involves a distance function $\beta^2(\mathbf{s}) = (\mathbf{s} - \boldsymbol{\mu})^T \Sigma_s^{-1} (\mathbf{s} - \boldsymbol{\mu})$, which can be used to measure the distance from the nominal value. If $\mathcal{A}_s = \{\mathbf{s} \in \Omega_s | \mathbf{c}(\mathbf{s}) \in \Omega_c\}$, where Ω_c indicates the acceptance region for the constraint function values, the worst-case point s_i^{wc} can be defined as

$$s_i^{wc} = \text{argmax}_{s \in \partial \mathcal{A}_{c_i}} \phi(s) \quad = \quad \text{argmin}_{s \in \partial \mathcal{A}_{c_i}} \beta^2(s)$$

In the worst-case analysis in[13] first linearization with respect to θ is done, and a θ_{WC} is derived. Next, this value is used in the linearization with respect to x, and a x_{WC} is determined. This process is iteratively repeated in a Gauss-Seidel-like manner. For several problems this algorithm gives reasonable results and yield estimations (as is confirmed by our own experiments). However, the algorithm appears not to be robust on the above constraint function c, which gives way for further research. Also, because of linearizations, normality of distributions is an essential ingredient.

As IC technologies scale down to finer feature sizes, it becomes increasingly difficult to control the relative process variations, especially when introducing new technologies. Hence large variations can be observed. Furthermore, these variations are different for each chip fabrication location. Depending on the physical design properties of the circuit, the designer can introduce additional correlations to the given distributions, targeting improved circuit performance. Note that with the introduction of the s parameters, one must add additional constraints to the problem formulation. These are called yield constraints and their targets are expressed in terms of probabilistic quantities, typically the number of standard deviations from the mean value.

With fundamentally increasing variation magnitudes, non-normality of performance distributions must be taken into account. The APEX algorithm[12] addresses this point. A response surface model of f is built that is quadratic in the process variations. The method relies (yet) on the explicit calculation of the stochastic moments of f and on AWE-techniques (with their drawbacks) to efficiently approximate the transfer function. Here more robust model order reduction techniques may be applied.

A design that is capable of achieving the performance targets under the given constraints, across all so-called environmental corner conditions can be called robust. Robustness of a design starts to play an increasingly important role in modern circuit design, as design margins are decreasing and process variability is increasing. The use of traditional nominal-design optimization techniques is falling short because they tend to give seemingly good results, but these results collapse under process variations. Therefore, automation of the robust design challenge is in high demand.

5. A Robust Design Example

We now describe a simplified but realistic example of robust design. The purpose of this example is to get a better idea of the intricacies and challenges of optimization applied to circuit design.

The design is a high-frequency divide-by-two circuit fabricated in a state-of-the-art Complementary Metal-Oxide Semiconductor (CMOS) technology. The primary target of the optimization job is to increase the maximum input frequency at which the circuit still correctly divides by two, i.e. the output frequency is half the input frequency across all specified environmental conditions. Denoting with $\omega(v) = \text{freq}(v)$, the frequency of a (scalar) signal $v = v(\mathbf{x})$, that depends on parameters x_j, the problem can in terms of (1) be formulated as follows:

$$\begin{aligned} &\text{minimize} && -\omega(v_{in}(\mathbf{x})), \\ &\text{subject to} && \omega(v_{out}(\mathbf{x}))/\omega(v_{in}(\mathbf{x})) - 0.5 \leq 0 \\ &&& 0.5 - \omega(v_{out}(\mathbf{x}))/\omega(v_{in}(\mathbf{x})) \leq 0 \\ &&& a_j \leq x_j \leq b_j, \qquad j = 1, \ldots, n. \end{aligned} \tag{12}$$

In this formulation, the x_j are coupled to the widths of selected transistors in the divider circuit, where typically the number of selected transistors is at least $2n$ to enforce so-called "matched pairs" of transistors. The lengths of the transistors are kept fixed at a reasonably small value guided by design experience.

The first practical issue that comes into play is how to "measure" the frequency $\omega(v)$. As the circuit exhibits a non-linear large-signal behavior, we need to use a transient analysis to measure its response. This calls for a robust measurement expression that *always* gives a meaningful result, even if the simulation does not converge. Devising robust and accurate simulator expressions is a non-trivial task that requires design and simulation tool knowledge.

Because we also want to have a *robust* solution that is inherently insensitive to process variations, we typically resort to using the well-known Monte Carlo (MC) analysis (placed as a loop around the former transient analysis). To limit the overall simulation and optimization time, a limit of ten MC trials per circuit evaluation with a fixed set of p_j values was enforced. Ten MC trials are not sufficient to guarantee so-called sign-off accuracy, but it appears to be sufficient for typical optimization purposes.

Another important aspect of applied optimization is the fact that the original objective(s) and constraints typically need to be transformed and tuned into a set of new equations that give more directional information to the optimization algorithm; if this process is done well, most algorithms will usually give better results in fewer iterations. This is also the case for this design example. We add two more constraints to prevent getting an output voltage swing that is too small to be properly detectable by the succeeding circuit. Design-specific knowledge and feedback from the circuit behavior during initial optimization runs is normally used to refine the optimization setup further.

The complete optimization problem is:

$$\begin{aligned}
\text{minimize} \quad & -\omega(v_{in}(\mathbf{x}, \mathbf{s}, \boldsymbol{\theta})), && (13)\\
\text{subject to} \quad & \omega(v_{out}(\mathbf{x}, \mathbf{s}, \boldsymbol{\theta}))/\omega(v_{in}(\mathbf{x}, \mathbf{s}, \boldsymbol{\theta})) \leq 0.5 && (14)\\
& \omega(v_{out}(\mathbf{x}, \mathbf{s}, \boldsymbol{\theta}))/\omega(v_{in}(\mathbf{x}, \mathbf{s}, \boldsymbol{\theta})) \geq 0.5 && (15)\\
& \min_t \left(v_{out}(\mathbf{x}, \mathbf{s}, \boldsymbol{\theta})(t)\right) \geq 0.9 && (16)\\
& \max_t \left(v_{out}(\mathbf{x}, \mathbf{s}, \boldsymbol{\theta})(t)\right) \leq 0.1 && (17)\\
& a_j \leq x_j \leq b_j, \;\; j = 1, \ldots, n.
\end{aligned}$$

Solving this problem with the optimization algorithm described in this paper required about 200 iterations, depending on the stopping criterion and other tuning parameters. In each iteration the algorithm evaluates the circuit performance by running a 10-trial MC simulation in which a transient analysis is embedded. The

overall throughput time is the number of iterations times the time required for a full circuit evaluation; approximately 15 hours on a contemporary Linux system with a single CPU. From this information one can directly derive the average CPU time per 10-trial MC simulation: about 5 minutes. Note also that the time required for an MC simulation can become prohibitive quite quickly. This has motivated recent research in this area with the goal of finding alternative methods to efficiently simulate performance variability under process parameter variability, e.g.[12]
Table 1 shows the optimization results in tabular format. The information on the optimization variables x_j is given in Table 2. To illustrate the increase in robustness of the divider circuit we show the results of a 100-trial MC simulation before and after optimization. Fig. 1(a) shows the unoptimized circuit performance at the maximized input frequency of 17.1GHz (from 10.5GHz). All graphs represent histograms; in typewriter order the graphs show bin count versus output frequency, safeguarded frequency ratio (with voltage thresholds), raw frequency ratio, minimum "high" output voltage, maximum "low" output voltage, and average power dissipation. Fig. 1(b) shows the performance of the divider circuit after optimization. From these plots we can draw the following conclusions:

Table 1. Optimization results (all values in μm)

Specification	Target	Initial	Optimized
(13)	maximize	10.5 GHz	17.1 GHz
(14) + (15)	= 0.5 V	[0.4999994, 0.5000011]	[0.4999991, 0.5000001]
(16)	> 0.9 V	[1.01077, 1.018599]	[0.9612433, 1.014029]
(17)	< 0.1 V	[-0.017398, -0.011710]	[-0.016031,-0.011626]
power dissipation		1.25 mW, [1.14, 1.41]	1.36 mW, [1.21, 1.51]

Table 2. Optimization variables data (in μm)

OV	Initial	$[a_j, b_j]$	Final
x_1	2.5	[0.5, 7.5]	1.187
x_2	1.0	[0.5, 3.0]	1.195
x_3	4.0	[0.5, 12]	5.974
x_4	2.5	[0.5, 7.5]	1.638
x_5	4.0	[0.5, 12]	5.436
x_6	10.0	[0.5, 20]	14.13

- The performance of the divider circuit has been improved by increasing the maximum input frequency by more than 60%, with high robustness;

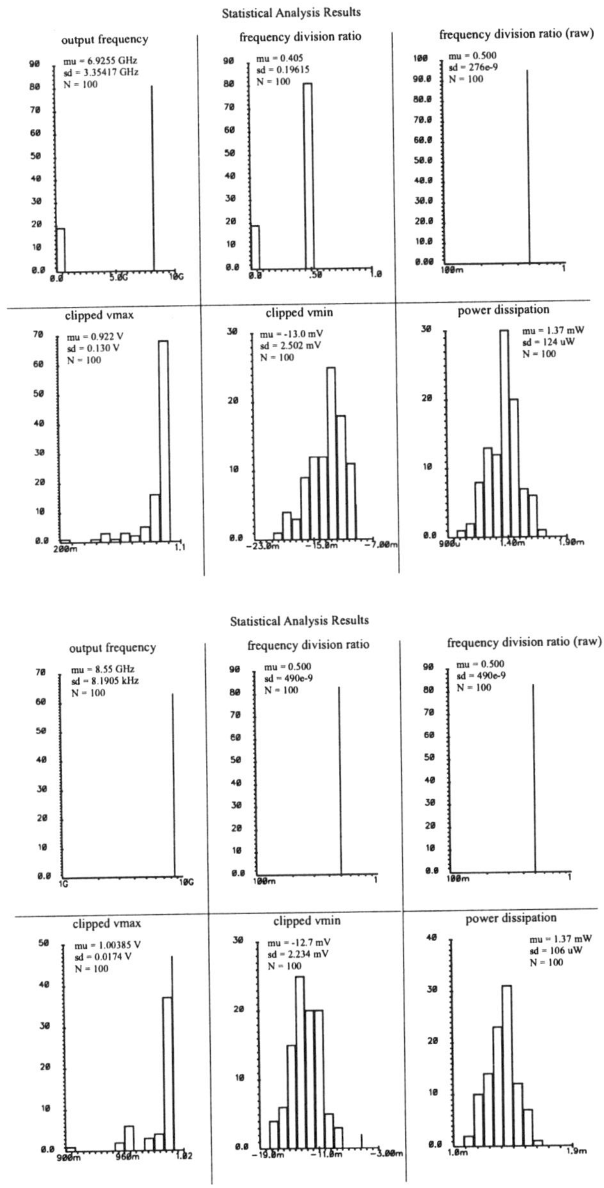

Fig. 1. (a) The divide-by-two circuit operating at 17.1GHz before optimization (top); and (b) the circuit operating at 17.1GHz after optimization (bottom).

- The optimized divider circuit can operate equally well at a maximized input frequency of 17.1GHz, with only a relatively small increase in power dissipation compared to 10.5GHz operation;
- Attaining increased robustness did not deteriorate any other metric.

References

1. J.S. Arora, M.W. Haug: *Methods for optimization of nonlinear problems with discrete variables: A review*, Struct. Opt., 8, pp. 69–85, 1994.
2. A.R. Conn, P.K. Coulman, R.A. Haring, G.L. Morrill, C. Visweswariah, C.W. Wu: *JiffyTune: cicuit optimization using time-domain sensitivities*, IEEE Trans. Computer-Aided Design, 17, pp. 1292–1309, 1998.
3. G. Di Pillo, L. Palagi: *Nonlinear programming: introduction, unconstrained and constrained optimization*, In: P.M. Pardalos, M.G.C. Resende (Eds): *Handbook of applied optimization*, Oxford Univ. Press, pp. 263–298, 2002.
4. C. Elster, A. Neumaier: *A grid algorithm for bound constrained optimization of noisy functions*, IMA J. Numer. Anal., Vol. 15, pp. 585–608, 1995.
5. A.A. Groenwold, N. Stander, J.A. Snyman: *A pseudo-discrete rounding method for structural optimization*, Struct. Opt., 11, pp. 218–227, 1996.
6. T. Heijmen, C. Lin, J. ter Maten, M. Kole: *Augmented Lagrangian algorithm for optimizing analog circuit design*, in: A. Buigis, R. Ciegis, A.D. Fit: *Progress in industrial mathematics at ECMI 2002*, Mathematics in Industry 5, Springer, pp. 179–184, 2003.
7. R. Hooke, T.A. Jeeves: *Direct search solution of numerical and statistical problems*, J. of the ACM, 8, pp. 212–229, 1961.
8. Z. Ilievski, H. Xu, A. Verhoeven, E.J.W. ter Maten, W.H.A. Schilders, R.M.M. Mattheij: *Adjoint Transient Sensitivity Analysis in Circuit Simulation*, Presented at Scientific Computing in Electrical Engineering, Sinaia, Romania, 2006.
9. D.R. Jones, M. Schonlau, W.J. Welch: *Efficient global optimization of expensive black-box functions*, J. of Global Optimization, Vol. 13, pp. 455–492, 1998.
10. T.G. Kolda, V. Torcson: *On the convergence of asynchronous parallel pattern search*, SIAM J. Optim, Vol. 14-4, pp. 939–964, 2004.
11. R.M. Lewis, V. Torczon: *A globally convergent augmented lagrangian pattern search algorithm for optimization with general constraints and simple bounds*, SIAM J. Optim., 12-4, pp. 1075–1089, 2002.
12. X. Li, J. Le, P. Gopalakrishnan, L.T. Pileggi: *Asymptotic probability extraction for non-normal distributions of circuit performance*, Proc. ICCAD 2004.
13. M. Pronath, H. Graeb, K. Antreich: *On parametric test design for analog integrated circuits considering error in measurement and stimulus*, In: K. Antreich, R. Bulirsch, A. Gilg, P. Rentrop (Eds.): *Modeling, simulation and optimization of integrated circuits*, In. Series of Numer. Maths., Vol. 146, pp. 283–301, 2003.
14. S. Rajeev, C.S. Krishnamoorthy: *Discrete optimization of structures using genetic algorithms* J. of Struct. Engineering, 118-5, pp. 1233–1250, 1992.
15. J.F. Rodriguez, J.E. Renaud, L.T. Watson: *Convergence of trust region augmented Lagrangian methods using variable fidelity approximation data*, Structural Optim., Vol. 15, pp. 141–156, 1998.
16. M.J. Sasena: *Flexibility and efficiency enhancements for constrained global design optimization with Kriging approximations*, PhD-Thesis, Univ. of Michigan, 2002.
17. Y. Young Soon, K. Gi Hwa: *Stochastic search techniques for global optimization of structures* In C.K. Choi, H. Sugimoto, C.B. Yun (Eds): *Proceedings of the Korea-Japan Joint Seminar on Structural Optimization, Seoul, Korea*, pp. 87–97, 1992.

TRANSMISSION PHENOMENA ACROSS HIGHLY CONDUCTIVE INTERFACES

L. TERESI

Dipartimento di Strutture,
Università degli Studi Roma Tre,
Via Corrado Segre 6, I-00146 Roma, Italy
E-mail: teresi@uniroma3.it

E. VACCA

Dipartimento di Metodi e Modelli Matematici per la Tecnologia e la Società,
Università degli Studi di Roma "La Sapienza",
Via Antonio Scarpa 16, I-00161 Roma, Italy
E-mail: vacca@dmmm.uniroma1.it

We deal with transmission phenomena across highly conductive interfaces, investigating the limit behavior of solutions for second order transmission problems. In particular, we investigate via finite element method the influence of thermal conductivity on the heat flux flowing into an interface.

Keywords: Heat transmission phenomena; highly conductive interfaces: Dirac heat flow.

1. Introduction

A bulk material property like thermal conductivity may differ by many orders of magnitude, for example, from the $\approx 2 \times 10^3$ Watt per meter per Kelvin of the perfect crystalline structure of diamond, to the $\approx 4 \times 10^{-2}$ Watt per meter per Kelvin of insulating fiberglass. Thus, for problems involving some materials whose bulk conductivities differ highly, it is important to investigate transmission phenomena in the highly conductive ones, even if their bulk volumes are negligible with respect to the insulating ones.

Moreover, several phenomena in nature concern irregular interfaces: for instance current flows through rough electrodes in electrochemistry or steady-state diffusion processes across irregular physiological membranes with finite permeability (see Ref. [4,13]). Thus fractal and prefractal geometries appear to be the natural mathematical tools to model those interfaces.

The aim of this paper is to present some numerical results on transmission phenomena across highly conductive interfaces. We deal with a model second order transmission problem which was firstly introduced by H. Pham Huy and E. Sanchez Palencia in [12] for a three-dimensional domain having a flat interface.

An analogous problem for the two-dimensional case has been solved by M.R. Lancia and M.A. Vivaldi [8] by proving existence, uniqueness and regularity results for the variational solution. Then, the same authors consider the transmission problem involving prefractal and fractal interface of von Koch type; they prove existence, uniqueness and regularity results for the variational solution for a prefractal interface in [8] and for a fractal one in [6,7]. Moreover, they analyze the asymptotic behavior of the solutions of prefractal transmission problems by considering them as approximations to the limit fractal problem (see Ref. [9]).

The finite element approximation that we provide in this paper concerns the steady-state heat diffusion treated in [8] and involves a highly conductive prefractal interface of von Koch type. More precisely, we consider a sequence of prefractal transmission problems $\{(P_n)\}_{n\in\mathbb{N}_0}$, where we denote by (P_n) the problem posed over the domain having the n-th prefractal curve K_n as interface (thus K_0 is a flat interface). The transmission condition imposed on the interface models the presence of a *highly conductive* material that, despite its negligible thickness, is the seat of a non negligible heat flow.

2. Physics of heat transmission

We consider the classical problem of heat conduction, that we pose for rigid bodies whose dimensions differ greatly; we begin with a simple, yet non trivial example, consisting of two squares in contact through a conductive 1D interface. In particular, we aim at investigating how heat flow is affected by the ratio between the conductivity coefficients of a 2D and a 1D conductor in contact each other.

2.1. *Balance laws*

Let us consider the 2D square $\Omega = [0,1]\times[0,1]$ (Fig. 1, left), whose boundary has constant temperature $u = 0$, and with an homogeneous bulk heat source q. Denoting by Γ the heat flux, by n the outward normal to the boundary $\partial\Omega$, by dA and ds the area and length measure respectively, we may write

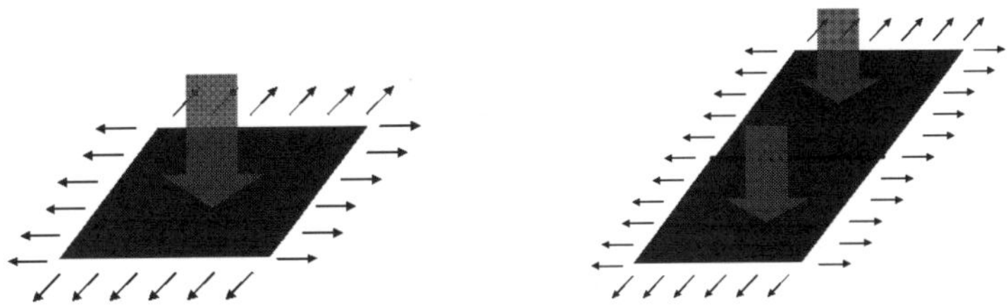

Fig. 1. Pictorial view of heat conduction problems; large gray arrows represent the external source, black arrows the outward flux.

the following balance law

$$\int_{\Omega} q\, dA = \int_{\partial\Omega} \Gamma \cdot n\, ds\,, \tag{1}$$

stating that the heat entering in the system–the left term–equals that flowing out from it–the right term. The total heat is $q\,A$ with A the area of the square; it is obvious from the symmetry of the problem that from each side exits the same amount of heat $q\,A/4$. Now, let us consider two squares $\Omega_u = [0,1] \times [0,1]$ and $\Omega_l = [0,1] \times [-1,0]$, sharing a same edge $[0,1] \times \{0\}$, having a constant temperature $u = 0$ on $\partial(\Omega_u \cup \Omega_l)$ and an homogeneous bulk heat source q (Fig. 1, right); under these conditions, balance law (1) holds, provided $\Omega = \Omega_u \cup \Omega_l$. Circumstances change completely if the common edge is a 1D conductive interface, that we denote by $K_0 := [0,1] \times \{0\}$. In such a case, part of the outward flux from Ω_u and Ω_l constitutes a bulk source for K_0; heat can flows along the interface (see exploded view of Fig. 2), and another balance equation must be considered. Let us denote by Γ_u, Γ_l the heat flux in the upper and the lower domain respectively, by n_u, n_l the respective outward normal and by Γ_c the heat flux along the interface; we may write the following integral balance equations

$$\textit{2D balance:} \int_{\Omega} (-\Gamma \cdot \nabla\tilde{u} + q \cdot \tilde{u})\, dA = 0, \ \forall\, \tilde{u} \,\text{test on}\, \Omega\,, \tag{2}$$

$$\textit{1D balance:} \int_{K_0} \Big(-\Gamma_c \cdot \nabla_t \tilde{u} + (\Gamma_u \cdot n_u + \Gamma_l \cdot n_l)\, \tilde{u}\Big)\, ds = 0, \ \forall\, \tilde{u} \,\text{test on}\, \Omega\,, \tag{3}$$

where ∇_t is the tangential gradient on K_0, that is, $\nabla_t\, u := \nabla(u|_{K_0})$. Let us note that the unique heat source for the whole system is q; then, heat can flow out from the system across $\partial\Omega$–as heat flow per unit length, and across ∂K_0, consisting of the two end points–as Dirac heat flows.

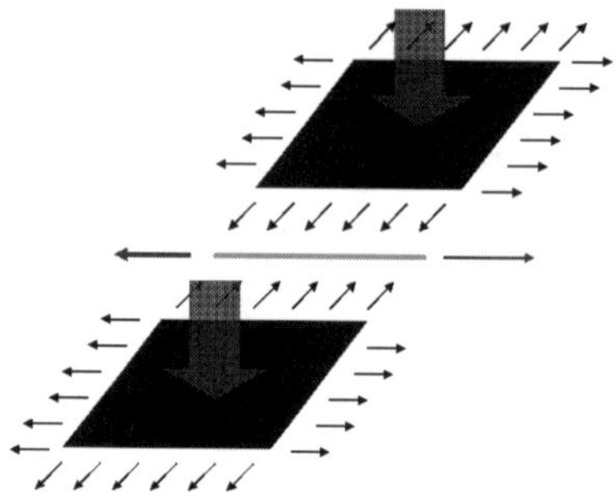

Fig. 2. Exploded view of heat conduction in two bodies connected through a conductive flat interface.

2.2. *Constitutive relations*

Flux is related to the temperature via a constitutive relation; in our case, we have

$$\Gamma = k\,\nabla u\,, \mathrm{on}\,\Omega\,; \quad \Gamma_c = k_c\,\nabla u\,, \mathrm{on}\,K_0\,. \tag{4}$$

Let a square bracket $[\cdot]$ denote the physical dimension of a quantity, and let L, P and θ denote measure of length, power and temperature, respectively; we have

$$\begin{array}{ll} [q] = P/dA = P/L^2 & \text{power per unit area,} \\ [\Gamma] = P/ds = P/L & \text{power per unit lenght,} \\ [k] = P/dA\,/(\theta/L) = P/\theta & \text{conductivity (2D),} \\ [\Gamma_c] = P & \text{power,} \\ [k_c] = P/(\theta/L) = P\,L/\theta & \text{conductivity (1D);} \end{array} \tag{5}$$

thus, the ratio between 1D and 2D conductivities has the physical dimension of a length: $[k_c/k] = L$. Equation (3) can be written exploiting constitutive relations as

$$\int_{K_0} \Big(-\frac{k_c}{k}\,\nabla_t u\ \cdot \nabla_t \tilde{u} + (\nabla u \cdot n_u + \nabla u \cdot n_l)\,\tilde{u} \Big)\,ds = 0,\ \forall\,\tilde{u}\,\text{test on}\,\Omega\,. \tag{6}$$

Let us assume k be a fixed positive real number throughout the whole paper. Previous equation shows how the ratio $[k_c/k] = L$ can influence the heat flow in the system; for $k_c = 0$, the interface is neutral and heat flows out entirely from $\partial\Omega$; for $k_c > 0$, heat flows along the interface, and two concentrated heat fluxes appear on the two points constituting the boundary of K_0; when $k_c \to \infty$, the Dirac heat fluxes amount as $q\,A/4$, that is, the interface drains the whole heat that would have escaped from the two sides of the squares it connects (as if the squares were separated),

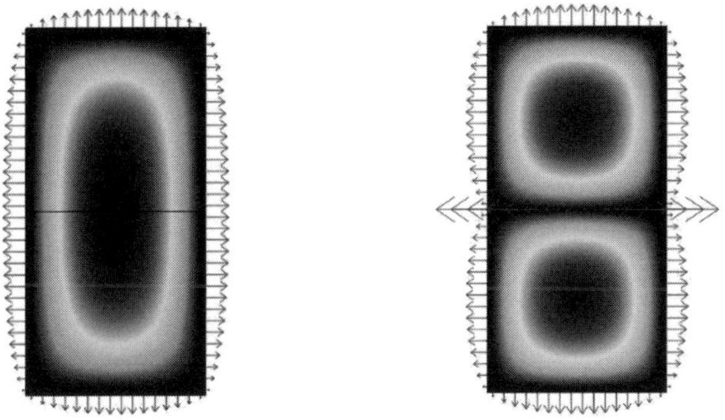

Fig. 3. Solution of heat conduction with a neutral (left) and highly conductive interface (right); colormap denotes the temperature field, arrows the boundary flux

and drive it to the boundary points, (Fig. 3); moreover, temperature on the interface tends to zero.

3. The Von Koch curve

We describe an iterative construction of the von-Koch fractal curve, which is an example of "*nested fractal*" according to the definition introduced by T. Lindstrøm in [10]. Each iteration generates a step of the construction, that is a von Koch prefractal curve. Let K_0 be defined as in Section 2. Let K_1 be the curve obtained by dividing K_0 in three equal parts, removing the central segment and replacing it by the other two sides of the equilateral triangle based on the removed segment. The prefractal curve K_2 can be constructed by applying the same procedure to each of the segments of the curve K_1. Iterating this procedure, we construct a sequence of prefractal polygonal curves K_n which tends in the Hausdorff metric to the limit "von Koch fractal curve" K, as n tends to infinity. We show in Fig. 4 some prefractal curves and we refer to the books by K. Falconer [3] and by B.B. Mandelbrot [11] for more details on fractal geometry and its applications. The same construction holds for any line segment K_0. We observe that the von-Koch curve is a *self-similar fractal* as the usual definition given by J.E. Hutchinson in [5]. Roughly speaking, this means that every part of the von Koch fractal curve is similar to the whole or to some copies of it. T. Lindstrøm subsequently gave in [10] an alternative construction of the von Koch curve proving so that K is also a "*nested fractal*". He defined a sequence of "nested sets" $\{V^n\}_{n\in\mathbb{N}_0}$, where V^n is the set of vertices of K_n, and he proved that $V^n \subset V^{n+1}$, $\forall n \in \mathbb{N}_0$. The fractal curve K can be seen

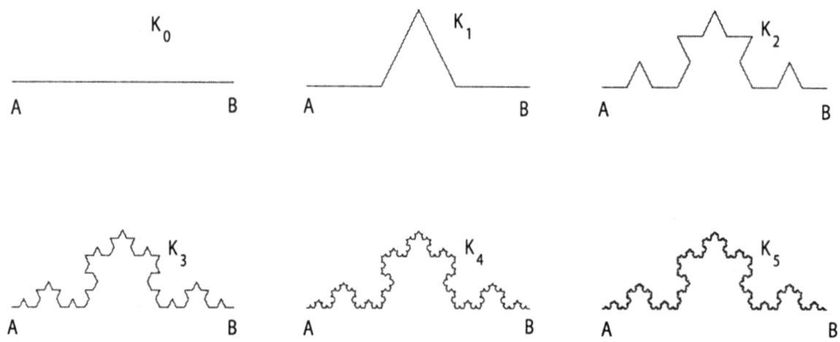

Fig. 4. The prefractal curves $K_0, \ldots, K_5$.

as the closure of the infinite union of all V^n, with respect to the Hausdorff metric. The following plots (see Fig. 5) show that just after five iterations, K_n has more than one thousand segments and its length is four times greater than the initial one. After fifteen iterations, there are more than 10^9 segments and the curve length has increased seventy times; actually, the limit fractal curve K has infinite length, and its Hausdorff dimension is equal to $d_f = \ln 4/\ln 3$ (see Ref. [3]).

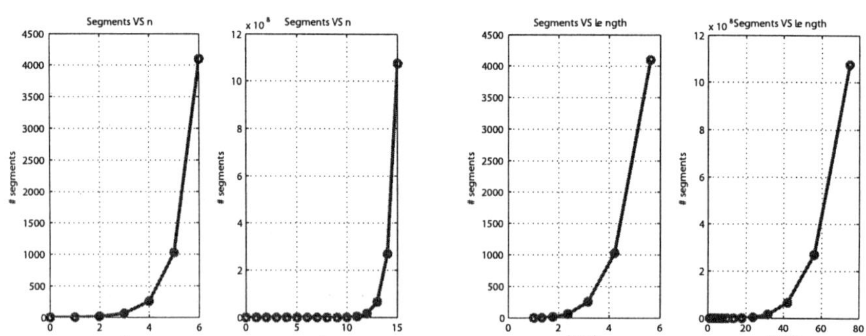

Fig. 5. Number of segments and the length of K_n, as n tends to infinity.

4. A highly conductive transmission model problem

In this Section we describe the model second order transmission problem related to the integral balance equations (2) and (3). It was firstly treated by H. Pham Huy and E. Sanchez Palencia in [12] in a 3D domain having a flat interface. Then, M.R. Lancia and M.A. Vivaldi analyzed the same problem in a 2D domain in three different cases in accordance with the interface's geometry (see Ref. [6,8]). The transmission condition imposed on interface is called of second order due to the presence of a tangential laplacian. This condition is used to account for the physics of highly conductive materials with vanishing thickness, modeled as flat interfaces where intrinsic one-dimensional diffusion phenomena occur.

4.1. *Flat interface*

We consider $\Omega = \Omega_u \cup \Omega_l \cup K_0$ and the interface K_0 as defined in Section 2. We denote by $L^2(\Omega)$ the usual Lebesgue space, by $H_0^s(\Omega)$, $s \in \mathbb{R}$, the usual Sobolev spaces and by $C^0(\overline{\Omega})$ the space of continuous functions (for details see [14]). Let Γ, Γ_c be defined as in (4), the heat source q a function in $L^2(\Omega)$, and $[u] = u|_{\Omega_u} - u|_{\Omega_l}$ the jump of u across K_0. Balance laws (2) and (3) may be interpreted as a 0-th transmission problem, or a *flat transmission problem*; they can be stated in differential form as follows

$$(P_0) \begin{cases} \operatorname{div}(\Gamma) + q = 0 & \text{in } \Omega_u \cup \Omega_l\,, \\ \operatorname{div}(\Gamma_c) + [\Gamma \cdot n] = 0 & \text{on } K_0\,, \\ [u] = 0 & \text{across } K_0\,, \\ u = 0 & \text{on } \partial\Omega\,, \end{cases} \tag{7}$$

We observe that $[\Gamma \cdot n] = k\left[\frac{\partial u}{\partial n}\right]$, and the jump of normal derivative is $\left[\frac{\partial u}{\partial n}\right] = \left(\frac{\partial(u|_{\Omega_u})}{\partial n_u} + \frac{\partial(u|_{\Omega_l})}{\partial n_l}\right)$. Using constitutive relations, the second order transmission condition $(7)_{ii}$ can be reformulated as

$$-\frac{k_c}{k}\,\Delta_t\, u = \left[\frac{\partial u}{\partial n}\right] \qquad \text{on } K_0\,, \tag{8}$$

where Δ_t is the tangential laplacian and the characteristic length $\left[\frac{k_c}{k}\right] = L_c$ is set equals to the length of K_0. The jump of normal derivative is a *source term* for the tangential laplacian that can generate heat flow inside K_0. We denote by

$$V(\Omega, K_0) := \{v \in H_0^1(\Omega) \,:\, v|_{K_0} \in H_0^1(K_0)\}, \tag{9}$$

the "test functional space" endowed with the natural graph norm

$$\| v \|_{V(\Omega,K_0)} := \left\{ \int_\Omega |v|^2 + \int_{K_0} |v|^2 \right\}^{1/2}. \tag{10}$$

The integral representation of balance laws (2) and (3) becomes now

$$\int_{\Omega} \Gamma \cdot \nabla \tilde{u} \, dA + \int_{K_0} \Gamma_c \cdot \nabla_t \tilde{u} \, ds = \int_{\Omega} q \cdot \tilde{u} \, dA, \quad \forall \tilde{u} \in V(\Omega, K_0) \tag{11}$$

and it suggests the following definition of a *flat energy form* for transmission problems across a flat conductive interface

$$a_0(u, \tilde{u}) = \int_{\Omega} \nabla u \cdot \nabla \tilde{u} \, dA + \frac{k_c}{k} \int_{K_0} \nabla_t u \cdot \nabla_t \tilde{u} \, ds , \quad \forall \tilde{u} \in V(\Omega, K_0) \tag{12}$$

Proposition 4.1. *For every q in $L^2(\Omega)$, the problem (P_0) is equivalent to the following well-posed variational problem: find u in $V(\Omega, K_0)$ such that*

$$a_0(u, \tilde{u}) = \int_{\Omega} q \, \tilde{u} \, dA , \quad \forall \tilde{u} \in V(\Omega, K_0). \tag{13}$$

There exists one and only one solution u of (13) such that $u \in C^0(\overline{\Omega})$, $u|_{\Omega_u} \in H^2(\Omega_u)$, $u|_{\Omega_l} \in H^2(\Omega_l)$ and $u|_{K_0} \in H^{\frac{5}{2}}(K_0)$.

Proof. For the proof of existence and uniqueness see Theorem 3.1 in [12] dealing with the more general 3D case. For the proof of regularity results see Ref. [8,14].

4.2. *Prefractal interface*

We consider $\Omega = [0,1] \times [-1,1]$ having as interface a prefractal curve K_n with endpoints $(0,0)$ and $(1,0)$. K_n splits Ω in two domains, the upper domain Ω_u and the lower one Ω_l, respectively. Thus $\Omega = \Omega_u \cup \Omega_l \cup K_n$. We denote by $H^{2,\alpha}(\Omega)$, α in $\mathbb{R}$ and $0 < \alpha < 1$, the usual weighted Sobolev spaces (for definition see [2]). With the same notations given in Subsection 4.1, for each n in $\mathbb{N}$, we can state the n-th differential problem

$$(P_n) \begin{cases} \operatorname{div}(\Gamma) + q = 0 & \text{in } \Omega_u \cup \Omega_l \,, \\ \operatorname{div}(\Gamma_c) + [\Gamma \cdot n] = 0 & \text{on } K_n \,, \\ [u] = 0 & \text{across } K_n \,, \\ u = 0 & \text{on } \partial\Omega \,, \end{cases} \tag{14}$$

which is called the n-th transmission problem or *n-th prefractal transmission problem.* Analogously, the second order transmission condition $(14)_{ii}$ rewrites as

$$-\frac{k_{cn}}{k} \Delta_t u = \left[\frac{\partial u}{\partial n}\right] \qquad \text{on } K_n \,, \tag{15}$$

with the characteristic length $L_c = L_{cn} = (\frac{4}{3})^n$ assumed equal to the length of K_n. Denoting by $V(\Omega, K_n)$ the "test functional space" analogous to (9), the integral representation of balance laws is given by

$$\int_{\Omega} \Gamma \cdot \nabla \tilde{u}\, dA + \int_{K_n} \Gamma_c \cdot \nabla_t \tilde{u}\, ds = \int_{\Omega} q \cdot \tilde{u}\, dA, \quad \forall \tilde{u} \in V(\Omega, K_n). \tag{16}$$

The *prefractal energy form* for transmission problems across a prefractal conductive interface is defined as follows

$$a_n(u, \tilde{u}) = \int_{\Omega} \nabla u \cdot \nabla \tilde{u}\, dA + \frac{k_{cn}}{k} \int_{K_n} \nabla_t u \cdot \nabla_t \tilde{u}\, ds, \quad \forall \tilde{u} \in V(\Omega, K_n). \tag{17}$$

Proposition 4.2. *For every q in $L^2(\Omega)$, the problem (P_n) is equivalent to the following well-posed variational problem: find u in $V(\Omega, K_n)$ such that*

$$a_n(u, \tilde{u}) = \int_{\Omega} q\, \tilde{u}\, dA, \quad \forall \tilde{u} \in V(\Omega, K_n). \tag{18}$$

There exists one and only one solution u of (18) such that $u \in C^0(\overline{\Omega})$, $u|_{\Omega_u} \in H^{2,\alpha_u}(\Omega_u)$, $\alpha_u > \frac{2}{5}$, $u|_{\Omega_l} \in H^{2,\alpha_l}(\Omega_l)$, $\alpha_l > \frac{1}{4}$ and $u|_{K_n} \in H^2(K_n)$.

Proof. For the proof see [8,14].
Moreover, for the variational solution of (18) it holds a quasi-optimal error estimate of order one obtained by using linear finite element over adaptively refined meshes (see Ref. [1,2]).

4.3. *Fractal interface*

For the sake of completeness, we cite the analytical treatment given by M.R. Lancia in [6] for the analogous fractal problem (P). Let $\Omega = \Omega_u \cup \Omega_l \cup K$ be the same 2D domain having the von Koch curve K as interface. Redefining suitably some analytical objects, such as the laplacian on K, we can state the following fractal differential problem

$$(P) \begin{cases} \operatorname{div}(\Gamma) + q = 0 & \text{in } \Omega_u \cup \Omega_l\,, \\ \operatorname{div}(\Gamma_c) + [\Gamma \cdot n] = 0 & \text{on } K\,, \\ [u] = 0 & \text{across } K\,, \\ u = 0 & \text{on } \partial\Omega\,, \end{cases} \tag{19}$$

which is called the limit transmission problem or *fractal transmission problem*. The *fractal energy form* is defined, on a suitable "test functional space" $V(\Omega, K)$, as follows

$$a(u, \tilde{u}) = \int_{\Omega} \nabla u \cdot \nabla \tilde{u}\, dA + \frac{k_{\infty}}{k} E(u|_K, \tilde{u}|_K), \quad \forall \tilde{u} \in V(\Omega, K), \tag{20}$$

where k_∞ is the fractal conductivity and $E(\cdot,\cdot)$ is the energy on fractal curve. For definition, existence, uniqueness and regularity results see Ref. [6].

4.4. *Asymptotic behaviour*

The asymptotic behavior as $n \to \infty$ for the solution of (18) has been dealt with by M.R. Lancia and M.A. Vivaldi in [9]. The idea is to regard to the problem (P_n) as the n-th prefractal problem approximating the limit fractal problem (P), for $n \to \infty$, provided a suitably choosen conductivity k_{cn}. Next Proposition follows from M-convergence in $L^2(\Omega)$ of the *approximating energy forms* $a_n(u, \tilde{u})$ to the *limit energy form* $a(u, \tilde{u})$.

Proposition 4.3. *Let us set* $k_{cn} = k_\infty \cdot L_{cn}$ *for all* n *in* $\mathbb{N}_0$. *Let* u_n *and* u *be the solution of the problem* (P_n) *and* (P), *respectively. The following asymptotic convergence results hold*

- $u_n \to u$ *strongly in* $H_0^1(\Omega)$;
- $u_n|_K \to u|_K$ *pointwise on* $\bigcup_{n=1}^{\infty} V^n$.

5. Numerical tests

We solve problem (P_n) via finite element method, up to $n = 6$; two different codes have been used, a commercial finite element code [15], using unstructured meshes, and a code written by E. Vacca in [14] for this specific problem, that uses linear finite element over adaptively refined meshes, with the aim of a quasi-optimal error estimate of order one. In accordance

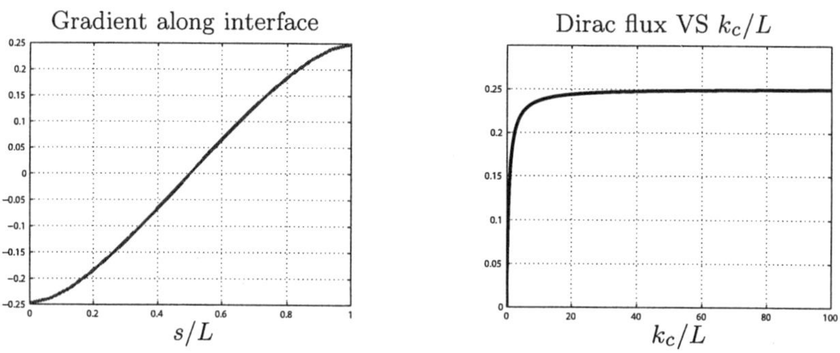

Fig. 6. Plots of $\nabla_t u(x, 0)$, x in $[0, 1]$, for the problem (P_0) (left) and Dirac heat flux $\Gamma_c(1, 0)$ (right).

with some remarks asserted in Subsection 2.2, we show in Fig. 3 two plots

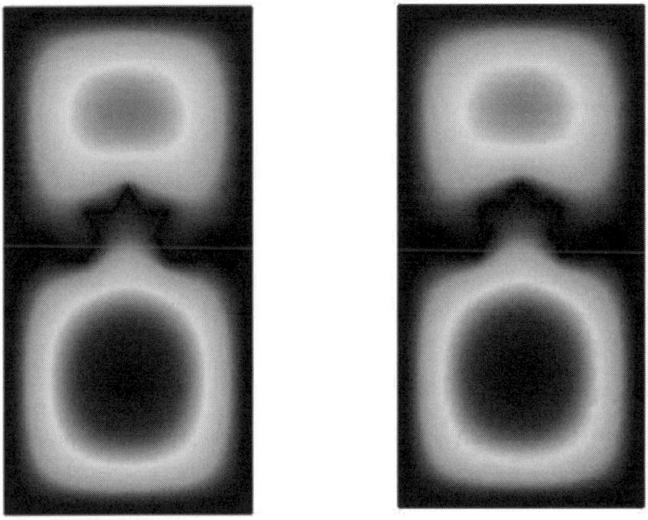

Fig. 7. Plots of temperature u for prefractal problem (P_2) (left) and $(P6)$ (right) with $k_\infty = 50$.

of temperature u solution of the flat transmission problem (P_0) for different values of k_c; more precisely $k_c = 0$ (left) and $k_c > 0$ (right). Actually, we show in Fig. 6 the graph of Dirac heat flux, at the right endpoint of K_0, $\Gamma_c(1,0) = k_c \cdot \nabla_t u(1,0)$, versus conductivity k_c. We note that just for $k_c = 20$ the flux value is nearly the limit value $\Gamma_c(1,0) = 0.25$ for $k_c \to \infty$.

Analogously, we show in Fig. 7 two plots of temperature u solution of the prefractal transmission problem (P_2) (left) and (P_6) (right) for $k_{cn} = k_\infty\,(4/3)^n$, $n = 2, 6$, coherently with Proposition 4.3 on the asymptotic behavior.

We note that temperature is nearly zero in a small neighborhood of the prefractal interface, even for a finite value of k_{cn}. As expected, the amount of heat which flows through the conductive interface K_n, growths very fast with k_{cn}, according to the previous asymptotic law. Finally, we show in Fig. 8 some graphs of Dirac heat flux, at the right endpoint of K_n, $\Gamma_{cn}(1,0)$, for the conductivity $k_{cn} = k_\infty\,(4/3)^n$ and for $n = 0, \cdots, 6$. The diagram on the right (see Fig. 8) compares the values of heat flux $\Gamma_{cn}(1,0)$, having fixed k_∞ equal to 100, for prefractal transmission problems (P_0), $(P_1), \cdots, (P_6)$. It seems that Dirac heat flux at the right endpoint converges quite fast to a limit value.

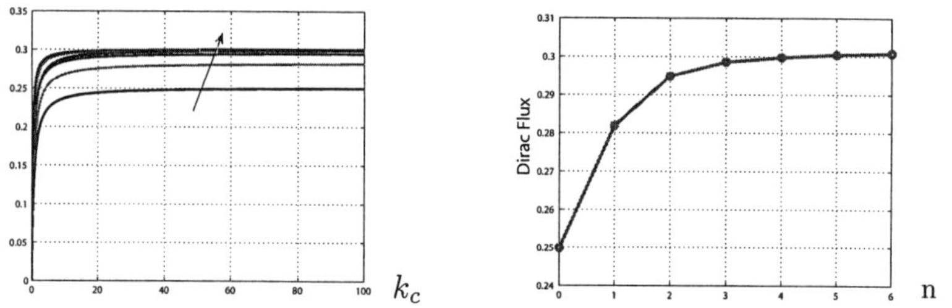

Fig. 8. Dirac heat flux $\Gamma_{cn}(1,0)$ at $n = 0, \cdots, 6$, versus k_c; arrow denotes increasing n (left); asymptotic values of Dirac heat fluxes for $P_0, ...P_6$ (right).

References

1. BAGNERINI P., BUFFA A., VACCA E. "Finite elements for a prefractal transmission problem", Comptes Rendus Mathematique (2006), Vol. 342, Issue 3, pp 211-214.
2. BAGNERINI P., BUFFA A., VACCA E. "Galerkin Method fo Highly Conductive Prefractal Layers", preprint del Dipartimento Me.Mo.Mat., Universitá "La Sapienza di Roma, 2/2006.
3. FALCONER K. "The geometry of fractal sets", 2nd edition Cambridge University Press, Cambridge 1990.
4. FILOCHE M., SAPOVAL B. *Transfer across random versus Deterministic Fractal Interfaces*, Phys. Rev. Lett., 84(2000), p. 5776-5779.
5. HUTCHINSON J.E. *Fractals and selfsimilarity*, Indiana Univ. Math. J.,30(1981), p. 713-747.
6. LANCIA M.R. *A transmission problem with a fractal interface*, Z. Anal. und Ihre Anwend., 21(2002), p. 113-133.
7. LANCIA M.R. *On some second order transmission problems*, Arab. J. Sci. Eng. Sect., 29 N.2C(2004), p. 85-100.
8. LANCIA M.R., VIVALDI M.A. *On the regularity of the solutions for transmission problems*, Adv. Math. Sc. Appl., 12(2002), p. 455-466.
9. LANCIA M.R., VIVALDI M.A. *Asymptotic convergence of transmission energy forms*, Adv. Math. Sc. Appl., 13(2003), p. 315-341.
10. LINDSTRØM T. *Brownian motion on nested fractals*, Memoirs Amer. Math. Soc., (420) 83(1990).
11. MANDELBROT B.B. "The fractal geometry of nature", Freeman, San Francisco, 1982.
12. PHAM HUY H., SANCHEZ-PALENCIA E. *Phénomènes des transmission à travers des couches minces de conductivité élevée*, J. Math. Anal. Appl., 47 (1974), p. 284-309.
13. SAPOVAL B. *General formulation of laplacian transfer across irregular surfaces*, Phys. Rev. Lett., 73(1994), p. 3314-3316.
14. VACCA E. *Galerkin Approximation for Highly Conductive Layers*, Ph.D.-Thesis, Dipartimento di Me.Mo.Mat., Universitá "La Sapienza di Roma, a.a. 2004-05.
15. COMSOL MULTIPHYSICS, release 3.2a; http://www.comsol.com.

SOME EXACT FORMULAS FOR THE POST-GELATION MASS OF THE COAGULATION EQUATION WITH PRODUCT KERNEL

H. VAN ROESSEL* and M. SHIRVANI

Department of Mathematical and Statistical Sciences, University of Alberta, Edmonton, Alberta, T6G 2G1, Canada
** E-mail: henry.vanroessel@ualberta.ca*
E-mail: mazi.shirvani@ualberta.ca
www.ualberta.ca

It is well known that solutions of the coagulation equation do not conserve mass if the coagulation kernel grows too rapidly. The phenomenon whereby conservation of mass breaks down in finite time is known as gelation and is physically interpreted as being caused by the appearance of an infinite "gel" or "superparticle." In this paper we discuss the post-gelation behaviour of the coagulation equation with product kernel. Several exact formulas for post gelation mass are given.

Keywords: coagulation, gelation, product kernel

1. Introduction

The general coagulation equation may be written as

$$\frac{\partial c}{\partial t}(\lambda,t) = \frac{1}{2}\int_0^\lambda K(\lambda-\mu,\mu)c(\lambda-\mu,t)c(\mu,t)\,d\mu - c(\lambda,t)\int_0^\infty K(\lambda,\mu)c(\mu,t)\,d\mu, \tag{1}$$

$$c(\lambda,0) = c_0(\lambda).$$

Equation (1) represents the evolution of particles $c(\lambda,t)$ of size $\lambda \geqslant 0$ at time $t \geqslant 0$ undergoing a change in size governed by reaction kernel K. A physical interpretation of the terms in (1) can be found in Melzak [6] or the review article by Drake [2].

Global existence and uniqueness of solutions to (1) has been proven by Aizenman and Bak [1] for constant kernels with suitable conditions imposed on the initial particle distribution c_0. For the product kernel $K(\lambda,\mu) = \lambda\mu$,

McLeod[5] has shown, again for a suitably restricted initial particle distribution, that a mass conserving solution exists for some finite time. That solutions in this case need not conserve mass for all time was demonstrated by Stewart [7]. The phenomenon whereby conservation of mass breaks down in finite time is known as gelation and is physically interpreted as being caused by the appearance of an infinite "gel" or "superparticle" (Ernst et al. [3]).

We consider the coagulation equation with the separable bilinear kernel

$$K(\lambda, \mu) = \theta(\lambda)\theta(\mu),$$

where $\theta(\lambda) = \alpha + \beta\lambda$, $\alpha, \beta \geqslant 0$. This kernel contains both the constant kernel and product kernel as special cases. For this kernel it is well known that gelation occurs in finite time if and only if $\beta > 0$.

2. Formulation as a PDE

The approach we employ is to transform Eq. (1) into a first order quasi-linear PDE using Laplace transforms. Let

$$u(x,t) := \int_0^\infty e^{-\lambda x}\theta(\lambda)c(\lambda,t)\,d\lambda; \tag{2}$$

$$N(t) := \int_0^\infty \theta(\lambda)c(\lambda,t)\,d\lambda; \tag{3}$$

$$h(x) := \int_0^\infty e^{-\lambda x}\theta(\lambda)c_0(\lambda)\,d\lambda. \tag{4}$$

Multiply Eq. (1) by $e^{-\lambda x}$ and integrate to obtain the following first-order PDE for the transform variable u:

$$\frac{\partial u}{\partial t} + \beta(u - N(t))\frac{\partial u}{\partial x} = \frac{\alpha}{2}u^2 - \alpha N(t)u, \qquad x,t > 0, \tag{5}$$

$$u(x,0) = h(x), \qquad x \geqslant 0, \tag{6}$$

$$u(0,t) = N(t), \qquad t \geqslant 0, \tag{7}$$

where (7), which comes from (2) and (3), is in effect a compatibility condition for u along $x = 0$. This initial value problem for u is unusual in that the function $N(t)$ in (5) and (7) is itself unknown and must be determined as part of the solution.

In principle one now proceeds to solve the PDE (using the method of characteristics or otherwise), then take the inverse Laplace transform to get the solution to the original coagulation equation. The problem, however, is that it is almost impossible to get an explicit solution to the PDE, let alone being able to compute its inverse Laplace transform. The only case for which an explicit solution has to date been found is for the pure product kernel ($\alpha = 0$) with mono-disperse initial conditions (see Ernst, et al. [3]).

The characteristic equations for (5) and (6) are:

$$\frac{dx}{dt} = \beta(z - N(t)), \qquad x\big|_{t=0} = \xi \geqslant 0, \tag{8}$$

$$\frac{dz}{dt} = \frac{\alpha}{2}z^2 - \alpha N(t)z, \qquad z\big|_{t=0} = h(\xi). \tag{9}$$

The equation for z is a Bernoulli equation which is easily solved. Once done the equation for x is then easily integrated. The result is:

$$x = X(\xi,t) = \xi - \frac{2\beta}{\alpha}\ln[1-\frac{\alpha}{2}h(\xi)w(t)] + \frac{\beta}{\alpha}\ln[w'(t)], \tag{10}$$

$$z = Z(\xi,t) = \frac{h(\xi)w'(t)}{1-\frac{\alpha}{2}h(\xi)w(t)}, \tag{11}$$

where

$$w(t) := \int_0^t e^{-\alpha\int_0^\tau N(\zeta)\,d\zeta}\,d\tau. \tag{12}$$

Thus, we have the solution to the characteristic equations in terms of N. However, from condition (7) we get an equation for N by setting $x = 0$ in (5) to get:

$$N'(t) = -\frac{\alpha}{2}N^2(t), \quad N(0) = h(0). \tag{13}$$

Of course, this equation can not be expected to hold if a shock occurs in (5) since $\partial u/\partial x$ would become unbounded there. Eq. (13) is only valid prior to the occurrence of a shock. Another equation is required to determine N in the post-shock regime.

Pre-shock solution

Prior to the occurrence of a shock N is governed by Eq. (13). Therefore

$$N(t) = \frac{h(0)}{1+\frac{\alpha}{2}h(0)t}, \quad \text{for} \quad t\in[0,T_0), \tag{14}$$

where the shock time T_0 has yet to be determined. Equations (10) and (11) become:

$$X(\xi,t) = \xi - \frac{2\beta}{\alpha}\ln[1+\frac{\alpha}{2}(h(0)-h(\xi))t], \qquad t\in[0,T_0), \tag{15}$$

$$Z(\xi,t) = \frac{h(\xi)}{(1+\frac{\alpha}{2}h(0)t)[1+\frac{\alpha}{2}(h(0)-h(\xi))t]}. \tag{16}$$

In principle at least, we can then invert

$$x = X(\xi, t) \qquad \text{to obtain} \qquad \xi = \widehat{\xi}(x,t)$$

and the solution to (5) is then

$$u(x,t) = Z(\widehat{\xi}(x,t),t), \quad \text{for} \quad t \in [0, T_0). \tag{17}$$

Shock-time

A shock occurs if a derivative of $\partial u/\partial x$ becomes infinite, which corresponds to $\partial X/\partial\xi = 0$. Differentiating (15) yields

$$\frac{\partial X}{\partial \xi}(\xi,t) = 1 + \frac{\beta h'(\xi)t}{1+\frac{\alpha}{2}(h(0)-h(\xi))t}.$$

Thus

$$\frac{\partial X}{\partial \xi}(\xi,t) = 0 \quad \Longrightarrow \quad t = \frac{-1}{\frac{\alpha}{2}(h(0)-h(\xi)) + \beta h'(\xi)} =: T(\xi).$$

Using the fact that h is completely monotonic, one can easily verify that T is an increasing function so that the shock-time is given by:

$$T_0 := \inf_{\xi \geqslant 0} T(\xi) = T(0) = -\frac{1}{\beta h'(0)}. \tag{18}$$

Post-shock solution

In order to use the solutions (10) and (11) to the characteristic equations for (5) in the post-shock regime we need an expression for N, or equivalently w, valid for $t \geqslant T_0$. Suppose we have such an N. Then it must satisfy

$$N(t) = u(0,t) = Z(\widehat{\xi}(0,t),t) = Z(\xi_0(t),t), \quad \text{for all} \quad t \in [0,\infty),$$

where

$$\xi_0(t) := \widehat{\xi}(0,t).$$

Thus, N is known once ξ_0 is known. The function ξ_0 is given implicitly by

$$X(\xi_0(t),t) \equiv 0. \tag{19}$$

Differentiation with respect to t yields

$$\frac{\partial X}{\partial \xi}(\xi_0(t),t)\,\xi_0'(t) + \frac{\partial X}{\partial t}(\xi_0(t),t) = 0.$$

Then, using (8) we obtain

$$\frac{\partial X}{\partial t}(\xi_0(t),t) = \beta(Z(\xi_0(t),t) - N(t)) = 0.$$

Thus, we conclude

$$\frac{\partial X}{\partial \xi}(\xi_0(t),t)\cdot \xi_0'(t)=0, \quad \text{for all} \quad t\in[0,\infty).$$

Prior to the shock, for $t \in [0, T_0)$, we have

$$\frac{\partial X}{\partial \xi}\neq 0 \quad \Longrightarrow \quad \xi_0'(t)=0 \quad \Longrightarrow \quad \xi_0(t)\equiv 0.$$

Post-shock, for $t \geqslant T_0$, we have $\partial X/\partial\xi = 0$, which means that ξ_0 satisfies both Eq. (19) and

$$\frac{\partial X}{\partial \xi}(\xi_0(t),t)=0. \tag{20}$$

With ξ_0 required to satisfy both (19) and (20), it would seem that the problem is over determined. But this is not the case since, for $t \geqslant T_0$, N, or equivalently w, is also unknown. This becomes more clear if we write (10) and (11) as follows:

$$X(\xi,t)=F(\xi,w(t),w'(t)),$$
$$Z(\xi,t)=G(\xi,w(t),w'(t)),$$

where

$$F(\xi,\zeta_1,\zeta_2):=\xi-\frac{2\beta}{\alpha}\ln[1-\frac{\alpha}{2}h(\xi)\zeta_1]+\frac{\beta}{\alpha}\ln\zeta_2, \tag{21}$$

$$G(\xi,\zeta_1,\zeta_2):=\frac{h(\xi)\zeta_2}{1-\frac{\alpha}{2}h(\xi)\zeta_1}. \tag{22}$$

Then ξ_0 and w are determined from the system

$$\begin{aligned} &F(\xi_0(t),w(t),w'(t))=0,\\ &\frac{\partial F}{\partial \xi}(\xi_0(t),w(t),w'(t))=0. \end{aligned} \tag{23}$$

One can easily verify that for $t\in[0,T_0)$ we have $\xi_0(t)\equiv 0$ and system (23) reduces to

$$F(0,w(t),w'(t))=0,$$

from which we recover w, and hence N, obtained earlier in Eq. (14).

Writing Eq. (23) explicitly we have

$$\xi_0-\frac{2\beta}{\alpha}\ln(1-\frac{\alpha}{2}h(\xi_0)w)+\frac{\beta}{\alpha}\ln w'=0 \tag{24}$$

$$1+\frac{\beta h'(\xi_0)w}{1-\frac{\alpha}{2}h(\xi_0)w}=0. \tag{25}$$

From (25) we can solve for w:

$$w(t) = \frac{1}{\frac{\alpha}{2} h(\xi_0(t)) - \beta h'(\xi_0(t))}. \tag{26}$$

Insertion of (26) into (24) leads to:

$$\xi_0' = \frac{\beta^2 h'^2(\xi_0)}{\beta h''(\xi_0) - \frac{\alpha}{2} h'(\xi_0)} e^{-\alpha \xi_0/\beta}, \qquad \xi_0(T_0) = 0. \tag{27}$$

3. Post-Shock Mass

If we multiply (1) by $\lambda e^{-\lambda x}$ and integrate, we get

$$\int_0^\infty \lambda e^{-\lambda x} \frac{\partial c}{\partial t}(\lambda, t)\, d\lambda = \frac{\partial u}{\partial x}(x,t)[N(t) - u(x,t)].$$

Thus, the mass satisfies

$$\frac{dM}{dt}(t) = \lim_{x \to 0+} \frac{\partial u}{\partial x}(x,t)[N(t) - u(x,t)], \qquad M(0) = h(0). \tag{28}$$

But $u(x,t) = Z(\widehat{\xi}(x,t), t)$, so

$$\frac{\partial u}{\partial x} = \frac{\partial Z}{\partial \xi}(\widehat{\xi}, t) \frac{\partial \widehat{\xi}}{\partial x} = \frac{\frac{\partial Z}{\partial \xi}(\widehat{\xi}, t)}{\frac{\partial X}{\partial \xi}(\widehat{\xi}, t)},$$

thus

$$\frac{dM}{dt}(t) = \lim_{x \to 0+} \frac{\frac{\partial Z}{\partial \xi}(\widehat{\xi}, t)}{\frac{\partial X}{\partial \xi}(\widehat{\xi}, t)} [N(t) - Z(\widehat{\xi}(x,t), t)].$$

Using the definition of F and G this becomes

$$\frac{dM}{dt}(t) = \lim_{x \to 0+} \frac{N(t) - G(\widehat{\xi}(x,t), w(t), w'(t))}{\frac{\partial F}{\partial \xi}(\widehat{\xi}(x,t), w(t), w'(t))} \frac{\partial G}{\partial \xi}(\widehat{\xi}(x,t), w(t), w'(t))$$

Using l'Hopital's rule we get

$$\lim_{x\to 0^+} \frac{N(t) - G(\widehat{\xi}(x,t), w(t), w'(t))}{\dfrac{\partial F}{\partial \xi}(\widehat{\xi}(x,t), w(t), w'(t))} = \lim_{x\to 0^+} \frac{-\dfrac{\partial G}{\partial \xi}(\widehat{\xi}(x,t), w(t), w'(t)) \cdot \dfrac{\partial \widehat{\xi}}{\partial x}}{\dfrac{\partial^2 F}{\partial \xi^2}(\widehat{\xi}(x,t), w(t), w'(t)) \cdot \dfrac{\partial \widehat{\xi}}{\partial x}}$$

$$= \frac{-\dfrac{\partial G}{\partial \xi}(\xi_0(t), w(t), w'(t))}{\dfrac{\partial^2 F}{\partial \xi^2}(\xi_0(t), w(t), w'(t))}.$$

Thus

$$M'(t) = -\frac{\left(\dfrac{\partial G}{\partial \xi}(\xi_0(t), w(t), w'(t))\right)^2}{\dfrac{\partial^2 F}{\partial \xi^2}(\xi_0(t), w(t), w'(t))}.$$

From the earlier expressions for F and G we get

$$M'(t) = \frac{1}{\beta} h'(\xi_0) e^{-\frac{\alpha}{\beta}\xi_0}\, \xi_0', \qquad M(T_0) = M_0. \tag{29}$$

Therefore, using the fact that $\xi_0(t) \equiv 0$ for $0 \leqslant t \leqslant T_0$, we may summarize as follows:

$$M(t) = M_0 + \frac{1}{\beta}\int_0^{\xi_0(t)} h'(\zeta) e^{-\frac{\alpha}{\beta}\zeta}\, d\zeta, \qquad \forall t \geqslant 0, \tag{30}$$

where $\xi_0(t) \equiv 0$ for $t \in [0, T_0)$ and satisfies the initial value problem (27) for $t \in [T_0, \infty)$, where T_0 is given by (18).

What is interesting about Eq. (30) is that it is an explicit formula for the mass for all $t > 0$, depending only on the solution of the initial value problem (27).

Special Case: $K(\lambda, \mu) = \lambda\mu$.

When $\alpha = 0$ and $\beta = 1$, Eq. (27) reduces to

$$\xi_0' = \frac{h'^2(\xi_0)}{h''(\xi_0)}, \qquad \xi_0(T_0) = 0,$$

from which we recover the formulas obtained by Ernst, et al. [3], namely

$$h'(\xi_0(t)) = -\frac{1}{t}, \tag{31}$$

which leads to

$$M(t) = h(\xi_0(t)). \tag{32}$$

4. Examples

One utility of having exact formulas is in validating numerical schemes. We give several examples where the mass can be computed explicitly.

Example 1.

For initial particle concentration $c_0(\lambda) = \lambda^p \delta(\lambda - q)$, where $p \geqslant 0$ and $q > 0$ are constant, we get

$$h(x) = (\alpha + \beta q) q^p e^{-qx},$$

which leads to a gelation time of $T_0 = [\beta(\alpha + \beta q) q^{p+1}]^{-1}$, an initial mass of $M_0 = q^{p+1}$ and

$$M(t) = \begin{cases} M_0, & t \in [0, T_0), \\ \frac{(\alpha + 2\beta q) M_0}{2(\alpha + \beta q)\frac{t}{T_0} - \alpha}, & t \in [T_0, \infty). \end{cases} \quad \blacksquare$$

The remainder of the examples are for the pure product kernel $K(\lambda, \mu) = \lambda\mu$. For these cases ξ_0 and the mass M are given by Eqs. (31).

Example 2.

For initial particle concentration $c_0(\lambda) = \lambda^p e^{-q\lambda}$, where $p > -2$ and $q > 0$ are constant, we get

$$h(x) = \frac{\Gamma(p+2)}{(q+x)^{p+2}},$$

which leads to

$$M(t) = \begin{cases} M_0, & t \in [0, T_0), \\ M_0 \left(\frac{T_0}{t}\right)^{\frac{p+2}{p+3}}, & t \in [T_0, \infty), \end{cases} \tag{33}$$

where

$$T_0 = \frac{q^{p+3}}{\Gamma(p+3)}, \qquad M_0 = \frac{\Gamma(p+2)}{q^{p+2}}. \quad \blacksquare$$

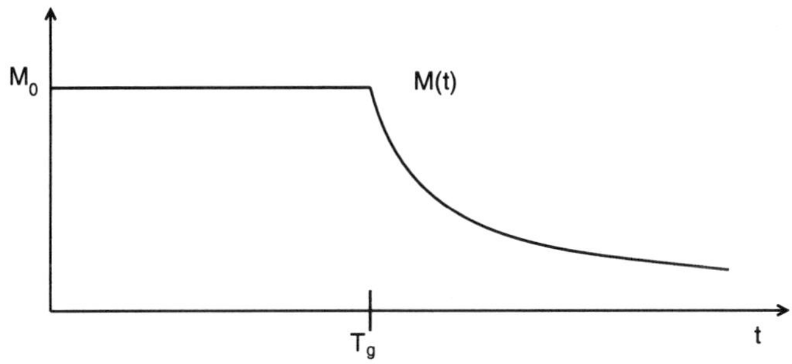

Fig. 1. A schematic of mass $M(t)$

For both of the previous examples, the mass M has a discontinuity in the derivative at the gelation point $t = T_0$, as illustrated in Figure 1. It is possible to construct examples for which the derivative of M is continuous at the gelation point. This is done by choosing an initial particle concentration c_0 that has its first two moments finite, but its third moment infinite. Two such example follow.

Example 3.

For initial particle concentration

$$c_0(\lambda) = \frac{e^{-\frac{1}{4\lambda}}}{2\sqrt{\pi}\lambda^{\frac{7}{2}}},$$

we get $h(x) = 2(\sqrt{x} + 1)\, e^{-\sqrt{x}}$ and $T_0 = 1$, and the mass given by

$$M(t) = \begin{cases} 2, & t \in [0, 1), \\ \dfrac{2(1 + \ln t)}{t} & t \in [1, \infty). \end{cases} \quad \blacksquare \tag{34}$$

Example 4.

One can easily verify that

$$h(x) = \sqrt{(x^2 + 2x)} - x + \tan^{-1}\left(\frac{1}{\sqrt{(x^2 + 2x)}}\right) - 1$$

is completely monotonic and that $h''(0)$ is infinite but that both $h(0)$ and $h'(0)$ are finite. The symbolic mathematical manipulator Maple will easily compute the inverse Laplace transform of this yielding

$$c_0(\lambda) = \frac{e^{-\lambda}}{\lambda^2}[\lambda I_0(\lambda) - I_1(\lambda) - \frac{1}{6}\lambda^3 F(\frac{3}{2}, 2, \frac{5}{2}, \frac{\lambda^2}{4})],$$

where I_0 and I_1 are modified Bessel functions of the first kind, and F is the hypergeometric function. This leads to

$$M(t) = \begin{cases} \frac{\pi}{2} - 1, & t \in [0, 1), \\ \tan^{-1}(\frac{\sqrt{2t-1}}{t-1}) - \frac{1}{\sqrt{2t-1}}, & t \in [1, \infty). \end{cases} \qquad (35)$$

A schematic of M is given in Figure 2. ■

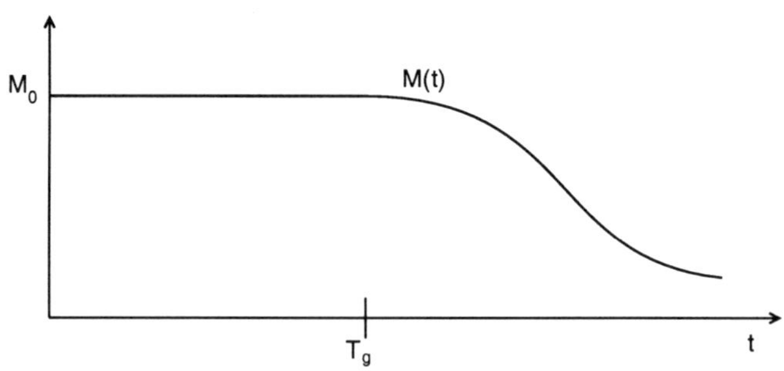

Fig. 2. A schematic of mass $M(t)$ with continuous derivative.

Examples with instantaneous gelation can also be constructed.

Example 5.

One can easily verify that

$$h(x) = 1 + \frac{x}{2} - \frac{\sqrt{x^2 + 4x}}{2}$$

is completely monotonic and that $h'(0)$ is infinite so that $T_0 = 0$. The symbolic mathematical manipulator Maple will easily compute the inverse Laplace transform of this yielding

$$c_0(\lambda) = \frac{e^{-2\lambda}}{\lambda^2} I_1(2\lambda).$$

This leads to

$$M(t) = \frac{1}{\sqrt{1+t}}, \qquad t \in [0, \infty). \quad ■ \qquad (36)$$

Example 6.

For initial particle concentration $c_0(\lambda) = \lambda^{-2}\mathrm{erfc}(-\frac{1}{\sqrt{\lambda}})$, we get

$$h(x) = e^{-2\sqrt{x}},$$

which leads to

$$M(t) = e^{-W(2t)}, \qquad t \in [0, \infty),$$

where the Lambert W function is defined implicitly by $W(t)e^W(t) = t$. ■

Example 7.

For initial particle concentration $c_0(\lambda) = \dfrac{e^{-a\lambda} - e^{-b\lambda}}{\lambda^2}$, where $b > a > 0$, we get

$$h(x) = \ln\left(\frac{x+b}{x+a}\right),$$

which leads to

$$M(t) = \begin{cases} \ln(b/a), & t \in [0, T_0), \\ \ln\left(\dfrac{\sqrt{1+4t/(b-a)}+1}{\sqrt{1+4t/(b-a)}-1}\right), & t \in [T_0, \infty), \end{cases}$$

where

$$T_0 = \frac{ab}{b-a}. \qquad ■$$

Acknowledgements

The authors would like to acknowledge financial support from Natural Sciences and Engineering Research Council of Canada.

REFERENCES

(1) Aizenman, M., and Bak, T. A. Convergence to equilibrium in a system of reacting polymers. *Comm. Math. Phys. 65* (1979), 203–230.
(2) Drake, R. L. A general mathematical survey of the coagulation equation. In *Topics in Current Aerosol Research 3 (Part 2)*, G. Hidy and J. R. Brock, Eds. Pergamon Press, 1972.
(3) Ernst, M. H., Ziff, R. M., and Hendriks, E. M. Coagulation processes with a phase transition. *J. Coll. Interf. Sci. 97*, 1 (1984), 266–277.

(4) LEYVRAZ, F. Scaling theory and exactly solved models in the kinetics of irreversible aggregation. *Physics Reports 383* (2003), 95–212.

(5) MCLEOD, J. B. On the scalar transport equation. *Proc. London. Math. Soc. 3*, 14 (1964), 445–458.

(6) MELZAK, Z. A. A scalar transport equation. *Trans. Amer. Math. Soc. 85* (1957), 547–560.

(7) STEWART, I. W. On the coagulation–fragmentation equation. *J. Appl. Math. Phys. (ZAMP) 41* (1990), 917–924.

(8) VAN ROESSEL, H. J., AND SHIRVANI, M. A formula for the post-gelation mass of a coagulation equation with a separable bilinear kernel. *Physica D. 222* (2006), 29–36.

MOTIF DISCOVERY FIXING MISMATCH POSITIONS

M. ZANTONI and A. POLICRITI
Dipartimento di Matematica e Informatica
University of Udine, Italy
E-mail: {zantoni,policriti}@dimi.uniud.it

E. DALLA and C. SCHNEIDER
Laboratorio Nazionale CIB
Area Science Park - Trieste, Italy
E-mail: {dalla,schneider}@lncib.it

Motif discovery abstracts many problems encountered during the analysis of biological sequence data, where sequences correspond to nucleotides or protein molecules and motifs represent short functionally important patterns. In this work we focus on a new computational approach to the problem of *Transcription Factor Binding Sites* (TFBS) identification. The task is to search for genomic motifs responsible for the binding of transcription factors to promoters and other regulative elements, the major event underlying gene expression control. We tackled the problem by designing data structures and algorithms for solving a consensus problem under an *error fixed layout* hypothesis, that seems to be biologically plausible and makes the task computationally treatable.

Keywords: Approximate string matching, Motif discovery, Transcription Factor Binding Sites

1. Introduction

The discovery/identification of short strings occurring approximately in a set of longer strings/sequences is one of the major tasks in today's computational biology. In this work we refer to these short strings as ***motifs***. In our initial setting the notion of "approximate occurrence" means that motifs must match a segment of (each) sequence with at most some specified number of mismatches.[1–4]

More specifically, in this paper we focus our attention on the problem of, given a set of strings, finding a substring common to (a significant portion of) the strings in the input set, allowing a *fixed layout* for mismatches in our output, motivated by biological/physical observations. Our general strategy to tackle the problem is based on the introduction of a data structure encoding *a la* Karp-Rabin substrings of the strings in the input set. This idea, together with the fixed layout error assumption and the observation that layout mismatches can be ordered, allows us a quick and efficient motif indexing and storing, based on an incremental construction highly reducing the overall number of basic operations. An important byproduct of this approach —currently under investigation— is the possibility of extending our method to the more challenging case (in general intractable) of the *Closest (Sub)String Problem.*

2. The mathematics of the problem

For a string s over an alphabet Σ, we use a standard notation: $|s|$ denotes the length of s, $s[i]$ is the *i-th* character of the string s, $s[i \dots i+\ell-1]$ is the substring of ℓ characters starting from $s[i]$, and $s \trianglelefteq t$ denotes the fact that s is a substring of t. The Hamming distance between two strings of the same length is the number of symbols that disagree. Let $\mathcal{F} = \{s_1, \dots, s_m\}$ be a set of strings over an alphabet Σ such that $|s_i| \leq n$, $1 \leq i \leq m$, and integers ℓ and d such that $0 \leq d < \ell \leq n$.

2.1. *The consensus problem*

Finding a *consensus* sequence, representing the best approximation of all the similar results that have been obtained, is crucial in order to recover useful information from the examined (biological) sequences. Formally, we need substring that is similar to many substrings of the strings in the input set. We consider the Hamming distance d_H to define the concept of "similarity" among substrings. The problem of deciding whether this substring exists is NP-hard; practical solutions are thus non-trivial. Moreover, in general, we are concerned with more than simply the decision problem, hence we may have to produce an output of exponential size. Specifically we are interested in the *closest substring* problem (CSSP):

> **CSSP(ℓ,d):** given $\mathcal{F}$, ℓ and d, find a string p of length ℓ, and m substrings $s'_i \trianglelefteq s_i$, such that $d_h(p, s'_i) \leq d$.

CSSP is NP-complete (even with the further restriction to a binary alphabet) and remains so even for the special case of the *closest string* problem (CSP), where the string p that we search for is of same length as the input strings.

In practical applications it is useful to introduce a further parameter q (the *quorum*) to the CSSP. q will represent a lower-bound to the number of input strings admitting a substring "close" to the consensus.

2.2. *Approaches*

Initially, guided by the needs of genomic research, optimization[5] and statistical[6,7] approaches were used to give solutions for the CSSP. The problem had been previously studied because of its connection with the area of coding theory, where it was proved to be NP-hard.[8] The CSSP models the more general situation where the strings that must be compared do not have the same length, and one wants to find just parts of the string that are similar. For example, the usefulness of this approach in genomic research can be seen when performing cross-species sequences comparisons, where the dataset is made of orthologous regions that can share some common features while having different length due to differences in the genomic structure.

Enumerating substring: a popular technique for finding motifs is to enumeratively test all strings over the sequence alphabet having length equal to the desired motif length. Enumerative algorithms produce all possible motifs for a set of sequences. This allows us to evaluate, according to other criteria, the discovered motifs that possess a certain combinatorial property. For this reason, enumerative algorithms can provide input to other algorithms that filter motifs based on other properties. We are concerned with more than simply the decision, so we may have to produce output of exponential size. This most naive form of search introduces a factor of $\Omega(|\Sigma|^{\ell})$ into the time complexity. The benefit of this type of enumeration is that it requires space bounded by a linear function of the size of the input. New ground was broken when Sagot[9] introduced a different approach that enumerates only those strings that are potential motifs, letting information from the sequences guide the enumeration. The method of Sagot has a time complexity of $\mathcal{O}(\ell m^2 nN)$, a space complexity of $\mathcal{O}(\ell m^2 n)$. Furthermore, the algorithm is designed with a "quorum" parameter, so that a motif is only required to be common to some $q \leq m$ of the sequences.

Optimization: for many string comparison problems, it is possible to find what is called an approximation algorithm. An algorithm $\mathcal{A}$ for a minimization problem is said to be an approximation algorithm if the worst solution returned by $\mathcal{A}$ is not greater than ε times the objective solution, for $\varepsilon > 1$. Here, ε is called the approximation guarantee of the proposed algorithm. A vast literature on approximation algorithms has been developed in the last decade.[10,11] Work has been done in determining efficient approximation algorithms for CSSP and CSP. Both problems admit a polynomial time approximation scheme (PTAS) although in both cases the exponent of the polynomial bounding the running time depends on the goodness of the approximation, they are not EPTASs (Efficient PTAS).[12] In terms of parameterized complexity, the main results for the CSSP is that it cannot be solved in polynomial time, even when the distance parameter is fixed. This is expressed, in terms of parameterized complexity theory, by showing that the CSSP is in the class W[1]-hard.

Parametric: for CSP as well as for CSSP it is natural and important to study their parameterized complexity. Considering that the number m of strings or the distance d are comparatively small in many practical situations, it is important to know whether the problems are tractable when these parameters are fixed. CSP was recently shown to be linear time fixed parameter tractable for parameter d, for a parameter function bounded by d^d, and also linear time fixed parameter tractable for parameter m, but in this case the parametric complexity contribution is much less encouraging. The parameterized complexity of CSSP has remained open for a long time. Niedermeier showed[13] that CSSP with parameter m is W[1]-hard even for a binary alphabet. That is, the problem is fixed parameter intractable unless FPT = W[1], the parametric analog of P = NP.

Local multiple alignment: multiple local alignment of nucleotide sequences is useful for finding conserved motifs, like TFBS, or greater regions putatively corresponding to entire promoters. The local multiple alignment problem (also known as the *general consensus patterns problem*) consists, given a set $\mathcal{F} = \{s_1, s_2, \ldots, s_m\}$ of m strings, in locating a substring at fixed length ℓ from each string in the set, so that the *score* determined from the set of substrings is optimal (called "Holy Grail" by Gusfield[14]).

Initially, guided by the needs of genomic research, statistical approaches were used to give solutions for the CSSP. The problem had been previously studied because of its connection with the area of coding theory, where it was proved to be NP-hard.[8] The CSSP models the more general situation where the strings that must be compared do not have the same length, and one wants to find just parts of the string that are similar. We can formulate the problem as follows: given a set $\mathcal{F} = \{s_1, s_2, \ldots, s_m\}$ of strings, and an integer ℓ, find a substring t_i of length ℓ from each s_i, maximizing the score of

$(t_1, \ldots, t_m)$, where $(t_1, \ldots, t_m)$ is a local multiple alignment, in according to a particular scoring scheme, such as Information Content, Sum of Paris, and so on. Although multiple local alignment seems to be easier than the CSSP, it is NP-hard under each of these scoring schemes. In addition, multiple local alignment is APX-hard under the average information content[15,16] scoring: it implies that unless P=NP, there is no polynomial time algorithm whose worst case approximation error can be arbitrarily small (precisely, a polynomial time approximation scheme).

Results[6] suggest that the scoring schemes greatly influence the approximability and thus, should be considered as an important factor in approximation algorithms.

3. TFBS detection

We begin this section recalling a classic in string manipulation which turns out to be the basic ingredient of our approach: the Karp-Rabin encoding.

The algorithm proposed by Karp and Rabin,[17] solves the pattern discovery problem on the exact string matching background. This algorithm assumes that we can efficiently shift a vector of bits and that we can efficiently perform arithmetical operations on integers. To take advantage of this assumptions, we can see a string like an integer, mapping each character of Σ in a digit using a function μ. For example, in the DNA context, $\Sigma = \{A, C, G, T\}$ can be mapped to $\Sigma' = \{0, 1, 2, 3\}$.

Definition 3.1. For a text string $T$of length n, let T_r denote the $\ell-$length substring of T starting at character r. We can now define the following function:

$$f(T_1) = \sum_{i=1}^{\ell} |\Sigma|^{\ell-i} \cdot \mu(T[i])$$

$$f(T_r) = |\Sigma| \cdot (f(T_{r-1}) - |\Sigma|^{\ell-1} \cdot \mu(T[r-1]) + \mu(T[r+\ell-1])$$

The following theorem holds.

Theorem 3.1. *There is an occurrence of a pattern P starting at position r of T if and only if $f(P) = f(T_r)$.*

Moreover, Karp and Rabin introduced the *randomized fingerprint* method,[17] that preserves the spirit of the above numerical approach, but allows us to deal with larger numbers in an extremely efficient way. It is a *randomized* method because introduces a probability of error, where the probability that a false match occurs can be bounded.[18] The fingerprint function is useful to deal with long patterns as it allows us to obtain reasonably small encoding numbers.

The fingerprint function is useful to deal with long patterns as it allows us to obtain reasonably small encoding numbers. In this work, we use only the first part of the algorithm idea, because the TFBS discovery problem leads to handle substring short enough to save us from using fingerprint functions. Moreover, working with approximate string matching, it is not simple to introduce an efficient hashing function. We are working out a suitable hashing function to finalize the full-fledged version of the algorithm.

3.1. *Our proposal*

TFBS are typically short ($\sim 5-12$ bp), and considerable sequence variation between functional binding sites is tolerated by most TFs. With rare exceptions, our understanding of biological properties is insufficient to enable the creation of effective computational methods. Interesting considerations about the limits of algorithmic approaches to this problem have been described by Pevzner.[19]

When using a solution to the above problem in a biological framework like the identification of known and putative TFBS, some considerations must be done in order to simplify the analysis. First of all, if an *ab-initio* search has to be done, it is useful to consider that, in general, transcription factors that regulate the expression of a group of genes involved in a given biological process tend to bind these genes' promoters in the same region. Therefore the dimension of the sequences taken into account for comparison can be greatly reduced.[20] The second useful assumption is that, given a motif (TFBS), only some of its characters (nucleotides) are important for the binding of the transcription factor. In fact, when transcription factors interact together chemically and physically it is important their position with respect to the DNA helix, and therefore with respect to nucleotides to which they can bind.[21–23] This feature, in general, simplifies and reduces the number of consensus TFBS to be generated and analyzed algorithmically.[24,25] Thus, we propose an algorithm based on the concept of localized nucleotide mutation: a protein can "tolerate" one or more mutations (wildcards) in a binding site, but always in the same positions (fixed layout error). Formally:

Definition 3.2. A solution for the *fixed-layout* (ℓ, d, q)*-consensus problem* over $\mathcal{F}$ is S if and only if

- $S \subseteq \Sigma^\ell$
- for all $s' \in S$, there exists $s_i \in \mathcal{F}$ such that $s' \trianglelefteq s_i$
- $|\{i|\exists s' \in S, s' \trianglelefteq s_i\}| \geq q$
- there exists a set of indexes $\mathcal{E} = \{i_1, \ldots, i_d\}$, $1 \leq i_1 < \cdots < i_d \leq \ell$, such that for all s'_i and s'_j in S, $s'_i[k] \neq s'_j[k] \Rightarrow k \in el$.

We assume to produce in output all the positions of the common subsequences of length ℓ with d errors discovered in at least q sequences, with the addition of the constraint that the errors, if occur, are in the same position (the layout).

3.2. *The algorithm ScanPro*

We classify error layouts into two classes: *basic* and *shifted.* The algorithm, as in the approach by Karp and Rabin, exploits the relation between these classes to perform a cost-effective encoding of substrings according to all possible layouts.

A generic error layout $\mathcal{E} = \{i_1, \ldots, i_d\}$, such that $1 \leq i_1 < \cdots < i_d \leq \ell$, is the set of the d positions where an error may occur during a comparison between strings. Without loss of results, we assume that i_1 is strictly greater than 1, that is no error can be in the first position; in this way there are $T = \binom{\ell-1}{d}$ error layouts called $EL = \{el_1, \ldots, el_T\}$.

Basic layouts, $\mathcal{B}$, are characterized by having $i_d = \ell$ and are $\binom{\ell-2}{d-1}$ overall. *Shifted layouts* relative to a given basic layout $\mathcal{B}_x$, with $x \in \{1, \ldots, \binom{\ell-2}{d-1}\}$, are denoted by $\mathcal{H}_{x,j}$ and look like $\{i_1 - j, \ldots, i_d - j\}$, with $i_1 - j \geq 2$.

For a given error layout, the function $fl : (\Sigma^\ell \times EL \rightarrow \mathbb{N})$ ($\mathbb{N}$ is the set of natural numbers) gives the encoding of a string of ℓ characters. For each string $s \in \mathcal{F}$ and each basic layout $\mathcal{B}_x = \{i_1, \ldots, i_d\}$, we have to encode all possible $n - \ell + 1$ substrigs of length ℓ in s, thus performing $\mathcal{O}(n \cdot \ell)$ operations. Below we will show that this step can, in fact, be done recursively at a lower cost.

Considering a shifted layout $\mathcal{H}_{x,j} = \{i_1 - j, \ldots, i_d - j\}$, with $j \in \{1, \ldots, i_1 - 2\}$, we obtain the encoding of the substring $s[i, \ldots, i + \ell - 1]$ for the given layout by taking the encoding of $s[i - 1, \ldots, i + \ell - 2]$ for the error layout $\mathcal{H}_{x,j-1}$, performing a left shift and adding the value relative to $s[i + \ell - 1]$, making $\mathcal{O}(n)$ operations only. Since $|\Sigma| = 4$, we map the alphabet on $\mathbb{Z}_4$ (the integers modulo 4) using only 2 bits for each character. Hence, the function fl maps an ℓ-characters-long string to a $2(\ell - d)$ bits integer, so any shift involves 2 bits only. The implementation of the algorithm (whose code is written in C and is available upon request) exploits this compact binary representation to improve performances. This idea was suggested by the *Shift-Or* algorithm.[26]

The function fl returns a value used as index of an array: in each position c of this array there is a pointer to a $T \times (m+1)$ matrix M_c. $M_k(i,j) \neq 0$ if in s_j there exists a substring w such that $|w| = \ell$ and $fl(w, fl_i) = k$. In column $m+1$ we store the number of different strings $s \in \mathcal{F}$ in which we can find the motif corresponding to each row. This value is incremented when an element is put in a null position in the relative row. If this last value is greater than the quorum, the motif is a solution for the *fixed-layout problem* and we retrieve all positions of the occurrences using a modified generalized suffix tree[14] able to manage *don't care* positions. Using these occurrences, we are able to generate a consensus c_k, according to a consensus string model (i.e. the majority string). In practice, to obtain a most biologically informative consensus we use the IUPAC alphabet.

We implemented our algorithm with a faster treatment of the basic layouts, observing that basic layouts of dimension d are easily obtainable from shifted layouts of dimension $d-1$. Iterating this argument, it is not difficult to show that we can reach the previous results in time $\mathcal{O}(mdl^{d-1}n)$. The space needed ($\mathcal{O}(|\Sigma|^{l-d}l^{d-1}m)$) depends on the number of strings in input but not on their length. It is possible to reduce the space needed at least of a factor m worsening time complexity. Furthermore, with our implementation, this algorithm is completely on-line and parallelizable.

As a matter of fact, additional information is gathered by our algorithm while building the data-structure, allowing subsequent improvements such as the bottom-up construction of a graph similar to the one introduced by Pevzner.[27] We used this *graph-extension* feature in our tests.

3.3. *Experiments and results*

In this section, to prove our algorithm on significant data-sets, we will show both the results of ScanPro on artificially generated data statistically coherent with real profile matrices and on real datasets. All these datasets, together with the code and further results, are available upon request.

Profile matrices and fixed layout: in order to confirm the biological relevance of our assumption, in this section we discuss the statistical relatedness between (real) *profile matrices* and the fixed error layout.

Given a collection of known binding sites, it is possible to generate a *profile*: a representation of those sites that can be used to search new sequences and reliably predict where additional similar binding sites occur. The mathematical background of profile models used for describing TF binding properties was extensively reviewed elsewhere.[4,28] In brief, a profile consists of a matrix tabulating observed nucleotides occurrences in each position of the protein-DNA interface, typically counted from an alignment of known sites. The rows of the matrix, in general, correspond to the four letters of the DNA alphabet, while the columns represent consecutive positions of the DNA sequence.

Profiles are converted to log-scaled position weight matrices (PWMs) in order to evaluate possible binding sites in an input sequence. A *weight matrix* is a two-dimensional table of numbers reflecting the base preferences of a protein along its cognate DNA binding site. A weight matrix assigns a match score to any oligonucleotide sequence of the corresponding length, which is usually defined as the sum of the base weights at each position. There are several approaches to derive the parameters of a weight matrix characterizing the sequences'specificity of a DNA-binding protein. We used the method[4] that exploits existing profiles.

The Homo Sapiens profile collection we used is drawn from the JASPAR database, an open-access, non-redundant collection of profiles.[29] We inspected known profiles to further confirm that our biological assumption of localized substitution was substantial. Then we used related weight matrices to score results obtained from a blind search using ScanPro.

First of all, given a profile, we computed a consensus sequence containing wildcards representing the binding site or, if it was too long, a considerable subset. In each position there is a nucleotide, if its frequency is greater or equal to 75%, otherwise a wildcard. The consensus sequence did not match with all binding site occurrences that filled the matrix but we showed that the consensus matched with a large number of these. Considering the worst condition, we had a 6-10 nucleotides-long consensus sequence containing 1-3 wildcards that matched with at least 30% of the possible occurrences of the binding sites. In most cases this rate was over 50%, up to 90%.

Considering average conditions this rate grows up increasing the conservation, this meaning that ScanPro would have been able to extract the motif correspondent to the consensus sequence and matching occurrences.

Simulated data: we considered 36 out of 49 Jaspar Homo Sapiens matrices related to as many transcription factors. Since in many cases the exact binding sites used to build a given matrix are not specified, for each profile, hence for each corresponding motif, we produced several (up to 10) different sets of occurrences of the motif "compatible" with the matrix: each set is a collection of occurrences of the related motif that could have generate the matrix. For each of these sets, we "implanted" in m (up to 100) random sequences of length n (up to 1000) one or more of the generated occurrences, guaranteeing to each one same probability to occur. The discussion in the previous section between profile matrices and fixed error layout, suggests us to run our algorithm with different parameters of length and quorum. Exploiting statistical considerations on the quorum,[34] we run our algorithm to find patterns that occur in at least half of the sequences. Then, we searched the patterns reported by the algorithm with no restrictions on the location of the mismatches taking advantage of the *graph-extension* feature. Using this technique, we obtained in output the expected motif and much of the corresponding implanted occurrences, from 80% up to 100%. Furthermore, using the related weight matrix to evaluate our results, false positives included in the solution-set obtained a high score.

Real data: we tested ScanPro on 4 datasets composed by real sequences created using the 500bp/1000bp upstream gene regions suggested by the bibliography. Encouraging results were obtained with **HIF-1**, **E2F1**, **MEF2A**, and **PPARgamma** TFBS datasets. An extended inspection of the results obtained with ScanPro on this and similar datasets is in progress, including further biological verifications and the drawing up of a more extensive work. Biological evaluation results are going to be submitted elsewhere.

Performances: experience suggests that most common parameters values for the motif discovery problem go from 8 to 12 for the substring length and 2 to 3 for wildcards. In these conditions ScanPro takes few seconds to run on 2.0GHz AMD processor and needs at most 2GB of RAM on experiments of the above size. Other tests were performed on datasets of 30 sequences 50k nucleotides with quorum at 50%. The most demanding combination of parameters required less than 10 minutes. Time requested by the filtering steps, like statistical evaluation or Ensembl cross comparison, is considered as external overhead.

4. Conclusions and further work

Predicting promoter families or extracting consensi for them is an important biological problem. The approach we are proposing in this paper tries to exploit biological considerations on the problem in order to reduce it to a tractable one and is designed to gather as much information as possible from the computation of indexes associated to substrings. The most significant goal consists in finding the best constraints that allow us to compute the consensus sequence for a set of strings. We are investigating on the properties of the *graph-extension* feature and its use to create consensus sequences for set of putative TFBS.

Tests show that our implementation, the algorithm ScanPro, spends from seconds to few minutes running using parameter ℓ (the length of the motif) less than 16. The applicability of the full-fledged Karp-Rabin technique, using a hashing function (computing non injective indexes of substrings with fixed-layout errors) similar to the deterministic sampling presented by Vishkin,[35] could allow to deal with longer motifs. We are investigating on a our own full-fledged version with a randomized behavior similar to the Karp-Rabin's one.

Many recent papers have stressed the importance of considering the correct combination and precise spatial organization between regulatory sites for both predicting such sites and extracting consensi from them. The order and relative distances of the binding sites in DNA are therefore not unrelated constraints. Another direction of study consists in encoding and manipulating strings of motifs instead of strings of nucleotides. This could allow to adapt our tool to find higher level structures like modules or gene enhancers.

The idea of the fixed layout leaded us to consider different applications. At present, we are investigating the estimation of the repetitiveness in DNA sequences. The description of algorithms, data structures, and experiments made on contigs of grape-vine's DNA have been topic of a master's thesis.[36]

Bibliography

1. A. Apostolico and Z. Galil (eds.), *Combinatorial Algorithms on Words*, NATO ASI Series, Series F: Computer and System Science, Vol. 12 (Springer-Verlag, 1985).
2. B. Brejova, C. D. Marco, T. Vinar, S. Hidalgo, G. Holguin and C. Patten, Finding patterns in biological sequences (2000).
3. M. Li, B. Ma and L. Wang, *Proceedings of the 30th ACM Symposium of Theory of Computing (STOC'99)* , 473 (1999).
4. G. Stormo, *Bioinformatics* **16**, 16 (2000).
5. P. Horton, *Proc. Pacific Symp. Biocomputing (PSB)* , 368 (1996).
6. T. Akutsu, H. Arimura and S. Shimozono, *RECOMB* (2000).
7. C. Lawrence and A. Reilly, *PROTEINS: Structure, Function, and Genetics* **7**, 41 (1990).
8. M. Frances and A. Litman, *Theor. Comput. Syst.* **30**, 113 (1997).

9. M. Sagot, *Lecture Notes in Computer Science* **1380**, 111 (1998).
10. D. Hochbaum, *Approximation Algorithms for NP-hard Problems.* (PWS Publishing, 1996).
11. V. Vazirani, *Approximation Algorithms* (Springer-Verlag, 2001).
12. M. Cesati and L. Trevisan, *Information Processing Letters* **64**, 165 (1997).
13. M. Fellows, J. Gramm and R. Niedermeier, *Lecture Notes in Computer Science* , 262 (2002).
14. D. Gusfield, *Algorithms on Strings, Trees and Sequences: Computer Science and Computational Biology* (Cambridge University Press, 1997).
15. S. Arora, C. Lund, R. Motwani, M. Sudan and M. Szegedy, *Proceedings of the 33rd IEEE Symposium Foundations of Computer Science* , 14 (1992).
16. G. Ausiello, P. Crescenzi, G. Gambosi, V. Kann, S. Spaccamela and M.Protasi, *Complexity and Approximation. Combinatorial Optimization Problems and their Approximability Properties* (Springer-Verlag, 1999).
17. R. Karp and M. Rabin, *IBM J. Res. Dev.* **31**, 249 (1987).
18. R. Motwani and P. Raghavan, *Randomized Algorithms* (Cambridge University Press, 1995).
19. E. Eskin and P. Pevzner, *Bioinformatics* **18**, 354 (2002).
20. E. Segal, N. Friedman, N. Kaminski, A. Regev and D. Koller, *Nature Genetics* **37**, Suppl:S38(June 2005).
21. V. J. Makeev, A. P. Lifanov, A. G. Nazina and D. A. Papatsenko, *Nucleic Acids Research* **31**, 6016(October 2003).
22. A. Moses, D. Chiang, M. Kellis, E. Lander and M. Eisen, *BMC Evolutionary Biology* **3**, p. 19 (2003).
23. G. Terai and T. Takagi, *Bioinformatics* **20**, 1119(May 2004).
24. S. Cawley *et al.*, *Cell* **116**, 499(February 20 2004).
25. O. Hallikas, K. Palin, N. Sinjushina, R. Rautiainen, J. Partanen, E. Ukkonen and J. Taipale, *Cell* **124**, 47(January 2006).
26. R. Baeza-Yates and G. Gonnet, *Communications of the ACM* **35**, 74 (1992).
27. P. A. Pevzner and S. H. Sze, *8th International Conference on Intelligent Systems for Molecular Biology (ISMB 2000), San Diego, California* , 269.
28. W. Wasserman and W. Krivan, *Naturwissenschaften* **90**, 156 (2003).
29. A. Sandelin, W. Alkema, P. Engstrom, W. Wasserman and B. Lenhard, *Nucleic Acids Research* **32**, D91 (2004).
30. J. Buhler and M. Tompa, *Journal of Computational Biology* **9**, 225 (2002).
31. J. Hughes, P. Estep, S. Tavazoie and G. Church, *Journal of Molecular Biology* **10**, 1205(March 2000).
32. U. Keich and P. Pevzner, *Bioinformatics* **18**, 1374 (2002).
33. X. Zhao, H. Huang and T. Speed, *In Proceedings of RECOMB* , 68 (2004).
34. G. Pavesi, G. Mauri and G. Pesole, *Bioinformatics* **17**, S207 (2001).
35. U. Vishkin, Deterministic sampling - a new technique for fast pattern matching, in *STOC '90: Proceedings of the 22nd annual ACM symposium on Theory of computing*, (ACM Press, New York, NY, USA, 1990).
36. F. Vezzi, Algoritmi e strutture dati per l'analisi di ripetitività in stringhe di DNA, Master's thesis, University of Udine (2006).

AUTHOR INDEX

Series on Advances in Mathematics for Applied Sciences

Aims and Scope

This Series reports on new developments in mathematical research relating to methods, qualitative and numerical analysis, mathematical modeling in the applied and the technological sciences. Contributions related to constitutive theories, fluid dynamics, kinetic and transport theories, solid mechanics, system theory and mathematical methods for the applications are welcomed.

This Series includes books, lecture notes, proceedings, collections of research papers. Monograph collections on specialized topics of current interest are particularly encouraged. Both the proceedings and monograph collections will generally be edited by a Guest editor.

High quality, novelty of the content and potential for the applications to modern problems in applied science will be the guidelines for the selection of the content of this series.

Instructions for Authors

Submission of proposals should be addressed to the editors-in-charge or to any member of the editorial board. In the latter, the authors should also notify the proposal to one of the editors-in-charge. Acceptance of books and lecture notes will generally be based on the description of the general content and scope of the book or lecture notes as well as on sample of the parts judged to be more significantly by the authors.

Acceptance of proceedings will be based on relevance of the topics and of the lecturers contributing to the volume.

Acceptance of monograph collections will be based on relevance of the subject and of the authors contributing to the volume.

Authors are urged, in order to avoid re-typing, not to begin the final preparation of the text until they received the publisher's guidelines. They will receive from World Scientific the instructions for preparing camera-ready manuscript.

SERIES ON ADVANCES IN MATHEMATICS FOR APPLIED SCIENCES

Vol. 17 The Fokker–Planck Equation for Stochastic Dynamical Systems and Its Explicit Steady State Solution
by C. Soize

Vol. 18 Calculus of Variation, Homogenization and Continuum Mechanics
eds. G. Bouchitté et al.

Vol. 19 A Concise Guide to Semigroups and Evolution Equations
by A. Belleni-Morante

Vol. 20 Global Controllability and Stabilization of Nonlinear Systems
by S. Nikitin

Vol. 21 High Accuracy Non-Centered Compact Difference Schemes for Fluid Dynamics Applications
by A. I. Tolstykh

Vol. 22 Advances in Kinetic Theory and Computing: Selected Papers
ed. B. Perthame

Vol. 23 Waves and Stability in Continuous Media
eds. S. Rionero and T. Ruggeri

Vol. 24 Impulsive Differential Equations with a Small Parameter
by D. Bainov and V. Covachev

Vol. 25 Mathematical Models and Methods of Localized Interaction Theory
by A. I. Bunimovich and A. V. Dubinskii

Vol. 26 Recent Advances in Elasticity, Viscoelasticity and Inelasticity
ed. K. R. Rajagopal

Vol. 27 Nonstandard Methods for Stochastic Fluid Mechanics
by M. Capinski and N. J. Cutland

Vol. 28 Impulsive Differential Equations: Asymptotic Properties of the Solutions
by D. Bainov and P. Simeonov

Vol. 29 The Method of Maximum Entropy
by H. Gzyl

Vol. 30 Lectures on Probability and Second Order Random Fields
by D. B. Hernández

Vol. 31 Parallel and Distributed Signal and Image Integration Problems
eds. R. N. Madan et al.

Vol. 32 On the Way to Understanding The Time Phenomenon: The Constructions of Time in Natural Science — Part 1. Interdisciplinary Time Studies
ed. A. P. Levich

Vol. 33 Lecture Notes on the Mathematical Theory of the Boltzmann Equation
ed. N. Bellomo

SERIES ON ADVANCES IN MATHEMATICS FOR APPLIED SCIENCES

Vol. 34 Singularly Perturbed Evolution Equations with Applications to Kinetic Theory
by J. R. Mika and J. Banasiak

Vol. 35 Mechanics of Mixtures
by K. R. Rajagopal and L. Tao

Vol. 36 Dynamical Mechanical Systems Under Random Impulses
by R. Iwankiewicz

Vol. 37 Oscillations in Planar Dynamic Systems
by R. E. Mickens

Vol. 38 Mathematical Problems in Elasticity
ed. R. Russo

Vol. 39 On the Way to Understanding the Time Phenomenon: The Constructions of Time in Natural Science — Part 2. The "Active" Properties of Time According to N. A. Kozyrev
ed. A. P. Levich

Vol. 40 Advanced Mathematical Tools in Metrology II
eds. P. Ciarlini et al.

Vol. 41 Mathematical Methods in Electromagnetism Linear Theory and Applications
by M. Cessenat

Vol. 42 Constrained Control Problems of Discrete Processes
by V. N. Phat

Vol. 43 Motor Vehicle Dynamics: Modeling and Simulation
by G. Genta

Vol. 44 Microscopic Theory of Condensation in Gases and Plasma
by A. L. Itkin and E. G. Kolesnichenko

Vol. 45 Advanced Mathematical Tools in Metrology III
eds. P. Cianlini et al.

Vol. 46 Mathematical Topics in Neutron Transport Theory — New Aspects
by M. Mokhtar-Kharroubi

Vol. 47 Theory of the Navier–Stokes Equations
eds. J. G. Heywood et al.

Vol. 48 Advances in Nonlinear Partial Differential Equations and Stochastics
eds. S. Kawashima and T. Yanagisawa

Vol. 49 Propagation and Reflection of Shock Waves
by F. V. Shugaev and L. S. Shtemenko

SERIES ON ADVANCES IN MATHEMATICS FOR APPLIED SCIENCES

Vol. 50 Homogenization
eds. V. Berdichevsky, V. Jikov and G. Papanicolaou

Vol. 51 Lecture Notes on the Mathematical Theory of Generalized Boltzmann Models
by N. Bellomo and M. Lo Schiavo

Vol. 52 Plates, Laminates and Shells: Asymptotic Analysis and Homogenization
by T. Lewiński and J. J. Telega

Vol. 53 Advanced Mathematical and Computational Tools in Metrology IV
eds. P. Ciarlini et al.

Vol. 54 Differential Models and Neutral Systems for Controlling the Wealth of Nations
by E. N. Chukwu

Vol. 55 Mesomechanical Constitutive Modeling
by V. Kafka

Vol. 56 High-Dimensional Nonlinear Diffusion Stochastic Processes — Modelling for Engineering Applications
by Y. Mamontov and M. Willander

Vol. 57 Advanced Mathematical and Computational Tools in Metrology V
eds. P. Ciarlini et al.

Vol. 58 Mechanical and Thermodynamical Modeling of Fluid Interfaces
by R. Gatignol and R. Prud'homme

Vol. 59 Numerical Methods for Viscosity Solutions and Applications
eds. M. Falcone and Ch. Makridakis

Vol. 60 Stability and Time-Optimal Control of Hereditary Systems — With Application to the Economic Dynamics of the US (2nd Edition)
by E. N. Chukwu

Vol. 61 Evolution Equations and Approximations
by K. Ito and F. Kappel

Vol. 62 Mathematical Models and Methods for Smart Materials
eds. M. Fabrizio, B. Lazzari and A. Morro

Vol. 63 Lecture Notes on the Discretization of the Boltzmann Equation
eds. N. Bellomo and R. Gatignol

Vol. 64 Generalized Kinetic Models in Applied Sciences — Lecture Notes on Mathematical Problems
by L. Arlotti et al.

SERIES ON ADVANCES IN MATHEMATICS FOR APPLIED SCIENCES

Vol. 65 Mathematical Methods for the Natural and Engineering Sciences
by R. E. Mickens

Vol. 66 Advanced Mathematical and Computational Tools in Metrology VI
eds. P. Ciarlini et al.

Vol. 67 Computational Methods for PDE in Mechanics
by B. D'Acunto

Vol. 68 Differential Equations, Bifurcations, and Chaos in Economics
by W. B. Zhang

Vol. 69 Applied and Industrial Mathematics in Italy
eds. M. Primicerio, R. Spigler and V. Valente

Vol. 70 Multigroup Equations for the Description of the Particle Transport in Semiconductors
by M. Galler

Vol. 71 Dissipative Phase Transitions
eds. P. Colli, N. Kenmochi and J. Sprekels

Vol. 72 Advanced Mathematical and Computational Tools in Metrology VII
eds. P. Ciarlini et al.

Vol. 73 Introduction to Computational Neurobiology and Clustering
by B. Tirozzi, D. Bianchi and E. Ferraro

Vol. 74 Wavelet and Wave Analysis as Applied to Materials with Micro or Nanostructure
by C. Cattani and J. Rushchitsky

Vol. 75 Applied and Industrial Mathematics in Italy II
eds. V. Cutello et al.